NORTH-HOLLAND
PERSONAL LIBRARY

SIMULATION OF LIQUIDS AND SOLIDS

SIMULATION OF LIQUIDS AND SOLIDS

Molecular Dynamics and Monte Carlo Methods in Statistical Mechanics

Editors

Giovanni CICCOTTI

Università di Roma "La Sapienza", Italy

Daan FRENKEL

FOM Institute for Atomic and Molecular Physics,
Amsterdam, The Netherlands

Ian R. McDONALD

University of Cambridge, UK

NORTH-HOLLAND
AMSTERDAM · OXFORD · NEW YORK · TOKYO

ISBN: 0 444 87062 8 (hardbound edition)
0 444 87061 x (paperback edition)
First edition: 1987
Reprinted: 1990

Published by:

North-Holland
Elsevier Science Publishers B.V.
P.O. Box 211
1000 AE Amsterdam
The Netherlands

Sole distributors for the U.S.A. and Canada:

Elsevier Science Publishing Company, Inc.
655 Avenue of the Americas
New York, N.Y. 10010
U.S.A.

Library of Congress Cataloging-in-Publication Data

Simulation of liquids and solids.

(North-Holland personal library)
Bibliography: p.
1. Liquids - Mathematical models. 2. Liquids - data processing. 3. Solids - Mathematical models. 4. Solids - Data processing. 5. Statistical mechanics. 6. Molecular dynamics. 7. Monte Carlo method. I. Ciccotti, Giovanni. II. Frenkel, Daan, 1948- . III. McDonald, Ian R. (Ian Ranald). IV. Series.
QC145.2.S54 1987 530.4′2′0724 87-21982
ISBN 0-444-87062-8
ISBN 0-444-87061-x (pbk.)

Printed in The Netherlands

Preface

This book is a collection of reprints of papers on the computer simulation of statistical-mechanical systems, together with some interlinking and explanatory material of our own. A total of 49 titles are included; inevitably, many excellent papers have had to be omitted, but we hope that the selection we have made will provide the reader with a balanced view of the history and methodology of the subject. We wish to express our gratitude to the authors for allowing us to reproduce their work and for supplying us with corrections and other, new information, much of which has been incorporated into the text and notes. We thank Bill Hoover and John Valleau for their valuable and patiently given advice, and Joost Kircz of North-Holland for his help in realizing the project. We are grateful, finally, for the financial support provided by the Netherlands Organization for the Advancement of Pure Research (ZWO) and the Consiglio Nazionale delle Richerche (CNR) under the ZWO–CNR exchange scheme, NATO grant no. 1865, and the CEE Programma di Gemellaggio.

Giovanni Ciccotti

Daan Frenkel

Ian R. McDonald

Contents

Preface v
Contents vii

Introduction 1

I. Early Papers 4

I.1. Equation of State Calculations by Fast Computing Machines, N. Metropolis, A.W. Rosenbluth, M.N. Rosenbluth, A.H. Teller and E. Teller
J. Chem. Phys. 21 (1953) 1087–1092 10
Notes to Reprint I.1 15

I.2. Monte Carlo Equation of State of Molecules Interacting with the Lennard–Jones Potential. I. A Supercritical Isotherm at About Twice the Critical Temperature, W.W. Wood and F.R. Parker
J. Chem. Phys. 27 (1957) 720–733 17
Notes to Reprint I.2 31

I.3. Preliminary Results from a Recalculation of the Monte Carlo Equation of State of Hard Spheres, W.W. Wood and J.D. Jacobson
J. Chem. Phys. 27 (1957) 1207–1208 32

I.4. Phase Transition for a Hard Sphere System, B.J. Alder and T.E. Wainwright
J. Chem. Phys. 27 (1957) 1208–1209 33
Note to Reprint I.4 34

I.5. Dynamics of Radiation Damage, J.B. Gibson, A.N. Goland, M. Milgram and G.H. Vineyard
Phys. Rev. 120 (1960) 1229–1253 35
Notes to Reprint I.5 59

I.6. Correlation in the Motion of Atoms in Liquid Argon, A.

Rahman
Phys. Rev. 136 (1964) A405–A411 60
Notes to Reprint I.6 66

I.7. Computer "Experiments" on Classical Fluids. I. Thermodynamical Properties of Lennard–Jones Molecules, L. Verlet
Phys. Rev. 159 (1967) 98–103 68
Notes to Reprint I.7 73

I.8. Structure of Water: A Monte Carlo Calculation, J.A. Barker and R.O. Watts
Chem. Phys. Lett. 3 (1969) 144–145 74
Notes to Reprint I.8 75

I.9. Decay of the Velocity Autocorrelation Function, B.J. Alder and T.E. Wainwright
Phys. Rev. A 1 (1970) 18–21 76
Notes to Reprint I.9 79

II. Free Energies and Phase Equilibria 80

II.1. Phase Transition in Elastic Disks, B.J. Alder and T.E. Wainwright
Phys. Rev. 127 (1962) 359–361 85
Notes to Reprint II.1 87

II.2. Use of Computer Experiments to Locate the Melting Transition and Calculate the Entropy in the Solid Phase, W.G. Hoover and F.H. Ree
J. Chem. Phys. 47 (1967) 4873–4878 88
Notes to Reprint II.2 93

II.3. Phase Transitions of the Lennard–Jones System, J.P. Hansen and L. Verlet
Phys. Rev. 184 (1969) 151–161 95
Notes to Reprint II.3 105

II.4. New Monte Carlo Method to Compute the Free Energy of Arbitrary Solids. Application to the FCC and HCP Phases of Hard Spheres, D. Frenkel and A.J.C. Ladd
J. Chem. Phys. 81 (1984) 3188–3193 107
Notes to Reprint II.4 113

II.5. Calculation of the Entropy of Liquid Chlorine and Bromine by Computer Simulation, S. Romano and K. Singer
Mol. Phys. 37 (1979) 1765–1772 114
Notes to Reprint II.5 121

II.6. Efficient Estimation of Free Energy Differences from Monte Carlo Data, C.H. Bennett

J. Comput. Phys. 22 (1976) 245–268 123
Notes to Reprint II.6 146
II.7. Nonphysical Sampling Distributions in Monte Carlo Free-Energy Estimation: Umbrella Sampling, G.M. Torrie and J.P. Valleau
J. Comput. Phys. 23 (1977) 187–199 147
Note to Reprint II.7 159

III. Transport and Non-Equilibrium Molecular Dynamics 160

III.1. Studies in Molecular Dynamics. VIII. The Transport Coefficients for a Hard-Sphere Fluid, B.J. Alder, D.M. Gass and T.E. Wainwright
J. Chem. Phys. 53 (1970) 3813–3826 167
Notes to Reprint III.1 181
III.2. Computer "Experiments" on Classical Fluids. IV. Transport Properties and Time-Correlation Functions of the Lennard–Jones Liquid near Its Triple Point, D. Levesque, L. Verlet and J. Kürkijarvi
Phys. Rev. A 7 (1973) 1690–1700 182
Notes to Reprint III.2 192
III.3. Statistical Error due to Finite Time Averaging in Computer Experiments, R. Zwanzig and N.K. Ailawadi
Phys. Rev. 182 (1969) 280–283 193
Note to Reprint III.3 196
III.4. Velocity -Inversion and Irreversibility in a Dilute Gas of Hard Disks, J. Orban and A. Bellemans
Phys. Lett. A 24 (1967) 620–621 197
III.5. "Thought-Experiments" by Molecular Dynamics, G. Ciccotti, G. Jacucci and I.R. McDonald
J. Stat. Phys. 21 (1979) 1–22 199
Notes to Reprint III.5 220
III.6. Determination of the Shear Viscosity of Atomic Liquids by Non-Equilibrium Molecular Dynamics, K. Singer, J.V.L. Singer and D. Fincham
Mol. Phys. 40 (1980) 515–519 221
Note to Reprint III.6 226
III.7. Lennard–Jones Triple-Point Bulk and Shear Viscosities. Green–Kubo Theory, Hamiltonian Mechanics, and Nonequilibrium Molecular Dynamics, W.G. Hoover, D.J. Evans, R.B. Hickman, A.J.C. Ladd, W.T. Ashurst and B. Moran
Phys. Rev. A 22 (1980) 1690–1697 227

Note to Reprint III.7 234

III.8. Nonequilibrium Molecular Dynamics via Gauss' Principle of Least Constraint, D.J. Evans, W.G. Hoover, B.H. Failor, B. Moran and A.J.C. Ladd
Phys. Rev. A 28 (1983) 1016–1021 235
Notes to Reprint III.8 240

III.9. Classical Response Theory in the Heisenberg Picture, B.L. Holian and D.J. Evans
J. Chem. Phys. 83 (1985) 3560–3566 241
Notes to Reprint III.9 247

IV. Other Ensembles 248

IV.1. *NpT*-ensemble Monte Carlo Calculations for Binary Liquid Mixtures, I.R. McDonald
Mol. Phys. 23 (1972) 41–58 253
Note to Reprint IV.1 271

IV.2. Primitive Model Electrolytes. I. Grand Canonical Monte Carlo Computations, J.P. Valleau and L.K. Cohen
J. Chem. Phys. 72 (1980) 5935–5941 272
Notes to Reprint IV.2 278

IV.3. Molecular Dynamics Simulations at Constant Pressure and/or Temperature, H.C. Andersen
J. Chem. Phys. 72 (1980) 2384–2393 279

IV.4. Polymorphic Transitions in Single Crystals: A New Molecular Dynamics Method, M. Parrinello and A. Rahman
J. Appl. Phys. 52 (1981) 7182–7190 289
Notes to Reprint IV.4 297

IV.5. A Unified Formulation of the Constant Temperature Molecular Dynamics Methods, S. Nosé
J. Chem. Phys. 81 (1984) 511–519 298
Notes to Reprint IV.5 306

IV.6. Canonical Dynamics: Equilibrium Phase-Space Distributions, W.G. Hoover
Phys. Rev. A 31 (1985) 1695–1697 308
Note to Reprint IV.6 310

IV.7. New High-Pressure Phase of Solid ^{4}He is BCC, D. Levesque, J.J. Weis and M.L. Klein
Phys. Rev. Lett. 51 (1983) 670–673 311

IV.8. Stability of the High-Pressure Body-Centered-Cubic Phase of Helium, D. Frenkel

Phys. Rev. Lett. 56 (1986) 858–860 315
Note to Reprint IV.8 318
IV.9. Ensemble Dependence of Fluctuations with Application to Machine Computations, J.L. Lebowitz, J.K. Percus and L. Verlet
Phys. Rev. 153 (1967) 250–254 319
Notes to Reprint IV.9 324

V. Molecular and Ionic Systems 325

V.1. Molecular Dynamics Study of Liquid Water, A. Rahman and F.H. Stillinger
J. Chem. Phys. 55 (1971) 3336–3359 330
Notes to Reprint V.1 354
V.2. Molecular Dynamics of Rigid Systems in Cartesian Coordinates. A General Formulation, G. Ciccotti, M. Ferrario and J.-P. Ryckaert
Mol. Phys. 47 (1982) 1253–1264 355
Notes to Reprint V.2 366
V.3. Singularity Free Algorithm for Molecular Dynamics Simulation of Rigid Polyatomics, D.J. Evans and S. Murad
Mol. Phys. 34 (1977) 327–331 367
Notes to Reprint V.3 371
V.4. Introduction of Andersen's Demon in the Molecular Dynamics of Systems with Constraints, J.P. Ryckaert and G. Ciccotti
J. Chem. Phys. 78 (1983) 7368–7374 373
Notes to Reprint V.4 379
V.5. Constant Pressure Molecular Dynamics for Molecular Systems, S. Nosé and M.L. Klein
Mol. Phys. 50 (1983) 1055–1076 380
Notes to Reprint V.5 401
V.6. Statistical Mechanics of Dense Ionized Matter. IV. Density and Charge Fluctuations in a Simple Molten Salt, J.P. Hansen and I.R. McDonald
Phys. Rev. A 11 (1975) 2111–2123 402
Notes to Reprint V.6 414
V.7. Dipole Moment Fluctuation Formulas in Computer Simulations of Polar Systems, M. Neumann
Mol. Phys. 50 (1983) 841–858 415
Notes to Reprint V.7 432

VI. Trends and Prospects 434

VI.1. Molecular Dynamics Simulations of the Incommensurate Phase of Krypton on Graphite Using More than 100,000 Atoms, F.F. Abraham, W.E. Rudge, D.J. Auerbach and S.W. Koch
Phys. Rev. Lett. 52 (1984) 445–448 440
Note to Reprint VI.1 443

VI.2. Faceting at the Silicon (100) Crystal-Melt Interface: Theory and Experiment, U. Landman, W.D. Luedtke, R.N. Barnett, C.L. Cleveland, M.W. Ribarsky, E. Arnold, S. Ramesh, H. Baumgart, A. Martinez and B. Khan
Phys. Rev. Lett. 56 (1986) 155–158 444

VI.3. Comparison of the Molecular Dynamics Method and the Direct Simulation Monte Carlo Technique for Flows around Simple Geometries, E. Meiburg
Phys. Fluids 29 (1986) 3107–3113 448
Note to Reprint VI.3 455

VI.4. Dynamics of the A + BC Reaction in Solution, J.P. Bergsma, P.M. Edelsten, B.J. Gertner, K.R. Huber, J.R. Reimers, K.R. Wilson, S.M. Wu and J.T. Hynes
Chem. Phys. Lett. 123 (1986) 394–398 456
Notes to Reprint VI.4 460

VI.5. Active Site Dynamics of Ribonuclease, A.T. Brünger, C.L. Brooks III and M. Karplus
Proc. Natl. Acad. Sci. USA 82 (1985) 8458–8462 461

VI.6. Study of Electron Solvation in Liquid Ammonia Using Quantum Path Integral Monte Carlo Calculations, M. Sprik, R.W. Impey and M.L. Klein
J. Chem. Phys. 83 (1985) 5802–5809 466

VI.7. Path-Integral Computation of the Low-Temperature Properties of Liquid ^{4}He, D.M. Ceperley and E.L. Pollock
Phys. Rev. Lett. 56 (1986) 351–354 474
Note to Reprint VI.7 477

VI.8. Unified Approach for Molecular Dynamics and Density-Functional Theory, R. Car and M. Parrinello
Phys. Rev. Lett. 55 (1985) 2471–2474 478
Note to Reprint VI.8 481

Introduction

The increasing use of computer simulation has in recent years transformed the manner in which the classical many-body problem is treated. The resulting shift in emphasis away from purely analytical methods has had an impact in two ways. First, the stimulus provided by the results of simulations has led to unprecedented advances in areas that are the traditional preserve of classical statistical mechanics, notably in liquid-state theory. Secondly, in conjunction with the dramatic developments that have occurred in computer technology over the same period, the use of simulation has greatly extended the range of problems to which the methods of statistical mechanics can usefully be applied. Interest in simulation has therefore spread beyond the confines of statistical physics to include groups such as physical chemists, solid-state physicists, materials scientists and, increasingly, biochemists and biophysicists.

The techniques used in the simulations are of two types: molecular dynamics and Monte Carlo. Their common feature is the fact that they are based on a molecular description of the system of interest; the main ingredient in each case is the law that describes the interactions between the constituent particles, be they atoms, molecules or ions. The different versions of the Monte Carlo method are schemes for sampling from a probability distribution appropriate to one of the ensembles of equilibrium statistical mechanics. In molecular dynamics, the particles are allocated initial coordinates and momenta, and their subsequent trajectories are mapped out by integration of the classical equations of motion. Observable properties of the system are then obtained as time averages over the trajectories. Hence molecular dynamics, at least in its conventional form, represents a realization of Boltzmann's approach to statistical mechanics, whereas the Monte Carlo method is rooted in Gibbs' formulation of the problem. Unlike Monte Carlo, molecular dynamics can be used in the study of time-dependent processes. To that extent, molecular dynamics is the more powerful technique, but the Monte Carlo method has its own advantages and is often simpler to apply.

The purpose of this book is to draw together some of the key papers in what is now a large and rapidly expanding literature; to keep the material within reasonable bounds, we include nothing on the simulation of spin systems and other lattice models. Chapter I contains a number of pioneering papers and records the early successes that established the place of simulation as a major investigative tool in condensed-matter physics. Chapters II to V consist largely of papers in which significant advances in methodology are described, but a number of illustrative applications are also included. Chapter II deals with the problem of phase equilibrium and the calculation of free energies. Chapter III is concerned with transport phenomena and with the class of methods known as "non-equilibrium" molecular dynamics. Chapter IV takes up the question of how to sample from ensembles other than the familiar canonical and microcanonical ones. Chapter V discusses some of the special problems that arise in the study of molecular and ionic systems. Finally, Chapter 6 is made up of a number of recent, mostly brief papers that emphasize how the subject is broadening and possibly give some clues as to how it may develop in the future. There is a short introduction to each chapter that serves – in part – to motivate the choice of papers, and some supplementary information, updating of references and corrections of misprints are contained in the notes attached to the individual papers. Despite the accent on methods the book is not intended as a manual for users, nor does it attempt a state-of-the-art survey of the field. These two needs are well catered for already, the former by the book of Allen and Tildesley [1] and the latter by the published proceedings [2] * of a recent International School on simulation, together with a variety of books [3] and specialist review articles [4]. The aim is rather to present a selection of reprints that both illuminates the historical development of the subject and provides easy access to material that is relevant to today's practitioner.

References

[1] M.P. Allen and D.J. Tildesley, Computer Simulation of Liquids (Clarendon Press, Oxford, 1987).

[2] G. Ciccotti and W.G. Hoover, eds, Molecular Dynamics Simulation of Statistical-Mechanical Systems (North-Holland, Amsterdam, 1986).

[3] P. Lykos, ed., Computer Modeling of Matter (American Chemical Society, Washington, 1978).
K. Binder, ed., Monte Carlo Methods in Statistical Physics (Springer-Verlag, Berlin, 1979).
R.W. Hockney and J.W. Eastwood, Computer Simulation Using Particles (McGraw-Hill, New York, 1981).
C.R.A. Catlow and W.C. Mackrodt, eds, Computer Simulation of Solids (Springer-Verlag, Berlin, 1982).
D. Nicholson and N.G. Parsonage, Computer Simulation and the Statistical Mechanics of Adsorption (Academic Press, London, 1982).
M.H. Kalos, ed., Monte Carlo Methods in Quantum Problems (Reidel, Dordrecht, 1984).

* This book contains a record of the 97th International School of Physics "Enrico Fermi", held in Varenna in July–August 1985.

O.G. Mouritsen, Computer Studies of Phase Transitions and Critical Phenomena (Springer-Verlag, Berlin, 1984).
K. Binder, ed., Applications of the Monte Carlo Method in Statistical Physics (Springer-Verlag, Berlin, 1984).
M.H. Kalos and P.A. Whitlock, Monte Carlo Methods, Vol. I: Basics (Wiley, New York, 1986).
D.W. Heerman, Computer Simulation Methods in Theoretical Physics (Springer-Verlag, Berlin, 1986).
W.G. Hoover, Molecular Dynamics (Springer-Verlag, Berlin, 1986).
[4] J.A. McCammon and M. Karplus, Simulation of Protein Dynamics, Ann. Rev. Phys. Chem. 31 (1980) 29.
D. Frenkel, Intermolecular Spectroscopy and Computer Simulations, in: Intermolecular Spectroscopy and Dynamical Properties of Dense Systems, ed. J. van Kranendonk, (North-Holland, Amsterdam, 1980).
D. Frenkel and J.P. McTague, Computer Simulations of Freezing and Supercooled Liquids, Ann. Rev. Phys. Chem. 31 (1980) 491.
B.J. Alder and E.L. Pollock, Simulation of Polar and Polarizable Fluids, Ann. Rev. Phys. Chem. 32 (1981) 311.
C.A. Angell, J.H.R. Clarke and L.V. Woodcock, Interaction Potentials and Glass Formation: A Survey of Computer Experiments, Adv. Chem. Phys. 48 (1981) 397.
W.G. Hoover, Nonequilibrium Molecular Dynamics, Ann. Rev. Phys. Chem. 34 (1983) 103.
J.A. McCammon, Protein Dynamics, Rep. Prog. Phys. 47 (1984) 1.
D.J. Evans and G.P. Morriss, Non-Newtonian Molecular Dynamics, Comput. Phys. Rep. 1, (1984) 297.
A. Baumgärtner, Simulation of Polymer Motion, Ann. Rev. Phys. Chem. 35 (1984) 419.
M.L. Klein, Computer Simulation Studies of Solids, Ann. Rev. Phys. Chem. 36 (1985) 525.
E. Dickinson, Brownian Dynamics with Hydrodynamic Interactions: The Application to Protein Diffusional Problems, Chem. Soc. Rev. 14 (1985) 421.
B.J. Alder, D.M. Ceperley and E.L. Pollock, Quantum Mechanical Simulation of Liquids, Acc. Chem. Res. 18 (1985) 265.
D. Fincham and D.M. Heyes, Recent Advances in Molecular-Dynamics Computer Simulation, Adv. Chem. Phys. 63 (1985) 493.
D.J. Evans and W.G. Hoover, Flows Far From Equilibrium via Molecular Dynamics, Ann. Rev. Fluid Mech. 18 (1986) 243.
G. Ciccotti and J.P. Ryckaert, Molecular Dynamics Simulation of Rigid Molecules, Comput. Phys. Rep. 4 (1986) 345.
F.F. Abraham, Computational Statistical Mechanics: Methodology, Applications and Supercomputing, Adv. Phys. 35 (1986) 1.

CHAPTER I

Early Papers

The reprints gathered together in this chapter comprise a selection of the most important early papers on the simulation of classical many-body systems. In later parts of the book, articles are discussed primarily in the context of specific computational problems. We begin, however, by taking a more historical point of view. The papers included here showed both that computer simulations were feasible and, equally importantly, that they were capable of yielding new results that could not be obtained by other, existing means.

It was not accidental that the earliest simulations were of simple dense fluids. The theoretical techniques already available for the prediction of properties of solids and dilute gases were ill-suited to the study of liquids. Indeed, it was only with the arrival of computer "experiments" that it became possible even to test theories of the liquid state in a satisfactory way. It has taken much longer for computer simulation to become an accepted tool in the study of crystalline solids; not surprisingly, it has generally been applied to those problems in solid-state physics that can be treated only approximately, if at all, by conventional, lattice-dynamics techniques. The first computer "experiment" on a model of a solid was the simulation by Fermi, Pasta and Ulam [1] * of the dynamics of a linear, anharmonic chain. The results were described in a 1955 internal report of the Los Alamos Scientific Laboratory. This work has had only a limited impact on equilibrium statistical mechanics, but has played a significant role in subsequent developments in ergodic theory and non-linear dynamics. An interesting account of the early history of computational physics, including both the Fermi–Pasta–Ulam work and the first Monte Carlo simulations, can be found in a paper by Anderson [2], written for a meeting held at Los Alamos in 1985 in honour of N. Metropolis.

Reprint I.1. The appearance of the paper by Metropolis, Rosenbluth, Rosenbluth, Teller and Teller marked the birth of computer simulation as a statistical-mechanical technique. The paper is famous for its introduction of the Monte Carlo method for the study of systems of interacting particles and

* A discussion of the Fermi paper in the context of molecular dynamics simulation can be found in the paper of Hoover and Ashurst [1].

for its proposal of a specific form of "importance sampling" that is still used in most present-day Monte Carlo programs. The latter also holds true of the periodic boundary conditions that are briefly described in the paper. Wood [3] has written as follows of the early Monte Carlo work: "When the Los Angeles MANIAC computer became operational in March 1952, Metropolis was interested in having as broad a spectrum of problems as possible tried on the new machine, in order to evaluate its logical structure and to demonstrate the capabilities of the machine. The classical statistical mechanical N-body problem via Monte Carlo techniques was one of the first problems, done by Metropolis in collaboration with the Tellers and the Rosenbluths, and led to the development of what is now known as the Metropolis Monte Carlo method. Metropolis also recalls that J.E. Mayer had frequently urged the importance of this problem, and the general idea of a computational approach, in conversations at Chicago..." The application described in the paper is to a system of hard disks in two dimensions, but this is now of only historical interest. The equation of state of hard disks has subsequently [4] been obtained with much higher accuracy, while the relatively short runs that were made were inadequate to reveal the hard-disk melting transition later discovered by Alder and Wainwright (Reprint II.1).

Reprint I.2. The results of the first successful Monte Carlo simulation of a "realistic" model of an atomic fluid were published by Wood and Parker in 1957. The system studied was a three-dimensional Lennard–Jones fluid, and the results were compared with experimental equation-of-state data for dense fluid argon. The paper contains a wealth of information on practical aspects of Monte Carlo simulations of systems with continuous interparticle potentials. The choice of the Lennard–Jones (6–12) potential was dictated partly by considerations of computing speed, but the good agreement that was found between simulation and experiment has contributed to the later popularity of "Lennard–Jonesium" as a model of the liquefied noble gases. The results are also notable for the evidence they give of a solid–fluid phase transition at high densities. As in the work of Alder and Wainwright on hard spheres (Reprint I.4), it was only about ten years later that the precise location of the transition was established (Hansen and Verlet, Reprint II.3).

One of the important technical questions discussed in the paper is the procedure used to truncate a continuous potential in a system with periodic boundary conditions. Most workers continue to use the "minimum distance image method" described below eq. (15) in the text, combined with truncation of the potential at a distance r_c that is less than half the length of the periodic cell. Long-range corrections are subsequently added by integration over a uniform distribution for distances $r > r_c$. Other practical matters discussed include the computation of the heat capacity from fluctuations in the potential energy, the calculation of the pressure from the virial equation, and a method of obtaining the radial distribution function.

Reprints I.3 and I.4. These two papers appeared together in the Journal of Chemical Physics in 1957. The article by Alder and Wainwright was the first report in the open literature of a molecular dynamics simulation (a paper by the same authors for the 1956 International Symposium on Statistical Mechanical Theory of Transport Processes was presented earlier but published later [5]). It established the status of molecular dynamics as a powerful computational approach by solving what was one of the outstanding problems in statistical mechanics in the late 1950's, namely the question of whether or not a fluid–solid transition occurs in a system of hard spheres. The data obtained by Alder and Wainwright can be found in fig. 1 of the accompanying paper by Wood and Jacobson, where Monte Carlo simulations of the same model system are described. The fact that the two sets of results were in agreement with each other was an important finding, since the equivalence * of the two methods of simulation was not obvious at the time. Indeed, there was some evidence to the contrary, since the molecular dynamics results were in conflict with earlier Monte Carlo data of Rosenbluth and Rosenbluth [7]. The reason for this discrepancy is probably that the systems studied by the Rosenbluths were not sufficiently well equilibrated.

The observation of a hard-sphere melting transition triggered off an intense theoretical effort aimed at providing quantitative predictions for the density at which solid and fluid coexist. In the case of hard spheres, the problem was solved [8] by computer simulation in 1968, but reasonably accurate analytical theories of melting have become available only in the last few years (see, for example, ref. [9]).

Reprint I.5. Unlike most of the other reprints in this book, the paper by Gibson, Goland, Milgram and Vineyard is not concerned with simulation in a statistical-mechanical context. Inclusion of the paper here is nonetheless justified, partly because of its pioneering character and partly because the methods used are very close in spirit to those later adopted in molecular dynamics calculations for Lennard–Jones and similar fluids. The incentive for attempting a simulation of radiation damage is much the same as for equilibrium simulations: the problem is too complex to be tackled exclusively by analytical methods. We quote from some autobiographical remarks of G.H. Vineyard [10]: "Somewhere the idea came up that a computer might be applied to follow in more detail what actually goes on in radiation damage

* The effect of the constraint of fixed total momentum in molecular dynamics calculations has been analyzed by Hover and Alder, and Erpenbeck and Wood [6].

cascades. We got into quite an argument, some maintaining that it wasn't possible to do this on a computer, others that it wasn't necessary. John Fisher insisted that the job could be done well enough by hand, and was then goaded into promising to demonstrate. He went off to his room to work. Next morning he asked for a little more time, promising to send me the results as soon as he got home. After about two weeks, not having heard from him, I called and he admitted that he had given up. This stimulated me to think further about how to get a high-speed computer into the game..." The resulting article describes the simulation of radiation damage in copper; calculations were made for a crystallite consisting of 500 copper atoms embedded in energy-absorbing boundaries. The central-difference algorithm used to generate the atomic trajectories is very similar to the one later developed by Verlet (Reprint I.7). In its implementation, however, use was made of the special nature of the radiation-damage process. Perhaps for this reason, the paper is rarely referred to in the literature of statistical-mechanical simulations, but it had an influence on early work in the field, including that of Rahman (Reprint I.6) (see Note I.6.1).

Reprint I.6. The results of the first molecular dynamics simulation of a "realistic" model fluid were described in a paper by Rahman that appeared in 1964. The system studied consisted of 864 Lennard–Jones particles under conditions appropriate to liquid argon at 84.4 K and 1.374 g cm^{-3}. Rahman's work has played a very important role in convincing experimentalists of the value of computer simulations. What the paper shows is that molecular dynamics can be used to gain a better understanding of experiments (e.g. neutron scattering) on liquid argon and, more generally, of the microscopic properties of real fluids. The impact that the paper has had may be judged by the fact that a special meeting was held at Argonne National Laboratory in 1984 to celebrate the twentieth anniversary of its publication.

Reprint I.7. Verlet's molecular dynamics simulation of Lennard–Jones "argon" was directly inspired by the work of Rahman (Reprint I.6) and has itself served as the inspiration for much of the later progress in the field. At the technical level, the paper is important for the introduction of two computational devices that have been used repeatedly in later simulations. The first is the so-called Verlet algorithm (eq. (4) in the text) for the integration of the equations of motion. This simple yet very stable algorithm has been widely used for both atomic and molecular liquids and has shown itself to be at least the equal of other, more complicated schemes. A second device that has endured is the use of "neighbour lists" in the evaluation of the interparticle forces; this leads to appreciable savings in computing time, particularly for large systems. In addition, the paper contains a systematic comparison, over a wide range of temperature and density, between the equilibrium properties of real argon and the corresponding data for the Lennard–Jones fluid. The altogether remarkable agreement that is found, together with the earlier results of Wood and Parker (Reprint I.2), has contributed greatly to the present status of the Lennard–Jones fluid as the classic model of a simple liquid.

Reprint I.8. Given the success achieved by Wood and Parker, Rahman and Verlet (Reprints I.2, I.6 and I.7) in modelling the behaviour of simple atomic fluids, the obvious next step was the extension to molecular systems. This article by Barker and Watts on the Monte Carlo simulation of liquid water was the first attempt to use such methods to obtain structural information about a real molecular liquid. The potential used in the work was based on experimental gas-phase data; although that particular model has now been superseded, the practice of taking semi-empirical, two-body, effective potentials as input to a simulation is one that has persisted. It is worth remembering that at the time that the paper of Barker and Watts appeared it was by no means clear that a simple potential of this type was capable of reproducing correctly the unusual structural features of real water.

Reprint I.9. The paper by Alder and Wainwright does not contain any innovations in technique, nor does it extend the scope of computer simulation to more complex systems. Its importance lies instead in the fact that it represents the culmination of a programme of work by the same authors that has led to a complete overhaul of the kinetic theory of dense fluids. The discovery of an algebraic long-time tail in the velocity autocorrelation functions of hard disks (in $d=2$ dimensions) and hard spheres ($d=3$) was totally unexpected. Alder and Wainwright show in their paper that the $t^{-d/2}$ tails are hydrodynamic in origin and can be understood in terms of the transient response of the hydrodynamic flow field to the motion of a particle. The fact that macroscopic, hydrodynamic arguments could be used on a molecular scale was itself a surprise. The discovery of the hydrodynamic long-time tails, later observed in other computer "experiments", has stimulated the development of new theoretical methods for the description of transport in dense fluids; a recent comparison between modern theories and the results of computer simulations can be found in an article by Erpenbeck and Wood [11].

References

[1] E. Fermi, J.G. Pasta and S.M. Ulam, Studies of Non-Linear Problems, LASL Rep. LA-1940 (1955).
E. Amaldi, H.L. Anderson, E. Persico, F. Rasetti, C.S. Smith, A. Wattenberg and E. Segré, eds, Collected Works of Enrico Fermi, Vol. II (Chicago, University of Chicago Press, 1965) pp. 978–88.

W.G. Hoover and W.T. Ashurst, Adv. Theor. Chem. 1 (1975) 1.
[2] H.L. Anderson, J. Stat. Phys. 43 (1986) 731.
[3] W.W. Wood, in: Molecular Dynamics Simulation of Statistical-Mechanical Systems, eds G. Ciccotti and W.G. Hoover (North-Holland, Amsterdam, 1986) p. 3.
[4] J.J. Erpenbeck and M. Luban, Phys. Rev. A 32 (1985) 2920.
[5] B.J. Alder and T.E. Wainwright, in: Proceedings of the International Symposium on Statistical Mechanical Theory of Transport Processes, Brussels 1956, ed. I. Prigogine (Interscience, New York, 1958).
[6] W.W. Wood, in: Physics of Simple Liquids, eds H.N.V. Temperley, J.S. Rowlinson and G.S. Rushbrooke (North-Holland, Amsterdam, 1968).
W.G. Hoover and B.J. Alder, J. Chem. Phys. 46 (1967) 686.
J.J. Erpenbeck and W.W. Wood, in: Statistical Mechanics. Part B, ed. B.J. Berne (Plenum Press, New York, 1977).
[7] M.N. Rosenbluth and A.W. Rosenbluth, J. Chem. Phys. 22 (1954) 881.
[8] W.G. Hoover and F.H. Ree, J. Chem. Phys. 47 (1968) 4873.
[9] T.V. Ramakrishnan and M. Yussouff, Phys. Rev. B 19 (1979) 2775.
M. Baus, Mol. Phys. 50 (1983) 543.
[10] G.H. Vineyard, in: Interatomic Potentials and Simulation of Lattice Defects, eds P.C. Gehlen, J.R. Beeler and R.I. Jaffe (Plenum Press, New York, 1972).
[11] J.J. Erpenbeck and W.W. Wood, Phys. Rev. A 32 (1985) 412.

THE JOURNAL OF CHEMICAL PHYSICS VOLUME 21, NUMBER 6 JUNE, 1953

Equation of State Calculations by Fast Computing Machines

NICHOLAS METROPOLIS, ARIANNA W. ROSENBLUTH, MARSHALL N. ROSENBLUTH, AND AUGUSTA H. TELLER,
Los Alamos Scientific Laboratory, Los Alamos, New Mexico

AND

EDWARD TELLER,* *Department of Physics, University of Chicago, Chicago, Illinois*
(Received March 6, 1953)

A general method, suitable for fast computing machines, for investigating such properties as equations of state for substances consisting of interacting individual molecules is described. The method consists of a modified Monte Carlo integration over configuration space. Results for the two-dimensional rigid-sphere system have been obtained on the Los Alamos MANIAC and are presented here. These results are compared to the free volume equation of state and to a four-term virial coefficient expansion.

I. INTRODUCTION

THE purpose of this paper is to describe a general method, suitable for fast electronic computing machines, of calculating the properties of any substance which may be considered as composed of interacting individual molecules. Classical statistics is assumed, only two-body forces are considered, and the potential field of a molecule is assumed spherically symmetric. These are the usual assumptions made in theories of liquids. Subject to the above assumptions, the method is not restricted to any range of temperature or density. This paper will also present results of a preliminary two-dimensional calculation for the rigid-sphere system. Work on the two-dimensional case with a Lennard-Jones potential is in progress and will be reported in a later paper. Also, the problem in three dimensions is being investigated.

* Now at the Radiation Laboratory of the University of California, Livermore, California.

II. THE GENERAL METHOD FOR AN ARBITRARY POTENTIAL BETWEEN THE PARTICLES

In order to reduce the problem to a feasible size for numerical work, we can, of course, consider only a finite number of particles. This number N may be as high as several hundred. Our system consists of a square† containing N particles. In order to minimize the surface effects we suppose the complete substance to be periodic, consisting of many such squares, each square containing N particles in the same configuration. Thus we define d_{AB}, the minimum distance between particles A and B, as the shortest distance between A and any of the particles B, of which there is one in each of the squares which comprise the complete substance. If we have a potential which falls off rapidly with distance, there will be at most one of the distances AB which can make a substantial contribution; hence we need consider only the minimum distance d_{AB}.

† We will use the two-dimensional nomenclature here since it is easier to visualize. The extension to three dimensions is obvious.

Our method in this respect is similar to the cell method except that our cells contain several hundred particles instead of one. One would think that such a sample would be quite adequate for describing any one-phase system. We do find, however, that in two-phase systems the surface between the phases makes quite a perturbation. Also, statistical fluctuations may be sizable.

If we know the positions of the N particles in the square, we can easily calculate, for example, the potential energy of the system,

$$E=\tfrac{1}{2}\sum_{i=1}^{N}\sum_{\substack{j=1\\ i\neq j}}^{N} V(d_{ij}). \tag{1}$$

(Here V is the potential between molecules, and d_{ij} is the minimum distance between particles i and j as defined above.)

In order to calculate the properties of our system we use the canonical ensemble. So, to calculate the equilibrium value of any quantity of interest F,

$$\bar{F}=\left[\int F\exp(-E/kT)d^{2N}pd^{2N}q\right]\Big/\left[\int \exp(-E/kT)d^{2N}pd^{2N}q\right], \tag{2}$$

where $(d^{2n}pd^{2n}q)$ is a volume element in the $4N$-dimensional phase space. Moreover, since forces between particles are velocity-independent, the momentum integrals may be separated off, and we need perform only the integration over the $2N$-dimensional configuration space. It is evidently impractical to carry out a several hundred-dimensional integral by the usual numerical methods, so we resort to the Monte Carlo method.‡ The Monte Carlo method for many-dimensional integrals consists simply of integrating over a random sampling of points instead of over a regular array of points.

Thus the most naive method of carrying out the integration would be to put each of the N particles at a random position in the square (this defines a random point in the $2N$-dimensional configuration space), then calculate the energy of the system according to Eq. (1), and give this configuration a weight $\exp(-E/kT)$. This method, however, is not practical for close-packed configurations, since with high probability we choose a configuration where $\exp(-E/kT)$ is very small; hence a configuration of very low weight. So the method we employ is actually a modified Monte Carlo scheme, where, instead of choosing configurations randomly, then weighting them with $\exp(-E/kT)$, we choose configurations with a probability $\exp(-E/kT)$ and weight them evenly.

This we do as follows: We place the N particles in any configuration, for example, in a regular lattice. Then we move each of the particles in succession according to the following prescription:

$$\begin{aligned} X&\rightarrow X+\alpha\xi_1\\ Y&\rightarrow Y+\alpha\xi_2, \end{aligned} \tag{3}$$

where α is the maximum allowed displacement, which for the sake of this argument is arbitrary, and ξ_1 and ξ_2 are random numbers§ between (-1) and 1. Then, after we move a particle, it is equally likely to be anywhere within a square of side 2α centered about its original position. (In accord with the periodicity assumption, if the indicated move would put the particle outside the square, this only means that it re-enters the square from the opposite side.)

We then calculate the change in energy of the system ΔE, which is caused by the move. If $\Delta E<0$, i.e., if the move would bring the system to a state of lower energy, we allow the move and put the particle in its new position. If $\Delta E>0$, we allow the move with probability $\exp(-\Delta E/kT)$; i.e., we take a random number ξ_3 between 0 and 1, and if $\xi_3<\exp(-\Delta E/kT)$, we move the particle to its new position. If $\xi_3>\exp(-\Delta E/kT)$, we return it to its old position. Then, whether the move has been allowed or not, i.e., whether we are in a different configuration or in the original configuration, we consider that we are in a new configuration for the purpose of taking our averages. So

$$\bar{F}=(1/M)\sum_{j=1}^{M}F_j, \tag{4}$$

where F_j is the value of the property F of the system after the jth move is carried out according to the complete prescription above. Having attempted to move a particle we proceed similarly with the next one.

We now prove that the method outlined above does choose configurations with a probability $\exp(-E/kT)$. Since a particle is allowed to move to any point within a square of side 2α with a finite probability, it is clear that a large enough number of moves will enable it to reach any point in the complete square.|| Since this is true of all particles, we may reach any point in configuration space. Hence, the method is ergodic.

Next consider a very large ensemble of systems. Suppose for simplicity that there are only a finite number of states¶ of the system, and that ν_r is the number of

‡ This method has been proposed independently by J. E. Mayer and by S. Ulam. Mayer suggested the method as a tool to deal with the problem of the liquid state, while Ulam proposed it as a procedure of general usefulness. B. Alder, J. Kirkwood, S. Frankel, and V. Lewinson discussed an application very similar to ours.

§ It might be mentioned that the random numbers that we used were generated by the middle square process. That is, if ξ^n is an m digit random number, then a new random number ξ_{n+1} is given as the middle m digits of the complete $2m$ digit square of ξ_n.

|| In practice it is, of course, not necessary to make enough moves to allow a particle to diffuse evenly throughout the system since configuration space is symmetric with respect to interchange of particles.

¶ A state here means a given point in configuration space.

Notes on p. 15

systems of the ensemble in state r. What we must prove is that after many moves the ensemble tends to a distribution

$$\nu_r \propto \exp(-E_r/kT).$$

Now let us make a move in all the systems of our ensemble. Let the *a priori* probability that the move will carry a system in state r to state s be P_{rs}. [By the *a priori* probability we mean the probability before discriminating on $\exp(-\Delta E/kT)$.] First, it is clear that $P_{rs}=P_{sr}$, since according to the way our game is played a particle is equally likely to be moved anywhere within a square of side 2α centered about its original position. Thus, if states r and s differ from each other only by the position of the particle moved and if these positions are within each other's squares, the transition probabilities are equal; otherwise they are zero. Assume $E_r > E_s$. Then the number of systems moving from state r to state s will be simply $\nu_r P_{rs}$, since all moves to a state of lower energy are allowed. The number moving from s to r will be $\nu_s P_{sr} \exp(-(E_r-E_s)/kT)$, since here we must weigh by the exponential factor. Thus the net number of systems moving from s to r is

$$P_{rs}(\nu_s \exp(-(E_r-E_s)/kT)-\nu_r). \tag{5}$$

So we see that between any two states r and s, if

$$(\nu_r/\nu_s) > [\exp(-E_r/kT)/\exp(-E_s/kT)], \tag{6}$$

on the average more systems move from state r to state s. We have seen already that the method is ergodic; i.e., that any state can be reached from any other, albeit in several moves. These two facts mean that our ensemble must approach the canonical distribution. It is, incidentally, clear from the above derivation that after a forbidden move we must count again the initial configuration. Not to do this would correspond in the above case to removing from the ensemble those systems which tried to move from s to r and were forbidden. This would unjustifiably reduce the number in state s relative to r.

The above argument does not, of course, specify how rapidly the canonical distribution is approached. It may be mentioned in this connection that the maximum displacement α must be chosen with some care; if too large, most moves will be forbidden, and if too small, the configuration will not change enough. In either case it will then take longer to come to equilibrium.

For the rigid-sphere case, the game of chance on $\exp(-\Delta E/kT)$ is, of course, not necessary since ΔE is either zero or infinity. The particles are moved, one at a time, according to Eq. (3). If a sphere, after such a move, happens to overlap another sphere, we return it to its original position.

III. SPECIALIZATION TO RIGID SPHERES IN TWO DIMENSIONS

A. The Equation of State

The virial theorem of Clausius can be used to give an equation of state in terms of $\bar{n}$, the average density of other particles at the surface of a particle. Let $\mathbf{X}_i^{(\text{tot})}$ and $\mathbf{X}_i^{(\text{int})}$ represent the total and the internal force, respectively, acting on particle i, at a position $\mathbf{r}_i$. Then the virial theorem can be written

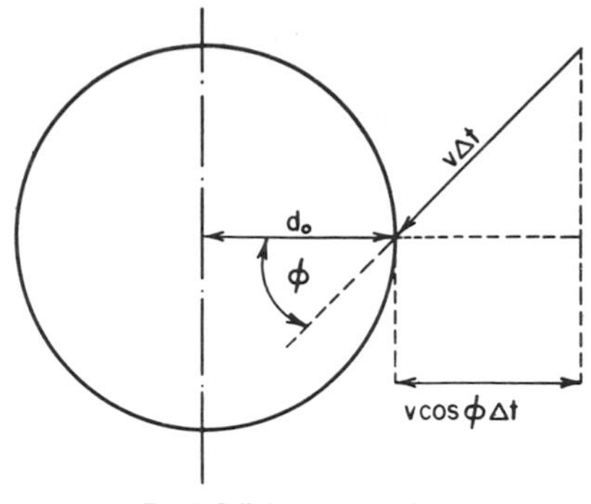

FIG. 1. Collisions of rigid spheres.

$$\langle \sum_i \mathbf{X}_i^{(\text{tot})}\cdot \mathbf{r}_i\rangle_{\text{Av}} = 2PA + \langle \sum_i \mathbf{X}_i^{(\text{int})}\cdot \mathbf{r}_i\rangle_{\text{Av}} = 2E_{\text{kin}}. \tag{7}$$

Here P is the pressure, A the area, and E_{kin} the total kinetic energy,

$$E_{\text{kin}} = Nm\bar{v}^2/2$$

of the system of N particles.

Consider the collisions of the spheres for convenience as represented by those of a particle of radius d_0, twice the radius of the actual spheres, surrounded by $\bar{n}$ point particles per unit area. Those surrounding particles in an area of $2\pi d_0 v \cos\phi \Delta t$, traveling with velocity v at an angle ϕ with the radius vector, collide with the central particle provided $|\phi| < \pi/2$. (See Fig. 1.) Assuming elastic recoil, they each exert an average force during the time Δt on the central particle of

$$2mv \cos\phi/\Delta t.$$

One can see that all ϕ's are equally probable, since for any velocity-independent potential between particles the velocity distribution will just be Maxwellian, hence isotropic. The total force acting on the central particle, averaged over ϕ, over time, and over velocity, is

$$\bar{F}_i = m\bar{v}^2 \pi d_0 \bar{n}. \tag{8}$$

The sum

$$\langle \sum_i \mathbf{X}_i^{(\text{int})}\cdot \mathbf{r}_i\rangle_{\text{Av}}$$

is

$$-\tfrac{1}{2}\sum_i\{\sum_{\substack{j\\ i\neq j}} r_{ij}F_{ij}\},$$

with F_{ij} the magnitude of the force between two particles and r_{ij} the distance between them. We see that

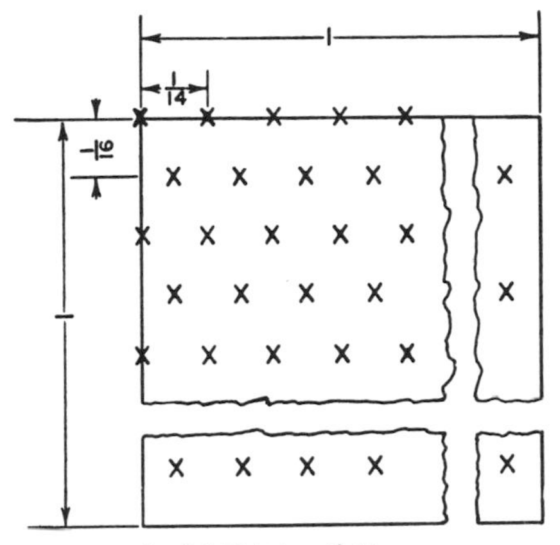

FIG. 2. Initial trigonal lattice.

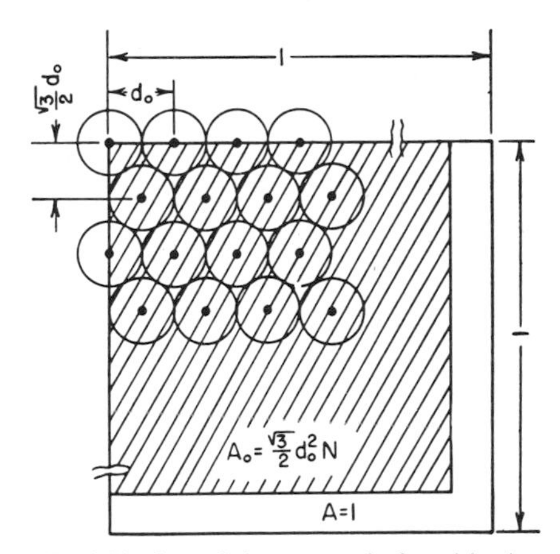

FIG. 3. The close-packed arrangement for determining A_0.

$r_{ij} = d_0$ and $\sum_j F_{ij}$ is given by Eq. (8), so we have

$$\langle \sum_i \mathbf{X}_i^{(\text{int})} \cdot \mathbf{r}_i \rangle_{\text{Av}} = -(N m \bar{v}^2 / 2) \pi d_0^2 \bar{n}. \quad (9)$$

Substitution of (9) into (7) and replacement of $(N/2) m\bar{v}^2$ by E_{kin} gives finally

$$PA = E_{\text{kin}}(1 + \pi d_0^2 \bar{n}/2) \equiv NkT(1 + \pi d_0^2 \bar{n}/2). \quad (10)$$

This equation shows that a determination of the one quantity $\bar{n}$, according to Eq. (4) as a function of A, the area, is sufficient to determine the equation of state for the rigid spheres.

B. The Actual Calculation of $\bar{n}$

We set up the calculation on a system composed of $N = 224$ particles ($i = 0, 1 \cdots 223$) placed inside a square of unit side and unit area. The particles were arranged initially in a trigonal lattice of fourteen particles per row by sixteen particles per column, alternate rows being displaced relative to each other as shown in Fig. 2. This arrangement gives each particle six nearest neighbors at approximately equal distances of $d = 1/14$ from it.

Instead of performing the calculation for various areas A and for a fixed distance d_0, we shall solve the equivalent problem of leaving $A = 1$ fixed and changing d_0. We denote by A_0 the area the particles occupy in close-packed arrangement (see Fig. 3). For numerical convenience we defined an auxiliary parameter ν, which we varied from zero to seven, and in terms of which the ratio (A/A_0) and the forbidden distance d_0 are defined as follows:

$$d_0 = d(1 - 2^{\nu - 8}), \quad d = (1/14), \quad (11a)$$

$$(A/A_0) = 1/(3^{\frac{1}{2}} d_0^2 N/2) = 1/0.98974329(1 - 2^{\nu - 8})^2. \quad (11b)$$

The unit cell is a parallelogram with interior angle 60°, side d_0, and altitude $3^{\frac{1}{2}} d_0/2$ in the close-packed system.

Every configuration reached by proceeding according to the method of the preceding section was analyzed in terms of a radial distribution function $N(r^2)$. We chose a $K > 1$ for each ν and divided the area between πd_0^2 and $K^2 \pi d_0^2$ into sixty-four zones of equal area ΔA^2,

$$\Delta A^2 = (K^2 - 1)\pi d_0^2/64.$$

We then had the machine calculate for each configuration the number of pairs of particles N_m ($m = 1, 2, \cdots 64$) separated by distances r which satisfy

$$(m-1)\Delta A^2 + \pi d_0^2 < \pi r^2 \leqslant m \Delta A^2 + \pi d_0^2. \quad (12)$$

The N_m were averaged over successive configurations according to Eq. (4), and after every sixteen cycles (a cycle consists of moving every particle once) were extrapolated back to $r^2 = d_0^2$ to obtain $N_{\frac{1}{2}}$. This $N_{\frac{1}{2}}$ differs from $\bar{n}$ in Eq. (10) by a constant factor depending on N and K.

The quantity K was chosen for each ν to give reasonable statistics for the N_m. It would, of course, have been possible by choosing fairly large K's, with perhaps a larger number of zones, to obtain $N(r^2)$ at large distances. The oscillatory behavior of $N(r^2)$ at large distances is of some interest. However, the time per cycle goes up fairly rapidly with K and with the number of zones in the distance analysis. For this reason only the behavior of $N(r^2)$ in the neighborhood of d_0^2 was investigated.

The maximum displacement α of Eq. (3) was set to $(d - d_0)$. About half the moves in a cycle were forbidden by this choice, and the initial approach to equilibrium from the regular lattice was fairly rapid.

Notes on p. 15

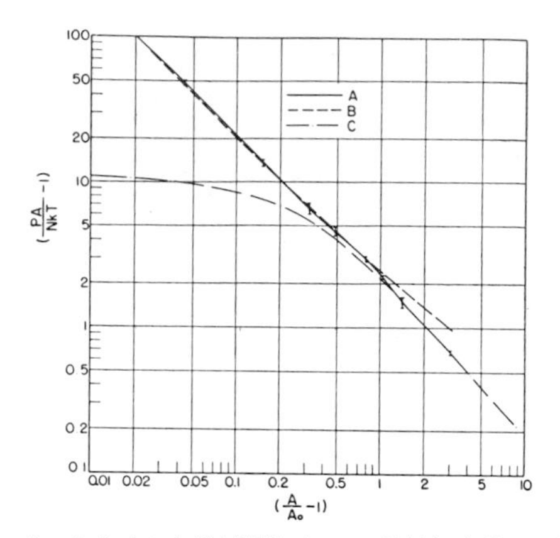

FIG. 4. A plot of $(PA/NkT)-1$ *versus* $(A/A_0)-1$. Curve A (solid line) gives the results of this paper. Curves B and C (dashed and dot-dashed lines) give the results of the free volume theory and of the first four virial coefficients, respectively.

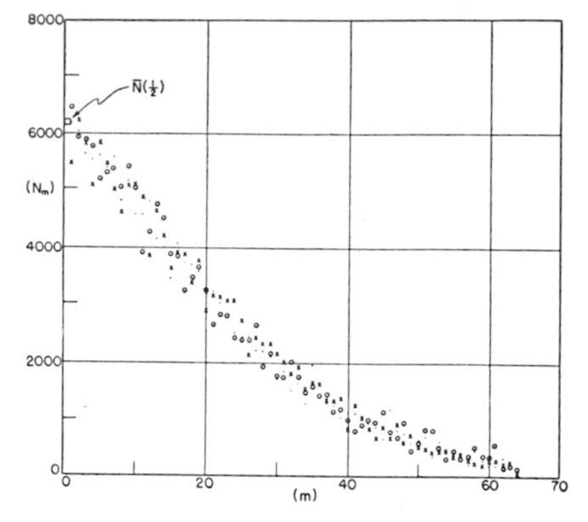

FIG. 5. The radial distribution function N_m for $\nu=5$, $(A/A_0)=1.31966$, $K=1.5$. The average of the extrapolated values of $N_{\frac{1}{2}}$ in $\bar{N}_{\frac{1}{2}}=6301$. The resultant value of $(PA/NkT)-1$ is $64\bar{N}_{\frac{1}{2}}/N^2(K^2-1)$ or 6.43. Values after 16 cycles, ●; after 32, ×; and after 48, ○.

IV. NUMERICAL RESULTS FOR RIGID SPHERES IN TWO DIMENSIONS

We first ran for something less than sixteen cycles in order to get rid of the effects of the initial regular configuration on the averages. Then about forty-eight to sixty-four cycles were run at

$$\nu=2,\ 4,\ 5,\ 5.5,\ 6,\ 6.25,\ 6.5,\ \text{and}\ 7.$$

Also, a smaller amount of data was obtained at $\nu=0$, 1, and 3. The time per cycle on the Los Alamos MANIAC is approximately three minutes, and a given point on the pressure curve was obtained in four to five hours of running. Figure 4 shows $(PA/NkT)-1$ *versus* $(A/A_0)-1$ on a log-log scale from our results (curve A), compared to the free volume equation of Wood[1] (curve B) and to the curve given by the first four virial coefficients (curve C). The last two virial coefficients were obtained by straightforward Monte Carlo integration on the MANIAC (see Sec. V). It is seen that the agreement between curves A and B at small areas and between curves A and C at large areas is good. Deviation from the free volume theory begins with a fairly sudden break at $\nu=6(A/A_0\simeq 1.8)$.

A sample plot of the radial distribution function for $\nu=5$ is given in Fig. 5. The various types of points represent values after sixteen, thirty-two, and forty-eight cycles. For $\nu=5$, least-square fits with a straight line to the first sixteen N_m values were made, giving extrapolated values of $N_{\frac{1}{2}}^{(1)}=6367$, $N_{\frac{1}{2}}^{(2)}=6160$, and $N_{\frac{1}{2}}^{(3)}=6377$. The average of these three was used in constructing PA/NkT. In general, least-square fits of the first sixteen to twenty N_m's by means of a parabola, or, where it seemed suitable, a straight line, were made. The errors indicated in Fig. 4 are the root-mean-square deviations for the three or four $N_{\frac{1}{2}}$ values. Our average error seemed to be about 3 percent.

Table I gives the results of our calculations in numerical form. The columns are ν, A/A_0, $(PA/NkT)-1$, and, for comparison purposes, $(PA/NkT-1)$ for the free volume theory and for the first four coefficients in the virial coefficient expansion, in that order, and finally PA_0/NkT from our results.

[1] William W. Wood, J. Chem. Phys. **20**, 1334 (1952).

V. THE VIRIAL COEFFICIENT EXPANSION

One can show[2] that

$$(PA/NkT)-1=C_1(A_0/A)+C_2(A_0/A)^2+C_3(A_0/A)^3+C_4(A_0/A)^4+0(A_0/A)^5,$$

$$C_1=\pi/3^{\frac{1}{2}},\quad C_2=4\pi^2A_{3,3}/9,$$

$$C_3=\pi^3(6A_{4,5}-3A_{4,4}-A_{4,6})/3^{\frac{3}{2}},$$

$$C_4=(8\pi^3/135)\cdot[12A_{5,5}-60A_{5,6}'-10A_{5,6}''+30A_{5,7}'+60A_{5,7}''+10A_{5,7}'''-30A_{5,8}'-15A_{5,8}''+10A_{5,9}-A_{5,10}].\quad (13)$$

TABLE I. Results of this calculation for $(PA/NkT)-1=X_1$ compared to the free volume theory (X_2) and the four-term virial expansion (X_3). Also (PA_0/NkT) from our calculations.

ν	(A/A_0)	X_1	X_2	X_3	(PA_0/NkT)
2	1.04269	49.17	47.35	9.77	48.11
4	1.14957	13.95	13.85	7.55	13.01
5	1.31966	6.43	6.72	5.35	5.63
5.5	1.4909	4.41	4.53	4.02	3.63
6	1.7962	2.929	2.939	2.680	2.187
6.25	2.04616	2.186	2.323	2.065	1.557
6.5	2.41751	1.486	1.802	1.514	1.028
7	4.04145	0.6766	0.990	0.667	0.4149

[2] J. E. Mayer and M. G. Mayer, *Statistical Mechanics* (John Wiley and Sons, Inc., New York, 1940), pp. 277–291.

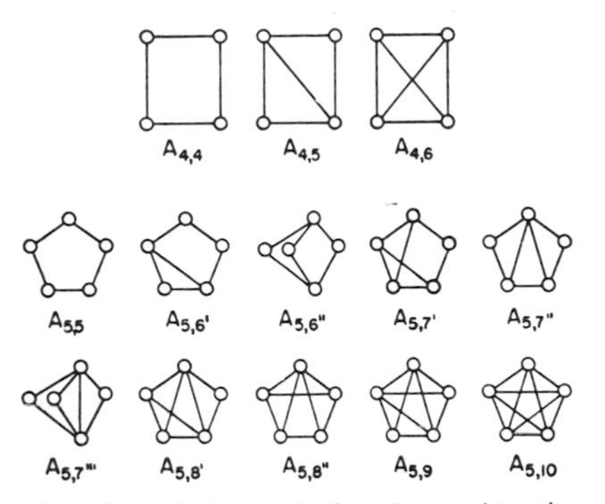

FIG. 6. Schematic diagrams for the various area integrals.

The coefficients $A_{i,k}$ are cluster integrals over configuration space of i particles, with k bonds between them. In our problem a bond is established if the two particles overlap. The cluster integral is the volume of configuration space for which the appropriate bonds are established. If k bonds can be distributed over the i particles in two or more different ways without destroying the irreducibility of the integrals, the separate cases are distinguished by primes. For example, A_{33} is given schematically by the diagram

and mathematically as follows: if we define $f(r_{ij})$ by

$$f(r_{ij})=1 \quad \text{if} \quad r_{ij}<d,$$
$$f(r_{ij})=0 \quad \text{if} \quad r_{ij}>d,$$

then

$$A_{3,3}=\frac{1}{\pi^2 d^4}\int\cdots\int dx_1dx_2dx_3dy_1dy_2dy_3(f_{12}f_{23}f_{31}).$$

The schematics for the remaining integrals are indicated in Fig. 6.

The coefficients $A_{3,3}$, $A_{4,4}$, and $A_{4,5}$ were calculated algebraically, the remainder numerically by Monte Carlo integration. That is, for $A_{5,5}$ for example, particle 1 was placed at the origin, and particles 2, 3, 4, and 5 were put down at random, subject to $f_{12}=f_{23}=f_{34}=f_{15}=1$. The number of trials for which $f_{45}=1$, divided by the total number of trials, is just $A_{5,5}$.

The data on $A_{4,6}$ is quite reliable. We obtained

$$A_{4,6}/A_{4,4}=0.752(\pm 0.002).$$

However, because of the relatively large positive and negative terms in C_4 of Eq. (13), the coefficient C_4, being a small difference, is less accurate. We obtained

$$C_4=8\pi^3(0.585)/135 \quad (\pm\sim 5 \text{ percent}).$$

Our final formula is

$$(PA/NkT)-1=1.813799(A_0/A)+2.57269(A_0/A)^2+3.179(A_0/A)^3+3.38(A_0/A)^4+0(A_0/A)^5. \quad (14)$$

This formula is plotted in curve C of Fig. 4 and tabulated for some values of (A/A_0) in column 5 of Table I. It is seen in Fig. 4 that the curves agrees very well with our calculated equation of state for $(A/A_0)>2.5$. In this region both the possible error in our last virial coefficients and the contribution of succeeding terms in the expansion are quite small (less than our probable statistical error) so that the virial expansion should be accurate.

VI. CONCLUSION

The method of Monte Carlo integrations over configuration space seems to be a feasible approach to statistical mechanical problems which are as yet not analytically soluble. At least for a single-phase system a sample of several hundred particles seems sufficient. In the case of two-dimensional rigid spheres, runs made with 56 particles and with 224 particles agreed within statistical error. For a computing time of a few hours with presently available electronic computers, it seems possible to obtain the pressure for a given volume and temperature to an accuracy of a few percent.

In the case of two-dimensional rigid spheres our results are in agreement with the free volume approximation for $A/A_0<1.8$ and with a five-term virial expansion for $A/A_0>2.5$. There is no indication of a phase transition.

Work is now in progress for a system of particles with Lennard-Jones type interactions and for three-dimensional rigid spheres.

Notes to Reprint I.1

1. The Monte Carlo calculations referred to in the first note on p. 1088 were later published (Alder et al., J. Chem. Phys. 23 (1955) 417).

2. The optimal choice of acceptance rate for Monte Carlo trial moves depends both on the model under investigation (hard-core or continuous interactions) and the nature of the computer program (scalar or vector code). On a scalar computer, the "best" acceptance rate for hard-core systems is typically a factor two to three times smaller than for systems with continuous interactions. The reason is the fact that in a hard-core model a trial move can be rejected as soon as an overlap is detected; rejected moves therefore require less time, on average, than accepted

ones. For continuous potentials, by contrast, all interactions have to be evaluated before a move can be accepted or rejected; thus accepted and rejected moves take equal amounts of computer time. On a vector computer, the situation is different. The structure of such machines makes it efficient to compute all interactions in a single, vectorized loop, and to decide only at the end whether the move is to be accepted or not. In this case, accepted and rejected moves always require the same time.

THE JOURNAL OF CHEMICAL PHYSICS VOLUME 27, NUMBER 3 SEPTEMBER, 1957

Monte Carlo Equation of State of Molecules Interacting with the Lennard-Jones Potential. I. A Supercritical Isotherm at about Twice the Critical Temperature*

W. W. WOOD AND F. R. PARKER†
Los Alamos Scientific Laboratory, Los Alamos, New Mexico
(Received May 2, 1957)

Values obtained by Monte Carlo calculations are reported for the compressibility factor, excess internal energy, excess constant-volume heat capacity, and the radial distribution function of Lennard-Jones (12,6) molecules at the reduced temperature $kT/\epsilon^*=2.74$, and at thirteen volumes between $v/v^*=0.75$ and 7.5. (v is the molar volume; $v^*=2^{-\frac{1}{2}}N_0r^{*3}$; N_0 is Avogadro's number; ϵ^* is the depth, and r^* the radius of the Lennard-Jones potential well.) The results are compared with the experimental observations of Michels (~150–2000 atmos) and Bridgman (~2000–15 000 atmos) on argon at 55°C, using Michels' second virial coefficient values for the potential parameters. Close agreement with Michels is found, but significant disagreement with Bridgman. The Monte Carlo calculations display the fluid-solid transition; the transition pressure and the volume and enthalpy increments are not precisely determined. The Lennard-Jones-Devonshire cell theory gives results which disagree throughout the fluid phase, but agree on the solid branch of the isotherm. Limited comparisons with the Kirkwood-Born-Green results indicate that the superposition approximation yields useful results at least up to $v/v^*=2.5$.

INTRODUCTION

THE Monte Carlo method for obtaining the equation of state of a system of interacting particles was devised by Metropolis *et al.*[1] and applied by them to the case of hard spheres in two dimensions and to hard spheres in three dimensions by Rosenbluth and Rosenbluth.[2] The latter investigation also presented qualitative results for Lennard-Jones molecules in two dimensions. Here we shall present quantitative results for three-dimensional molecules interacting in pairs according to the Lennard-Jones (LJ) potential

$$u_{\mathrm{LJ}}(r)=\epsilon^*\left[\left(\frac{r}{r^*}\right)^{-12}-2\left(\frac{r}{r^*}\right)^{-6}\right]; \qquad (1)$$

$u_{\mathrm{LJ}}(r)$ is the potential energy of interaction of two molecules at the distance r. This paper will discuss an isotherm at about twice the critical temperature, specifically one for which the reduced temperature $\theta=kT/\epsilon^*$ has the value 2.74. The range of the reduced volume $\tau=v/v^*$ is from 0.75 to 7.5; see the abstract for the definition of these quantities. Additional results for the isotherms $\theta=1$, 5, 20, and 100 are being prepared for publication.

This program, which has been carried out on IBM type 701 and 704 electronic calculators, was undertaken in order to establish the feasibility of the Monte Carlo method for nonsingular potentials such as the Lennard-Jones potential, as contrasted with the hard-sphere potential in which there is no long-range interaction. In addition, two other objectives have determined the course of our investigation. It is of course desirable to compare the Monte Carlo results with those of other theories, such as the Lennard-Jones and Devonshire cell theory[3] and the Kirkwood-Born-Green theory in the superposition approximation.[4,5] Also, it seemed of interest to compare the calculated results with experimental observations of the equation of state of molecules for which a simple potential would be reasonably realistic. The Lennard-Jones potential was thought to be a reasonable compromise, inasmuch as most of the existing statistical mechanical theories have been

* Work performed under the auspices of the U. S. Atomic Energy Commission.

† Present address: Department of Mathematics, University of Wisconsin, Madison, Wisconsin.

[1] Metropolis, Rosenbluth, Rosenbluth, Teller, and Teller, J. Chem. Phys. **21**, 1087 (1953).

[2] M. N. Rosenbluth and A. W. Rosenbluth, J. Chem. Phys. **22**, 881 (1954).

[3] See, e.g., Wentorf, Buehler, Hirschfelder, and Curtiss, J. Chem. Phys. **18**, 1484 (1950). The numerical values used here were calculated by Dr. Wildon Fickett of this Laboratory (LASL), from the Lennard-Jones-Devonshire theory as modified by Wentorf *et al.*

[4] Kirkwood, Lewinson, and Alder, J. Chem. Phys. **20**, 929 (1952).

[5] Zwanzig, Kirkwood, Stripp, and Oppenheim, J. Chem. Phys. **21**, 1268 (1953). The values used here include the correction for the hard core, but not the empirical scale factor c introduced by Zwanzig *et al.*

applied to it, and since it is reasonably realistic. In addition, it has the advantage of being inherently faster (as regards calculator time) than some possibly more realistic alternatives such as the exp-six potential.

The most reasonable choices of experimental comparison fluids are clearly the rare gases, of which argon has been the most studied. Since it was our intention to establish internally (insofar as possible) the feasibility of the Monte Carlo method as a valid statistical mechanical procedure for potentials of the LJ type, the comparison with experiment is viewed as being mostly a test of the assumptions regarding the intermolecular potential. This being the case, *we have introduced no adjustable parameters*. The values of the potential constants for argon have been taken throughout as $\epsilon^*/k = 119.76$°K, $r^* = 3.822$ A, $v^* = 23.79$ cm³/mole, as determined by Michels[6] from second virial coefficient data. Using these values, the reduced temperature $\theta = 2.74$ corresponds to a temperature of 55°C for argon, and was so chosen in order to permit comparison with Bridgman's[7] high pressure data at this temperature. This temperature is also in the region of Michels'[6] medium pressure data.

In the following parts of this paper, we first sketch a somewhat more elaborate proof than given in reference 1 of the equivalence of the general Monte Carlo method and the classical-mechanical petite canonical ensemble of Gibbs, using the theory of Markov chains. Next we discuss the approximations which must be introduced into the Monte Carlo method in order to adapt it to the capabilities of even the fastest modern calculators. Then follow the presentation of the present results and comparisons with the experimental data for argon and with the other theories.

THEORY

The Monte Carlo method (throughout this paper, we will mean by this, the Metropolis-Rosenbluth-Teller method) is a means of estimating the configurational or "excess" contributions to the thermodynamic properties of a classical statistical mechanical system; the momenta make their usual classical contributions. Applied to a system of N such molecules confined in a volume V at temperature T, the method consists in the generation of a Markov chain with constant transition probabilities,[8] in which the states of the chain are points in the $3N$-dimensional configuration space of the system. It is convenient in the theoretical discussion to consider this space to be subdivided into a sufficiently large number S of cells, such that the positions of the N molecules are specified by giving the single number associated with the cell into which its representative point falls. This simplification permits us to utilize the theory of discrete chains; it is physically reasonable that a sufficiently fine subdivision of configuration space should give results indistinguishable from a continuum of possible states, and in any case such a subdivision is inherent in numerical calculation, the fineness depending on the number of binary bits used to represent the position coordinates of the molecules. Thus, a single integer $k = 1, 2 \cdots S$ suffices to specify a state in the chain; associated with each such integer are other variables such as U_k, the potential energy of configuration k, etc. In order to avoid circumlocution, we will adopt the usual terminology in which successive states of the chain are said to occur at successive instants of time. It must be emphasized that no *physical* time is involved; there are no molecular velocities involved. The succession of configurations at consecutive instants of "time" is of course intimately related to calculating machine time, as the calculator generates the chain of configuration states one after the other by the procedure to be described.

The object of the Monte Carlo method is to generate a Markov chain in which asymptotically (indefinitely longer chain) each state k recurs with a frequency proportional to the Boltzmann factor $\exp(-U_k/kT)$ for that state. Then the average over the chain of any function of the configuration state, such as U_k, in which each occurrence of any state is given equal weight, will converge to the corresponding petite canonical ensemble average of the same quantity as the chain length increases. In order for a Markov chain to have this property it is sufficient[8] that the fundamental one-step transition probabilities satisfy certain conditions. Let $p_{jk} \equiv p_{jk}^{(1)}$ be the conditional probability that if the system be in state j at time t, it will at time $t+1$ be in state k. This matrix of transition probabilities is independent of the time t, and in conjunction with the specification of the state at $t=0$ uniquely defines a particular Markov process. The p_{jk} must satisfy the normalization conditions

$$\sum_{k=1}^{S} p_{jk} = 1, \quad j = 1, 2 \cdots S. \tag{2}$$

Higher order, or multistep, transition probabilities can be defined, such as the probability $p_{jk}^{(2)}$ that if the system be in state j at time t, then it will be in state k at time $t+2$; this probability is given by

$$p_{jk}^{(2)} = \sum_{k'=1}^{S} p_{jk'}^{(1)} p_{k'k}^{(1)}; \tag{3}$$

and analogously

$$p_{jk}^{(n)} = \sum_{k'=1}^{S} P_{jk'}^{(n-1)} p_{k'k}^{(1)}, \tag{4}$$

for the probability after n steps. It is the asymptotic behavior of $p_{jk}^{(n)}$ as n increases indefinitely in which we are interested. The following theorem from the theory of Markov chains[8] concerns this limiting behavior. Provided that all states are *ergodic* and in the

[6] Michels, Wijker, and Wijker, Physica **15**, 627 (1949).
[7] P. W. Bridgman, Proc. Am. Acad. Arts Sci. **70**, 1 (1935).
[8] W. Feller, *Probability Theory and Its Applications* (John Wiley and Sons, Inc., New York, 1950), Chap. 15.

same *class*, the limits

$$\lim_{n\to\infty} p_{jk}^{(n)} = u_k, \quad k=1, 2\cdots S \tag{5}$$

exist for all j and are independent of j; furthermore

$$u_k > 0, \quad k=1, 2\cdots S \tag{6}$$

and

$$\sum_{k=1}^{S} u_k = 1. \tag{7}$$

Also, the limits u_k satisfy the system of linear equations

$$u_k = \sum_{j=1}^{S} u_j p_{jk}, \quad k=1, 2\cdots S; \tag{8}$$

in fact, the limits u_k are uniquely determined by (7) and (8). An ergodic state is a nonperiodic state with finite mean recurrence time (see reference 8 for a complete discussion). For our case in which only a finite number of discrete states is possible, the assumption that all states are members of the same ergodic class will be fulfilled if for any pair j, k there exists a nonvanishing multistep probability of finite order n; that is, there exists a n such that $p_{jk}^{(n)} > 0$. Speaking less formally it is required that any state may follow any other state, not necessarily immediately.

We wish to invert the usual process; that is, we inquire what set of probabilities p_{jk} will converge to previously known set of u_k, namely,

$$u_k = c\exp(-U_k/kT), \quad k=1, 2\cdots S, \tag{9}$$

where c is a normalization constant. As far as (8) is concerned, we note that it will be satisfied identically if the p_{jk} satisfy the condition of microscopic reversibility,

$$u_j p_{jk} = u_k p_{kj}, \quad \text{all } j, k, \tag{10}$$

provided the normalization condition (2) holds. Therefore, if the p_{jk} satisfy (2) and (10) and also the ergodicity condition, then the Markov chain converges to the petite canonical ensemble, in the sense that the various states tend to occur with the frequencies indicated by the Boltzmann factors. The convergence is ultimately independent of the initial state. These are just the conditions formulated earlier.[1,2]

Equation (10) evidently does not determine a unique set p_{jk}. We now proceed to discuss the transition probabilities which we actually use, and afterwards we will consider the ergodic question. The transition probabilities used in most of the present calculations[9] are given by

$$\left.\begin{aligned} p_{jk} &= A_{jk}, && \text{if } U_k \leqslant U_j, \\ &= A_{jk}\exp[-(U_k-U_j)/kT], && \text{if } U_k > U_j, \end{aligned}\right\} k\neq j, \tag{11}$$

$$p_{jj} = 1 - \sum_{k\neq j} p_{jk},$$

where

$$A_{jk} = \frac{1}{8N\delta^3}, \quad \text{if } x_j^{(\alpha,r)} = x_k^{(\alpha,r)},$$

and

$$\begin{aligned} &|x_j^{(\alpha,r')} - x_k^{(\alpha,r')}| < \delta; \\ &r = 1, 2\cdots r'-1, r'+1, \cdots N; \\ &r' = 1, 2\cdots N; \\ &\alpha = 1, 2, 3; \\ &\quad = 0, \text{ otherwise.} \end{aligned} \tag{12}$$

Here $x_j^{(\alpha,r)}$ is the αth, $\alpha=1, 2, 3$, Cartesian coordinate of molecule r in configuration j, etc. In words, the single-step transition probability is nonzero only for configurations which differ in the position of a single molecule r', which may be any one of the N molecules. Furthermore, the differences of the three coordinates of r' between the two configurations must each be less than the fixed parameter δ. Transition probabilities between configurations satisfying this condition (which is symmetrical in the configurations) are determined in unsymmetrical fashion by their potential energies, according to (11). It is readily seen that (2) and (10) are satisfied independently of the value of δ; thus, the Markov chain averages converge to the petite canonical ensemble averages independently of the value of this parameter. However, the latter will be expected to influence the rate of convergence with increasing chain length, since it in part determines the rate of effective configurational "motion."

There remains to be examined the ergodic character of the Markov chain defined by (11) and (12). In the first place, if the potential energy function is finite for all configurations except possibly for those in which two or more molecules have exactly the same positions, then there will be a nonvanishing probability for moving from any configuration to any other in a finite time; a lower bound for this probability, and a corresponding minimum time, can readily be found by simply moving molecules one at a time from their positions in the first configuration to their positions in the second configuration, in each case by the minimum number of time intervals permitted by the parameter δ. Furthermore, a nonvanishing lower bound for the probability of the same succession of configurations occurring in reverse order can be obtained. Thus we conclude, in the case of potential energy functions of this kind, that the chain defined by (11) and (12) is ergodic, and therefore all requirements are met for equivalence of

[9] It should be pointed out that this formulation of p_{jk}, which was suggested by Dr. Marshall Rosenbluth, differs from that used in the previous work.[1,2] In the latter investigations r' in Eq. (12) was determined uniquely by the "time" position of configuration j in the chain, according to the relation $r'=t$ modulo N and an initially assigned ordering of the molecules. A few of our chains were also generated by this procedure, which corresponds to a Markov process with time-dependent transition probabilities. We have not studied the corresponding convergence properties except by a few empirical comparisons of chain generated by both methods, which are discussed later.

Notes on p. 31

the Monte Carlo procedure and the petite canonical ensemble.

For potential energy functions which have infinities over extended regions of phase space, such as is the case for "hard-sphere" or "billiard-ball" molecules, the situation is more complicated. Here it is possible that at sufficiently high densities there may be configurations between which it may be impossible to pass. Even if this is the case we note that if a transition (possibly multistep) is possible from configuration j to configuration k, then the reverse transition from k to j is also possible. Thus the total set of states break up into separate classes, all of which are ergodic, but between which no transitions can occur. If this is the case then the asymptotic behavior will be different for initial states belonging to different ergodic classes. However the corresponding dynamical physical system would presumably have the same compartmentalization of phase space imposed on it, so that this question is really related to the general question of accessibility in statistical mechanics.

The most serious question of this nature then seems to be what we may call the quasi-ergodic problem. That is, the possibility that in some situation of interest configuration space may be divided into two or more regions, all making appreciable contributions to the proper statistical mechanical phase integral, and which are formally members of the same ergodic class, but between which the n-step transition probabilities, though nonzero, are very small except for very large n (the basis of comparison being of course the conveniently attainable chain length). If this situation should arise, the Monte Carlo method will clearly be in danger of giving fallacious results. Clearly this difficulty, if it exists, will tend to be most serious in fairly dense systems. Comparison of the calculated results with experiment can of course be helpful in deciding the question, but the issue will tend to be obscured by uncertainties in the proper potential energy expressions. In addition, the possibility can be tested in a limited way by making perturbations on the initial configuration (see below), and of course by extending the chains to as great lengths as possible.

PERIODIC BOUNDARY CONDITION

Calculator speed seriously limits the number of molecules N which may be considered, as was noted in the earlier work.[1,2] The greater computational complexity of the LJ potential, especially the long-range character of the interaction, has limited the present investigation to systems of 32 and 108 molecules. In order to minimize surface effects with so small a number of molecules, we have used the same periodic boundary condition used in the earlier work.[1,2] This consists in filling three-dimensional space by repetitions of the fundamental cell of volume V, each containing N molecules in the same relative positions as occupied by the fundamental set of N molecules in the fundamental cell in the particular configuration being examined. Figure 1 illustrates this procedure in two dimensions with a system of four molecules. Thus only the motion of the N molecules of the fundamental set need be followed by the calculator, while at the same time the system is somewhat representative of the essentially infinite extent characteristic of ordinary thermodynamic systems. Use of the periodic boundary condition implies that the shape of the Monte Carlo cell of volume V containing the N molecules must be such as to fill space under successive unit translations. This requirement makes the cell similar to the familiar unit cell of crystallography, except the latter is ordinarily chosen to be the smallest of various possible alternatives, which may not be convenient for the Monte Carlo calculation. In the present work this cell is a cube containing 32 or 108 molecules; it may be considered to be made up of cubical arrays of the smallest cubical unit cell, containing 4 molecules, which will generate a face-centered cubic lattice. The most obvious result of imposing the periodic boundary condition is of course the omission of fluctuations in the number of molecules contained in volumes V and greater. A concomitant possibility is a limitation on the ergodic behavior compared to that to be expected from a macroscopic system. Consider, for example, the possibility of two molecules in a quite dense hard sphere system exchanging positions. At close packings this might be impossible when 32 or 108 particles are constrained to be in the fixed volume V. It might occur in a larger system without such a restriction, for instance during a fluctuation in which the number of molecules in the same volume V is reduced.

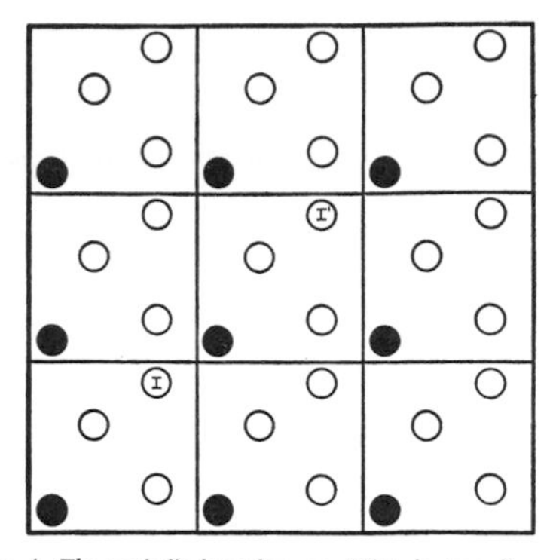

FIG. 1. The periodic boundary condition in two dimensions. The central square represents the fundamental Monte Carlo cell, the others are replicas of it. One of the molecules in the fundamental cell and its eight images are shown as solid circles.

Although not strictly necessary it is obviously desirable, if we hope to obtain useful results for condensed systems (certainly for crystalline systems), to choose the cell shape and its population N so as to permit the periodic boundary condition to generate a perfect lattice from a suitably regular arrangement of the N molecules in the fundamental cell. We note that a cell shape and a population N which are appropriate for one lattice (e.g., the face-centered cubic) will not in general generate a lattice of different type (e.g., hexagonal). In this way the imposition of the periodic boundary condition further limits the ergodic character of the resulting molecular system compared to that of a macroscopic system, since one lattice arrangement cannot be generated from the other. Thus, with the presently feasible number of molecules, we cannot expect the method to give a correct account of the relative importance of alternative lattice arrangements in the structure of a condensed phase, at least not at distances greater than the cell dimensions. In principle, independent calculations based on a face-centered cubic lattice on the one hand, and a hexagonal lattice on the other hand, might resolve the question of the relative stability of the two arrangements of LJ molecules; in actuality the differences seem much too small to be resolved from the inherent statistical fluctuations.

It will be noticed also that the introduction of this approximation results in a certain similarity to the cell theories of the fluid state, the distinguishing feature being of course the large number of molecules per cell. It should be remarked that the boundary condition for motion in the adjacent cells is different from that usually introduced in the cell theories. Other ways of characterizing the effects of the periodic boundary condition are as an incomplete elimination of surface effects, distortion of long wavelength lattice vibrations, or distortion of large clusters. In view of the fair success of the simple cell theories, and on intuitive grounds, it seems likely that in spite of such defects the method should produce useful approximations to the thermodynamic properties of systems of LJ molecules. Empirically, one can attempt to assess its effect by comparing the results obtained with different N, as is done below for the cases of 32 and 108 molecules in the previously described cube. A rather limited investigation of the effect of a different cell shape (one appropriate for the hexagonal lattice) has been made on another isotherm, and will be reported in a future paper. Briefly, although significant differences in the outer part of the radial distribution function were found, the thermodynamic functions did not differ within the statistical fluctuations.

EVALUATION OF THE ENERGY SUMMATION

Although use of the periodic boundary condition allows us to approximate the behavior of an infinite sample, further specifications are required concerning the evaluation of the potential energy of any resulting configuration. If the results of the calculation are to be representative of a macroscopic system of Lennard-Jones molecules, the long-range character of the potential (combined with the small size of the systems which can be studied) requires that the energy sum include more than just the interactions of the N fundamental molecules. It is well known that the lattice energy sum for particles interacting according to the inverse sixth power is slowly convergent. For a face-centered cubic lattice, the contributions to the potential energy per molecule of interactions beyond (and not including) neighbors of order n can be expressed in the form

$$U^{(n)}/N=\epsilon^*(A^{(n)}\tau^{-4}-B^{(n)}\tau^{-2}), \tag{13}$$

with the coefficients given in Table I; the values for $n=0$ are the complete sums including all interactions. The slow convergence of the attractive contribution is evident. For later use and for comparison with these values we also calculate the corresponding potential energy resulting from interaction of each molecule with a uniform distribution *outside* a shell of radius r, which is given by

$$U'(r)/N=2\pi\sqrt{2}\epsilon^*(\tfrac{1}{9}x^{-9}\tau^{-4}-\tfrac{2}{3}x^{-3}\tau^{-2}), \tag{14}$$

where $x=r/a$, a being the nearest neighbor separation. This expression is often used for estimating the previously defined lattice sums by setting $r=r^{(n)}$, where $r^{(n)}$ is chosen to be the radius of a sphere which, if filled at the macroscopic number density, contains as many molecules as are in all shells of the lattice of order n or less. With this correspondence and with

TABLE I. Summation coefficients for the Lennard-Jones potential.

n	$A^{(n)}$	$B^{(n)}$	$a^{(n)}$	$b^{(n)}$	$x^{(n)}$
0	6.0659	14.454	...	...	...
1	0.0659	2.454	0.0934	2.700	1.300
2	0.0191	1.704	0.0299	1.847	1.475
3	0.0026	0.815	0.0026	0.816	1.936
4	0.0011	0.627	0.0012	0.638	2.100
5	0.0004	0.435	0.0004	0.444	2.37

$$\frac{U'(r^{(n)})}{N}=\epsilon^*(a^{(n)}\tau^{-4}-b^{(n)}\tau^{-2}), \tag{15}$$

we obtain the values shown in Table I. We note that $b^{(n)}$ and $a^{(n)}$ rather quickly become useful approximations to $B^{(n)}$ and $A^{(n)}$. This comparison forms the basis for the method of correction to be described later.

Considerations of machine speed obviously require some limitation on the number of interactions to be included in calculating the potential energy of any configuration. The most severe truncation, which is the fastest method and the one we have used most, is that used in the earlier work.[1,2] It may be characterized as the minimum-image distance method, and consists in considering for inclusion in the sum all pair interactions among the fundamental set of N molecules, but for each pair taking as interaction distance the smallest distance between any images of the two molecules. In the ex-

Notes on p. 31

ample of Fig. 1, the interaction between the solid circle and circle I' in the central square would be computed as that between the solid circle in the central square with circle I in the lower left square. Only one such image interaction is calculated for each of the $\frac{1}{2}N(N-1)$ pairs. If the results from this convention are compared with a complete summation of all interactions without restriction (except for use of the periodic boundary condition), it is readily seen that the minimum image distance method includes all interactions at separations less than one-half the cell edge, but for longer range interactions includes a smaller and smaller fraction, decreasing to none beyond the cell diagonal. In terms of the radial distribution function, at relatively low densities it has its normal appearance in the range of smaller distances, but in the range from 0.5 to 1.73 times the cell edge decreases smoothly from values near one to zero. At higher densities, where the true radial distribution function begins to consist of a series of disconnected peaks, the minimum distance convention gives peaks in the range 0.5 to 1.73 times the cell edge which may be highly asymmetric, and of course much reduced in amplitude. For a face-centered cubic lattice, with 32 molecules the edge of the Monte Carlo unit cell is $2\sqrt{2}a$. Thus the minimum-distance convention begins to distort the true structure at $\sqrt{2}a$, which happens to coincide with the second shell of neighbors. With 108 molecules, the Monte Carlo cell edge is $3\sqrt{2}a$, and distortion begins between the 4th and 5th shells.

Since an appreciable increase in calculating speed can be obtained by excluding interactions which fall in the range of distortion (that is, between 0.5 and 1.73 times the cell edge) and since it seemed possible that excluding them entirely might introduce less error than including some of them in unsymmetrical fashion, the majority of the investigations reported here (speaking of those in which the minimum image distance convention was used) excluded all interactions at distances greater than half the cell edge. A few calculations in which this procedure was compared with inclusion of all minimum image distance interactions showed no significant differences in the radial distribution functions at distances less than half the cell edge. With this convention and a 32-molecule system, the 704 calculator generates about 19 000 configurations per hour. With 108 molecules the rate is about 6500 configurations per hour.

Inspection of Table I will suggest that the problem of estimating the contribution of the excluded interactions can become quite serious for the 32-molecule system, as will be discussed below. We have found that a useful procedure in this connection is a so-called "augmented" summation convention, in which *all* interactions between any molecule of the fundamental cell and all the other molecules and their images are included providing they are at distances less than an appropriately chosen maximum. This procedure was suggested by the fact that it will give correct results in the extremes of a static lattice on one hand, and a uniform ideal fluid on the other. Test calculations comparing the minimum image distance convention for 108 molecules and this augmented convention for 32 molecules but including all interactions within the "distortionless" range of the larger system yielded radial distribution functions and thermodynamic functions which were not appreciably different. Under these conditions the augmented convention yields about 6300 configurations per hour. This is in some respects an improvement, timewise, over the 108-molecule-minimum-distance method, since a large number of configurations *per molecule* is undoubtedly very desirable in order to sample configuration space well. On the other hand the larger systems have smaller over-all fluctuations, so that the short-term statistical scatter is smaller and the average over a fixed total number of configurations correspondingly better determined. Also as earlier discussed, certain transition probabilities may be smaller than desirable in the smaller system because of the more severe restraints. Thus the augmented convention is useful only as a means of correcting for truncation of the energy summations of the minimum-image-distance 32-molecule chains.

FURTHER DETAILS OF CALCULATION

The procedure by which the calculator is caused to develop a chain of configurations in accordance with (11) and (12) is briefly as follows. Let the configuration existing at time t be called configuration j. One of the N molecules is selected at random (but uniformly) by use of pseudo-random numbers generated by the calculator; call this molecule r'. Again, by use of pseudo-random numbers, the molecule r' is given a new position in which each of its coordinates differs randomly and uniformly on the interval $(-\delta,\delta)$ from its value in configuration j. Designate the resulting configuration, which differs from j only in the position of molecule r', as configuration k. The potential energies U_j and U_k are then computed and compared. If U_k is less than U_j the configuration at time $t+1$ is taken to be configuration k, and the process begins over again. If the opposite is the case, the exponential appearing in (11) is calculated and compared with another pseudo-random number uniform on the interval (0,1). If the exponential is the smaller, the configuration at $t+1$ is j; if larger, it is k; in either case, the process then repeats.

In practice it is convenient to take the side of the Monte Carlo cell as unit of length. The desired reduced volume τ then determines the magnitude of r^* in these units, while the desired reduced temperature θ becomes a multiplicative factor in the exponential Boltzmann factors, which in practice is absorbed into the scaling of the energy expressions. Little use is made of the floating-point arithmetic operations of the 704. Account must be taken of the indefinite increase of the LJ potential as the separation distance decreases. In the present work, the potential as given by (1) has been modified as follows:

$$u(r)=\infty, \quad r<r_1, \qquad =u_{LJ}(r), \quad r\geqslant r_1. \tag{16}$$

The cutoff point r_1 is set at sufficiently high interaction energies so that their omission cannot be expected to appreciably change the thermodynamic properties of the system. We determined r_1 from the relation

$$u_{LJ}(r_1)=u_{LJ}(a)+40kT. \tag{17}$$

This has apparently been satisfactory, since when it is used the first few subdivisions of the cumulative distribution function (see below) have always been empty.

The total potential energy of a configuration is taken, using the assumption of additivity of pair potentials, as

$$U=\tfrac{1}{2}\sum_{ij}{}' u(r_{ij}), \tag{18}$$

where r_{ij} is the distance between molecules i and j, and where the interpretation of the summation sign has been discussed above. Then the petite ensemble gives the excess molar internal energy E' as

$$E'/RT=\bar{U}/NkT, \tag{19}$$

the excess (constant-volume) molar heat capacity C_v' as

$$\frac{C_v'}{R}=N\left[\left\langle\left(\frac{U}{NkT}\right)^2\right\rangle_{Av}-\left(\frac{\bar{U}}{NkT}\right)^2\right], \tag{20}$$

and the compressibility factor as

$$pv/RT=1-\left(\frac{\bar{\phi}}{3NkT}\right), \tag{21}$$

where

$$\phi=\tfrac{1}{2}\sum_{ij}{}' r_{ij}\frac{du(r_{ij})}{dr_{ij}}. \tag{22}$$

The bars in these formulas indicate the petite canonical ensemble averages or by the previously shown equivalence, the Monte Carlo or Markov chain averages. The codes used cause these quantities to be calculated as the Markov chain is developed, except that the earlier codes did not evaluate the heat capacity as given by (20). For convenience in monitoring the problem the calculator produces, in addition to the grand average of U/NkT used in (19), a sequence of subaverages of U/NkT each over a fixed number of consecutive configurations of the chain. The codes also classify each r_{ij} for every configuration into a suitable number of classes or counters in equal intervals Δr_{ij}^2 and maintain a continual count of the number in each counter. Step by step summation of the counters yields, after averaging over a sufficiently large number of configurations, the cumulative radial distribution function giving the average number $N(r)$ molecules inside a sphere of radius r, excluding the reference molecule at the center. Differentiation of this cumulative distribution function gives the usual radial distribution function:

$$g(r)=\frac{v}{4\pi N_0 r^2}\frac{dN(r)}{dr}. \tag{23}$$

The required numerical differentiation has usually been done using centered two or four-point difference formulas. There is of course statistical fluctuation in $N(r)$ which is magnified upon differentiation, and in addition we have sometimes not used a sufficient number of counters under the steeply rising inside of the first peak to obtain good detail on this portion of g.

The parameter δ appearing in (12), controlling the maximum difference between successive configurations, as previously mentioned affects the rate of convergence but not the finally convergent results. We have not studied in detail the effect of varying δ. In practice we have chosen δ so that successive configurations in the chain are identical approximately one-half the time. A satisfactory estimate has been the smaller of the two values $(a-r_1)/2$ or the rms amplitude of vibration calculated from the nearest-neighbor, harmonic-oscillator approximation to the Lennard-Jones-Devonshire cell theory. The pseudo-random numbers were obtained as appropriate portions of 70 bit numbers generated by the middle square process.

Except where otherwise mentioned, the initial configuration of the Markov chains has been taken as a regular face-centered cubic lattice arrangement. As discussed before, the convergence of the process should be independent of this choice, but we may reasonably expect to obtain more rapid convergence by omitting from the average the first few early configurations which are closely associated with the initial regular arrangement.

CORRECTION OF RESULTS

As mentioned previously, interactions at distances greater than those included in the Monte Carlo summation contribute importantly to the thermodynamic functions, and corrections for their omission must be applied to the direct results if they are adequately to represent the behavior expected of a macroscopic system of Lennard-Jones molecules. The procedure which we have used is as follows. From the calculated cumulative distribution function we choose the largest distance r_c in the "distortionless" region (r_c less than one-half the cell edge unless the augmented convention is used) at which $N(r_c)$ is equal to the sum of an integral number of neighbor shells of the f.c.c. lattice; for $N=32$ this sum is 12, for $N=108$ it is 54. All interactions at distances greater than r_c are removed from the Monte Carlo averages, by use of the calculated radial distri-

Notes on p. 31

bution function and the relations

$$\frac{E'}{RT}=\frac{2\pi N_0}{vkT}\int u(r)g(r)r^2dr, \tag{24}$$

$$\frac{pv}{RT}-1=\frac{2\pi N_0}{3vkT}\int\frac{du(r)}{dr}g(r)r^3dr. \tag{25}$$

Next, two different estimates are made for the contributions of interactions at distances beyond r_c: one, called the "lattice correction," is calculated from (13) (and the analogous expression for pv/RT); the second, called the "fluid correction," is calculated from (14) (and its analog for pv/RT) with $r=r_c$. The discrepancy between these two values is taken as a measure of the systematic uncertainty in the corrected results, unless the appearance of the radial distribution function clearly indicates a preference for one or the other. The appearance also influences, of course, the decision as to whether to use one or the other pairs of values, or some combination of them, to correct the Monte Carlo results.

When the two estimates of the correction are quite different, and there is no clear preference for one or the other, as happens for the $N=32$ systems at the smaller reduced volumes, we cannot obtain a reliable correction without a better estimate of $g(r)$ for $r>r_c$. In this case we obtain the required information from a corresponding $N=108$ or "augmented" 32 chain, with use of the correction procedure just described based on the larger r_c for these systems. Examples of these various methods of correction will be found below. No correction of the calculated C_v'/R values has been attempted, since this would require the triplet correlation function.

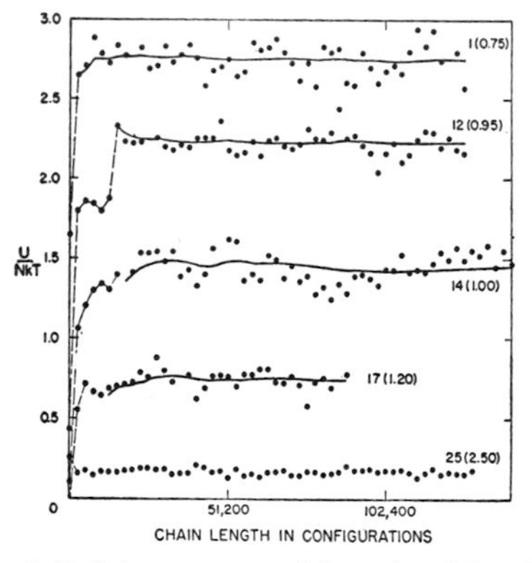

FIG. 2. Typical convergence and fluctuation of the reduced internal energy obtained with 32-molecule Monte Carlo chains at various reduced volumes. The first number in the label for each curve is the chain number from Table II; the second number, in parentheses, is the reduced volume. The solid curves show the variation of the cumulative average with increasing chain length. The circles represent the average over only those configurations generated since the preceding circle. The abscissa should be multiplied by 4 for chain 12. The origin for the ordinates is arbitrary and different for each volume. For chain 25 the solid curve is omitted to avoid confusion.

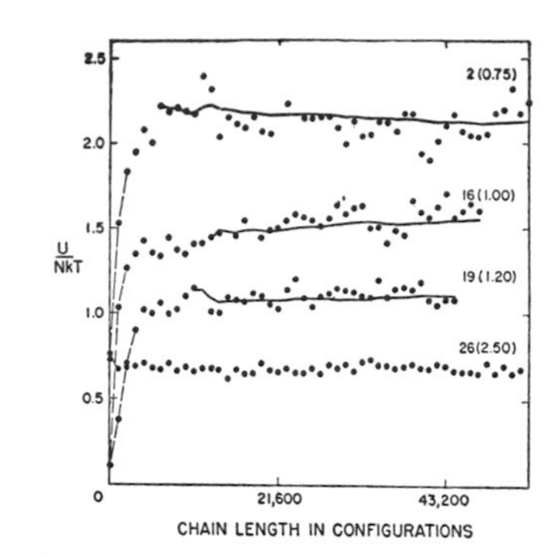

FIG. 3. Typical convergence and fluctuation of the reduced internal energy obtained with 108-molecule Monte Carlo chains at various reduced volumes. See the legend of Fig. 2. The abscissas should be multiplied by 4 for chain 16 and by 2 for chain 19. The origin for the ordinate scale is arbitrary, and different for each volume. For chain 26 the solid curve is omitted to avoid confusion.

DISCUSSION OF RESULTS

The thermodynamic results obtained from the 31 chains calculated on this isotherm are shown in Table II. The extreme values given in columns 5, 6, and 7 are given primarily as indications of the over-all trend of the cumulative averages, and also as relatively crude measures of the statistical reliability of the reported results. On the whole, the agreement of the results from different chains tends to be compatible with the tabulated extremes. At all except the lowest densities, the average giving the excess heat capacity converges very slowly; this may indicate a comparatively slow convergence of the triplet correlation function, or may simply be due to the usual difficulty in estimating a variance by sampling techniques. The convergence of most of the 108-molecule chains is poor, due in part to the longer time required, and in part to our tendency to regard these chains as useful mostly for their radial distribution functions between $x=1.3$ and 2.1. In this region of the radial distribution function, convergence seems to be quite rapid; it is the first peak which equilibrates rather slowly. The two longest 108-molecule chains, at reduced volumes 1.0 and 0.85, show a slight tendency to give higher values of energy and pressure than the corresponding 32-molecule chains, but the differences cannot be said to be significant.

The best empirical comparison of the results of the two types of Markov processes[9] is among the four chains at reduced volume 1.00, where the agreement is seen to be quite good.

Graphs showing the convergence of the average

TABLE II. Equation of state results obtained by Monte Carlo calculation on the Lennard-Jones isotherm $kT/\epsilon^*=2.74$.

Chain number[a]	Reduced volume v/v^*	Number of molecules[b] N	Number of configurations in averages[c] (in thousands; approx)	Reduced excess internal energy (corrected)[d] E'/RT	Compressibility factor (corrected)[d] pv/RT	Reduced excess heat capacity[d] C_v'/R	x_c	Corrections[e] E'/RT	Corrections[e] pv/RT
1 (R)	0.75	32	128	−1.04(−0.01+0.04)	16.47(−0.04+0.13)	1.42(−0.04+0.12)	1.211	−1.55(−1.52;−1.96)	−2.91(−2.88;−3.51)
2 (R)		108	49	−0.97(−0.01+0.05)	16.77(−0.05+0.19)	1.40(−0.25+0.14)	2.102	LF(−0.41;−0.41)	LF(−0.81;−0.82)
3 (R)	0.80	32	256	−1.54(−0.01+0.00)	12.31(−0.03+0.02)	1.25(−0.07+0.14)	1.215	−1.38(−1.34;−1.73)	−2.61(−2.56;−3.16)
4 (R)		32 (A)	77	−1.51(−0.04+0.01)	12.44(−0.16+0.04)	1.28(−0.00+0.14)	2.103	LF(−0.36;−0.36)	LF(−0.71;−0.72)
5 (R)	0.85	32	256	−1.78(−0.00+0.05)	9.57(−0.00+0.22)	1.17(−0.04+0.05)	1.221	−1.24(−1.19;−1.53)	−2.36(−2.30;−2.83)
6 (R)		32 (A)	77	−1.75(−0.01+0.03)	9.72(−0.04+0.13)	1.24(−0.03+0.13)	2.101	LF(−0.32;−0.32)	LF(−0.63;−0.64)
7 (R)		32 (0.96)	204	−1.78(−0.01+0.01)	9.56(−0.04+0.06)	1.28(−0.11+0.02)	1.221	−1.24(−1.19;−1.53)	−2.36(−2.30;−2.83)
8 (R)		32 (0.92)	204	−1.76(−0.01+0.00)	9.68(−0.04+0.03)	1.21(−0.01+0.20)	1.221	−1.24(−1.19;−1.53)	−2.36(−2.30;−2.83)
9 (R)		108	186	−1.73(−0.00+0.03)	9.84(−0.00+0.16)	1.26(−0.03+0.23)	2.094	LF(−0.32;−0.32)	LF(−0.63;−0.64)
10 (R)	0.90	32	261	−1.86(−0.01+0.02)	7.81(−0.03+0.11)	1.30(−0.06+0.02)	1.229	−1.12(−1.07;−1.35)	−2.13(−2.06;−2.53)
11 (R)		32 (A)	77	−1.90(−0.00+0.01)	7.70(−0.02+0.04)	1.11(−0.07+0.06)	2.099	LF(−0.28;−0.29)	F(−0.56;−0.57)
12 (R)	0.95	32	463	−1.50(−0.00+0.02)	8.47(−0.01+0.09)	1.10(−0.03+0.03)	1.262	−1.01(−0.96;−1.14)	−1.96(−1.87;−2.17)
13 (S)	1.00	32	55	−1.60(−0.01+0.06)	7.03(−0.05+0.30)	...	...	...	...
14 (R)		32	128 (15)	−1.60(−0.02+0.04)	6.97(−0.12+0.19)	...	1.270	−0.92(−0.87;−1.01)	−1.78(−1.70;−1.94)
15 (R)		32 (0.95)	46	−1.62(−0.02+0.05)	6.90(−0.10+0.22)	0.85(−0.00+0.13)	1.270	−0.92(−0.87;−1.01)	−1.78(−1.70;−1.94)
16 (S)		108	140 (41)	−1.59(−0.03+0.01)	7.06(−0.18+0.06)	...	2.092	LF(−0.23;−0.24)	LF(−0.46;−0.47)
17 (S)	1.20	32	78 (13)	−1.57(−0.00+0.02)	4.08(−0.01+0.13)	...	1.290	−0.65(−0.61;−0.68)	−1.28(−1.20;−1.33)
18 (S)		32 (0.95)	49	−1.58(−0.00+0.02)	4.04(−0.01+0.13)	0.64(−0.01+0.05)	1.290	−0.65(−0.61;−0.68)	−1.28(−1.20;−1.33)
19 (S)		108	69 (22)	−1.59(−0.03+0.01)	4.01(−0.17+0.03)	...	2.101	F(−0.16;−0.16)	F(−0.32;−0.32)
20 (R)	1.40	32	58 (13)	−1.46(−0.01+0.01)	2.66(−0.02+0.07)	...	1.294	F(−0.45;−0.50)	F(−0.89;−0.98)
21 (R)	1.80	32	66 (13)	−1.17(−0.01+0.00)	1.64(−0.03+0.01)	...	1.302	F(−0.27;−0.30)	F(−0.54;−0.59)
22 (S)		108	58 (22)	−1.16(−0.02+0.00)	1.76(−0.13+0.00)	...	2.102	F(−0.07;−0.07)	F(−0.14;−0.14)
23 (R)	2.50	32	96 (53)	−0.86(−0.00+0.00)	1.13(−0.06+0.01)	0.21(−0.03+0.01)	1.296	F(−0.14;−0.16)	F(−0.28;−0.31)
24 (R)		32	29	−0.85(−0.01+0.00)	1.20(−0.08+0.00)	0.18(−0.04+0.00)	1.296	F(−0.14;−0.16)	F(−0.28;−0.31)
25 (R)		32	128	−0.85(−0.00+0.01)	1.18(−0.00+0.03)	0.19(−0.02+0.00)	1.294	F(−0.14;−0.16)	F(−0.28;−0.32)
26 (R)		108	54	−0.87(−0.01+0.00)	1.16(−0.09+0.01)	0.19(−0.04+0.00)	1.933	F(−0.05;−0.05)	F(−0.10;−0.10)
27[f] (R)		32	64	−0.98(−0.01+0.00)	1.46(−0.02+0.01)	0.03(−0.00+0.01)	1.298	F(−0.14;−0.16)	F(−0.28;−0.31)
28 (R)	3.50	32	115	−0.61(−0.01+0.00)	1.03(−0.01+0.01)	0.14(−0.00+0.01)	1.287	F(−0.07;−0.08)	F(−0.14;−0.16)
29 (R)	5.00	32	38	−0.43(−0.00+0.01)	0.98(−0.01+0.01)	0.10(−0.00+0.00)	1.282	F(−0.04;−0.04)	F(−0.07;−0.08)
30 (R)		108	27	−0.44(−0.00+0.01)	0.98(−0.00+0.02)	0.11(−0.01+0.00)	2.100	F(−0.01;−0.01)	F(−0.02;−0.02)
31 (R)	7.00	32	115	−0.29(−0.00+0.00)	0.97(−0.01+0.00)	0.08(−0.00+0.01)	1.280	F(−0.02;−0.02)	F(−0.04;−0.04)

[a] The chain numbers are used for reference in the figures. The letters in parentheses refer to the method of generating the chain: R indicates that successive configurations are generated by a random choice of the molecule to be moved, as indicated in Eq. (12). S indicates that the molecules were moved sequentially, according to an initially assigned ordering, as in the earlier calculations of references 1 and 2.

[b] The letter A indicates that the augmented summation convention described in the text was used. Where a fraction f is given in parentheses, the initial f.c.c. lattice was slightly compressed into one corner of the Monte Carlo cell so as to reduce the intermolecular distances to f times their uncompressed values.

[c] The numbers in parentheses give the number of configurations (in thousands) utilized in determining the radial distribution function, in those cases where it differs from the number utilized in obtaining the thermodynamic quantities.

[d] The numbers in parentheses give the extreme values (with reference to the quoted average) of the Monte Carlo averages over the last three-fourths of the chain.

[e] Beyond the reduced cutoff distance $x_c=r_c/a$, the contributions to E'/RT and pv/RT were estimated by approximated radial distribution functions. The first number in parentheses corresponds to a perfect lattice, the second to the ideal fluid ($g=1$). LF (or F) preceding the parentheses indicates that the unweighted average of these two extremes was used (or the fluid value). Where numbers precede the parentheses, they are the corrections obtained from the radial distributions calculated from the chains at the same reduced volume, but with either 108 molecules or with the augmented summation convention. For chain 12 the corrections were obtained by weighted interpolation between the extreme values, the weight factors being themselves interpolated from their values at the adjacent volumes, as determined from the chains with larger N. For chain 13, for which no radial distribution function was calculated, the corrections were estimated from chains 14 and 15.

[f] Chain 27 was calculated using the modified Lennard-Jones potential of reference 4.

internal energy, as well as giving a qualitative picture of the fluctuation of successive averages over a small fixed number of configurations, are given in Fig. 2 for representative 32-molecule chains, and in Fig. 3 for 108-molecule chains. The behavior of chain 1 in Fig. 2 is representative of the other chains at reduced volumes less than 0.95. Except for the greater fluctuation, corresponding to the expected greater excess heat capacity, it will be noticed that the behavior is qualitatively similar to that at the larger volumes, exemplified by chains 17 and 25. The sudden jump occurring in the energy of chain 12 (Fig. 2), after it apparently began to level off at a lower value, was our first indication of the apparent transition to be discussed below. Previous to obtaining this result, we had been accustomed to terminate the chains at about the point where the jump occurs in chain 12 (note that in Fig. 2 the abscissas are to be multiplied by 4 for this chain). When we observed this jump we lengthened chain 12 to the extent shown; since there seems to be no tendency to return to the low initial level, the low-lying first configurations were omitted from the reported averages. Furthermore, in order to guard against missing similar jumps at other volumes, the other chain lengths were also appreciably increased. At the slightly higher reduced volume 1.00, we have chain 14 (Fig. 2), which displays a more gradual buildup period followed by a

Notes on p. 31

rather long period oscillation with more rapid fluctuations superimposed.

The 108-molecule chains in Fig. 3 show less stabilized averages, as expected from both their generally somewhat shorter absolute chain lengths, and the much shorter length in terms of configurations per molecule. This lack of convergence is particularly evident for chains 2 and 16. The lack of a long period oscillation in chain 16 (whose reduced volume is the same as that of chain 14) is possibly significant in terms of the absolute lengths of both chains, but not if the relevant comparison is in terms of chain length per molecule, as may well be the case.

Columns 8–10 of Table II concern the corrections mentioned earlier. Besides giving the actual corrections which were applied to the directly calculated results to obtain the values reported in columns 6–8, the extreme values obtained from a perfect lattice arrangement of the distant molecules on the one hand, and a uniform distribution on the other, are also displayed. It will be noticed that over the whole range of reduced volumes reported here, the two extreme estimates are essentially identical for the 108-molecule chains, well within the statistical precision of the directly calculated contributions. Furthermore, for the 108-molecule chains the tabulated value of $x_c = r_c/a$, obtained as described earlier, in each case agrees closely with the corresponding value in Table I.[10] These observations indicate that there should be little systematic error in the corrections applied to these chains. Thus we believe that the 108-molecule results have as principal source of uncertainty their relatively short length and correspondingly poorer statistical convergence.

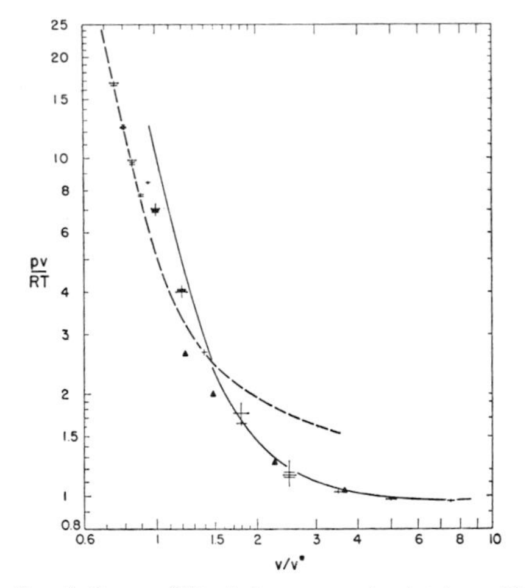

FIG. 4. Compressibility factor *versus* reduced volume. The horizontal bars represent the positions of the Monte Carlo averages; those bearing the small dots at their ends represent results from 108-molecule systems. The vertical extent of the symbols covers the range obtained by subtracting from and adding to each average the difference of largest magnitude shown for it in column 6 of Table II. The upper solid curve represents Bridgman's measurements on argon; the lower one, Michels' measurements. The dashed curve is the Lennard-Jones-Devonshire cell theory isotherm. The triangles are the superposition integral equation results of reference 5.

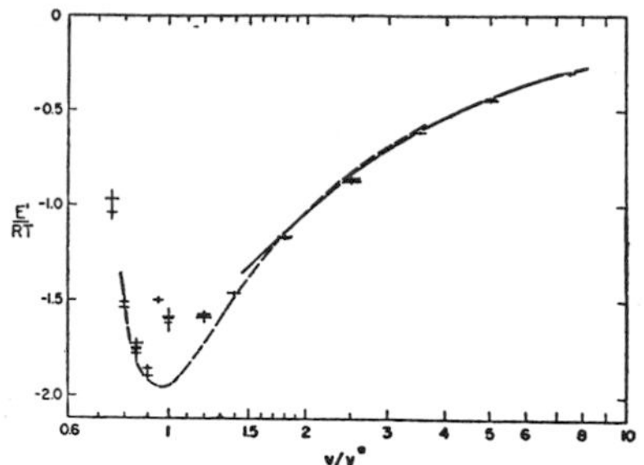

FIG. 5. Reduced excess internal energy *versus* reduced volume. See the legend for Fig. 4. The solid curve represents Michels' experimental results; the dashed curve, the Lennard-Jones-Devonshire cell theory.

For the 32-molecule chains, where the directly calculated contributions are used only through the first shell of 12 neighbors, Table II shows that the necessary correction is poorly bracketed by the two extremes at the higher densities. Here, therefore, we have followed the previously described procedure of estimating the necessary corrections from chains having a longer range of intermolecular interaction. It is believed that this procedure is quite reliable as an estimate of the excluded interactions. It is to be emphasized, of course, that these corrections cannot be expected to result in complete agreement between 32-molecule and 108-molecule chains. They are concerned with the gross effects of truncating the energy sums, and simply supply estimates of the omitted interactions. They cannot account for the likelihood that inclusion of the missing contributions would alter to some extent the distribution of the directly calculated interactions. The reasonably good agreement shown in Table II encourages us to believe that such effects are reasonably small.

In Figs. 4, 5, and 6 the values of the thermodynamic functions are compared with the experimental p-v-T data for argon obtained by Michels[6] and Bridgman,[7] as well as the thermodynamic functions derived from Michels' data. In both cases 55°C isotherms were used, and conversion to reduced volume v/v^* as independent variable was made with use of the previously mentioned value of v^*. In the figures comparison is also made with the two theoretical equations of state mentioned earlier, the Lennard-Jones-Devonshire cell theory[3] and the

[10] The value of x_c shown in Table II for chain 26 corresponds to a cumulative distribution function equal to 42 (three complete f.c.c. shells), and also agrees well with the corresponding value in Table I.

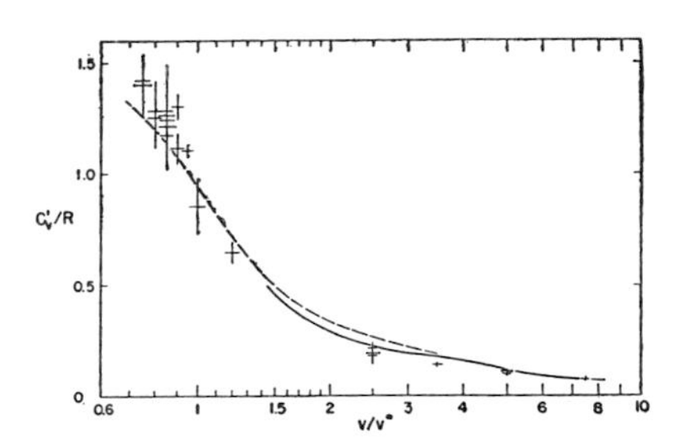

FIG. 6. Reduced excess constant-volume heat capacity *versus* reduced volume. See the legend for Fig. 4. The solid curve represents Michels' experimental results; the dashed curve, the values obtained from the Lennard-Jones-Devonshire cell theory.

Kirkwood-Born-Green theory.[4] With the cell theory, we have of course used the same Lennard-Jones potential as in the Monte Carlo calculations reported here. The Kirkwood-Born-Green equations were solved[4] for a modified Lennard-Jones potential which satisfies (15) with $r_1 = 2^{-1/6}r^*$, the hard core beginning at the crossover point of the ordinary LJ potential. Zwanzig *et al.*[5] attempted a rough correction to make the results applicable to the LJ potential; these are the points shown in Fig. 4.

In discussing Figs. 4–6 it is convenient to distinguish three reduced volume intervals according to the available experimental information. In the range 1.5 to 10, we have Michels' extensive p-v-T data and the derived thermodynamic functions. Between 0.95 and 1.5 we have Bridgman's p-v values, but no thermodynamic functions. In the range 0.75 to 0.95 there is no experimental information. Throughout Michels' region the Monte Carlo compressibility factors (Fig. 4) are in essential agreement with the experimental values. At the largest volumes there is a tendency for the calculated values to be slightly low, by a percent or so. In the higher density portion of Michels' range the deviations are larger, about 5%, but the scatter in the Monte Carlo results is quite comparable. The pv/RT values of Zwanzig *et al.*[5] are seen to approach the experimental results at low density, but tend to become too low as the density increases. Those of the cell theory are of course quite poor in this region. The excess internal energy (Fig. 5) obtained from the Monte Carlo calculations tends to be slightly low by amounts close to the limits of significance at Michels' high density limit, the agreement improving as the density decreases. Such agreement is not extremely significant, since even rather crude approximations such as the cell theory give quite reasonable values of the energy in this range in which their errors in pressure are very large. Thus the divergent trend at higher densities may very well be significant. The Monte Carlo and experimental heat capacities shown in Fig. 6 are also in reasonably good agreement. Here again, however, so crude a treatment as the ordinary cell theory gives comparably good results.

Bridgman's experimental results begin where those of Michels' end, with a small discontinuity of about 5% in the compressibility factor; there is also a noticeable discontinuity in the slopes of the two curves at this point. In the range of Bridgman's volumes, the Monte Carlo results give much lower values of pv/RT, amounting to 30% and more at the highest pressures. We do not wish to minimize this disagreement, since it is perfectly possible that either the LJ pair potential itself or the assumption of additivity of pair potentials may be responsible. However, two points may be worth mentioning. First, the Monte Carlo points seem to represent the more natural continuation of Michels' curve. Secondly, comparison of more recent p-v-T data[11,12] for nitrogen with those of Bridgman reported in the same paper[7] as his argon data shows comparable differences. Benedict's data was obtained in Bridgman's laboratory, and Professor Bridgman[13] has stated that he believes this more recent data to be better, and that the argon data may well be no more reliable than the earlier results for nitrogen. Thus, the significance of the disagreement in this region remains an open question, until the p-v-T behavior of argon is carefully reinvestigated. In this region of volumes (0.95 to 1.5 in v/v^*), the cell theory gives too small values for pv/RT. No experimental values of internal energy and heat capacity are available in the Bridgman region. The Monte Carlo energies in this region deviate strongly from those of the cell theory, passing through a smaller (in absolute value) minimum at a larger volume than the latter. The heat capacity results are quite scant, and begin to show large scatter, but are compatible with the apparent trend of Michels' data as are those of the cell theory also.

At still smaller volumes than reached by Bridgman we observed the very interesting behavior shown in Figs. 4 and 5, where the points for reduced volumes 0.90 and less are seen to lie on an essentially different curve from those at the larger volumes. A number of considerations have led us to identify the region of the break in the curve with the first-order transition between fluid and solid (crystalline) phases; the qualitative appearance of Figs. 4 and 5 corresponds to this behavior. To discuss first the corresponding behavior of argon molecules, the melting curve has been investigated by Bridgman[7] and by Robinson,[14] but has not been carried up to the temperature of our isotherm. Extrapolation of Robinson's results by means of the empirical Simon relation between p and T along the coexistence line indicates that the freezing pressure of argon at this temperature should be about 13 500 atmos. Bridgman's measurements are the only ones

[11] M. Benedict, J. Am. Chem. Soc. **59**, 2233 (1937).
[12] D. S. Tsiklis, Doklady Akad. Nauk USSR **79**, 289 (1951).
[13] P. W. Bridgman (private communication).
[14] D. W. Robinson, Proc. Roy Soc. (London) **A225**, 393 (1954).

Notes on p. 31

giving the volume change upon fusion. A rough extrapolation suggests a volume increment of somewhat less than 0.8 cm^3/mole, which would be an increment of about 0.03 on our reduced volume scale. Over the range of roughly 1000 to 6000 atmos, Bridgman obtained (from his p, v, T measurements by use of the Clapeyron equation) an essentially constant enthalpy of fusion of about 260 cal/mole. If the same value obtains at 328°K, $\Delta H/RT$ would be approximately 0.40.

Turning to the Monte Carlo results in the vicinity of the apparent transition, several observations are in order. First and foremost, the small number of molecules, the periodic boundary condition, and the formal equivalence of the Monte Carlo method to the *petite* canonical ensemble, will all have an effect on the results to be expected across a first-order phase transition.[15] Secondly, we may very well expect the Monte Carlo process to be slowly convergent, with possible appearance of metastable plateaus of long chain length, in such a transition region (chain 12). Thus a detailed investigation of such a region might well be extraordinarily time consuming, particularly since it would undoubtedly be essential to investigate the dependence of the results on the number of molecules. Thirdly, we have too few points to permit even a Maxwell equal-area determination of the coexistent phases of the 32-molecule Monte Carlo system. Thus, although the existence of an apparently first-order transition seems indicated by our results, the parameters characterizing it are very crudely determined. Furthermore, the dependence of the transition on the number of molecules has not been investigated as well as would be desirable, since little confidence could be placed on anything but chains much longer than any obtained to date with 108 molecules. Thus, it has not seemed worthwhile to generate 108-molecule chains at the two points on the apparent boundary of the transition region ($v/v^*=0.90$ and 0.95). At the two next adjacent volumes (0.85 and 1.0) we do have our two longest 108-molecule investigations. In both cases the resulting pressures and energies are slightly higher than the corresponding 32-molecule results, but the convergence of the larger systems in both cases leaves something to be desired, and we hestitate to attribute significance to the differences. In any case, the qualitative nature of our results is not changed. It should be pointed out that the existence of a break in the curves would be indicated even if the results at $v/v^*=0.90$ and 0.95 were ignored, especially in the case of the internal energy curve. In order to check the stability of our results in this vicinity with respect to perturbation of the initial configurations, we generated chains 7, 8, 15, and 18, in which the ordinary regular initial array was slightly compressed into one corner of the Monte Carlo cell. In the case of chain 8 this compression was sufficient to raise the initial potential energy noticeably above the average value. In all cases the results are in essential agreement with the normally started chains. Thus, the transition behavior seems not to be an artifact introduced by any of the approximations which we can investigate. It is to be noted, as far as the ergodic question is concerned, that the usual regular initial configuration and the compressed one just mentioned obviously are members of the same class, so that such agreement as obtained should be expected.

With due regard for the limitations mentioned in the previous paragraph, it is still of interest to compare the rough values of the transition parameters obtained by Monte Carlo with those to be expected for argon. The values of pv^*/RT are 8.6 at $v/v^*=0.90$, and 8.9 at $v/v^*=0.95$. Thus, we have an apparent region of mechanical instability, but on the borderline of statistical significance. If the previously mentioned parameters for argon are used to convert to the physical pressure scale, these two points correspond to a pressure of about 10 000 atmos, rather lower than the expected freezing pressure for argon at this temperature. The difference of 0.05 in the reduced volumes is somewhat larger than the value 0.03 estimated from Bridgman's data. The difference in melting pressures may be significant; another way of stating it is that Robinson's data indicate quite clearly that under a pressure of 10 000 atmos argon should melt at about 270°K.

If the difference in E'/RT at $v/v^*=0.90$ and 0.95 is taken as $\Delta E/RT$ for the transition, and the values $p=10\,000$ atmos and $\Delta v/v^*=0.05$ are used, we find $\Delta H/RT=1.0$. This is two and one-half times the value extrapolated from Bridgman's measurements at 6000 atmos. The interpretation of these differences in melting parameters is not clear, if they really represent convergent behavior of the 32-molecule system. They could possibly be due to the small number of molecules used in our calculations. But they could well be due to departures of the argon potential energy from the additive Lennard-Jones form; it may be expected that high-pressure melting phenomena will be quite sensitive to the details of the repulsive potential.

The heat capacity results in this region are too scattered to allow much interpretation; there is a slight tendency to lie above the cell theory curve. There is no sign of the separated high segment of the curve to be expected in the coexistence region,[16] but this can hardly be surprising in view of uncertainty as to the exact position of this region, as well as the unanswered questions involving the necessary length of chain, etc.

The sudden improvement in agreement of the cell theory with our Monte Carlo results on the solid side of the transition is very striking. This may be associated with another striking observation, which increases our confidence in the reality of the observed transition. The calculator codes used do not study in detail the

[15] T. L. Hill, *Statistical Mechanics* (McGraw-Hill Book Company, Inc., New York, 1956), Appendix 9.

[16] See, for example, the experimental measurements by A. Michels and J. Strijland, Physica **18**, 613 (1952), of C_v for CO_2 in the liquid-vapor coexistence region.

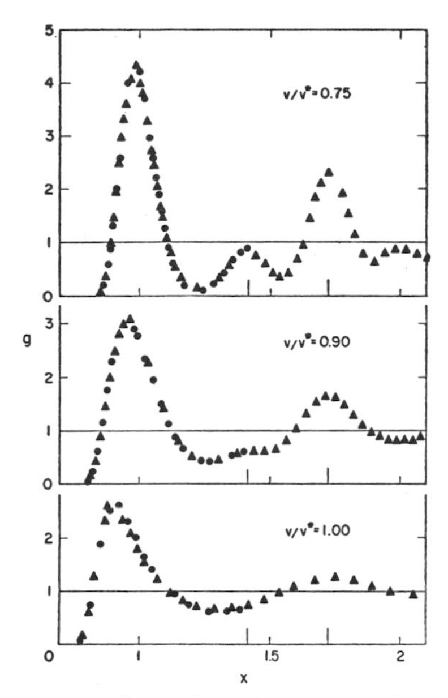

FIG. 7. The radial distribution function g *versus* the reduced distance $x=r/a$, in the neighborhood of the apparent transition. The circles were obtained from the 32-molecule chains 1, 10, and 15; the triangles, from the 108-molecule (or "augmented" 32) chains 2, 11, and 16.

diffusion of the molecules through the Monte Carlo cell. However, we compared the positions and arrangements of the molecules in the final configurations of chains 1, 3, 5, 10, 12, and 14 with the original regular lattice structures. For the chains on the solid branch (volume 0.90 and less), the relative positions of all the molecules with respect to one another were the same as at the start except for more or less haphazard displacements from a regular lattice small compared to the distance between nearest neighbors. (All molecules had in some cases moved through the cell essentially in unison, subject to the periodic boundary condition, as might be expected if the system as a whole is describing a sort of symmetric random walk, all molecules being essentially in step in formation.) This is in great contrast with the observations at volumes 0.95 and 1.0, where the final arrangement was in both cases completely disordered with molecules appearing in random fashion at positions well removed from where they began.

The appearance of the radial distribution functions on the two sides of the transition region is shown in Fig. 7.[17] There seems to be no abrupt change in their qualitative appearance as the transition region is crossed, the peaks simply becoming increasingly well defined as the density increases. Radial distribution functions at larger volumes are shown in Fig. 8, where they are also compared in two instances with the results of the Kirkwood-Born-Green[4] theory. It must again be emphasized, however, that the latters' radial distribution functions correspond to the previously mentioned modified LJ potential. In order to determine how much of the discrepancy is due to this difference in the intermolecular potential, and how much to introduction of the superposition approximation, chain 27 was generated in which we used the same modified potential in a Monte Carlo calculation. In Fig. 9 these Monte Carlo and superposition radial distribution functions are compared; much of the discrepancy has been removed by using the same potential function. The superposition values for E'/RT and pv/RT are, respectively, -0.99 and 1.31. Thus there is essential agreement in the values of the energy, and a residual difference of about 10% in the compressibility factor, some of which however may be due to uncertainty in the precision of our single Monte Carlo chain for these conditions. Thus at this fairly large volume the superposition approximation is seen to give fairly good results.

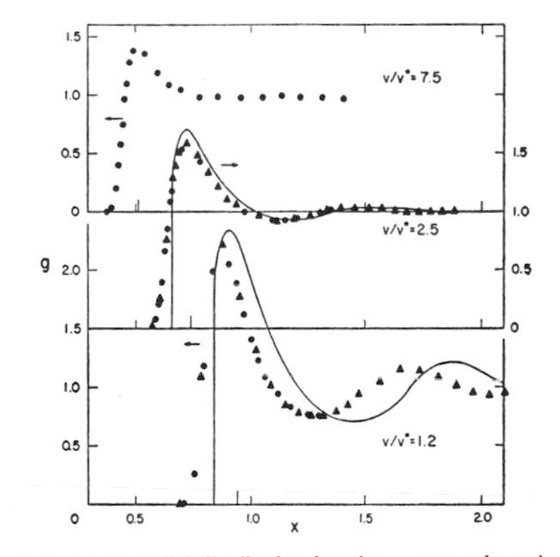

FIG. 8. The radial distribution function g *versus* the reduced distance $x=r/a$, in the fluid region. The circles are from the 32-molecule chains 31, 25, and 18; the triangles, from 108-molecule chains 26 and 19. The curves are the superposition integral equation results for the modified Lennard-Jones potential (reference 4).

[17] A set of tables of the cumulative distribution functions obtained from 29 of the 31 chains is available in limited number from one of the authors (W. W. Wood). They may also be obtained as Document No. 5330 upon remitting \$5.00 for a photoprint or \$2.25 for a 35-mm microfilm copy to Chief, Photoduplication Service, ADI Auxiliary Publications Project, Library of Congress, Washington 25, D. C. The tabulations consist of values of $N(x)$ at mostly equally spaced intervals in x^2.

Notes on p. 31

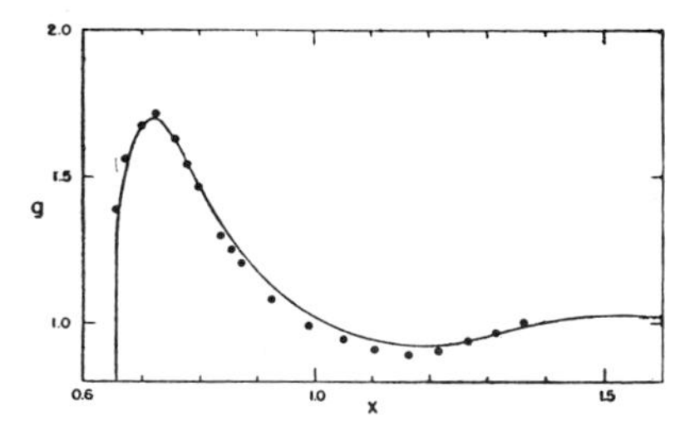

FIG. 9. Comparison of Monte Carlo and superposition-integral equation radial distribution functions for the same modified Lennard-Jones potential. The points are from chain 27; the curve from the tabulations of reference 4. The reduced volume v/v^* is 2.5.

CONCLUSIONS

The essential agreement of our calculations with Michels' data suggests that in this region effects of deviations from the Lennard-Jones pair potential and additivity of pair potentials are fairly small. A reinvestigation of the experimental behavior of argon in the Bridgman region would be very useful for a further assessment of the magnitude of these effects. At volumes removed from the transition region our 32-molecule chains are intentionally much longer (by perhaps a factor of 3 or 4) than needed to obtain useful results, in order to establish with some certainty that the unusual behavior observed is really peculiar to that region. On the other hand, it seems likely that even the 32-molecule chains cannot practically be made long enough to study the detailed behavior in the transition region, and a study of the effect of the number of molecules seems completely out of question. Thus, we shall await the availability of calculators of considerably higher speed before undertaking these investigations. Even outside the transition region, except at the lowest densities, it would be desirable to have 108-molecule chains several times longer than those reported here in order to make a precise assessment of the effect of the number of molecules. The present results are, however, certainly compatible with the hypothesis that these are small.

The simple cell theory seems to be unsatisfactory in the entire fluid region, but is surprisingly good in the crystalline region. The present results should be useful in assessing the effect of various elaborations of the cell theory in the fluid region. The superposition approximation used with the (otherwise rigorous) Kirkwood-Born-Green theory seems likely to be useful on this isotherm at least for volumes greater than 2.5.

As this material was being prepared for publication, we learned of the results for the equation of state of hard spheres obtained by Alder and Wainwright[18] from a numerical integration of the elementary equations of motion. In the range of reduced volumes from 1.5 to 2.0, Alder and Wainwright found substantial discrepancies between their results and the Monte Carlo results reported by the Rosenbluths.[2] We are currently repeating the Monte Carlo investigation, and have already found the earlier Monte Carlo results in the region in question to be wrong, apparently because of insufficient chain length. This investigation, as well as the other Lennard-Jones isotherms previously mentioned, will be the subject of further papers.

ACKNOWLEDGMENTS

It is a pleasure to express our gratitude for several helpful discussions with Professors John G. Kirkwood and Joseph O. Hirschfelder, and especially with Dr. Marshall N. Rosenbluth. Much of the calculator operation was supervised by Mr. J. T. Mann and Mr. W. H. Lane, to whom we are very grateful. We would also like to thank Mr. Donald D. Fitts who performed the necessary interpolations of the Kirkwood-Born-Green tabulations to obtain the radial distribution functions at our values of reduced volume.

[18] B. J. Alder and T. Wainwright, Proc. I.U.P.A.P. Symposium on Statistical Mechanical Theory of Transport Processes, Brussels, 1956 (to be published).

Notes to Reprint I.2

1. The following typographical errors have been pointed out to us by W.W. Wood. (a) On p. 721, right-hand column, line 22: "indefinitely longer" should read "indefinitely long". (b) In eq. (4), $P_{jk'}^{(n-1)}$ should be lower case. (c) In the penultimate line of Note 9 (p. 722), "chain" should read "chains". (d) The last line of eq. (12) should read "$A_{jk} = 0$, otherwise".

2. According to Dr. Wood, the "future paper" referred to in the closing lines of the paragraph on periodic boundary conditions (p. 724) was never written.

3. It is sometimes preferable to use periodic cells that are more nearly spherical than the usual, cubic box. The "most spherical", space-filling unit cell that can be used is a truncated octahedron (Adams, Chem. Phys. Lett. 62 (1979) 329). The replacement of cubic by non-cubic periodic boundary conditions tends to slow down a simulation. The net result may nevertheless be a saving of computer time because the alternative, in general, is to simulate a larger system. Specific advantages of "almost spherical" boundary conditions are that (a) for the same number of particles, the radial distribution function can be computed over larger separations; (b) angle-dependent, three-body correlations are less distorted, and (c) simulations of large molecules (e.g. proteins) in solution require fewer solvent molecules per macromolecule.

4. The "lattice summation" method described in the text (p. 724) is nowadays rarely used.

5. The thermodynamic properties of liquid rather than supercritical Lennard–Jones "argon" were later studied systematically by Monte Carlo (McDonald and Singer, Disc. Faraday Soc. 43 (1967) 40) and by molecular dynamics (Verlet, Reprint I.7).

Preliminary Results from a Recalculation of the Monte Carlo Equation of State of Hard Spheres*

W. W. Wood and J. D. Jacobson
Los Alamos Scientific Laboratory, Los Alamos, New Mexico
(Received August 15, 1957)

THE disagreement between the hard sphere equation of state obtained by Rosenbluth and Rosenbluth[1] using the Monte Carlo method[2] and that reported in the accompanying paper by Alder and Wainwright[3] using detailed molecular dynamics led us to repeat the Monte Carlo investigation. Preliminary results for 32 molecules with cubical periodic boundary conditions[1,2] are shown in Fig. 1 along with Alder and Wainwright's[3] results with which there is rather good agreement. The previous Monte Carlo calculations[1] at reduced volumes v/v_0 (v_0=close-packed volume) from about 1.5 to 2.0 are in error due to inadequate chain length to detect the behavior described below; the difficulty was aggravated by the concentration of effort on the system of 256 molecules, which requires considerably longer computing time.

The present calculations have been made on IBM Type 704 calculators and use the same method as the earlier work[1,2] except that the molecules are "moved" in random rather than ordered sequence.[4]

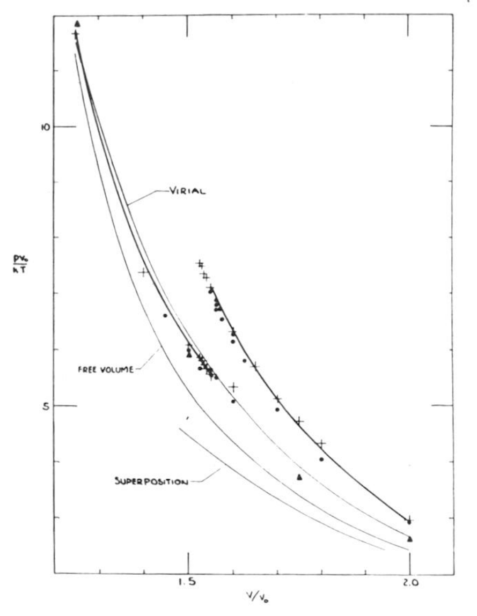

Fig. 1. The equation of state of hard spheres. The heavy solid curve represents Alder and Wainwright's[3] 108 molecule results; +, their 32 molecule results. ● and ▲ represent the present and previous[1] Monte Carlo results. Virial = five term virial expression.[1] Superposition = reference 5.

The equation of state is given by $pv/kT=1-2\pi\sqrt{2}v_0g(\sigma)/3v$, where $g(\sigma)$ is the radial distribution function at the collision diameter σ. For v/v_0 between 1.55 and 1.6 the Markov chains developed by the Monte Carlo method indicate that the set of configuration states is divided into two classes characterized by different central values of $g(\sigma)$. Transitions between classes take place only rarely; the Markov chains display configurational relaxation. A typical chain started from a regular f.c.c. lattice fluctuates for a variable length of time (which tends to increase with density) in the low $g(\sigma)$ class, then jumps rather suddenly to the high $g(\sigma)$ class. Calculator periods as long as 10–30 hours may sample only 0–3 such interclass transitions, so that the over-all canonical average is very poorly estimated. As a consequence we have averaged the low and high $g(\sigma)$ classes separately to obtain two values of pv_0/kT at each v/v_0 in this range, as shown in Fig. 1. The presence of only two classes of states is indicated by agreement of $g(\sigma)$ values obtained from different chains at the same v/v_0, and also those obtained when a single chain reenters a class; furthermore, the pv_0/kT values attributed to the two classes in Fig. 1 vary reasonably smoothly with v/v_0.

The manner in which complete canonical averaging would connect the two separate, overlapping branches of the equation of state is at present undetermined. A first-order phase transition is, however, strongly suggested. Of some interest in this connection is the fact that the high $g(\sigma)$ states seem to be characterized by relatively free diffusion, while in the low $g(\sigma)$ states diffusion is much restricted.

The conjecture[1] that some high-order virial coefficients might be negative is not necessarily supported by the present results, since only to the left of the apparent transition do the latter give lower pressures than the five-term virial expression.

Some further investigation for both 32 molecule and larger systems will be made on the present calculators, but a satisfactory determination of the detailed behavior in the apparent transition region will require higher speed equipment. The possibility that a similar phenomenon for hard spheres in two dimensions may have been missed in the original Monte Carlo calculations[1] will also be investigated.

* Work performed under the auspices of the U. S. Atomic Energy Commission.

[1] M. N. Rosenbluth and A. W. Rosenbluth, J. Chem. Phys. **22**, 881 (1954).
[2] Metropolis, Rosenbluth, Rosenbluth, Teller, and Teller, J. Chem. Phys. **21**, 1087 (1953).
[3] B. J. Alder and T. Wainwright, J. Chem. Phys. **27**, 1208 (1957).
[4] W. W. Wood and F. R. Parker, J. Chem. Phys. **27**, 720 (1957). This paper discusses the Monte Carlo method in some detail, as well as giving computational results for Lennard-Jones molecules.
[5] Kirkwood, Maun, and Alder, J. Chem. Phys. **18**, 1040 (1950).

Phase Transition for a Hard Sphere System

B. J. Alder and T. E. Wainwright
University of California Radiation Laboratory, Livermore, California
(Received August 12, 1957)

A CALCULATION of molecular dynamic motion has been designed principally to study the relaxations accompanying various nonequilibrium phenomena. The method consists of solving exactly (to the number of significant figures carried) the simultaneous classical equations of motion of several hundred particles by means of fast electronic computors. Some of the details as they relate to hard spheres and to particles having square well potentials of attraction have been described.[1,2] The method has been used also to calculate equilibrium properties, particularly the equation of state of hard spheres where differences with previous Monte Carlo[3] results appeared.

The calculation treats a system of particles in a rectangular box with periodic boundary conditions.[4] Initially, the particles are in an ordered lattice with velocities of equal magnitude but with random orientations. After a very short initial run[1,2] the system reached the Maxwell-Boltzmann velocity distribution so that the pressure could thereafter be evaluated directly by means of the virial theorem, that is by the rate of change of the momentum of the colliding particles.[1,2] The pressure has also been evaluated from the radial distribution function.[5] Agreement between the two methods is within the accuracy of the calculation.

A 32-particle system in a cube and initially in a face-centered cubic lattice proceeded at about 300 collisions an hour on the UNIVAC. For comparison a 96-particle system in a rectangular box and initially in a hexagonal arrangement has been calculated, however only at high densities so far. No differences in the pressures can be detected. It became apparent that some long runs were necessary at intermediate densities, accordingly the IBM-704 was utilized where, for 32 particles, an hour is required for 7000 collisions. Larger systems of 108, 256, and 500 particles can also conveniently be handled; in an hour 2000, 1000, and 500 collisions, respectively, can be calculated. The results for 256 and 500 particles are not now presented due to inadequate statistics.

The equation of state shown in Fig. 1 of the accompanying paper[6] for 32 and 108 particles is for the intermediate region of density, where disagreement was found with the previous Monte Carlo results. The volume, v, is given relative to the volume of close packing, v_0. Plotted also are the more extended Monte Carlo results; the agreement between these three systems is within the present accuracy of the pressure determination. This agreement provides an interesting confirmation of the postulates of statistical mechanics for this system.

Figure 1 of the accompanying paper shows two separate and overlapping branches. In the overlapping region the system can, at a given density, exist in two states with considerably different pressures. As the calculation proceeds the pressure is seen to jump suddenly from one level to the other. A study of the positions of the particles reveals that as long as the system stays on the lower branch of the curve the particles are all confined to the narrow region in space determined by their neighbors, while on the upper branch of the curve the particles have acquired enough freedom to exchange with the surrounding particles. Since the spheres are originally in ordered positions, the system starts out on the lower branch; the first jump to the upper branch can require very many collisions. The trend, as expected, is that at higher densities more collisions are necessary for the first transition, however, there are large deviations. At $v/v_0=1.60$, 5000 collisions were required; at 1.55, 25 000; while at 1.54 only 400; at 1.535, 7000; at 1.53, 75 000; and at 1.525, 95 000. Runs in excess of 200 000 collisions at v/v_0 of 1.55 and 1.53 have not shown any return to the lower branch, while at 1.525 the system has returned several times, however only for relatively few collisions. The lowest density at which the system did not jump to the upper curve is at 1.50, however the run extends only to 50 000 collisions and at that density it might take very many collisions before the appropriate fluctuation occurs for a molecule to escape from its neighborhood. For comparison, the first jump for 108 particles occurred for $v/v_0=1.55$ and 1.60 at about 2000 collisions. This is fewer collisions per particle than for the smaller system and is indicative of larger possible density fluctuations in larger systems. Apparently, the reason these jumps occur rather than the system's taking on some mean pressure is that it is not possible to have the two states simultaneously in equilibrium in such a small sample. The fact that the 108 and 32 particles systems give identical branches indicates that the effect of the periodic boundary is not serious even for 32 particles. It can be shown that the first few virial coefficients for a finite system of N particles with periodic boundary conditions have corrections of order $1/N$.

Clearly, it is presently impossible to connect these two branches, though it may be possible to do this with a faster machine by averaging the pressure over a very long run which includes many jumps instead of averaging the branches separately. In the transition region the effect of the number of particles might appear in the average time the system spends in each branch, such that for an infinite system a horizontal line might appear, characteristic of the first-order transition which is strongly indicated by the present results. Such a transition was suggested earlier by the superposition theory at a $v/v_0=1.48$.

We gratefully acknowledge Mrs. Shirley Campbell's and Mary Shephard's help with the coding problems.

[1] Proceedings of the International Union of Pure and Applied Physics on "Statistical mechanical theory of transport properties," Brussels (1956).
[2] Symposium on the "Many body problem" in New York (1957).
[3] M. N. Rosenbluth and A. W. Rosenbluth, J. Chem. Phys. **22**, 881 (1954).
[4] Alder, Frankel, and Levinson, J. Chem. Phys. **23**, 417 (1955).
[5] Kirkwood, Maun, and Alder, J. Chem. Phys. **18**, 1040 (1950).
[6] W. W. Wood and J. D. Jacobson, J. Chem. Phys. **27**, 1207 (1957).

Note to Reprint I.4

A full description of the molecular dynamics method used in this paper can be found in (Alder and Wainwright, J. Chem. Phys. 31 (1959) 459).

Dynamics of Radiation Damage*

J. B. GIBSON, A. N. GOLAND,† M. MILGRAM, AND G. H. VINEYARD
Brookhaven National Laboratory, Upton, New York
(Received July 14, 1960)

Radiation damage events at low and moderate energies (up to 400 ev) are studied by machine calculations in a model representing copper. Orbits of knock-on atoms are found and the resulting damaged configurations are observed to consist of interstitials and vacancies. Thresholds for producing permanently displaced atoms (i.e., interstitials) are about 25 ev in the ⟨100⟩ direction, 25 to 30 ev in the ⟨110⟩ direction, and around 85 ev in the ⟨111⟩ direction. Collision chains in the ⟨100⟩ and ⟨110⟩ directions are prominent; at low energies the chains focus, at higher energies they defocus. Above threshold, the chains transport matter, as well as energy, and produce an interstitial at a distance. The range of ⟨110⟩ chains has been studied in detail. Localized vibrational modes associated with interstitials, agitations qualitatively like thermal spikes, ring annealing processes, and a higher energy process somewhat like a displacement spike have been observed. Replacements have been found to be very numerous.

The configurations of various static defects have also been studied in this model. The interstitial is found to reside in a "split" configuration, sharing a lattice site with another atom. The crowdion is found not to be stable, and Frenkel pairs are stable only beyond minimum separations, which are found to be very much dependent on orientation.

1. INTRODUCTION

THE initial event in the damaging of a crystal lattice by high-energy radiation is the sudden transfer of a rather large amount of kinetic energy (10 to perhaps 10^5 ev) to a single atom. The energized atom then ploughs through the lattice knocking other atoms from their sites and leaving a damaged region behind. From a theoretical standpoint this damaging event is a complex many-body problem, and it has been treated in the past only by making drastic approximations.[1] Generally it has been considered as a cascade of independent, two body collisions between knock-on atoms and stationary atoms. The knock-on atoms have been assumed to move freely between collisions. The stationary atoms have been assumed to behave as though randomly located, and their binding in the lattice has been taken into account by the very much simplified assumption that they will be displaced and enter the group of freely moving knock-ons if and only if endowed with energy above a certain threshold, generally in the neighborhood of 25 ev. On this cascade model the damage is predicted to be a set of interstitial atoms and an equal number of vacant lattice sites, distributed randomly over a small region. Other models have been proposed in which many-body effects are given prominence. Thermal spike and displacement spike models are of this character. In the former, the region around the site of a knock-on is assumed to behave as if suddenly heated, and its subsequent cooling is treated by the classical laws of heat conduction in a homogeneous medium. In the displacement spike models, qualitative arguments about the character of damage are advanced on the assumption that a kind of miniature "explosion" occurs around the site of the knock-on. These models are difficult to harmonize with one another, and each has obvious shortcomings. Patchwork attempts at improving the models in individual details have not yet been very impressive.

In the last few years a number of sophisticated radiation damage experiments have been made. In the most notable of these highly purified metals have been

* Work supported by the U. S. Atomic Energy Commission.
† Guest Scientist from Ordnance Materials Research Office, Watertown, Massachusetts.
[1] For reviews see F. Seitz and J. S. Koehler, in *Solid-State Physics*, edited by F. Seitz and D. Turnbull (Academic Press, Inc., New York, 1956), Vol. 2, p. 305; also G. J. Dienes and G. H. Vineyard, *Radiation Effects in Solids* (Interscience Publishers, Inc., New York, 1957).

bombarded in a variety of ways at very low temperatures, and the recovery of the specimens induced by careful annealing has been studied.[2-6] Such experiments have shown that damage under even the simplest conditions still has a complex character, and controversies over its nature have increased, rather than decreased, in vigor.[7-10]

For these reasons a more realistic calculation of some typical damage processes is highly desirable. It has seemed to us that analytical methods are inadequate and that numerical treatment with the aid of a high-speed computing machine is required.[11] This paper is the first full length report on results to date.[12] Our procedure is to consider a crystallite containing a reasonably large number of atoms which interact with realistic forces. Atoms on the surface of the crystallite are supplied with extra forces simulating the reaction of atoms outside, as though the crystallite were embedded in an infinite crystalline matrix. A radiation damage event starts with all atoms on their lattice sites and all but one at rest. That one atom is initially endowed with arbitrary kinetic energy and direction of motion, as though it had just been struck by a bombarding particle. A high speed computer then integrates the classical equations of motion for the set of atoms, showing how the initially energized atom (the knock-on) transfers energy to neighboring atoms, how the dynamic stages evolve, and how the kinetic energy finally dies away and the atoms of the set come to rest in a damaged configuration. A series of "runs" are made, corresponding to a representative variety of initial conditions.

The computer program can also be used to study the stability, energy, equilibrium configuration, and other properties of lattice defects permitted by the model. One guesses the positions of the atoms in the defect and uses these as initial conditions, with zero initial velocities (actually a very small kinetic energy may be imparted to one atom to spoil the symmetries of the starting configuration). The computing machine then shows how the lattice relaxes, and with dissipation of energy at the boundary of the crystallite, aided, when desired, by artificial damping, an equilibrium configuration is eventually reached. Concurrently with the study of dynamic damage events a study of point defects has been undertaken. These are of interest in themselves, they give an independent check on the adequacy of the model, and they assist in interpreting the dynamic results.

All of the calculations made to date are for metallic copper. This material has been chosen because it is a reasonably simple metal and because more radiation damage experiments at low temperatures have been performed on it than on any other substance. It has seemed advisable to treat this one material very thoroughly before extending the calculations to other substances. The effect of a finite temperature during bombardment could, in principle, be put into the calculations by supplying small initial agitations to all the atoms. Doing this properly would require averaging over a large set of initial agitations and would enormously extend the number of computations required. Most of the annealing effects can, we feel, be sufficiently well estimated by calculating activation energies for migration of the defects and applying the theory of absolute rate processes in solids. Thus all calculations made to date have been for bombardments at a temperature of absolute zero.

A further important limitation of the calculations must be pointed out. Because of the speed and the size of memory of the computing machine available, the fundamental crystallite dealt with has been of modest size. Most of the computations reported were on crystallites containing about 500 atoms; a few computations have been done more recently on crystallites of about 1000 atoms. The forces applied to the atoms bounding the crystallite are admittedly somewhat inaccurate, and this renders the calculations less certain when boundary atoms receive large displacements during the crucial stages of a run. To avoid this it is necessary to limit the kinetic energy of the knock-on atom. Our most energetic events have been at 400 ev, and the majority of runs have been at energies an order of magnitude lower than this. The calculations are thus pre-eminently concerned with threshold and near-threshold events; many inferences for the high-energy events typical of reactor neutron bombardment can be drawn, but we are not yet able to extend the machine calculations to cover these fully.

A detailed discussion of the model and the force laws

[2] J. K. Redman, T. S. Noggle, R. R. Coltman, and T. H. Blewitt, Bull. Am. Phys. Soc. **1**, 130 (1956); R. R. Coltman, T. H. Blewitt, C. E. Klabunde, and J. K. Redman, Bull. Am. Phys. Soc. **4**, 135 (1959); T. H. Blewitt, *Symposium on Vacancies and Point Defects, Harwell* (1958).

[3] H. G. Cooper, J. S. Koehler, and J. W. Marx, Phys. Rev. **97**, 599 (1955); G. D. Magnuson, W. Palmer, and J. S. Koehler, Phys. Rev. **109**, 1990 (1958).

[4] J. W. Corbett, J. M. Denney, M. D. Fisk, and R. M. Walker, Phys. Rev. **108**, 954 (1957); J. W. Corbett, R. B. Smith, and R. M. Walker, Phys. Rev. **114**, 1452 (1959); **114**, 1450 (1959).

[5] J. W. Corbett and R. M. Walker, Phys. Rev. **115**, 67 (1959).

[6] C. J. Meechan and A. Sosin, Phys. Rev. **113**, 422 (1958).

[7] A. Seeger, *Proceedings of the Second United Nations International Conference on Peaceful Uses of Atomic Energy, Geneva, 1958* (United Nations, Geneva, 1958), Paper No. 998.

[8] *Proceedings of the Lattice Defects in Noble Metals Conference*, edited by J. A. Brinkman, J. Meechan and A. Sosin, North American Aviation Report NAA-SR-3250 (Office of Technical Services, Department of Commerce, Washington, D. C.).

[9] National Academy of Sciences Report, *Perspectives in Materials Research* (to be published).

[10] C. J. Meechan, A. Sosin, and J. A. Brinkman, Phys. Rev. **120**, 411 (1960)

[11] Calculations of the threshold energy for producing a permanently displaced atom have been made by H. B. Huntington [Phys. Rev. **93**, 1414 (1954)], for copper, and by W. Kohn [Phys. Rev. **94**, 1409 (1954)], for germanium. These treatments necessarily relied on numerous assumptions.

[12] Brief preliminary reports have been given in the following places: J. Appl. Phys. **30**, 1322 (and cover) (1959). In this report [100] should be replaced by [010]; G. H. Vineyard, J. B. Gibson, A. N. Goland, and M. Milgram, Bull. Am. Phys. Soc. **5**, 26 (1960), and the next three abstracts; Brookhaven National Laboratory, Annual Report, July 1, 1959 (Office of Technical Services, Department of Commerce, Washington, D. C.), pp. 12–14.

is given in Sec. 2. Section 3 outlines the scheme of integration and the computational procedure. Section 4 describes the results of some static calculations on the defects supported in the model. Those defects are described that are necessary to an understanding of the dynamic results; a more complete report on static calculations will be given in a future paper. Section 5 reports the principal dynamic results achieved to date, and Sec. 6 summarizes the conclusions reached. In the Appendix, there is a table of all the dynamic events that have been run.

2. THE MODEL

All computations have been made on a model designed to represent metallic copper. The atoms are allowed to interact with two-body, central repulsive forces. For these a Born-Mayer form is assumed, the interaction energy of a pair of atoms at separation r being

$$\varphi = Be^{-\beta r}. \tag{1}$$

This interaction describes the repulsion of atoms at close approach. The choice of the constants in this law will be discussed below. A cohesive tendency is also needed, and for this a constant inward force is applied to each atom on the boundary of the crystallite. In the equilibrium configuration this force just balances the Born-Mayer repulsions of neighboring atoms. The equilibrium configuration, of course, is a face-centered cubic array with the normal lattice spacing of copper. Since all crystallites considered are rectangular parallelepipeds, for an atom in a face the surface force is normal to the face, for an atom in an edge the force is normal to the edge (along $\langle 110 \rangle$) and for an atom in a corner it is along the inwardly directed cube diagonal. In any distortion involving small displacement of surface atoms these surface forces give an increment of total binding energy proportional to the increment of volume of the crystallite. The forces can thus represent any binding energy that is a function only of volume of the crystallite and which varies at the right rate with volume to equilibrate the Born-Mayer repulsions. In a monovalent metal the conduction electrons are the major source of binding, and their cohesive energy is, to a certain approximation, dependent only on volume. Thus the constant surface forces employed here represent, in first approximation, the cohesive effect of the conduction electrons. This combination of two-body repulsions plus constant surface forces is easy to apply in machine computation, it gives a crystallite which, at equilibrium, has no distortions near the surface, and it would seem to be at least as faithful to the forces in a real metal as any two-body force law with a repulsive core and attractive tail (Morse potentials, Lennard-Jones potentials, etc.) that can be devised. Since it is not a purely central-force model, it does not require the Cauchy relation for the elastic constants.

Since the crystallite is supposed to behave as a set of atoms on the interior of an infinite perfect crystal, it is necessary to have additional forces on the surface atoms to represent the reaction forces of atoms beyond the surface caused by any displacement of atoms in the microcrystallite. For small displacements an elegant expression for these reactions can be written in terms of a Green's function and an integral over the history of the motions in the crystallite. It does not seem feasible to use this expression in an actual calculation, however, both because of the difficulty of finding the Green's function explicitly and because of added requirements on storage of information during the computation. Instead, the additional surface forces were simply taken to be a spring force, proportional to the displacement of the surface atom, and a viscous force, proportional to the velocity of the surface atom. These are only approximations to the true reaction forces, but with judicious choice of the spring and viscosity constants, they are thought to be adequate for the accuracy required. The spring forces represent the tendency of material just outside the crystallite to resist slow or static deformation of the crystallite by a system of forces proportional to the deformation.

The spring constants were arrived at by the approximate arguments that follow: The crystallite is first replaced by a sphere of equal volume, embedded in an infinite homogeneous, elastic medium with isotropic elastic properties. If the sphere is expanded from radius R to $R+\delta R$ the equations of elasticity show that a pressure P acts on the surface of the sphere, and

$$P = (4\mu/R)\delta R, \tag{2}$$

where μ is the shear modulus of the medium. The effective normal spring force on each atom at the surface of the crystallite is found by dividing P by the number of atoms per unit area in a cube face. The shear modulus is taken to be c_{44} as determined from the Born-Mayer potential φ used in Eq. (6) below. The effective normal spring force per atom, so determined, is proportional to δR. The normal part of the effective spring constant, k_n, is then taken to be this force for unit displacement δR. Static tangential displacements of a surface atom are assumed to be resisted by a tangential force, given by a tangential effective spring constant, k_t. To calculate k_t we consider a long right circular cylinder of radius R, embedded in an infinite homogeneous medium with isotropic elastic properties. If this cylinder is rotated about its axis so that the tangential displacement of a point on its surface is δR, a shear stress is set up with magnitude $\mu\delta R/R$ at the surface. Distributing this equally over surface atoms in the same manner as before, one finds that the tangential displacement of each atom in the surface is opposed by a force proportional to that displacement. Thus one finds $k_t = \frac{1}{4}k_n$.

These spring constants are smaller by a factor of the order of a/R (a is the lattice constant) than the constants that would be obtained by holding the atoms beyond the surface fixed while displacing a surface atom. Physically, this factor allows for the tendency of atoms beyond the surface to move in cooperation with the motion of a surface atom and thus to oppose its motion less strongly than if they were fixed.

The spring forces on surface atoms are conservative, and it is essential to have surface dissipations, which will allow the large energies introduced into the crystallite by a primary knock-on to disappear. If disturbances reach the surface with small amplitude they can be analyzed there into harmonic plane waves, and if the surface were treated rigorously it would absorb these waves with no reflection. Using only viscous damping on the surface it is not possible to absorb all waves perfectly. As the best compromise we ask that a normally incident wave of selected frequency be absorbed as well as possible. We have chosen the viscosity coefficient for motion of a surface atom perpendicular to the surface so as to give the maximum possible absorption for a normally incident plane wave of longitudinal polarization and of half the maximum frequency of such waves. Also we have chosen the viscosity coefficient for motion of a surface atom parallel to the surface so as to give maximum possible absorption for a normally incident wave of transverse polarization and of half the maximum frequency of such waves. The criterion for maximum absorption is easy to find since a plane wave normally incident on a plane boundary presents a one dimensional problem. Under the present conditions it is found that the reflection coefficient (for power) at the mid-frequency is about 0.02; at the maximum frequency and at zero frequency the coefficient is unity, but drops steeply from these extremes and averages about 0.28 over the entire range of frequencies.

The above account has shown how the two spring force constants and two viscosity constants for each atom in the flat faces of the crystallites were chosen. The atoms in the corners and edges of the crystallite were given spring and viscosity constants derived from these by the rather obvious device of superimposing the constants associated with each face in which the atom was simultaneously resident.

These crude criteria for choosing force constants are best justified by experience with the calculations. Results have not proved very sensitive to boundary effects, as is demonstrated by cases where the same dynamic event has been run twice, starting at a different point in the crystallite, and where the same event has been run both in a large and in a small crystallite. Comments on these happenings will be made subsequently.

Choice of a size and shape for the microcrystallite is a matter of balancing the need to keep strong action away from the boundaries against the increased computing time required by a larger set of atoms. Except for a few special trials all runs to date have been made in three fundamental microcrystallites, which will be referred to as sets A, B, and C. All were rectangular parallelepipeds bounded by {100} planes. Set A was made up of 5×4×4 unit cells, and contained 446 atoms. Set B was 2×6×7 unit cells, and contained 488 atoms. Set C was 2×9×10 unit cells and contained 998 atoms. To standardize the descriptions an origin of a cartesian coordinate system will always be located in a corner of the set, and the x, y, and z directions will always be the same. All lengths are measured in such units that the cubic unit cell of copper has length 2. The range of starting positions of atoms in the various sets are then given in Table I.

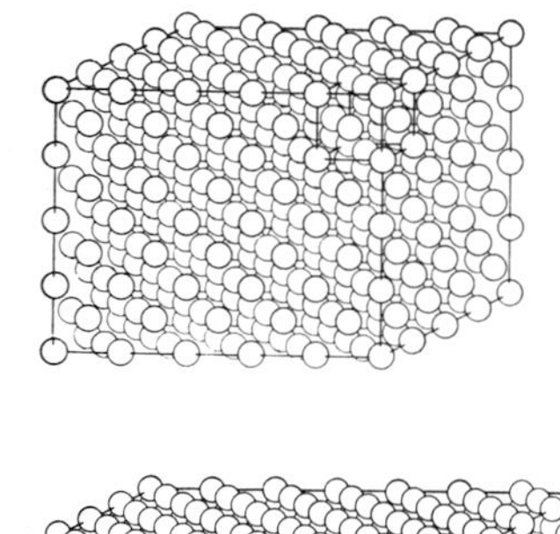

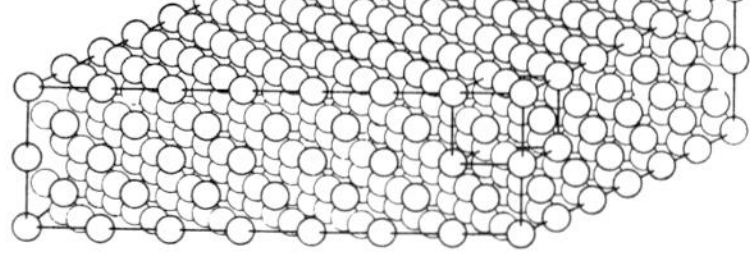

FIG. 1. Two of the sets of atoms used in the calculations. Set A is above, Set B is below.

Sets A and B are pictured in Fig. 1. Set A has been used for a variety of shots at odd directions. Sets B and C have been used only for shots with initial knock-on velocity in the (100) plane. The initial knock-on was always in the plane $x=2$, which appears to be sufficiently far away from the boundaries $x=0$ and $x=4$, and the larger y and z dimensions of these sets allowed more energetic events to be contained.

The Born-Mayer form of repulsive potential, Eq. (1), was chosen largely after the lead of Huntington[13,14] and Seitz[13] in work on point defects and self diffusion in face centered cubic metals. It is admittedly an approximation, but it is hoped that it may be an adequate approximation over the range of distances important to the present problem if the constants B and β are properly chosen. At very close approach, potentials of interaction can be established on theoretical grounds, and at separations near the equilibrium separation in the crystal some information is available from considerations of elastic constants and from atom-atom scattering experiments that have been conducted in gases. The radiation damage problem, unfortunately, demands knowledge of the potential at intermediate separations, where no reliable information exists. We have attempted to bridge this gap in the following way: We have three Born-Mayer potentials, which will be referred to simply by number, all of which give a moderately good account of the elastic constants and their variation with pressure when employed near equilibrium separations, and which are plausible as extrapolations to large separations of the theoretical repulsive potentials at small separation. The difference between the three potentials shows most strongly in the threshold energy for permanent displacement of an atom by irradiation, and the choice among the three is ultimately made by comparison between the calculated and measured threshold energy.

TABLE I. Characteristics of fundamental sets.

Set	Range of initial atomic positions	Number of atoms
A	$0\leq x\leq 10$, $0\leq y\leq 8$, $0\leq z\leq 8$	446
B	$0\leq x\leq 4$, $0\leq y\leq 12$, $0\leq z\leq 14$	488
C	$0\leq x\leq 4$, $0\leq y\leq 18$, $0\leq z\leq 20$	998

[13] H. B. Huntington and F. Seitz, Phys. Rev. **61**, 315 (1942).
[14] H. B. Huntington, Phys. Rev. **91**, 1092 (1953).

TABLE II. Constants in the Born-Mayer potentials employed [see Eq. (3)].

Potential	A (ev)	ρ
1	0.0392	16.97
2	0.0510	13.00
3	0.1004	10.34

Figure 2 shows various curves of repulsive potential energy between a pair of copper atoms plotted against the separation of the pair. At separations smaller than 0.1 A the screened Coulomb potential suggested by Bohr,[1] $\varphi=Z^2e^2r^{-1}\exp(-r/\alpha)$, where $\alpha=a_0 2^{-\frac{1}{2}}Z^{-\frac{1}{3}}$, Z being the atomic number, e the charge of the electron, and a_0 the Bohr radius, is a good representation. At larger separations the Bohr potential is undoubtedly too small.[15] Theoretical potentials which should be better than the Bohr potential at moderately small separations and which agree with the Bohr potential at very small separations have been found by Abrahamson,[16] using the Thomas-Fermi and the Thomas-Fermi-Dirac approximations (labeled TF and TFD, respectively, in Fig. 2). The TFD curve is probably the more accurate of the two. Both of these become unreliable at about 1 A, and so the curves are terminated a little beyond this point. The three Born-Mayer potentials employed in the present work are the three straight lines, labeled Pot. 1, Pot. 2, and Pot. 3. Potentials 1 and 2 are close to those suggested by Huntington for copper. For energies in the range 1- to 100-ev potential 2 represents a smaller atom than potential 1. Potential 3 was chosen arbitrarily to give the same bulk modulus as potential 1 and to give, at intermediate energies, the smallest atom of the three. It is seen from Fig. 2 that any of the three potentials might be joined to the TFD curve between 100 and 1000 ev by moderate alterations in the range 10 to 100 ev, although this would require the least alteration if done with potential 2, and would produce a more complex curve if done with potential 1. Allowing for a considerable uncertainty in the TFD result, no one of the potentials 1 to 3 is immediately ruled out for the low and moderate energy range, although potential 2 looks best. It will be shown subsequently that all three potentials give qualitatively similar results, both for the static configurations of lattice defects and for dynamic damage events, but that the threshold energies for producing a permanently displaced atom are very different for the three, being too high for potential 1, too low for potential 3, and approximately

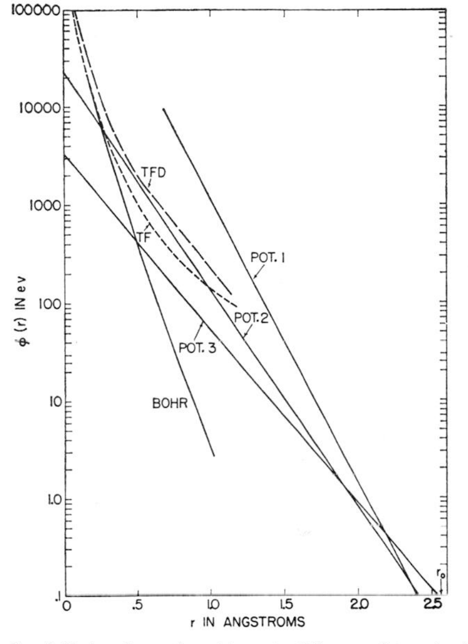

FIG. 2. Various forms of repulsive potential energy for a pair of copper atoms. Potentials 1, 2, and 3 were used in the calculations. r_0 is the equilibrium separation in the crystal.

right for potential 2. The majority of our calculations have been made with potential 2.

To specify the potentials it is convenient to recast Eq. (1) into a commonly used form,

$$\varphi=A\exp[-\rho(r-r_0)/r_0], \tag{3}$$

where r_0 is the near neighbor distance at zero pressure and absolute zero of temperature. Taking r_0 for copper to be 2.551 A, the constants A and ρ for the potentials employed in the present work are given in Table II.

The contribution of the repulsive force to the elastic moduli can be computed from the formulas[17]

$$c_{11}'=2^{\frac{1}{2}}r_0^{-1}[\varphi''+r_0^{-1}\varphi'], \tag{4}$$

$$c_{12}'=2^{-\frac{1}{2}}r_0^{-1}[\varphi''-5r_0^{-1}\varphi'], \tag{5}$$

$$c_{44}'=2^{-\frac{1}{2}}r_0^{-1}[\varphi''+3r_0^{-1}\varphi'], \tag{6}$$

$$B=(c_{11}+2c_{12})/3, \tag{7}$$

where primes on φ denote derivatives, evaluated at $r=r_0$, and only contributions from nearest neighbors have been considered. B is the bulk modulus. The complete elastic constants are the above expressions augmented by contributions from the conduction

[15] See J. A. Brinkman, J. Appl. Phys. **25**, 961 (1954). Measurements of ranges of knock-on atoms also confirm this conclusion. See R. A. Schmitt and R. A. Sharp, Phys. Rev. Letters **1**, 445 (1958), and D. K. Holmes and G. Leibfried, J. Appl. Phys. **31**, 1046 (1960).

[16] A. A. Abrahamson, thesis, New York University, 1960 (unpublished); A. A. Abrahamson, R. D. Hatcher, and G. H. Vineyard, Bull. Am. Phys. Soc. **5**, 231 (1960).

[17] H. B. Huntington, in *Solid-State Physics*, edited by F. Seitz and D. Turnbull (Academic Press, Inc., New York, 1958), Vol. 7, p. 213.

Notes on p. 59

TABLE III. Elastic moduli[a] (units 10^{11} dynes/cm^2).

	c_{11}	c_{12}	c_{44}	B
Potential 1	14.5	10.0	6.3	11.5
Potential 2	10.9	8.1	4.5	9.0
Potential 3	13.2	10.9	5.2	11.6
Experiment, 0°K	17.6	12.5	8.2	14.2

[a] First three rows list Born-Mayer part of moduli, as given by Eqs. (4)–(7). Experimental values from W. C. Overton, Jr., and J. Gaffney, Phys. Rev. 98, 969 (1955).

electrons and from Coulomb interactions of the ion cores.[17] The simplest possible estimate of the electronic contribution treats the conduction electrons as free and neglects the strain dependence of the energy of the bottom of the conduction band. In this way a contribution to the bulk modulus and to c_{11} and c_{12} of 6.4×10^{11} dynes/cm^2 is predicted. If an effective mass for electrons of $1.47m_0$ is assumed, this contribution drops to 4.3×10^{11} dynes/cm^2. The free electron contribution to these moduli must, in our model, be allotted to the surface forces. The third elastic modulus, c_{44}, is not directly affected by the Fermi energy of conduction electrons; Fuchs calculated that for copper the electrostatic interaction of the ions with the conduction electrons should contribute 2.6×10^{11} dynes/cm^2 to c_{44}. Table III lists the values of the contributions to elastic moduli from the various Born-Mayer potentials employed here, and also the experimental elastic moduli for copper at 0°K.

Table III shows that an electron contribution to the bulk modulus of about 2.6×10^{11} dynes/cm^2 is needed with potentials 1 and 3 and about 5.2×10^{11} dynes/cm^2 is needed with potential 2. These contributions would also bring c_{11} and c_{12} approximately to the experimental values. Electron contributions of this magnitude are reasonable, in view of estimates mentioned above, but exact values are difficult to establish. This contribution could be brought into our model by additional spring forces on the boundary. The spring forces already employed act in this direction, but are too small to contribute appreciably to the effective elastic constants. The dynamic stages of a damage event clearly would not be affected by such changes; experience with the computations leads us to believe that the static results also would not be very much affected, since most of the distortion around interstitials and vacancies is large near the defect, but very small at the boundary of the crystallite. More work is needed, however, to establish this point fully.

The value of c_{44} that comes from the Born-Mayer forces also is small compared with the experimental value, and an electrostatic effect such as that calculated by Fuchs is needed. This shortcoming of our model would not be expected to affect the dynamic stages of damage events appreciably, but it may affect the stability of lattice defects. Because of this and also because of the lack of any other special energies associated with electron rearrangement near lattice defects, our model may not give correct stability and energy for all defects. The argument for pursuing calculations in great detail with this model is that it will provide one complete and self-consistent picture of both the damage process and the ensuing lattice disruption. Corrections to the model can then be made as a second approximation.

3. METHOD OF COMPUTATION

A. Integrating the Equations of Motion

Solving a large number of coupled differential equations is time consuming, even on a high speed computer. Since the force law employed is only an approximation to the true force law, it was deemed not worthwhile to strive for extremely high accuracy in the integration scheme; instead, a simple central difference procedure was used which gives reasonable accuracy along with reasonable speed.

Let the ith atomic coordinate at time t be $x_i(t)$ and let the associated velocity be $v_i(t)$, where $i=1,2,\cdots N$ and N is three times the number of atoms in the crystallite. The force in the ith degree of freedom depends, in general, on the positions of all atoms. In the case that x_i refers to a boundary atom the force depends in addition on the velocity in the ith degree of freedom (because of the viscous damping). Thus the force may always be written $F_i[x_1(t),\cdots x_N(t);\, v_i(t)]$. Letting m be the mass of an atom, the classical equations of motion of the system are

$$\dot{v}_i(t)=m^{-1}F_i[x_1(t),\cdots x_N(t);\, v_i(t)], \qquad (8)$$

$$\dot{x}_i(t)=v_i(t), \quad i=1,2,\cdots N. \qquad (9)$$

Our procedure is to replace time derivatives by finite differences with arbitrary interval Δt; coordinates are defined on integer steps and velocities on half integer steps:

$$\dot{v}_i(t)\cong[v_i(t+\Delta t/2)-v_i(t-\Delta t/2)]/\Delta t, \qquad (10)$$

$$\dot{x}_i(t+\Delta t/2)\cong[x_i(t+\Delta t)-x_i(t)]/\Delta t. \qquad (11)$$

In Eq. (9) t is replaced by $t+\Delta t/2$, then (10) and (11) are inserted in (8) and (9). Rearrangement gives[18]

$$v_i(t+\Delta t/2)\cong v_i(t-\Delta t/2)+\Delta t m^{-1}F_i[x_1(t),\cdots x_N(t);\, \times v_i(t-\Delta t/2)], \qquad (12)$$

$$x_i(t+\Delta t)\cong x_i(t)+\Delta t v_i(t+\Delta t/2), \quad i=1,2,\cdots N. \qquad (13)$$

Starting with a complete set of positions $x_i(t)$ at arbitrary time t, and corresponding velocities $v_i(t-\Delta t/2)$, the machine essentially employs (12) to compute the new velocities $v_i(t+\Delta t/2)$ and (13) to compute new coordinates $x_i(t+\Delta t)$. The process is then iterated to generate coordinates at $t+2\Delta t$, $t+3\Delta t$, etc.,

[18] There is a minor inconsistency in that the viscous force in (12) is computed from $v_i(t-\Delta t/2)$. This introduces no appreciable error.

together with the corresponding velocities. The optimum size for Δt depends on the maximum velocity of any atom. Thus in the early stages of a calculation Δt is small; after the velocities of moving atoms have diminished, Δt may be increased to hasten the computation. The calculation is stopped whenever the configuration is judged to have stabilized sufficiently. Considerations governing the choice of Δt are discussed below. The program itself will be described in detail in a forthcoming Brookhaven National Laboratory report.

The coordinates and velocities of all atoms at alternate time steps are stored on magnetic tape and can be printed out as desired. The positions of selected atoms can also be displayed as dots on a cathode-ray screen. Displays are presented sequentially and multiflash pictures, such as Figs. 14 and 15, can be taken on stationary film; by advancing the film after each display, moving pictures have been made.

B. Units Used in the Calculation

For convenience the unit of length was chosen to be one-half a cubic cell edge of Cu,

$$l_0 = 1.804\times10^{-8}\text{ cm},$$

the unit of time was such that a 1000-ev copper atom would have unit velocity,

$$t_0 = 3.273\times10^{-15}\text{ sec},$$

and the unit of energy was one electron volt,

$$E_0 = 1.602\times10^{-12}\text{ erg}.$$

From these one obtains the unit velocity (v_0) as 5.512×10^6 cm/sec and the unit of mass $m_0 = E_0/v_0^2 = 5.275\times10^{-26}$ g, which is 1/2000 the mass of a copper atom.

C. Energy Checks

The central difference scheme employed in the present calculation leads to a rigorous conservation law, analagous to the energy conservation principle of classical mechanics and reducing to the latter when $\Delta t\to 0$. To demonstrate this, one first notes, from two applications of Eq. (13), that

$$v_i(t+\Delta t/2)+v_i(t-\Delta t/2) = (\Delta t)^{-1}[x_i(t+\Delta t)-x_i(t-\Delta t)]. \quad (14)$$

Rewriting Eq. (12), one has

$$m[v_i(t+\Delta t/2)-v_i(t-\Delta t/2)] = \Delta t F_i[x_1(t),\cdots x_N(t);\, v_i(t-\Delta t/2)]. \quad (15)$$

Multiplying Eq. (14) by Eq. (15), one finds

$$(m/2)[v_i^2(t+\Delta t/2)-v_i^2(t-\Delta t/2)] = F_i[x_1(t),\cdots x_N(t);\, v_i(t-\Delta t/2)] \times\tfrac{1}{2}[x_i(t+\Delta t)-x_i(t-\Delta t)]. \quad (16)$$

This relation shows that the increase of kinetic energy in the ith degree of freedom in one time step is rigorously equal to an effective work done during that time step. The effective work is seen to be the product of the force at time t by an averaged displacement $(\frac{1}{2})[x_i(t+\Delta t)-x_i(t-\Delta t)]$. This displacement may also be written $(\frac{1}{2})[x_i(t+\Delta t)+x_i(t)]-(\frac{1}{2})[x_i(t)+x_i(t-\Delta t)]$, which shows that it is the increment in average position associated with time t. Equation (16) may be written for each time step, summed over the time steps from starting time 0 to time $T=M\Delta t$, and summed again over all degrees of freedom. This gives the master conservation law

$$K(T)-K(0) = -[\Phi(T)-\Phi(0)]-D(T). \quad (17)$$

Here $K(T)$ is the total kinetic energy of the system at time T,

$$K(T) = (m/2)\sum_{i=1}^{N} v_i^2(T+\Delta t/2). \quad (18)$$

$\Phi(0)-\Phi(T)$ is a version in finite differences of the work done on the system by all conservative forces, and $D(T)$ is essentially the dissipative work in the interval 0 to T. Φ might be termed a pseudopotential. If the force F_i is divided into a conservative part F_i^c (the Born-Mayer plus the spring forces) and a dissipative part F_i^d (the viscous force), one has

$$\Phi(T)-\Phi(0) = -\sum_{i=1}^{N}\sum_{\mu=1}^{M} F_i^c[x_1(\mu\Delta t),\cdots x_N(\mu\Delta t)] \times\frac{[x_i(\mu\Delta t+\Delta t)-x_i(\mu\Delta t-\Delta t)]}{2}, \quad (19)$$

and

$$D(T) = -\sum_{i=1}^{N}\sum_{\mu=1}^{M} F_i^d[v_i(\mu\Delta t-\Delta t/2)] \times\frac{[x_i(\mu\Delta t+\Delta t)-x_i(\mu\Delta t-\Delta t)]}{2}. \quad (20)$$

If $\Delta t\to 0$ $\Phi(T)$ approaches the classical potential energy $V(x)$, where $-\partial V(x)/\partial x_i = F_i^c$, and x is the coordinate set at time T.

The following use has been made of the conservation law (17). The kinetic energy K, and the cumulative dissipation (to surface viscosity) D, are computed at each time step. Equation (17) is then employed to compute the pseudopotential Φ. The classical potential $V(x)$ is also computed directly from its analytic form and compared with Φ at each stage. If the difference is less than a preset tolerance the machine proceeds automatically to the next time step; if the difference exceeds the tolerance, the machine repeats the calculation of the present time step. If the difference is now within tolerance the machine proceeds, if it is again outside tolerance the machine stops. A large class

Notes on p. 59

of possible machine errors will cause a discrepancy beyond tolerance, and in the automatic repetition of the step the error may be rectified. Truncation errors (inadequacy of the finite difference approximation) also will cause a discrepancy between Φ and V, and thus the energy check monitors errors of this kind as well. Tolerance is usually kept at about 1.0 ev when Φ and V are in the vicinity of 125 ev (sets A and B). It is felt that this is sufficient accuracy. When the truncation error proves to be larger than this, the calculation is restarted with a smaller value of Δt. All of the energies are printed out on line for each time step; this shows the progress of the calculations, and also gives, at the end, the stored energy of the defect configuration produced.

D. Choice of Δt

The choice of an optimum interval Δt is a matter of some delicacy. Analytic solutions of the finite difference equations for one-dimensional problems with certain simplified potentials can be found, and these give useful insight. In general the force must not vary by too large a fraction of itself in one time step. This means that in an actual calculation, the pair of atoms with strongest interaction anywhere in the system places the most severe demand on Δt, and our choice of Δt has generally been governed by the energy in the strongest interaction. During stages of vibration at small amplitude, on the other hand, the system is behaving as a set of coupled oscillators. The analytic solution of the simple harmonic oscillator problem shows that Δt must be small compared with the period of the oscillator, and we have interpreted this to mean that the coupled oscillator problem requires Δt to be small compared with the shortest normal period of the system.

Ultimately, the only reliable check on Δt is afforded by repeating a calculation with a smaller Δt and comparing the results. This has been done on selected test problems. The problem shown in Fig. 8 (25 ev in the (100) plane, 15° away from [010]) was run with three different time steps, 2, $\frac{1}{2}$, and $\frac{1}{8}$, and the first collision was checked in detail for each. This collision takes place between times 4 and 8, which is only two time steps in the case of the largest Δt. Positions and velocities during the collision differed only in the fourth place in the case of $\Delta t=\frac{1}{2}$ and $\frac{1}{8}$. The calculation with $\Delta t=2$ differed, at most, in the second place from the results with $\Delta t=\frac{1}{8}$. It was concluded that for this collision $\Delta t=\frac{1}{2}$ was small enough, reduction to $\Delta t=\frac{1}{8}$ being unnecessary. It is surprising that $\Delta t=2$, sampling the collision in only two steps, gave an error of but 1% in position. The energy transfer to the atom struck in the collision was about 3% higher for $\Delta t=2$ than for $\Delta t=\frac{1}{8}$. Following experience of this kind, the following procedures were developed: Problems with initial energies around 25 ev were regularly started with $\Delta t=1$, those with initial energy around 100 ev were started with $\Delta t=\frac{1}{2}$, and those with initial energies around 400 ev were started with $\Delta t=\frac{1}{4}$. After energy checks had been built into the program, these values of Δt were reduced by a factor of two so that the larger discrepancies in energy occurring only during collisions would not stop the program.

As the collision cascade proceeds the energy of a moving atom is divided about in half at each collision. As this energy decreases Δt may be increased. The upper limit for Δt is determined by the frequency of the localized modes of the split interstitials. This is perhaps 2.5 times the Debye frequency. The Debye period is about 70 time units, the period of the localized mode is about 24. In order to describe this mode there should be several time steps per quarter cycle, and thus $\Delta t\cong 2$ is the maximum. All other modes are below the Debye frequency and permit $\Delta t\cong 6$. This largest time step has been used in some static problems.

E. Computational Speed

With 500 atoms the computing time required for one time step is about one minute on the IBM 704. The program is such that this time is proportional to the number of atoms. Total time to run a problem to quiescence varies a great deal. In some problems the end point is reasonably certain after about an hour. More ambitious problems, in which the settling down is followed in detail, require considerably more time. The history of such a problem will be described, to give an idea of fairly typical procedures. The problem is shown in Fig. 13, and was a shot at 35 ev, in the (100) plane, directed 1° away from [010], in set B (488 atoms). From $t=0$ to 23, Δt was taken as $\frac{1}{2}$ (46 steps). Δt was then increased to 1 for $t=23$ to 82 (59 steps). Next Δt was increased to 2 and the problem was run from $t=82$ to $t=182$, (50 steps). At this time the motions were very small; Δt was increased to 6 and the run was continued to $t=566$ (64 steps). The total was 219 steps, or a little over three and one half hours of machine time. Miscellaneous operations such as setting up tapes, taking edits, and making movies might require one half to one hour more.

F. Reliability

Errors may be divided into truncation error, round off error and machine errors. Truncation errors were discussed in the section on the choice of Δt, machine errors in the energy check section. Round off error would not seem to be important, since the equivalent of eight decimal places is carried by the machine. This would make truncation error much more important, by several orders of magnitude. Before energy checks were built into the system results were checked for physical reasonableness. Most problems contained some symmetry, and errors were detected by the loss of this property. Several problems were also run twice to check reliability. Even with our built in energy check, energies, positions, and velocities are checked for reasonableness whenever possible. It is believed that all large errors have been detected by these means.

4. STATIC RESULTS

It is desirable to know the configuration and stability of various lattice defects that can be housed in our model. Accordingly a number of "static" calculations have been run. In these the equilibrium configuration of a defect is estimated from simple considerations, and the atoms are given these coordinates at the beginning. All the atoms are started from rest, except that in cases where the configuration has symmetry one atom is given a very small initial velocity in such a way as to spoil the symmetry without introducing appreciable kinetic energy. False equilibria corresponding to "dead center" positions are thus avoided. The machine calculates the motions of the atoms from these initial conditions, until a static, equilibrium configuration is reached. Artificial damping (in which the kinetic energies of all atoms are set equal to zero each time the total kinetic energy reaches a maximum) is usually employed to hasten the attainment of equilibrium.

A full report on the static results will be presented in a future paper. In this section enough information will be given to aid in interpreting the dynamic calculations.

The vacancy seems entirely normal.[19] All three potentials have been employed, and in the case of potentials 2 and 3 the calculations have been run long enough to come very close to equilibrium. The behavior is qualitatively similar in all three cases, with the amount of relaxation being largest in the case of potential 3 and least with potential 1. The nearest neighbors relax radially inward by a small amount—about 1.5% of the equilibrium distance, $\sqrt{2}$, in the case of potential 1; 2.5% in the case of potential 2; and 3.2% in the case of potential 3. The second neighbors, and more distant neighbors in or near the cubic axial directions, relax slightly outward. In the case of second neighbors, the percentage outward relaxation is about one twentieth the inward relaxation of near neighbors. Such apparently anomalous relaxation has already been found by others[20] and is easily understood by considering the geometry of the lattice. An immediate consequence is that the strain field at a distance from the vacancy cannot be very well fitted to the field of a point singularity in an *isotropic* elastic continuum. A cubic elastic continuum is required and the outward relaxation along cubic directions can be considered as a manifestation of the anisotropic character of the medium.[21]

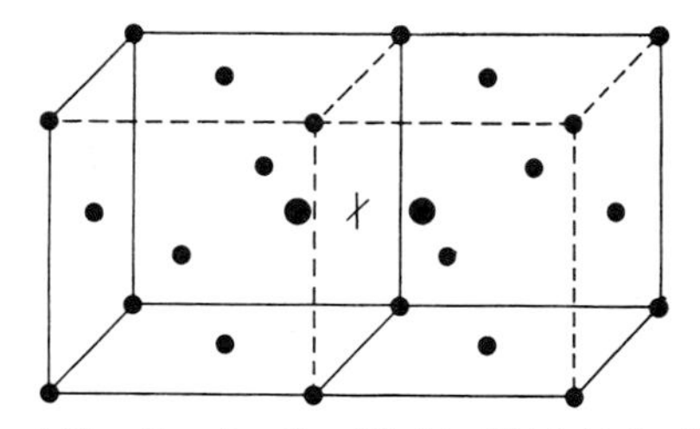

FIG. 3. The split configuration of the interstitial that is found to be stable. Relaxation of neighbors not shown.

The interstitial has been investigated carefully with potential 2, and is found not to reside at the center of the cubic unit cell. Instead, the interstitial has what may be termed a split configuration, in which it shares a lattice site symmetrically with another atom, the axis of the pair being along a cubic axis of the lattice. Figure 3 shows a split interstitial in the face-centered lattice. For potential 2 the separation of the two atoms is very nearly 1.2 (in units in which the lattice constant is 2). The possibility of this configuration of the interstitial in copper was pointed out by Huntington and Seitz,[12] although its stability was not settled at that time. More recently, Johnson *et al.*[22] have also demonstrated its stability in a lattice model rather similar to ours. It should be noted that there are three possible orientations of this interstitial on each lattice site; its symmetry is only tetragonal, and it should thus give rise to resonant anelastic effects.

The stability of this interstitial has been demonstrated in two calculations. First an interstitial was set up in the cube center, with relaxations of its neighbors, according to our first estimate of the stable position of the interstitial. The machine calculation showed that this atom rapidly moved away from the cube center toward a neighboring atom, in a direction determined by minor asymmetries in the starting conditions, and settled down in the split configuration with this atom. Later, as a check, the split configuration, with minor perturbations, was set up as an initial condition, and a long machine run was made. This demonstrated the complete stability of the split configuration and gave accurate values of the relaxations of surrounding atoms. As will be seen in the following section, a number of dynamic events have also produced interstitials (see, for example, Figs. 12 and 13), and in all cases these are seen to settle down in the split configuration.

[19] Our results give no support to the "relaxion" picture of the vacancy which has been put forward by N. H. Nachtrieb and G. S. Handler, Acta Met. 2, 797 (1954), and N. H. Nachtrieb, H. A. Resing, and S. A. Rice, J. Chem. Phys. 31, 135 (1959).

[20] H. Kanzaki, J. Phys. Chem. Solids 2, 24 (1957); G. L. Hall, J. Phys. Chem. Solids 3, 210 (1957); A. Seeger and E. Mann, J. Phys. Chem. Solids 12, 326 (1960); L. A. Girifalco and V. G. Weizer, J. Phys. Chem. Solids 12, 260 (1960).

[21] The writers are indebted to Dr. E. Kröner and Dr. A. Seeger for information on strain fields of point sources in a cubic elastic medium. Quantitative comparisons with Kröner's solutions will be given in a future paper.

[22] R. A. Johnson, G. H. Goedecke, E. Brown, and H. B. Huntington, Bull. Am. Phys. Soc. 5, 181 (1960).

Notes on p. 59

The foregoing calculations do not demonstrate fully that there are no other stable configurations of the interstitial, and indeed experimental evidence has occasionally been interpreted as requiring that the crowdion be stable in copper. We have tested the stability of the crowdion in our model by two static calculations, both employing potential 2. In both of these a crowdion was formed by inserting an interstitial atom in a ⟨110⟩ line and moving three atoms on either side of the inserted atom outward along the line by diminishing amounts. A total of eight neighbors on adjoining ⟨110⟩ lines were also relaxed away from the interstitial, in each case, to obtain the lowest energy configuration possible. In one calculation the extra atom was inserted half way between two neighboring lattice sites and the relaxation along the line was made symmetrical about this point. This might be called a space-centered crowdion. In the second calculation, which might be called a site-centered crowdion, the extra atom was placed on one side of a lattice site and the pattern of relaxation was made symmetrical about this site. Both crowdions proved to be unstable, and decayed into a split interstitial by simple rotation of a pair of atoms near the center. The decay occurred rather slowly, however, which demonstrates that the potential energy is fairly flat near these crowdion configurations. We conclude that the crowdion is not stable in our model; nevertheless, rather modest changes in the force laws might make it stable.

It is also necessary to know which Frenkel pairs are stable. Accordingly a series of static runs on Frenkel pairs at various separations were made with potential 2. The results for pairs in the {100} plane are shown in Fig. 4. Here the split interstitial is shown at a fixed position in the lower left corner of the figure. Lattice sites around this interstitial at which a vacancy yielded a stable Frenkel pair (by actual calculations) are indicated by S. Sites for the vacancy which yielded an unstable Frenkel pair are indicated by U. All sites inside the dotted line are unstable, all sites outside it are stable. It is seen that a surprisingly large separation of the pair is needed to produce stability, particularly for a pair on a close packed line. The size of this region of instability has obvious implications for near-threshold damage events, and also for the annealing process in which a migrating interstitial recombines with a vacancy.

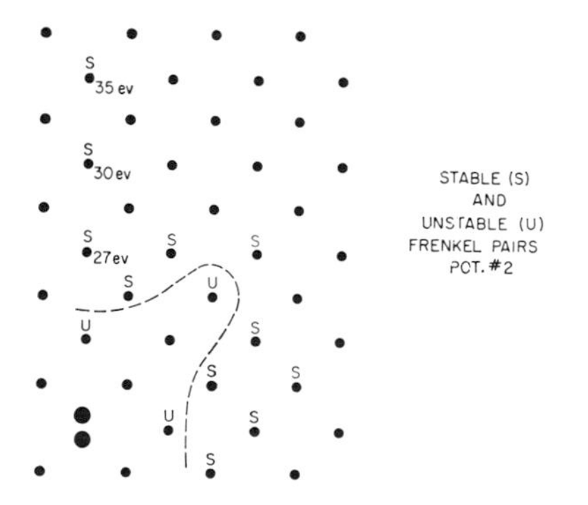

Fig. 4. Stability of Frenkel pairs in {100} plane of copper. Split interstitial is at lower left. Dotted line separates stable from unstable sites for a vacancy. Approximate threshold energies for dynamic production of three particular pairs are indicated.

Energies associated with various point defects are also under investigation, and a very brief account will be given here. Let W_v be the work needed to take an atom reversibly from a normal lattice site to a distant point out of the crystal. Let W_I be the work needed to take an atom from this distant point and to insert it, reversibly, into a perfect crystal, forming a split interstitial. In our model, W_v is given by the potential energy of the perfect crystallite, minus the potential energy of the crystallite containing a thoroughly relaxed vacancy. Similarly W_I is the potential energy of the crystallite containing a relaxed split interstitial, minus the potential energy of the perfect crystallite. W_v and W_I include the potential energies of the surface forces. For potential 2, static runs give

$$W_v = -0.71 \text{ ev},$$
$$W_I = 3.38 \text{ ev}.$$

To appreciate the significance of these numbers, one must consider that the thermodynamic energy of formation of a vacancy, E_v, and energy of formation of an interstitial, E_I, are related to W_v and W_I by

$$E_v = W_v - W_s,$$
$$E_I = W_I + W_s,$$

where W_s is the work needed to take an atom reversibly from an average site on the surface of the crystal to a remote point outside the crystal. E_v and E_I are the energies that determine the equilibrium concentrations of vacancies and interstitials, in the familiar way. Since our model does not give a proper account of a free surface it is not convenient to compute W_s directly. Instead one can consider the work needed to disassemble the entire crystal, after cutting off the surface forces at a suitable finite range. This work, per atom, will also be an estimate of the sublimation energy, W_s; such considerations show that, in our model, W_s is negative. Reasonable values of W_s give reasonably good values of E_v and E_I, although the latter tends to lie lower than expected.

The energy of formation of a separated Frenkel pair, E_F, is given by

$$E_F = E_v + E_I = W_v + W_I = 3.38 - 0.71 = 2.67 \text{ ev}.$$

This is lower than values indicated by some earlier

calculations,[23] but is not entirely outside reasonable limits.

The divacancy and trivacancy have been studied in a preliminary way with potential 2, and have configurations generally consistent with findings of other investigators.[24,25] The binding energy of the divacancy, against separation into isolated vacancies, is about 0.06 ev in our model, and of the trivacancy, against separation into three isolated vacancies, is about 0.5 ev. These binding energies are lower than those reported in the literature.[24,25] Work on these energies, on activation energies for migration, and on other clusters of point defects is continuing.

5. DYNAMIC RESULTS

We come now to the chief results of this work, dynamic problems corresponding to real radiation damage events. To date about 45 distinct calculations leading to usable results have been run. These have involved all three of the repulsive force laws (see Fig. 2), a rather wide variety of knock-on directions, and a number of knock-on kinetic energies up to 400 ev. In this section, a representative selection of the dynamic events will be discussed. These events have been chosen to give a good general idea of what has been learned, without attempting to discuss details of every event run. In the Appendix Table IV lists all the events run successfully, and Fig. 5 is a diagrammatic presentation of all of the events in which potential 2 was used for the repulsive force (for discussion see below). As will be seen the dynamic events give strong preference to

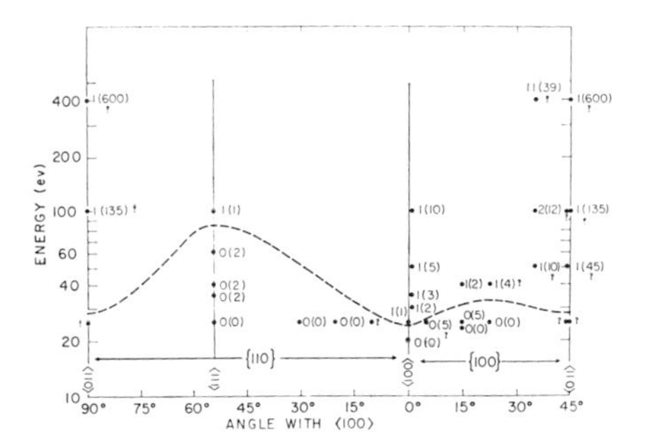

FIG. 5. Diagram showing all dynamic events calculated with potential 2 (see also Appendix). A dot is shown for each event, and indicates kinetic energy and direction of knock-on atom. First figure attached gives number of stable Frenkel pairs created, figure in parenthesis gives number of replacements. Dotted line is estimated threshold for creation of at least one stable Frenkel pair.

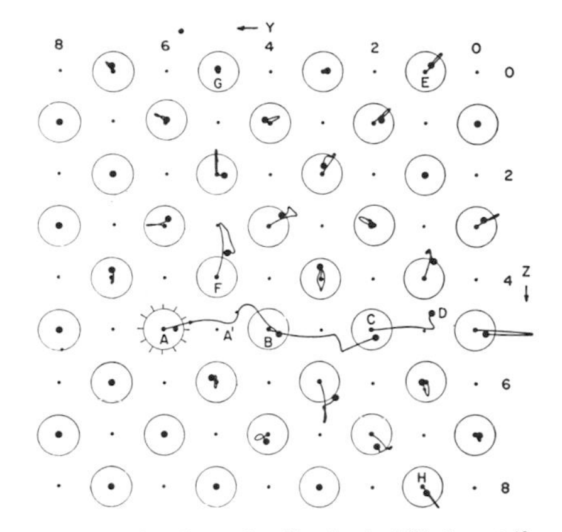

FIG. 6. Atomic orbits produced by shot in (100) plane at 40 ev. Knock-on was at A and was directed 15° above $-y$ axis. Large circles give initial positions of atoms in plane; small dots are initial positions in plane below. Vacancy is created at A, split interstitial at D. Run to time 99. (Run No. 12).

potential 2 among the three forms tried. Consequently this one has been used for most of the calculations.

A. Description of Events

Figure 6 shows the trajectories resulting in the y-z plane when an atom (A) is set in motion with 40 ev of kinetic energy in a direction in this plane making an angle of 15° with the $-y$ axis. The initial positions of atoms in the planes immediately above and below the plane of major action are shown by small black dots. Large open circles show the atoms in the plane of major action at time 0, large black dots show the positions of these atoms at time 99 (one time unit is 3.27×10^{-15} sec). The large open circles give the sizes of the atoms, as determined by the distance of closest approach in a head-on collision between a 40-ev atom and a stationary atom. Atoms for which no trajectory is shown suffered negligible displacements. Replacement collisions can be seen at B and C, a vacancy is left at A, and an interstitial is formed at D. This appears to be the usual split configuration (see Fig. 3) in which the atom at $y=1$, $z=4$ is displaced upward from its lattice site, the site being shared by D. Also notable are the chains of strongly focused collisions along ⟨110⟩ and ⟨100⟩ directions, including the chains AD, AE, FG, BH, etc. The ⟨110⟩ focusing is essentially that predicted by Silsbee[26] (see below), but the ⟨100⟩ focusing occurs only because of the influence of neighboring lines of atoms, and had not been anticipated. Atoms along all lines other than AD are in the process of relaxing back to the vicinity of their original positions. Although the relaxation has not

[23] Huntington,[13] found 5 to 6 ev for E_F. Radiation damage experiments have yielded 5.4 ev[6] and also much lower values.[2]

[24] J. H. Bartlett and G. J. Dienes, Phys. Rev. **89**, 848 (1953).

[25] A. C. Damask, G. J. Dienes, and V. G. Weizer, Phys. Rev. **113**, 781 (1959).

[26] R. H. Silsbee, J. Appl. Phys. **28**, 1246 (1957).

Notes on p. 59

TABLE IV. Table of dynamic events calculated.

Run number	Potential (see Fig. 2)	Direction of knock-on	Kinetic energy of knock-on (ev)	Set (see Table I)	Coordinates of knock-on atom[a]	Final time[b]	Number of replacements	Number of displacements (Frenkel pairs)	Remarks[c]
4	1	[100]	25	*A*	2,4,4	23	0	0	Below threshold. Two knock-ons in this run.
4	1	$22\frac{1}{2}$° from $[\bar{1}00]$[d]	25	*A*	8,4,4	23	0	0	Below threshold. Two knock-ons in this run.
2	1	[111]	25	*A*	4,4,4	18	0	0	Below threshold. See Fig. 20.
14	3	[111]	20	*A*	2,2,2	92	0	0	Atom does not reach cube center. Below threshold. Two knock-ons in this run. See Fig. 19.
15	3	[111]	23	*A*	2,2,2	258	0	0	Atom does not reach cube center. Below threshold. Two knock-ons in this run. See Fig. 19.
15	3	$[\bar{1}\bar{1}1]$	27	*A*	8,6,2	258	0	0	Atom reaches cube center but returns. Below threshold. Two knock-ons in this run. See Fig. 19.
17	3	$[\bar{1}\bar{1}1]$	27	*A*	8,6,2	130	0	0	Same as 27 volt atom in run No. 15. Confirms run 15.
14	3	$[\bar{1}\bar{1}1]$	30	*A*	8,6,2	92	2	0	Struck atom goes to cube center. Static calculations indicate it is unstable and will decay by 2 replacements. Two knock-ons in this run.
16	3	$[\bar{1}\bar{1}1]$	30	*A*	8,6,2	200	2	0	Same as 30-volt atom in run No. 14, but carried to longer time. Confirms No. 14. See Fig. 19.
18	3	[111]	100	*A*	2,2,2	45	2	1	At end of run, interstitial is in cube center at 7,7,7. Must make split interstitial in this cell. Vacancy left at 2,2,2. $\Delta t=4$, which is rather large.
39	2	[100]	20	*A*	2,4,4	175	0	0	Temporarily forms split interstitial around 4,4,4, oriented along [100]. This is unstable and reverts to original configuration.
30	2	[010]	25	*B*	2,2,6	380	1 (est)	1 (est)	Appears to be making interstitial at 2,6,6, oriented along [010]. Vacancy is at 2,2,6. Stability is not quite certain. Machine error about $t=180$.
46	2	[100]	25	*A*	2,4,4	47	1 (est)	1 (est)	Confirms No. 30 as far as it goes.
54	2	1° from [010][d]	30	*B*	2,2,6	317	2	1	Makes split interstitial at 2,8,6, oriented along [010]. Vacancy left at 2,2,6. See Fig. 12.
50	2	1° from [010][d]	35	*B*	2,2,6	566	3	1	Makes split interstitial at 2,10,6 oriented along [010]. Vacancy left at 2,2,6. See Fig. 13.
72	2	1° from [001][d]	50	*B*	2,5,1	75	5	1	Should make split interstitial either at 2,5,13, or 2,5,15 with vacancy at 2,5,1. See Fig. 14.
73	2	1° from [001][d]	100	*B*	2,5,1	42	10 (est)	1 (est)	Defocusing chain gives maximum KE of 59 ev to atom at 2,5,11. Should make interstitial at about 2,5,21 with vacancy at 2,5,1. See Fig. 15.
31	2	5° from [010][d]	25	*B*	2,2,6	250	5 (est)	0 (est)	At end has made split interstitial at 2,5,5, oriented along [001]. May decompose by 5-membered ring mechanism. (See No. 25.)
91	2	15° from [010][d]	23	*B*	2,2,6	199	0	0	Below threshold.
25	2	15° from [010][d]	25	*B*	2,2,6	646	5	0	Forms split interstitial at 2,5,7, vacancy at 2,2,6, then decomposes by 5-membered ring mechanism. $\Delta t=2$ up to $t=442$. See Figs. 8–11.
35	2	15° from [010][d]	25	*B*	2,2,6	22	5 (est)	0 (est)	$\Delta t=0.125$, for check on truncation error in No. 25.

[a] Coordinates are relative to origin at corner of set, with axes oriented as in Table I.
[b] In units of 3.273×10^{-16} sec.
[c] Estimated positions of some defects are outside atomic set used.
[d] In (100) plane.
[e] In $(01\bar{1})$ plane.

TABLE IV. *Continued.*

Run number	Potential (see Fig. 2)	Direction of knock-on	Kinetic energy of knock-on (ev)	Set (see Table I)	Coordinates of knock-on atom[a]	Final time[b]	Number of replacements	Number of displacements (Frenkel pairs)	Remarks[c]
35A	2	15° from [010][d]	25	*B*	2,2,6	51	5 (est)	0 (est)	$\Delta t = 0.5$, for check on truncation error in No. 25.
12	2	15° from [0$\bar{1}$0][d]	40	*A*	3,6,5	99	2	1	Interstitial formed at 3,1,4, oriented along [001]. Vacancy at 3,6,5. Two knock-ons in this run. See Fig. 6.
32	2	22½° from [010][d]	25	*B*	2,2,6	114	0	0	Below threshold.
12	2	22½° from [010][d]	40	*A*	7,2,3	99	4 (est)	1 (est)	Interstitial estimated to be forming at about 7,7,8, orientation uncertain. Vacancy at 7,2,3. Two knock-ons in this run. See Fig. 7.
40	2	[011]	25	*B*	2,2,2	44	?	?	Cannot run long enough to determine whether above threshold. See Fig. 16.
44	2	[011]	100	*B*	2,2,2	27	135 (est)	1 (est)	Dynamic crowdion. See Fig. 17.
45	2	[011]	400	*B*	2,2,2	14	600 (est)	1 (est)	Dynamic crowdion. See Fig. 18.
74	2	1° from [011][d]	25	*B*	2,2,2	78	?	?	Cannot run long enough to determine whether above threshold.
75	2	1° from [011][d]	50	*B*	2,2,2	54	45 (est)	1 (est)	Dynamic crowdion. See No. 95.
95	2	Continuation of No. 75		*B*	2,1,1 and near neighbors	46	...	...	Displacements and velocities of all near neighbors of atom 2,11,11 of No. 75, at time 36 in that problem, were used as starting conditions for atom 2,1,1 and near neighbors in this problem.
76	2	1° from [011][d]	100	*B*	2,2,2	33	?	?	Collision chain is defocusing. Atom at 2,10,10 acquires 59 ev.
71	2	10° from [011][d]	50	*B*	2,2,2	92	10 (est)	1	Interstitial estimated to form at about 2,13,13. Vacancy is at 2,2,2. See Fig. 22.
62	2	10° from [011][d]	100	*B*	2,2,2	130	11 (est)	2 (est)	Interstitials appear to be formed at about 2,11,11 and 2,0,10, vacancies at 2,2,2 and 2,4,6. See No. 96.
96	2	10° from [011][d]	100	*C*	2,6,8	703	12	2	Repeat of No. 62 in larger set. Interstitials form and stabilize at 2,14,18 (along [001]) and 2,3,17 (along [010]). See Figs. 23 and 24.
63	2	10° from [011][d]	400	*B*	2,2,2	15			Estimates of end point are very crude because large energies reach boundary. See No. 97.
97	2	10° from [011][d]	400	*C*	2,6,8	45	39 (est)	11 (est)	Repeat of No. 63 in larger set. See Fig. 25.
10	2	[$\bar{1}\bar{1}$1]	25	*A*	8,6,2	78	0	0	Below threshold. Note dual excitation. See Fig. 20.
34	2	[111]	35	*A*	2,2,2	31	2 (est)	0	Struck atom goes to cube center at 3,3,3. Decay in two membered ring is expected.
10	2	[111]	40	*A*	2,2,2	78	2 (est)	0 (est)	Struck atom goes to cube center at 3,3,3. Decay by two membered ring is expected. Note dual excitation. See also No. 33 and Fig. 20.
33	2	[111]	40	*A*	2,2,2	66	2 (est)	0 (est)	Single knock-on, to check No. 10. Gave same behavior of 40-ev knock-on as No. 10.
69	2	[111]	60	*A*	2,2,2	141	2 (est)	0 (est)	Struck atom is near 3,3,3 at end of run. This is expected to decay by two membered ring. See Fig. 21.
70	2	[111]	100	*A*	2,2,2	49	1	1	Vacancy left at 2,2,2. Interstitial will be near 5,5,5 but exact location is uncertain. See Fig. 21.
42	2	20° from [100][e]	25	*A*	2,4,4	86	0	0	Below threshold.
43	2	30° from [100][e]	25	*A*	2,4,4	70	0	0	Below threshold.

Notes on p. 59

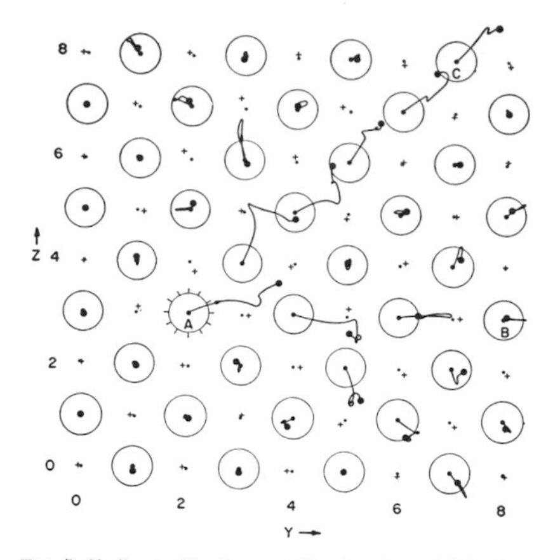

FIG. 7. Similar to Fig. 6 except that knock-on (A) is directed $22\frac{1}{2}°$ above y axis. Small crosses show positions of atoms in plane below at end of calculation (time 99). Vacancy is left at A, interstitial is estimated to form near C (Run No. 12).

been entirely completed by time 99, experience with this and other events convinces us that the further relaxation will not change the topology of the final configuration from that which is evident in Fig. 6. The net result of this event is two replacements and one Frenkel pair. Although velocities are not given in Fig. 6, it should be noted that a great range of velocities occurs. The original knock-on, which had 40 ev at A, has slowed to 19 ev just before A' (where it is in nearly head-on collision with B), and has dropped to only 0.1 ev just beyond the point A'.

Figure 7 shows an event similar to that in Fig. 6, but different in detail. Atom A was initially projected in the y-z plane with a kinetic energy of 40 ev but at an angle of $22\frac{1}{2}°$ with the y axis. The large circles and large black dots indicate initial positions and positions at time 99, respectively, as in Fig. 6. The small black dots indicate initial positions of atoms in the planes just above and below, the small crosses indicate positions in these adjacent planes at time 99. A focusing chain AB is again seen, but atoms in this chain are returning to their original sites, while a chain of replacements occurs in the diagonal direction AC, and an interstitial is being formed somewhere along this chain, most likely at the site $y=7$, $z=8$ (but possibly at $y=6$, $z=7$). The orientation of this interstitial is in doubt, and cannot be reliably determined by running the calculation longer because the displacement of the boundary atom C has become large and assumptions employed in the boundary forces are brought into question. A vacancy is clearly left at A, one Frenkel pair has been created, and four replacements appear to have occurred. The disturbance in adjacent planes, as shown by the crosses in the figure, is surprisingly slight. This is the case with all our events in which the initial velocity lies in a {100} plane. Comparing Figs. 6 and 7 it is noteworthy how much difference in location of the interstitial is caused by a small change in the direction of initial motion. The qualitative similarity of focusing tendencies in the two figures is also evident. More will be said below about focusing.

The preceding events were well above the threshold for production of lattice defects. An event very near threshold (actually, as it turns out, just below threshold) will be discussed next. An atom at 2,2,6, was given 25 ev of kinetic energy in a direction 15° from the y axis and lying in the y-z plane. Figure 8 shows the trajectories of atoms in this plane (the plane $x=2$) from time 0 to time 128. In this and in the next two figures, trajectories are indicated by dotted lines. Successive dots are separated by two units in time, so speeds along the trajectories can be estimated from the spacing of the dots. The struck atom was initially at the site indicated by the large circle. Figure 9 shows the same plane of atoms between times 130 and 254, and Fig. 10 shows the plane between times 256 and 380. In Fig. 8 a replacement is seen to occur at 2,4,6 and this atom is forced into the split interstitial position around the site 2,5,7. Prominent focusing chains are seen to branch off along several close packed lines. These transport energy, but not matter, away from the scene of initial action. In Fig. 9 the split interstitial is better established and the organized motion in the focused chains has begun to disperse. In Fig. 10 the kinetic energy has died away further and the vacancy and interstitial are clearly evident. Because of the localized vibrational modes associated with the vacancy and interstitial, particularly with the latter, there is still an appreciable amount of kinetic energy in and near the interstitial.

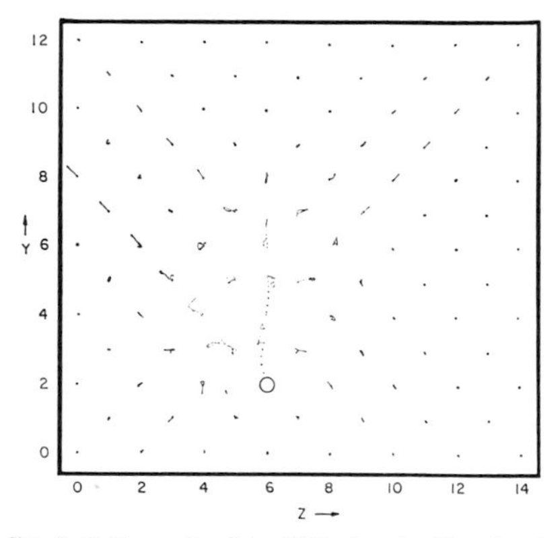

FIG. 8. Orbits produced in (100) plane by 25-ev knock-on (large circle) directed 15° away from y axis. Atomic positions at intervals of 2 units in time are shown. Orbits are shown from time 0 to 128 (Run No. 25).

As has been noted before this particular Frenkel pair is stable in our model, but very precariously so, and when the event we have been discussing was carried to still longer times, the kinetic energy retained near the defect caused a self-annealing. This occurred by means of a ring interchange involving five atoms and the five sites 2,2,6; 2,4,6; 2,5,7; 2,4,8; and 2,3,7; the orbits of these atoms during the complete run from time 0 to 646 are shown in Fig. 11. Times at various points along the orbits are indicated by numbers in the figure. The configuration evident in Fig. 10 is beginning to drift between time 300 and time 400. Between 400 and 500 the two members of the split interstitial are developing a more pronounced motion, and shortly after time 500 the two atoms at 2,4,8 and 2,3,7 are pushed along so that the former replaces the latter and the latter falls into the vacancy. By time 646 the atoms are very near their new lattice sites, from which they will obviously not escape again. The results of this shot are thus 5 replacements and no permanent displacements.

Inspection of Fig. 11 shows how a truly cooperative motion of several atoms can be involved in some annealing processes. It is also plausible that localized vibrational modes associated with a defect can play an important role in trapping a portion of the initial kinetic energy near the defect, creating a local hot spot which cools more slowly than in a perfect lattice and thus enhances self-annealing. Clearly it is a matter of considerable delicacy to establish exactly what defect is created by each near-threshold event.

A series of shots with initial velocity vector near a cube axis will be considered next. In Run No. 39, the atom at 2,4,4 was given 20 ev directed along [100]. This formed a temporary split interstitial on the sites at 4,4,4, leaving a vacancy at 2,4,4. This Frenkel pair is unstable and collapses back to the original lattice after a short time. Run No. 30 was similar except that the initial energy was 25 ev. Here a split interstitial is formed 4 units away from the vacancy, a Frenkel pair that has been proved in static runs to be stable. In the dynamic run a machine error occurred at about time 180, so that

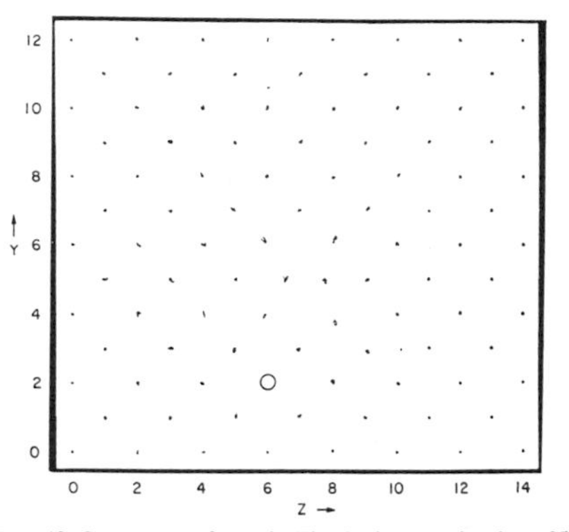

FIG. 10. Same event shown in Fig. 8, time running from 256 to 380. Temporary Frenkel pair has been produced. Note quasi-thermal agitation in vicinity of pair (Run No. 25).

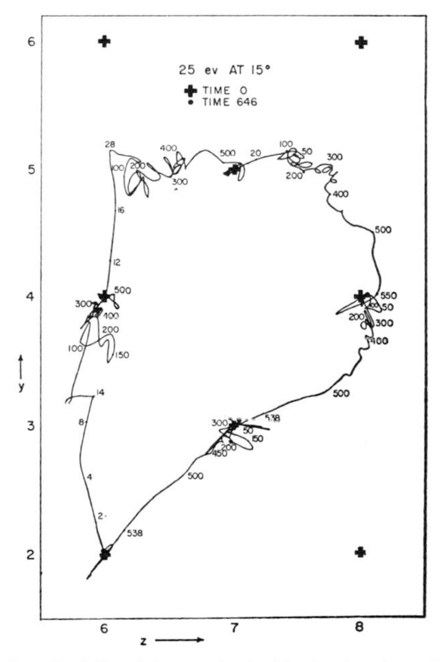

FIG. 11. Orbits of 5 atoms involved in ring interchange by which Frenkel pair seen in Fig. 10 spontaneously annealed after time 400. Figures along orbits indicate times (Run No. 25).

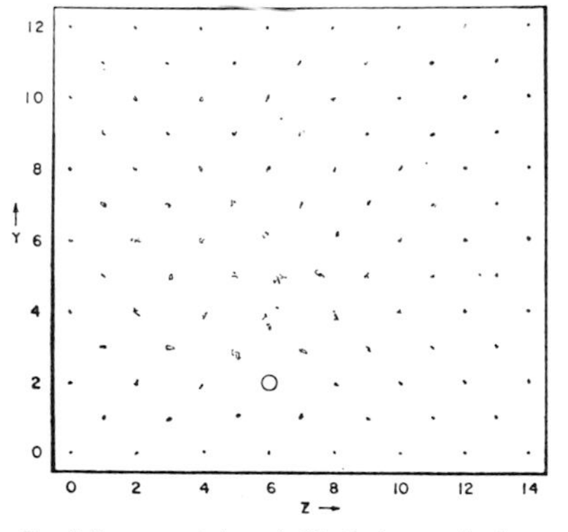

FIG. 9. Same event shown in Fig. 8, time running from 130 to 254. (Run No. 25).

Notes on p. 59

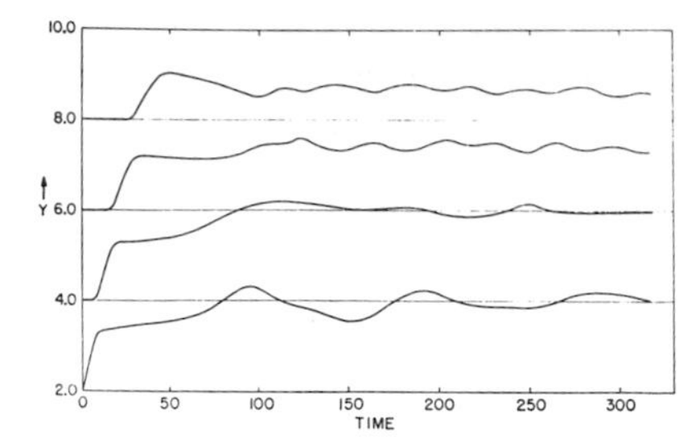

FIG. 12. y coordinates vs time for line of atoms. Knock-on (bottom curve) shot at 30 ev, 1° away from y axis. Split interstitial is created at $y=8$ (Run No. 54).

it was not possible to be quite sure that this configuration would not anneal itself before quieting down completely. The threshold for this direction is thus estimated to be very near 25 ev with our potential 2. Run No. 54 was at 30 ev, with atom 2,2,6 directed initially 1° away from the y axis in the plane $x=2$. This seems clearly to make a Frenkel pair after 2 replacements, the vacancy being left at 2,2,6 and the split interstitial, oriented along y, being centered on 2, 8, 6. Run No. 50 was exactly like No. 54 except that the initial energy was 35 ev. This time there were 3 replacements and the interstitial formed at 2,10,6. Figure 12 shows the y coordinate vs time for the replaced and displaced atoms in the 30-ev run (No. 54) and Fig. 13 shows the same information for the 35-ev run (No. 50). In Fig. 12 the localized antiphase vibration of the two members of the split interstitial is clearly evident after time 120. The period of this vibration is about 33 time units, which means that its frequency is about twice the Debye frequency for copper. Assisted focusing down the cubic axis is clearly evident in these runs, and Figs. 12 and 13 may be considered to picture the transport of both matter and energy in ⟨100⟩ chains. Similar action along ⟨110⟩ has been termed a "dynamic crowdion;"[8] and it is seen that the dynamic crowdion, if this is to be the name adopted, can indeed act along the ⟨100⟩ axes as well. The energy loss in the ⟨100⟩ case is at a greater rate than in ⟨110⟩. In runs No. 50 and No. 54 (see, e.g., the maximum slopes of the successive curves in Figs. 12 and 13) and also in runs No. 72 and No. 73 at higher energies, the attenuation of energy in well focused ⟨100⟩ chains occurs at the rate of 7 to 8 ev per collision. Atoms other than those in the direct path of the knock-ons in these events are not moved very far and return to the vicinity of their original sites.

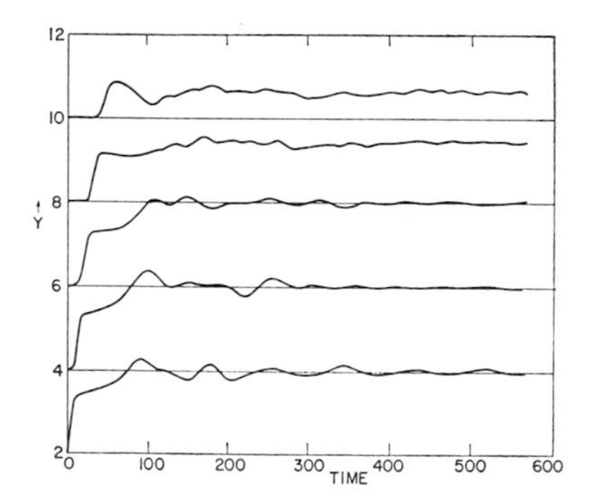

FIG. 13. y coordinates vs time for line of atoms. Knock-on (bottom curve) shot at 35 ev, 1° away from y axis. Split interstitial is created at $y=10$ (Run No. 50).

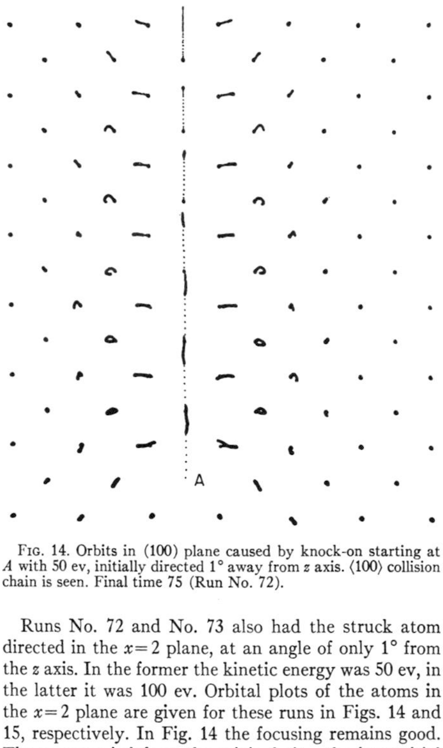

FIG. 14. Orbits in (100) plane caused by knock-on starting at A with 50 ev, initially directed 1° away from z axis. ⟨100⟩ collision chain is seen. Final time 75 (Run No. 72).

Runs No. 72 and No. 73 also had the struck atom directed in the $x=2$ plane, at an angle of only 1° from the z axis. In the former the kinetic energy was 50 ev, in the latter it was 100 ev. Orbital plots of the atoms in the $x=2$ plane are given for these runs in Figs. 14 and 15, respectively. In Fig. 14 the focusing remains good. The vacancy is left at the original site, the interstitial is projected to a point outside the original set of atoms. One estimates that the interstitial would be formed at either 2,5,13, or 2,5,15, which means 6 or 7 replacements. In Fig. 15 a pronounced early defocusing is visible (associated with the higher energy). It is estimated that about 10 replacements would occur and the interstitial would be left at about 2,5,21.

A series of shots in or near the close-packed direction

⟨110⟩ have also been made. At low energies pronounced focusing in this direction has been found, at high energies defocusing occurs, followed by focusing when the chain has lost enough energy. When directed close to ⟨110⟩ these chains lose energy only very slowly, at a rate of about 2/3 ev per collision for energies from 3 to several hundred ev. Consequently their range is so long that they cannot be stopped inside our set of atoms. Contrary to early expectations, the threshold energy for producing permanent displacements in ⟨110⟩ is rather low. Because of the difficulty with the long range of these chains, we have not been able to obtain an accurate value for this threshold, but the best estimate is that,

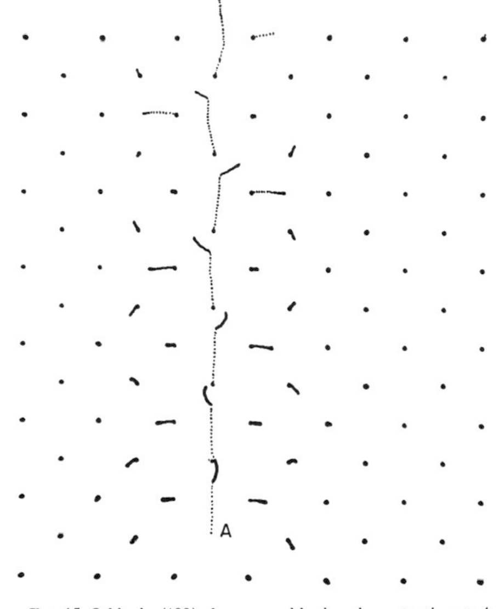

FIG. 15. Orbits in (100) plane caused by knock-on starting at A with 100 ev, initially directed 1° away from z axis. ⟨100⟩ collision chain with pronounced defocusing is seen. Final time 42 (Run No. 73).

for potential 2, it is less than 35 ev and probably is in the neighborhood of 25 ev. Since the ⟨100⟩ threshold is also around 25 ev, this means that both of these directions are important in near-threshold bombardments. Also it is clear that the interstitial produced by a ⟨110⟩ chain will be far from the beginning of the chain—at least 10 atomic spacings near threshold and as much as 150 spacings at 100 ev. Fig. 16 gives the z-displacements versus time for successive atoms in a chain initiated at 25 ev exactly along [011]. Each atom moves just slightly past the midpoint ($\Delta z = 1/2$) between it and its neighbor, before being brought to rest, and the relaxation thereafter is extremely slow. The first atom, unlike

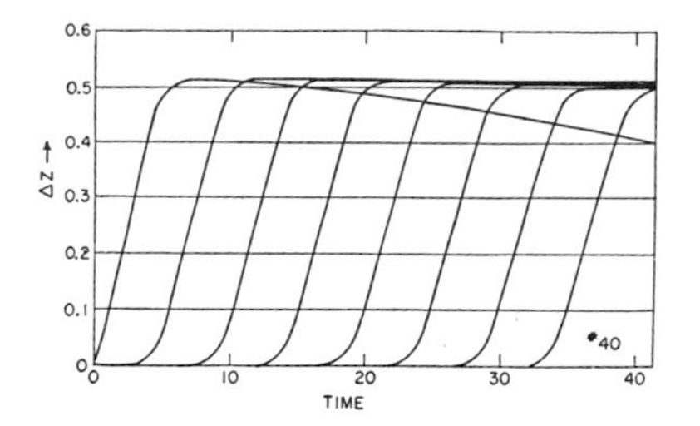

FIG. 16. Δz vs time for series of atoms in [011] chain. Knock-on, starting at time 0, had 25-ev kinetic energy, directed along [011] (Run No. 40).

the others, relaxes back toward its original site and will clearly return there. The anomalous behavior of this atom is explained by the fact that the atom preceding it has not moved much, and thus an unbalanced restoring force is supplied by the second atom in the chain. The second and neighboring atoms are lingering near saddle points, and it is impossible to tell for certain from this calculation whether their relaxation will finally be forward or back. In any case, a slightly higher initial energy should insure the forward relaxation, and if this occurs an interstitial must be produced some distance down the chain and a vacancy will be left at the site of the *second* atom in the chain. In some other events that have been run, it would appear that the vacancy may even form at the site of the third atom in the chain.

A higher energy event which clearly does produce a permanently displaced atom is shown in Fig. 17. This event is exactly like that of Fig. 16 except that the initial atom was given a kinetic energy of 100 ev. Now each atom moves well past the midpoint in its first strong collision, and subsequent relaxation is proceeding in a forward direction. This time the vacancy is being formed at the site of the first atom. The interstitial should be formed about 150 atomic distances away. Figure 18 shows the same thing again at a still higher initial energy, 400 ev. Results are much like the 100-ev case, with a more pronounced tendency for the entire

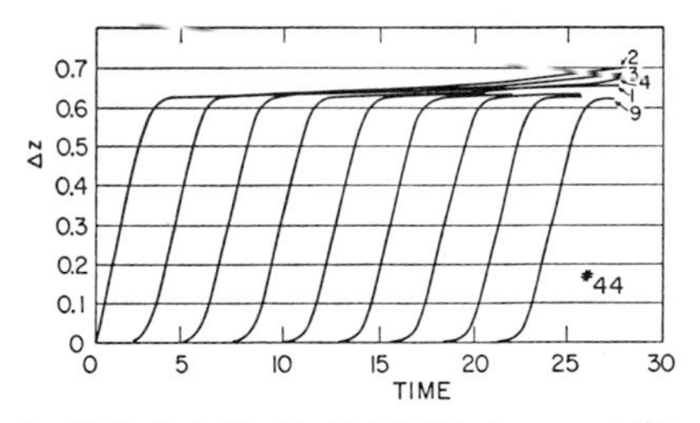

FIG. 17. Similar to Fig. 16, with initial kinetic energy of 100 ev. Curves belonging to consecutive atoms are identified by numbers at upper right (Run No. 44).

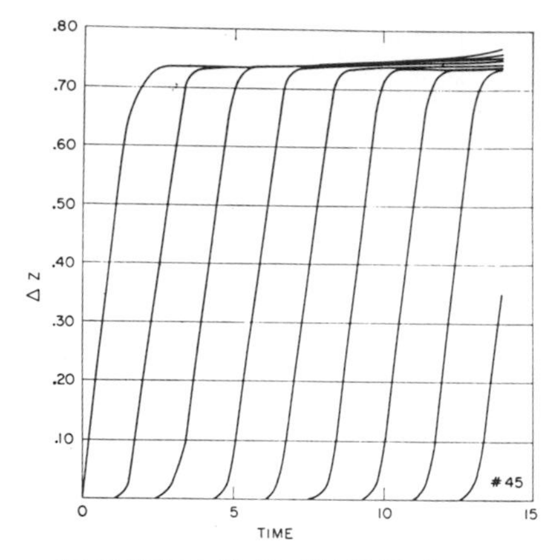

FIG. 18. Similar to Fig. 16, with initial kinetic energy of 400 ev (Run No. 45).

chain to move forward one step. In all three of these figures the high degree of preservation of energy down the chain is evident. Action of the dynamic crowdion is thus seen to occur in ⟨110⟩ as well as in ⟨100⟩.

A number of shots along ⟨111⟩ have been made. All three repulsive potentials have been used, and here the most extensive comparison of potentials two and three exists. In all cases the chains of displacements occurred in the ⟨111⟩ line, with considerable energy transferred to adjoining atoms and carried off in other chains. With enough energy the struck atom penetrates a triangle of near neighbors and lodges temporarily in the center of a unit cube. With more energy it replaces the atom at the far corner of the cube, causing this atom to lodge temporarily in the center of the next cube. With still more energy two replacements occur and the interstitial first appears at the center of the third cube. In all cases this cube center position is known, from separate static calculations, to be unstable, although these dynamic runs were not continued long enough to show it. The instability will ultimately (in times of about 200 to 400) cause the interstitial at the cube center to move toward one of the six nearest lattice sites, forming a split interstitial on that site. In case the interstitial is produced in the center of the first cube, collapse of the Frenkel defect by a ring-type self-annealing mechanism must ultimately occur. In this process two replacements will occur and no defect will remain.

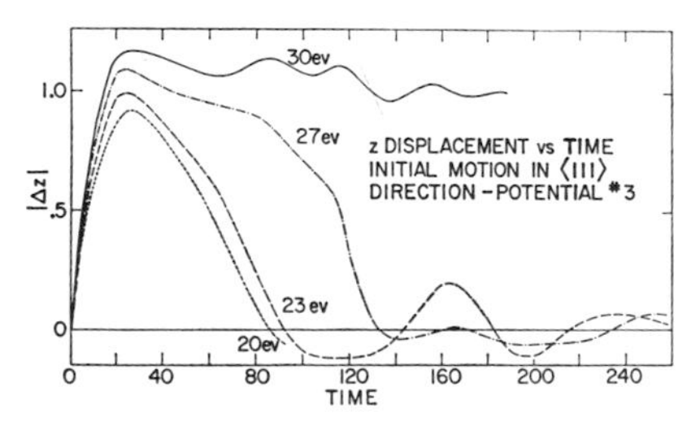

FIG. 19. z displacement vs time for knock-on directed along ⟨111⟩ at various energies. Potential 3 (Runs Nos. 14, 15, and 17).

Figure 19 shows the z component of displacement (Δz) of the struck atom for a series of shots along ⟨111⟩ with interatomic potential number 3. Initial energies of 20, 23, 27, and 30 ev were used. The triangle of near neighbors is at $\Delta z=2/3$; the cube center is at $\Delta z=1$. Beyond the cube center is a second triangle of atoms at $\Delta z=4/3$. The 20 and 23 volt shots pass a little beyond the first triangle, spreading these atoms out and also pushing them slightly ahead. The atoms of the triangle then push the struck atom back to its place of origin. With 30 ev the struck atom goes far enough through the triangle to allow the triangle to close in behind it, confining it near the center of the cube for the duration of the calculation, 200 time units. From a second calculation started quasi-statically it is known that this Frenkel pair is unstable, and will decay by a replacement process in another 100 to 200 time units. The 27-ev shot almost locks into the cube center, but drops back after time 100.

The 30-volt shot in Fig. 19 was useful for testing the adequacy of the boundary conditions employed. The struck atom was at 8,6,2 in atomic set A (see Fig. 1), and was projected in the direction $[\bar{1}\bar{1}1]$. This location is symmetric with respect to the y and z faces of the set, but not with respect to the x faces, since the x dimension of the set was 10, the y and z dimensions 8. When Δx and Δy are plotted against time and compared with Δz, it is found that Δx, Δy, and Δz remain the same to high accuracy (1 part in 10^4) until time 130, when fluctuating differences commence to occur, primarily in Δx. The velocity of longitudinal waves in this model is about 0.08 (our units). The time required for a longitudinal plane wave to travel from the struck atom to the faces $y=0$ and $z=8$, be reflected, and return to the struck atom is thus very close to the time when asymmetric behavior sets in, and this reinforces the belief that reflection at the boundaries is indeed responsible. The resulting asymmetry is slight, however, remaining less than 1%, which argues that reflections are not a serious disturbance to the calculations.

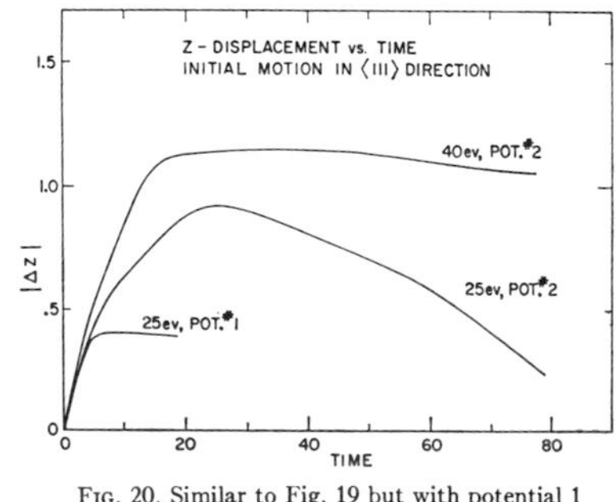

FIG. 20. Similar to Fig. 19 but with potential 1 (run No. 2) and potential 2 (run No. 10).

Figure 20 shows the z-component of displacement for more shots along ⟨111⟩, first 25 volts with interatomic potential number 1, then 25 and 40 volts with interatomic potential number 2. In the first of these the struck atom does not even penetrate the triangle of near neighbors, and is clearly returning directly to its site, even though the calculation was run only a very short time. The effect of increasing the atomic size is evident—the threshold for reaching the center of the cube has become much higher. The shot at 25 ev with potential 2 is very much like that at 20 ev with potential 3 (Fig. 19), and is also below threshold. That at 40 ev with potential 2 is like that at 30 ev with potential 3.

Two higher energy shots are shown in Fig. 21. Both involve interatomic potential 2. In both cases the struck atom was 2,2,2 and its initial motion was along [111]. Δz is plotted against time for atoms 2,2,2, 4,4,4, and 6,6,6. In one shot (dotted lines in Fig. 21) the initial kinetic energy was 60 ev, in the second (solid lines) it was 100 ev. It is seen that 60 ev is only enough to project the interstitial into the nearest cube-center position, whence it must decay by the previously mentioned mechanism, resulting in two replacements and no displacements. With 100 ev, an interstitial is created temporarily at the center of the second cube (5,5,5). In times beyond the end of the run this should produce an interstitial in the split configuration, whose exact location will be determined by any slight departures from symmetry; each of the six possible locations is known from static calculations to be stable. The threshold for production of a permanently displaced atom by a shot in the direction ⟨111⟩ (with interatomic potential 2) is thus seen to be between 60 and 100 ev, a value notably higher than for the directions ⟨100⟩ and ⟨110⟩.

The next three figures show results of some shots above threshold in a direction well away from symmetry axes, namely 10° away from [011] in the plane $x=2$. In the first of these the atomic set B was used, in the last two the large set C was used (see Table I). Orbital plots of the plane $x=2$ are shown. In all cases the interatomic potential was number 2. In Fig. 22 the initial kinetic energy was 50 ev. A collision chain is seen, which leaves a vacancy at its beginning (marked A) and is about to produce an interstitial after perhaps 10 replacements. The transition from a defocused to a focused condition is evident in this chain. Figure 23 shows a 100-ev shot (No. 96) which, because of the large set in which it was run is completely contained. Time runs up to about 200 units in this figure. The knock-on atom was at V_1. Two vacancies are created, at V_1 and V_2, and

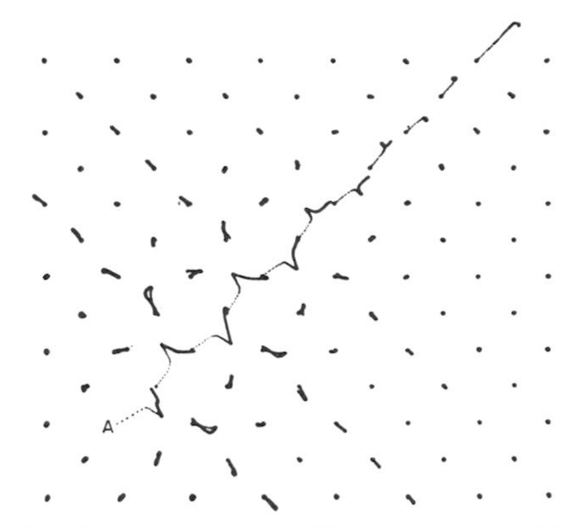

FIG. 22. Shot in (100) plane at 50 ev, 10° away from [011]. Orbits in (100) plane, to time 92, are shown. Knock-on started at A (Run No. 71).

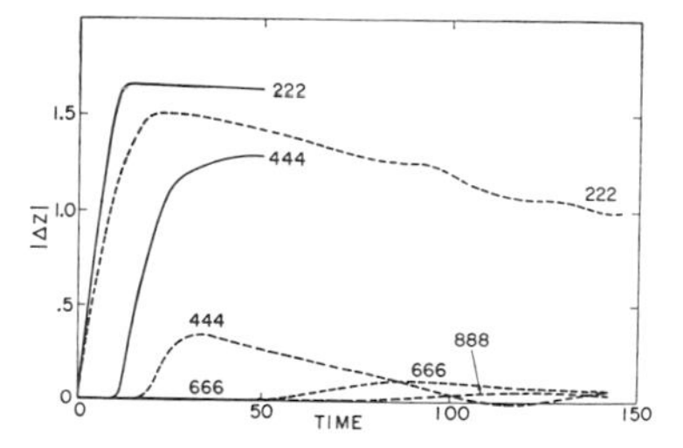

FIG. 21. Two ⟨111⟩ shots at higher energies with potential 2. Δz vs time is shown for several atoms. Dotted lines, knock-on energy 60 ev (Run No. 69); solid line, knock-on energy 100 ev (Run No. 70).

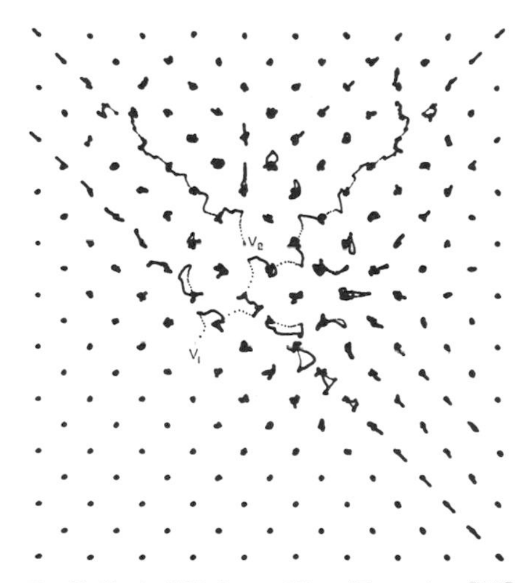

FIG. 23. Shot in (100) plane at 100 ev, 10° away from [011]. Orbits in plane to time 200 (approximately) are shown. Knock-on started at V_1 (Run No. 96).

Notes on p. 59

FIG. 24. Same event shown in Fig. 23, orbits from time 200 to 703. Two vacancies, V_1 and V_2, and two stable interstitials, I_1, and I_2 have formed.

two interstitials, at the somewhat distant points I_1 and I_2 (see Fig. 24). Twelve replacements are seen. Figure 24 shows the same plane from the time where Fig. 23 was terminated up to time 704. The defect pattern is now well established and the atomic vibrations are all quite restricted. Vacancies and interstitials are marked by the same symbols as in Fig. 23. Figure 25(a) (Run No. 97) shows orbits produced by a 400-ev knock-on atom, directed initially 10° away from [011]. The knock-on atom started at K and goes to K'. This shot runs to time 45, at which time large motions have reached the boundary and the configuration is still rather far from equilibrium.

The 400-volt event in Fig. 25 really exceeds the capabilities of our present computing methods. It is presented, however, as a suggestive example of intermediate energy damage events. By looking at the energies of key orbits at the end of this run and drawing on experience gained with lower energy shots, it is possible to estimate the final configuration. This estimate is no more than a plausible guess, and many of its particular features are likely to be revised when more powerful computing methods become available. The general character of the damage may, however, be correctly assessed. The action remaining at the end of the calculation is analyzed into a number of collision chains. ⟨100⟩ chains are still active at A, B, C, D, E, F, G, and H. Looking at the kinetic energies at these points and using the rule that a focused ⟨100⟩ chain loses about 7 ev per step, one estimates that 8 interstitials would eventually be formed at sites outside the fundamental set, as indicated in Fig. 25(b). In addition 3 interstitials appear to be forming inside the set, at sites also indicated in Fig. 25(b). A total of 11 vacancies must also have been produced; the sites of some of these are obvious, others are found by extrapolation, and these locations are indicated by open circles in the figure. About 39 replacements are estimated to occur. It is quite possible that some of these closely spaced vacan-

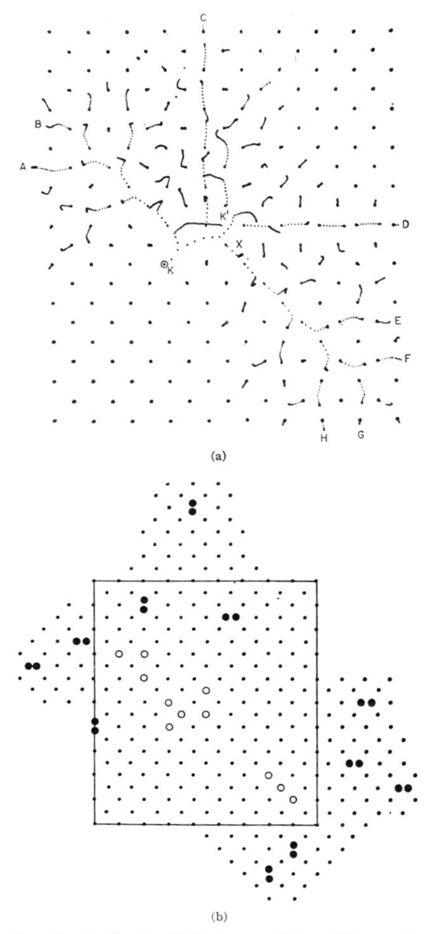

FIG. 25. (a) Shot in (100) plane at 400 ev, 10° away from [011]. Orbits in plane to time 45 are shown. Knock-on started at K, moves to K'. At end of run collision chains $A,B,\cdots H$ are still active (Run No. 97). (b) Estimated array of 11 vacancies (circles) and 11 interstitials (double dots) that could result from shot in Fig. 25(a). Set used in Fig. 25(a) indicated by rectangle. Indicated vacancy arrangement may not be stable.

cies will immediately rearrange themselves (as, for example, in the case of a closely spaced trivacancy[25]), but it is not possible to make any reliable statements about this yet.

The configuration of vacancies and interstitials produced is noteworthy for several reasons. The vacancies are near the site of the original knock-on, the interstitials are farther away. ⟨110⟩ collision chains have not played as prominent a role in this event as in some of the lower energy events discussed earlier. This can be attributed to the fact that the energies here are far above the focusing limit for ⟨110⟩ chains, causing these to spray out into ⟨100⟩ chains which are near or below their focusing threshold. An especially clear example is the chain which starts at X and moves toward the lower right corner of the figure. The kinetic energy at X is 125 ev. The action along this line is reminiscent in many ways of Brinkman's displacement spikes,[1] although there is nothing like the melting and turbulent mixing that he predicted. Note especially the 3 vacancies in a line and the 5 interstitials outside them. It seems clear that, at this energy, almost every kind of stable cluster of vacancies and interstitials will be produced, at least when there is even the slightest annealing.

Shots 96 and 97 had the same starting velocities in the large set C as shots 62 and 63, respectively, had in the smaller set B (see Appendix). The latter two shots could not be run to completion, because of boundary limitations, (No. 62 ran very nearly to completion) but the individual orbits in 62 coincided almost exactly with the corresponding orbits in 96, and those in 63 agreed very closely with the corresponding orbits in 97. This gave further evidence that boundary conditions are not seriously disturbing our results.

B. Collision Chains

One of the most striking features of the orbital plots reported here is the strong tendency of energy to propagate along two preferred lines of atoms, the close packed ⟨110⟩ lines, and the cubic ⟨100⟩ lines. As mentioned before, the ⟨110⟩ effect was anticipated by Silsbee, who first pointed out that focusing occurs in an isolated, uniformly spaced straight line of hard spheres. Figure 26 shows such a line. If the first sphere is projected toward the second at an angle θ_1 with the line of centers, the second will be driven away at an angle θ_2, given by

$$\theta_2 = \sin^{-1}[(S/D \sin\theta_1] - \theta_1, \tag{21}$$

where S is the separation of centers and D is the diameter of a sphere. If the spheres are sufficiently closely spaced θ_2 will be less than θ_1, and in general θ_{i+1} will be less the θ_i. If θ_1 is small, Eq. (21) reduces to

$$\theta_2 = (S/D - 1)\theta_1, \tag{22}$$

and focusing occurs if $S/D < 2$. For atoms that are soft spheres, a first approximation is obtained if an equi-

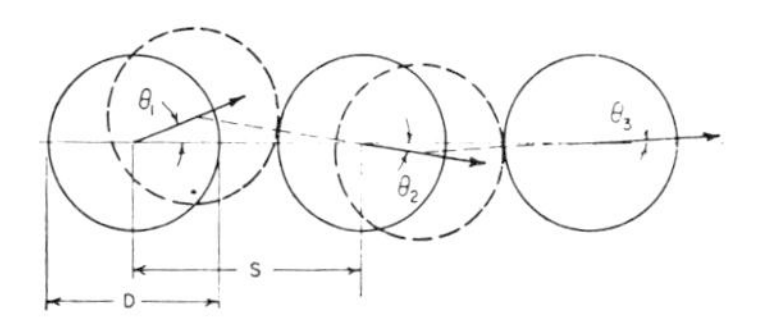

FIG. 26. Hard sphere collision chain.

valent hard-sphere diameter is defined, equal to the distance of closest approach of the atoms in a head-on collision. Considering a moving atom with kinetic energy E to be in collision with a stationary atom of equal mass, and using the exponential repulsive potential

$$\varphi = Be^{-\beta r},$$

one finds the hard-sphere diameter for energy E to be

$$D = (1/\beta) \ln 2B/E. \tag{23}$$

The strength of the focusing can be conveniently described by a focusing parameter Λ defined as the ratio of angles θ_2 and θ_1:

$$\Lambda = \theta_2/\theta_1. \tag{24}$$

For small angles, one has from (22)

$$\Lambda = (S/D) - 1, \tag{25}$$

with D given by Eq. (23).

In a more realistic model, as employed in the present calculations, the row of atoms is not isolated, but is embedded in adjoining rows of atoms. Also a moving atom is in continuous interaction with its next neighbor in the line, and it is not possible to make a rigorous separation of the collision into before and after stages: the moving atom pursues a curved trajectory, losing speed continuously, and the struck atom moves away on a curved trajectory as it gradually picks up speed. It is thus of considerable interest that the calculations produce collision sequences having such close qualitative resemblances to the Silsbee chains. In order to check more closely on the resemblance, and also in order to see if the focused chains observed can be easily characterized so that complex damage events can be resolved into simple elements, a quantitative study has been made of the chains appearing in our calculations. A chain was characterized in the ith stage ($i = 1,2,3,\ldots$) by a kinetic energy E_i and by the angle θ_i between the axis of the chain and a tangent to the orbit of the moving atom. E_i was always chosen as the kinetic energy of one atom, at the point of its maximum kinetic energy, and θ_i was also defined at this point[27]; this seemed to be the best compromise between the require-

[27] θ_i was taken to be the angle between the axis of the chain and the velocity vector of the moving atom at its point of maximum kinetic energy. This is nearly, but not quite, the same as the angle between the axis of the chain and a line running from the original site of the atom to its point of maximum kinetic energy.

Notes on p. 59

ments that the preceding collision be ended and the next collision not yet begun. If the angle in the chain at its next stage is θ_{i+1}, it is convenient to define a focusing parameter $\Lambda(E_i)$ as

$$\Lambda(E_i)=\theta_{i+1}/\theta_i. \tag{26}$$

Examining the major ⟨110⟩ focusing chains occurring in all of our calculations with interatomic potential 2, and limiting attention to cases where θ_i was less than 20° and did not belong to a boundary atom, values of $\Lambda(E)$ for a large variety of energies E were found. These points are plotted in Fig. 27, and are seen to lie on a rather well defined curve. Very little dependence of Λ on θ was observed, and most of the scatter of the points in the figure can be attributed to the somewhat arbitrary attempt to characterize chains initiated in a variety of ways by only two parameters, θ and E, and also, to a minor extent, to truncation error. From Fig. 27 it is seen that $\Lambda=1$ at $E\cong 30$ ev, so that chains above 30 ev are defocused, chains below 30 ev are focused. A defocused chain increases its angle and causes a more rapid loss of energy. In ⟨110⟩ chains, the energy lost at each stage is found to be approximately

$$\Delta E_i=\tfrac{2}{3}(\text{ev})+E\sin^2(\theta_i+\theta_{i+1}). \tag{27}$$

As the angle increases, the attrition of energy increases until the chain drops into the focusing range. Its angle then rapidly approaches zero and the chain continues for a distance determined by the first term in Eq. (27). This term arises because even the perfectly focused chain must force its way between neighbors and lose some energy to them. The value 2/3 ev per step was found to be a fairly good approximation for repulsive potential 2 at small angles and for chain energy E between about 3 and 400 ev.

Using Eq. (23) for D in Eq. (25) one has what may be called the modified hard sphere approximation. This result is also plotted in Fig. 27, again for repulsive potential 2, and is seen to overestimate the true degree of focusing.

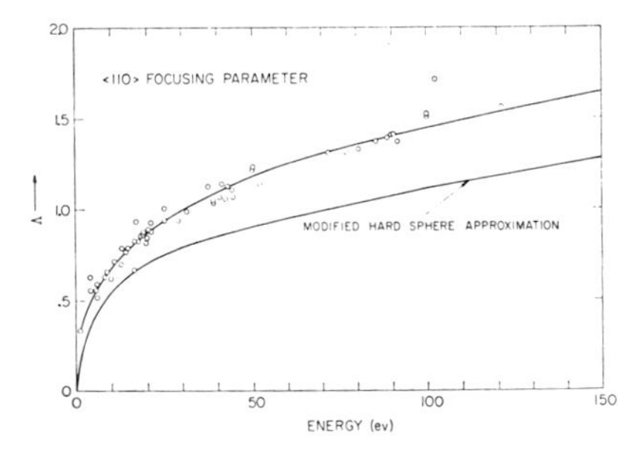

FIG. 27. Focusing parameter $\Lambda=\theta_2/\theta_1$, as found from ⟨110⟩ chains in various runs (open circles).

The number of replacements occurring in a chain that starts with energy E_1 and angle θ_1 can now be calculated. Taking $\Lambda(E_1)$ from the solid line through the "experimental" points in Fig. 27, one finds the second angle in the chain, θ_2, from $\theta_2=\Lambda(E_1)\theta_1$. Then using Eq. (27) to find the energy loss ΔE_1 one has the second energy in the chain

$$E_2=E_1-\Delta E_1.$$

The process is now repeated, starting with E_2 and θ_2, to find energy and angle in the third stage of the chain; by iteration, energy and angle at each successive stage are found. From the dynamic events run to date it is estimated that a well-focused chain produces an interstitial (and thus ceases to transport matter) when its energy falls to about 3 or 4 ev. This energy is subject to rather wide limits of error, in the present stage of our computa-

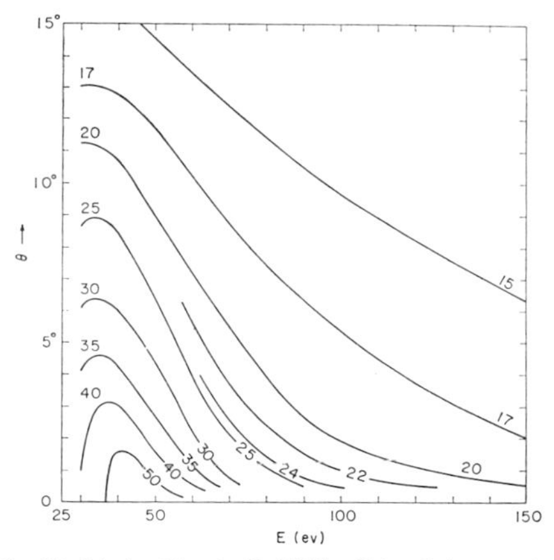

FIG. 28. Calculated lengths N of ⟨110⟩ collision chains started at various angles and energies. See Sec. 5B.

tions, but will be assumed to be 3.5 ev for the present purpose. A quantity $N(E,\theta)$ is defined as the number of collisions required for the energy of a chain that starts at energy E and angle θ (assuming $E\gtrsim 30$ ev and $|\theta|<15°$) to drop to 3.5 ev. This quantity, $N(E,\theta)$, is the number of replacements in the chain, and it also is the distance (in atomic spacings) between the vacancy at the start of the chain and the interstitial at its end. Contours of constant $N(E,\theta)$ are shown on a plane of E and θ in Fig. 28. It is seen that the length of a chain initiated at low energies (around 50 ev) diminishes rather slowly as θ increases, while the length of a higher energy chain drops very rapidly as θ increases. Along the line $\theta=0$, $N(E,\theta)=\tfrac{3}{2}(E-3.5)$. This predicts that a 100-volt chain with $\theta=0$ travels 146 atomic spaces, while a 100-volt chain with $\theta=1°$ travels only 22 spaces. It is obvious that the considerations leading to Fig. 28

are quite crude, but it is felt that the figure represents a better approximation than earlier work based on a hard-sphere model. The contours in Fig. 28 are terminated at 30 ev because the assumptions in their derivation do not apply to chains initiated at energies lower than this.

It is noteworthy that an appreciable fraction of the total energy in a ⟨110⟩ chain resides, at any time, as potential energy of compressed bonds in the chain. This potential energy is a minimum when the kinetic energy of the atom at the center of a pulse is a maximum, and reaches a maximum one half cycle later. In the chain initiated at 25 ev (No. 40) the minimum of potential energy is about 3 ev and the maximum is about 10 ev; in the chain initiated at 100 ev (No. 44) the minimum is about 6 ev, the maximum about 45 ev; in the chain initiated at 400 ev (No. 45) the minimum is about 12 ev, the maximum about 195. The average potential energy is approximately the mean of the minimum and the maximum. This behavior is very different from that of the chain of hard spheres, where the average potential energy is zero. Also the storage of potential energy is associated with a drop in the maximum kinetic energy between the initial stage and the second stage, and Eq. (27) does not apply to the energy loss in the initial stage. The struck atom must supply the chain's potential energy from kinetic energy, and thus the maximum kinetic energy of the second atom (when $\theta=0$) is less than the knock-on kinetic energy by approximately 2/3 ev plus the minimum value of the potential energy of the chain.

The ⟨100⟩ lines are more widely spaced than the ⟨110⟩ lines by a factor $\sqrt{2}$. A modified hard-sphere theory for these lines would predict defocusing at all energies above 5.5 ev. Much stronger focusing effects are actually observed in these lines, and examination of Fig. 6 shows that this occurs because of the confining action exerted by neighboring lines: it is quite insufficient to consider a ⟨100⟩ line to be isolated. Focusing in a variety of ⟨100⟩ chains in our calculations (all for interatomic potential 2) was examined, and characterized at each stage by a parameter Λ, defined exactly as for the ⟨110⟩ chains [Eq. (26)]. The results are presented in Fig. 29. Again there is some scatter of points, attributable to the same causes. Focusing occurs, in general, when the kinetic energy is less than about 40 ev, and defocusing occurs at energies above this. As energy increases above the focusing threshold, Λ grows more rapidly in the ⟨100⟩ case than in the ⟨110⟩ case. The angles θ range up to 20° for the events represented in Fig. 29, but the majority of angles are below 3°. As with ⟨110⟩ chains, no systematic dependence of Λ on θ could be found within the range examined.

All of the chains represented in Figs. 27 and 29 lay in {100} planes.[28] Since ⟨110⟩ is a twofold axis, chains

[28] The plane in which a collision chain is said to lie is defined by the axis of the chain and the velocity of an atom at the center chain.

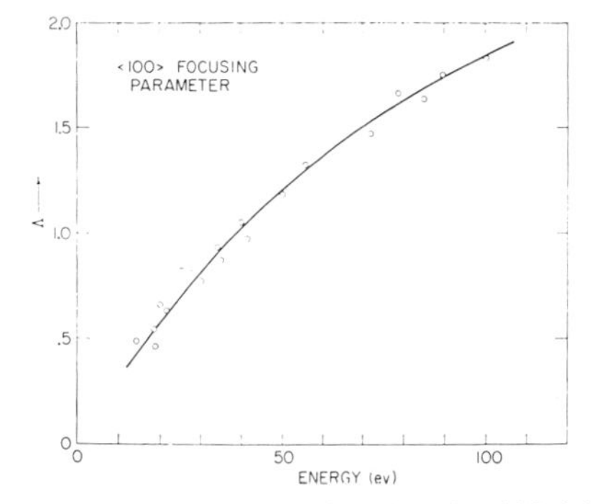

FIG. 29. Focusing parameter $\Lambda=\theta_2/\theta_1$, as found from ⟨100⟩ chains in various runs (open circles).

along this axis but not lying in a {100} plane need not behave exactly like those determined. However, for small and moderate angles θ (up to perhaps 15°) ⟨110⟩ chains are highly independent of their surroundings, and thus all ⟨110⟩ chains at small angles, regardless of the plane in which they lie, should focus about as indicated in Fig. 27. The ⟨100⟩ chains, as has been pointed out, are not independent of their surroundings, indeed require them for focusing. However, the ⟨100⟩ axes have fourfold rotational symmetry. This requires that, to second order in θ, a ⟨100⟩ chain is also independent of the orientation of the plane in which it lies.

A simple theory of assisted focusing can be constructed using the impulse approximation to treat the glancing collision with the ⟨110⟩ neighbor and a modified hard-sphere model for the nearly head-on collision with the ⟨100⟩ neighbor down the chain. This has been done by the writers and independently by Thompson and Nelson.[29] The impulse approximation gives a focusing parameter which rises with energy very much like the curve of Fig. 29, but which is too low at every point. In our version of the impulse approximation the threshold energy, below which ⟨100⟩ focusing would be expected, is 84 ev, and in the form of Thompson and Nelson it is 74 ev, in contrast with the value 39 ev found from the present machine calculations. A large part of the discrepancy can be blamed on failure of the impulse method at these rather low energies. A more accurate analytical treatment would appear to be rather complicated.

C. Number of Defects Produced

Let us return now to Fig. 5 where the results of all shots made to date with interatomic potential 2 are represented as points in a plane. Plotted vertically is the initial kinetic energy of the shot, and horizontally the angle between the initial velocity of the knock-on and

[29] We are indebted to Dr. Thompson for informing us of his results before publication.

Notes on p. 59

the ⟨100⟩ axis. In the left portion of the figure are cases with initial velocities lying in the {110} plane, and in the right portion are cases with initial velocities in the {100} plane. Each shot is represented by a point and attached to most points are two numbers. The first number gives the number of atoms permanently displaced to interstitial positions by the shot; the second figure (in parentheses) gives the number of atomic replacements occurring in the shot. In some cases one or both of these numbers had to be estimated by a substantial extrapolation, and in such cases a question mark is affixed. In four cases no extrapolation sufficiently reliable to report could be made. In these no numbers are given. Points on the left and right edges of the figure are the same, both edges being ⟨110⟩ axes. The dotted line gives an estimate, from these points and from general considerations, of the threshold energy, at each angle, for producing a permanent displacement. The ⟨110⟩ threshold is assumed, somewhat arbitrarily, to be 28 ev, and the ⟨100⟩ threshold is shown as 24 ev. Lattice symmetry requires this line to have zero slope at ⟨110⟩, ⟨111⟩, and ⟨100⟩. The thresholds on these axes are fairly well determined, but considerable uncertainty prevails in several regions.

It is clear from Fig. 5 that for most directions the threshold for producing a replacement lies well below the threshold for displacement, and the number of replacements is generally much larger than the number of displacements. From the data presented, some remarks can be made about the important quantity $p(E)$, defined as the probability of producing at least one displacement, starting with a knock-on of kinetic energy E and of random direction. The probability $p(E)$ rises from zero at 24 ev, and becomes unity at about 85 ev. From considerations of the symmetry of the fcc lattice, one can show that the curve of $p(E)$ versus E must commence with a finite slope and must reach 1 with a finite slope. It must also have at least three more discontinuities of slope between these discontinuities, in the lowest of which three the slope increases discontinuously with increasing E and in the highest of which three the slope decreases discontinuously with increasing E. The effects of quantum mechanics and thermal vibrations, however, probably would round off these discontinuities and make them unobservable. A straight line rising from zero to one in the interval $E=24$ ev to $E=85$ ev should be a reasonably good approximation to $p(E)$ as indicated by the present calculations with potential 2.

Of more direct interest is the average number of permanent displacements $\nu(E)$ produced by a knock-on atom with kinetic energy E and random direction. The results presented in Fig. 5 allow the conclusion that $\nu(E)$ equals 0 for E less than 24 ev, and rises at about this point with a finite slope, not reaching unity until E is considerably above 24. Unfortunately the calculations completed to date provide too small a sample of directions and energies of the knock-on to give much of a curve of $\nu(E)$. From the fact that two displacements were not produced in any shot below 100 ev, and at this energy in only one out of 5 examples, it is plausible that $\nu(E)$ reaches 1 between 80 and 90 ev. The curve of $\nu(E)$ versus E will contain all the discontinuities of slope of $p(E)$ plus many more at higher energies. Again it is questionable whether these would exist in a model taking account of thermal motions and quantum effects.

6. SUMMARY AND CONCLUSIONS

The calculations presented here give a more intimate view of radiation damage events at low and moderate energies in a face-centered cubic metal than has been obtained before. It should be remembered, however, that all results are based on a simple model of metallic copper which is plausible, but whose accuracy has not been finally established. With this reservation, the following conclusions concerning radiation damage and lattice defects in copper have been reached:

1. Damage at low energies consists of vacancies and interstitials. This point is only confirmation of what has been commonly supposed.
2. Vacancies are of the conventional character, but interstitials reside in the split configuration (Fig. 3); no other configuration of the interstitial has been found to be stable.
3. The regular arrangement of atoms on a lattice has an important influence on the character of damage events. Collision chains occur in both ⟨110⟩ and ⟨100⟩ directions, propagating with especially low loss of energy in the former direction, as anticipated by Silsbee.[26] Chains in ⟨110⟩ focus at kinetic energies below approximately 30 ev, chains in ⟨100⟩ focus below 40 ev; all chains defocus at higher energies. These thresholds are surprisingly low.
4. A chain with energy above 25 or 30 ev carries matter, as well as energy, somewhat in the fashion of the "dynamic crowdion,"[8] and produces an interstitial atom near its terminus. The ranges of various ⟨110⟩ chains have been estimated (Fig. 28).[30]
5. Because of the "dynamic crowdion" action interstitials tend to be produced at a distance from the site of a primary knock-on, while the vacancies, having no mechanism of propagation, remain behind in fairly compact groups. At moderate energies a variety of clusters of vacancies, and possibly more complex configurations resulting from the collapse or rearrangement of such clusters, can be expected. The present calculations have not yet been able to follow such rearrangements in detail. The question of the existence of amorphous zones at the site of a damage event, as suggested by Seeger,[7,8] is not yet settled.
6. Another result of the collision chains is the production of many more replacements than displacements. In

[30] See also G. Leibfried, J. Appl. Phys. 30, 1388 (1959); 31, 117 (1960). Our conclusions about focusing chains are somewhat different from Leibfried's.

compounds or alloys of nearly homogeneous mass this effect would produce many more disordered atoms than displaced atoms.[31]

7. The threshold energy for producing a single Frenkel pair is lowest (about 25 ev) in or near ⟨100⟩ and probably is almost as low in or near ⟨110⟩. The threshold is much higher, probably 85 ev, around ⟨111⟩. Experiments on the directional dependence of the threshold are clearly indicated.

8. The closest Frenkel pairs are not stable, in the present model, and pairs along ⟨110⟩ directions must be separated to 4th neighbor positions in order to be stable (see Fig. 4). These conclusions are probably rather sensitive to the details of the force law employed.

9. Knock-ons with energy near threshold produce a variety of Frenkel pairs. The explanation advanced by Corbett, Smith, and Walker[4] for the substages that they observed in the lowest temperature annealing of electron irradiated copper are consistent with results reported here, except that the interstitial is not in the position assumed by Corbett *et al.*, and the present calculations are not far enough advanced to identify particular Frenkel pairs with all of the particular annealing substages. In further work is it hoped that such identification, which would constitute a sensitive check on the model, can be made. Corbett and Walker[32] have also studied the effect on the annealing spectrum of varying the bombardment energy, and find that lowering the bombardment energy appears to decrease the number of distant pairs relative to the number of closer pairs by a surprisingly small amount. This phenomenon finds ready explanation in the calculations reported here—the ⟨110⟩ threshold appears to be very little, if any, above the ⟨100⟩ threshold, and yet ⟨110⟩ displacement events, through dynamic crowdion action produce interstitials at a considerable distance from the vacancy, while ⟨100⟩ displacement events produce the interstitial relatively close by. Varying the maximum energy of knock-ons from 115 ev to 37 ev, as Corbett and Walker have done, would thus produce rather little change in the relative number of distant (⟨110⟩ type) and close (⟨100⟩) type Frenkel pairs.

10. Agitations following damage events of moderate energy are seen to bear some resemblance to thermal spikes (see Figs. 9 and 10), but the transport of energy is far from isotropic, as would be predicted by thermal-spike models in a cubic material. Localized vibrational modes associated with interstitials are prominently excited, and retain their energy longer than other modes. Localized annealing appears to be promoted by the excitation that lingers in these modes.

11. It would appear that these calculations have proved the feasibility of simulating events of radiation damage by mathematical models on high speed computers. Limitations on the size of the set of atoms that can be treated are still a matter of concern, and practical means of increasing this size are under study. Further checks and improvements on the force laws are needed. Work in these areas is continuing.

[31] G. H. Kinchin and R. S. Pease [*Reports on Progress in Physics* (The Physical Society, London, 1955), Vol. 18, p. 1] suggested the importance of replacements, although their treatment bore little resemblance to the results of our calculations.

[32] J. W. Corbett and R. M. Walker, Phys. Rev. **115**, 67 (1959).

7. ACKNOWLEDGMENTS

We are indebted to M. E. Rose, S. S. Rideout, and Mrs. R. E. Larsen for help in arranging for machine computations; to H. B. Huntington and E. Brown for information on their calculations and a number of suggestions; to J. F. Garfield and R. J. Walton for resourceful aid with photographic problems; and to Miss B. Garnier for dedicated assistance in all phases of the work.

Notes to Reprint I.5

1. Details of later developments in the computer simulation of radiation damage are given in an article by Vineyard (in: Interatomic Potentials and Simulation of Lattice Defects, eds P.C. Gehlen, J.R. Beeler and R.I. Jaffe, (Plenum Press, New York, 1972)).

2. The central-difference algorithm given by eqs. (12) and (13) has recently been used in a molecular dynamics simulation of fluid nitrogen (Johnson et al., J. Chem. Phys. 80 (1984) 1279). The trajectories that are generated are in fact identical to those obtained with the Verlet algorithm (Reprint I.7 and Note I.6.2); see, for example, the article by Berendsen (in: Molecular Dynamics Simulation of Statistical-Mechanical Systems, eds G. Ciccotti and W.G. Hoover (North-Holland, Amsterdam, 1986)).

Correlations in the Motion of Atoms in Liquid Argon*

A. RAHMAN
Argonne National Laboratory, Argonne, Illinois
(Received 6 May 1964)

A system of 864 particles interacting with a Lennard-Jones potential and obeying classical equations of motion has been studied on a digital computer (CDC 3600) to simulate molecular dynamics in liquid argon at 94.4°K and a density of 1.374 g cm^{-3}. The pair-correlation function and the constant of self-diffusion are found to agree well with experiment; the latter is 15% lower than the experimental value. The spectrum of the velocity autocorrelation function shows a broad maximum in the frequency region $\omega=0.25(k_BT/\hbar)$. The shape of the Van Hove function $G_s(r,t)$ attains a maximum departure from a Gaussian at about $t=3.0\times10^{-12}$ sec and becomes a Gaussian again at about 10^{-11} sec. The Van Hove function $G_d(r,t)$ has been compared with the convolution approximation of Vineyard, showing that this approximation gives a too rapid decay of $G_d(r,t)$ with time. A delayed-convolution approximation has been suggested which gives a better fit with $G_d(r,t)$; this delayed convolution makes $G_d(r,t)$ decay as t^4 at short times and as t at long times.

I. INTRODUCTION

IN recent years considerable use has been made of large digital computers to study various aspects of molecular dynamics in solids, liquids, and gases.[1] The following is a description of a computer experiment on liquid argon (using the CDC 3600) to study the space and time dependence of two-body correlations which determine the manner in which slow neutrons are inelastically scattered from the liquid. If neutron scattering data of unlimited accuracy and completeness was available, then the kind of work presented here would serve the useful though unexciting purpose of confirming the results already obtained with neutrons. At present, however, the situation is that theorists are trying to build models for these two-body dynamical correlations to account for the observed neutron spectra; the current interest in the work presented here is thus to throw some light on the validity of these models, and to suggest the manner in which some improvements can be made.

The calculations presented here are based on the assumption that classical dynamics with a two-body central-force interaction can give a reasonable description of the motion of atoms in liquid argon. For practical reasons, further assumptions have to be made, namely, the interaction potential has to be truncated beyond a certain range, the number of particles in the assembly has to be kept rather small, and suitable boundary conditions have to be imposed on the assembly. Finally, the equations of motion have to be solved as a set of difference equations, thus involving a certain increment of time to go from one set of positions and velocities to the next. The details will be set forth in the next section. At the end of the paper a brief mention will be made of checks on the validity of these assumptions. The results presented in this paper are confined mainly to one pair of values of the temperature and the density of the system, namely, 94.4°K and 1.374 g cm^{-3}. A less exhaustive study, at 130°K and 1.16 g cm^{-3}, is mentioned briefly at the end.

II. METHOD OF COMPUTATION

The calculations reported here were based on the following ingredients.

Particles with mass $39.95\times1.6747\times10^{-24}$ g (the mass of an argon atom) were assumed to interact in pairs according to the potential $V(r)=4\epsilon\{(\sigma/r)^{12}-(\sigma/r)^6\}$, $\epsilon/k_B=120$°K, $\sigma=3.4$ Å, r being the distance between the particles. This interaction was assumed to extend up to a range $R=2.25\sigma$, so that a particle interacts with all particles situated within a sphere of that radius; $V(2^{1/6}\sigma)=-\epsilon$ is the minimum of $V(r)$ and at $r=R$, $V\sim-0.03\epsilon$.

864 such particles were placed in arbitrary positions in a cubical box of side $L=10.229\sigma$, thus providing a density of 1.374 g cm^{-3}. Periodic boundary conditions were imposed, so that at any given moment a particle with coordinates x, y, z inside the real box implied the presence of 26 periodic images with coordinates obtained by adding or subtracting L from each Cartesian coordinate. The density was conserved because when a particle moves out across one face of the cube another moves in across the opposite face.

The particles were then allowed to move, and their motions were calculated using a set of difference equations with a time increment of 10^{-14} sec. The details have been given in an Appendix. The positions and velocities obtained at successive moments were recorded on magnetic tape for later analysis. The only quantity monitored during the progress of the calculation was the mean-square velocity of the particles expressed in temperature units,

$$T=\frac{M}{3Nk_B}\sum_{i=1}^{N}\mathbf{v}_i^2,$$

where $N=864$. In the initial stages of the calculation, if T was not in the region of temperature (90°K) at which the system was to be studied, all velocities were

* Based on work performed under the auspices of the U. S. Atomic Energy Commission.

[1] J. R. Beeler, Jr., in *Physics of Many-Particle Systems*, edited by E. Meeron (Gordon and Beach Publishers, Inc., New York, 1964).

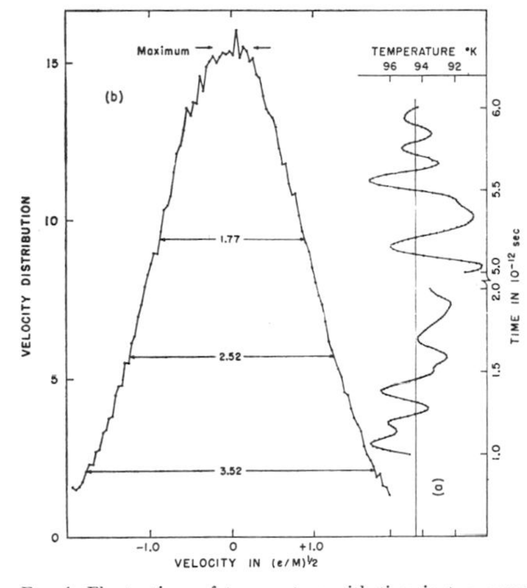

FIG. 1. Fluctuations of temperature with time in two sample regions (curve a); distribution of velocities is shown as curve b; widths of the distribution are shown at $e^{-1/2}$, e^{-1}, and e^{-2} of maximum.

stepped up or down by a constant factor and the system again left to follow its course.

At the completion of one such "experiment," the tape containing the record of positions and velocities was analyzed for the time-independent and time-dependent correlations. For the former, the information at each time can be analyzed without reference to the information at other times, and the correlations calculated at different times can be assembled into one ensemble average. For time-dependent correlations, any moment can be considered as the time origin, and again an ensemble average can be made with a succession of time origins.

The time-independent correlations investigated were the distribution of velocities and the pair distribution function $g(r)$; if $n(r)$ particles are situated at a distance between r and $r+\Delta r$ from a given particle we have

$$g(r)=(V/N)[n(r)/4\pi r^2\Delta r].$$

The time-dependent correlations investigated were: (i) The mean values of the even powers of the displacements $\langle r^{2n}\rangle$, given by

$$\langle r^{2n}\rangle=\frac{1}{N}\sum_{i=1}^{N}[\mathbf{r}_i(t)-\mathbf{r}_i(0)]^{2n},\quad n=1,2,3,4.$$

We define a function[2] $G_s(r,t)$ which gives the probability of a particle attaining a displacement r in time t. We then have

$$\langle r^{2n}\rangle=\int r^{2n}G_s(r,t)d\mathbf{r}.$$

TABLE I. Mean temperature and the rms deviation after ν increments of time have been calculated. The value of the increment $=10^{-14}$ sec.

ν	$\bar{T}$ (°K) for Steps 1 to ν	$(\langle T^2\rangle_{\mathrm{av}}-\bar{T}^2)^{1/2}/\bar{T}$
100	94.64	0.0167
200	94.47	0.0161
300	94.55	0.0158
400	94.55	0.0155
500	94.67	0.0160
600	94.51	0.0170
700	94.43	0.0170
780	94.45	0.0165

(ii) The velocity autocorrelation function, $\langle\mathbf{v}(0)\cdot\mathbf{v}(t)\rangle$, given by

$$\langle\mathbf{v}(0)\cdot\mathbf{v}(t)\rangle=\frac{1}{N}\sum_{i=1}^{N}\mathbf{v}_i(0)\cdot\mathbf{v}_i(t).$$

(iii) The time-dependent pair correlation function[2] $G_d(r,t)$; if at time t, $n(r,t)$ particles are situated at a distance between r and $r+\Delta r$ from the position which was occupied by a certain atom at $t=0$ then we define

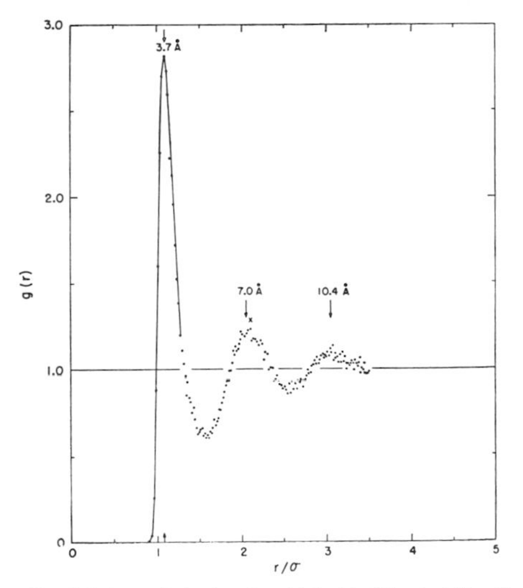

FIG. 2. Pair-correlation function obtained in this calculation at 94.4°K and 1.374 gcm^{-3}. The Fourier transform of this function has peaks at $\kappa\sigma=6.8$, 12.5, 18.5, 24.8.

[2] The functions G_s and G_d defined here are closely related to but not identical with, the Van Hove functions [L. Van Hove, Phys. Rev. **95**, 249 (1954)] G_s and G_d; for a discussion of this relationship see R. Aamodt, K. M. Case, M. Rosenbaum, and P. F. Zweifel [Phys. Rev. **126**, 1165 (1962)] and A. Rahman [Phys. Rev. **130**, 1334 (1963)].

Notes on p. 66

this function as

$$G_d(r,t)=\frac{V}{N}\frac{n(r,t)}{4\pi r^2\Delta r}.$$

The suffix d, for "distinct," indicates that in this counting process the particle originally at the origin is excluded. We may remark here that $G_d(r,t)$ gives the time decay of the pair correlation, $g(r)$, which is identical with $G_d(r,0)$.

III. RESULTS

Figure 1(a) shows the fluctuation of temperature with the passage of time. The figure shows only two sample regions extending from step 100 to step 200 and from step 500 to step 600. A more complete analysis is given in Table I, which shows the mean the rms deviation relative to the mean.

The table shows that the mean remains steady, and we have adopted the value 94.4°K as the temperature of the system.

Figure 1(b) shows the distribution of velocities and the widths w_1, w_2, w_3 of the distribution at heights of $e^{-1/2}$, e^{-1}, and e^{-2} of the maximum. In a Maxwellian distribution we should have $w_2=2(2k_BT/\epsilon)^{1/2}$ [if the velocities are expressed in units of $(\epsilon/M)^{1/2}$] and $w_1=w_2/2^{1/2}=w_3/2$. At $T=94.4$°K, the numerical values of w_1, w_2, w_3 should be 1.78, 2.51, and 3.55, whereas in Fig. 1(b) they are 1.77, 2.52, and 3.52, respectively.

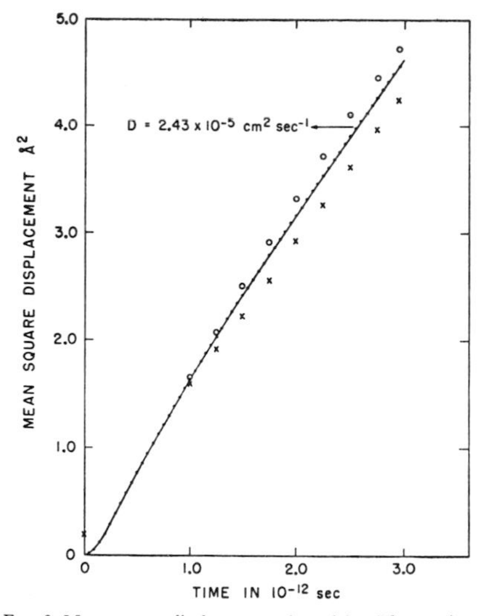

FIG. 3. Mean-square displacement of particles. The continuous curve is the mean of a set of 64 curves; the two members of the set which have *maximum* departures from the mean are shown as circles and as crosses. The asymptotic form of the continuous curve is $6Dt+C$, with D as shown on the figure and $C=0.2$ Å².

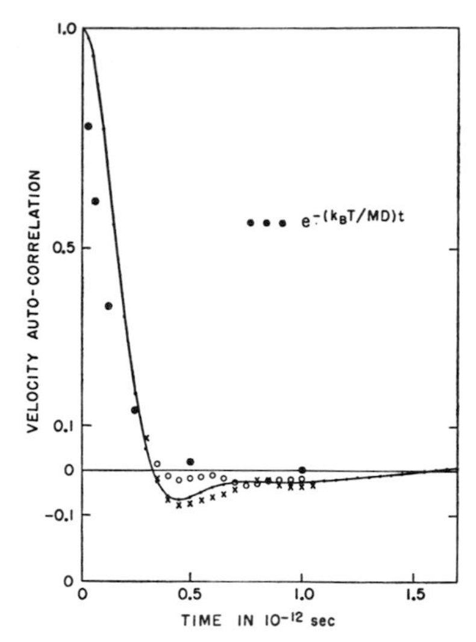

FIG. 4. The velocity autocorrelation function. The Langevin-type exponential function is also shown. The continuous curve, the circles, and the crosses correspond to the curves shown in Fig. 3.

Figure 2 shows $g(r)$ the pair distribution function. Using x rays, Eisenstein and Gingrich[3] have obtained $g(r)$ (at 91.8°K and 1.8-atm pressure) and the agreement with the $g(r)$ shown in Fig. 2 is quite satisfactory. To get a further check we have also calculated the transform of our $g(r)$, namely, the function

$$\gamma(\kappa)=\int_0^\infty \frac{\sin\kappa r}{\kappa r}\left\{\frac{N}{V}[g(r)-1]\right\}4\pi r^2dr.$$

The transform has peaks at $\kappa\sigma=6.8$, 12.5, 18.5, 24.8, whereas the peaks in the x-ray scattering[3] occur at $\kappa\sigma=6.8$, 12.3, 18.4, 24.4, respectively.

Figure 3 shows the mean-square displacement $\langle r^2\rangle$ obtained by averaging over an ensemble of 64 curves with as many different origins of time. Two *extreme* members of the set are also shown in Fig. 3 to exhibit the degree to which individual members of the set differ from their average. One can thus say that $\langle r^2\rangle$ written equivalently as $\langle r^2\rangle=(1/N)\sum(\mathbf{r}_i(t_0+t)-\mathbf{r}_i(t_0))^2$ is independent of the origin t_0, as it should be for a system in equilibrium.

From Fig. 3, it is seen that the asymptotic behavior $6Dt+C$ of $\langle r^2\rangle$ is already achieved at about $t\sim10^{-12}$ sec. At $t=2.5\times10^{-12}$ sec, its value is 3.9 Å² so that the rms displacement at that time is only about half the first-neighbor distance (3.7 Å). Thus, even after 2.5×10^{-12} sec we would expect that the identity of the first

[3] A. Eisenstein and N. S. Gingrich, Phys. Rev. **62**, 261 (1942).

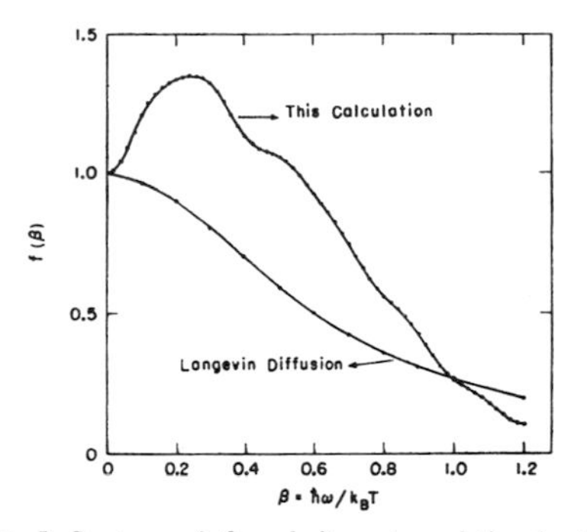

FIG. 5. Spectrum of the velocity autocorrelation function. The Lorentzian spectrum of a Langevin-type correlation is also shown.

neighbors is not completely lost. We shall see a more quantitative indication of this fact further below.

From the slope of the linear part of the curve for $\langle r^2\rangle$ one finds the diffusion constant D to be 2.43×10^{-5} $cm^2\ sec^{-1}$; the temperature of our system is 94.4°K and the density is 1.374 g cm^{-3}; the experimental value of Naghizadeh and Rice,[4] for argon at 90°K and 1.374 g cm^{-3} is also 2.43×10^{-5} $cm^2\ sec^{-1}$. The agreement thus is quite good.

Figure 4 shows the velocity autocorrelation function, $\langle \mathbf{v}(0)\cdot\mathbf{v}(t)\rangle$, normalized to unity at $t=0$ by dividing by $\langle \mathbf{v}^2\rangle$. Notice that the correlation becomes negative at $t=0.33\times10^{-12}$ sec and remains essentially negative as it goes to zero. In this respect it is radically different from the Langevin type of velocity autocorrelation, namely, $\exp(-k_BTt/MD)$, which is also shown in Fig. 4. A more illuminating way of exhibiting this qualitative difference is to consider the Fourier transform of the correlation, defined as

$$f(\omega)=\frac{kT}{MD}\int_0^\infty \frac{\langle \mathbf{v}(0)\cdot\mathbf{v}(t)\rangle}{\langle \mathbf{v}^2\rangle}\cos\omega t dt\,,$$

so that $f(0)=1.0$. Writing $\beta=\hbar\omega/k_BT$, $\lambda=\hbar/MD$, and $u=tk_BT/\hbar$, we get

$$f(\beta)=\lambda\int_0^\infty \frac{\langle \mathbf{v}(0)\cdot\mathbf{v}(u)\rangle}{\langle \mathbf{v}^2\rangle}\cos\beta u du\,.$$

Figure 5 shows $f(\beta)$ obtained from the correlation shown in Fig. 4; it has a broad maximum at about $\beta=0.25$. The transform of a Langevin-type correlation is a Lorentzian $\lambda^2/(\lambda^2+\beta^2)$ which is also shown in Fig. 5.

The time-dependent pair-correlation function, $G_d(r,t)$, was calculated for values of t ranging from 0 to 3.0×10^{-12} sec at intervals of 0.1×10^{-12} sec. It is shown in Fig. 6(a) for $t=10^{-12}$ sec and in Fig. 6(b) for $t=2.5\times10^{-12}$ sec. [$G_d(r,t=0)$ is the static pair distribution, $g(r)$, shown in Fig. 2.] It is seen that even at $t=2.5\times10^{-12}$ sec the remnants of the first-neighbor shell in $g(r)$ are visible.

This remark about the persistence of short-range correlations with the passage of time is relevant if one tries to describe the behavior of the liquid as quasi-crystalline. Whereas the increase of $\langle r^2\rangle$ with time (see Fig. 3) shown no such behavior, the function $\langle \mathbf{v}(0)\cdot\mathbf{v}(t)\rangle$ does show such a behavior, through that of its transform $f(\beta)$ which has a maximum reminiscent of the maximum in the frequency spectrum of a solid; moreover, the short-range order in the arrangement of an atom and its neighbors also shows a certain degree of permanence which is reminiscent of the permanent correlation existing in a solid.

IV. NON-GAUSSIAN BEHAVIOR OF $G_s(r,t)$

If $G_s(r,t)$ has the Gaussian form, $[4\pi\rho(t)]^{-3/2}\times\exp[-r^2/4\rho(t)]$, one has the following relations:

$$\langle r^2\rangle=6\rho(t)\,,$$
$$\langle r^{2n}\rangle=C_n\langle r^2\rangle^n\,,$$
$$C_n=1\times3\times5\times7\cdots(2n+1)/3^n\,,$$

for $n=1, 2, 3, \cdots$. Thus, a departure of $G_s(r,t)$ from a Gaussian form can be expressed in terms of the functions, $\alpha_n(t)$, defined by

$$\alpha_n(t)=(\langle r^{2n}\rangle/C_n\langle r^2\rangle^n)-1\,.$$

For a non-Gaussian $G_s(r,t)$ the α_n, $n=2, 3, \cdots$ will not vanish.

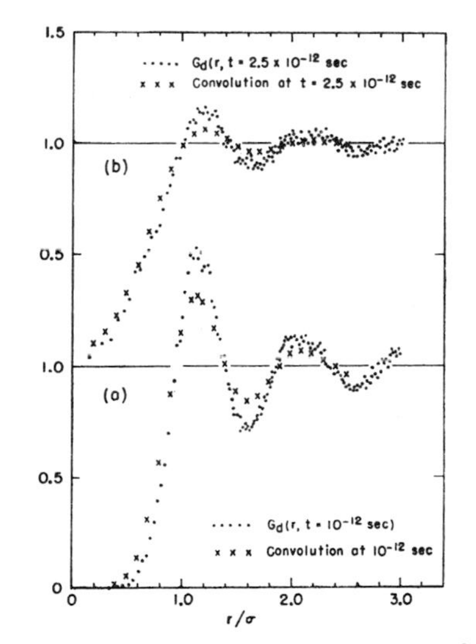

FIG. 6. Time-dependent pair-correlation function $G_d(r,t)$ shown at two values of t. The convolution approximation of Vineyard (Ref. 6) gives a too rapid decay of G_d.

[4] J. Naghizadeh and S. A. Rice, J. Chem. Phys. 36, 2710 (1962).

Notes on p. 66

In Fig. 7 we have shown α_2, α_3, and α_4. Since the values are all positive, we conclude that G_s goes to zero with increasing r more slowly than a Gaussian. The flatness of the curves near the origin reflects the Maxwellian distribution of velocities, because at short times $\langle r^{2n}\rangle$ tends to $\langle v^{2n}\rangle t^{2n}$.

As $t\rightarrow\infty$ the non-Gaussian behavior of $G_s(r,t)$ should disappear. Figure 7 shows that $\alpha_2, \alpha_3, \alpha_4$ start to decrease after 3.0×10^{-12} sec. By extrapolating the curves to the right, one can roughly put down 10^{-11} sec as the time when G_s becomes Gaussian again. At $t=10^{-11}$ sec, the value of $\langle r^2\rangle^{1/2}$ is 3.8 Å, and this is very nearly equal to the first-neighbor distance of 3.7 Å (Fig. 2).

The non-Gaussian behavior of $G_s(r,t)$ can be expressed alternatively by expanding the function in a series of $\mathrm{He}_{2n}(x)$, the even Hermite polynominals[5]; it is straightforward to show that the coefficients in the expansion,

$$G_s(r,t)=[4\pi\rho(t)]^{-3/2}\exp[-r^2/4\rho(t)]\times\{1+b_6(t)\mathrm{He}_6(\alpha r)+b_8(t)\mathrm{He}_8(\alpha r)+\cdots\},$$

with $\alpha^2=3/\langle r^2\rangle$, are given in terms of the $\alpha_n(t)$ by

$$6b_6=(3/4!)\alpha_2(t),$$

$$b_6+8b_8=(3\times5/6!)\alpha_3(t)-(3/2\times4!)\alpha_2(t),$$

$$b_8+10b_{10}=(3\times5\times7/8!)\alpha_4(t)-(3\times5/2\times6!)\alpha_3(t)+(3/2^2\times2\times4!)\alpha_2(t),$$

etc. To illustrate the situation let us substitute the values of α_2, α_3, α_4 at $t=2.5\times10^{-12}$ sec (see Fig. 7). Putting $\alpha_2=0.13$, $\alpha_3=0.40$, $\alpha_4=0.83$ we get $b_6=0.0027$, $b_8=-0.0003$, and $b_{10}=0.00003$, showing that the first few terms in the expansion above give a good description of the non-Gaussian behavior of G_s. In fact the values of $\alpha_2, \alpha_3, \alpha_4$ are such that we essentially have $b_6+8b_8\approx0$ and $b_8+10b_{10}\approx0$.

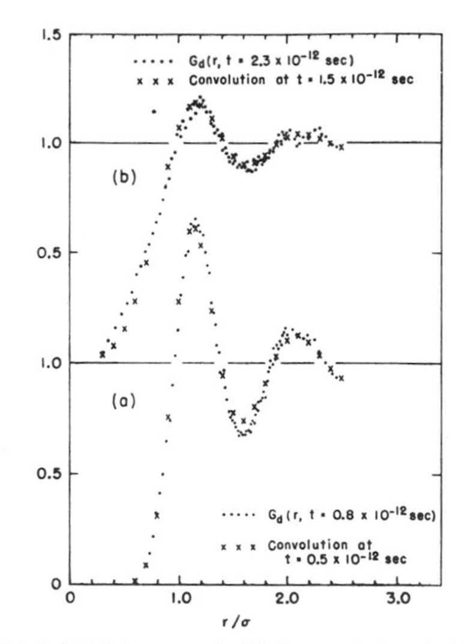

FIG. 8. $G_d(r,t)$ is compared with the convolution of $g(r)$ and $G_s(r,t')$ with $t'<t$, for two pairs of values of t and t' showing the extent to which such a delayed convolution improves the Vineyard approximation (Fig. 6).

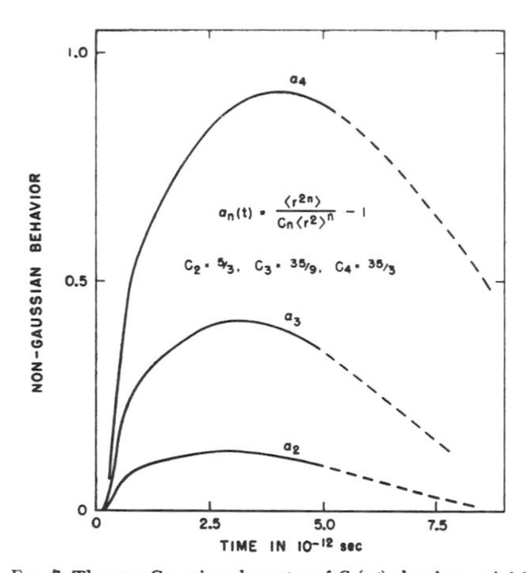

FIG. 7. The non-Gaussian character of $G_s(r,t)$ showing an initial Gaussian behavior lasting about 0.15×10^{-12} sec and, on extrapolating to the right, a return to a Gaussian form at about 10^{-11} sec. Maximum departure of $\langle r^4\rangle$ from its Gaussian value is only about 13%.

V. THE CONVOLUTION APPROXIMATION OF VINEYARD

To describe the time dependence of $G_d(r,t)$, Vineyard[6] has suggested an approximation which makes G_d a convolution between $g(r)$ and $G_s(r,t)$. Following Vineyard one first writes the formal equality

$$G_d(\mathbf{r},t)=\int g(\mathbf{r}')H(\mathbf{r}-\mathbf{r}',t)d\mathbf{r}',$$

where $H(\mathbf{r}-\mathbf{r}',t)$ is the probability that the particle at $\mathbf{r}'$ travels to $\mathbf{r}$ in time t, given that another particle was situated at the origin at $t=0$.

Vineyard's approximation consists in putting $H=G_s$ in the above equation. However, the motions of particles in the first shell are strongly correlated with the occupation of the origin by another particle at $t=0$, and Vineyard's approximation overlooks this fact. In other words, the approximation leads to a too rapid decay of $g(r)$. This is shown in Fig. 6 where the actual $G_d(r,t)$ and the Vineyard approximation are compared at $t=10^{-12}$ sec, and at $t=2.5\times10^{-12}$ sec.

[5] *Tables of Integral Transforms, Bateman Manuscript Project*, edited by H. Erdelyi (McGraw-Hill Book Company, Inc., New York, 1954).

[6] G. H. Vineyard, Phys. Rev. 110, 999 (1958).

From this it follows that the Vineyard approximation might be improved by delaying the convolution in the following way. We write

$$G_d(\mathbf{r},t)=\int g(\mathbf{r}')G_s(\mathbf{r}-\mathbf{r}',t')d\mathbf{r}',$$

where the delayed time $t'(t)$ is always earlier than t. When t is small we should have $t'\to t^2$, so that at small times G_d starts decaying as t^4 as it should and not as t^2 as in the Vineyard approximation. When t is large we should have $t'\to t$.

By matching G_d obtained by a convolution at time t' with the actual G_d at time t, one finds the following pairs (t',t) in units of 10^{-12} sec: (0.2,0.4), (0.5,0.8), (1.0,1.6), (1.5,2.3), (2.0,2.9), (2.5,3.5). Figure 8 shows two examples of how a convolution at $t'<t$ fits the G_d at t. This can be described as a functional relation between t' and t, and the following is suggested as a simple one-parameter function:

$$t'=t-\tau[1-\exp(-t/\tau)-(t^2/\tau^2)\exp(-t^2/\tau^2)].$$

With $\tau=1.0\times10^{-12}$ sec one gets the pairs of values (0.21,0.4), (0.59,0.8), (1.0,1.6), (1.4,2.3), (2.0,2.9), (2.5,3.5).

There are two points to be clearly stated here. Firstly, a delayed convolution will certainly be an improvement over the Vineyard approximation; secondly, in the light of the results obtained in our calculations an empirical functional form for the delay has been suggested involving just one parameter τ.

If we denote the Fourier transform of $G_d(r,t)$ by $F_d(\kappa,t)$ and of $G_s(r,t)$ by $F_s(\kappa,t)$, the delayed convolution gives $\gamma(\kappa)F_s(\kappa,t')$ as an approximation for $F_d(\kappa,t)$ instead of the Vineyard approximation $\gamma(\kappa)F_s(\kappa,t)$. The extent to which this gives an improvement is being investigated.

VI. CONCLUSIONS

A classical 864-body problem with a truncated two-body interaction of the Lennard-Jones type, with periodic boundary conditions is, by itself, a problem of interest, in which case the assumptions involved reduce simply to the assumptions in solving the set of differential equations as a set of difference equations.

The question of identifying such a system with a physical system like liquid argon is very difficult to answer on the basis of the limited amount of information presented in this paper. Firstly, the value of the diffusion constant obtained here is in good agreement with the observed value; this is some justification for saying that the time-dependent mean-square displacement $\langle r^2\rangle$ obtained here is correct; in that case the non-Gaussian behavior of $G_s(r,t)$ shown above should also be dependable. Secondly, the function $g(r)$ we have calculated is in good agreement with the observed pair-distribution function and the $G_d(r,t)$ obtained here differs from that obtained with the Vineyard approximation in the right direction; this has enabled us to suggest an improvement over the Vineyard approximation which can be checked by using neutron scattering data.

A more stringent test for the validity of a model for self-diffusion is the dependence of the diffusion constant on temperature. A calculation of the type described above at 130°K and 1.16 g cm^{-3} gave a diffusion constant $D=5.67\times10^{-5}$ cm^2 sec^{-1}. The experimental value of Naghizadeh and Rice[4] at 120°K and 1.16 g cm^{-3} is $D=6.06\times10^{-5}$ cm^2 sec^{-1}. Thus the variation of D with temperature and density is also in fairly good agreement with the variation measured in the laboratory. It should be noticed, however, that our calculated values are in both cases lower than those measured in the laboratory by about 20%. Calculations are now being made to check if this discrepancy can be reduced by allowing for a softer repulsive part in the interaction potential.

ACKNOWLEDGMENTS

The author wishes to thank Dr. Lester Guttman for many useful discussions during the progress of the work and for invaluable help in the preparation of this paper. The unrelenting patience and cooperation of the scheduling and operating personnel of the CDC 3600 at Argonne are most thankfully acknowledged.

APPENDIX

If x_i and v_i are the components of the position and velocity of the particle i in any direction we have

$$dx_i/dt=v_i\,, \tag{1}$$

$$\frac{dv_i}{dt}=a_i=24\frac{\epsilon}{M}\sum_{j\neq i}\frac{x_i-x_j}{r_{ij}^2}\left\{2\left(\frac{\sigma}{r_{ij}}\right)^{12}-\left(\frac{\sigma}{r_{ij}}\right)^6\right\}. \tag{2}$$

Taking σ as the unit of length and $(\epsilon/M)^{1/2}$ as that of velocity and using dimensionless variables ξ, η, u, ρ, and α for x, v, t, r, and a we have

$$d\xi_i/du=\eta_i\,, \tag{3}$$

$$\frac{d\eta_i}{du}=\alpha_i=24\sum_{j\neq i}\frac{\xi_i-\xi_j}{\rho_{ij}^2}\left\{\frac{2}{\rho_{ij}^{12}}-\frac{1}{\rho_{ij}^6}\right\}. \tag{4}$$

Corresponding to an interval $\Delta t=10^{-14}$ sec we have an interval $\Delta u=10^{-14}(\epsilon/M)^{1/2}(1/\sigma)$.

Let us assume that we are given the positions $\xi_i^{(n-1)}$ at time u_{n-1} and the positions, velocities and accelerations $\xi_i^{(n)}$, $\eta_i^{(n)}$ and $\alpha_i^{(n)}$ at time $u_n=u_{n-1}+\Delta u$.

Using a predictor formula for positions $\bar{\xi}$ at time u_{n+1}, we have

$$\bar{\xi}_i^{(n+1)}=\xi_i^{(n-1)}+2\Delta u\eta_i^{(n)}.$$

With these we get predicted accelerations $\bar{\alpha}_i^{(n+1)}$ using Eq. (4). Using these we get the new positions and velocities

$$\eta_i^{(n+1)}=\eta_i^{(n)}+\tfrac{1}{2}\Delta u(\bar{\alpha}_i^{(n+1)}+\alpha_i^{(n)}),$$
$$\xi_i^{(n+1)}=\xi_i^{(n)}+\tfrac{1}{2}\Delta u(\eta_i^{(n+1)}+\eta_i^{(n)}).$$

Notes on p. 66

This process can be repeated until the predicted and corrected values of $\xi_i^{(n+1)}$ differ by less than a prescribed value. However, the procedure adopted was to make trial runs on the system of 864 particles with one and with two repetitions of this predictor-corrector procedure. A comparison of the results in terms of the correlations discussed in this paper showed no observable difference. As a further check, the motion of a diatomic system was calculated with one and with two repetitions of this procedure. The two particles were initially at a distance $\rho_{12}=1.9$ and were allowed to oscillate; their positions at 2000 successive intervals Δu were recorded covering a little over three periods of oscillation and the following is a summary of the results to show the degree to which the approximations involved in using the difference equations affect the motion.

(a) At the end of three successive oscillations the separations were: 1.8958, 1.8932, 1.8890, when the predictor-corrector procedure was used only once and 1.9018, 1.9016, and 1.9044 when it was used twice, thus giving improved results.

(b) The distance of closest approach was successively 1.0039, 1.0040, 1.0041 in the first case and 1.0038, 1.0038, 1.0038 in the other.

(c) The mean-square velocity in °K while going through the minimum of the potential was 36.65, 36.61, 36.60, 36.59, 36.59, 36.54, in one case and 36.65, 36.67, 36.67, 36.68, 36.67, 36.70, in the other.

(d) The period of oscillation was (in units of 10^{-12} sec) 6.27, 6.22, 6.17, in one case, and 6.31, 6.32, 6.33 in the other.

This gives an idea of the errors involved in using the difference equations given above. The results given in the paper were all obtained in a run with two passes through the predictor-corrector procedure.

There are five factors which determine the time for computing one step Δu, namely, N, R, the number of predictor-corrector cycles, the manner of writing the program, and the computer used. For $N=864, R=2.25\sigma$, using floating point arithmetic each cycle takes 45 sec on the CDC-3600 computer. For $N=250$, $R=2.0\sigma$, using fixed point arithmetic each cycle takes 40 sec on the IBM-704 machine. For the most time consuming part the program was written in machine language and in FORTRAN for the rest.

Notes to Reprint I.6

1. Reference [1] discusses the earlier computer simulations of Wood, Alder and Vineyard.

2. The algorithm adopted in this paper was not used in later work. The great majority of present-day molecular dynamics calculations make use either of a Gear predictor–corrector method or of a central-difference algorithm in the form given by Verlet (Reprint I.7). The requirements that a good algorithm must meet are that (a) it should be stable, (b) it should not require excessive storage, and (c) the (time-consuming) calculation of the intermolecular forces should be carried out only once per time step. In the nth-order Gear algorithm, the positions and their derivatives (up to order n) at time t, i.e. $\{q_0(t), q_1(t), \ldots, q_n(t)\}$, where

$$q_m(t) \equiv (\delta t^m/m!)(\mathrm{d}^m q_0(t)/\mathrm{d}t^m)$$

are used to predict the corresponding set $\{p_0(t+\delta t), p_1(t+\delta t), \ldots, p_n(t+\delta t)\}$ at time $t+\delta t$. This part of the algorithm uses a simple Taylor expansion:

$$\begin{aligned} p_0(t+\delta t) &= q_0 + q_1 + \cdots + q_n \\ p_1(t+\delta t) &= q_1 + 2q_2 + \cdots + nq_n \\ &\vdots \\ p_n(t+\delta t) &= q_n. \end{aligned}$$

The forces corresponding to the predicted coordinates $p_0(t+\delta t)$ are then calculated, leading to the corrected accelerations $c_2(t+\delta t)$. The predicted accelerations $p_2(t+\delta t)$ will, in general, be different from $c_2(t+\delta t)$, and the difference is used to obtain an improved ("corrected") estimate for the coordinates and derivatives at time $t+\delta t$ from

$$c_m(t+\delta t) = p_m(t+\delta t) + f_{m2}^n[c_2(t+\delta t) - p_2(t+\delta t)].$$

The coefficients f_{m2}^n are chosen such that the resulting algorithm combines accuracy with stability. For example, the coefficients in a fifth-order predictor–corrector scheme for the integration of a second-order differential equation are

$$f_{02}^5 = 3/16, \qquad f_{12}^5 = 251/360, \qquad f_{22}^5 = 1,$$
$$f_{32}^5 = 11/18, \qquad f_{42}^5 = 1/6, \qquad f_{52}^5 = 1/60.$$

The Verlet algorithm, at first sight, is much cruder than the Gear method, yet for a wide class of applications it turns out to be scarcely inferior, and in some cases even superior, to the predictor–corrector schemes (see the article by Berendsen quoted in Note I.5.2). The Verlet algorithm is easily derived by writing two Taylor expansions for the particle coordinates:

$$q(t+\delta t) = q(t) + (\mathrm{d}q/\mathrm{d}t)\ \delta t + (\mathrm{d}^2q/\mathrm{d}t^2)(\delta t^2/2!) + \cdots$$
$$q(t-\delta t) = q(t) - (\mathrm{d}q/\mathrm{d}t)\ \delta t + (\mathrm{d}^2q/\mathrm{d}t^2)(\delta t^2/2!) + \cdots$$

Subtraction of the two expansions gives

$$q(t+\delta t) = 2q(t) - q(t-\delta t) + (\mathrm{d}^2q/\mathrm{d}t^2)\ \delta t^2 + \mathrm{O}(\delta t^4).$$

The Verlet algorithm requires less storage than a fifth-order predictor–corrector scheme. This is an important advantage for large system sizes.

PHYSICAL REVIEW VOLUME 159, NUMBER 1 5 JULY 1967

Computer "Experiments" on Classical Fluids. I. Thermodynamical Properties of Lennard-Jones Molecules*

LOUP VERLET†
Belfer Graduate School of Science, Yeshiva University, New York, New York
(Received 30 January 1967)

The equation of motion of a system of 864 particles interacting through a Lennard-Jones potential has been integrated for various values of the temperature and density, relative, generally, to a fluid state. The equilibrium properties have been calculated and are shown to agree very well with the corresponding properties of argon. It is concluded that, to a good approximation, the equilibrium state of argon can be described through a two-body potential.

I. INTRODUCTION

THE "exact" machine computations relative to classical fluids have several aims: It is possible to realize "experiments" in which the intermolecular forces are known; approximative theories can thus be unambiguously tested and some guidelines are provided to build such theories whenever they do not exist. The comparison of the results of such computations with real experiments is the best way to obtain insight into the interaction between molecules in the high-density states.

The Monte Carlo method initiated by the Los Alamos group[1] is a first example of these "exact" methods. It amounts to a direct computation of the integrals involved in the canonical averages. It is easy to carry out, with the inconvenience, however, of providing no information on the time properties of the system.

The dynamics of an isolated system can also be considered and used to calculate time averages and time-dependent properties. The case of hard spheres and hard spheres surrounded by a square well has been extensively studied by Alder *et al.*[2] In the case of a two-body interaction simulating more closely the interaction between the molecules, it is possible to integrate directly the equation of the motions of about a thousand particles, as brilliantly demonstrated by Rahman.[3] The present paper presents some of the results which have been obtained, using a technique inspired by Rahman's work, for a system of 864 particles interacting through a Lennard-Jones potential.

In Sec. II we give some technical details on the method which we use; in particular, we describe a bookkeeping device that cuts the computing time by a factor of the order of 10.

In Sec. III we give and discuss the results obtained for the pressure, the internal energy, the high-frequency elastic moduli, and the isotopic separation factor. These results, summarized in Table I, are sufficiently numerous to allow a comparison on the whole density range of the fluid state and on a wide temperature range which essentially excludes the extremely high temperatures. The over-all agreement is surprisingly good. It appears that the many-body forces, if they are at all important, behave so as to realize an effective interaction which is state independent to a good approximation.

The correlation functions are described and discussed in a separate paper.[4] The formalism necessary to express the fluctuations in the microcanonical ensemble was discussed recently.[5] It can be applied to calculate the derivatives of the thermodynamic functions (e.g., the specific heat and $\partial p/\partial\rho$) in terms of fluctuations averaged over time. The results are not very precise and will only be presented as an illustration of these theoretical considerations.[6]

The results on the time-dependent properties will be reported later.

II. DESCRIPTION OF THE COMPUTER EXPERIMENTS

We consider a system of 864 particles, enclosed in a cube of side L, with periodic boundary conditions interacting through a two-body potential of the Lennard-Jones type

$$V(r)=4((\sigma/r)^{12}-(\sigma/r)^6). \tag{1}$$

This potential is cut at $r_v=2.5\sigma$ in most of our experiments, or, in some of them at $r_v=3.3\sigma$. The problem is to integrate the equation of motion

$$m\frac{d^2\mathbf{r}_i}{dt^2}=\sum_{j\neq i}\mathbf{f}(r_{ij}). \tag{2}$$

We choose the following units: The lengths are expressed in units of σ ($\sigma=3.405$ Å for argon), and the energies in units of ϵ ($\epsilon=119.8$°K for argon).[7] The thermodynamic quantities will thus be measured in the

* Supported by the U. S. Air Force Office of Scientific Research Grant No. 508-66.

† Permanent address: Faculté des Sciences, Laboratoire de Physique Théorique et Hautes Energies, Bâtiment 211, 91-Orsay.

[1] W. W. Wood and F. R. Parker, J. Chem. Phys. **27**, 720 (1957).

[2] B. J. Alder and T. E. Wainwright, J. Chem. Phys. **33**, 1439 (1960).

[3] A. Rahman, Phys. Rev. **136**, A405 (1964).

[4] L. Verlet (to be published).

[5] J. L. Lebowitz and J. K. Percus, Phys. Rev. **124**, 1673 (1961).

[6] J. L. Lebowitz, J. K. Percus, and L. Verlet, Phys. Rev. **153**, 250 (1967).

[7] A. Michels and H. Wijker, Physica **15**, 627 (1949).

usual "reduced" units. The time unit is chosen so that $m=48\epsilon\sigma^{-2}$; it turns out to be, for argon, equal to 3×10^{-13} sec. This time is of the order of the kinetic relaxation times of the system in the case considered in this paper. With this in mind, we have, for the force acting on particle i in the x direction,

$$f_x(r_{ij})=m(x_i-x_j)(r_{ij}^{-14}-0.5r_{ij}^{-8}). \quad (3)$$

To integrate (2), we use the very simple algorithm

$$\mathbf{r}_i(t+h)=-\mathbf{r}_i(t-h)+2\mathbf{r}_i(t)+\sum_{j\neq i}\mathbf{f}(r_{ij}(t))h^2, \quad (4)$$

where h is the time increment which we take equal to 0.032. This is practically the value chosen by Rahman (i.e., 10^{-14} sec in the case of argon). We have checked that this time increment is adequate and even superfluously small in most cases. For instance, for $T=1.38$, $\rho=0.55$ (i.e., temperature just above critical, density almost twice critical), we have performed two integrations up to the time $t=4$. In one case we have taken $h=0.032$, in the other $h=0.016$, with the same initial conditions. The difference in position at time t is typically of the order of 0.001 and the difference in the thermodynamic quantities, at the same final time, amounts to 1/10 000. In this kind of calculation, most of the time is spent in computing the force. If no special devices are introduced, we must, at each step, compute $\frac{1}{2}N(N-1)$ terms, most of which turn out to be zero. We introduce the following bookkeeping device[8] which cuts the computing time by a factor of the order of 10: Every nth step, we compute all the $\frac{1}{2}N(N-1)$ distances, and, given a particle i, we make a table of all the particles which are within a distance r_M of that particle. Then, for the next $n-1$ steps in time, we take into account only the particles in the tables. There is no error as long as r_M is sufficiently larger than r_v so that no particle outside the table traverses the "skin" of depth r_M-r_v and gets into the range r_v of the potential. The feasibility of such a procedure can be easily appreciated by giving some orders of magnitude: Let $\bar{v}$ be the root-mean-square velocity in our units, it is typically of the order of 0.3; if this is so, no error is made as long as

$$r_M-r_v\lesssim n\bar{v}h. \quad (5)$$

If, for instance $n=16$, $n\bar{v}h=0.15$. By choosing $r_M=3.3$ for $r_v=2.5$, the condition (5) is largely met, and at the same time the "skin" depth stays reasonably small. We have checked, by following some systems for several hundred steps in time, that no difference at all was observed when n was reduced. Moreover, the conservation of the total energy and of the total velocity, which stays of the order of 10^{-7}, is a guarantee of the soundness of the whole procedure.

[8] Some time-saving tricks have been considered before: See Ref. 2 and, for the hard-sphere case, A. Rotenberg [New York University Report No. NYO-1480-3, 1964 (unpublished)].

With this device, the time spent for an integration step at the density 0.45 is about 12 sec on the UNIVAC 1107 of the Faculté des Sciences, Orsay, where the first calculations were made, and ten times less (with careful machine coding of the time-consuming subroutines) on the CDC 6600 of New York University, where the greatest part of the results reported here were obtained.

With that machine, a typical "experiment" takes about 1 h. It goes as follows: The positions are initially taken, in general, at the nodes of a face-centered-cubic lattice which has the desired density, and the velocities are chosen at random with a Gaussian probability law. Three hundred steps in time are sufficient, in general, to reach equilibrium. The computation is then carried on for 1200 steps in time (this corresponds, for argon, to 1.2×10^{-11} sec). The main part of the computation concerns the study of equilibrium quantities (thermodynamic functions: temperature, pressure, internal energy, specific heat, etc.; time-independent correlation functions) and of nonequilibrium quantities (velocity autocorrelation function, elastic constants, viscosities, heat conductivity, etc.). The necessary technical details will be given when these results are reported.

III. THERMODYNAMIC QUANTITIES

A. Temperature

At each step in time the velocities are calculated simply by the formula

$$\mathbf{v}_i(t)=[\mathbf{r}_i(t+h)-\mathbf{r}_i(t-h)]/2h. \quad (6)$$

The temperature is, at time t, $\frac{2}{3}$ of the kinetic energy, in our units

$$T=16\sum_i v_i^2/N. \quad (7)$$

The error entailed by the use of (6) is of the order of 1/1000. This error is of no consequence, except that it gives rise to small irregularities in the total energy which should be otherwise strictly constant. The temperature, averaged over the time, is affected by a statistical error of the order of 0.004.

B. Pressure

The pressure is calculated from the virial theorem

$$\frac{p}{\rho kT}=1-\frac{1}{6NkT}\left\langle\sum_i\sum_{j>i}r_{ij}\frac{\partial v_{ij}}{\partial r_{ij}}\right\rangle-\frac{\rho}{6kT}\int_{r_v}^{\infty}r\frac{\partial v}{\partial r}g(\mathbf{r})d\mathbf{r}. \quad (8)$$

The second term of (8) is the time average of the virial. The last term is a correction term which takes into account the effect on the pressure of the tail of the potential which has been neglected in the dynamics.

The influence on the main term of (8) of the tail of the potential, which has been neglected in the dynamics can be appreciated by considering that cutoff tail as a

Notes on p. 73

TABLE I. List of thermodynamical results: ρ is the particle density, and T the temperature; $\beta p/\rho$ is the compressibility factor, Ω_0^2 is equal to $\langle\nabla^2 V\rangle/m$, K_∞ and G_∞ are the infinite-frequency bulk and shear moduli, respectively, K_∞/G_∞ their ratio, and r_v the distance at which the potential is cut. Reduced units are used throughout.

ρ	T	$\beta p/\rho$	U^i	Ω_0^2	K_∞	G_∞	K_∞/G_∞	r_v
0.88	1.095	3.48	−5.66	23.3	58.2	32.1	1.81	2.5
0.88	0.94	2.72	−5.84	21.4	52.4	29.8	1.76	2.5
0.88	0.591	−0.18	−6.53	17.3	42.5	26.2	1.62	2.5
0.85	2.889	4.36	−4.25	35.1	90.7	44.5	2.04	2.5
0.85	2.202	4.20	−4.76	30.9	77.4	39.3	1.97	2.5
0.85	1.214	3.06	−5.60	22.4	54.7	30.3	1.81	2.5
0.85	1.128	2.78	−5.69	21.5	52.2	29.3	1.78	2.5
0.85	0.880	1.64	−5.94	18.8	45.0	26.4	1.70	2.5
0.85	0.782	0.98	−6.04	17.8	42.1	25.3	1.66	2.5
0.85	0.786	0.99	−6.05	17.8	42.2	25.4	1.67	2.5
0.85	0.760	0.78	−6.07	17.5	41.4	25.0	1.66	2.5
0.85	0.719	0.36	−6.12	17.0	39.9	24.4	1.63	2.5
0.85	0.658	−0.20	−6.19	16.4	38.3	23.8	1.61	2.5
0.85	0.591	−1.20	−6.26	15.5	35.7	22.7	1.57	2.5
0.75	2.849	3.10	−4.07	26.8	59.3	30.2	1.96	2.5
0.75	1.304	1.61	−5.02	⋯	36.1	20.9	1.73	2.5
0.75	1.069	0.90	−5.19	15.4	32.0	19.3	1.65	2.5
0.75	1.071	0.89	−5.17	15.4	31.6	19.1	1.66	3.3
0.75	0.881	−0.12	−5.31	14.1	27.8	17.5	1.56	2.5
0.75	0.827	−0.54	−5.38	13.2	26.9	17.2	1.53	2.5
0.65	2.557	2.14	−3.78	⋯	35.8	19.3	1.85	3.3
0.65	1.585	1.25	−4.23	⋯	25.5	15.0	1.70	2.5
0.65	1.036	−0.11	−4.52	11.5	19.7	12.7	1.55	3.3
0.65	0.900	−0.74	−4.61	10.4	17.8	11.9	1.50	2.5
0.55	2.645	1.63	−3.24	⋯	23.0	12.7	1.80	2.5
0.5426	3.26	1.86	−3.00	⋯	26.1	13.9	1.88	2.5
0.5426	1.404	0.57	−3.63	⋯	14.2	9.0	1.58	2.5
0.5426	1.326	0.42	−3.66	⋯	13.9	8.9	1.56	3.5
0.5	1.36	3.40	−3.38	8.5	11.4	7.4	1.54	2.3
0.45	4.625	1.68	−2.22	14.9	21.4	11.2	1.91	2.5
0.45	2.935	1.38	−2.60	11.0	14.8	8.3	1.79	2.5
0.45	1.744	0.74	−2.90	8.20	11.0	6.7	1.64	2.5
0.45	1.764	0.76	−2.89	8.20	11.0	6.7	1.64	2.5
0.45	1.710	0.74	−2.95	8.20	11.0	6.7	1.64	2.5
0.45	1.552	0.75	−2.98	7.7	10.1	6.4	1.58	2.5
0.4	1.462	0.41	−2.72	⋯	7.27	4.77	1.53	2.5
0.4	1.424	0.38	−2.73	⋯	7.22	4.75	1.52	3.3
0.35	1.620	0.58	−2.31	5.8	5.70	3.72	1.54	2.5
0.35	1.118	0.40	−2.21	⋯	4.92	3.31	1.49	2.5

weak long-range perturbation.[9] The effect on the main term of (8) is quite small: It amounts to 0.006 in the one case studied ($\rho=0.75$, $T=1.05$). It is smaller than the statistical error which is generally of the order of 0.01 when 1200 steps in time are considered. The main error comes certainly from the correction term which is very large at high density and low temperature where it is of the order of 1 for $r_v=2.5$.

The replacement of $g(r)$ by 1 in the correction term leads, for $r_v=2.5$, to an error which may reach 0.05 at the highest density and lowest temperature considered here. Such a replacement is, however, not necessary as the extrapolated values of $g(r)$[4] may be used.

As a whole, we believe that the over-all error on $p/\rho kT$ is probably of the order of 0.01 around the critical density and may reach 0.05 in the high-density and low-temperature region.

The results so far obtained have been gathered in Table I. Fig. 1 shows $p/\rho kT$ for some isochores at high

[9] J. Lebowitz (private communication).

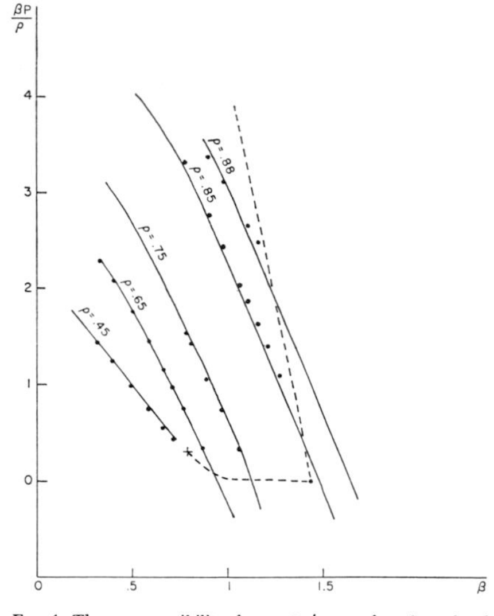

FIG. 1. The compressibility factor $\beta p/\rho$ as a function of β for the isochores $\rho=0.88$, 0.85, 0.75, 0.65, and 0.45 (solid lines), as compared with the experimental data for argon (Refs. 10–12) (dots). The dashed curves correspond to the gas-liquid and solid-fluid coexistence lines.

density ($\rho=0.88$, 0.85, 0.75, 0.65, and 0.45), and the comparison of the equation of state determined by various experimental groups in Amsterdam,[7,10] Toronto,[11] and Louvain.[12]

The over-all agreement between "theory" and experiment is surprisingly good: It appears that the Lennard-Jones potential is a quite satisfactory interaction as far as the equilibrium properties of argon are concerned. It is to be noted that if, instead of argon, xenon were chosen for the comparison, the agreement would not be so good. For instance, with, for $T=1.35$, $\rho=0.75$, the value of $\beta p/\rho$ from molecular dynamics is 0.86; the same quantity is equal to 0.86 in argon. For xenon (with the reduction parameters $\epsilon/k=225.3$°K, $\sigma=4.07$ Å), the value 1.05 is obtained.[10]

On Fig. 2 are represented three isotherms: One at high temperature ($T=2.74$) where comparison can be made with the results obtained a long time ago by Wood and Parker,[2] and also with a Lennard-Jones potential through the Monte Carlo technique; the low-temperature isotherm $T=1$; and the isotherm $T=1.35$, which is also an isotherm for which Monte Carlo com-

[10] J. M. H. Levelt, Physica **26**, 361 (1960).

[11] W. Van Witzenburg, University of Toronto dissertation (unpublished); copy available in Dissertation Abstracts **25**, 1268 (1964).

[12] A. Van Itterbeck, O. Verbeke, and K. Staes, Physica **29**, 742 (1963).

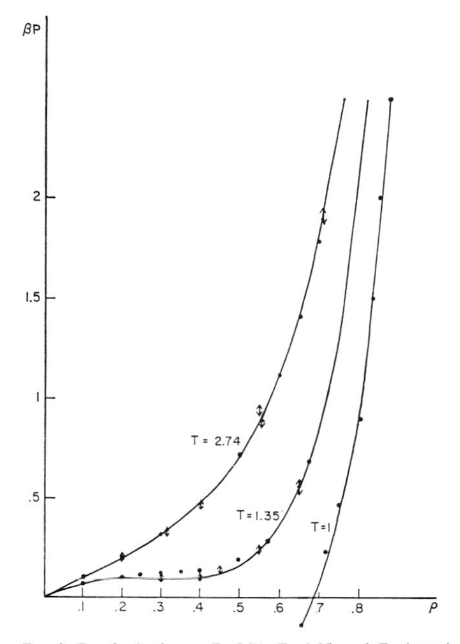

FIG. 2. For the isotherms $T=2.74$, $T=1.35$, and $T=1$, βp is represented as a function of ρ. The low-density parts of the curves were obtained with the help of the PY and PY II equations (Ref. 13). The arrows show the results of Monte Carlo calculations (Refs. 1, 13). The dots are the experimental results in argon (Refs. 10–12).

putations have been made.[13] That isotherm will be used below to examine the question of the critical constants.

For low values of the pressure, our computation may describe metastable states: The periodic boundary prevents the occurrence of inhomogeneities of a size larger than the side of the box, so that the formation of a two-phase system is very much hindered. The low pressure part of the isochores at $\rho=0.85$ and 0.88 has been obtained by "cooling" a liquid configuration: That is, we start from a spatial distribution relative to the liquid state, and a Gaussian initial velocity distribution. Equilibrium configurations corresponding to supercooled liquids may thus be reached. We have also started from solid-like spatial configurations, with again Gaussian velocity distributions. The temperature T_0 corresponding to this initial velocity is progressively raised until "melting" occurs. Our criterion[14] for melting is based on the evolution in time of $\rho_k=\sum_i \cos kx_i$: There k corresponds to the smallest vector of the reciprocal lattice, i.e., $k=4\pi/a$, where a is the side of unit cell of the original fcc lattice, which, in the x direction extends from 0 to L. For a solid, ρ_k is of the order of N. It oscillates around 0 with an amplitude of the order of $\sqrt{N}$ for a liquid. If we admit that the metastable states of the solid are short-lived (as compared to the total time of the computation, i.e. 1.2×10^{-11} sec for argon), we can find by trial and error the melting temperature. For $\rho=0.85$ a solid configuration is still observed for $T=0.695$. It corresponds to a value of the initial temperature $T_0=1.308$. For $T_0=1.366$, melting does occur. Experimentally (see Fig. 1) melting occurs at $T=0.704$. For $\rho=0.88$, we have a solid for $T=0.825$; this corresponds to $T_0=1.545$. With $T_0=1.67$, melting is reached. Experimentally, melting takes place for $T=0.82$ at that density. The agreement with experiment is seen to be good.

[13] D. Levesque and L. Verlet, Physica (to be published).
[14] P. Nozières (private communication).

C. Critical Constants

Because of computational errors, the precise determination of the critical constants is difficult. We believe, however, that a careful examination of the data warrants the following discussion and conclusions.

(a) The extrapolation of the isochores towards the critical temperature yields results which are in good agreement with the Monte Carlo results of Levesque and Verlet[13] (108 particles, about 300 000 configurations). Those results, for $T=1.35$, are shown in Fig. 2 and in Table II.

(b) The PY II equation leads at $\rho=0.45$ and $\rho=0.5$ to values of $\beta p/\rho$ which are definitely too low. The trend to give too low values for the quantities at intermediate density is already noticeable at $\rho=0.4$, as may be seen from Table III, although one is at the border of computational errors. If the values yielded by the PY II equation are corrected by 0.005 at $\rho=0.35$, and by 0.02 at $\rho=0.4$—and this seems to be the most that can be expected—the following critical constants are obtained: $T_c=1.32$, $\rho_c=0.32$, and $\beta_c p_c/\rho_c=0.3$, instead of those obtained from the PY II equation, $T_c=1.36$, $\rho_c=0.36$, and $\beta_c p_c/\rho_c=0.36$. We expect the critical constants to be situated somewhere between those limits.

TABLE II. Values of $\beta p/\rho$ for the isotherm $T=1.35$ obtained from the PY and PY II equations (Ref. 13), Monte Carlo computations (M.C.) (Ref. 13), extrapolation of the molecular dynamics (M.D.) results, 5-term virial series (Ref. 15), its Padé approximant, and experiment on argon (Ref. 10).

ρ	PY	PY II	M.C.	M.D.	5 Vir.	Padé	Exp.
0.2	0.510	0.50			0.506	0.505	0.527
0.3	0.396	0.36	0.36±0.03		0.377	0.374	0.404
0.35	0.376	0.30		0.32±0.02	0.350	0.343	0.368
0.4	0.386	0.27	0.28±0.03	0.29±0.02	0.351	0.338	0.349
0.45	0.434	0.25	0.33±0.03	0.30±0.02	0.382	0.359	0.353
0.5	0.532	0.30		0.33±0.03	0.446	0.406	0.388
0.55	0.692	0.45	0.44±0.03	0.42±0.02	0.545	0.481	0.471
0.65	1.256	1.22	0.86±0.05	0.82±0.03	0.860	0.709	0.863

Notes on p. 73

TABLE III. Values of $\beta p/\rho$, of the inverse compressibility (Ref. 4), and of the interaction part of the integral as given by the molecular dynamics (M.D.), the PY II, and the PY equation for $T=1.46$, $\rho=0.4$.

	M.D.	PY II	PY
$\beta p/\rho$	0.41±0.01	0.40	0.51
$\beta\partial p/\partial\rho$	0.54	0.53	0.49
U^i	−2.72±0.01	−2.72	−2.71

(c) The five virial series[15] gives values of $\beta p/\rho$ which are clearly too high by 0.015 at $\rho=0.3$, by 0.05 at $\rho=0.35$, and by 0.06 at $\rho=0.4$. The difference then diminishes and changes sign at high density. The Padé approximant shows the same behavior but a little less accentuated. We conclude that one or several virial coefficients after the fifth one must be negative, and that some higher ones must be positive again.

The critical point deduced by Barker *et al.*[15] from their series is situated at $T_c=1.29$, $\rho_c=0.26$, and $\beta_c p_c/\rho_c=0.35$. The critical density is quite low, a consequence of the too low pressures at intermediate densities.

(d) Table IV summarizes the situation: There a comparison with experiment on argon is made, which shows that there exists a discrepancy of at least 5% on the critical temperature. This raises the following problem: It is known nowadays[16,17] that the low-density data are not well fitted by a Lennard-Jones potential; a potential with a deeper bowl and a weaker tail, such as the recently determined Kihara potential,[17] seems much more appropriate. It would seem natural to guess that using such a potential will improve the situation in the critical region. The reverse is true: Levesque and Vieillard-Baron,[18] using the PY II equation, find a critical temperature of 1.42 (with the same units as before). The blame for this discrepancy may be put on many-body forces. Even at density around the critical region, a relatively large amount of those forces would be needed. It appears that these many-body forces should then manifest themselves in the form of an interaction which is clearly state dependent at high density; and this seems to be contradicted by the excellence of the fit obtained in the present paper.

TABLE IV. Critical constants obtained from the PY II equation (Ref. 13), from the PY II equation corrected as explained in Sec. III.A, from the 5-term virial series (Ref. 15) and from experiment (Ref. 10).

	T_c	ρ_c	$\beta_c p_c/\rho_c$
PY II	1.36±0.04	0.31±0.03	0.36±0.03
PY II Maximum correction	1.32	0.31	0.30
5-term virial	1.29	0.26	0.352
Experiment	1.26	0.297	0.316

D. Internal Energies

The part of the internal energy due to the interaction U^i is calculated as the time average of the sum of the interparticle interaction energies. A tail correction is included, as for the pressure and the energy; the over-all error appears to be of the order of 0.02. The agreement between our results and the (not too extensive) experimental data[10] is satisfactory, although the calculated internal energies are a little larger in magnitude than the experimental one. For instance, for $\rho=0.45$, at $T=2.935$, $U_{\text{calc}}^i=-2.60$, $U_{\text{exp}}^i=-2.49$; at the same density, for $T=1.764$, $U_{\text{calc}}^i=-2.89$, $U_{\text{exp}}^i=-2.84$. At higher densities, the same trend is noticeable, for $\rho=0.5426$, $T=1.404$, $U_{\text{calc}}^i=-3.63$, $U_{\text{exp}}^i=-3.55$. It appears, roughly speaking, that the Lennard-Jones potential is slightly too deep.

E. High-Frequency Elastic Moduli

Zwanzig and Mountain[19] have shown that in simple fluids the infinite-frequency bulk modulus K_∞ and the infinite-frequency shear modulus G_∞ can be expressed in terms of the radial distribution function; and that, furthermore, in the case of the Lennard-Jones potential, the necessary integrals can be expressed in terms of the pressure and of the internal energy, so that G_∞ and K_∞ are simply given by the following expressions:

$$G_\infty=3p-2\rho T-(24/5)\rho U^i, \tag{9}$$

$$K_\infty=(5/3)G_\infty+2(p-\rho T). \tag{10}$$

Results for K_∞, G_∞, and the ratio K_∞/G_∞ are given in Table I. When the temperature is low, the internal-energy effects predominate and K_∞/G_∞ is near (5/3), the value obtained from Cauchy condition for solids.

We may compare the "exact results" given in Table I with those obtained by Zwanzig and Mountain who used (9) and (10), and the experimental data of Levelt,[10] supposing that the interaction between argon atoms is exactly the Lennard-Jones potential. The good agreement between those results and ours shows again—although this proof is not independent of the preceding ones—that the Lennard-Jones potential is appropriate to describe argon in the temperature and density domain considered in the present paper.

F. Isotopic Separation Factor

The separation faction in a liquid-gas mixture of isotopes of argon can be expressed[20] in terms of the

[15] J. A. Barker, P. J. Leonard, and A. Pompe, J. Chem. Phys. **44**, 4266 (1966).
[16] E. A. Guggenheim and M. C. McGlasham, Proc. Roy. Soc. (London) **225A**, 456 (1960); R. J. Munn and J. Alder, J. Chem. Phys. **43**, 3998 (1965); J. Rowlinson, Discussions Faraday Soc. **40**, 19 (1965).
[17] J. A. Barker, W. Fock, and F. Smith, Phys. Fluids **7**, 429 (1964).
[18] D. Levesque and J. Vieillard-Baron (to be published).
[19] R. Zwanzig and R. D. Mountain, J. Chem. Phys. **43**, 4464 (1965).
[20] G. Casanova, A. Levi, and N. Terzi, Physica **30**, 937 (1964).

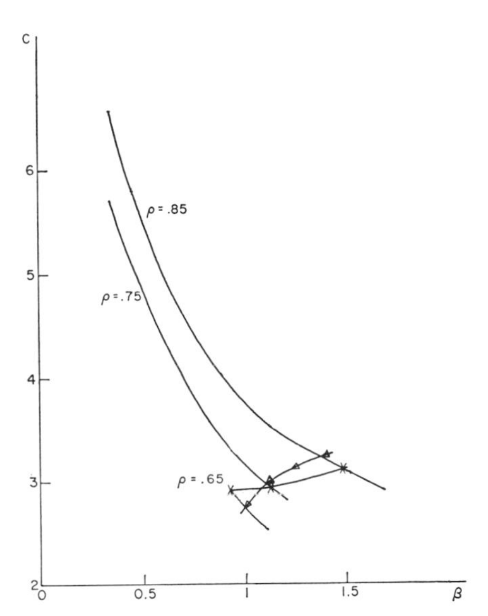

FIG. 3. The solid lines show the quantity $C=\Omega_0^2/2\pi\rho m = \langle\Delta^2 V\rangle/96\pi\rho$ for the isochores $\rho=0.85$, $\rho=0.75$, and $\rho=0.65$, as a function of β. The crosses represent the intercepts of those isochores with the experimental liquid-gas coexistence line. It should be compared with the isotopic separation factor in argon (Ref. 20) (triangles).

average value of the Laplacian of the potential $\langle\nabla^2 V(r)\rangle$. The quantity $\Omega_0^2=(1/M)\langle\nabla^2 V(r)\rangle$, which is also important in the time development of the system, has been calculated, and the values we obtained are shown in Table I. The quantity $C=\Omega_0^2/2\pi\rho$ is plotted for several isochores in Fig. 3. We have used experimental information (pressure-versus-density curve for the coexistence curve, as shown on Fig. 1), to locate the temperature of the coexistence line on the isochores. It may be seen easily, however, that this introduces very little arbitrariness. A good fit of the experiment is obtained, as seen from Fig. 3.

IV. CONCLUSIONS

We have shown, using Rahman's work as a starting point, how it is possible to integrate the equations of motion of about a thousand particles in a relatively easy way. The first application of this tool is the thermodynamic study of a fluid composed of atoms interacting through a Lennard-Jones potential. The striking result of this study is the over-all agreement between the results thus obtained and the thermodynamics of real argon. It is likely that those results can still be slightly improved and that a two-body potential can fit the experimental data with a large degree of success.

ACKNOWLEDGMENTS

The author is very grateful for the hospitality of Yeshiva University where most of this work was done. It is a particular pleasure to thank Professor Joel L. Lebowitz; numerous long discussions have been both inspiring and stimulating in the series of studies whose results we have begun to describe. We are grateful also to D. Levesque, J. Vieillard-Baron, and D. Schiff for help and discussions. Last, but not least, we must acknowledge the profuse computing opportunities and the warm welcome in the Courant Institute of New York University, where the largest part of the computations were performed, and where the author benefitted from interesting and helpful discussions with Professor J. Percus and Dr. M. Kalos.

Notes to Reprint I.7

1. This article is the first in a series of papers on the properties of Lennard–Jones "argon", the fourth and last of which (on collective dynamical properties) is included in Chapter III (Reprint III.2). The others in the series are (Verlet, Phys. Rev. 165 (1968) 201) (structural properties) and (Levesque and Verlet, Phys. Rev. A 2 (1970) 2514) (single-particle dynamics).

2. For a discussion of algorithms other than eq. (4), see Note I.6.2.

3. Other bookkeeping devices have been devised that, in some circumstances, are more efficient than Verlet's neighbour-list method (Hockney and Eastwood, Computer Simulation Using Particles (McGraw-Hill, New York, 1981)). In its original form, Verlet's method is ill suited to vector computers, but vectorizable versions have subsequently been developed (Fincham and Ralston, Comput. Phys. Commun. 23 (1981) 127; van Gunsteren et al., J. Comput. Chem. 5 (1984) 272).

4. The results of a more recent and extensive computer simulation study of the thermodynamic properties of the Lennard–Jones fluid can be found in a paper by Nicholas et al. (Mol. Phys. 37 (1979) 1429).

STRUCTURE OF WATER; A MONTE CARLO CALCULATION *

J. A. BARKER and R. O. WATTS
Departments of Applied Mathematics and Physics, University of Waterloo, Waterloo, Ontario, Canada

Received 10 February 1969

An *a priori* calculation of the energy, specific heat and radial distribution function of liquid water at 25°C is made using the Monte Carlo technique and an intermolecular pair potential determined by Rowlinson from the properties of ice and steam. Agreement with experiment is sufficiently good to demonstrate the feasibility of this approach to water.

The structure of liquid water presents important and controversial questions; for recent reviews see Kavanau [1] and Conway [2]. Spectroscopic and structural evidence [3,4] appears to favour 'continuum' as against 'mixture' pictures, but this is not generally accepted [5]. Simple liquids are now well understood as a result of Monte Carlo and molecular dynamics calculations [6-8], advances in perturbation theory [9] and increased knowledge of intermolecular forces [10,11]. Here we describe a Monte Carlo calculation for water at 25°C using an orientation dependent additive pair potential energy function derived by Rowlinson [12] from second virial coefficients and the lattice energy and spacing of ice. This interaction consists of a) the electrostatic interaction of four charges on one molecule with four on another, b) a Lennard-Jones 6:12 function of the distance between the charge centres of the molecules, with parameters $\epsilon = 0.707$ kcal/mole, $\sigma = 2.725$ Å. The charges are $+0.3278e$ at $0.5844\boldsymbol{a} \pm 0.7616\boldsymbol{b}$ and $-0.3278e$ at $\pm 0.2539\boldsymbol{c}$; e is the electronic charge and $\boldsymbol{a}$, $\boldsymbol{b}$, $\boldsymbol{c}$ are orthogonal unit (Å) vectors fixed in the molecule. The origin is the oxygen nucleus and the charge centre is at $0.2922\boldsymbol{a}$. We imposed a hard-sphere cut-off at 2Å and neglected interactions beyond 6.2Å. The hard-sphere cut-off is necessary to remove the physically unreal configurations in which charges of opposite sign on different molecules coincide. Apart from these unreal configurations the Lennard-Jones potential effectively prevents molecules from approaching closer than about 2.2 Å, so that the cut-off has no other effect on the results.

We considered 64 molecules in a cube with periodic boundary conditions at the experimental density at 25°C. We generated configurations as follows: (i) select a molecule at random, (ii) select displacements δ_x, δ_y, δ_z, each uniformly distributed on $(-\frac{1}{2}\Delta, \frac{1}{2}\Delta)$, (iii) select the x, y or z axis at random, (iv) select an angle θ uniformly distributed on $(-\psi, \psi)$, (v) calculate the change in potential energy $\delta\varphi$ on displacing the chosen molecule by $(\delta_x, \delta_y, \delta_z)$ and rotating it through θ about the chosen axis, (vi) select a number u uniformly distributed on (0, 1), (vii) if $\exp(-\delta\phi/kT) < u$ take the next configuration identical with the previous configuration, otherwise take it as the configuration with the chosen molecule moved and rotated as indicated. These rules satisfy the conditions of reversibility and accessibility [3] required to ensure that averaging over long chains approaches classical canonical averaging, with weighting of configurations proportional to $\exp(-\phi/kT)$. Thus we ignored quantum effects, which are probably not large in water at 25°C.

We started the molecules in a cubic ice structure at 2980 °K and reduced the temperature by 298 °K after each 500 configurations until it reached 298 °K. The angle ψ was fixed at 10° and Δ was chosen so that about half the attempted moves were actually made, the resulting value being 0.5 Å. A further 120 000 configurations were generated to equilibrate the system; there was some drift in the energy at the beginning of this stage but this was absent at the end. We then generated 110 000 configurations and averaged the potential energy ϕ, its square

* This research has been supported by the National Research Council and the Department of Energy, Mines and Resources, Canada, and by the US Department of the Interior, Office of Saline Water.

ϕ^2 (to permit estimation of the specific heat C_v) and the radial distribution function $g(R)$.

After adding kinetic energy contributions ($3NkT$ for the energy, $3Nk$ for C_v) we found for the thermodynamic energy U (referred to separated molecules) the values -8.38 kcal/mole after 54 000 configurations and -8.36 kcal/mole after 110 000 configurations, compared with the experimental [13] value -8.12 kcal/mole. For C_v we found 21.2 cal/deg. mole after 54 000 configurations, 20.5 cal/deg. mole after 110 000 compared with the experimental value 18 cal/deg. mole. Considering that our model contains no adjustable parameter, since the potential was determined from the properties of ice and steam, these results are strikingly good. The calculated radial distribution is compared with the experimental function of Narten, Danford and Levy [14] in fig. 1. The calculated number of neighbours within 3.5 Å is near 6.4, compared with the experimental value 5.1; in part this difference is due to the fact that the potential function apparently permits molecules to approach too closely. The agreement with experiment for the radial distribution function is not outstanding but we believe that these results are sufficiently good to establish the feasilibity of this approach to water. In later work we plan to calculate the pressure to estimate the volumetric behaviour. and also to improve the potential function by studying a wide range of properties of water and the various modifications of ice.

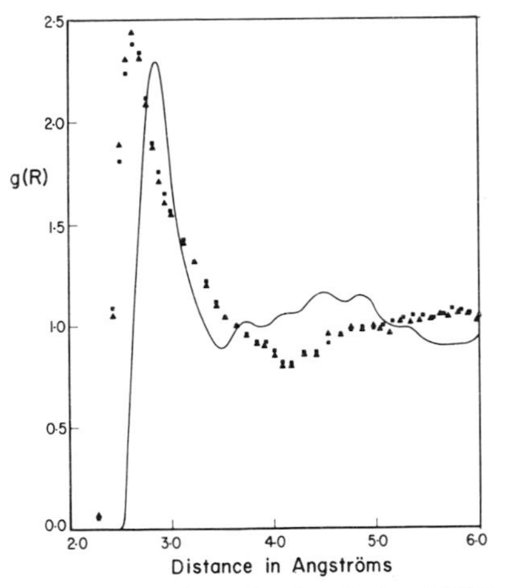

Fig. 1. Radial distribution function of water at 25°C. Solid line, experiment [14]; ▲ calculated, 54 000 configurations: ■ calculated, 110 000 configurations.

We acknowledge helpful discussions with Professor D. Henderson.

REFERENCES

[1] J. Lee Kavanau, Water and Solute-Water Interactions (Holden-Day, Inc., 1964).
[2] B. E. Conway, Ann. Revs. Phys. Chem.17 (1966) 481.
[3] M. Falk and T. A. Ford, Can. J. Chem. 44 (1966) 1699; 46 (1968) 3579.
[4] E. Whalley, Ann. Rev. Phys. Chem. 18 (1967) 205.
[5] G. E. Walrafen, J. Chem. Phys. 48 (1968) 244.
[6] W. W. Wood and F. R. Parker, J. Chem. Phys. 27 (1957) 720.
[7] I. R. McDonald and K. Singer. Disc. Faraday Soc. 43 (1967) 40.
[8] L. Verlet, Phys. Rev. 159 (1967) 98; 165 (1968) 201.
[9] J. A. Barker and D. Henderson, J. Chem. Phys. 47 (1967) 4714.
[10] J. A. Barker and A. Pompe, Austral. J. Chem. 21 (1968) 1683.
[11] J. A. Barker. D. Henderson and W. R. Smith, Phys. Rev. Letters 21 (1968) 134.
[12] J. S. Rowlinson, Trans. Faraday Soc. 47 (1951) 120.
[13] N. E. Dorsey, Properties of Ordinary Water-Substance, A. C. S. Monograph No. 18 (1940).
[14] A. H. Narten, M. D. Danford and H. A. Levy. Disc. Faraday Soc. 43 (1967) 97.

Notes to Reprint I.8

1. For a follow-up on the simulation of water, and a discussion of the potentials used, see the notes to Reprint V.1.

2. The Monte Carlo sampling of the orientations of a rigid molecule poses no special difficulty, as the paper shows. On the other hand, the choice of sampling scheme for the internal degrees of freedom of partially flexible molecules requires some care (see, for example, Fixman, Proc. Natl. Acad. Sci. USA 71 (1974) 3050). Perhaps for this reason, most authors prefer the use of the molecular dynamics method for the study of non-rigid molecules.

PHYSICAL REVIEW A VOLUME 1, NUMBER 1 JANUARY 1970

Decay of the Velocity Autocorrelation Function*

B. J. Alder and T. E. Wainwright
Lawrence Radiation Laboratory, University of California, Livermore, California 94550
(Received 10 July 1969)

Molecular-dynamic studies of the behavior of the diffusion coefficient after a long time s have shown that the velocity autocorrelation function decays as s^{-1} for hard disks and as $s^{-3/2}$ for hard spheres, at least at intermediate fluid densities. A hydrodynamic similarity solution of the decay in velocity of an initially moving volume element in an otherwise stationary compressible viscous fluid agrees with a decay of $(\eta s)^{-d/2}$, where η is the viscosity and d is the dimensionality of the system. The slow decay, which would lead to a divergent diffusion coefficient in two dimensions, is caused by a vortex flow pattern which has been quantitatively compared for the hydrodynamic and molecular-dynamic calculations.

A previous study[1] of the diffusion coefficient has shown that the velocity autocorrelation function has a long positive tail, indicating a surprising persistence of velocities. Subsequently,[2] the collective nature of this persistence was established by the observation that the value of the diffusion coefficient depends strongly on the number of particles, particularly in two dimensions where the results did not seem to converge as larger systems were investigated. Finally, by studying the velocity correlation between a molecule and its neighborhood, a vortex flow pattern was found on a microscopic scale which could qualitatively explain the tail. Since the persistence of the vortex flow is long compared to the mean collision time, it is natural to ask whether a hydrodynamic model could calculate such vortex motion, and hence, the behavior of the velocity autocorrelation function for long times. This paper addresses itself to that question.

In such a hydrodynamic model, a fluid is imagined to be at rest except that a small volume element is given an initial velocity. A compression wave develops in front of this region and a rarefaction wave to the rear. When the sound waves have separated, the residual flow is in the form of a double vortex in two dimensions, or a vortex ring in three dimensions. At late times, the circulatory flow approaches that of an incompressible fluid, and hence the velocity decays solely due to the influence of the shear viscosity. It should be emphasized at the outset that this hydrodynamic model differs conceptually from the Stokes-Einstein model which also relates the diffusion coefficient to the viscosity. In that model, a sphere representing a molecule is assumed to slow down adiabatically in a viscous fluid; that is, the retarding force at each instant of time is assumed to be the steady-state value, which is proportional to the velocity, so that the velocity decays exponentially. In the model described here, on the other hand, a transient solution of the Navier-Stokes equation is carried out to find the long-time behavior of the initially moving volume element.

The initially moving volume element is made equal in size to the average volume per molecule and is given a velocity comparable to the root-mean-square molecular velocity. The subsequent motion of the fluid is then calculated by direct numerical integration of the Navier-Stokes equation.[3] Both Eulerian and Lagrangian formulations have been used successfully. A comparison of the flow pattern between the hydrodynamic and molecular-dynamic calculation at a fairly late time is given in Fig. 1. The nearly quantitative agreement obtained lends credence to the applicability of the model. The values for the viscosity η and $y = Pv/NkT-1$ used in the hydrodynamic calculation were obtained from molecular-dynamic calculations at the same density. The comparison in Fig. 1 was made at a fairly late time, so that the flow pattern had approached the hydrodynamic regime, but not so late that there was any interference from the sound waves coming over the periodic border of the finite system, nor so late that the velocity had decayed to such a small value as to prevent an accurate determination. A correction of $1/N-1$ has been added to the velocity autocorrelation function as calculated by molecular dynamics. This is because whenever a given molecule has a velocity v, the average velocity of the other molecules is $-v/N-1$ in a system of N molecules where momentum is conserved.

Figure 2 illustrates, in the case of hard spheres, the agreement of the autocorrelation function $\rho(s)$ from the molecular-dynamic calculation with that from the hydrodynamic calculation. The latter is simply the velocity at the center of the flow pattern divided by the initial velocity. The two disagree at short times, as might be expected. The molecular-dynamic velocity autocorrelation function shows an initial exponential decay lasting for a few mean collision times s. However, at times great-

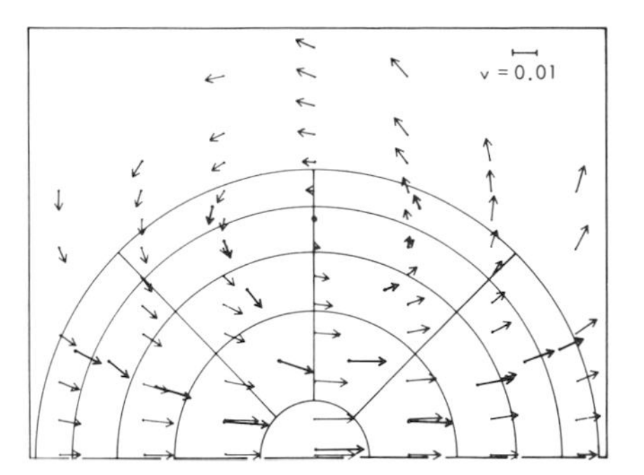

FIG. 1. Statistically averaged velocity field around a central disk from molecular dynamics (heavy arrows) compared to that given by the hydrodynamic model (light arrows). Because of symmetry only half the plane is shown. The scale of distance is indicated by the size of the central disk as shown by the smallest half-circle. The sizes of the other four concentric circles have been determined so as to include roughly six neighboring particles each. These semicircles have been partitioned further into four parts, as indicated by the lines, so as to have a measure of direction relative to the velocity vector of the central particle at zero time. The size of the arrows indicates the magnitude of the velocity (the scale of velocity is indicated as 0.01 of the initial velocity in the upper right-hand corner) and the direction of the arrow is determined by the parallel and perpendicular components of the velocity (relative to that of the central particle initially) averaged over all the particles in that section at a particular time. The arrow is hence drawn at the center of the section. A correction of $1/N$-1 has been added to the parallel component. The comparison is made at 9.9 collision times where the molecular-dynamic and hydrodynamic velocity autocorrelations begin to nearly agree, as seen on the graph by the velocity vectors of the central particle. (See also Fig. 3.) In the molecular-dynamics run, 224 hard disks were used at an area relative to close packing of 2. For the hydrodynamic run, the conditions are given in Table I.

er than about 10 mean collision times, both calculations show a decay like $s^{-3/2}$.

Figure 3 illustrates the same agreement at various densities in the case of hard disks where the decay is like s^{-1}. Figure 3 shows furthermore that the $1/N-1$ correction brings into agreement the velocity autocorrelation functions calculated in molecular-dynamic systems of various sizes and that the long-time behavior of the hydrodynamic solution is independent of the initial velocity. It was also found that the long-time hydrodynamic solution does not depend upon the bulk viscosity. Heat conductivity was not included in the hydrodynamic calculation. The late-time kinks seen in Fig. 3 in the velocity autocorrelation functions calculated for 504 particle systems are caused by the arrival of sound waves from the periodic images. The arrival time of these interferences can be predicted by the hydrodynamical model.

A simple analysis of the hydrodynamical model

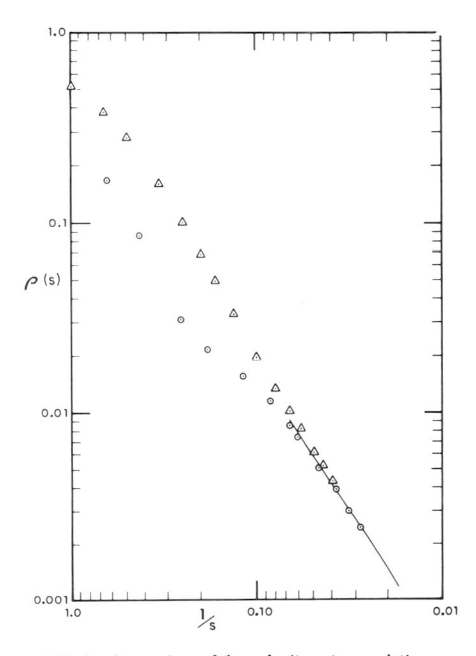

FIG. 2. Comparison of the velocity autocorrelation function $\rho(s)$ as a function of time (in terms of mean collision times s) between the hydrodynamic model (circles) and a 500-hard-sphere molecular-dynamic calculation (triangles) at a volume relative to close packing of 3 on a log-log plot. The straight line is drawn with a slope corresponding to $s^{-3/2}$. To the molecular dynamics $\rho(s)$ a correction of $1/N$-1 has been added. Furthermore, the function has only been graphed up to the time where serious interference between neighboring periodically repeated systems is indicated. In the hydrodynamic calculation the viscosity predicted by the Enskog theory has been used while the molecular-dynamic calculations indicate a 2% larger value. A value of pv/NkT of 3.03 was employed, and the initial velocity of the fluid volume element was normalized to unity for comparison purposes. If the initially moving cylindrical region is made to have the same volume as that corresponding to the volume per particle in the molecular system, a distance and hence a time scale can be obtained to make the above comparison.

Notes on p. 79

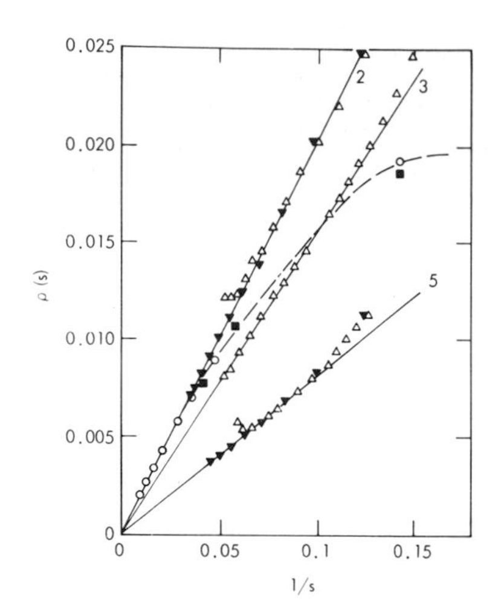

FIG. 3. The decay of the velocity autocorrelation function at large times for hard disks at three densities: $A/A_0 = 2, 3$, and 5. The closed and open triangles refer to molecular-dynamic runs of 986 and 504 particles, respectively. A $1/N-1$ correction to the molecular-dynamic results has been applied. At A/A_0 of 2 and 5 the 504-particle results include the initial deviations due to the interference of neighboring cells at the boundary while all other results have not been plotted beyond the point where serious interference is indicated. The dashed line represents the results of a hydrodynamic run at A/A_0 of 2 (see Table I for conditions) in which the initially moving square area element was given two different velocities, the root-mean-square molecular velocity (squares) and that $\frac{1}{14}$th as large (circles).

shows that a similarity solution exists for the circulatory flow at late times. The linear dimensions of the flow pattern increase at $(\nu s)^{1/2}$ and, since total momentum is conserved, the velocity decays as $(\nu s)^{-d/2}$, where ν is the kinematic viscosity (η divided by the density) and d is the dimensionality of the system. This result verifies the observed behavior.

Table I lists the values of the decay constants α, found by molecular dynamics in two dimensions, where $\rho(s) = \alpha s^{-1}$. Since the hydrodynamic model predicts $\rho(s)$ is proportional to $(\nu s)^{-1}$, $\alpha_H = \alpha\eta s/\eta_0 s_0 = \alpha\eta/\eta_0 y$ should be a constant; the collision rate being proportional to y. The zero subscript indicates the low-density Boltzmann values for the reference system. The value of α_H is π^{-1}, as accurately as it can be determined from the numerical hydrodynamic calculations at a number of different densities and also according to the analytic asymptotic solution. This solution does, however, depend on the empirically supported assumption that the two sound waves carry off $\frac{1}{2}$ of the original momentum, independent of all parameters involved in the calculation, the remaining half being involved in the vortex flow.

The slight remaining density dependence of α_H and its small disagreement with the hydrodynamic solution can be ascribed to the unrealistic nature of the hydrodynamical model. The hydrodynamic flow will not carry the molecule appreciably away from the center of the vortex pattern; but, in fact, in an actual system a molecule has a density-dependent probability of diffusing away from the center. At low densities, particularly, intermolecular diffusion can carry molecules away from the center to a distance comparable with the size of the vortex pattern. To account approximately for the diffusive motion, the vortex flow pattern has to be sampled over a spreading Gaussian distribution representing the probability that the molecule has moved away from the center. This argument leads to the following correction factor F:

$$F = \frac{\int \exp(-r^2 D_0/D_E s)\exp(-r^2\eta_0/\eta s) r\,dr}{\int \exp(-r^2 D_0/D_E s) r\,dr}$$

$$= \frac{\eta/\eta_0}{D_E/D_0 + \eta/\eta_0} .$$

The Enskog value of the diffusion coefficient D_E is used because it is intended to describe only the diffusion of the molecules among its neighbors and not the collective motion of the neighborhood for which the hydrodynamic model is used.

A comparison of the last two columns of Table I shows that this correction factor F accounts for the density dependence of α_H to within a few percent, that is, within the accuracy of the determina-

TABLE I. Values of the decay coefficient α.

A/A_0	α	y	η/η_0	D_E/D_0	$\alpha_H = \alpha\eta/\eta_0 y$	$\alpha_H\pi$	F
2	0.206	2.42	3.39	0.375	0.29	0.91	0.90
3	0.157	1.08	1.66	0.560	0.24	0.76	0.75
5	0.082	0.50	1.29	0.725	0.21	0.66	0.64

tion of α and η/η_0. The above argument leads to the prediction that in the low-density limit $F = \frac{1}{2}$. Thus, the velocity autocorrelation function in two dimensions decays as s^{-1} at any finite density, leading to a divergent diffusion coefficient at any nonzero density. This result is in contradiction to previous theories on the density expansion of the diffusion coefficient away from the low-density limit. The study of the late-time autocorrelation function at very low densities by molecular dynamics is unfortunately very difficult since the system must be so large that a molecule undergoes many collisions before a sound wave travels across the size of the system.

The hydrodynamic model, as discussed so far, cannot reverse the velocity of the region initially in motion, and thus cannot reproduce the negative part of the velocity autocorrelation found at high densities. This deficiency can be remedied at least qualitatively by the inclusion of visco-elastic forces in the Navier-Stokes equations. These forces can be obtained from the autocorrelations of the elements of the stress tensor as calculated by molecular dynamics. A trial calculation at $A/A_0 = 1.4$ has shown that negative autocorrelation functions can be obtained in this way, but that at very late times, in agreement with molecular dynamics, the function becomes again positive and decays like s^{-1}.

We wish to thank E. D. Giroux and J. A. Viecelli for invaluable help with the hydrodynamic calculations and M. A. Mansigh similarly with the molecular-dynamic calculations.

---

*Work performed under the auspices of the U.S. Atomic Energy Commission.

[1] B. J. Alder and T. E. Wainwright, Phys. Rev. Letters 18, 988 (1967).

[2] B. J. Alder and T. E. Wainwright, J. Phys. Soc. Japan, Suppl. 26, 267 (1968).

[3] M. L. Wilkins, in Methods in Computational Physics (Academic Press Inc., New York, 1964), Vol. 3, p. 211.

Notes to Reprint I.9

1. A recent survey of long-time tails in computer simulations has been given by Alder (in: Molecular Dynamics Simulation of Statistical-Mechanical Systems, eds G. Ciccotti and W.G. Hoover (North-Holland, Amsterdam, 1986)).

2. A long-time tail was first observed in the velocity autocorrelation function of a system with a continuous interparticle potential (a truncated Lennard–Jones interaction) by Levesque and Ashurst (Phys. Rev. Lett. 33 (1974) 277).

CHAPTER II

Free Energies and Phase Equilibria

The methods described in the previous chapter can be used to obtain equilibrium averages of "mechanical" properties of classical many-body systems. The term "mechanical" refers here to quantities that can be expressed as functions of the coordinates and momenta of the particles: examples are the potential energy and instantaneous pressure. By contrast, it is not in general possible to obtain directly from a simulation information on "thermal" properties, i.e., quantities that depend on the total phase-space volume accessible to the system: examples include the entropy, Helmholtz free energy and chemical potential(s). Knowledge of the latter is particularly important in situations where two or more phases may coexist.

In order to understand the problems involved in the calculation, for example, of the free energy, it is useful to recall how this quantity is obtained in real experiments. In the real world, as in simulations, free energies are not directly measurable. The procedure commonly used in determining the free energy at a particular state point is first to find a reversible path that links the state point of interest to a state of known free energy, such as the ideal gas or the low-temperature, harmonic crystal. The change in free energy along the reversible path is then calculated by thermodynamic integration based on the relations

$$\left(\frac{\partial A}{\partial V}\right)_{N,T} = -P, \qquad \left(\frac{\partial(A/T)}{\partial(1/T)}\right)_{N,V} = E,$$

where A is the Helmholtz free energy, V the volume, P the pressure, T the temperature and E the internal energy of the system. In some cases, the same procedure can be used in computer "experiments", but the method often fails in situations where the path of thermodynamic integration crosses a first-order phase transition. The reason for the failure is the fact that in the relatively small, periodic systems studied by computer simulation, two-phase coexistence is difficult to achieve, especially when the transition involves a solid phase. What is much more commonly found in practice is that one phase is supercooled or superheated to a point at which it becomes mechanically unstable and transforms irreversibly to the other. The earliest attempts to study first-order phase transitions by simulation, including those described in Reprints I.2, I.3 and I.4, suffered from precisely this problem. The first simulation that was successful in locating the coexistence point of a first-order transition is described in the paper of Alder and Wainwright (Reprint II.1) on the melting of a two-dimensional system of hard disks. The method used involves a thermodynamic integration through the two-phase region. Because of the small sample size, the pressure along an isotherm in the two-phase region is not constant, but instead displays a van der Waals-like loop; the origin of the loop is discussed in some detail by Mayer and Wood [1]. The densities of the coexisting phases are computed via a Maxwell equal-areas construction on the quasi-van der Waals loop.

The method adopted by Alder and Wainwright is straightforward in principle, but requires very long simulations and large samples. In addition, for three-dimensional systems, the hysteresis effects associated with the solid–fluid transition are so severe that other methods are needed in order to locate the melting point. The first technique designed for this purpose is described in the paper by Hoover and Ree (Reprint II.2). The idea behind the method is as follows. Thermodynamic integration from dilute gas to solid is troublesome because the path is not everywhere reversible. This can be remedied by carrying out the integration along an artificial path from a reference state of known free energy to the solid under consideration. The task, therefore, is to find a path that is free of hysteresis. The particular choice made by Hoover and Ree was to take as reference system a dilute, single-occupancy, lattice gas that can be compressed reversibly to give the solid phase. In their paper, the single-occupancy cell method is tested on the only model for which the melting transition had already been located, namely the hard-disk system studied earlier by Alder and Wainwright (Reprint II.1), but Hoover and Ree [2] were subsequently able to use the same approach to determine the freezing point of the hard-sphere fluid. The method of Hoover and Ree was later applied to the calculation of the melting line of the Lennard–Jones solid by Hansen and Verlet (Reprint II.3), who also devised a similar technique for the study of liquid–vapour coexistence. The resulting phase diagram of "Lennard–Jonesium" was found to be in good agreement with the experimentally known phase diagram of argon.

Although the single-occupancy cell method has a general applicability, it suffers from certain practical drawbacks. First, the thermodynamic-integration path, even though reversible, apparently does cross a second-order phase transition (or even, in three dimensions, a weakly first-order one [3]); this manifests itself as a cusp-like feature in the equation of state. Integration of the pressure through such a transition requires some care. Secondly, although the method can, in principle, be extended to molecular systems, its implementation in such cases is cumbersome and, to our knowledge, has not yet been attempted.

Another method to determine the absolute free energy of solid phases was developed by Hoover et al. [4]. This is applicable to any solid phase that can be cooled or compressed reversibly to a crystalline state that is effectively harmonic. The free energy of a harmonic crystal can be calculated by standard methods of lattice dynamics; the free energy of the solid at higher temperatures and/or lower densities is then obtained by straightforward thermodynamic integration. Clearly the method is useful only if a nearly harmonic state can be reached. It will therefore fail for hard-core systems and for solid phases that are strongly anharmonic throughout their range of (mechanical) stability. In such situations, a thermodynamic-integration scheme devised by Frenkel and Ladd (Reprint II.4) can be useful. Along the integration path, the potential-energy function U of the system is made to transform smoothly from the form corresponding to the solid under consideration to that of an Einstein crystal of the same structure. The free energy A_0 of the Einstein crystal can be evaluated analytically and A_1, that of the solid of interest, is obtained from

$$A_1 - A_0 = \int_0^1 (\, \mathrm{d}A/\mathrm{d}\lambda) \, \mathrm{d}\lambda = \int_0^1 U(\lambda) \, \mathrm{d}\lambda,$$

where $U(\lambda)$ is the generalized potential-energy function. The dependence of $U(\lambda)$ on the parameter λ is such that $U(\lambda = 0)$ corresponds to the Einstein crystal and $U(\lambda = 1)$ to the solid under consideration. The method was originally developed for hard spheres but was later generalized to both non-spherical molecules [5] and continuous potentials (see Reprint IV.8). There are also many other cases in which an artifical thermodynamic integration has been used to measure a free energy *difference*. Illustrative examples can be found in refs. [6] (calculation of solubilities) and [7] (free energies of hydration).

The methods outlined above are all rather time consuming insofar as they require several simulations in order to obtain the free energy (or chemical potential) at a given state point. Under certain conditions, however, it is possible to compute the chemical potential of a given species (atomic or molecular) from the results of a single simulation. One possibility is to use the "particle-insertion" method first formulated by Widom [8]. Another is to carry out a simulation in the grand canonical ensemble, in which chemical potential is an independent variable. Grand canonical simulations, which were pioneered by Norman and Filinov [9], Adams [10], and Rowley et al. [11], are discussed in Chapter IV. Here we concentrate on the particle-insertion technique. The method is based on the relation between the excess chemical potential μ_{ex} and the average of the Boltzmann factor associated with the insertion of a test particle at a random point in the fluid. This relation has the form

$$\mu_{\mathrm{ex}} = -kT \log \langle \exp(-u/kT) \rangle,$$

where μ_{ex} is the difference between the chemical potential of the species of interest and that of an ideal gas at the same temperature and density, and u is the interaction energy of the test particle with all particles already present in the system. The quantity $\exp(-u/kT)$ may be thought of as the probability of acceptance of a Monte Carlo trial move that attempts to add a particle at a random point in the system, though it must be stressed that no such move is ever carried out in practice. When the average "acceptance rate" of particle moves becomes very small (u/kT almost always much larger than unity), the statistical error in μ_{ex} becomes such that the method is no longer useful. That situation is typical of dense systems of particles for which the short-range interactions are strongly repulsive. For this reason, the method is ill-suited to the study of solid–fluid coexistence. It has been used, however, in the investigation of phase equilibria that occur at lower densities, including the liquid–vapour [12] and isotropic–nematic [13] transitions. The particle-insertion method was first implemented for atomic systems by Adams [10] and generalized to molecular fluids by Romano and Singer (Reprint II.5).

The early versions of the particle-insertion method were not particularly efficient, partly for the reasons already explained. Considerable improvement was achieved by Shing and Gubbins [14], who proposed a scheme that employs a form of "umbrella sampling" (see below) and also considers both particle insertion and particle removal. In its latter aspect, the Shing–Gubbins approach is a special case of the "overlapping-distribution" method suggested by Bennett (Reprint II.6) for the measurement of free-energy differences between two "similar" systems. In this context, "similar" systems are systems with hamiltonians $\mathscr{H}_0$ and $\mathscr{H}_1$, say, for which there is a significant degree of overlap of the accessible volumes in configuration space. Let U_1 be the potential energy of a given configuration of system 1 and let U_0 be the potential energy for the same configuration of system 0. Bennett shows that the free-energy difference $A_1 - A_0$ is related to the ratio $p_1(U_1 - U_0)/p_0(U_1 - U_0)$ by

$$p_1(x)/p_0(x) = \exp[(A_1 - A_0 - x)/kT]$$

where $p_0(x)$ (or $p_1(x)$) is the probability density for the potential-energy difference $x = U_1 - U_0$ for a canonical distribution of states of system 0 (or 1). Another scheme described by Bennett is one called the "acceptance-ratio" method, the basis of which is a relation between the free-energy difference and averages of the Fermi function $f(x) = (1 + e^x)^{-1}$ over the distributions $p_0(x)$ and $p_1(x)$:

$$\exp[(A_1 - A_0)/kT] = (\langle f(C - x)\rangle_1 / \langle f(x - C)\rangle_0) \exp C$$

Here C is a constant that can be chosen so as to minimize the estimated error in $A_1 - A_0$. Both of Bennett's methods have been applied to a wide range of problems. See, for example ref. [15]. In most cases studied, the acceptance-ratio approach appears to be the more accurate of the two. The same ideas can also be exploited in the calculation of free-energy differences between systems that are not "similar" in the sense explained above. This is achieved by defining a number of systems with hamiltonians $\mathscr{H}_i$, intermediate between $\mathscr{H}_0$ and $\mathscr{H}_1$, such that the configuration spaces of two successive systems again have a significant overlap. "Multistage-sampling" methods of simpler type had earlier been developed by several authors [16,17].

Yet another method for the computation of free-energy differences is to work with a single, artificial hamiltonian designed in such a way that the configuration space accessible to the system overlaps as much as possible with the parts accessible to the systems of physical interest. The earliest example of this "umbrella-sampling" approach is contained in a paper by Torrie and Valleau [18]; the technique itself is explained in detail in Reprint II.7. The advantage of the method is the fact that a judicious choice of the "non-physical" sampling distribution can greatly reduce the number of intermediate stages required to compute the relative free energies, or any other thermodynamic properties, of a whole class of "physical" systems.

The calculation of free energies by computer simulation is by no means a solved problem. At present, much effort is being expended on the development of efficient techniques for obtaining information about chemical potentials in complex systems, including molecular solids, multicomponent mixtures and quantum fluids.

References

[1] J.E. Mayer and W.W. Wood, J. Chem. Phys. 42 (1965) 4268.
[2] W.G. Hoover and F.H. Ree, J. Chem. Phys. 47 (1968) 4873.
[3] H. Ogura, H. Matsuda, T. Ogawa, N. Ogita and A. Ueda, Prog. Theor. Phys. 58 (1977) 419.
[4] W.G. Hoover, M. Ross, D. Henderson, J.A. Barker and B.C. Brown, J. Chem. Phys. 52 (1970) 4931.
[5] D. Frenkel and B. Mulder, Mol. Phys. 55 (1985) 1171.
[6] W.C. Swope and H.C. Andersen, J. Phys. Chem. 88 (1984) 6548.
[7] W.L. Jorgensen and C. Ravimohan, J. Chem. Phys. 83 (1985) 3050.
[8] B. Widom, J. Chem. Phys. 39 (1963) 2808.
[9] G.E. Norman and V.S. Filinov, High Temp. Res. USSR 7 (1969) 216.
[10] D.J. Adams, Mol. Phys. 28 (1974) 1241.
[11] L.A. Rowley, D. Nicholson and N.G. Parsonage, J. Comput. Phys. 17 (1975) 401.
[12] J.G. Powles, Mol. Phys. 41 (1980) 715.
J.G. Powles, W.A.B. Evans and N. Quirke, Mol. Phys. 46 (1982) 1347.
[13] D. Frenkel and R. Eppenga, Phys. Rev. Lett. 49 (1982) 1089.
[14] K.S. Shing and K.E. Gubbins, Mol. Phys. 46 (1982) 1109.
[15] A. Rahman and G. Jacucci, Nuovo Cimento D 4 (1984) 357 and references therein.
[16] I.R. McDonald and K. Singer, J. Chem. Phys. 50 (1969) 2308.
[17] J.P. Valleau and D.N. Card, J. Chem. Phys. 57 (1972) 5457.
[18] G.M. Torrie and J.P. Valleau, Chem. Phys. Lett. 28 (1974) 578.

PHYSICAL REVIEW VOLUME 127, NUMBER 2 JULY 15, 1962

Phase Transition in Elastic Disks*

B. J. ALDER AND T. E. WAINWRIGHT
University of California, Lawrence Radiation Laboratory, Livermore, California
(Received October 30, 1961)

The study of a two-dimensional system consisting of 870 hard-disk particles in the phase-transition region has shown that the isotherm has a van der Waals-like loop. The density change across the transition is about 4% and the corresponding entropy change is small.

A STUDY has been made of a two-dimensional system consisting of 870 hard-disk particles. Simultaneous motions of the particles have been calculated by means of an electronic computer as described previously.[1] The disks were again placed in a periodically repeated rectangular array. The computer program has been improved such that about 200 000 collisions per hour can be calculated by the LARC computer regardless of the number of particles in the system. This speed made it possible to follow large systems for several million collisions.

It became necessary to study larger systems in the phase transition region when for smaller ones in three dimensions, it did not seem to be possible for the two phases to exist together in equilibrium.[2,3] Even in the largest three-dimensional system investigated with the improved program (500 hard spheres), the particles were either all in the fluid phase or all in the crystalline phase. The system would typically remain in one phase for many collisions. The occasional shift from one phase to the other would be accompanied by a change of pressure. The equation of state was represented by two disconnected branches overlapping in the density range of the transition, since with the limited number of phase interchanges it was not possible to average the two branches.

Two-dimensional systems were then studied, since the number of particles required to form clusters of particles of one phase of any given diameter is less than in three dimensions. Thus, an 870 hard-disk system is effectively much larger than a 500 hard-sphere system. First, however, it was necessary to establish that small two-dimensional systems behave analogously to the three-dimensional systems. This is illustrated in Fig. 1 by the two disconnected branches drawn lightly through the triangular points for a 72-particle system. In that figure, the reduced pressure pA_0/NkT is plotted against the reduced area A/A_0, where A_0 is the area of the system at close packing. In the region of A/A_0 from 1.33 to 1.35 the system fluctuated infrequently between a high-pressure fluid branch and a low-pressure crystalline branch, while at A/A_0 of 1.31 and higher densities the solid phase was always stable.

For the larger 870-particle system, however, the two phases exist side by side. One piece of evidence for this coexistence is the cathode-ray tube pictures described earlier (see Fig. 2).[1] The trajectories of the particles plotted on the oscilloscope show regions where the particles are localized (crystallites) in between regions of mobile particles (fluid). Further evidence is the characteristically large pressure fluctuations in the phase transition region where two states can exist with almost equal probability. The extent of the fluctuations in a typical run of about 10 million collisions is obtained

* This work was performed under the auspices of the U. S. Atomic Energy Commission.

[1] B. J. Alder and T. E. Wainwright, J. Chem. Phys. 31, 459 (1959).

[2] B. J. Alder and T. E. Wainwright, J. Chem. Phys. 33, 1439 (1960).

[3] W. W. Wood, R. R. Parker, and J. P. Jacobson, Suppl. Nuovo cimento 9, 133 (1958).

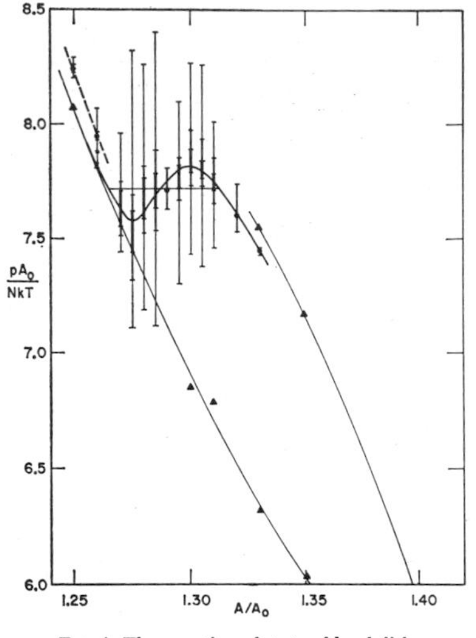

FIG. 1. The equation of state of hard disks in the phase transition region.

by breaking the run into intervals of 50 thousand collisions (about 100 collisions for each particle) and determining the average pressure in each interval. The light vertical lines at various densities (Fig. 1) extend from the maximum to the minimum pressure found among these intervals. As can be seen, the fluctuations are much larger in the phase transition region than in the pure solid ($A/A_0<1.26$) and pure fluid ($A/A_0>1.33$) phases and they are also larger in the middle of the phase transition region than near the ends. The medium vertical lines in Fig. 1 indicate the middle range of the fluctuations, that is, $\frac{1}{4}$ of the intervals show pressures above the top of the vertical line and $\frac{1}{4}$ show pressures below the bottom of the line. For the shorter runs at A/A_0 of 1.32 and 1.29 only this medium vertical line could be drawn.

The heavy vertical lines in Fig. 1 indicate the estimated accuracy of the average pressure determination at each density. These estimates are made by comparing several runs of 10 million collisions each with various starting conditions. The comparison was typically within 1% at a few selected densities. The fact that the pressures calculated with varying starting conditions agree rather closely is the only indication that phase space has been adequately sampled. Four different starting conditions have been used: (1) all the particles are located in lattice positions with only one particle in motion; (2) all particles are in lattice positions and in motion with randomly selected velocities; (3) the starting configuration for a run at one density is taken from an instantaneous configuration at a lower density, having been effectively changed by increasing the diameter of the particles; and (4) the same as (3) except a higher density configuration is used as an initial low-density one.

The smooth curve drawn through the heavy vertical sections in Fig. 1 clearly shows a van der Waals loop-like behavior for the equilibrium state of a finite system. To confirm this it was found that decreasing the density from A/A_0 of 1.29 to 1.30 by procedure (4) above, increased the average pressure. Similarly, increasing the density from A/A_0 of 1.285 to 1.280 and subsequently to 1.275 by procedure (3) decreased the average pressure in both cases, although, immediately after increasing the density, the pressure at 1.275 was higher for some millions of collisions. This shows that the density must be increased very slowly near the solid region or otherwise the particles will be locked into a disordered configuration. Thus, on increasing the density by procedure (3) from A/A_0 of 1.275 to 1.26 the crystalline region of phase space (lower value) became disconnected from the disordered region (upper value) as seen in Fig. 1, since the extremes of the pressure fluctuations no longer overlap. A further increase in density of the disordered or glass-like configuration at A/A_0 of 1.26 to 1.25 defines the dashed line in Fig. 1.

It has been shown that in an infinite system the isotherms for this system would always have to be of negative or at most zero slope.[4] Thus, the loop could not exist. The existence of a loop for finite systems probably derives from the fact that the constraint of constant density is imposed over a region occupied by however many particles are dealt with. This constraint restricts

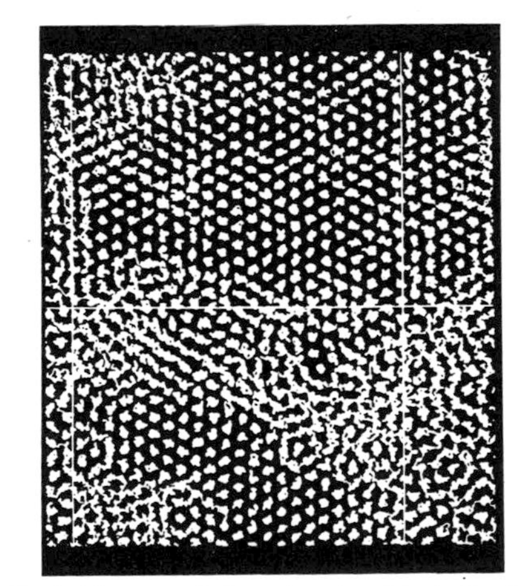

FIG. 2. The traces of the centers of particles in the phase-transition region showing fluid and crystalline regions. The horizontal and vertical lines represent an arbitrary grid.

[4] L. van Hove, Physica 15, 951 (1949).

the configurations which can be reached in a finite system, that is, for example, a fluid in equilibrium with crystallites of average size greater than the number of particles dealt with is impossible to achieve. This constraint for the small systems previously investigated resulted in stabilization of the predominant phase. Thus, the system was either all solid and when a rare fluctuation disordered enough of the system, it became completely fluid. For the 870-particle system the constraint again stabilizes the more abundant phase causing the pressure to be high on the fluid side and low on the crystal side. It thus seems that the phase separation which might occur in infinite systems is not complete in finite systems, since a sizable portion of the system lies in the fluid-crystal boundary region and this region is of intermediate density and evidently takes on more of the character of the predominant phase.

The horizontal line in Fig. 1 drawn at $pA_0/NkT=7.72$ and extending from A/A_0 of 1.266 to 1.312 corresponds to the usual "equal area" rule. If the phase transition for an infinite system is of first order at the pressure indicated by this straight line, then the resulting entropy change across the transition $\Delta S/Nk$ is $p\Delta A/NkT=0.36$. The change of entropy across the same density interval corresponding to the expansion of the one particle cell as calculated by the free volume theory is 0.30. This indicates that the change of communal entropy (0.06) across the transition is very much smaller than unity. This is hardly in accord with the view[5] that the difference between a dense fluid and a solid is one of the accessibility of the entire space in the fluid and localization of a molecule in a solid.

The complete equation of state and comparisons of it with the predictions of various theories will be the subject of further publications.

We are deeply indebted to Mary Ann Mansigh and Norman Hardy for their invaluable help in programming, and to Dr. Sidney Fernbach of the Livermore Computing Division for his cooperation.

[5] J. O. Hirschfelder, D. P. Stevenson, and H. Eyring, J. Chem. Phys. 5, 896 (1937); however, see also O. K. Rice, J. Chem. Phys. 6, 476 (1938).

Notes to Reprint II.1

1. An estimate of the melting point of hard disks in an infinite system was later obtained (Hoover and Ree, J. Chem. Phys. 49 (1968) 3609) by means of the single-occupancy cell method (see Reprint II.2). The melting pressure of the infinite system ($PA/NkT=8.08$) was found to be slightly higher than that calculated by Alder and Wainwright for a system of 870 disks ($PA/NkT=7.72$). Simulations on other systems support the notion that the periodic boundary conditions tend to stabilize the solid phase, thereby lowering the coexistence pressure.

2. More recent examples of the direct coexistence method include (Toxvaerd and Praestgaard, J. Chem. Phys. 67 (1977) 5291; Hiwatari et al., J. Chem. Phys. 68 (1978) 3401; Ladd and Woodcock, Mol. Phys. 36 (1978) 611; Ladd and Woodcock, Chem. Phys. Lett. 59 (1978) 271; Cape and Woodcock, J. Chem. Phys. 73 (1980) 2420; Ueda et al., J. Phys. Soc. Jpn. 50 (1981) 307).

THE JOURNAL
OF
CHEMICAL PHYSICS

VOLUME 47, NUMBER 12 15 DECEMBER 1967

Use of Computer Experiments to Locate the Melting Transition and Calculate the Entropy in the Solid Phase*

WILLIAM G. HOOVER AND FRANCIS H. REE
Lawrence Radiation Laboratory, University of California, Livermore, California

(Received 19 June 1967)

Modern computers can accurately simulate the behavior of idealized systems of several hundred particles, but they have trouble in studying the melting process in which small-system surface effects make the transition irreversible. It is here suggested that a thermodynamically reversible path linking the solid and fluid phases can be obtained by using a periodic "external field" to stabilize the solid phase at low density. The properties of the artificially stabilized solid at low density are studied theoretically, and two practical schemes are outlined for determining the melting parameters by using computer-calculated entropies.

I. INTRODUCTION

Monte Carlo[1,2] and molecular dynamic[3–5] calculations are two popular computer methods for obtaining thermodynamic data under conditions in which exact statistical–mechanical calculations are difficult. In *pure* phases, either fluid or solid, the computer techniques can measure pressure and energy to within a percent of limiting thermodynamic values. In or near density–temperature regions where two phases can coexist, the computer calculations become inefficient. Distortions due to boundary effects become important in a two-phase system. Both density and energy fluctuations become large. In addition, the decay time for such fluctuations increases by orders of magnitude so that convergent thermodynamic averages cannot easily be obtained.

Volume and temperature are the usual independent variables for Monte Carlo computer work; volume and total energy are the independent variables for dynamic calculations. In either case the pressure, calculated from the virial theorem, together with the energy or temperature averages can be put in the form of an equation of state relating E, T, P, and V. Such an equation of state can be measured accurately only for pure phases.

In order to calculate the conditions under which two phases coexist, the $ETPV$ relations for the pure phases are not enough; the difference between the *entropies* of the two phases is also required.[6] Calculating entropy differences between two different thermodynamic states requires an integration of $dS=(1/T)dE+(P/T)dV$ along a thermodynamically reversible path linking the two states. To connect the gas and liquid phases, one can use a path which avoids the two-phase region by going above the critical point in temperature. To connect the fluid and solid phases is not so easy. The melting transition persists even at high temperature, and unlike the gas–liquid transition, cannot be avoided by heating the system.

Computer "experiments" with systems of a few hundred particles show distortion in the melting–freezing transition.[2–4] If one slowly decreases the density of a solid-phase system, then the system will change suddenly, at constant volume, from solid to fluid. Often this change is *not* reversible; i.e., going the other way, compressing a fluid to higher densities, leads not to a crystalline solid, but to a glassy state instead (see Fig. 1). Assuming that at high density a solid phase, not a glass, is thermodynamically stable, failure to observe the solid can only be due to surface effects.[7]

---

* Work performed under the auspices of the U.S. Atomic Energy Commission.

[1] W. W. Wood and J. D. Jacobson, Proc. Joint Computer Conf. San Francisco, March 1959, 261 (1959).

[2] W. W. Wood, "Monte Carlo Calculations of the Equation of State of Systems of 12 and 48 Hard Circles," Los Alamos Sci. Lab. Rept. LA-2827, 1963.

[3] B. J. Alder and T. E. Wainwright, J. Chem. Phys. **33**, 1439 (1960).

[4] B. J. Alder and T. E. Wainwright, Phys. Rev. **127**, 359 (1962).

[5] A. Rahman, Phys. Rev. **136**, A405 (1964).

[6] See Gibbs' interesting discussion of this point in *The Collected Works of J. Willard Gibbs* (Yale University Press, New Haven, 1957), Vol. 1, p. 37.

[7] J. E. Mayer and W. W. Wood, J. Chem. Phys. **42**, 4268 (1965).

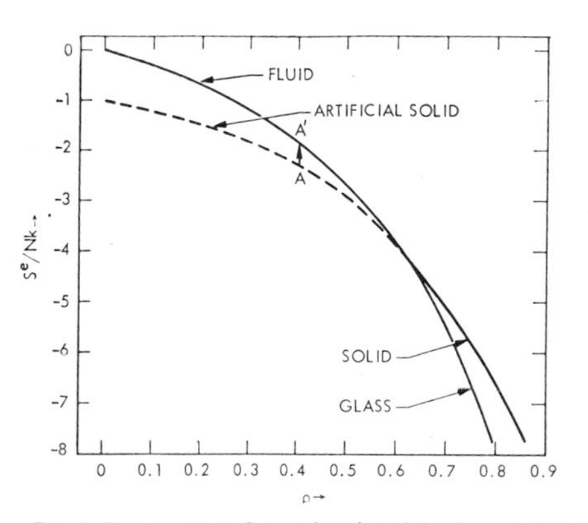

FIG. 1. Excess entropy S^e as a function of density ρ at fixed energy. For infinite thermodynamic systems, only the fluid and solid phases can be observed, and the two phases can coexist over a density range near the intersection of the two curves. In computer calculations with small systems, attempts to compress the fluid to solid-phase densities result in metastable glassy states. In the text, it is explained how generating the artificial solid states, indicated by dashes, can locate the melting transition. The properties of the artificial solid can be related to those of the real one either by following the entire dashed line or by converting the artificial solid to a fluid ($A \rightarrow A'$) by using an external field. The figure was drawn using approximate data for the hard-sphere system (the entropy excess was measured relative to an ideal gas at the same density and temperature, and volume units were chosen such that $\rho = 1$ at close packing).

It takes so much free energy to form a surface between two phases that even in a system containing 500 particles, a solid-phase nucleus cannot be formed.

This paper develops two ways to avoid the *discontinuous* change from solid to fluid. The first method is to prevent the melting transition from taking place at all by applying an external field, stabilizing the solid phase, at all densities. Each particle is held in a private cell; this artificial constraint can be thought of as the effect of an external field of infinite strength.[8]

The second method is to allow the melting transition to occur, but to force it to take place gradually and *reversibly*, instead of discontinuously, by a two-step process: first, using the infinite-strength external field already mentioned, the solid can be expanded to a reasonably low density; then the field strength can be gradually reduced to allow the system to "melt" in a reversible way. Both methods permit the exact calculation of the solid-phase entropy.

The three-dimensional hard-sphere system has been studied extensively by both the Monte Carlo and molecular-dynamic techniques,[9] and the pressure has been accurately determined in the pure-solid and pure-fluid phases. The hard-sphere model is of particular interest in the solid–fluid transition region, because the model gives a qualitative description of melting in real systems.[10] So far, however, computer attempts to accurately locate the two-phase region for hard spheres have failed. For the two phases to co-exist, it is necessary to consider larger systems (or possibly make much longer runs) than is now practical with present computers. Because of the difficulty in treating the interesting hard-sphere model, it is an appropriate test case for external-field calculations. In this paper we derive the basic equations needed for the hard-sphere calculation and evaluate the low-density limiting properties of the artificial hard-sphere solid. These latter results should prove valuable in extrapolating computer results from intermediate densities to the low-density limit. We expect to report on numerical work on the hard-sphere system at a later date.

II. ARTIFICIAL SOLID

The idea of locating the melting transition by studying a system which cannot melt is not so strange as it appears. By confining the center of each particle in an N-particle system to its own cell, of volume V/N, at *all* densities, the solid phase can be artificially extended to cover the entire density range. Particles in the artificial single-occupancy solid can collide both with the walls which confine each particle to its cell and with other nearby particles. At high density, particles are usually confined by their neighbors alone, rather than by cell walls. Each particle stays near the center of its cell, and the single-occupancy cell system faithfully represents the properties of a perfect solid. At low density, collisions with cell walls become appreciable—these collisions prevent the artificial solid from melting. That is, these collisions keep the particles ordered within the lattice of individual cells and prevent the diffusion throughout the entire system which is characteristic of particles in a normal fluid. Instead of melting, the cell system artificially continues solid-phase thermodynamic properties to low density.

At low-enough density all of the thermodynamic properties of either the constrained cell system or the unconstrained real system can be calculated exactly. We want to show how this low-density limit, coupled with computer-generated thermodynamic properties for the *artificial* cell system spanning the whole density range, can be used to calculate the entropy in the *real* solid phase.

To examine the differences between the artificial solid and the real system in a quantitative way, we first write down the configurational integrals which connect the thermodynamic properties of these systems with the microscopic potential-energy function $\Phi(\mathbf{r}^N)$. For the real system, the configurational inte-

[8] J. G. Kirkwood, J. Chem. Phys. **18**, 380 (1950), used a constrained system of this kind as the first step in his derivation of the Lennard-Jones–Devonshire cell theory.

[9] A summary of recent work will shortly appear in *The Physics of Simple Liquids*, H. N. V. Temperley, J. S. Rowlinson, and G. S. Rushbrooke, Eds. (North-Holland Publ. Co., Amsterdam, to be published).

[10] J. S. Rowlinson, Mol. Phys. **8**, 107 (1964).

Notes on p. 93

gral Q_N has the form

$$Q_N \equiv \exp[-(\bar{\Phi}/kT)+(S/k)]$$
$$\equiv (N!)^{-1}\int \exp\left(\frac{-\Phi(\mathbf{r}^N)}{kT}\right) d\mathbf{r}^N. \quad (1)$$

The configurational entropy S and the average potential energy $\bar{\Phi}$ can be evaluated from the configurational integral $Q_N(T)$. The integration in Eq. (1) is carried out over the coordinates of all N particles in the system. The *single-occupancy*[8,11] configurational integral Q_{so} for the artificial solid differs from Q_N; the potential energy in the single-occupancy system contains an extra cell-wall term Φ_{cw}, which constrains each particle to lie within its own cell. This extra term modifies both the average potential energy $\bar{\Phi}$ and the configurational entropy S. Using Δ to indicate the difference, real-system property minus artificial-solid property, the single-occupancy configurational integral Q_{so} has the form

$$Q_{so} \equiv \exp[-(\bar{\Phi}/kT)+(S/k)+(\Delta\bar{\Phi}/kT)-(\Delta S/k)]$$
$$\equiv \int \exp\left(-\frac{\Phi(\mathbf{r}^N)}{kT}\right)\exp\left(-\frac{\Phi_{cw}(\mathbf{r}^N)}{kT}\right) d\mathbf{r}^N. \quad (2)$$

In the integration in Eq. (2), each of the particles is confined by the cell-wall potential to a particular cell. Thus the factor of $(N!)^{-1}$ included in Eq. (1) has been automatically offset in Eq. (2) by restricting the integration to only one of the $N!$ different permutations of the N particles in N cells.

At low density both $\bar{\Phi}$ and $\Delta\bar{\Phi}$ approach zero; the two configurational integrals approach limits $Q_N \to (Ve/N)^N$ and $Q_{so} \to (V/N)^N$, and ΔS attains its maximum value $\Delta S \to Nk$. At high density the entropy difference, the so-called "communal entropy," goes to zero. Just how ΔS changes with density is not known in detail. The view that the entire communal entropy of Nk makes its appearance at melting, thus "explaining" the entropy of fusion, is by now obsolete.[4,11]

If the solid-phase configurational integral could itself be measured by a computer, then there would be no point in introducing the single-occupancy system. *In principle* the configurational integral could be calculated by taking random configurations of N particles in a volume V and averaging $\exp(-\Phi/kT)$ over these configurations. *In practice* one cannot measure configurational integrals at interesting densities, because almost all configurations of an N-particle system picked at random will have two or more particles so close together that the weight of that configuration is negligibly small.[12] The larger the system under consideration, the more serious this limitation becomes. It is, however, possible to determine accurate derivatives of Q with respect to external parameters. $(\partial \ln Q/\partial V)_T$ and $(\partial \ln Q/\partial T)_V$ are proportional to the average pressure P and the average potential energy $\bar{\Phi}$. Both averages can be measured in computer experiments. Then, knowing Q theoretically at low density, the derivative can be integrated to the solid-phase density of interest.

The derivative of Q with respect to V can be evaluated numerically by a direct application of the virial theorem.[2,3] For a D-dimensional system with a pairwise-additive potential energy $\Phi \equiv \sum \phi(\mathbf{r}_{ij})$, the virial theorem has the form

$$PV/NkT = N^{-1}(\partial \ln Q/\partial \ln V)_T$$
$$= 1 - [\langle \sum \mathbf{r}_{ij}\cdot\nabla\phi(\mathbf{r}_{ij})\rangle / \langle NDkT\rangle]. \quad (3)$$

All distinct pairs of particles $1 \leq i < j \leq N$ are included in the sum. The angle brackets correspond dynamically to a time average. For a Monte Carlo calculation, the time average is replaced by a configuration-space average. In the event that the forces are short range and the density is high, so that particles interact effectively only with z equivalent nearest neighbors, Eq. (3) can be simplified to

$$PV/NkT \doteq 1 - \tfrac{1}{2}z\langle \mathbf{r}_{12}\cdot\nabla\phi(\mathbf{r}_{12})\rangle / \langle DkT\rangle, \quad (4)$$

where z is the coordination number of the lattice, and Particles 1 and 2 occupy nearest-neighbor cells.

Since the artificial cell walls partitioning the system make no contribution to the pressure themselves, the single-occupancy configurational integral can be calculated from the pressure using either Monte Carlo or molecular dynamics. The constraint of singly occupied cells is taken into account by rejecting any Monte Carlo moves which would take the center of a particle outside its cell, or in the dynamic case, by reflecting the component of velocity normal to a cell wall whenever the center of a particle reaches the wall.

To compute the entropy of the artificially stabilized solid at any density, $\rho \equiv N/V$, simply compute the integral, at constant energy,

$$\frac{S(\rho)}{Nk} = \frac{S_o(\rho_o)}{Nk} + \int_\rho^{\rho_o} \frac{P}{(\rho^2 kT)} d\rho, \quad (5)$$

where S_o is the entropy at ρ_o, chosen low enough so that S_o can be calculated. This does *not* mean zero density. The low-density limit can be calculated analytically by performing an f-function expansion[13] of the integrand of Q_{so}. For nearest-neighbor interactions, one finds the result

$$Q_{so} = (V/N)^N \exp\left[\tfrac{1}{2}(Nz)\rho^2 \int d\mathbf{r}_1 \int d\mathbf{r}_2 f_{12} + \text{higher-order terms}\right], \quad (6)$$

[11] W. G. Hoover and B. J. Alder, J. Chem. Phys. **45**, 2361 (1966).

[12] E. Byckling, Physica **27**, 1030 (1961).

[13] This kind of expansion has been considered before by many authors, usually with the idea of calculating thermodynamic properties at liquid densities! See, for some hard-disk calculations, E. G. D. Cohen and B. C. Rethmeier, Physica **24**, 959 (1958).

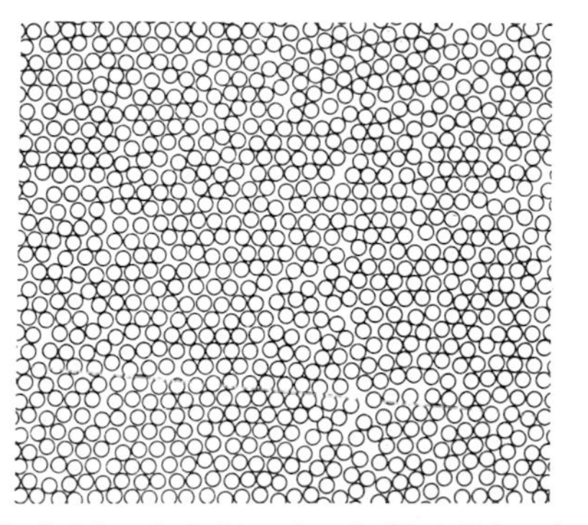

FIG. 2. A "snapshot" picture of a typical Monte Carlo configuration of 870 disks in the solid phase, expanded about 27% from close packing. The figure shows clearly the absence of long-range order in the two-dimensional solid phase. The displacement of the disks from their "lattice sites" is large. The configuration shown could not occur in the single-occupancy cell system. This figure was kindly furnished by W. W. Wood (see Ref. 9).

where Particles 1 and 2 occupy neighboring cells and the Mayer f-function f_{12} is equal to $\exp[-\phi(\mathbf{r}_{12})/kT]-1$. For a short-range potential function $\phi(\mathbf{r})$, the integral in Eq. (6) is at low density[14] proportional to the area of the cell wall separating Particles 1 and 2, $\sim(V/N)^{1-1/D}$ in D dimensions.

The procedure just outlined for evaluating the solid-phase entropy makes the assumption that the communal entropy $\Delta S=S-S_{so}$ is negligibly small in the solid phase. This is certainly valid provided that in the actual solid phase, particles remain localized within half a particle diameter of their most likely position at the cell center. The extent to which this localization prevails depends very strongly on the number of dimensions D. In one or two dimensions,[15] even at the highest densities, particles are not localized but instead move cooperatively back and forth over distances large with respect to the nearest-neighbor spacing. Thus the mean-squared displacement of a particle from its lattice site should diverge in one or two dimensions. Striking confirmation of the irregular structure and lack of long-range order, in the two-dimensional solid phase, can be seen in one of Wood's "snapshot" pictures of the computer-generated 870-disk solid, reproduced in Fig. 2. Of course, the fact that the mean-squared displacement diverges does not necessarily mean that thermodynamic properties calculated with a finite displacement (forced on a system by using cells) will be very different from those of a real system with a diverging displacement. In fact, thermodynamic properties are not very sensitive to these low-frequency motions. In the most extreme case, a one-dimensional hard-rod system, the mean-squared displacement is proportional to the number of particles N, and yet a comparison of Q_N and Q_{so} for this system shows that both the entropy and the pressure differences vanish near close packing.[11] The discrepancy is small at high density and must be much smaller in two or three dimensions; in our three-dimensional applications we will ignore it.

Since three-dimensional solids generally melt at a linear expansion of order 10% from an effective "close-packed" volume, the root-mean-squared displacement from the center of a cell should remain small with respect to the cell diameter even at melting; in three dimensions, the single-occupancy approximation should be accurate throughout the solid phase. Experimentally, the field-ion microscope pictures show that over periods long with respect to a molecular vibration time, particles are localized in three-dimensional solids.[16] Thus, the single-occupancy configurational integral, which can be connected to low-density thermodynamic properties, should realistically represent a three-dimensional system in the solid phase.

III. EXTERNAL FIELD OF VARIABLE STRENGTH

As an alternative to the evaluation of the single-occupancy configurational integral at many densities, one can instead connect the artificial solid phase with the fluid phase at some convenient sufficiently low density by using an external field of varying strength. In Fig. 1 the thermodynamic path joining Points A and A' could be used to make the solid-fluid connection. The "external field" is chosen so that when turned on at full strength, the system is forced into the single-occupancy configuration, with one particle per cell. At vanishing field strength the system behaves normally. We expect then to be able to go *reversibly* between the solid and fluid phases by varying the field strength.

To set up the field, imagine a cell structure of N cells superimposed on the volume V. To stabilize singly occupied cells, as opposed to empty or multiply occupied ones, we introduce an external field which furnishes an absorption energy, $-\epsilon$, per singly occupied cell. The absorption energy is added to the usual potential energy of the system $\Phi(\mathbf{r}^N)$. The total absorption energy for a configuration with ν singly occupied cells is $-\nu\epsilon$. Increasing the external field strength parameter ϵ increases the average number of singly occupied cells $\bar{\nu}$. For such a system the generalized configurational integral, depending on the field strength, can be written

$$Q_N(\epsilon)\equiv(N!)^{-1}\int\exp\left(-\frac{\Phi(\mathbf{r}^N)}{kT}\right)\exp\left(\frac{\epsilon\nu(\mathbf{r}^N)}{kT}\right)d\mathbf{r}^N. \tag{7}$$

As ϵ approaches zero, $Q_N(\epsilon)$ reduces to the uncon-

[14] At low density the integral reduces to, apart from a proportionality constant, a surface-tension integral considered and worked out for hard spheres, by A. Bellemans, Physica **28**, 493 (1962).

[15] L. D. Landau and E. M. Lifshitz, *Statistical Physics* (Pergamon Press, Inc., London, 1958), Sec. 125.

[16] R. H. Good, Jr., and E. W. Müller, *Encyclopedia of Physics*, S. Flügge, Ed. (Springer–Verlag, Berlin, 1956) Vol. 21, p. 176.

Notes on p. 93

strained configurational integral for the real system Q_N. In the opposite limit, for ϵ large, $Q_N(\epsilon)$ approaches $\exp(\epsilon N/kT)Q_{so}$ as all N cells become singly occupied. The two special cases, $\epsilon=0$ and $\epsilon\to\infty$, can be used to establish the useful identity

$$-\frac{\Delta\bar{\Phi}}{NkT}+\frac{\Delta S}{Nk}\equiv N^{-1}\ln\left(\frac{Q_N}{Q_{so}}\right)=(kT)^{-1}\int_0^\infty\left(\frac{N-\bar{\nu}}{N}\right)d\epsilon, \quad (8)$$

where $\bar{\nu}=\bar{\nu}(\epsilon)$ depends on the field strength and temperature. We expect that as long as the chosen density is not too high, $\bar{\nu}$ could be determined by computer calculations, and that the transition could be made to occur reversibly, even for a small system. It seems likely that there is a critical density (analogous to the magnetic Curie temperature) below which the transition from artificial solid to fluid is continuous and above which the transition is discontinuous. Assuming that at sufficiently low density the transition from solid to fluid takes place ***reversibly*** for small systems, one can calculate the entropy in the solid phase by a two-step process: first, the single-occupancy pressure is measured from solid-phase density down to a density low enough for the field-induced transition to take place reversibly; next, the external field is gradually reduced in strength, at fixed density, to zero. The entropy changes for the two steps, added together, give the total entropy difference between the initial solid-phase state and lower-density fluid. The density at which the transition becomes reversible would have to be determined empirically. If it turns out to be extremely low, say one-tenth of close packing, then there is no advantage in using a variable external field. If, however, the transition becomes reversible at about half of close packing, the amount of numerical calculation needed could be reduced by using a variable external field.

The field strength required to "freeze" a low-density fluid should be of the order kT and should decrease at higher density when the cells are more likely to be singly occupied even without the help of an external field. In the low-density limit, we can evaluate the configurational integral in the presence of an external field by the usual variational technique

$$Q_N(\epsilon)\to N!(V/N)^N\sum_{\{N_j\}}\exp(N_1\epsilon/kT)\prod_{j=0}^{\infty}[(j!)^{N_j}N_j!]^{-1}, \quad (9)$$

where the sum is over all sets of N_j satisfying the restrictions $\sum N_j=\sum jN_j=N$. (N_j is the number of cells containing exactly j particles. *Two* restrictions occur, because we set both the number of cells $\sum N_j$ and the number of particles $\sum jN_j$ equal to N, to make a single-occupancy system occur when the external field is strong.) The solution[17] of this problem, in the large-system low-density limit, is

$$Q_N(\epsilon)=(V/N)^N[\exp(\epsilon/kT)+e-1]^N,$$
$$N_j/N=\{(j!)[\exp(\epsilon/kT)+e-1]\}^{-1}, \quad \text{for } j\neq 1,$$
$$=\exp(\epsilon/kT)[\exp(\epsilon/kT)+e-1]^{-1}, \quad \text{for } j=1. \quad (10)$$

[17] The same mathematical problem has no doubt been solved many times before. For a mathematically more general case, see N. G. Van Kampen, Phys. Rev. **135**, A362 (1964).

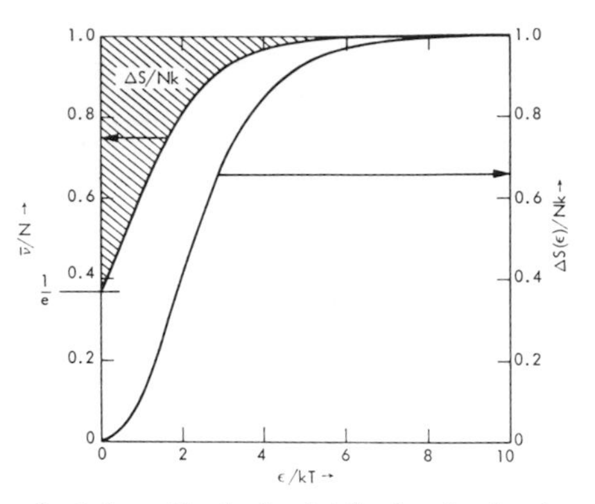

FIG. 3. An exact low-density calculation shows how the entropy difference $\Delta S(\epsilon)$ between an artificially stabilized solid and an ideal gas varies with the strength of the stabilizing external field. When the field is at full strength, all N cells are singly occupied. When the field is turned off, the artificial system reduces to the real one, and the fraction of cells singly occupied $\bar{\nu}/N$ is $1/e$. The shaded area is numerically equal to the communal entropy and illustrates the low-density limit of the integration indicated by Eq. (8) of the text.

In Fig. 3 both the entropy and the number of singly occupied cells are shown as functions of field strength. Notice that even at zero density a field strength of $4kT$ is sufficient to make over 96% of the cells singly occupied. At higher densities, weaker fields would be adequate.

In numerical applications it is reasonable to expect that the communal entropy, calculated in the way just outlined, could be determined with an error of order $0.01Nk$. In typical applications this would allow the phase transition pressure to be determined within about 1%.

IV. APPLICATION TO HARD SPHERES

To the extent that attractive forces can be idealized as forming a uniform background of negative energy and repulsive forces can be replaced by an energy-independent collision diameter, the hard-sphere system can be used to represent real systems. The hard-sphere system is probably the simplest which can reproduce melting qualitatively and is therefore of particular theoretical interest.

Because of its simplicity and intrinsic interest, the hard-sphere model has been extensively investigated by both the Monte Carlo and molecular-dynamic methods. So far, however, in systems containing up to 500 particles, it has not been possible to observe coexisting solid and fluid phases. At linear expansions of about 14% from close packing, the solid phase melts irreversibly[18] to the fluid, and attempts to compress the fluid back to solid-phase densities have always resulted in the generation of a metastable glassy phase.

The small size of the systems studied is responsible for these unsatisfactory results. The dependence of the transition on the number of particles has been estimated. The deviation from the large-system thermodynamic limit is reasonably small, of order $(\ln N)/N$,[19] but without a definite value for the transition pressure for some finite N, the extrapolation to infinite systems cannot be carried out. Instead, one must use entropy estimates from slowly converging theories[20] or use *ad hoc* assumptions[21] to connect the thermodynamic properties of the solid and fluid phases.

A fluid-phase equation of state for hard spheres, accurate over the whole density range, is already available.[22] Thus, the entropy of the single-occupancy hard-sphere system can be determined from Eq. (8) by measuring the change in the number of singly occupied cells with field strength. Then an integration along the single-occupancy isotherm ($TdS=PdV$ on a hard-sphere isotherm) would establish the entropy of the hard-sphere solid and allow the melting transition to be accurately located.

At the low-density limit, the configurational integral expansion indicated in Eq. (6) can be carried out analytically for hard spheres, at least through the first term. The integral corresponds to the average overlap of two spheres occupying neighboring cells and is the cell-system analog of the second virial coefficient.

For one-, two-, and three-dimensional hard spheres, respectively, one finds for the integral $\int d\mathbf{r}_1\int d\mathbf{r}_2 f_{12}$: $-\frac{1}{2}\sigma^2$, $-\frac{2}{9}\sqrt{3}\sigma^4 x^{-1/2}$, and $-(\pi\sqrt{2}\sigma^6/16)x^{-2/3}$, where σ is the sphere diameter, and x is the density divided by the density at close packing. Using the density expansion of the configurational integral Q_N, one can use these results to calculate the communal entropy in the low-density limit

$$\Delta S/Nk = 1-1.00000x-O(x^3), \quad \text{one dimension;}$$

$$= 1-1.81380x+1.53960x^{3/2}-O(x^2), \quad \text{two dimensions;}$$

$$= 1-2.96192x+3.33216x^{4/3}-O(x^{5/3}), \quad \text{three dimensions.} \qquad (11)$$

The low-density limiting cases, Eq. (11), coupled with Monte Carlo or dynamic values of the single-occupancy pressure from low density to solid-phase densities will make it possible to determine the transition pressure and densities of the coexisting phases.

An investigation of the two-dimensional hard-disk system probably would not yield any new thermodynamic information, because the 870-disk transition has already been located.[4] It would nevertheless be of some interest to investigate the hard-disk single-occupancy system in the solid phase to find out if the qualitative difference in mean-squared displacement between the real system and the artificial one causes noticeable differences between their thermodynamic properties.

V. SUMMARY

Two different ways have been suggested to locate the melting transition accurately by using computer-generated thermodynamic properties of an artificial single-occupancy solid. The first method is to calculate the pressure of the artificial solid over the whole density range and by integration to calculate the entropy of the solid at high density, where the entropy difference vanishes between the actual solid and the artificial one. The second method avoids the low-density part of the integration by connecting the artificial solid to the real system at fixed density by turning on an external field reversibly. Computer results for the hard-sphere system, augmented by the low-density limiting results described in this paper, should make it possible to accurately locate the hard-sphere melting transition and to determine how the communal entropy varies with density. If this program is successful it would be logical to try the same techniques for more general force laws and for more complicated solid–solid transitions.

[18] Some unpublished dynamic data, provided by B. J. Alder, indicate that a 500-sphere system "melts" at an expansion of 50% from close packing. The melting occurs in both the face-centered and hexagonal versions of the hard-sphere crystal. The transition is only "irreversible" in the context of times short enough for computer calculations, of course.

[19] W. G. Hoover and B. J. Alder, J. Chem. Phys. **46**, 686 (1967).

[20] F. H. Stillinger, Z. W. Salsburg, and R. L. Kornegay, J. Chem. Phys. **43**, 932 (1965).

[21] M. Ross and B. J. Alder, Phys. Rev. Letters **16**, 1077 (1966).

[22] Either the integral of the Padé approximant to the hard-sphere pressure, or for simplicity, a Padé approximant to the entropy itself can be used. The former approximant can be found in F. H. Ree and W. G. Hoover, J. Chem. Phys. **40**, 939 (1964).

Notes to Reprint II.2

1. A later paper by Hoover and Ree (J. Chem. Phys. 49 (1968) 3609) uses the single-occupancy cell method to locate the melting points of hard disks in two dimensions and hard spheres in three dimensions.

2. At low densities, the pressure of the single-occupancy lattice gas is proportional to a

fractional power of the density

$$(P - \rho kT)/\rho^2 = b_2 \rho^\alpha + \cdots$$

where $\alpha = 1/d$ (d is the dimensionality) for a space-filling lattice such as that described in this paper, and $\alpha = (d+1)/d$ for a lattice of touching, spherical shells (Hoover et al., J. Chem. Phys. 55 (1971) 1128). The quantity b_2 plays the role of a second virial coefficient; its value depends both on the interparticle potential and on the shape of the lattice cells.

PHYSICAL REVIEW VOLUME 184, NUMBER 1 5 AUGUST 1969

Phase Transitions of the Lennard-Jones System*

Jean-Pierre Hansen and Loup Verlet

Laboratoire de Physique Théorique et Hautes Energies, 91-Orsay, France†

(Received 19 February 1969)

Monte Carlo computations have been performed in order to determine the phase transitions of a system of particles interacting through a Lennard-Jones potential. The fluid-solid transition has been investigated using a method recently introduced by Hoover and Ree. For the liquid-gas transition a method has been devised which forces the system to remain always homogeneous. A comparison is made with experiment in the case of argon. An indirect determination of the phase transition of the hard-sphere gas is made which is essentially in agreement with the results of the more direct calculations.

I. INTRODUCTION

The present paper is devoted to the study of the gas-liquid and fluid-solid phase changes of a system of particles interacting through the Lennard-Jones potential

$$V(r) = 4\epsilon[(\sigma/r)^{12} - (\sigma/r)^6] \ . \tag{1}$$

In the one-phase region the thermodynamic properties of that system are nowadays rather well known: When the density is too high for using either the virial expansion or results from integral equations, they are obtained through Monte Carlo[1–6] or molecular dynamics[7,8] computations. These results are reviewed and discussed in Ref. 6. It has been shown furthermore[1,4,8] that the equilibrium properties of the system of Lennard-Jones atoms are very similar to those of argon if σ and ϵ are given those values which fit the second virial coefficient at not too low temperatures,[9] i.e., $\sigma = 3.405$ Å, $\epsilon/k = 119.8$°K.

If one attempts "computer experiments" in the neighborhood of a phase change, one meets serious difficulties. The tendency for separation into two phases entails large fluctuations, and there is a very slow approach to "equilibrium" in the computations. It also appears difficult to reach all the relevant regions of configuration space. As a result, few quantitative results have so far been obtained.

In the liquid-gas coexistence region, Wood has studied the isotherm $T = 1.0579$ (reduced units, i.e., $\sigma = \epsilon/k = 1$, are used throughout) for a system of 32 atoms.[3] Owing to its small size, the system does not separate into two phases at that temperature, and the fluctuation of the pressure should remain fairly small.

The pressure versus volume curve obtained by Wood shows a van der Waals loop in the transition region. The results are too few to allow a precise Maxwell equal area construction which would yield the transition pressure.

In the melting region, Wood and his collaborators have found in several cases that the pressure versus density curve is composed of two branches. One of these corresponds to the homogeneous solid state, the other to the pure fluid. For a 32-particle system, occasional jumps from one of these states to the other are observed in the transition region. These jumps are rare, however, so that an adequate sampling of the configuration space cannot be made. The transition pressures of the 32-particle system can therefore not be determined with any precision. For larger systems, such as the 864-particle system that we consider in the present paper, the only observed transitions are those from the fluid to the solid, and the transition pressure cannot be determined directly.

In order to locate the phase transitions we shall use methods which involve only homogeneous phases. The two-phase region with its difficulties will be completely avoided, at the price, however, of a more indirect approach.

In the condensation region, the 864-particle system exhibits large density fluctuations due to the tendency of the system to separate into regions of different densities. This problem was encountered earlier by Rotenberg[10] in his study of the system of 256 hard spheres each embedded in an attractive well. This author observed a van der Waals loop in the transition region, accompanied by such large fluctuations that no quantitative conclusions could be drawn. In order to avoid these difficulties, we shall devise a reversible isothermal path joining continuously the two physical one-phase states by constraining the system to remain homogeneous in the transition region. This is the spirit underlying Van Kampen's[11] solution of the condensation problem in the van der Waals limit. Practically, the homogeneity condition is met by subdividing the system into a certain number of boxes and, in the computer "experiments," setting upper and lower bounds to the number of atoms in each box. Large density fluctuations leading to a gradual phase separation are thus prevented and a reversible path joining the gas to the liquid phase can be constructed. The limitation in the density fluctuations is chosen so that the liquid and gas-phase thermodynamic properties are not perturbed. The pressure versus volume curve in the transition region turns out to be a van der Waals-like loop which runs smoothly into the gas and liquid isotherms. Integration of the pressure along this continuous isotherm yields the liquid-phase free energy and this in turn allows the determination of the transition data.

In Sec. 2 of this paper, we give the results of computation, using the method that we have outlined, for two subcritical isotherms of the Lennard-Jones fluid. The calculated data are the transition pressure, densities, and latent heat. The agreement with the same quantities for argon is quite good except in the critical region, where the machine computations are unrealistic in not allowing large density fluctuations, and for the gas properties at very low temperature, where the Lennard-Jones potential is known to be inadequate.[12]

After the free energy of the liquid has been calculated by the above procedure, one is left with the problem of computing the free energy of the solid phase in order to determine the melting transition. A method allowing the numerical computation of the solid-phase free energy has recently been proposed by Hoover and Ree[13]; their method consists in stabilizing the solid phase over the whole density range by confining each atom to its own cell of volume V/N (where N is the number of atoms in the total volume V). This scheme prevents the system from melting and provides a reversible path joining the density domain of the true solid to the low density region where the cell model free energy can be evaluated analytically. A simple integration of the pressure along an isotherm computed in this way yields the solid-state free energy. Since the free energy of the liquid is known, the transition data are easily determined.

In Sec. III, we give the results obtained using this method for the Lennard-Jones system. The thermodynamic properties of the artificial solid are computed along three different isotherms by the Monte Carlo method. These "exact" calculations have revealed a curious property of the cell model: In addition to the well-known "liquid-gas" transition of the cell model, there seems to exist a second-order phase transition at a density about 10% lower than the melting density. At very low temperatures, large fluctuations of the pressure occur and the method becomes inadequate. In that case we have solved the cell model for the Lennard-Jones potential deprived of its attractive tail, the density ranging up to that of the actual solid. The attractive part of the potential is then progressively turned on. A reversible path which makes it possible to determine the free energy of the solid is again available.

The transition data we have obtained, ranging from twice the critical temperature down to that of the triple point, are in very good agreement with those for argon. We show that along the solidification line of the Lennard-Jones fluid, the maximum of the structure factor takes the value 2.85 which is the value of the same quantity for a hard-sphere gas at solidification.

Barker and Henderson[14] have recently shown how a system of particles interacting through a repulsive potential can be replaced by hard spheres. This equivalence, checked by "exact" computations, enables us to use the results of Sec. III to obtain the transition data of the hard-sphere gas. These data are very close to those obtained more directly and precisely by Hoover and Ree[15] who have solved the cell model for the hard-sphere system.

II. LIQUID-GAS PHASE TRANSITION

We shall show in this section how an equation of state can be obtained for the Lennard-Jones fluid, not only for the pure one-phase states, but also in the coexistence region. Once the equation of state is known along an isotherm, the interaction part of the free energy is given by

$$\beta F_i/N = \int_0^{\rho} (\beta p/\rho' - 1)\,d\rho'/\rho'. \qquad (2)$$

To this quantity must be added the perfect gas contribution $\ln\rho - 1$ in order to have the total configurational part of the fluid's free energy F_l divided by NkT. From the free energy the transition densities will be obtained using the Maxwell double-tangent construction.

The equation of state in the liquid region was obtained for the reduced temperatures $T = 1.15$ and $T = 0.75$ by a standard Monte Carlo calculation.[3] A system of 864 particles with periodic boundary conditions was used. Between 3 and 10×10^5 configurations were generated at each volume. In the course of the computation we calculated the internal energy and the compressibility factor $\beta p/\rho$ by averaging the corresponding microscopic quantities. As the Lennard-Jones potential is cut off at $r = 2.5\,\sigma$, a correction is made as described in Ref. 8 to take into account the effect of the neglected tail on the thermodynamical quantities. The error on $\beta p/\rho$ is of the order of 0.02 at densities around critical and may reach 0.05 at densities around that of the triple point. The error on the internal energy is about twice smaller.

In the gas region, the equation of state can easily be obtained from the virial expansion; the densities, on the isotherms we have considered, remain sufficiently small so that using the five known virial coefficients,[16] we obtained a precise answer.

As we mentioned in the Introduction, the two-phase region requires more care. If we consider the 864-particle system with no constraint, it tends to separate into two phases. Owing to the rather large size of the system, this process takes a relatively long time. For instance at $T = 1.15$ and $\rho = 0.1$, after 10^6 configurations had been generated, the pressure and internal energy had not yet reached stable values, and the computation was given up. The same result was reached for the state $T = 0.75$, $\rho = 0.05$.

So as to obtain a faster convergence in the coexistence region, we force the system into an artificial single-phase state. In order to do so, we divide the volume into ν cubic cells of equal size and we require the number of particles in each cell to vary only between $\langle n\rangle - \delta n$ and $\langle n\rangle + \delta n$. Here $\langle n\rangle = 864/\nu$ is the average number of particles per cell and δn is a fixed number. Practically this constraint is realized in the following manner: At each Monte Carlo move, we ask if the particle under consideration tries to move outside of its cell. Should it do so, the move is prevented if it violates the constraint. The constraint parameters ν and δn are at our disposal. They must be chosen in such a way as to prevent the phase separation as well as possible without affecting the thermodynamical properties of the system in the physical, one-phase region.

If there is no constraint and if we are in a one-phase region, the standard deviation Δn to the average number of particles $\langle n\rangle$ in a cell is given by the well-known relation

$$\Delta n = \left(2\langle n\rangle \Big/ \beta\frac{\partial p}{\partial \rho}\right)^{1/2}, \qquad (3)$$

where $\beta\,\partial p/\partial\rho$ is the inverse compressibility for the thermodynamical state under consideration. We shall choose δn substantially larger than Δn as determined in the liquid region. The constraint should therefore have no influence for the liquid. We shall check *a posteriori* that this is indeed so, and that the properties of the gas phase are not modified either.

At $T = 1.15$, we choose $\nu = 27$ and $\delta n = 12$. For the lowest liquid density, we obtain, by numerically differentiating the computed equation of state, $\beta\,\partial p/\partial\rho = 2.4$. Δn is thus found to be 5.1, a value considerably smaller than δn. We have observed that during the computations which have actually been made in the liquid region, the constraint never operates and has therefore no measurable influence on the thermodynamics of the liquid phase.

In the gas region the constraint eliminates some possible configurations, but this has practically no influence on the equation of state. For instance for $\rho = 0.1$, which is a density high enough to be in the two-phase region, $\beta p/\rho$ is equal to 0.61 when computed by the Monte Carlo method with the constraint, and to 0.613 when calculated through the virial series.

For $T = 0.75$ the above-mentioned restriction proves insufficient; i.e., the computed thermal average fluctuates too much to allow a precise determination of the equation of state. Consequently we must take a stronger constraint. We choose $\nu = 64$ and $\delta n = 2.5$. At the lowest liquid density $\beta\partial p/\partial\rho = 12.3$, which leads to $\Delta n = 1.5$. We see that Δn is

Notes on p. 105

TABLE I. Compressibility factor and free energy per particle in the region of the liquid-gas transition for the isotherms $T=1.15$ and $T=0.75$.

ρ	$T=0.75$ $\beta p/\rho$	$T=0.75$ F_l	$T=1.15$ $\beta p/\rho$	$T=1.15$ F_l
0.02	0.829	−3.81	0.918	−5.75
0.06	0.504	−3.26	0.760	−4.70
0.1	0.234	−3.08	0.612	−4.24
0.15			0.470	−3.98
0.2	−0.292	−3.07	0.345	−3.84
0.3	−0.784	−3.23	0.124	−3.74
0.4	−1.201	−3.45	−0.090	−3.74
0.5	−1.688	−3.69	−0.130	−3.77
0.55			−0.075	−3.78
0.6	−2.052	−3.93	0.070	−3.78
0.65			0.306	−3.76
0.7	−1.705	−4.15		
0.75			1.165	−3.65
0.8	−0.531	−4.27		
0.84	0.371	−4.28		
0.85			2.860	−3.38
0.92			4.723	−3.03

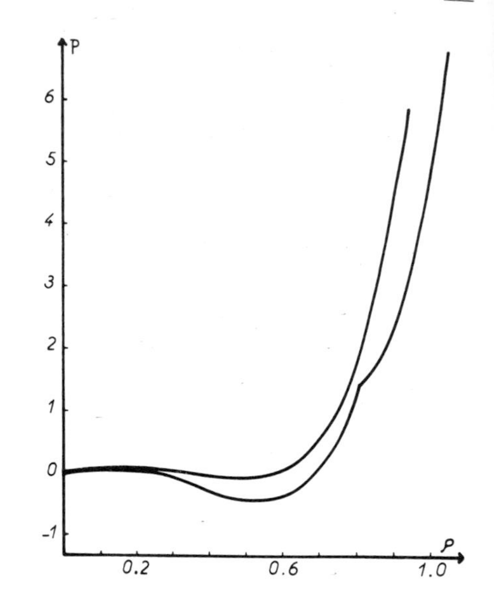

FIG. 1. Reduced pressure versus reduced density for the Lennard-Jones "homogenized" fluid (upper curve) and the corresponding cell-model (lower curve) at the reduced temperature $T=1.15$. Both isotherms exhibit a van der Waals loop and the cell-model isotherm exhibits an angular point around the reduced density $\rho=0.83$.

still somewhat smaller than δn. We have checked as above that the constraint, although it eliminates some configurations both in the liquid and the gas phase, has no practical effect on the thermodynamical quantities.

It should be noted that the constraint does not prevent some form of phase separation: at low densities some of the boxes tend to fill up to the maximum value, $\langle n\rangle + \delta n$, whereas the number of particles in other boxes decreases down to $\langle n\rangle - \delta n$. This separation is however rapid: $\beta p/\rho$ reaches its equilibrium value after 10^5 configurations. The internal energy does not stabilize so rapidly, but this matters little as we have no use for this quantity in the two-phase region.

Table I gives the results obtained for the compressibility factor $\beta p/\rho$ on the two isotherms $T=1.15$ and $T=0.75$. The error in the two-phase region is the same as in the one-phase domain: less than 0.01 for densities less than critical, around 0.02 for densities around 0.7, somewhat more when the density is high and the temperature is low. It may reach 0.05 on the point $\rho=0.85$, $T=0.75$. The $\beta p/\rho$ data along an isotherm can easily be fitted by a polynomial of order 5 or 6 in ρ. Using (2) the free energy can then be obtained. The configurational part of the free energy is also given in Table I.

On Figs. 1 and 2 the pressure is represented as a function of density. It is seen that the finite system exhibits a van der Waals loop. The double-tangent construction made on the free energy versus volume curve enables us to obtain the transition data. They are shown in Table II. A comparison is made with the experimental argon data.[17] The experimental value of the latent heat of vaporization was obtained through the Clapeyron equation.

The data of Table II together with the known critical constants for the Lennard-Jones fluid[5] $T_c=1.36$, $\rho_c=0.36$ are used to draw the curve shown in Fig. 3. It is seen that the coexistence curve for argon is flatter in the critical region ($T_c=1.26$ experimentally) than the one deduced from machine computation. The long-range density variations, responsible for the peculiar singularities characteristic of the critical point, cannot be included in the Monte-Carlo calculation. If we could remove these density fluctuations in real argon, the critical temperature would probably rise to about 1.34 as shown in Ref. 6. The coexistence curve would then very much resemble that determined for the Lennard-Jones fluid.

We also notice in Table II that at very low tem-

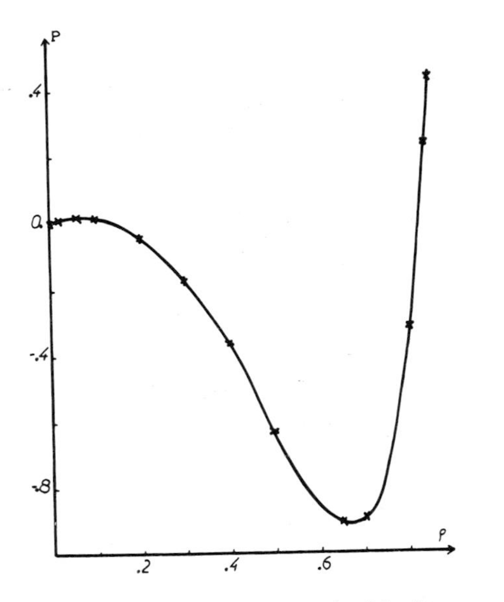

FIG. 2. Reduced pressure versus reduced density for the "homogenized" fluid at the reduced temperature $T = 0.75$.

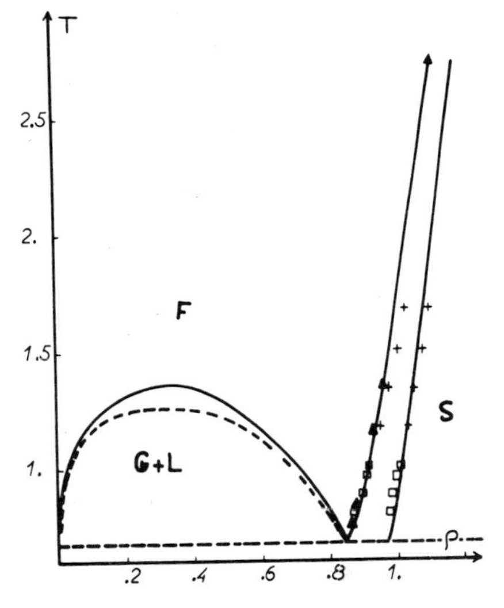

FIG. 3. Coexistence curve for the Lennard-Jones system (temperatures and densities in reduced units). The solid line gives our theoretical results. The broken line gives the experimental argon liquid-gas coexistence line taken from Michels *et al.*[17,22] The circles are experimental argon melting data taken from van Witzenburg and Stryland,[20] the crosses are experimental melting data taken from Crawford and Daniels.[21] The triangles indicate the crystallization densities according to the "law" stating that crystallisation occurs whenever $S(k_0)$ reaches the value 2.85.

peratures ($T = 0.75$) the transition density for the liquid branch shows a very good agreement between theory and experiment, but that there is a rather large disagreement in the case of the gas. This discrepancy is not surprising as it is well known that the properties of dilute argon at very low temperature are very poorly accounted for by the Lennard-Jones potential.[12] Part of the similar discrepancy for the gas at $T = 1.15$ may also be due to the same cause. In view of the fact that the transition pressure has been predicted quite accurately, it seems that the rather poor agreement for the latent heats is due mostly to the behavior of the low-density gas.

We have also given in Table II the results ob-

TABLE II. Liquid-gas transition data expressed in reduced units for the Lennard-Jones (LJ) fluid and for argon at the temperatures $T = 1.15$ and $T = 0.75$. Here $\mathcal{L}$ is the latent heat of melting.

		$T = 1.15$				$T = 0.75$			
		P	ρ_{gas}	ρ_{liquid}	$\mathcal{L}$	P	ρ_{gas}	ρ_{liquid}	$\mathcal{L}$
LJ	Monte Carlo	0.0597	0.073	0.606	4.34	0.0025	0.0035	0.825	6.62
	Eq. of state	0.0566	0.085	0.675	4.59	0.0034	0.0037	0.842	6.75
argon	Experiment	0.0664	0.093	0.579	3.73	0.0031	0.0047	0.818	5.44
	Eq. of state	0.0480	0.155	0.695	3.97	0.0033	0.0035	0.854	7.21

tained using the approximate equation of state of Ref. 6. We recall that these equations are supposed to hold for dense liquids and are quite poor for gases. The fit to the transition data is not very good. It is somewhat better in the Lennard-Jones case. This is probably due to the fact that the approximate equation of state for the Lennard-Jones fluid involves fewer adjustable parameters and more data with which to obtain them than is the case for argon.

III. MELTING TRANSITION

In order to determine the melting transition, we must calculate, in addition to the free energy of the fluid which has been computed in the preceding section, that of the solid phase. The method of Hoover and Ree,[13] the basis of which was described in the Introduction, has been used for this purpose. It amounts to solving the cell model "exactly" for densities ranging from very small values where a suitably modified cluster expansion just ceases to apply, up to those of the actual solid. The lattice structure of the Lennard-Jones solid has been chosen to be the same as that of real argon, i.e., fcc. The corresponding cells are dodecahedra with rhombic faces.[18]

The numerical solution of this cell model was carried out for three isotherms: $T=2.74$, 1.35, and 1.15. As an example of the equation of state yielded by the model, the isotherm for $T=1.15$ is given in Fig. 1.

The low-density behavior of the equation of state has been obtained by making a cluster expansion. If we consider a density sufficiently low so that only binary collisions occur and so that the dimension of the faces of the cells is much larger than the scale factor σ of the potential, the following expansion holds[19]:

$$\beta p/\rho = 1 + a_2' \rho^{4/3} + O(\rho^{5/3}), \tag{4}$$

$$\text{with } a_2' = -\tfrac{2}{3}\pi\gamma \int_0^\infty r^3 (e^{-\beta V(r)} - 1)\,dr, \tag{5}$$

where γ is a numerical factor characteristic of the cell geometry. In the case of the fcc structure, we have $\gamma = 3\times 2^{5/6}$.

At higher densities, the single occupancy constraint is added to the Monte Carlo procedure used for the fluid state: Starting from a configuration where all the atoms are at the center of their cells, the atoms are moved according to the usual Metropolis procedure: The only difference is that whenever a move takes the center of an atom outside its cell the corresponding configuration is rejected. The initial configurations for which thermodynamic equilibrium has not been reached are rejected. About 3.10^5 configurations are needed in order to obtain the same precision in the pressure as was obtained for the fluid. In the computation, five shells of particles were considered around each cell. A lattice sum over the more distant shells was made in order to take the effect of the tail of the potential into account. The compressibility factor $\beta p/\rho$ and the configurational internal energy U_i corresponding to the three temperatures $T=2.74$, 1.35, and 1.15 are given in Tables III, IV, and V, respectively. We also give the configurational part of the free energy $F_S(\rho)$ obtained by integration of the equation of state over the density:

$$\beta F_S(\rho)/N = \ln\rho + \int_0^\rho (\beta p/\rho' - 1)\,d\rho'/\rho'. \tag{6}$$

Equation (6) differs from (2) because of the distinguishability of the particles in the cell model.

For the isotherm $T=1.15$ we have calculated the mean-square deviation of the ith atom from the center of its cell

$$S^2 = \frac{1}{N}\sum_{i=1}^{N} \langle (\vec{r}_i - \vec{R}_i)^2 \rangle. \tag{7}$$

This quantity, given in Table V, is plotted as a function of the density in Fig. 4. We see that the curve consists of two very distinct branches presumably with a transition for a value of the density between 0.8 and 0.85. As may be verified in the course of the Monte Carlo computation, the low-density branch corresponds to a regime where the

TABLE III. Thermodynamic properties of the cell model for the isotherm $T=2.74$.

ρ	$\beta p/\rho$	U_i/N	F_S/N
0.05	0.949	−0.196	−8.32
0.1	0.895	−0.454	−6.57
0.2	0.803	−1.048	−4.97
0.3	0.785	−1.684	−4.09
0.4	0.851	−2.341	−3.44
0.5	1.081	−2.975	−2.86
0.6	1.486	−3.560	−2.24
0.7	2.178	−4.058	−1.49
0.8	3.21	−4.417	−0.51
0.9	4.68	−4.601	0.75
1.0	6.25	−4.77	2.32
1.05	6.99	−4.898	3.20
1.1	7.82	−4.991	4.14
1.125	8.47	−4.909	4.68
1.15	9.16	−4.816	5.18
1.2	10.65	−4.559	6.33
1.23	11.89	−4.21	7.12

TABLE IV. Thermodynamic properties of the cell model for the isotherm $T = 1.35$.

ρ	$\beta p/\rho$	U_i/N	F_s/N
0.025	0.937	−0.100	−5.03
0.05	0.851	−0.227	−4.19
0.1	0.671	−0.522	−3.47
0.15	0.477	−0.831	−3.15
0.2	0.311	−1.172	−3.00
0.25	0.140	−1.508	−2.93
0.3	0.027	−1.801	−2.91
0.35	−0.105	−2.222	−2.92
0.4	−0.186	−2.584	−2.95
0.45	−0.217	−2.943	−2.98
0.55	−0.131	−3.671	−3.03
0.6	0.07	−4.023	−3.03
0.7	0.68	−4.714	−2.97
0.8	1.79	−5.343	−2.76
0.85	2.72	−5.596	−2.57
0.9	3.34	−5.923	−2.33
0.95	3.74	−6.330	−2.07
1.	4.74	−6.611	−1.78
1.05	6.21	−6.779	−1.43
1.1	8.32	−6.78	−0.98
1.2	14.26	−6.355	0.14

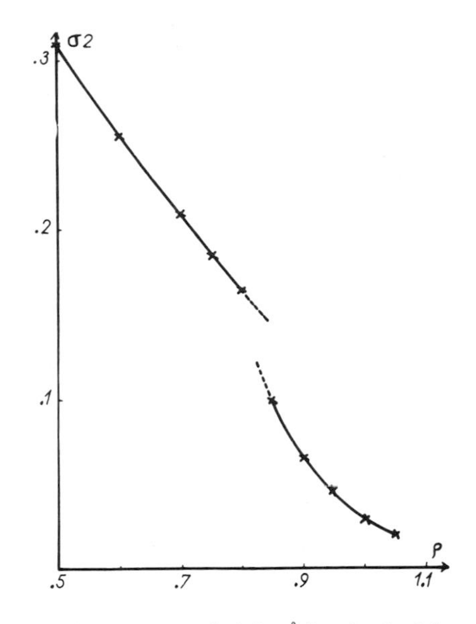

FIG. 4. Mean-square deviation s^2 (in reduced units) of an atom from its lattice site in the cell model versus reduced density for $T = 1.15$.

localization of the particles in the cell is brought about predominantly by the cell boundaries; for the high-density branch, the localization, more pronounced (as may be seen by considering the product $S_\rho{}^{-1/3}$), is caused by the neighbors and no longer by the cells.

It is most probable that this is the explanation of what seems to be an angular point in the pressure versus density curve for the temperature 1.15. If such an angular point really exists, it means that there is, in the cell model, a second-order transition. We believe that this is indeed the case, and we give the transition densities as

$$\text{for } T = 2.74, \quad \rho_t = 1.10 \pm 0.01;$$
$$T = 1.35, \quad \rho_t = 0.88 \pm 0.01;$$
$$T = 1.15, \quad \rho_t = 0.83 \pm 0.01.$$

TABLE V. Thermodynamic properties of the cell model and mean-square deviation from the center of the cell for the isotherm $T = 1.15$.

ρ	$\beta p/\rho$	U_i/N	F_s/N	s^2
0.05	0.820	−0.240	−3.60	
0.1	0.593	−0.538	−3.02	
0.2	0.156	−1.188	−2.72	
0.3	−0.283	−1.897	−2.75	0.48
0.4	−0.572	−2.635	−2.89	0.38
0.5	−0.678	−3.369	−3.05	0.31
0.6	−0.550	−4.110	−3.19	0.25
0.7	0.100	−4.818	−3.24	0.21
0.75	0.560	−5.162	−3.21	0.185
0.8	1.300	−5.476	−3.14	0.165
0.85	1.710	−5.865	−3.03	0.100
0.9	2.180	−6.258	−2.91	0.065
0.95	3.00	−6.592	−2.74	0.046
1.0	4.070	−6.896	−2.54	0.028
1.05	6.050	−7.015	−2.26	0.019

We shall see (Table VII) that these transitions occur at densities smaller than those of the melting point. Consequently this feature of the model has no physical consequence. For densities above those transitions, the isotherms are smooth continuations of those belonging to the solids with no constraint. In Fig. 1 we see, at low density, a van der Waals loop which is also present for $T = 1.35$. It is related to the well-known "liquid-gas" transition of the cell model.

For the temperature $T = 0.75$, which is near the

triple point, this method is no longer feasible. At this low temperature the number of steps needed before the pressure values stabilize becomes increasingly large and this is believed to be essentially due to the attractive part of the potential. For that isotherm we used a more indirect method. As in Refs. 14 and 6 we divided the potential into two parts: The repulsive part $u(r)$ of the Lennard-Jones potential and its attractive part $w(r)$ multiplied by a charging parameter λ. The following two-step process is considered: Up to some reference density $\rho_0 = 1.1$, located in the physical solid phase, only the repulsive part of the interaction is taken into account, in the "exact" solution of the cell model. Using (6) we obtain the free energy $F_S^0(\rho)$ of the system of particles interacting through $u(r)$ when the equation of state is known on the isotherm.

The second part of the process consists of turning on the attractive potential, at the density ρ_0. We thus obtain the free energy of the solid at the density ρ_0 through the easily derived relation

$$F_S(\rho_0) = F_S^0(\rho_0) + \int_0^1 \langle W \rangle_\lambda \, d\lambda , \tag{8}$$

where $\langle W \rangle$ is the total attractive interaction for the interaction $u(r) + \lambda w(r)$, averaged over the ensemble. This quantity can be calculated by the Monte Carlo method for several values of λ and the integral over λ can be performed. The λ dependence of $\langle W \rangle_\lambda$ is almost linear and it proves sufficient to perform the integral over λ with a step of 0.2.

Let $\langle W \rangle_\lambda$ be expanded in powers of λ. The above computation shows that the contribution to the integral in (8) of the λ-independent term is equal to -7.10; the term linear in λ yields: -0.38. The remainder of the series gives: 0.06. The series seems to converge almost as well as in the case of the liquid near the triple point.[6]

When the free energy of the real solid is known at ρ_0, it can be calculated at neighboring values of the density by using (6), once the equation of state has been determined through the Monte Carlo method. The results of the computations made on the isotherm $T = 0.75$ are given in Table VI. The isotherm shows, as in the case of the full Lennard-Jones potential, a branch point around $\rho = 0.95$.

The melting properties are given in Table VII. We give there, for the four temperatures at which the computation has been made, the melting pressures, the density of the solid at melting and that of the fluid at freezing, the volume change during the transition and the latent heat of fusion. These quantities are compared with similar quantities measured in the case of argon when they are available.[20,21] It is seen that the agreement is altogether surprisingly good. This confirms the excellence of the Lennard-Jones potential as an effective two-body potential for argon at high density.

With the data of Table VII the phase diagram of Fig. 3 for the Lennard-Jones system can be completed. It is seen again that the agreement with argon data for the fluid-solid transition is quite good. From Fig. 3, the triple point can be located. It corresponds to a temperature $T_T = 0.68 \pm 0.02$ and a density $\rho_T = 0.85 \pm 0.01$. These values are very near to those of argon[22]: $T_T = 0.70$, $\rho_T = 0.841$.

In Fig. 5 we give the curve of melting pressure versus temperature obtained using the data of Table VII. The experimental points are shown as crosses. The triangles represent the melting results due to Barker and Henderson.[23] Those points are obtained through the following approximations: The liquid-state properties are given by an approximate version of the λ expansion,[14] amply discussed in Refs. 6 and 24. This theory is known[14] to reproduce fairly well the thermodynamical properties of the Lennard-Jones fluid. Barker and Henderson use the free-volume theory for the solid: It yields values for the free energies that are not too different from the "exact" values, as shown in Table VIII where a comparison is made for the isotherm $T = 0.75$. The results obtained by Barker and Henderson should lie on the curve

TABLE VI. Thermodynamic results obtained for various values of the changing parameter λ on the isotherm $T = 0.75$.

λ	ρ	$\beta p/\rho$	U_i/N	$\beta F_S/N$	S^2
0	0.05	1.047	0.002	−2.96	
	0.1	1.121	0.004	−2.21	1.05
	0.2	1.314	0.012	−1.35	0.64
	0.3	1.652	0.024	−0.79	0.47
	0.4	2.070	0.041	−0.25	0.38
	0.5	2.675	0.063	0.25	0.31
	0.6	3.46	0.092	0.81	0.265
	0.7	4.54	0.137	1.43	0.222
	0.8	6.01	0.195	2.12	0.190
	0.9	8.23	0.280	2.94	0.150
	0.94	9.12	0.322	3.34	0.131
	1.0	9.32	0.334	3.90	0.055
	1.02	9.56	0.330	4.09	
	1.05	10.06	0.348	4.39	
	1.06	10.27	0.353	4.56	0.034
0	1.10	10.92	0.379	4.85	0.023
0.2	1.10	10.42	−1.122	2.85	0.020
0.4	1.10	9.69	−2.691	0.87	0.017
0.6	1.10	9.36	−4.280	−1.10	0.015
0.8	1.10	9.14	−5.903	−3.08	0.012
1.0	1.10	9.12	−7.540	−5.07	0.009
1.0	1.05	5.21	−7.530	−5.31	0.013
1.0	1.025	3.66	−7.467	−5.39	0.017
1.0	1.0	2.35	−7.370	−5.48	0.029

TABLE VII. Fluid-solid transition data expressed in reduced units for the Lennard-Jones (LJ) fluid and for argon at the temperatures $T = 2.74$, 1.35, 1.15, and 0.75. ΔV is the volume change at the transition. $\mathcal{L}$ is the latent heat of melting.

	T	P	ρ_{fluid}	ρ_{solid}	ΔV	$\mathcal{L}$
LJ	2.74	32.2	1.113	1.179	0.050	2.69
argon	2.74	37.4				2.34
LJ	1.35	9.00	0.964	1.053	0.087	1.88
argon	1.35	9.27	0.982	1.056	0.072	1.63
LJ	1.15	5.68	0.936	1.024	0.091	1.46
argon	1.15	6.09	0.947	1.028	0.082	1.44
LJ	0.75	0.67	0.875	0.973	0.135	1.31
argon	0.75	0.59	0.856	0.967	0.133	1.23

of Fig. 5. They are not very far off, although one should note that the pressure is plotted on a logarithmic scale.

We shall now give a relation between the crystallization of the Lennard-Jones fluid and that of a hard-sphere gas. The crystallization of a hard-sphere gas of diameter a and density ρ occurs, as we shall see in the next section, whenever the packing fraction $\eta = \pi \rho a^3/6$ reaches the value of 0.49. Let us define the structure factor in the usual way:

$$S(k) = \sum_{i,j} \frac{\langle \exp[i\vec{k} \cdot (\vec{r}_i - \vec{r}_j)] \rangle}{N} \tag{9}$$

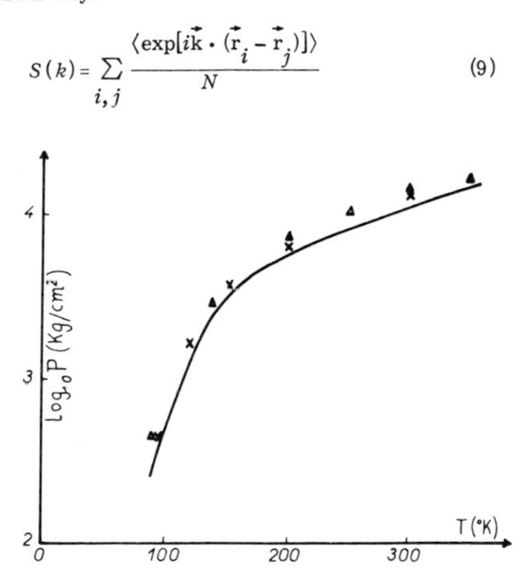

FIG. 5. Melting pressure versus temperature. The solid line gives our results for the Lennard-Jones system. The triangles are the theoretical results for the LJ system taken from Barker and Henderson[23]; the crosses give some experimental argon results taken from Crawford and Daniels[21] which are in close agreement with the results of van Witzenburg and Stryland.[20]

At solidification, the maximum value $S(k_0)$ of the exactly determined[25] structure factor takes the value 2.85. We recall the success of a slightly modified version[26] of the hard-sphere model of Ashcroft and Lekner[27] in explaining the structure factor of the Lennard-Jones fluid: To each state of the fluid is associated a hard-sphere gas of diameter $a(\rho, T)$ which is the only parameter of the theory. It is adjusted in such a way that $S(k_0)$ for the hard-sphere gas of packing fraction $\pi\rho a^3/6$ has the same value as that of the Lennard-Jones fluid at the density ρ and temperature T. The structure factor of the Lennard-Jones fluid is then very well reproduced by that of the hard-sphere gas. In view of that success it is tempting to go one step further and to associate the crystallization of the Lennard-Jones fluid with that of the underlying hard-sphere model which embodies the geometrical aspects of the problem. We then obtain a simple "law" of crystallization by stating that it should occur when $S(k_0)$ reaches the value 2.85. In order to check this hypothesis, we need to know the structure factor for various isotherms. This can be done with the method of Ref. 26, using the correlation functions obtained in that paper and those obtained for the four isotherms studied in the

TABLE VIII. Free energy per particle in the solid phase: comparison between the prediction from free volume theory and exact results.

	F_i/N	
ρ	free volume	exact
1.01	−3.89	−4.10
1.176	−2.99	−3.07

Notes on p. 105

present work. $S(k_0)$ obtained in this way is shown for various isotherms in Fig. 6. The transition densities are obtained as intercepts of those curves with the horizontal line $S(k_0)=2.85$. The results are shown as triangles on Fig. 3. They are seen to agree very well with the "exact" transition curve shown on that figure. The statement that solidification occurs when the maximum value of the structure factor reaches the value 2.85 is thus confirmed.

A comparison with experiment can be made in the case of argon near its triple point. The structure factor yielded by x-ray experiments[28] reaches a value compatible, within the experimental errors, with the value 2.85. It would be interesting to have data relative to xenon and krypton on the solidification line.

IV. HARD-SPHERE TRANSITION

The calculations we have just reported for the isotherm $T=0.75$, where the cell model was solved exactly for a system of particles interacting through the repulsive potential $u(r)$ can be used in order to obtain information concerning the fluid-solid transition of the hard-sphere system. The link between the two systems is provided by the work of Barker and Henderson.[14] These authors have shown that the system of particles interacting through $u(r)$ at the temperature T is equivalent, as far as the thermodynamics is concerned, to a hard-sphere gas of diameter d at the same density. The diameter d is given by

$$d=\int_0^\sigma dz\{1-\exp[-u(z)/kT]\}. \tag{10}$$

It turns out to be equal to 0.978 at $T=0.75$. The derivation holds both for the fluid and solid state. It is apparent that the theory is better when the density is low. It must obviously break down near the close packing of the equivalent hard-sphere system. Its range of validity can only be ascertained by direct computations. We know that for the highest densities considered in Table VI the cell model is equivalent to the real solid. We can thus compare the results with those obtained for the solid state.[29,30] For instance, let us consider the highest density $\rho=1.10$. It corresponds to a hard-sphere gas of a volume relative to that of the close packing of $V/V_0=1.373$. For the hard-sphere system we use the very recently published work of Alder, Hoover and Young[30]: By interpolating their results we obtain the value 10.9 for the compressibility factor. This should be compared with the value 10.92 of Table VI. The agreement is quite good for the other points (for $\rho\geq 1$) except for the point at $\rho=1.06$ for which the pressure is too high by about 1%.

We now are in a position to fix the tie line of the hard-sphere system: For the hard-sphere solid the free energies of Table VI are used for the density ρd^3. The equation of state of the hard-sphere gas is well known.[31,29] We have a convenient fit of the exact data by adding to the known seven-term virial series[32] for $p/\rho kT$ the correction[6]

$$1.6049(\rho d^3)^8+0.46142(\rho d^3)^{13}$$

We find by the double-tangent construction that the melting transition occurs at $V/V_0=1.371$ and the freezing transition at $V/V_0=1.513$, which corresponds to the value 0.49 for the packing fraction. Since this work was completed, Hoover and Ree have given[15] the solution of the cell model which was announced in their preceding paper.[13] They find that the tie line lies between $V/V_0=1.359$ and $V/V_0=1.500$. These are very close to the figures we give. The small discrepancy is probably due to the various numerical manipulations involved in both papers.

We have made a careful comparison of the data in order to see if we can get a little more information on the transition. We have made fits for the combined data of Hoover and Ree and those of Table VI excluding only our point at $\rho=1.06$ which is somewhat out of range. We obtain in this new revision of the data exactly the same transition data as above. It is situated between the values

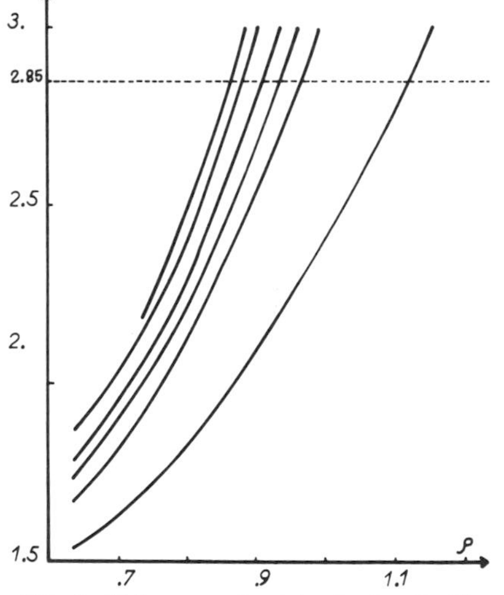

FIG. 6. $S(k_0)$ versus reduced density along the following isotherms (in reduced units): $T=0.75$ (upper curve), 0.833, 1., 1.15, 1.35, 2.74 (lower curve). Our empirical "law" states that crystallization takes place at the density where each of the curves intersects the horizontal line $S(k_0)=2.85$.

1.371 and 1.513 for V/V_0.

We then obtain the thermodynamic quantities characterizing the transition similar to those found by Hoover and Ree: The communal entropy at transition is equal to 0.12; the pressure of the tie line is given by $pV_0/NkT = 8.02$.

V. CONCLUSIONS

We have shown that the phase transitions of the Lennard-Jones fluid can be calculated, using methods where only homogeneous phases are considered. We plan in the near future to study the gas-solid transition at very high temperatures and to extend the same kind of methods to more complicated systems.

VI. ACKNOWLEDGMENTS

The authors are very grateful to Dominique Levesque for his large contribution to the present work and to John Valleau for revising the manuscript.

---

*This article is based on a thesis submitted by J.-P. Hansen for the degree of Docteur d'Etat ès Sciences Physiques at the Faculté des Sciences d'Orsay, Paris University (1969).

†Laboratoire associé au Centre National de la Recherche Scientifique.

[1]W. W. Wood and F. R. Parker, J. Chem. Phys. 27, 720 (1957).
[2]W. Fickett and W. W. Wood, Phys. Fluids 3, 204 (1960).
[3]W. W. Wood, Physics of Simple Liquids, edited by J. Rowlinson, G. S. Rushbrooke, and H. N. V. Temperley (North-Holland Publishing Co., Amsterdam, 1968), Chap. V.
[4]J. Mc Donald and K. Singer, J. Chem. Phys. 47, 4766 (1967).
[5]L. Verlet and D. Levesque, Physica 36, 245 (1967).
[6]D. Levesque and L. Verlet, Phys. Rev. 182, 304 (1969).
[7]A. Rahman, Phys. Rev. 136, A405 (1964).
[8]L. Verlet, Phys. Rev. 159, 98 (1967).
[9]A. Michels, H. Wijker, and H. K. Wijker, Physica 15, 627 (1949).
[10]A. Rotenberg, J. Chem. Phys. 47, 4873 (1967).
[11]N. G. Van Kampen, Phys. Rev. 135, A362 (1964).
[12]R. D. Weir, I. Wynn Jones, J. S. Rowlinson, and G. Saville, Trans. Faraday Soc. 63, 1320 (1967).
[13]W. G. Hoover and F. H. Ree, J. Chem. Phys. 47, 4873 (1967).
[14]J. A. Barker and D. Henderson, J. Chem. Phys. 47, 4714 (1967).
[15]W. G. Hoover and F. H. Ree, J. Chem. Phys. 49, 3609 (1968).
[16]J. A. Barker, P. J. Leonard, and A. Dombe, J. Chem. Phys. 44, 4206 (1966).
[17]A. Michels, J. M. H. Levelt, and W. de Graaff, Physica 24, 659 (1958).
[18]R. J. Buehler, R. H. Wentorf, J. O. Hirschfelder, and C. F. Curtiss, J. Chem. Phys. 19, 61 (1951).
[19]A. Bellemans, Physica 28, 493 (1962).
[20]W. van Witzenburg and J. C. Stryland, Can. J. Phys. 46, 811 (1968).
[21]R. K. Crawford and W. B. Daniels, Phys. Rev. Letters 21, 367 (1968).
[22]A. M. Clark, F. Din, J. Roob, A. Michels, A. Wasendar, and T. N. Zwietering, Physica 17, 876 (1951).
[23]D. Henderson and J. A. Barker, Mol. Phys. 14, 587 (1968).
[24]J. A. Barker and D. Henderson (report of work prior to publication).
[25]D. Schiff and L. Verlet (to be published).
[26]L. Verlet, Phys. Rev. 165, 201 (1968).
[27]N. W. Ashcroft and J. Lekner, Phys. Rev. 145, 83 (1966).
[28]N. Gingrich and C. W. Thomson, J. Chem. Phys. 36, 2398 (1962).
[29]A. Rotenberg, New York University, Report No. NYO 1480-3, 1964 (unpublished).
[30]D. J. Alder, W. G. Hoover, and D. A. Young, J. Chem. Phys. 49, 3688 (1968).
[31]B. J. Alder and T. E. Wainwright, J. Chem. Phys. 33, 1439 (1968), and results quoted in Ref. 32.
[32]F. H. Ree and W. G. Hoover, J. Chem. Phys. 40, 939 (1964).

Notes to Reprint II.3

1. In order to obtain the corresponding phase diagram for neon, quantum corrections must be taken into account (Hansen and Weis, Phys. Rev. 188 (1969) 314).

2. The liquid–gas coexistence curve of the (classical) Lennard–Jones fluid was later recalculated by Adams (Mol. Phys. 32 (1976) 647; 37 (1979) 211) by a grand canonical Monte Carlo method. The results obtained by Adams are in good agreement with those of Hansen and Verlet; the estimates that Adams gives for the critical parameters are $kT_c/\epsilon = 1.30 \pm 0.02$, $N\sigma^3/V_c = 0.33 \pm 0.03$, and $P_c\sigma^3/\epsilon = 0.13 \pm 0.02$. An alternative method to compute two-phases coexistence

properties has been recently developed by A.Z. Panagiotopoulos (Mol. Phys. 61 (1987)).

3. The freezing criterion proposed in this paper ($S(k_0) = 2.85$ for the liquid at coexistence) was later shown to hold for a large class of three-dimensional systems with spherically symmetric interactions (Hansen and Schiff, Mol. Phys. 25 (1973) 1281). The "universal" value of $S(k_0)$ at freezing plays an important part in modern density-functional theories of the liquid–solid transition. See, for example (Ramakrishnan and Yussouff, Phys. Rev. B 19 (1979) 2775; Baus, Mol. Phys. 50 (1983) 543).

New Monte Carlo method to compute the free energy of arbitrary solids. Application to the fcc and hcp phases of hard spheres

Daan Frenkel
Fysisch Laboratorium, Rijksuniversiteit Utrecht, P.O. Box 80000, 3508 TA Utrecht, The Netherlands
Anthony J. C. Ladd[a)]
Department of Applied Science, University of California at Davis, Davis, California 95616

(Received 14 March 1984; accepted 18 April 1984)

We present a new method to compute the absolute free energy of arbitrary solid phases by Monte Carlo simulation. The method is based on the construction of a reversible path from the solid phase under consideration to an Einstein crystal with the same crystallographic structure. As an application of the method we have recomputed the free energy of the fcc hard-sphere solid at melting. Our results agree well with the single occupancy cell results of Hoover and Ree. The major source of error is the nature of the extrapolation procedure to the thermodynamic limit. We have also computed the free energy difference between hcp and fcc hard-sphere solids at densities close to melting. We find that this free energy difference is not significantly different from zero: $-0.001 < \Delta F < 0.002$.

INTRODUCTION

In this paper we present a computer simulation method to determine the domain of thermodynamic stability of solid phases. The recent development by Parinello and Rahman[1] of a new molecular dynamics simulation technique have stimulated the use of computer simulations in investigating solid–solid phase transitions in model systems. The idea upon which the Parinello–Rahman method is based is that the fixed periodic boundary conditions employed in conventional MD simulations exclude the direct observation of solid–solid phase transitions, as the boundary conditions chosen to be compatible with one solid phase are, in general, incompatible with the other. Hence, fixed periodic boundary conditions tend to stabilize one solid phase well beyond its range of thermodynamic stability, and may easily overlook the existence of other more stable phases altogether. In the Parinello–Rahman method the shape of the periodic box is no longer fixed; shape and size of the periodic box are expressed in terms of variables which play the role of generalized coordinates in an extended Hamiltonian. The resulting equations of motion describe the "natural" time evolution of the shape and size of the periodic box under constant applied external pressure and zero applied stress. The Parinello–Rahman method provides a "reaction path" from one solid phase to the other as the boundary conditions adjust themselves to the favored solid structure. For this reason this method is now being used to map phase diagrams involving several solid phases (see e.g., Refs. 2 and 3). It should be noted however that the method does not provide a *reversible* route from one solid phase to the other; the solid–solid phase transformation takes place when the initial solid phase becomes mechanically unstable. The actual thermodynamic phase transition is bracketed by the width of the hysteresis region. In order to locate the thermodynamic phase transition precisely one needs information on the free energy of both solid phases.

Two methods have traditionally been used to obtain such information. Both methods rely on the construction of a reversible path from a state of known free energy to the solid phase under consideration. The first method is the single occupancy cell (SOC) method introduced by Hoover and Ree.[4] This method starts with a lattice gas with one particle per lattice cell. At high densities the centers of the lattice cells coincide with the average atomic positions in the unconstrained solid. Expanding this lattice uniformly leads to a dilute gas which has the same pressure as an ideal gas at the same density, and a free energy that can be evaluated exactly. The free energy of the lattice gas at high densities coincides with the free energy of the corresponding unconstrained solid, provided that the density is sufficiently high to ensure that the artificial cell walls have negligible effect on the particle displacements. The free energy of the solid is then obtained by computing

$$F_{\text{solid}}(V_2) - F_{\text{lattice gas}}(V_1) = -\int_{V_1}^{V_2} P(V)dV \qquad (1)$$

at constant temperature. This method was used by Hoover and Ree to obtain the absolute free energy of the hard-sphere and hard-disk solids.[4,5] The actual numerical integration of Eq. (1) may require evaluating the pressure at many state points because lattice-gas isotherms exhibit a cusp at the point where the nearest-neighbor interactions take over from the cell walls in constraining the particles. There is even some evidence that a weak first-order transition takes place at this point,[6] in which case, the supposedly reversible path linking the solid to the dilute lattice gas may not be quite reversible, after all.

A second method of computing the free energy of a solid phase is to start from the low-temperature harmonic solid, the free energy of which can be computed exactly. This method was first used by Hoover, Gray, and Johnson.[7] There are two factors limiting the applicability of the latter method. The first is that it only works for solids which are harmonic at low temperatures (and/or high densities). This excludes all systems with discontinuous intermolecular

[a)] Present address: Lawrence Livermore National Laboratory L-454, Livermore, California 94550.

forces, e.g., the hard-sphere solid. Moreover, solid phases that are mechanically unstable at low temperatures cannot be investigated by this method. A practical problem with the harmonic-lattice method is that for all but the simplest solids, and in particular for molecular solids, evaluating the harmonic lattice free energy involves lengthy computation.

METHOD

In this section we introduce a new method to compute the absolute free energy of a solid phase. Our approach is once again based on the construction of a reversible path to a state of known free energy. In this case the reference state is an Einstein crystal with the same structure as the solid under consideration. This reference state can be reached from the real solid by slowly switching on harmonic springs which bind the atoms to their lattice sites. As the Einstein solid is structurally identical to the initial solid, it is very likely that such a path will be free of phase transitions and hence reversible. The simplest way to transform a solid to an Einstein crystal is to add a term λV to the unperturbed Hamiltonian H_0, such that

$$H(\lambda) = H_0 + \lambda V = H_0 + \lambda \sum_{i=1}^{N} (\mathbf{r}_i - \mathbf{r}_i^0)^2, \tag{2}$$

where $\mathbf{r}_i^0$ is the lattice position of particle i. The derivative of the free energy of this system with respect to the coupling constant λ is given by

$$\frac{\partial F}{\partial \lambda} = -kT \frac{\partial}{\partial \lambda} \left\{ \ln \int \cdots \int \exp[-\beta (H_0 + \lambda V)] d\mathbf{q}^N \right\}$$
$$= \langle V \rangle_\lambda, \tag{3}$$

from which it follows that the free energy of the real crystal is related to the free energy of a crystal with spring constant λ by

$$F(\lambda = 0) = F(\lambda) - \int_0^\lambda \langle V \rangle_{\lambda'} d\lambda'. \tag{4}$$

At sufficiently high λ the free energy of the system reduces to that of an Einstein crystal

$$F(\lambda) = \Phi_0 - kT \ln(\pi kT/\lambda)^{3N/2} + C(T) + O(1/\lambda), \tag{5}$$

where Φ_0 is the potential energy of the corresponding static lattice. $C(T)$ is the kinetic contribution to the free energy which depends only on the temperature T. Of course, a rather high value of λ may be required before terms of order $O(1/\lambda)$ in Eq. 5 become negligible. In practice there is no need to go to very high values of λ as it is rather simple to evaluate the leading corrections to the free energy at finite λ. In some cases these corrections can be evaluated analytically, as in the case of hard spheres to be discussed below, but in the most general case the free energy difference between the ideal Einstein crystal and the Einstein crystal with intermolecular interactions (henceforth referred to as the interacting Einstein crystal) can be found numerically by performing a Monte Carlo simulation on the ideal Einstein crystal, and deriving the free energy of the interacting Einstein crystal by umbrella sampling.[8] Depending on the nature of the system studied it may be useful to parametrize the Hamiltonian in a different way. For instance, for systems with continuous intermolecular interactions one may switch on the spring constants while switching off the intermolecular interactions. In general, the Hamiltonian may depend on λ in a nonlinear fashion. Equation (4) then becomes

$$F(\lambda = 0) = F(\lambda) - \int_0^\lambda \left\langle \frac{\partial H(\lambda')}{\partial \lambda'} \right\rangle d\lambda'. \tag{4a}$$

By a suitable choice of the parametrization of $H(\lambda)$ one may achieve a situation where $|\langle \partial H/\partial \lambda' \rangle|$ is quite small for $0 \leqslant \lambda' \leqslant \lambda$. Such a parametrization corresponds to a situation where the free energy of the Einstein crystal is quite close to the free energy of the initial crystal. It is likely that such a parametrization will result in an improved accuracy of the free energy computation. This method is also applicable to solids containing defects, particularly grain boundaries. The reference lattice in this case is the fully relaxed lattice containing the defect.

APPLICATION TO THE HARD-SPHERE SOLID

In order to test the usefulness of the method described above we used it to compute the free energy of the fcc and hcp phases of the hard-sphere solid at two densities close to the solid–fluid coexistence point. We chose this particular system for two reasons: First, of all reliable numerical results on the free energy of the fcc phase of the hard sphere solid are available,[5] yet the calculations on which these results are based are by no means trivial; they involve the computation of a complete isotherm of the single-occupancy cell system. Secondly, the old question of the relative stability of the hcp and fcc phases of the hard-sphere solid at melting is, thus far, unresolved. There is some evidence that as the solid density approaches ρ_0, the density of closest packing, the fcc phase is the more stable,[9–11] but as the pressure of the dense hcp phase is slightly higher than the pressure of the fcc phase,[10] it is not obvious which phase is the more stable at the melting point. In the present study we have computed the free energy of both fcc and hcp solids at a reduced density $\rho/\rho_0 = 0.7360$ (i.e., the density of the solid at the solid–fluid coexistence point, according to Ref. 5) and at a slightly higher density $\rho/\rho_0 = 0.7778$. Each simulation consisted of ten runs, for different values of the spring constant λ. Every run consisted of at least 10^4 sweeps (i.e., 10^4 attempted moves per particle) excluding equilibration; many runs were appreciably longer. The values of λ at which the different runs were carried out were chosen as follows. For $\lambda > \lambda_{\max}$ ($\lambda_{\max} \simeq 632$ for $\rho/\rho_0 = 0.7360$, and $\lambda_{\max} \simeq 1775$ for $\rho/\rho_0 = 0.7778$) the free energy of the interacting Einstein crystal could be accurately approximated by an analytical expression based on a "virial-like" cluster expansion, described in the Appendix. Hence, the numerical simulations were limited to the interval $0 < \lambda < \lambda_{\max}$. The mean-square particle displacement $\langle r^2 \rangle$, which is the integrand in Eq. (4), depends strongly on λ. At high values of λ, $\langle r^2 \rangle \sim 1/\lambda$ whereas as $\lambda \to 0$, $\langle r^2 \rangle \to \langle r^2 \rangle_0$, the mean-square displacement of an atom around its lattice site in the normal hard-sphere solid. Clearly, the function $(\lambda + c)\langle r^2 \rangle$ varies much less over the interval $0 < \lambda < \lambda_{\max}$, if we choose $c \simeq kT\sigma^2/\langle r^2 \rangle_0$. We make use of this fact to transform the integral in Eq. (4) in such a way that the integrand becomes a slowly varying function of the integration variable. To this end we write the integral in Eq. (4), i.e., the free energy difference between the interacting Einstein crystal and the hard-sphere solid as

$$\Delta F = -\int_0^{\lambda_{max}} \langle r^2\rangle_\lambda (\lambda + c)\frac{d\lambda}{(\lambda + c)}$$

$$= -\int_{\ln(c)}^{\ln(\lambda_{max}+c)} \{\langle r^2\rangle_\lambda(\lambda + c)\}d\ln(\lambda + c). \quad (6)$$

Here the integrand is a very smooth function of $\ln(\lambda + c)$ [we chose $c = \exp(3.5)$], and the integral could be evaluated using a 10-point Gauss–Legendre quadrature.[12] Later tests indicated that no significant loss of accuracy resulted if a five-point quadrature was used. A typical simulation consisting of ten runs of 10^4 sweeps for a 108 particle system took about 20 min on an IBM 192 computer.

Simulations were carried out for a number of fcc and hcp crystals of different size and shape. In the simulation we kept the center of mass of the system fixed. Without this constraint the mean-square particle displacement would become of order L^2 (L = boxlength) as $\lambda \to 0$, in which case the integrand in Eq. (6) would be sharply peaked around $\lambda = 0$. This would have an adverse effect on the accuracy of the numerical integration. In contrast, with the center of mass fixed, $\langle r^2\rangle$ tends to $\langle r^2\rangle_0$, as $\lambda \to 0$, and no problems occur. The fixed center of mass MC was implemented as follows. The coordinates of all particles were expressed relative to the center of the periodic box. During a trial move a particle displacement over a distance $\Delta \mathbf{r}$ is attempted. As the intermolecular interactions depend only on relative distances the tests for particle overlaps can be carried out without knowledge of the position of the center of mass. In contrast, in order to compute $V = \lambda \sum_{i=1}^N (\mathbf{r}_i - \mathbf{r}_i^0)^2$, one needs to know the absolute position of each particle with respect to the reference lattice. The distance $\mathbf{r}_i - \mathbf{r}_i^0$ can be written as $\mathbf{r}_i^{(B)} - \mathbf{r}_i^{0(B)} - (\mathbf{R}_{CM}^{(B)} - \mathbf{R}_{CM}^{0(B)})$, where the superscript (B) indicates coordinates relative to the center of the periodic box; $\mathbf{R}_{CM}^{(B)}$ is the position of the center of mass in these units. In order to compute $\mathbf{r}_i - \mathbf{r}_i^0$ we have to keep track of the displacement $(\mathbf{R}_{CM}^{(B)} - \mathbf{R}_{CM}^{0(B)})$. This is a simple matter; every time a particle is moved from $\mathbf{r}^{(B)} \to \mathbf{r}^{(B)} + \Delta\mathbf{r}$, $\mathbf{R}_{CM}^{(B)}$ changes to $\mathbf{R}_{CM}^{(B)} + \Delta\mathbf{r}/N$. Note that changing the box coordinate of one particle implies changing the absolute coordinates of all particles in order to keep the center of mass fixed. Keeping the center of mass of the system fixed reduces the partition function by a factor V (in the limit $\lambda = 0$). Hence, the free energy per particle in the fixed center of mass solid is $(\ln V)/N$ higher than in the unconstrained system. We have corrected our data for this effect; all final results refer to hard-sphere solids with unconstrained center of mass. In order to compare fcc and hcp crystals of the same size and shape, we chose the shape of the periodic box such that the basal plane (x–y plane) was parallel to a set of close-packed planes [e.g., the fcc (111) plane]. The height of the box was chosen such that a multiple of six close-packed planes fitted in the box. For hcp and fcc crystal structures only the stacking of these planes differs: *ABABAB*... stacking for hcp and *ABCABC*... stacking for fcc. In addition, we performed a number of simulations at $\rho/\rho_0 = 0.7360$ with fcc crystals of different sizes in a cubic box with edges parallel to the crystallographic 100,010, and 001 axes. All Monte Carlo data are collected in Table I. In order to compute the free energy of an N–particle hard-sphere crystal we first evaluated the free energy of the interacting Einstein crystal with fixed center of mass, at λ_{max}. This free energy contains three contributions: (i) the free energy of the corresponding noninteracting Einstein crystal, (ii) the virial contribution described in the Appendix, and (iii) a small correction to the virial contribution, which is also described in the Appendix; for the simulations with $\rho/\rho_0 = 0.7778$ this correction turned out to be negligible. The free energy of the unconstrained hard-sphere solid is then obtained by adding ΔF [Eq. (6)], together with a term $(\ln V)/N$ to account for the contribution of the center of mass motion to the free energy of the solid. Finally we compute the excess free energy F_{ex}^N, i.e., the free energy difference between the N-particle hard-sphere solid and an ideal gas with the same number of particles at the same density. All different contributions have been collected in Table I. In order to obtain the free energies of the different solid phases in the thermodynamic limit, we have to extrapolate to infinite system size. This extrapolation is not straightforward because we observe that the excess free energy depends not just on the number of particles, but also on the box shape. For instance, there is no significant difference between the excess free energy of a 54 ($= 3\times3\times6$) particle system and a 108 ($= 3\times3\times12$) particle system. But the excess free energy of the latter system *does* differ significantly from the corresponding value for the 108 particle system in a *cubic* box. We observed, however, that the excess free energy is a reasonably smooth function of v_{max}^{-1}, where v_{max} is the volume of the largest cubic box that fits into the periodic box (see Fig. 1). For a cubic box v_{max} equals the total box volume, for all other shapes $v_{max} = (L_{min})^3$, where L_{min} is the shortest diameter of the box. Clearly for fixed box shape extrapolating as a function of v_{max} is equivalent to extrapolating as a function of N. We considered two types of extrapolation viz., $F_{ex}^\infty - F_{ex}^N \sim 1/v_{max}$ and $F_{ex}^\infty - F_{ex}^N \sim (\ln v_{max})/v_{max}$, which correspond to $1/N$ and $(\ln N)/N$ extrapolations for fixed box shape. The results for the different extrapolation procedures have been collected in Table II. In most cases we find that the $1/v_{max}$ extrapolation fits marginally better to the MC data than the $(\ln v_{max})/v_{max}$ extrapolation. In Table II we have also included the extrapolated values for the free energy difference $F_{ex}^\infty(\text{hcp}) - F_{ex}^\infty(\text{fcc})$.

Let us first look at the fcc data at $\rho/\rho_0 = 0.7360$. From the data shown in Table II it is clear that the nature of the extrapolation procedure is the major source of uncertainty in the final result. This is probably made worse by the fact that we have combined the results for a number of different box shapes. At $\rho/\rho_0 = 0.7778$, where we have studied systems with rather similar box shapes, the different extrapolations yield more consistent results. The present values for the excess free energy of the fcc solid at $\rho/\rho_0 = 0.7360$ agree well with the Hoover and Ree value[5]

$$F_{ex}^\infty(\text{fcc}) = 5.924(15).$$

Next we turn to the hcp solid at the same density. The extrapolated values for $F_{ex}^\infty(\text{hcp})$ have also been collected in

Notes on p. 113

TABLE I. Contributions to the excess free energy of an N-particle hard-sphere solid. N-number of particles, fcc/hcp—crystal structure of the solid under consideration. $N_x \times N_y$—number of atoms in a close-packed plane, N_z—number of stacked close-packed planes. Three simulations on the fcc solid at $\rho/\rho_0 = 0.7360$ were carried out in a cubic box, as indicated in the table. $F^N_{\rm Einst}$—free energy of noninteracting Einstein crystal with fixed center of mass and spring constant $\lambda_{\max}$ ($\lambda_{\max} = 632.026$ at $\rho/\rho_0 = 0.7360$, $\lambda_{\max} = 1774.927$ at $\rho/\rho_0 = 0.7778$). All free energies in this table are expressed per particle. $\Delta F_{\rm vir}$—virial correction to the free energy of an Einstein crystal [Eq. (A6)]. $\Delta F_{\rm corr}$—correction to virial correction (see the Appendix). $\Delta F_{\rm MC}$—Monte Carlo result for the free energy difference between a hard-sphere solid and an interacting Einstein crystal (spring constant $\lambda_{\max}$) at the same density, both systems with fixed center of mass. $F^N_{\rm i.d.gas}(\rho)$—free energy of an N-particle ideal gas at density ρ. $F^N_{\rm excess}$—free energy difference between an N-particle hard-sphere solid and an ideal gas with the same number of particles at the same density. $F^N_{\rm excess} = F^N_{\rm Einst} + \Delta F_{\rm vir} - \Delta F_{\rm corr} + \Delta F_{\rm MC} - F^N_{\rm i.d.gas}(\rho) - (\ln V)/N$. ρ/ρ_0—(density)/(density at close packing). The error in the last two digits is indicated in brackets.

N	Type	$N_x \times N_y \times N_z$	$F^N_{\rm Einst}$	$\Delta F_{\rm vir}$	$\Delta F_{\rm corr}$	$\Delta F_{\rm MC}$	$F^N_{\rm i.d.gas}(\rho)$	$F^N_{\rm excess}$	ρ/ρ_0
54	fcc	3×3 ×6	7.9198	0.0183	1.5×10^{-3}	−2.8929(39)	−0.9060	5.8766(39)	0.7360
54	hcp	3×3 ×6	7.9198	0.0183	2.2×10^{-3}	−2.8976(42)	−0.9060	5.8712(42)	0.7360
108	fcc	3×3 ×12	7.9477	0.0183	6.0×10^{-4}	−2.9722(08)	−0.9275	5.8799(08)	0.7360
108	hcp	3×3 ×12	7.9477	0.0183	1.2×10^{-3}	−2.9729(07)	−0.9275	5.8787(07)	0.7360
72	fcc	3×4 ×6	7.9349	0.0183	1.7×10^{-3}	−2.9190(28)	−0.9175	5.8912(28)	0.7360
72	hcp	3×4 ×6	7.9349	0.0183	2.5×10^{-3}	−2.9205(24)	−0.9175	5.8889(24)	0.7360
96	fcc	4×4 ×6	7.9447	0.0183	1.5×10^{-3}	−2.9364(30)	−0.9266	5.9047(30)	0.7360
96	hcp	4×4 ×6	7.9447	0.0183	1.5×10^{-3}	−2.9335(30)	−0.9266	5.9076(30)	0.7360
192	fcc	4×4 ×12	7.9559	0.0183	1.5×10^{-3}	−2.9841(30)	−0.9415	5.9030(30)	0.7360
192	hcp	4×4 ×12	7.9559	0.0183	1.5×10^{-3}	−2.9825(30)	−0.9415	5.9046(30)	0.7360
216	fcc	6×6 ×6	7.9568	0.0183	2.4×10^{-3}	−2.9754(10)	−0.9432	5.9159(10)	0.7360
216	hcp	6×6 ×6	7.9568	0.0183	8.0×10^{-4}	−2.9761(09)	−0.9432	5.9168(09)	0.7360
32	fcc	cubic box	7.8701	0.0183	8.0×10^{-4}	−2.7933(30)	−0.8771	5.8644(30)	0.7360
108	fcc	cubic box	7.9477	0.0183	7.0×10^{-4}	−2.9403(30)	−0.9298	5.9117(30)	0.7360
256	fcc	cubic box	7.9577	0.0183	1.7×10^{-3}	−2.9776(21)	−0.9455	5.9208(21)	0.7360
72	fcc	3×4 ×6	9.4623	0.0006	···	−3.7710(33)	−0.8622	6.4960(33)	0.7778
72	hcp	3×4 ×6	9.4623	0.0006	···	−3.7639(33)	−0.8622	6.5031(33)	0.7778
144	fcc	4×6 ×6	9.4909	0.0006	···	−3.8165(26)	−0.8811	6.5223(26)	0.7778
144	hcp	4×6 ×6	9.4909	0.0006	···	−3.8125(21)	−0.8811	6.5263(21)	0.7778
576	fcc	6×8 ×12	9.5052	0.0006	···	−3.8575(12)	−0.8976	6.5351(12)	0.7778
576	hcp	6×8 ×12	9.5052	0.0006	···	−3.8551(11)	−0.8976	6.5375(11)	0.7778
1152	fcc	8×12×12	9.5061	0.0006	···	−3.8637(15)	−0.9008	6.5375(15)	0.7778
1152	hcp	8×12×12	9.5061	0.0006	···	−3.8652(10)	−0.9008	6.5363(10)	0.7778

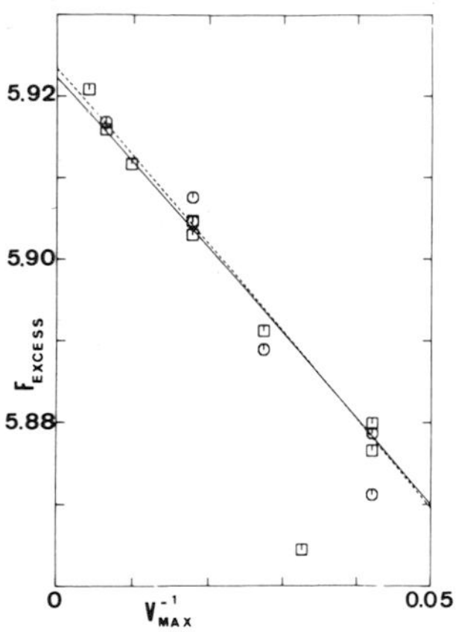

FIG. 1. System size dependence of the excess free energy of fcc (open squares) and hcp (open circles) hard-sphere solids at a reduced density $\rho/\rho_0 = 0.7360$. The excess free energy is in reduced units ($kT = 1$). The system size is characterized by the volume $v_{\max}$ of the largest cube that fits in the periodic box (see the text); the unit of volume is σ^3. Note that the excess free energy appears to be a smooth function of $v^{-1}_{\max}$ for all but the smallest system sizes even though this figure displays results of simulations on boxes of widely different shape. The drawn line is the best fit of $F^N_{\rm excess}$(fcc) to an expression of the form $F^N_{\rm ex} = F^{\infty}_{\rm ex} - a/v_{\max}$. The dashed line is the corresponding fit to the HCP data. The intercepts and regression coefficients for these and other fits have been collected in Table II.

Table II. Note that for both extrapolations the difference in excess free energy between hcp and fcc is very small: 0.0012(14) and 0.0015(15) for the $1/v_{\max}$ and $(\ln v_{\max})/v_{\max}$ extrapolations respectively. Of course, we can also *first* compute ΔF^N, the free energy difference between corresponding hcp and fcc systems, and *then* exrapolate to the thermody-

TABLE II. Estimate of the excess free energy (per particle) of the hard-sphere fcc and hcp solids at infinite system size. The nature of the $1/v_{\max}$ and $(\ln v_{\max})/v_{\max}$ extrapolations are discussed in the text. As a measure for the quality of the extrapolation procedure the regression coefficient R is also shown. $F^{\infty}_{\rm ex}({\rm hcp}) - F^{\infty}_{\rm ex}({\rm fcc})$—difference between the extrapolated excess free energies. $[F({\rm hcp})F({\rm fcc})]^{\infty}$—extrapolated free energy difference, i.e., $\lim_{N\to\infty} \Delta F^N$ (see the text). The error in the last two digits is indicated between brackets.

$\rho/\rho_0 = 0.7360$	$1/v_{\max}$ extrapolation	R	$(\ln v_{\max})/v_{\max}$ extrapolation	R
$F^{\infty}_{\rm ex}$(fcc)	5.9222(10)	0.97	5.9284(11)	0.97
$F^{\infty}_{\rm ex}$(hcp)	5.9234(10)	0.996	5.9299(12)	0.996
$F^{\infty}_{\rm ex}({\rm hcp}) - F^{\infty}_{\rm ex}({\rm fcc})$	0.0012(14)	···	0.0015(15)	···
$[F({\rm hcp}) - F({\rm fcc})]^{\infty}$	0.0017(14)	0.74	0.0021(17)	0.72
$\rho/\rho_0 = 0.7778$	$1/v_{\max}$ extrapolation	R	$(\ln v_{\max})/v_{\max}$ extrapolation	R
$F^{\infty}_{\rm ex}$(fcc)	6.5397(09)	0.97	6.5417(10)	0.94
$F^{\infty}_{\rm ex}$(hcp)	6.5404(11)	0.999	6.5436(13)	0.991
$F^{\infty}_{\rm ex}({\rm hcp}) - F^{\infty}_{\rm ex}({\rm fcc})$	0.0007(14)	···	0.0019(16)	···
$[F({\rm hcp}) - F({\rm fcc})]^{\infty}$	−0.0004(14)	0.79	−0.0011(16)	0.84

namic limit. This procedure is not quite equivalent to the previous one, because different data points have different statistical weights. Due to a cancellation of N-dependent contributions ΔF^N will be much less strongly N dependent than F_{ex}^N. If we assume that there is no systematic number dependence in ΔF^N we obtain $\Delta F^\infty = \langle \Delta F^N \rangle = 0.0001(8)$. Assuming a $1/v_{max}$ or a $(\ln v_{max})/v_{max}$ dependence of ΔF^N, yields $\Delta F^\infty = 0.0017(14)$ and $\Delta F^\infty = 0.0021(17)$, respectively. The extrapolated excess free energies of the hcp and fcc hard-sphere solid at $\rho/\rho_0 = 0.7778$ have been collected in the lower half of Table II. The corresponding differences between the hcp and fcc free energies are $F_{ex}^\infty(\text{hcp}) - F_{ex}^\infty(\text{fcc}) = 0.0007$ ($1/v_{max}$ extrapolation) and $F_{ex}^\infty(\text{hcp})$-$F_{ex}^\infty(\text{fcc}) = 0.0019(16)$ [$(\ln v_{max})/v_{max}$ extrapolation]. Once again we can perform subtraction and extrapolation in reverse order. In that case we obtain for ΔF^∞: 0.0012(11) (no number dependence), $-0.0004(14)$ ($1/v_{max}$) and $-0.0011(16)$ [$(\ln v_{max})/v_{max}$]. As before, the free energy difference is very small. At neither density is the free energy difference between hcp and fcc solids significantly different from zero. The upper bound on the free energy difference is once again largely determined by the uncertainty in the extrapolation procedure. If we assume that the procedure in which ΔF^N is extrapolated to its infinite-system value is least error prone (it involves the smallest number of assumptions), we conclude that the free energy difference between hcp and fcc hard-sphere solids close to melting is most likely in the interval $-0.001 < \Delta F^\infty < 0.002$, where we have lumped the results at both densities together. Note that the simulations at $\rho/\rho_0 = 0.7778$, which were performed over a wide range of system sizes (between $N = 72$ and $N = 1152$), suggest that there is a pronounced system size dependence of the free energy difference ΔF^N (see Fig. 2). We are not aware of other direct computations of the hcp–fcc free energy difference close to melting. However, several estimates exist for densities at or near close packing. For instance, Alder, Hoover, and Young[13] conclude that at close packing $|\Delta F^\infty| < 0.04$. In later work by Alder, Carter and Young[14] and Alder, Young, Mansigh, and Salzburg[10] the free energy difference between hcp and fcc at close packing is estimated to be $\Delta F^\infty = 0.002$. More recent work by Kratky[11] yields a much higher estimate of this free energy difference, viz, $\Delta F^\infty = 0.021(5)$ at $\rho/\rho_0 = 0.995$. This value is much larger than the earlier estimates, but fairly close to the free-volume prediction $\Delta F^\infty = 0.015(9)$. The present result is compatible with the results of Alder *et al.* but it appears to be incompatible with Kratky's value because if both the present results and the results of Ref. 11 were correct, this would imply that the pressure difference $\Delta P = P_{hcp} - P_{fcc}$ must be, on average, $\Delta P = 0.05$ between melting and close packing. This relatively large pressure difference is larger than the upper bound for ΔP that follows from the MD results of Refs. 10 and 13. It should be noted that in view of the very small free energy difference between hcp and fcc phases, the true state of lowest free energy at densities below close packing must contain stacking faults (i.e., stacking of the type *ABACBCABA*...). Of course, the contribution to the free energy due to this kind of disorder vanishes in the thermodynamic limit.

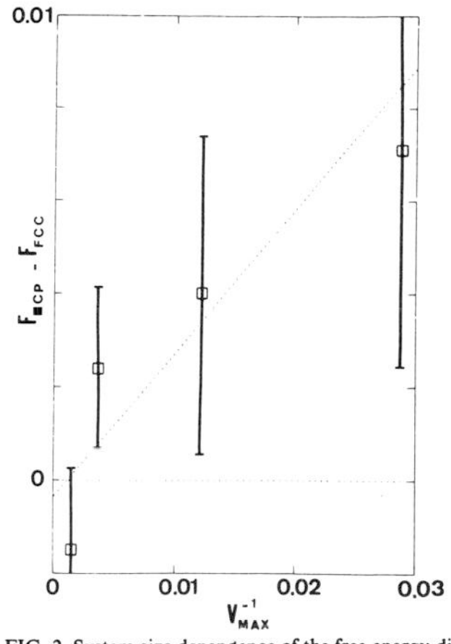

FIG. 2. System size dependence of the free energy difference between hcp and fcc hard-sphere solids of the same size and shape at a reduced density $\rho/\rho_0 = 0.7778$. Although the error bars on these data points are relatively large (but quite small in an absolute sense), the data suggest that the hcp–fcc free energy difference ΔF^N is strongly system size dependent. The dashed line is a fit to the MC data of the form $\Delta F^N = \Delta F^\infty - a'/v_{max}$. The intercept and regression coefficient for this fit are quoted in Table II. Units are as in Fig. 1.

CONCLUSION

In this paper we have developed a new method to compute the absolute free energy of arbitrary solid phases. The method is fast and accurate, and can also be applied to molecular solids[15] and solids containing defects. The basic idea is to constrict a reversible path between the solid under consideration and an Einstein crystal with the same structure. In the previous section we have demonstrated the conceptually simplest but numerically least sophisticated example of this method, namely one in which the spring constants are switched on while the intermolecular interations remain unaffected. For continuous intermolecular potentials the method may be expected to work much better if the intermolecular forces are swtiched off while the spring constant are being switched on, in such a way that the mean-square particle displacement remains approximately constant for all values of λ. Even for hard-core interactions it is easy to find a well behaved, nonlinear parametrization of the Hamiltonian which yields the unperturbed Hamiltonian at $\lambda = 0$ and the perfect Einstein crystal at $\lambda = \lambda_{max}$.

ACKNOWLEDGMENTS

This work was started at a CECAM workshop on transport in molecular fluids at Orsay, France in the summer of 1983. We thank Dr. Moser for creating the conditions which made this work possible; we thank the attendants of the workshop for many stimulating discussions.

APPENDIX

In this Appendix we derive an approximate expression for the free energy of an interacting Einstein crystal, which

Notes on p. 113

becomes exact at sufficiently high values of the spring constant λ. The configurational part of the partition function of the interacting Einstein crystal is of the following form:

$$Q(T;\lambda) = \int..\int \exp\Big[-\beta\lambda\sum_i(\Delta\mathbf{r}_i)^2\Big]\exp\Big[-\beta\sum_{i<j}u(r_{ij})\Big]d\mathbf{r}^N, \tag{A1}$$

where $\Delta\mathbf{r}_i = \mathbf{r}_i - \mathbf{r}_i^0$ ($\mathbf{r}_i^0$ is the lattice site of particle i), $\beta = 1/kT$ and $u(r_{ij})$ is the value of the pair potential of particles i and j at separation r_{ij}. Equation (A1) can be rewritten as

$$Q(T;\lambda) = Q_E(T;\lambda)\Big\langle\exp\Big[-\beta\sum_{i<j}u(r_{ij})\Big]\Big\rangle_E, \tag{A2}$$

where $Q_E(T;\lambda)$ is the partition function of the noninteracting Einstein crystal with spring constant λ. The subscript E in Eq. (A2) stands for averaging over all configurations of the noninteracting Einstein crystal. Such an average can, in principle, be carried out by Monte Carlo (umbrella sampling) but in the present case we use an expansion in cluster functions $f_{ij} = \{\exp[-\beta u(r_{ij})] - 1\}$:

$$\Big\langle\exp\Big[-\beta\sum_{i<j}u(r_{ij})\Big]\Big\rangle_E = \Big\langle 1 + \sum_{i<j}f_{ij} + \sum_{i<j}\sum_{k<l}f_{ij}f_{kl} + \cdots\Big\rangle_E. \tag{A3}$$

Note that for hard spheres $-\langle f_{ij}\rangle = \langle P^{ij}_{\text{overlap}}\rangle$, the probability that particles i and j in the noninteracting Einstein crystal are separated by a distance $|r_{ij}| < \sigma$. At high values of λ all $\langle f_{ij}\rangle$ for i and j not nearest neighbors become negligible, while $|\langle f^{\text{nn}}_{ij}\rangle| \ll 1$ (nn stands fo. "nearest neighbors"). In this limit we may approximate Eq. (A3) by

$$\Big\langle\exp\Big[-\beta\sum_{i<j}u(r_{ij})\Big]\Big\rangle_E \simeq (1 + \langle f^{\text{nn}}_{ij}\rangle)^{Nn/2}. \tag{A4}$$

In Eq. (A4) n stands for the number of nearest neighbors of a particle i (for both hcp and fcc solids $n = 12$). To the same level of approximation $\langle f^{\text{nn}}_{ij}\rangle$ can be evaluated by computing the probability of overlap of two isolated harmonically bound penetrating spheres of diameter σ at an average separation $a = |r_i^0 - r_j^0|$. This probability can be evaluated analytically. The result is

$$\langle P^{\text{nn}}_{\text{overlap}}\rangle_\lambda = \tfrac{1}{2}\{\text{erf}[(\beta\lambda/2)^{\frac12}(\sigma + a)] + \text{erf}[(\beta\lambda/2)^{\frac12}(\sigma - a)]\} - \{\exp[-\beta\lambda(\sigma - a)^2/2] - \exp[-\beta\lambda(\sigma + a)^2/2]\}/[(2\pi\beta\lambda)^{\frac12}a]. \tag{A5}$$

The expression which relates the free energy of the interacting Einstein crystal to that of a noninteracting Einstein crystal F_E then becomes

$$F(T;\lambda) = F_E(T;\lambda) - N(n/2)kT\ln(1 - \langle P^{\text{nn}}_{\text{overlap}}\rangle_\lambda). \tag{A6}$$

From this expression for $F(T;\lambda)$ we can derive an estimate for the mean-square particle displacement in the interacting Einstein crystal $\langle r^2\rangle_\lambda = -N^{-1}(\partial F/\partial\lambda)$:

$$\langle r^2\rangle_\lambda = \langle r^2\rangle_E - (\beta n/2)\frac{\{[\sigma a - \sigma^2 - (\beta\lambda)^{-1}]\exp[-\frac12\beta\lambda(a - \sigma)^2] + (\sigma a + \sigma^2 - (\beta\lambda)^{-1})\exp[\frac12\beta\lambda(a + \sigma]^2)\}}{2a(2\pi\beta\lambda)^{\frac12}(1 - \langle P^{\text{nn}}_{\text{overlap}}\rangle_\lambda)}. \tag{A7}$$

In Eq. (A7) $\langle r^2\rangle_E$ stands for the mean-square particle displacement in the nonintracting Einstein crystal. The second term on the right-hand side is the correction due to hard-core interactions. The above expression for the mean-square particle displacement in the interacting Einstein crystal can be compared directly to Monte Carlo results at high values of λ. In fact, our choice of λ_{max} [see Eq. (6)] was dictated by the requirement that the mean-square particle displacement at λ_{max} obeyed Eq. (A7). On closer scrutiny of the MC results we found that the MC results for the simulations at $\rho/\rho_0 = 0.7360$, $\lambda = 632.026$ still deviated slightly $[0(10^{-5})]$ from Eq. (A7). However, it was found that this difference $\Delta = \langle r\rangle^{\text{MC}}_\lambda - \langle r^2\rangle^{\text{vir}}_\lambda$ {the superscript "vir" stands for the "virial" expression [Eq. (A7)]} depends exponentially on λ: $\Delta = \alpha\exp(-\lambda/\delta)$ with $\alpha = 0(10^{-3})$ and $\delta = 0(10^2)$. As a consequence, Eq. (A6) overestimates the free energy of the interacting Einstein crystal by an amount $\Delta F_{\text{corr}} = \int_{\lambda_{\text{max}}}^\infty \Delta\, d\lambda = \alpha\lambda_{\text{max}}\exp(-\lambda_{\text{max}}/\delta)$. This correction, which turned out to be of the same order as the estimated error in the free-energy integration [Eq. (6)] was taken into account in the evaluation of the free energy of the hard-sphere solid. At $\rho/\rho_0 = 0.7778$, $\lambda_{\text{max}} = 1774.927$ ΔF_{corr} was negligible. Finally, we also corrected for the fact that in the Monte Carlo simulations we kept the center of mass of the system fixed. Hence, our reference state is not the normal Einstein crystal, but an Einstein crystal with fixed center of mass. The partition function of an Einstein crystal with fixed center of mass is given by

$$Q'_E(T;\lambda) = N^{-3/2}(\pi/\beta\lambda)^{3(N-1)/2}. \tag{A8}$$

Similarly, the state at $\lambda = 0$ in our simulation was not the normal hard-sphere solid, but a hard-sphere solid with fixed center of mass. The partition function of the latter system differs from the partition function of the former by a factor V^{-1}.

[1] M. Parinello and A. Rahman, Phys. Rev. Lett. **45**, 1196 (1980), J. Appl. Phys. **52**, 7182 (1981).
[2] D. Levesque, J.-J. Weis, and M. L. Klein, Phys. Rev. Lett. **51**, 670 (1983).
[3] M. Parinello, A. Rahman and P. Vashishta, Phys. Rev. Lett. **50**, 1073 (1983).
[4] W. G. Hoover and F. H. Ree, J. Chem. Phys. **47**, 4873 (1967).
[5] W. G. Hoover and F. H. Ree, J. Chem. Phys. **49**, 3609 (1968).
[6] H. Ogura, H. Matsuda, T. Ogawa, N. Ogita, and U. Ueda. Prog. Theor. Phys. **58**, 419 (1977).
[7] W. G. Hoover, S. C. Gray, and K. Johnson, J. Chem. Phys. **55**, 1128 (1971).
[8] J. P. Valleau and G. M. Torrie in *Statistical Mechanics A, Modern Theoretical Chemistry*, edited by B. J. Berne (Plenum, New York, 1977), Vol. 5, p. 178.
[9] D. A. Young and B. J. Alder, J. Chem. Phys. **60**, 1254 (1974).
[10] B. J. Alder, D. A. Young, M. R. Mansigh, and Z. W. Salzburg, J. Comp. Phys. **7**, 361 (1971).
[11] K. W. Kratky, Chem. Phys. **57**, 167 (1981).
[12] M. Abramowitz and I. A. Stegun, *Handbook of Mathematical Functions* (Dover, New York, 1970).
[13] B. J. Alder, W. G. Hoover and D. A. Young, J. Chem. Phys. **49**, 3688 (1968).
[14] B. J. Alder, B. P. Carter, and D. A. Young, Phys. Rev. **183**, 831 (1969).
[15] D. Frenkel and B. M. Mulder (to be published).

Notes to Reprint II.4

1. The method described in this paper was later extended to continuous potentials (see Reprint IV.8).

2. Reference [15] was later published as (Frenkel and Mulder, Mol. Phys. 55 (1985) 1171).

MOLECULAR PHYSICS, 1979, VOL. 37, No. 6, 1765–1772

Calculation of the entropy of liquid chlorine and bromine by computer simulation

by S. ROMANO†
Department of Chemistry, University of Southampton, U.K.
and K. SINGER
Department of Chemistry, Royal Holloway College,
Egham, Surrey, TW20 0EX, U.K.

(*Received* 29 *September* 1978)

The method of Widom [1] has been used to calculate the chemical potential of liquid chlorine and bromine in Monte Carlo simulations based on the two Lennard-Jones centres potential. The agreement between the computed and experimental entropies and vapour pressures over the range $T^*=1$ to 1·8 is satisfactory.

1. INTRODUCTION

The Monte Carlo (MC) and molecular dynamics (MD) methods have been used extensively to calculate respectively canonical and microcanonical ensemble averages of thermodynamic ((E, V, T), (p, V, T)) and structural properties from inter-particle pair potentials. Computations of quantities depending on the absolute value of the partition function (entropy, free energy, chemical potential) are more cumbersome and time-consuming and have therefore been reported less frequently. Current available methods can be broadly divided into direct and indirect ones. In the latter, the molar free energy of the condensed phase is related by a reversible expansion to that of the dilute phase for which the free energy can be evaluated by elementary statistical mechanics [2, 3]. Direct methods are the generation of a grand canonical ensemble, in which the chemical potential (μ) appears as a parameter [4–6], and the implementation of Widom's method [1], in which the absolute activity appears as (V/N) times the mean Boltzmann factor of a test particle, which is acted upon, but does not act upon the real particles. These and other methods e.g. umbrella sampling [7] whereby great economies can be made in covering the reversible path in indirect methods, have been reviewed [8, 9].

The purpose of this paper is to show (1) that the implementation of Widom's method in MC, and probably also MD calculations is practicable at not too high densities in the liquid range ; (2) that entropies and vapour pressures so calculated in conjunction with p, V, T and E, V, T data for the two Lennard-Jones centres model for liquid Cl_2 and Br_2 [10] yield rather satisfactory results along the co-existence line between $T^*=1$ and 1·8 (2·0). Determinations by computer simulation of absolute entropies for molecular fluids have to our knowledge not been reported.

† Present address : Montedison, G. Donegani Research Institute, 28100, Novara, Italy.

2. The method

Widom's result for the configurational chemical potential follows from

$$\mu_{\text{conf}}(V, T) = N_0 \left(\frac{\partial A_{\text{conf}}}{\partial N} \right)_{V,T} \simeq -N_0 kT \log \frac{Z_{\text{conf}}(N+1, V, T)}{Z_{\text{conf}}(N, V, T)} \tag{1}$$

and the identity

$$\frac{Z_{\text{conf}}(N+1, V, T)}{Z_{\text{conf}}(N, V, T)} = \frac{1}{N+1} \frac{\int \underset{V}{\cdots} \int \left[\exp - \frac{(U_{1,N}(\mathbf{r}', \mathbf{r}^N)\, d\mathbf{r}'}{kT} \right] \exp \left[-\frac{U(\mathbf{r}^N)}{kT} \right] d\mathbf{r}^N}{\int \underset{V}{\cdots} \int \exp \left[-\frac{U_N(\mathbf{r}^N)}{kT} \right] d\mathbf{r}^N}, \tag{2}$$

where $U_{1,N}(\mathbf{r}', \mathbf{r}^N)$ is the potential energy of the $(N+1)$th particle (at $\mathbf{r}'$) in the field of the remaining N particles, and $U_N(\mathbf{r}^N)$ is the potential energy of N particles interacting through the same pair potentials. In the numerator of (2) all coordinates $\mathbf{r}_i$ may be changed to $\mathbf{r}_i - \mathbf{r}'$, and the variable $\mathbf{r}'$ thereby be removed from the integrand. It follows that

$$\mu_{\text{conf}}(V, T) \simeq -N_0 kT \log \left(\frac{V}{N} \langle BFT \rangle \right), \tag{3}$$

where

$$\langle BFT \rangle = \langle \exp(-U_{1,N}/kT) \rangle$$

is the mean Boltzmann factor of a test particle, i.e. of a particle in the field of N other particles which are not affected by it. Inserting the remaining factors in partition function, one obtains for a diatomic fluid

$$-\frac{\mu}{N_0 kT} = \log \left\{ \left[\frac{2\pi(m_a + m_b)}{h^2} kT \right]^{3/2} \frac{T}{\sigma_{ab}\theta_r} [1 - \exp(-\theta_v/T)]^{-1} \right\} + \log \left(\frac{V}{N} \langle BFT \rangle \right), \tag{4}$$

where $\theta_r = h^2/(8\pi^2 \mu_{ab} r_{ab}{}^2 k)$, $\mu_{ab} = m_a m_b/(m_a + m_b)$ being the reduced mass, $\theta_v = h\nu/k$ the characteristic vibrational temperature, and σ_{ab} the effective symmetry number. When virtually only two isotopes are present, as for Cl_2 and Br_2, σ_{ab} is related to the mole fractions of the isotopes by $\sigma_{ab}{}^{-1} = 0{\cdot}5\,(x_1{}^2 + x_2{}^2) + 2x_1x_2$.

In the present calculations the following parameters were used

	ϵ/k/K	σ/m	r_{ab}/σ	$N_0 m_a$/(g/mol)	$N_0 m_b$/(g/mol)	x_1	θ_v/K
Cl_2	178·3	$3{\cdot}332 \times 10^{-10}$	0·63	35	37	0·7553	810
Br_2	257·2	$3{\cdot}538 \times 10^{-10}$	0·63	79	81	0·5054	470

The potential parameters are taken from reference [10] (where σ for Br_2 in table 9 is erroneously given as $2{\cdot}538 \times 10^{-10}$ m). Since these pair potentials for Cl_2 and Br_2 are conformal, the principle of corresponding states leads to

$$\mu^*(\epsilon_A, \sigma_A, V^*, T^*) = \mu^*(\epsilon_B, \sigma_B, V^*, T^*) - 3\log(\sigma_B/\sigma_A),$$
$$(V^* = V(\text{mol})/(N_0\sigma^3),\ T^* = kT/\epsilon,\ \mu^* = \mu/(N_0\epsilon)),$$

i.e. the same computer experiment yields the chemical potential of Cl_2 and Br_2.

3. Computational technique

The $\langle BFT \rangle$ is determined the course of conventional MC runs of 4×10^5 steps with sample cubes of appropriate size containing 256 molecules, and periodic boundary conditions. After every 100 steps the mean BFT for the current configuration is calculated according to the following procedure. The centres of mass of the test particles are placed on 125 sites of a simple $5 \times 5 \times 5$ cubic lattice which fills the sample cube. At each site the potential energy of the test particle in the field of the real MC particles is calculated for three mutually perpendicular orientations (X, Y, Z). Thus for the kth sampling

$$\langle \exp\left[-U_{1,N}/kT\right)\rangle_k = \frac{1}{375} \sum_{p=1}^{125} \sum_{\alpha=1}^{3} \exp\left[-U(\mathbf{r}'_p, \mathbf{u}'_\alpha, \{\mathbf{r}^N, \mathbf{u}^N\})_k\right], \tag{5}$$

where

$$U(\mathbf{r}'_p, \mathbf{u}'_\alpha, \{\mathbf{r}^N, \mathbf{u}^N\}_k) = \sum_{i=1}^{N} \phi(\mathbf{r}'_p, \mathbf{u}'_\alpha\,;\ \mathbf{r}_i^{(k)}, \mathbf{u}_i^{(k)}). \tag{6}$$

The right-hand side of (6) is the sum of pair potentials of the test molecule and the molecules in the kth configuration ($\mathbf{r}$ refers to positions, $\mathbf{u}$ to orientations). The left-hand side of (5) is averaged over the number M of sampled configurations :

$$\langle BFT \rangle = \frac{1}{M} \sum_{k=1}^{M} \langle \exp\left[-U_{1,N}/kT\right]\rangle_k. \tag{7}$$

The test-particle sampling does not affect the configurations generated by the MC process.

At high densities in the liquid range ($0{\cdot}45 \leqslant \rho^* \leqslant 0{\cdot}54$, $\rho^* = N\sigma^3/V(N)$) the Boltzmann factor of the test molecule is almost everywhere small, and statistical accuracy of log $\langle BFT \rangle$ is poor. At $\rho^* = 0{\cdot}4$ the statistical error is $\simeq 5$–10 per cent, and reliable estimate scan be obtained at and below this density. From the chemical potential (equation (4)) the entropy is obtained as

$$S/R = (E^* + p^*V^* - \mu^*)/T^*, \tag{8}$$

where $E^* = U^* + 2{\cdot}5T^* = E/N_0\epsilon$, $V^* = 1/\rho^*$, $p^* = p\sigma^3/\epsilon$, $\mu^* = \mu/(N_0\epsilon)$; E, V, S refer to 1 mole.

Expressions for the mean potential energy and the pressure in terms of the BFT are also given in reference [1] :

$$\langle U \rangle = \tfrac{1}{2} z V \langle U_{1,N} \exp\left(-U_{1,N}/kT\right)\rangle, \tag{9}$$

$$\langle p \rangle = NkT + \tfrac{1}{6} z \langle \Psi_{1,N} \exp\left(-U_{1,N}/kT\right)\rangle, \tag{10}$$

($z^{-1} = Z_{\text{conf}}(N+1)/[(N+1)Z_{\text{conf}}(N)]$; $\Psi_{1,N}$ = virial of the test particle).

The values so calculated are on the whole consistent with those obtained in the usual manner from ensemble averages, but statistically less accurate, as the following figures illustrate :

		equation (9)	equation (10)	ensemble average	
T^*	V^*	$-U^*$	p^*	$-U^*$	p^*
2·0	2·5	9·23 ± 0·23	0·3 ± 0·1	9·46 ± 0·01	0·26 ± 0·03
2·5	3·0	7·50 ± 0·15	0·47 ± 0·08	7·63 ± 0·01	0·50 ± 0·02

4. Results

The primary data for the evaluation of (8) are given in table 1. Because of the margins of error the S/R values on column 7 are not compatible with the relationships

$$S(V, T_2) - S(V, T_1) \simeq C_V \log (T_2/T_1), \quad (11\,a)$$

and

$$S(V_2, T) - S(V_1, T) = \int_{V_1}^{V_2} \left(\frac{\partial p}{\partial T}\right)_V dV \simeq (V_2 - V_1)\left[\left(\frac{\partial p}{\partial T}\right)_{V_2} + \left(\frac{\partial p}{\partial T}\right)_{V_1}\right]/2 \quad (11\,b)$$

$((V_2 - V_1)/V_1 \ll 1)$. Column 8 of table 1 contains S/R values which are adjusted within the limits of error so as to make them very nearly compatible with (11 *a*) and (11 *b*) based on the MC data :

$V^* =$	2·5	2·75	3·0	3·2	3·4	3·6
$C_v/R =$	3·35	3·25	3·18	3·10	3·0	2·95
$(\partial p^*/\partial T^*)_{V^*} =$	2·2	1·75	1·20	1·0	0·9	0·8

These C_V/R and $(\partial p^*/\partial T^*)_{V^*}$ data are somewhat higher than the values reported in reference [10] for $V^* = 2{\cdot}5$, $T^* = 2{\cdot}0$ ($C_V/R = 3{\cdot}1$, $(\partial p^*/\partial T^*)_{V^*} = 1{\cdot}9$), but this difference has very little effect on the best estimates of S/R. The data in table 1 correspond to the rigid rotator model for Cl_2. $(S/R)_{Br_2} = (S/R)_{Cl_2} + 3{\cdot}338$ for this model.

Table 1. Monte Carlo data for the calculation of chemical potential and the entropy by Widom's method.

T^*	V^*	$-U^*$	p^*	$\log \langle BFT \rangle$	$-\mu^*$	S/R	S/R^{adj}
2·00	2·50	9·46 ± 0·01	0·26 ± 0·03	2·74 ± 0·33	41·12	18·66	18·54
2·50	2·50	9·15 ± 0·01	1·35 ± 0·05	0·54 ± 0·24	47·30	19·11	19·29
2·00	2·75	8·69 ± 0·02	0·08 ± 0·03	2·79 ± 0·18	41·42	18·97	19·13
2·25	2·75	8·42 ± 0·01	0·33 ± 0·03	2·40 ± 0·11	46·38	19·77	19·56
2·5	2·75	8·31 ± 0·01	0·83 ± 0·03	1·43 ± 0·17	49·77	19·91	19·92
1·75	3·00	8·29 ± 0·03	−0·29 ± 0·02	4·33 ± 0·08	38·52	19·27	19·10
2·00	3·00	7·93 ± 0·02	−0·08 ± 0·02	3·06 ± 0·10	42·13	19·48	19·43
2·25	3·00	7·76 ± 0·01	0·17 ± 0·02	2·28 ± 0·11	46·30	19·86	19·90
2·50	3·00	7·63 ± 0·01	0·50 ± 0·02	1·43 ± 0·05	49·98	20·04	20·15
2·25	3·20	7·25 ± 0·01	0·15 ± 0·02	2·35 ± 0·09	51·79	20·20	20·14
2·25	3·40	6·97 ± 0·02	0·12 ± 0·02	2·17 ± 0·06	51·49	20·18	20·26
2·25	3·60	6·50 ± 0·01	0·04 ± 0·01	2·25 ± 0·05	46·65	20·41	20·47
						(±0·25)	(±0·10)

From the best estimate of S/R ($T^* = 2{\cdot}0$, $V^* = 2{\cdot}5$) = 18·54 the value at low temperature and high density ($T^* = 1{\cdot}0$, $\rho^* = 0{\cdot}539$) corresponding to $p^* \simeq 0$, is obtained by means of (11 *a*) and (11 *b*) with the use of the C_V/R and $(\partial p^*/\partial T^*)_{V^*}$ values given above for $V^* = 2{\cdot}5$, and values taken from table 1 of reference [10] at higher densities ; linear interpolation is used where necessary. Integration along isochores to $T^*(p^*=0)$ alternate with isothermal integrations from one isochore to another :

(T^*, V^*)	(T^*, V^*)		$\Delta(S/R)$
$(2{\cdot}00, 2{\cdot}5)$	$\rightarrow(1{\cdot}882, 2{\cdot}5)$;	$(11\ a)$:	$3{\cdot}35 \log \dfrac{1{\cdot}882}{2{\cdot}00} = -0{\cdot}203$
$(1{\cdot}882, 2{\cdot}5)$	$\rightarrow(1{\cdot}882, 1/0{\cdot}448)$;	$(11\ b)$:	$-0{\cdot}268(2{\cdot}2+3{\cdot}1)/2 = -0{\cdot}710$
$(1{\cdot}882, 1/0{\cdot}448)$	$\rightarrow(1{\cdot}625, 1/0{\cdot}448)$;	$(11\ a)$:	$3{\cdot}50 \log \dfrac{1{\cdot}625}{1{\cdot}882} = -0{\cdot}514$
$(1{\cdot}625, 1{\cdot}0{\cdot}448)$	$\rightarrow(1{\cdot}625, 1/0{\cdot}492)$;	$(11\ b)$:	$-0{\cdot}200(3{\cdot}1+4{\cdot}6)/2 = -0{\cdot}770$
$(1{\cdot}625, 1/0{\cdot}492)$	$\rightarrow(1{\cdot}322, 1/0{\cdot}492)$;	$(11\ a)$:	$3{\cdot}79 \log \dfrac{1{\cdot}322}{1{\cdot}625} = -0{\cdot}782$
$(1{\cdot}322, 1/0{\cdot}492)$	$\rightarrow(1{\cdot}322, 1/0{\cdot}539)$;	$(11\ b)$:	$-0{\cdot}177(4{\cdot}6+6{\cdot}4)/2 = -0{\cdot}974$
$(1{\cdot}322, 1/0{\cdot}539)$	$\rightarrow(1{\cdot}000, 1/0{\cdot}539)$;	$(11\ a)$:	$4{\cdot}09 \log \dfrac{1{\cdot}000}{1{\cdot}322} = -1{\cdot}142$
			sum $-5{\cdot}095$

Table 2. Comparison between calculated and experimental entropies for liquid Cl_2 and Br_2.

T^*	$C_p(Cl_2)/R/$ K^{-1} (5)	$S_l(Cl_2)/R$ calculated	experimental (7)	(9)	$C_p(Br_2)/R$ (6)	$S_l(Br_2)/R$ calculated	experimental (9)	(10)
0·965(1)	7·83	13·24	12·99					
1·00	7·84	13·51	13·28	13·17	8·30	17·31		17·29
1034(2)	7·79	13·77	13·54	13·44	8·32	17·59	17·26	17·55
1·10	7·81	14·25	14·05	13·96	8·34	18·10	17·83	18·00
1·20	7·86	14·94	14·74	14·75	8·35	18·83	18·63	18·60
1·288(3)	7·91	15·56	15·24	15·27	8·36	19·56		19·23
1·34(4)	7·93	15·81	15·53	15·61	8·36	19·75		19·41
1·40	7·95	16·15		15·97	8·34	20·12		19·74
1·60	8·03	17·22		17·07	8·37	21·24		20·70
1·80	8·08	18·17		18·05	8·37	22·22		21·63
2·00	8·12	19·02		19·02				

(1) Experimental normal melting point of Cl_2.
(2) Experimental normal melting point of Br_2.
(3) Experimental normal boiling point of Br_2.
(4) Experimental normal boiling point of Cl_2.
(5) From table 8 (a) in reference 10 with added vibrational contribution ($\theta_v=810$ K).
(6) From table 9 (a) in reference 10 with added vibrational contribution ($\theta_v=470$ K).
(7) Giaque and Powell [11].
(8) Seshadri *et al.* [12].
(9) Hildenbrand *et al.* [13].
(10) Seshadri *et al.* [14].

Thus S/R (1·00, 1/0·539) = 18·54 − 5·09 = 13·45. At this temperature the vibrational contribution

$$-\log(1\text{-}\exp(-\theta_v/T) + (\theta_v/T)[\exp(\theta_v/T) - 1]^{-1} = 0{\cdot}06$$

and the estimate of the total value is 13·51. The difference between this value and that for Br_2 results from the translational–rotational contribution (+3·338) and the vibrational contribution (0·526), giving S/R (1·00, 1/0·539) = 17·31.

Calculated and experimental values of S/R are compared in table 2. The experimental values are those along the co-existence line. The calculated values are for the isobar $p = 0$, obtained by $\Delta S = \int_{T_1}^{T_2} C_p(T)\, dT$ from C_p data in tables 8(a) and 9 of reference [10], to which the vibrational contribution is added.

The estimated correction due to the fact that the vapour pressure is >0 is very small (−0·01, −0·03 and −0·08 at $T^* = 1{\cdot}6$, 1·8 and 2·0) and is neglected in table 2. For both Cl_2 and Br_2 the discrepancies are ~0·2−0·3, i.e. probably outside the computational margin of error, but quite satisfactory in view of the simplicity of the potential model.

Table 3. Comparison between calculated and experimental vapour pressures of liquid Cl_2.

T^*	$\frac{H_l^*}{T^*}$	$\frac{\mu_l^*}{T^*}$	$-\frac{(\mu_g^0)^*}{T^*}$	$\log\left(\frac{p\gamma}{\text{atm}}\right)$	log(p/atm) calculated	log(p/atm) experimental	experimental (5)
0·965(1)	12·29	25·53	21·54	−3·99	−3·99	−4·29(3)	
1·00	11·59	25·10	21·67	−3·43	−3·43	−3·72(3)	[−4·49]
1·10	9·83	24·08	22·01	−2·07	−2·07	−2·36(3)	−2·32
1·20	8·36	23·30	22·32	−0·98	−0·97	−1·25(3)	−1·22
1·30	7·12	22·69	22·61	−0·08	−0·06	−0·33(3)	−0·31
1·34(2)	6·67	22·48	22·71	0·23	0·26	0·00(3)	0·01
1·40	6·05	22·20	22·88	0·68	0·72	0·48(4)	0·45
1·60	4·28	21·49	23·37	1·88	1·97	1·66(4)	1·58
1·80	2·93	20·97	23·80	2·83	3·00	2·60(4)	2·60
2·00	1·83	20·79	24·30	3·51	3·78	3·33(4)	3·55

(1) Experimental triple point.
(2) Experimental normal boiling point.
(3) Giauque and Powell [11].
(4) Pellaton [15].
(5) Seshadri *et al.* [12].

The experimental and calculated vapour pressures are compared in tables 3 and 4. The latter are obtained via

$$\mu_l^*(p^*, T^*) = (\mu_g^0(T^*))^* + T^* \log\left(\frac{p\gamma}{\text{atm}}\right), \tag{12}$$

from data which are also given in tables 3 and 4. The left-hand side of (12) is obtained from

$$\frac{\mu_l^*(T^*, p^*=0)}{T^*} = \frac{H^*(T^*, p^*=0)}{T^*} - \frac{S_l(T^*, p^*=0)}{R}, \tag{13}$$

Table 4. Comparison between calculated experimental vapour pressures of liquid Br_2.

T^*	$-\frac{H_l^*}{T^*}$	$-\frac{\mu_l^*}{T^*}$	$-\frac{(\mu_g^0)^*}{T^*}$	$\log\left(\frac{p\gamma}{\text{atm}}\right)$	log (p/atm) calculated	experimental (3)
1·00	11·29	28·60	23·56	−3·24	−3·24	−3·40
1·034(1)	10·64	28·23	25·49	−2·74	−2·74	−2·79
1·10	9·51	27·61	25·73	−1·88	−1·88	−1·92
1·20	8·03	26·86	26·07	−0·79	−0·78	−0·75
1·288(2)	6·92	26·29	26·27	−0·02	0·00	0·00
1·3	6·71	26·19	26·28	+0·09	0·11	0·12
1·4	5·69	25·81	26·70	0·89	0·93	0·87
1·6	3·94	25·38	27·21	1·83	1·90	2·09
1·8	2·57	24·76	27·69	2·93	3·08	3·05
2·0	1·48	24·58	28·12	3·54	3·77	3·82

(1) Experimental triple point.
(2) Experimental normal boiling point.
(3) Seshadri *et al.* [14] (in close agreement with Scheffer and Voogt [16]).

where the reduced enthalpies can, with sufficient accuracy, be equated to 19·11 − 7·47 T^* ([10], table 3) + $(\theta_v/T)\,[\exp(\theta_v/T)-1]^{-1}$, and S_1/R is taken from table 2. The small pressure correction V^*p^*/T^* (=0·01, 0·03 and 0·06 at T^* = 1·6, 1·8 and 2·0) are included in column 3 of tables 3 and 4.

μ_g^0 is the chemical potential of the ideal gas at 1 atm, calculated by elementary statistical mechanics. The activity coefficient γ is estimated from the second virial coefficient $B(T)$ according to

$$\log(p\gamma) \simeq \log p - \frac{Bp}{RT},$$

which may be solved by iteration. The required $B(T)$ values were obtained from the following figures calculated for the Cl_2-model potential [17].

T^*	1·0	1·2	1·4	1·6	1·8	2·0
$-B(T)/\text{cm}^3$	750	505	370	285	225	185

The discrepancy between experimental and calculated values of log p is ~0·3 up to T^* = 1·6 and somewhat worse at higher temperatures. For Br_2 the agreement is, no doubt fortuitously, excellent.

5. ⟨*BFT*⟩ SAMPLED IN MOLECULAR DYNAMICS CALCULATIONS

Since the entropy of the microcanonical and canonical ensembles agree in the limit of large N, one would expect that Widom's method could be implemented in MD as well as in MC calculations. To test this the potential energy of M test particles was calculated after every other time in two MD runs :

	Number of time steps	M	log ⟨BFT⟩
$T^*=2\cdot25$, $V^*=3\cdot2$	7200	3×3^3	$1\cdot8\pm0\cdot2$
	10000	3×5^3	$2\cdot26\pm0\cdot08$

The corresponding MC result (table 1) is $2{\cdot}35 \pm 0{\cdot}09$. The average of the MD values weighted according to the number of sample points (in space and time) is $2{\cdot}20 \pm 0{\cdot}10$, in fair agreement with the MC value. These results suggest that Widom's method can also be implemented in MD calculations, but that the statistical convergence is slow.

We wish to thank Dr. C. S. Murthy for the permission to use his unpublished data on $B(T)$ for the Cl_2 model. We also acknowledge the generous allocation of computing time by the University of London Computer Centre.

References

[1] Widom, B., 1963, *J. chem. Phys.*, **39**, 2808.
[2] Hoover, W. G., and Ree, F. H., 1968, *J. chem. Phys.*, **49**, 3609.
[3] Hansen, J. P., and Verlet, L., 1969, *Phys. Rev.*, **184**, 151
[4] Norman, G. S., and Filinov, V. S., 1969, *High Temp. Res. U.S.S.R.*, **7**, 216.
[5] Adams, D. J., 1974, *Molec. Phys.*, **28**, 1241 ; 1975, *Ibid.*, **29**, 307.
[6] Rowley, L. A., Nicholson, D., and Parsonage, N. G., 1975, *J. comput. Phys.*, **17**, 401.
[7] Torrie, G. M., and Valleau, J. P., 1974, *Chem. Phys. Lett.*, **28**, 578.
[8] Valleau, J. P., and Whittington, S. G., 1977, *Modern Theoretical Chemistry*, Vol. 5, edited by B. J. Berne (Plenum), chapter 4.
[9] Valleau, J. P., and Torrie, G. M., 1977, *Modern Theoretical Chemistry*, Vol. 5, edited by B. J. Berne (Plenum), chapter 5.
[10] Singer, K., Taylor, A., and Singer, J. V. L., 1977, *Molec. Phys.*, **33**, 1757.
[11] Giauque, W. F., and Powell, T. M., 1939, *J. Am. chem. Soc.*, **61**, 1970.
[12] Seshadri, D. M., Vishwanath, D. S., and Kuloor, N. R., 1966, *J. Indian Inst. Sci.*, **48**, 38.
[13] Hildenbrand, D. L., Kramer, W. R., McDonald, R. A., and Stull, D. R., 1958, *J. Am. chem. Soc.*, **80**, 4129.
[14] Seshadri, D. M., Viswanath, D. S., and Kuloor, N. R., 1970, *Indian J. Technol.*, **8**, 191.
[15] Pellaton, M., 1915, *J. Chim. phys.*, **13**, 926.
[16] Scheffer, F. E., and Voogt, M., 1926, *Recl. Trav. chim. Pays-Bas Belg.*, **45**, 21.
[17] Murthy, C. S. (unpublished).

Notes to Reprint II.5

1. On p. 1771, "log $p - Bp/RT$" should read "log $p + Bp/RT$".

2. An early method that resembles the particle-insertion technique is described in a paper by Byckling (Physica 27 (1961) 1030).

3. The first systematic application of the particle-insertion method can be found in a paper by Adams (Mol. Phys. 28 (1974) 1241). The same paper also contains a description of the closely related technique of grand canonical Monte Carlo simulations, which we discuss in Chapter V. In a later paper (Mol. Phys. 29 (1975) 307), Adams notes that the quality of his earlier results suffered from the use of a poor random-number generator.

4. At high densities, the "acceptance rate" of the (imaginary) particle insertions becomes very low, leading to large statistical errors in the chemical potential. Several authors have attempted to overcome this problem by use of biased sampling schemes in which the test particle is preferentially inserted in a cavity (Shing and Gubbins, Mol. Phys. 43 (1981) 717; Fixman, J. Chem. Phys. 78 (1983) 4223). Another improvement of the method is that due to Shing and Gubbins (Mol. Phys. 46 (1982) 1109; 49 (1983) 1121) in which trial insertions are combined with trial

removals. The Shing–Gubbins modification is similar in spirit to the "overlapping-distribution" method of Bennett (Reprint II.6); its use leads in some cases to better estimates of the chemical potential and, invariably, to an improved estimate of the statistical error.

5. The liquid–vapour coexistence curves of chlorine and bromine were calculated later (Powles, Mol. Phys. 41 (1980) 715).

6. The correct implementation (see Section 5) of the particle-insertion method in microcanonical molecular dynamics simulations requires some care. The problem is discussed by Frenkel (in: Molecular Dynamics Simulation of Statistical-Mechanical Systems, eds G. Ciccotti and W.G. Hoover (North-Holland, Amsterdam, 1986)).

Reprinted from JOURNAL OF COMPUTATIONAL PHYSICS
All Rights Reserved by Academic Press, New York and London

Vol. 22, No. 2, October 1976
Printed in Belgium

Efficient Estimation of Free Energy Differences from Monte Carlo Data

CHARLES H. BENNETT

IBM Thomas J. Watson Research Center, Yorktown Heights, New York 10598

Received February 13, 1976; accepted May 3, 1976

Near-optimal strategies are developed for estimating the free energy difference between two canonical ensembles, given a Metropolis-type Monte Carlo program for sampling each one. The estimation strategy depends on the extent of overlap between the two ensembles, on the smoothness of the density-of-states as a function of the difference potential, and on the relative Monte Carlo sampling costs, per statistically independent data point. The best estimate of the free energy difference is usually obtained by dividing the available computer time approximately equally between the two ensembles; its efficiency (variance $\times$ computer time)$^{-1}$ is never less, and may be several orders of magnitude greater, than that obtained by sampling only one ensemble, as is done in perturbation theory.

I. INTRODUCTION

A well-known deficiency of the Monte Carlo [1, 2] and molecular dynamics [3] methods, commonly used to study the thermodynamic properties of classical systems having 10^2 to 10^4 degrees of freedom, is their inability to calculate quantities such as the entropy or free energy, which cannot be expressed as canonical or microcanonical ensemble averages. In general, the free energy of a Monte Carlo (MC) or molecular dynamics (MD) system can be determined only by a procedure analogous to calorimetry, i.e., by establishing a reversible path between the system of interest and some reference system of known free energy. "Computer calorimetry" has a considerable advantage over laboratory calorimetry in that the reference system may differ from the system of interest not only in its thermodynamic state variables but also in its Hamiltonian, thereby making possible a much wider variety of reference systems and reversible paths. Often the path between an analytically tractable reference system and the system of ultimate physical interest will include one or more intermediate systems. These may be interesting in their own right (e.g., the hard sphere fluid), or they may be special systems, important only as calorimetric stepping stones, whose Hamiltonians contain artificial terms designed to stabilize the system against phase transitions [4, 5], induce favorable importance weighting [6, 7], or otherwise enhance the system's efficiency as a computational tool [8–10].

Whether the calorimetric path has one step or many, one eventually faces the statistical problem of extracting from the available data the best estimate of the free energy differences between consecutive systems. Specializing the question somewhat, one might inquire what is the best estimate one can make of the free energy difference between two MC systems (i.e., two canonical ensembles on the same configuration space), given a finite sample of each ensemble. Section II of this paper derives the "acceptance ratio estimator," a near-optimal solution of this estimation problem, based only on the data in the two ensemble samples (a special case is the estimation of the free energy difference between two ensembles using data from only one of them; however, it will be argued that it is usually preferable to gather data from both ensembles). The efficiency of the acceptance ratio estimator is proportional to the degree of overlap between the two ensembles.

Section III presents a related method, the "interpolation method," which yields an improved free energy estimate under an additional assumption that is often physically justified, namely, that the density of states in each ensemble is a smooth function of the difference potential. When this assumption is justified, the interpolation method can yield a good free energy estimate even when the overlap between the two ensembles is negligible. On the other hand, when the two ensembles neither overlap nor satisfy this smoothness assumption, no method of statistical analysis can yield a good estimate of the free energy difference, and one must collect additional MC data from one or more ensembles intermediate between the two originally considered.

Section IV compares the present methods with older methods of MC free energy estimation, viz, numerical integration of a derivative of the free energy, perturbation theory, and previous overlap methods. This section also discusses some of the problems of designing and sampling intermediate ensembles.

II. The Acceptance Ratio Method

IIa. *Acceptance Probabilities and Configurational Integrals*

In this section the acceptance ratio method, to be developed more rigorously in Sections IIb and IIc, will be discussed from a physical and qualitative point of view.

In most classical systems of interest the kinetic part of the canonical partition function is trivially calculable; hence the problem of finding the free energy of a given (N, T, V) macrostate reduces to that of evaluating the canonical configurational integral

$$Q = \int \exp[-U(q_1 \cdots q_N)]\, dq_1 \cdots dq_N . \tag{1}$$

Here $U = \Phi/\mathrm{kT}$ is the temperature-scaled potential energy, a function of the system's N configurational degrees of freedom, $q_1, q_2, \ldots, q_N$. It is convenient to allow U sometimes to take on the "value" plus infinity (but never minus infinity) so that external constraints, such as those that define the system's volume and shape, may be incorporated directly in the potential function and Q may be defined, as above, by an unbounded integral. With these conventions *any* nonsingular probability density $\rho(\mathbf{q})$ may be viewed as a canonical ensemble density, determined by a potential of the form $U(\mathbf{q}) = \text{const} - \ln \rho(\mathbf{q})$.

Existing methods are incapable, in general, of evaluating integrals of the form (1), because the dominant contribution typically comes from a small but intricately shaped portion of configuration space; however it is not difficult to derive useful formulas for the *ratio* between two such integrals, defined by two *different* potential functions, U_0 and U_1, acting on the *same* configuration space $\{(q_1 \cdots q_N)\}$. Equation (4), for example, to be derived presently, expresses the ratio Q_0/Q_1 as a ratio of canonical averages involving the "Metropolis" function, $M(x) = \min\{1, \exp(-x)\}$. The Metropolis function, because it has the property $M(x)/M(-x) = \exp(-x)$, is used in the standard Monte Carlo algorithm [1, 2] to assign Boltzmann-weighted acceptance probabilities to trial moves, a move that would change the (temperature-scaled) energy by ΔU being accepted with probability $M(\Delta U)$. Here, however, we consider an unorthodox kind of trial move—one that keeps the same configuration $(q_1 \cdots q_N)$, but switches the potential function from U_0 to U_1 or vice-versa. For each configuration, the acceptance probabilities for such a pair of trial moves must satisfy the relation

$$M(U_1 - U_0)\exp(-U_0) = M(U_0 - U_1)\exp(-U_1). \tag{2}$$

Integrating this identity over all of configuration space and multiplying by the trivial factors Q_0/Q_0 and Q_1/Q_1, one obtains:

$$Q_0 \frac{\int M(U_1 - U_0)\exp(-U_0)\,dq_1 \cdots dq_N}{Q_0}$$

$$= Q_1 \frac{\int M(U_0 - U_1)\exp(-U_1)\,dq_1 \cdots dq_N}{Q_1}. \tag{3}$$

The quotients on both sides can be recognized as canonical averages, i.e., quantities that can be measured during ordinary MC runs on systems 0 and 1 respectively. Representing these averages by the conventional angle brackets, one obtains the desired result:

$$\frac{Q_0}{Q_1} = \frac{\langle M(U_0 - U_1)\rangle_1}{\langle M(U_1 - U_0)\rangle_0}. \tag{4}$$

Notes on p. 146

The physical meaning of this formula is that a Monte Carlo calculation that included potential-switching trial moves (in a fixed ratio to ordinary, configuration-changing trial moves) would distribute configurations between the unknown U_1 and the reference U_0 system in the ratio of their configurational integrals. The potential-switching moves need not actually be carried out, however, since the desired ratio can be estimated more accurately simply by taking the indicated averages over separately-generated samples of the U_0 and U_1 ensembles.

Before proceeding further, a few general remarks on the scope and limitations of the acceptance ratio method are in order. The requirement that the two systems be defined by potentials acting on the same configuration space is not a serious limitation, since, for most pairs of macrostates one might care to compare, a rather trivial transformation of the coordinates (e.g., a dilation or shear) suffices to make the two configuration spaces congruent. It is possible even to compare systems with a different number of degrees of freedom (as in the MC simulation of a grand canonical ensemble); the lower-order system is simply given one or more dummy coordinates, whose contribution to Q can later be factored out and computed analytically.

Most special ensembles used in Monte-Carlo work can be expressed as canonical ensembles by appropriate definition of the potential function U. The (N, T, P) ensemble, for example, can be represented [2] by making the volume a coordinate and the pressure a parameter of U. Importance-weighted ensembles [6, 7] can be viewed as canonical ensembles defined by U functions containing additive terms designed to concentrate the probability density in desired portions of configuration space.

The only important practical limitation on the method is that *both* mean acceptance probabilities (i.e., both averages in Eq. (4)) must be large enough to be determined with reasonable statistical accuracy in a Monte-Carlo run of reasonable duration. If only one of the acceptance probabilities is too small, it can be increased, at the expense of the other, by shifting the origin of one of the potential functions by an additive constant. Simultaneous smallness of both probabilities indicates that there is insufficient overlap between the U_0 and U_1 ensembles, and, in order to obtain a good estimate of Q_1, one must either:

1. Find a new reference potential which exhibits greater overlap with U_1.

2. Perform additional MC calculations under one or more intermediate potentials, so as to form an overlapping chain between U_0 and U_1.

3. Use curve-fitting methods, to be discussed in Section III, to interpolate between the U_0 and U_1 ensembles, thereby obtaining a good estimate of the free energy difference in spite of the lack of overlap.

Although Eq. (4) is not strictly correct for the (N, E, V) or $(N, E, V$, linear momentum) ensembles sampled by molecular dynamics, in practice it often can be used with molecular dynamics data, owing to the close similarity (except near phase transitions) of the configurational distributions in the various ensembles, for systems having more than a few degrees of freedom. Since temperature is not an independent variable in a constant-energy ensemble, the temperatures used in defining the temperature-scaled potentials U_0 and U_1 would have to be taken from time averages of the kinetic energy.

An exact microcanonical analog of Eq. (4) exists for the somewhat special case of two systems at the same energy E whose Hamiltonians, H_0 and H_1, have equal "soft" parts, i.e., H_0 and H_1 are equal wherever neither is infinite. For such a pair of systems the ratio of the microcanonical phase integrals is given by

$$\frac{\int \delta(H_0 - E)\, dq^N\, dp^N}{\int \delta(H_1 - E)\, dq^N\, dp^N} = \exp(S_0 - S_1)/k = \frac{[M(H_0 - H_1)]_1}{[M(H_1 - H_0)]_0}, \tag{5}$$

with square brackets here denoting microcanonical phase averages. The numerator and denominator of Eq. (5) have a very simple interpretation: e.g., $[M(H_1 - H_0)]_0$ is the fraction of points on the H_0 energy surface that also lie on the H_1 energy surface. The formulation of a more general microcanonical analog of Eq. (4) is frustrated by the fact that, for a general pair of Hamiltonians, H_0 and H_1, the two energy surfaces would have an intersection of zero measure.

Returning to the canonical ensemble, it may be noted that Eq. (4) is not the most general formula for Q_0/Q_1 as a ratio of canonical averages. A more general formula results if one includes in both the numerator and denominator an arbitrary weighting function. Let $W(q_1 \cdots q_N)$ be any everywhere-finite function of the coordinates. It then follows easily that

$$\frac{Q_0}{Q_1} = \frac{Q_0 \int W \exp(-U_0 - U_1)\, dq^N}{Q_1 \int W \exp(-U_1 - U_0)\, dq^N} = \frac{\langle W \exp(-U_0)\rangle_1}{\langle W \exp(-U_1)\rangle_0}. \tag{6}$$

Note that configurations having infinite energy under either U_0 or U_1 or both make no contribution to Eq. (6) so long as W is finite; henceforth, W will by convention be set equal to zero for all such configurations.

Most previous direct or overlap methods for estimating free energy (to be reviewed in Section IV) can be viewed as special cases of Eq. (6), with particular forms of the potentials U_0 and U_1 and the weight function W. Equation (4), for example, corresponds to the choice $W = \exp(+\min\{U_0, U_1\})$. The next section (IIb) shows, by some rather lengthy statistical arguments, that the optimized estimator of Q_0/Q_1 as a ratio of canonical averages differs from Eq. (4) in two respects: (1) the Fermi function, $f(x) = 1/(1 + \exp(+x))$, is used instead of the Metropolis function; and (2) the origin of one of the potential functions is shifted so as to (roughly) equalize the two acceptance probabilities.

It is also shown that, when one is free to vary the amount of computer time spent sampling the two ensembles, roughly equal time should be devoted to each.

IIb. *Optimized Acceptance Ratio Estimator—Large Sample Regime*

Optimization of the free energy estimate is most easily carried out in the limit of large sample sizes. Let the available data consist of n_0 statistically independent configurations from the U_0 ensemble and n_1 from the U_1 ensemble, and let this data be used in Eq. (6) to obtain a finite-sample estimate of the reduced free energy difference $\Delta A = A_1 - A_0 = \ln(Q_0/Q_1)$. For sufficiently large sample sizes the error of this estimate will be nearly Gaussian, and its expected squre will be

$$\text{Expectation of } (\Delta A\text{est} - \Delta A)^2$$

$$\approx \frac{\langle W^2 \exp(-2U_1)\rangle_0}{n_0[\langle W \exp(-U_1)\rangle_0]^2} + \frac{\langle W^2 \exp(-2U_0)\rangle_1}{n_1[\langle W \exp(-U_0)\rangle_1]^2} - \frac{1}{n_0} - \frac{1}{n_1}$$

$$= \frac{\int ((Q_0/n_0) \exp(-U_1) + (Q_1/n_1) \exp(-U_0))\, W^2 \exp(-U_0 - U_1)\, dq^N}{[\int W \exp(-U_0 - U_1)\, dq^N]^2}$$

$$- (1/n_0) - (1/n_1). \tag{7}$$

By making the integral in the numerator stationary with respect to a variation of W at constant value of the integral in the denominator, the optimum W function is found:

$$W(q_1 \cdots q_N) = \text{const} \times \left(\frac{Q_0}{n_0} \exp(-U_1) + \frac{Q_1}{n_1} \exp(-U_0)\right)^{-1}. \tag{8}$$

Substituting this into Eq. (6) yields

$$\frac{Q_0}{Q_1} = \frac{\langle f(U_0 - U_1 + C)\rangle_1}{\langle f(U_1 - U_0 - C)\rangle_0} \exp(+C), \tag{9a}$$

where

$$C = \ln \frac{Q_0 n_1}{Q_1 n_0} \tag{9b}$$

and f denotes the Fermi function $f(x) = 1/(1 + \exp(+x))$. Equation (9a) is true for any value of the shift constant C, but the particular value specified by Eq. (9b) minimizes the expected square error (Eq. (7)) when the canonical averages are evaluated by finite sample means, with sample sizes n_0 and n_1.

The magnitude, σ^2, of this minimum square error can be found by taking the variance of Eq. (9a), or by substituting Eq. (8) into (7); σ^2 can be conveniently expressed in terms of n_0, n_1, and the normalized configuration-space density functions ρ_0 and ρ_1:

$$\sigma^2 = \frac{\langle f^2\rangle_0 - \langle f\rangle_0^2}{n_0\langle f\rangle_0^2} + \frac{\langle f^2\rangle_1 - \langle f\rangle_1^2}{n_0\langle f\rangle_0^2} \tag{10a}$$

$$= \left(\int \frac{n_0 n_1 \rho_0 \rho_1}{n_0\rho_0 + n_1\rho_1}\, dq^N\right)^{-1} - \frac{n_0 + n_1}{n_0 n_1}. \tag{10b}$$

In Eq. (10a) the argument of f is understood to be $(U_1 - U_0 - C)$ or $(U_0 - U_1 + C)$ in the 0 and 1 expectations, respectively, with $C = \ln(Q_0 n_1/Q_1 n_0)$ as specified by Eq. (9b). In Eq. (10b), $\rho(q_1 \cdots q_N)$ denotes the density $(1/Q)\exp[-U(q_1 \cdots q_N)]$. Since σ^2 is a monotonically decreasing function of both n_0 and n_1, it follows that for some $\bar{n}$, lying between n_0 and n_1,

$$\sigma^2 = \frac{2}{\bar{n}}\left[\left(\int \frac{2\rho_0\rho_1}{\rho_0 + \rho_1}\, dq^N\right)^{-1} - 1\right]. \tag{11}$$

The integral in this equation is clearly a measure of the "overlap" between the two densities in configuration space. Equation (11) thus says that Q_0/Q_1 can be determined accurately as a ratio of canonical averages by (and only by) sampling a number of configurations greater than the reciprocal of the overlap between ρ_0 and ρ_1.

The optimized formula for Q_0/Q_1 (Eq. (9a)) differs from that derived earlier (Eq. (4)) only in the use of the Fermi function in place of the Metropolis function, and in the shifting of the origin of one of the potentials by an additive constant C. Figure 1 shows both the Fermi and Metropolis functions along with a typical probability density for values of their argument x, the change in energy accompanying a potential-switching move (i.e., $x = U_0 - U_1 + C$ under U_1, and

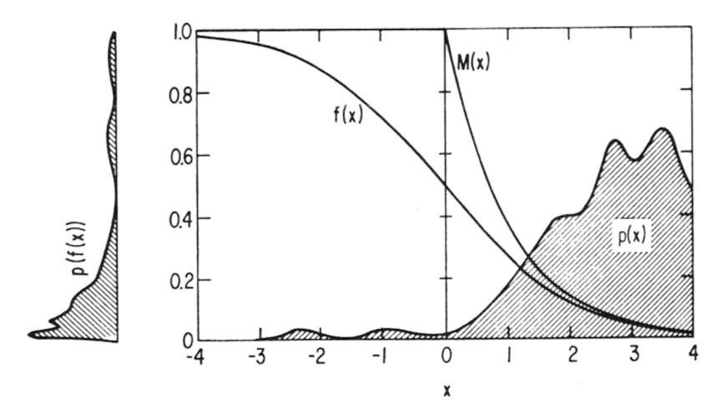

FIG. 1. The Fermi function, $f(x) = 1/[1 + \exp(+x)]$, and the Metropolis function, $M(x) = \min\{1, \exp(-x)\}$, are shown along with a typical probability density, $p(x)$, for their argument. Plotted on the left is a typical probability density for *values* of the Fermi function.

$U_1 - U_0 - C$ under U_0). When the shift constant is properly chosen (Eq. (9b)), most potential-switching moves (like most trial moves in an ordinary MC calculation) will result in an increase in energy and hence will lie on the positive "tail" of the f (or M) function. The advantage of the Fermi function in estimating free energy differences lies in its having a softer shoulder than the Metropolis function. This narrows the distribution of acceptance probabilities $p(f(x))$, and makes possible a more accurate estimation of the ensemble average acceptance probability from a given body of data. The Metropolis function, on the other hand, is the better acceptance function to use in the ordinary MC algorithm for *generating* new configurations, because here one seeks to maximize the acceptance probability itself, without regard to its variance. The shifting of the energy origin by C serves to maximize the number of configurations falling near the soft shoulder of the f function, while minimizing the number falling far out on its tail. It should perhaps be pointed out that in the special case of two potentials whose soft parts are identical, Eq. (9a) becomes equivalent to Eq. (4) and yields no better estimate of Q_0/Q_1.

In practice, of course, one cannot determine the optimum shift constant C exactly, because it depends on the unknown quantity Q_0/Q_1; however a value sufficiently close to the optimum can be found by adjusting C until Eqs. (9a) and (9b) become self-consistent for the given body of data. This estimation procedure can be expressed conveniently as a pair of simultaneous equations in $\Delta A\text{est}$ and C:

$$\Delta A\text{est} = \ln \frac{\sum_1 \{f(U_0 - U_1 + C)\}}{\sum_0 \{f(U_1 - U_0 - C)\}} + C - \ln(n_1/n_0) \tag{12a}$$

$$\Delta A\text{est} = C - \ln(n_1/n_0). \tag{12b}$$

Equation (12a) (the finite-sample analog of Eq. (9a)) estimates the free energy difference in terms of explicit sums over the n_1 configurations comprising the U_1 ensemble sample and the n_0 comprising the U_0 ensemble sample; Eq. (12b) (or equivalently $\sum_1 = \sum_0$) is the self-consistency criterion for selecting C.

The large-sample regime assumed in Eqs. (9)–(12) may now be expressed as a condition on the sums $\sum_0$ and $\sum_1$: namely, that for some range of C-values about the true C of Eq. (9b), these sums differ relatively little from their respective expectations, $n_0\langle\ \rangle_0$ and $n_1\langle\ \rangle_1$. Under this condition the self-consistent procedure yields essentially an optimum estimate of ΔA, differing from the true value by a quantity of order σ. This follows from the fact that the first term on the right-side of Eq. (12a) is a monotonically decreasing function of C with a slope of nearly -1 and a value, for the correct C of Eq. (9b), within about σ of zero. One may be sure of being in the large-sample regime whenever both $\sum_1$ and $\sum_0$ are large compared to unity, because the terms comprising the sums are statistically independent and all lie between zero and one (the large sample condition is thus equivalent to the condition discussed in connection with Eqs. (10) and (11), viz, n_0 and n_1 must be great enough to adequately sample the region of overlap between ρ_0 and ρ_1).

Figure 2 shows a representative graphical solution of Eqs. (12a) and (12b) for four sets of simulated Monte-Carlo data drawn from the same pair of ensembles (the sample sizes $n_0 = n_1 = 10^6$, lay in the large sample regime, with $\Sigma_0 \approx \Sigma_1 \approx 200$). Note that the straight line of Eq. (12b) cuts through a region where the four curves of (12a) differ least from each other and from the true value of ΔA. In the large sample regime the standard error of the estimate ΔA est will be less than ± 1, and can be computed in the usual manner by solving Eqs. (12a) and (12b) for several large, independent bodies of data, as was done in Fig. 2.

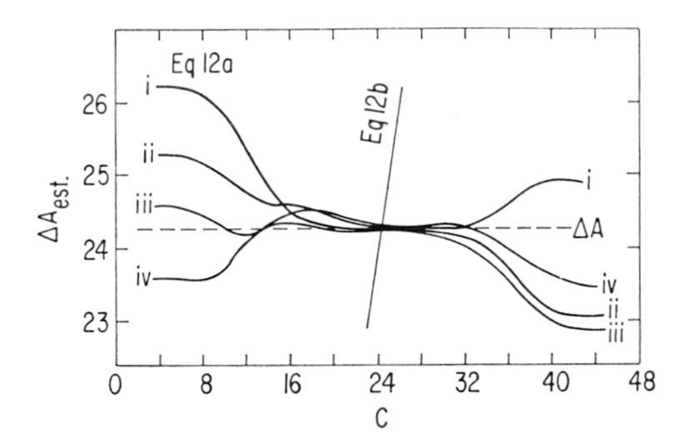

FIG. 2. Acceptance ratio estimate of the free energy difference between two MC ensembles in the large-sample regime by simultaneous graphical solution of Eqs. (12a) and (12b). For a description of the two ensembles, and the method of sampling them, see the Appendix. The four curves (i–iv) plot the right side of Eq. (12a) as a function of the shift constant C for four independent sets of data, each consisting of 10^6 points randomly chosen from the 0 ensemble and 10^6 from the 1 ensemble. The slanting straight line is Eq. (12b), while the dashed horizontal line gives the true free energy difference ΔA. The mean and standard error of the four finite-sample estimates are 24.290 ± 0.017; these are in satisfactory agreement with the true free energy difference, 24.268, and the error $\sigma = \pm 0.021$ predicted by Eq. (10b) for sample sizes $n_0 = n_1 = 4 \times 10^6$.

Equations (12a) and (12b) represent an estimation strategy optimized with respect to a given pair of large samples, with fixed sizes n_0 and n_1. We now consider the allocation of computer time *between* the two ensembles if the sample sizes are not fixed beforehand, but are free to be chosen so as to minimize σ^2 with respect to n_1/n_0 at a constant total cost in computer time. Let us assume that the time required to compute each (statistically-independent) data point is $\not{c}_0$ in the 0 ensemble and $\not{c}_1$ in the 1 ensemble, so that the total computing time is $n_0\not{c}_0 + n_1\not{c}_1$. A crude but effective rule for choosing n_1/n_0 is simply to allocate equal time to the two ensembles, i.e.,

$$n_1/n_0 = \not{c}_0/\not{c}_1 . \tag{13}$$

Notes on p. 146

The estimation efficiency, $1/[(n_0 \not{c}_0 + n_1 \not{c}_1)\,\sigma^2]$, resulting from this equal-time allocation is at least half as great as that resulting from any other allocation. This follows from the fact that σ^2 is a monotonically decreasing function of both n_0 and n_1; (i.e., even the best allocation of 1 hour of computer time *between* the two ensembles yields no better estimate than would obtained by devoting a full hour to *each* ensemble). In the special case $\not{c}_0 = \not{c}_1 = 1$, the equal-time estimation efficiency is approximately one fourth the overlap integral (cf. Eq. (11)).

The equal time rule gives a sufficiently good n_1/n_0 ratio for most practical situations; however, for the sake of elegance, the true optimum ratio can be expressed in terms of the variance of the Fermi functions by solving the variational equation

$$\not{c}_1(d\sigma^2/dn_0) = \not{c}_0(d\sigma^2/dn_1). \tag{14}$$

Explicitly differentiating Eq. (10a) with respect to n_0 and n_1, one obtains

$$-\frac{\not{c}_1(\langle f^2\rangle_0 - \langle f\rangle_0^2)}{n_0^{\;2}\langle f\rangle_0^2} = -\frac{\not{c}_0(\langle f^2\rangle_1 - \langle f\rangle_1^2)}{n_1^{\;2}\langle f\rangle_1^2}, \tag{15}$$

with the argument of f understood to be $U_1 - U_0 - C$ in the 0 expectations and $U_0 - U_1 + C$ in the 1 expectations. One might worry about the implicit n-dependence that the various Fermi expectations have by virtue of the n-dependent shift constant C, defined in Eq. (9b). However, these implicit n-dependences have no effect on the *derivatives* of σ^2, because Eq. (9b) is itself the solution to the variational condition $\partial\sigma^2/\partial C = 0$. Equations (9a) and (9b) also cause the denominators on the two sides of Eq. (15) to be equal. Thus Eq. (15) reduces to an equation in one unknown, defining an optimum value for the shift constant C:

$$\not{c}_1 \,\mathrm{Var}_0[f(U_1 - U_0 - C)] = \not{c}_0 \,\mathrm{Var}_1[f(U_0 - U_1 + C)], \tag{16}$$

with Var_0 and Var_1 denoting the absolute variances. These variances, of course, cannot be estimated with much precision from finite samples of the ensembles, but by adjusting the sample sizes until Eqs. (12a) and (12b) can be solved self-consistently for the same value of C as Eq. (16), one might obtain some improvement over the equal time strategy, particularly in cases where the optimum time ratio is far from 1: 1.

As noted earlier, the efficiency of estimating ΔA is at least half-optimal, and not very sensitive to the n_1/n_0 ratio, in the neighborhood of $n_1/n_0 = \not{c}_0/\not{c}_1$ (the equal time rule). This can be seen in Fig. 3, which plots the log estimation efficiency versus the log sampling ratio for the pair of model ensembles considered earlier. Although the optimum n_1/n_0 ratio is 1.8, the estimation efficiency at $n_1/n_0 = 1$ is almost (99.7 %) as good. (At first it might appear that by making the costs $\not{c}_0$ and $\not{c}_1$ very disparate, the optimum time ratio could be displaced far from unity;

however this is not so. If, for example $¢_1/¢_0$ is changed from 1 to 10^{-4}, the optimum n_1/n_0 ratio is indeed greatly increased as one would expect; but the optimum *time* ratio, $n_1¢_1/n_0¢_0$, changes only from 1.8 to 1.3)

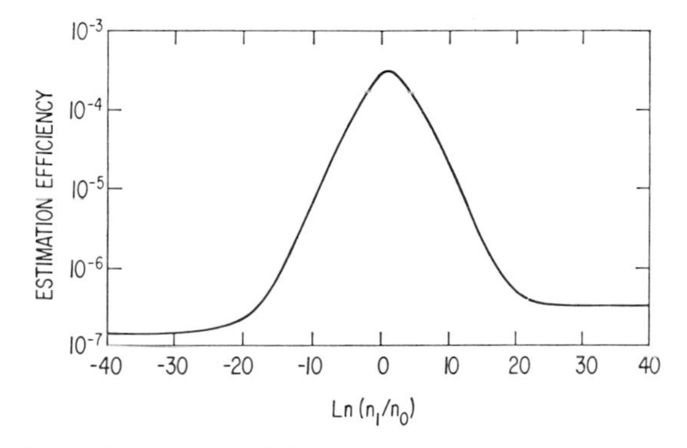

FIG. 3. Dependence of estimation efficiency, $1/[(n_0 + n_1)\sigma^2]$, on the n_1/n_0 ratio for the model ensembles described in the Appendix, assuming equal sampling costs ($¢_1 = ¢_0$) in the two ensembles. The horizontal wings of the curve indicate the rather poor efficiency with which A_1-A_0 can be estimated when only one ensemble is sampled, as in infinite-order perturbation theory. The efficiency curve was calculated by exact evaluation of Eq. (10b) over the rather trivial configuration space of the model ensembles.

Figure 3 also shows that the estimation efficiency can become very bad if one flouts the equal time rule by sampling only one ensemble. The poor efficiency results from the fact that, when only one ensemble is sampled (say the 0 ensemble), the acceptance ratio estimator reduces to the average of a pure exponential (cf. Eq. (22)), whose variance can be expressed in terms of the densities ρ_0 and ρ_1 as $[\int (\rho_1^2/\rho_0)\, dq^N - 1]/n_0$. This expression is less transparent than the formula (Eq. (11)) relating the variance of the two-ensemble estimate to the overlap integral; however, its qualitative meaning is that an accurate one-ensemble estimate requires that the sampled ensemble include *all* important configurations of the other ensemble. A good two-ensemble estimate, on the other hand, requires only that each ensemble include *some* important configurations of the other ensemble.

IIc. *Acceptance Ratio Estimates in the Small Sample Regime*

The treatment so far has been limited to the large sample regime, in which both sums in Eq. (12a) can be made simultaneously greater than unity. Unfortunately, in many cases of interest, the overlap between the two ensembles is so slight that even with the largest practical n_0 and n_1 this condition cannot be met. We shall now show that even in this small sample regime Eq. (12a) can yield a useful estimate of ΔA, though the error bounds will be greater than ± 1 and can no longer be estimated from the spread among independent estimates. When either sum in (12a) (say $\sum_1$) is small compared to unity, its most serious source of statistical error becomes the possibility that some important class of configurations, whose total ensemble probability is low ($1/n_1$ or less) but whose f values are high (near unity at worst), may not be sampled at all. Such failures to sample could cause either sum to *underestimate* its expectation by a quantity of order unity (corresponding errors due to *over*-sampling of high-f configurations could also occur, but, owing to the convexity of the log function, their effect on Eq. (12a) would be much less). In order to find bounds on the possible failure-to-sample errors, and hence on ΔA, we take advantage of the fact that $\sum_1$ is a monotonically decreasing function of C, while $\sum_0$ is monotonically increasing. Therefore, by decreasing C to that value, C_1, for which $\sum_1$ becomes equal to unity, we can make all the failure-to-sample errors appear in the denominator of Eq. (12a) and obtain a value, $\Delta A\text{est}+$, which may overestimate, but is unlikely to seriously under-

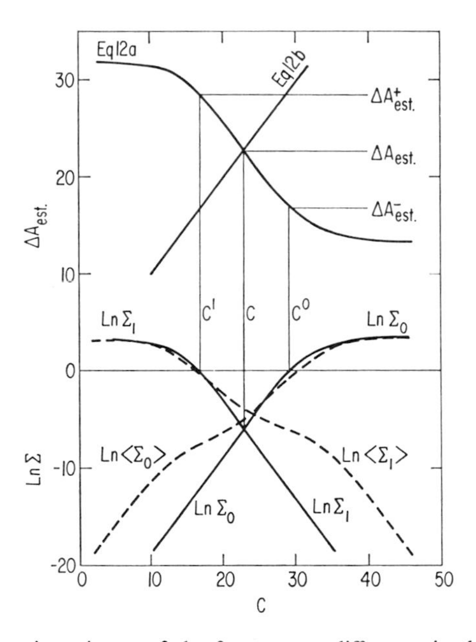

FIG. 4. Acceptance ratio estimate of the free energy difference in the small-sample regime. The upper pair of curves show the construction of the optimum estimate $\Delta A\text{est}$ by graphical solution of Eqs. (12a) and (12b), from two small samples ($n_0 \approx n_1 \approx 20$) of the pair of ensembles described in the Appendix. The lower pair of solid curves show the construction of the upper and lower estimates, $\Delta A\text{est}^+$ and $\Delta A\text{est}-$, by Eqs. (17a) and (17b). The dashed curves show the log expectations, $\ln \langle \Sigma_0 \rangle = \ln[n_0 \langle f(U_1 - U_0 - C)\rangle_0]$ and $\ln\langle \Sigma_1 \rangle = \ln[n_1 \langle f(U_0 - U_1 + C)\rangle_1]$, which are closely approximated by the observed log sums as long as the latter are greater than unity, but are poorly approximated at C values for which the log sums are less than unity.

estimate, the true free energy difference ΔA. The construction of this upper estimate and the corresponding lower estimate, $\Delta A\text{est}-$, is shown in Fig. 4. The computation of error bounds in the small sample regime may be summarized:

$$\Delta A\text{est}- = \text{R12a}(C_0) \lesssim \Delta A \lesssim \text{R12a}(C_1) = \Delta A\text{est}+, \tag{17a}$$

where R12a(C) denotes the right side of Eq. (12a), and C_0 and C_1 are defined by

$$\sum_0 \{f(U_1 - U_0 - C_0)\} = 1 = \sum_1 \{f(U_0 - U_1 + C_1)\}. \tag{17b}$$

Parenthetically it is interesting to note that the right side of Eq. (12a), which would be independent of C in the large-sample limit, here decreases with a slope of about -1 throughout the range where both sums are much smaller than unity. This is necessarily so because, under these conditions, neither sum can receive contributions except from the nearly exponential positive tail of the f function. The self-consistent estimate ΔAest obtained by solving Eq. (12a) with (12b) therefore lies about midway between the upper and lower bounds computed by Eq. (17). It is still the best estimate of ΔA in the sense that it equalizes the damage that would be done by a unit failure-to-sample error in the numerator or the denominator of Eq. (12a).

IId. *Practical Considerations in Using Acceptance Ratio Methods*

In using Eqs. (12) and (17) with real MC data account must be taken of the fact that successively generated configurations of a Markov chain are not statistically independent, but on the contrary highly correlated. Each of the sums, $\sum_0$ and $\sum_1$, and the numbers n_0 and n_1, must therefore refer to a subset of configurations, chosen from the chain so infrequently as to be uncorrelated, or else be defined in terms of the whole chain as follows:

$$\sum = \tau^{-1} \sum_{\text{wc}}; \qquad n = \tau^{-1} n_{\text{wc}}. \tag{18}$$

Here $\sum_{\text{wc}}$ denotes a sum over the whole chain, having n_{wc} configurations, and τ is an empirically estimated autocorrelation time of the Markov chain with respect to values of the f function (This autocorrelation time can be defined as the large k limit of the quantity $k \cdot \text{Var}[f^{(k)}]/\text{Var}[f]$, where $f^{(k)}$ denotes the mean of k consecutive f values. Clearly τ can be estimated accurately only if it is considerably shorter than the total chain length n_{wc}). The cost, ¢, of a statistically independent data point, discussed in connection with Eqs. (13)–(16), would be similarly defined as τ times the computer time required to make one MC move.

Another practical note: although the evaluation of the sums in Eq. (12a) could in principle be done after the MC data (typically a Markov chain of several million configurations) had been generated, it is inconvenient to store all this data. A better approach would be during the run to accumulate values of $\sum_0$ and $\sum_1$, using a mesh of C values sufficiently fine to permit accurate graphical solution of Eqs. (12a) and (12b) at the end of the run. Alternatively, one could store a pair of histograms, $h_0(\Delta U)$ and $h_1(\Delta U)$, of the values of the difference potential $\Delta U = U_1 - U_0$, observed while sampling the 0 and 1 ensembles, respectively. The interval width of the histograms need be no smaller than the desired precision of estimating ΔA, and they can be summed over easily at the end of the runs to evaluate Eq. (12a). Such histograms are also useful in their own right, in the interpolative method for estimating ΔA to be discussed in the next section.

The acceptance ratio method is at its best when the overlap between the two ensembles, as defined by the integral in Eq. (11), is not too small, e.g., in solid state vacancy calculations [8, 10] where the difference potential, $U_1 - U_0$, depends strongly on only a few atomic coordinates. It is more common for the difference potential to depend strongly on all the coordinates, resulting in an overlap many orders of magnitude less than unity. The two data histograms, $h_1(\Delta U)$ and $h_0(\Delta U)$ will then be separated by a wide gap (cf. Fig. 5) that cannot be filled in by any reasonable amount of additional sampling of either ensemble, and the acceptance ratio method (Eq. (17) in particular) will yield only the rather crude conclusion that the true free energy difference ΔA is somewhere between $\max\{\Delta U\}_1$ and $\min\{\Delta U\}_0$.

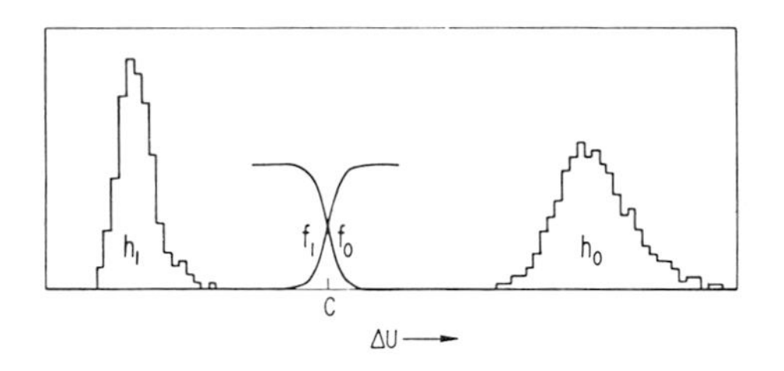

FIG. 5. Histograms of values of $\Delta U = U_1 - U_0$, h_0 representing a typical set of ΔU values sampled from the 0 ensemble, and h_1 a typical set sampled from the 1 ensemble. In the gap between the two histograms are a pair of complementary Fermi functions $f_0 = f(\Delta U - C)$ and $f_1 = f(C - \Delta U)$ whose origin can be shifted to the left or right, but whose widths are insufficient to achieve good overlap of f_0 with h_0, and f_1 with h_1, simultaneously.

The estimate of ΔA can be considerably improved if, as is often the case, the two histograms are sufficiently smooth to justify extrapolating them into the gap region of the ΔU spectrum, from which no data have been collected. The following section will deal with this method of estimating ΔA, which is no longer a pure acceptance ratio method, because it is based on the additional assumption that the ΔU spectrum is smooth even where no data have been collected.

When the smoothness assumption is not justified, i.e., when the data histograms are ragged as well as widely separated, an improved estimate of ΔA can be obtained by using the acceptance ratio method in a multistage manner, with data collected from a chain of intermediate ensembles extending from U_0 to U_1.

III. Interpolation or Curve-Fitting Method

This section concerns the estimation of ΔA from histograms of ΔU values, $h_0(\Delta U)$ and $h_1(\Delta U)$, which are smooth but may be separated by a gap wide compared to kT. This situation is likely to arise when the difference potential depends strongly but more or less equally on many coordinates, e.g., when U_0 and U_1 represent a condensed system of many identical atoms interacting via one pair potential in the U_0 system and another in the U_1 system. If one is willing to infer from thc histograms' smoothness that the reduced density-of-states functions $p_0(\Delta U)$ and $p_1(\Delta U)$, which the histograms approximate, extend smoothly into the gap region, one can obtain a much better estimate of ΔA than could be obtained from the acceptance ratio method alone.

This approach is less risky than it might first appear, and in fact is more akin to interpolation than to extrapolation, because p_0 and p_1 are not independent:

$$\frac{p_1(x)}{p_0(x)} \equiv \frac{\langle\delta(U_1 - U_0 - x)\rangle_1}{\langle\delta(U_1 - U_0 - x)\rangle_0} = \exp(\Delta A - x); \tag{19}$$

thus it is a matter of finding a single function (p_0, say) which fits both histograms, while satisfying the normalization constraints

$$\int_{-\infty}^{\infty} p_0(\Delta U)\, d\,\Delta U = 1, \tag{20a}$$

and

$$\int_{-\infty}^{\infty} p_1(\Delta U)\, d\,\Delta U \equiv \int_{-\infty}^{\infty} p_0(\Delta U)\exp(\Delta A - \Delta U)\, d\,\Delta U = 1. \tag{20b}$$

This can be done conveniently by expanding $\ln p_0$ as a polynomial in ΔU and performing a least-squares fit of the expansion coefficients, along with ΔA, to the histogram data, subject to the normalization constraints. The adequacy of the polynomial approximation, as well as the range of plausible ΔA values, can be judged by chi-square tests.

Estimation of ΔA also can be performed graphically (cf. Fig. 6), by plotting the two functions

$$-\tfrac{1}{2}\Delta U \qquad +\ln p_0\,, \tag{21a}$$

and

$$+\tfrac{1}{2}\Delta U \qquad +\ln p_1 \tag{21b}$$

versus ΔU on the same graph. Each function is plotted (solid curves) over the range of ΔU values for which it can be accurately estimated from the histogram data. By virtue of Eq. (19), the two functions are parallel, differing only by the unknown additive constant ΔA; hence, in order to estimate ΔA, one need only find a plausible parallel extrapolation (dashed curves) of the two functions into the range of ΔU values corresponding to the gap between the two histograms. Probably the easiest way to do this is to cut the graph in half vertically and slide the right half up or down until it can be smoothly joined onto the left half by some plausible extrapolation. The range of vertical shifts for which this can be done is then the range of plausible ΔA values. In Fig. 6 this range can be seen to be about $\Delta A = -85.5 \pm 2.5$, which is considerably narrower than the gap between the two histograms.

Clearly the interpolation method is at its best when the data histograms are smooth and significantly broader than kT (kT = 1 in the reduced units of ΔU),

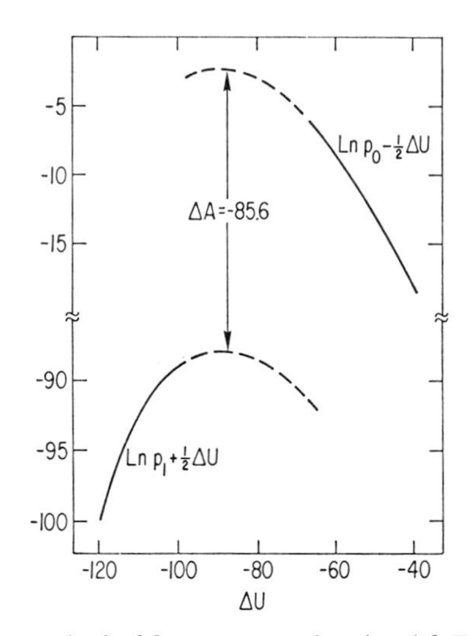

FIG. 6. The curve-fitting method of free energy estimation (cf. Eqs. (21a) and (21b)), applied to data of Valleau and Card [17]. The U_1 system here consisted of 200 charged (100+ and 100−) hard spheres at a fixed temperature and volume; the U_0 system was the same, except that the spheres were uncharged. The left and right solid curves in the present figure were obtained, respectively, from the left and right probability density curves of Valleau and Card [17, Fig. 2] (omitting the tails where the density was less than one tenth maximum) by: scaling the energy to units of kT, taking the logarithm, and adding $+\frac{1}{2}\Delta U$ and $-\frac{1}{2}\Delta U$, respectively. Their figure also includes a middle curve, corresponding to an intermediate ensemble which they used to bridge the gap between the two outer curves. With the help of this intermediate ensemble, they estimated ΔA to be -85.60 ± 0.58 (cf. [17, Table I]. As the present figure suggests, the outer curves are smooth enough to yield a fairly good estimate of ΔA (about -85.5 ± 2.5) by interpolation across the gap, without help from the intermediate ensemble.

and separated by a gap not much broader than the histograms themselves. Under these conditions much information about the shape of p_0 and p_1 in the overlap region can be inferred from data points lying far outside that region, data points which contribute hardly anything to the acceptance probabilities on which the method of the previous section is based. From this it can be seen that when two smooth histograms overlap very slightly (i.e., overlap integral in Eq. (11) of order $1/\bar{n}$), the interpolation method will still be an improvement over the straight acceptance ratio estimator. However, the more the two histograms overlap, the less important it becomes to guess the shape of the reduced density-of-states functions, p_0 and p_1, and, in the large-overlap limit (overlap integral ≈ 1), interpolation does not improve the estimate at all.

On the other hand, when the gap between h_0 and h_1 is excessively wide compared to the width of the histograms themselves, accurate interpolation becomes difficult, and it is best to use the interpolation method in a multistage manner, using MC data collected under one or more intermediate potentials, e.g., $U_\lambda = U_0 + \lambda(U_1 - U_0)$; $0 < \lambda < 1$. This linear form of the intermediate potential is convenient because it allows all the $\ln p_\lambda(\Delta U)$ data to be plotted on the same graph and fitted to the same polynomial, but it may be inferior to the more general intermediate potentials discussed in Section IVc.

IV. Discussion

In this section the acceptance ratio and interpolation methods of Sections II and III are compared with other methods of free energy estimation, viz, perturbation theory, numerical integration of a derivative of the free energy, and previous overlap methods.

IVa. *Perturbation Theory*

Perturbation theory [11, 12], which estimates the free energy of the U_1 system by extrapolation from the U_0 system, can be viewed as the limiting case of the acceptance ratio method in the absence of any data from the U_1 system. In this limit (i.e., $n_1 \to 0$) Eqs. (9a) and (9b) reduce to

$$A_1 - A_0 = -\ln\langle \exp(U_0 - U_1)\rangle_0, \tag{22}$$

an infinite-order perturbation formula [13] which is exact, provided there are no configurations for which U_0 is infinite but U_1 is finite. To obtain finite-order formulas, one assumes the potential U to depend on a continuous parameter

Notes on p. 146

λ in such a way that as λ is varied from 0 to 1, U passes smoothly from U_0 to U_1. The potential U_1 in Eq. (22) then can be expanded in a Taylor series about U_0:

$$\begin{aligned} A_1 - A_0 &= -\ln\langle \exp[-(\partial U/\partial\lambda) - (\partial^2 U/\partial\lambda^2)/2 - \cdots]\rangle_0 \\ &= \langle(\partial U/\partial\lambda)\rangle_0 \\ &\quad + \tfrac{1}{2}[\langle\partial^2 U/\partial\lambda^2\rangle_0 - \langle(\partial U/\partial\lambda)^2\rangle_0 + (\langle\partial U/\partial\lambda\rangle_0)^2] \\ &\quad + \cdots \\ &= \partial A/\partial\lambda + \tfrac{1}{2}\,\partial^2 A/\partial\lambda^2 + \cdots, \end{aligned} \tag{23}$$

where all the derivatives with respect to λ are evaluated in the U_0 ensemble, at $\lambda = 0$. The expansion is usually truncated at second order because the statistical uncertainty in measuring the higher derivatives, by a MC run of reasonable duration, is typically so great that they contribute only noise to the infinite-order formula (Eq. (22)). The simplest perturbation formula results by taking $U_\lambda = U_0 + \lambda\,\Delta U$, in which case $\partial^2 U/\partial\lambda^2 = 0$ and $\partial U/\partial\lambda = \Delta U$; Eq. (23) then expresses ΔA in terms of the mean and moments of ΔU, as measured in the reference system.

Alternatively, the acceptance ratio and curve fitting methods of Sections II and III may be viewed as double-ended, interpolative counterparts of ordinary extrapolative perturbation methods of infinite (Eq. (22)) and finite (Eq. 23)) order, respectively. The assumption of smoothness of the density-of-states function, on which the curve-fitting method is based, is then a less restrictive counterpart of the assumption that higher-order terms in Eq. (23) are negligible. Double-ended methods have the advantage of being able to set both upper and lower bounds on the free energy difference $A_1 - A_0$. The crudest of these bounds is the so-called Gibbs-Bogoliubov inequality [14],

$$\langle U_1 - U_0\rangle_1 \leqslant A_1 - A_0 \leqslant \langle U_1 - U_0\rangle_0\,, \tag{24}$$

which follows, via the convexity of the log function, from Eq. (22) and its analogue with U_0 and U_1 interchanged; more subtle bounds, e.g., Eq. (17), are discussed in Sections II and III.

In an ordinary single-ended perturbation treatment, on the other hand, no data is collected from the U_1 ensemble and only the right half of Eq. (24) can be used. One therefore cannot rule out the possibility of seriously overestimating $A_1 - A_0$ due to a failure to sample important configurations in the U_1 ensemble. Assurance that this kind of error has not occurred must come from specific knowledge of the potentials, or from independent confirmation of the properties of the U_1 system. The notably successful perturbation theory of liquids [12, 15, 16], whose reference system is the hard sphere fluid, was confirmed in this manner.

The perturbation theory of liquids also illustrates the chief strength of single-ended perturbation methods, namely, the possibility of using a single reference system to compute the properties of many different U_1 systems, without having to collect MC data on each of these separately. It should be noted, however, that when there is any doubt about the rapid convergence of an extrapolative perturbation from U_0 data, the estimate of $A_1 - A_0$ can be considerably improved, usually without much cost in computer time, by collecting a small amount of MC data on the U_1 system, then using a double-ended method. Indeed, whenever the second-order perturbation term differs significantly from zero, the two ensembles being compared are probably sufficiently different to warrant sampling both of them. The gain in estimation efficiency made possible by sampling both ensembles instead of only one is suggested in Fig. 3.

IVb. *Numerical Integration*

This method estimates the free energy difference $A_1 - A_0$ by numerically integrating the derivative $\partial A/\partial\lambda = \langle \partial U/\partial\lambda \rangle$, which is measured by equilibrium MC calculations at a mesh of values of the parameter λ between 0 and 1. The most commonly performed integration is of pressure versus volume [4, 5]; however, the method can be used to compute the free energy change attending any continuous deformation [8] of the potential, boundary conditions, or other parameters defining the MC macrostate. The integration method as ordinarily practiced is less than optimal because it ignores the information which each MC run provides about higher derivatives of the free energy (e.g., the isothermal compressibility); however, this information may be of poor statistical quality. Ideally each estimated derivative of A, at each mesh point, should be given a weight inversely proportional to its estimated standard error. Probably the easiest way to achieve this correct weighting of information is to use the acceptance ratio or interpolation methods in a multistage manner, to estimate $\int (\partial A/\partial\lambda)\, d\lambda$ between consecutive mesh points.

When information on higher derivatives is included it is clear that perturbation theory and the interpolation method of Section III are special cases of integration, with one or two mesh points, respectively. The number of mesh points actually needed depends on the smoothness of $\partial A/\partial\lambda$ as a function of λ, and on the ease and precision with which the derivatives can be estimated at each mesh point. In typical applications, where five to ten mesh points are used, numerical integration can determine $A_1 - A_0$ when the unknown and reference ensembles are too different to be compared by any method not using intermediate ensembles. On the other hand, when the 0 and 1 systems are similar enough to be compared directly, the generation of many intermediate ensembles, each of which must be allowed to equilibrate before representative data can be collected, is tedious and may be wasteful of computer time.

Notes on p. 146

IVc. *Previous Overlap Methods*

Most previous overlap methods for determining free energy differences can be regarded as special cases of the acceptance ratio method, with particular forms of the potentials U_0 and U_1 and of the weight function W in Eq. (6). Perhaps the most common special case is the comparison of a pair of systems, one of which is restricted to a subset of the configurations accessible to the other. In other words, the difference potential $\Delta U = U_1 - U_0$ is a hard function, taking on only the values zero and plus infinity. With such a pair of potentials, one of the acceptance probabilities in Eq. (4) becomes identically unity, while the other is simply the fraction of microstates in the less restricted ensemble belonging to the more restricted ensemble. In what was probably the first application of an overlap method to a realistic system, McDonald and Singer [9] sampled a nested set of ensembles, defined by a decreasing sequence of upper bounds on the total energy of a gaseous Lennard–Jones system, and obtained the unnormalized density of states as a function of energy over a wide range of energies, from which they were able to compute the thermodynamic properties of the gas over a wide range of temperature.

Such nested sets of hard constraints can be used to restrict a MC system to any desired region of configuration space, and to determine the spontaneous probability of occupancy of that region in the absence of the constraints. In a molecular dynamics calculation, analogous constraint terms in the Hamiltonian can be used to sample trajectories passing through an arbitrary region of phase space, and to estimate the spontaneous frequency of such passages in an unconstrained system [10].

Apparently the first overlap calculation of a free energy difference between two systems whose potentials had differing "soft" parts was that of Valleau and Card [17]. These authors, interested in determining the thermodynamic properties of a fluid of charged hard spheres as a function of temperature, compared systems whose U differed by a constant factor, i.e., systems having the same unscaled potential but different temperatures. Their somewhat complicated procedure for for estimating ΔA (cf. [17, Appendix]) can be recognized as accomplishing the same result, with somewhat less statistical efficiency, as the Fermi-function weighting used in the acceptance ratio method of Section II. By emphasizing the importance of the density-of-states functions, $p_0(\Delta U)$ and $p_1(\Delta U)$, these authors adumbrated the interpolation method of Section III.

In the same paper Valleau and Card pointed out the possibility of using a specially tailored bridging ensemble, designed to have significant overlap with both the unknown and reference ensembles, in place of the many intermediate ensembles ordinarily used in the numerical integration method. In principle it is always possible to define such a bridging ensemble, no matter how different U_0 and U_1 may be. This may be done, for example, by defining the bridging potential

U_B, as an appropriately weighted log mean exponential of a sequence of overlapping potentials U_λ, extending between U_0 and U_1,

$$U_B = -\ln \sum_{\lambda=0}^{1} w_\lambda \exp(-U_\lambda), \tag{25}$$

the discrete weights w_λ being chosen to approximate $\exp(A_\lambda)$. By using a continuous weight function $w(\lambda)$ one can even define a bridging ensemble whose density of states is perfectly flat over the entire relevant interval of the ΔU spectrum, but to guess such a weight function would be tantamount to guessing the function $p_0(\Delta U)$ over the same interval. Torrie, Valleau, and Bain [6] used discretely weighted bridging ensembles to pass between an unconstrained hard sphere fluid and a single occupancy fluid (no two particles allowed in the same Wigner–Seitz cell) in a few sampling stages, thereby avoiding the long pressure–volume integration by which the communal entropy is usually estimated. Unfortunately, the bridging ensembles exhibited such long autocorrelation times (cf. Eq. (18)) that the hoped-for gain in computational efficiency was not realized. The bridging system, in other words, diffused much too slowly between the part of configuration space overlapping with U_0 and the part overlapping with U_1. In later work, Torrie and Valleau [7] obtained much better performance using continuously weighted bridging ensembles to estimate the free energy of a 32 particle Lennard–Jones fluid relative to that of the corresponding purely repulsive inverse twelfth power system. The difference in diffusion rates obtained in these two experiments probably is due in part to the superiority of a continuous weighting, but may also reflect the many-body character of the structural rearrangements involved in accommodating to the single-occupancy constraint. In difficult cases like the communal entropy calculation, some improvement in diffusion through a given bridging ensemble might be obtained by modifying the MC transition algorithm used to sample it (infinitely many transition algorithms, with different rates of diffusion through configuration space, can be used to sample the same canonical ensemble). So far almost all MC calculations have used simple transition algorithms in which only one particle is moved at a time, and trial moves are symmetrically distributed in direction. More efficient diffusion in the desired direction might be obtained by making trial moves preferentially parallel and antiparallel to the local gradient of the difference potential.

Overlap methods using a bridging potential of the form of Eq. (25) bear a certain similarity to numerical integration. Under a bridging potential, the system diffuses freely back and forth between the U_0 and U_1 parts of configuration space; in numerical integration, a series of intermediate MC runs is made, and the system is forced to diffuse, by the increment in the integration parameter, as each successive run equilibrates. Given a particular MC transition algorithm, which limits the

Notes on p. 146

diffusion rate, numerical integration and bridging ensembles may be equally efficient statistically, if the higher-derivative information provided by the integration runs is taken into account.

V. Conclusion

The problem of free energy estimation can be broken into three parts:

1. what reference and (possibly) intermediate ensembles to use;
2. what MC transition algorithms to use for sampling the ensembles; and
3. how best to estimate the free energy from the resulting data.

The acceptance ratio and interpolation methods (developed in Sections II and III, respectively) offer a fairly complete solution to the third subproblem, viz, optimally estimating the free energy difference between two canonical ensembles given a finite MC sample of each (or, more generally, given MC routines able to sample each at some fixed cost per statistically independent data point). A good estimate can be arrived at if the ensembles being compared

1. exhibit significant overlap, allowing the acceptance ratio method of Section II to be used; or

2. are sufficiently similar that the density of states in each ensemble is a smooth function of $\Delta U = U_1 - U_0$, allowing the interpolation method of Section III to be used.

In either case it is dangerous not to sample both ensembles, unless one is known to include all important configurations of the other. When the two ensembles neither overlap nor satisfy the above smoothness condition, an accurate estimate of the free energy cannot be made without gathering additional MC data from one or more intermediate ensembles.

The first two subproblems are much less well understood than the third. The choice of a reference ensemble is more a matter of physics than of statistics, and it is probably best made on an individual, empirical basis. On the other hand the problems that arise in designing efficient bridging ensembles and transition algorithms to sample them appears to be the manifestation, in Monte Carlo work, of a general difficulty in the numerical simulation of systems with many degrees of freedom—the problem of moving efficiently through a complicated, labyrinthine configuration space. The problem arises whether one wishes to study the system dynamically (where it makes itself felt as a disparity of time scale between the phenomena of interest and the time step needed to integrate the equations of motion [18]), or statistically (as in Monte Carlo work), or merely by seeking the global energy minimum in a space filled with steep-sided curving valleys, saddle points, and spurious local minima [19, 20]. Judging from results in these other fields, the problem may be partly alleviated by transition algorithms that make intelligent use of local anisotropy information, but some complicated systems will remain intrinsically sluggish and hard to simulate.

APPENDIX: MODEL ENSEMBLES

The acceptance ratio estimators of Sections IIb and IIc were tested using a pair of simple model ensembles, defined on a discrete configuration space having only 23 states. Because of the trivial configuration space the ensembles could be sampled by a simple Poisson routine (rather than by a Markov chain) and the free energy difference, Fermi expectations, and all other quantities of interest could be calculated exactly, for convenient comparison with the estimates under study. These estimates were obtained by applying the estimators (e.g., Eqs. (12a) and 12b)) to finite random samples of the two ensembles. The defining properties of the ensembles are given in Table I.

TABLE I

State	ΔU	$-\ln p_0$	$-\ln p_1$	State	ΔU	$-\ln p_0$	$-\ln p_1$
a	2	30.352	8.084	*m*	26	7.352	9.084
b	4	26.352	6.084	*n*	28	5.352	9.084
c	6	22.352	4.084	*o*	30	4.352	10.084
d	8	18.352	2.084	*p*	32	2.352	10.084
e	10	15.352	1.084	*q*	34	1.352	11.084
f	12	13.352	1.084	*r*	36	1.352	13.084
g	14	12.352	2.084	*s*	38	1.352	15.084
h	16	11.352	3.084	*t*	40	2.352	18.084
i	18	11.352	5.084	*u*	42	4.352	22.084
j	20	10.352	6.084	*v*	44	6.352	26.084
k	22	9.352	7.084	*w*	46	8.352	30.084
l	24	8.352	8.084				

$A_1 - A_0$ is 24.268; overlap (integral in Eq. (11)) is 1.2×10^{-3}.

ACKNOWLEDGMENTS

I first learned of overlap methods, and of the statistical difficulty of estimating free energies from machine data, while working with Berni Alder and Mary Ann Mansigh in 1966–70. Discussions with Aneesur Rahman, John Barker, and Betty Flehinger, and particularly reading Valleau and Card's 1972 paper, stimulated my thinking.

Notes on p. 146

REFERENCES

1. M. METROPOLIS, A. W. ROSENBLUTH, M. N. ROSENBLUTH, A. H. TELLER, AND E. TELLER, *J. Chem. Phys.* **21** (1953), 1087.
2. W. W. WOOD, *in* "The Physics of Simple Liquids," (H. N. V. Temperley, J. S. Rowlinson, and G. S. Rushbrooke, Eds.), North-Holland, Amsterdam, 1968.
3. B. J. ALDER AND T. E. WAINWRIGHT, *J. Chem. Phys.* **31** (1959), 459; **33** (1960), 1439.
4. W. G. HOOVER AND F. H. REE, *J. Chem. Phys.* **47** (1967), 4873; **49** (1968), 3609.
5. J.-P. HANSEN AND L. VERLET, *Phys. Rev.* **184** (1969), 151.
6. G. TORRIE, J. P. VALLEAU, AND A. BAIN, *J. Chem. Phys.* **58** (1973), 5479.
7. G. TORRIE AND J. P. VALLEAU, *Chem. Phys. Lett.* **28** (1974), 578; *J. Comput. Phys.*, to be published.
8. D. R. SQUIRE AND W. G. HOOVER, *J. Chem. Phys.* **50** (1969), 701.
9. I. R. MCDONALD AND K. SINGER, *J. Chem. Phys.* **47** (1967), 4766.
10. C. H. BENNETT, *in* "Diffusion in Solids: Recent Developments" edited (A. S. Nowick and J. J. Burton, Eds.), Academic Press, New York, 1975.
11. R. W. ZWANZIG, *J. Chem. Phys.* **22** (1954), 1420.
12. W. R. SMITH, *in* "Statistical Mechanics 1" (K. Singer, Ed.), Chem. Soc. Publ., London, 1973.
13. Z. W. SALSBURG, J. D. JACOBSON, W. FICKETT, AND W. W. WOOD, *J. Chem. Phys.* **30** (1959), 65.
14. W. R. SMITH, *in* "Statistical Mechanics 1" (K. Singer, Ed.), pp. 98–99. Chem. Soc. Publ., London, 1973.
15. J. A. BARKER AND D. HENDERSON, *J. Chem. Phys.* **47** (1967), 2856 and 4714.
16. D. LEVESQUE AND L. VERLET, *Phys. Rev.* **182** (1969), 307.
17. J. P. VALLEAU AND D. N. CARD, *J. Chem. Phys.* **57** (1972), 5457.
18. C. H. BENNETT, *J. Comput. Phys.* **19** (1975), 267.
19. R. FLETCHER AND M. J. D. POWELL, *Comput. J.* **6** (1963), 163.
20. M. LEVITT AND A. WARSHEL, *Nature* **253** (1975), 694.

Notes to Reprint II.6

1. Examples of the use of the techniques described in this paper can be found in (Clark and Lal, J. Phys. A 11 (1978) L11; Quirke and Jacucci, Mol. Phys. 45 (1982) 823; Rahman and Jacucci, Nuovo Cimento D 4 (1984) 357).

2. The extension of the methods of this paper to molecular dynamics simulations is discussed in the article by Frenkel referred to in Note II.5.6.

JOURNAL OF COMPUTATIONAL PHYSICS 23, 187–199 (1977)

Nonphysical Sampling Distributions in Monte Carlo Free-Energy Estimation: Umbrella Sampling

G. M. TORRIE AND J. P. VALLEAU

Lash Miller Chemical Laboratories, University of Toronto, Toronto, Ontario, Canada

Received May 7, 1976; revised June 16, 1976

The free energy difference between a model system and some reference system can easily be written as an ensemble average, but the conventional Monte Carlo methods of obtaining such averages are inadequate for the free-energy case. That is because the Boltzmann-weighted sampling distribution ordinarily used is extremely inefficient for the purpose. This paper describes the use of arbitrary sampling distributions chosen to facilitate such estimates. The methods have been tested successfully on the Lennard–Jones system over a wide range of temperature and density, including the gas–liquid coexistence region, and are found to be extremely powerful and economical.

1. INTRODUCTION

Use of the Monte Carlo method of Metropolis *et al.* [1] to estimate averages for model systems is nowadays a relatively routine matter. This is suitable for mechanical properties such as the pressure or internal energy. On the other hand, statistical properties such as the entropy and free energy, because they cannot be expressed as ensemble averages, have not been so easily accessible. The conventional technique has been numerical integration, following the Monte Carlo determination of some derivative of the free energy at a series of state points connecting the state or system of interest to one with a known free energy. This somewhat cumbersome method is least efficient or altogether unworkable when the system undergoes a phase transition, because of the difficulty of defining a path of integration on which the necessary ensemble averages can be reliably measured, though it is in precisely such cases that free-energy estimates would be most useful. Recently [2] we described a generalization of the method of Valleau and co-workers [3–5] for measuring free-energy differences which overcomes such difficulties for the case of the liquid–gas transition of a Lennard–Jones fluid. The free-energy difference between the Lennard–Jones fluid and a soft-sphere fluid was determined at a series of densities on an isotherm below the Lennard–Jones critical temperature by sampling on an arbitrary distribution designed to explore in a single Monte Carlo experiment the parts of configuration space relevant both to the Lennard–Jones fluid and to the reference soft-sphere fluid. In this paper we demonstrate the feasibility of extending such techniques to explore systematically large regions of a phase diagram, applying them to the Lennard–Jones system in a wide range of temperature and pressure including part of the gas–liquid coexistence region.

ISSN 0021–9991

2. Outline of the Method

The free-energy difference between the "system of interest," with internal energy $U(\mathbf{q}^N)$ at temperature T, and a reference system, with internal energy $U_0(\mathbf{q}^N)$ at temperature T_0, is easily expressed as an ensemble average

$$\frac{A}{kT} - \frac{A_0}{kT_0} = -\ln \frac{\int \exp[(-U/kT) + (U_0/kT_0)] \exp(-U_0/kT_0)\, d\mathbf{q}^N}{\int \exp(-U_0/kT_0)\, d\mathbf{q}^N}$$

$$= -\ln \left\langle \exp\left(-\frac{U}{kT} + \frac{U_0}{kT_0}\right)\right\rangle_0$$

$$= -\ln\langle \exp(-\Delta U^*)\rangle_0, \tag{1}$$

where $\langle\ \rangle_0$ denotes an average over a canonical ensemble of reference systems and U^* is the reduced energy U/kT. Here we will briefly review the qualitative features of the Monte Carlo sampling schemes used to obtain an accurate estimate of the right-hand side of (1); a more complete and formal description may be found in [2].

In practice it is more useful to regard the average in (1) as a one-dimensional integral over ΔU^*, i.e.,

$$\frac{A}{kT} - \frac{A_0}{kT_0} = -\ln \int_{-\infty}^{\infty} f_0(\Delta U^*) \exp(-\Delta U^*) d\,\Delta U^*, \tag{2}$$

where $f_0(\Delta U^*)$ is the probability density of ΔU^* in the reference fluid, and therefore in a conventional Monte Carlo experiment on that fluid.

In order to determine accurately the right-hand side of (2) such a Monte Carlo experiment would evidently have to produce good estimates of the values of $f_0(\Delta U^*)$ for that range of ΔU^* over which the product $f_0(\Delta U^*) \exp(-\Delta U^*)$ takes on its largest values. The corresponding region of configuration space is in fact that which would normally be sampled by a conventional Monte Carlo experiment *not* on the reference system but on the "system of interest" itself. This is easy to see for, if $f(\Delta U^*)$ is the probability density of ΔU^* in such an experiment,

$$f(\Delta U^*) = f_0(\Delta U^*) \exp(-\Delta U^*)\, Q_0/Q, \tag{3}$$

where Q and Q_0 are the configurational integrals of the "system of interest" and the reference system, respectively. This is of limited usefulness since, without knowledge of the free-energy difference being sought, Q_0/Q is unknown, and the measurement of $f(\Delta U^*)$ (along with Eq. (3)) can give only *relative* values of $f_0(\Delta U^*)$ in the region where its *absolute* value is required. On the other hand, the range of ΔU^* over which a conventional *reference-system* experiment yields *absolute* values of $f_0(\Delta U^*)$ will not be adequate to evaluate the right-hand side of Eq. (2). Whether or not a reference system could be chosen so that the ranges of ΔU^* sampled by the two experiments overlapped, thus allowing a proper normalization of $f_0(\Delta U^*)$ throughout [3, 6], it is clear from the sharply peaked distributions of ΔU^* obtained in such experiments that conventional Boltzmann sampling is *not* an efficient way to explore the relevant part of configuration space.

Instead a Markov chain may be generated having a limiting distribution $\pi(\mathbf{q}'^N)$ which differs from the Boltzmann distribution for either system. We write it for convenience in the form

$$\pi(\mathbf{q}'^N) = \frac{w(\mathbf{q}'^N)\exp(-U_0(\mathbf{q}'^N)/kT_0)}{\int w(\mathbf{q}^N)\exp(-U_0(\mathbf{q}^N)/kT_0)\,d\mathbf{q}^N}, \tag{4}$$

where $w(\mathbf{q}^N) = W(\Delta U^*)$ is a weighting function chosen to favor those configurations with values of ΔU^* important to the integral in (2). Provided that $W(\Delta U^*)$ is such that the resulting Monte Carlo experiment continues as well to sample adequately those parts of configuration space that would be sampled by a Boltzmann-weighted experiment on the reference system, the unbiased ensemble average of any function $\theta(\mathbf{q}^N)$ can be recovered from the results of the π-sampling experiment according to

$$\begin{aligned}\langle\theta\rangle_0 &= \frac{\int (\theta/w)w\exp(-U_0/kT_0)\,d\mathbf{q}^N}{\int (1/w)w\exp(-U_0/kT_0)\,d\mathbf{q}^N}\\ &= \frac{\langle\theta/w\rangle_w}{\langle 1/w\rangle_w},\end{aligned} \tag{5}$$

where $\langle\ \rangle_w$ denotes an average over the distribution (4). Similarly $f_0(\Delta U^*)$ can be recovered from $f_w(\Delta U^*)$, the probability density of ΔU^* in the "biased" Monte Carlo experiment based on (4),

$$f_0(\Delta U^*) = \frac{f_w(\Delta U^*)/W(\Delta U^*)}{\langle 1/w\rangle_w}. \tag{6}$$

By trial and error $W(\Delta U^*)$ is adjusted until $f_w(\Delta U^*)$ is as wide and uniform as possible; the more rapidly varying $f_0(\Delta U^*)$ is then determined over this same wide range using (6).

Obviously it is required that the sampling distribution π specified by W should cover simultaneously the regions of configuration space relevant to two or more physical systems. We call this "umbrella sampling."

In the calculations described in the following sections weighting functions were used which brought about sampling of a range of ΔU^* up to three times that of a conventional Monte Carlo experiment, allowing accurate determination of values of $f_0(\Delta U^*)$ as small as 10^{-8}. Where this type of gain is still not sufficient to sample the entire range of ΔU^* values additional (equally powerful) "umbrella-sampling" experiments can be carried out with different weighting functions exploring successively overlapping ranges of ΔU^*. (Satisfactory weighting functions $W(\Delta U^*)$ are easy to find for such ranges of ΔU^*, and convergence of the runs was rapid. Evidently one could in principle always use only a *single* Monte Carlo run, choosing the sampling distribution to cover the *whole* of the relevant part of configuration space. However, for very wide-ranging distributions the choice of a successful $W(\Delta U^*)$ becomes tedious; we found it more convenient to use more modest overlapping umbrella distributions as described.)

To make maximum use of the information thus gained on $f_0(\Delta U^*)$ over a wide range of ΔU^*, it often proves useful to consider scaling the reduced energy difference

Note on p. 159

between the two systems by a strength parameter α. The information obtained on $f_0(\Delta U^*)$ then suffices to estimate

$$\left(\frac{A}{kT}\right)_\alpha - \frac{A_0}{kT_0} = -\ln \int f_0(\Delta U^*) \exp(-\alpha\, \Delta U^*) d\, \Delta U^* \tag{7}$$

for all α between 0 and 1 simultaneously. The physical interpretation of these "intermediate" systems depends on the particular systems and energy difference under consideration and is more conveniently discussed in the context of the various types of calculations described in the following section.

3. Application to a Model System

In order to test the effectiveness of these methods we have applied them to the Lennard–Jones fluid. We were especially interested in using the techniques to investigate a phase transition region, since this is notoriously difficult using conventional methods. At the same time there are some reliable earlier results for this system, using the more cumbersome conventional techniques, and these afford direct tests of the method.

The investigations were carried out in two stages.

(a) *Altering the Force Law*

We first determined the free-energy difference between a Lennard–Jones fluid, with internal energy

$$U = 4\epsilon \sum_{i<j} [(\sigma/r_{ij})^{12} - (\sigma/r_{ij})^6], \tag{8}$$

and an inverse-twelve "soft-sphere" fluid [7] at the same temperature, with internal energy

$$U_0 = 4\epsilon \sum_{i<j} (\sigma/r_{ij})^{12}. \tag{9}$$

This was done at seven densities on the supercritical isotherm $kT/\epsilon = 2.74$. The relevant energy difference is then simply

$$\Delta U^* = (U_6/kT) = (-4\epsilon/kT) \sum_{i<j} (\sigma/r_{ij})^6 \tag{10}$$

so that

$$(\Delta A/NkT) = -(1/N) \ln\langle \exp(-U_a/kT)\rangle_0\,, \tag{11}$$

and the umbrella sampling (4) is carried out using a weighting function which favors soft-sphere configurations with large negative values of U_6. These calculations are similar in all respects to those previously reported [2] for a subcritical isotherm and will not be described in more detail here. We note, however, that at the higher densities on the supercritical isotherm the "soft-sphere" reference system and the Lennard–

Jones system have sufficiently similar configurations that a *single* umbrella-sampling experiment is powerful enough to determine ΔA. This is illustrated in Fig. 1 for the highest density studied. The solid line is $f_w(U_6)$, the probability density of U_6 that resulted from the umbrella sampling; the dotted line is $f_0(U_6)$, the unbiased probability of U_6 for a soft-sphere fluid as obtained by reweighting $f_w(U_6)$ according to Eq. (6); the broken line is the function $f_0(U_6)\exp(-U_6/kT)$ normalized to unity, also obtained by reweighting $f_w(U_6)$. The weighting function used to carry out this particular experi-

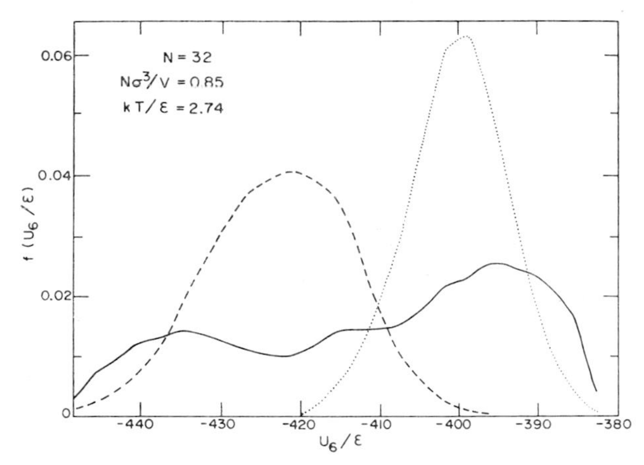

FIG. 1. Probability density functions for U_6 in a 32-particle soft-sphere fluid at $kT/\epsilon = 2.74$, $N\sigma^3/V = 0.85$. Solid line, f_w, the biased probability density. Dotted line, f_0, the unbiased probability density obtained by reweighting f_w. Broken line, relative values of $f_0(U_6/\epsilon)\exp(-U_6/kT)$ normalized to unity.

TABLE I

Numerical Weights Used for an Umbrella-Sampling Experiment for 32 Soft Spheres at $N\sigma^3/V = 0.85$, $kT/\epsilon = 2.74$[a]

U_6/ϵ	$W(U_6/\epsilon)$	U_6/ϵ	$W(U_6/\epsilon)$
$\leqslant$ −444.3	1,500,000	−411.5	2.60
−441.6	400,000	−408.7	1.75
−438.9	100,000	−406.0	1.25
−436.1	25,000	−403.3	1.15
−433.4	6,600	−400.5	1.00
−430.7	1,700	−397.8	1.24
−427.9	470	−395.0	1.70
−425.2	150	−392.3	2.50
−422.4	50	−389.9	4.30
−419.7	22	−386.8	8.50
−417.0	10	$\geqslant$ −384.1	16.00
−414.2	5		

[a] Linear interpolation was used to determine W for energies lying between table entries.

Note on p. 159

ment is shown in Table I. The free-energy differences and the resulting configurational free energy

$$A_c = -kT \ln(Q/N!\, \sigma^{3N}) \tag{12}$$

obtained by adding to ΔA the soft-sphere free energies of Hoover *et al.* [7] are shown in Table II.

TABLE II

Configurational Helmholtz Free Energy for the Lennard–Jones Fluid on the Isotherm $kT/\epsilon = 2.74$, Obtained from Eq. (11)

N	$N\sigma^3/V$	$(\Delta A/N\epsilon)$	$(A_c/N\epsilon)$
32	0.500	-5.90 ± 0.015	-4.45
108	0.500	-5.95 ± 0.01	-4.50
32	0.600	-7.62 ± 0.015	-3.64
32	0.700	-9.51 ± 0.015	-2.68
32	0.750	-10.54 ± 0.015	-2.11_5
32	0.800	-11.69 ± 0.015	-1.56_5
108	0.800	-11.68 ± 0.01	-1.557
32	0.835	-12.49 ± 0.015	-1.124
32	0.850	-12.85 ± 0.015	-0.890

For these calculations the "intermediate" systems of Eq. (7) would be ones in which the attractive part of the energy is scaled by the parameter α:

$$U_\alpha = 4\epsilon \sum_{i<j} \{(\sigma/r_{ij})^{12} - \alpha(\sigma/r_{ij})^6\}. \tag{13}$$

We have not reweighted the Monte Carlo data to derive any results for this somewhat unusual model, though such an approach could be used to obtain information on terms in a perturbation theory expansion of the Lennard–Jones free energy about that of the inverse-twelve soft-sphere system. A much more useful interpretation of this type of reweighting to intermediate systems can be made for the kind of calculations described in the following section.

(b) *Scaling the Temperature*

Whenever the reference system has the same internal energy function as the system of interest, Eq. (1) takes the simple form

$$\frac{A(T)}{kT} = \frac{A(T_0)}{kT_0} - \ln \left\langle \exp\left[-U\left(\frac{1}{kT} - \frac{1}{kT_0}\right)\right]\right\rangle_0, \tag{14}$$

and the most convenient way to write (2) is

$$\frac{A(T)}{kT} = \frac{A(T_0)}{kT_0} - \ln \int f_0(U) \exp\left[-U\left(\frac{1}{kT} - \frac{1}{kT_0}\right)\right] dU. \tag{15}$$

This is very powerful, because the "intermediate" systems that result from multiplying $-U$ in the exponent by a smaller number (cf. $\alpha < 1$ in (7)) can now be interpreted as those with temperatures *between* T and T_0. A single sampling of $f_0(U)$ can therefore give the free energy over a whole range of temperatures.

As reference systems we used the high-temperature Lennard–Jones systems at each of the seven densities in Table II. Umbrella sampling was used to measure $f_0(U)$ over a range extending to progressively lower energies, using additional sampling stages if necessary. For each density, the lowest energies so sampled will determine the lowest value of the temperature for which A can be reliably determined using (15). In addition to A, the average energy and specific heat can be calculated for any intermediate temperature, e.g.,

$$\langle U \rangle_T = \frac{\int U f_0(U) \exp\{-U[(1/kT) - (1/kT_0)]\}\, dU}{\int f_0(U) \exp\{-U[(1/kT) - (1/kT_0)]\}\, dU} \tag{16}$$

Mean values of quantities which are *not* functions of U can also be obtained in a similar way provided that $\bar{\theta}(U)$, the average value of θ for a fixed value of the energy is recorded for all U during the Monte Carlo run; then

$$\langle \theta \rangle_T = \frac{\int_{-\infty}^{\infty} f_0(U)\, \bar{\theta}(U) \exp\{-U[(1/kT) - (1/kT_0)]\}\, dU}{\int_{-\infty}^{\infty} f_0(U) \exp\{-U[(1/kT) - (1/kT_0)]\}\, dU} \tag{17}$$

Such ideas were first proposed and successfully carried out by McDonald and Singer [8, 9] who attempted relatively small temperature changes ($\leqslant 15$ %) in the data from a single Boltzmann-weighted experiment. They become powerful in the present context because of the much larger energy ranges that can be spanned by the nonphysical umbrella samples. The results of a typical two-stage experiment for 32 Lennard–Jones particles near the triple-point density are shown in Fig. 2. The solid lines show

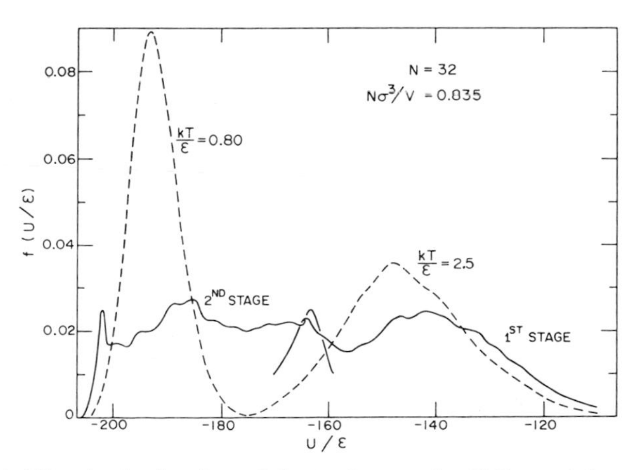

FIG. 2. Probability density functions of the total energy for 32 Lennard–Jones particles from umbrella sampling.

the energy distributions resulting from the two umbrella-sampling experiments while the broken lines are two examples of the reweighting of these data to obtain the energy distribution at various temperatures. This is typical of the possible gain—the energy and free energy of the system at fixed density in the temperature range $0.7 \leqslant kT/\epsilon \leqslant 2.8$, i.e., from the triple-point temperature to twice the critical temperature, are determined from only *two* umbrella-sampling experiments. For a system of 108 particles, where the Boltzmann distributions are much narrower relative to the total energy range, the same information can still be obtained with four umbrella experiments.

(c) *Corroboration of the Results*

In Table III some values of the free energy calculated from umbrella sampling are compared with previous Monte Carlo measurements by Hansen [10, 11] and by Levesque and Verlet [12], based on integration of pressures calculated for an 864-

TABLE III

Configurational Free Energy of the Lennard-Jones Fluid Obtained by Umbrella Sampling[a]

kT/ϵ	$N\sigma^3/V$	N	$A_c/N\epsilon$	Method[b]	Previous Monte Carlo results
2.74	0.80	32 108	−1.565 −1.556	I	−1.56[c]
1.35	0.50	32	−3.791	II	−3.85[d]
1.35	0.80	32 108	−3.236 −3.229	II	−3.25[d]
1.15	0.75	32	−3.633 −3.638	I II	−3.65[d]
0.75	0.50	32	−3.657	II	−3.69[e]
0.75	0.60	32	−3.913	II	−3.93[e]
0.75	0.80	32 108	−4.265 −4.265	II	−4.27[e]
1.16	0.835	32	−3.444	II	−3.83[f]
0.902	0.835	32	−3.966	II	−3.92[f]
0.81	0.835	32	−4.168	II	−4.10[f]

[a] The error usually quoted for the free energies of the final column obtained by thermodynamic integration is 0.01–$0.03N\epsilon$. The uncertainty in the present free energies includes this error for the reference system plus an uncertainty of $0.01N\epsilon$ in ΔA.

[b] Method I, calculated using Eq. (11); method II, calculated using Eq. (15).

[c] = [10].

[d] = [12].

[e] = [11].

[f] = [13].

particle system. The table also compares our results with estimates by Gosling and Singer [13] based on an intuitive "free-volume" interpretation of acceptance ratios in Monte Carlo runs. The quoted results of the umbrella-sampling methods are only a few values selected for these comparisons: data are of course available for the densities studied at *all* temperatures in the range $0.7 \leqslant kT/\epsilon \leqslant 2.8$.

Over the whole range of temperature and density the present results agree with those of the pressure integrations to within the combined statistical uncertainties. This is so whether A_c is determined by relating it directly to a soft-sphere reference system or by temperature reweighting to relate it to a high-temperature Lennard–Jones fluid using (15). At $kT/\epsilon = 1.15$, $N\sigma^3/V = 0.75$, both of these umbrella-sampling methods have been used independently, and agree excellently with each other and with Hansen's results.

Most of the calculations reported here are for a relatively small system of 32 particles. Consequently the overall agreement with the thermodynamic integration results for 864 particles seems to confirm our earlier conjecture [2] that the N-dependence of free-energy *differences* between dense systems is very mild. For example, calculations by the present methods for 32 and 108 particles at $N\sigma^3/V = 0.8$ (cf. Table III) show no statistically significant N-dependence. This is very pleasing since it means that good free-energy estimates can be made very economically, where there exist good data for a suitable reference system.

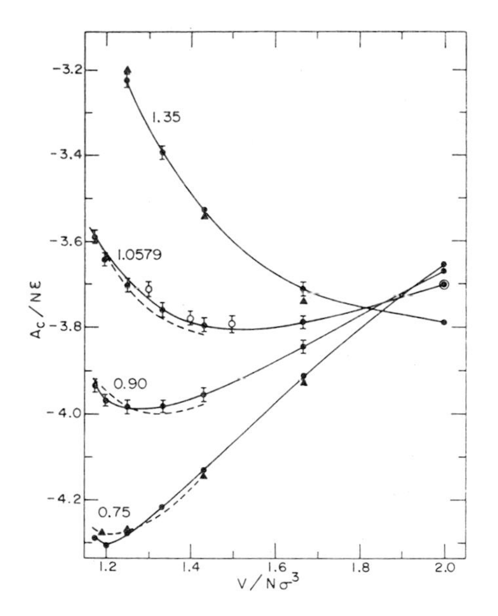

FIG. 3. Configurational Helmholtz free energy of the Lennard–Jones fluid in the vicinity of the liquid–gas coexistence curve. The open circles are free-energy estimates based on Eq. (11), the closed circles are based on Eq. (15). The triangles are the results of Verlet and co-workers [10–12], the broken curves the fitted equation of McDonald and Singer [14].

Note on p. 159

Figure 3 shows further comparisons between some free-energy results obtained by umbrella-sampling and those of thermodynamic integration results of the Orsay group [10–12] (triangles) and of McDonald and Singer [14] (broken lines [15]) in the vicinity of the gas–liquid coexistence curve. The open circles are the present results for direct soft-sphere to Lennard–Jones free-energy estimates using (11), and include some of the results we reported earlier [2], the closed circles are results of the temperature-reweighting procedure, Eq. (15). The solid lines are simply visual aids connecting isothermal free-energy estimates of this work, and the error bars show one standard deviation of the mean of ΔA. The agreement of the two umbrella-sampling routes to A_c on the isotherm $kT/\epsilon = 1.0579$ is particularly gratifying, as the original starting points were two quite different soft-sphere systems, one at $kT/\epsilon = 2.74$, the other at $kT/\epsilon = 1.0579$.

Although portions of the three low-temperature isotherms of Fig. 3 are well within the liquid–gas coexistence region and, when taken together with the low-density virial series, would clearly violate the stability criterion $(\partial^2 A/\partial V^2) \geqslant 0$, no associated convergence problems were encountered in the umbrella-sampling experiments. Apparently the supercritical Lennard–Jones fluid is as suitable as the soft-sphere fluid proved to be [2] for use as a reference state from which umbrella-sampling is able to explore successfully the unstable regime.

This close agreement between the present free energies and those obtained by thermodynamic integration is to be contrasted with some rather large discrepancies in the case of the free energies [16] resulting from the method for estimating entropies proposed by Gosling and Singer [13] (cf. Table III). For example, the disagreement of $0.4N\epsilon$ at $kT/\epsilon = 1.16$ corresponds to an error of 20 % in the nonideal entropy (the quantity actually estimated by their technique) and is equivalent to an overestimation of the configurational integral of a 108-particle system by a factor of 10^{16}. It appears that that technique cannot be relied upon to give quantitative results for the entropy and free energy.

TABLE IV

Unbiased Mean Internal Energy for the Lennard–Jones Fluid

$N\sigma^3/V$	kT/ϵ	$\langle U_{LJ}/N\epsilon\rangle_0$	
		Umbrella[a]	Conventional[b]
0.50	1.35	−3.32	−3.37 (a)
0.70	1.35	−4.66	−4.68 (a)
0.75	0.827	−5.38	−5.38 (b)
0.75	0.977	−5.26	−5.244 (c)
0.75	1.304	−5.01	−5.02 (b)
0.80	0.092	−5.65	−5.656 (c)
0.80	1.06	−5.51	−5.507 (c)

[a] Umbrella sampling: Obtained by reweighting the results of umbrella sampling using Eq. (16).
[b] Conventional: (a) Monte Carlo [12]; (b) Molecular dynamics [17]; (c) Monte Carlo [14].

Finally, in Table IV, the mean internal energy U of the Lennard–Jones fluid as determined by reweighting the results of umbrella-sampling experiments according to Eq. (16) is compared with conventionally determined values [12, 14, 17]. The generally good agreement is best for the higher densities, for which neglect of the density fluctuations caused by use of the small 32-particle system would be expected to have the least effect. The standard deviation of the mean for the energies obtained by reweighting data from umbrella-sampling Monte Carlo experiments, such as those in Table IV, is about $0.02N\epsilon$ for a run of $3\text{–}5 \times 10^5$ configurations, somewhat higher than for a Boltzmann-weighted experiment of similar length. These standard deviations are estimated in the customary way [18] by treating averages over sequences of m steps within the entire Monte Carlo run of M steps as M/m independent estimates to which the appropriate statistical formulas may be applied. The determinations of the value of m constituting an independent sample must be treated with more than usual care when recovering unbiased averages from umbrella-sampling experiments. In Eq. (5), for example, the values of the numerator and denominator will be strongly correlated for short blocks of configurations. As a result the mean value calculated as

$$\langle \theta \rangle_0 = \sum_{i=1}^{M/m} \langle \theta / w \rangle_i / \langle 1/w \rangle_i \tag{18}$$

will in general depend on m, where $\langle \ \rangle_i$ is an average over the ith block of m configurations. In practice the appropriate block size has been determined in each case by increasing m until no systematic trends could be detected in either the average given by (18) or in the standard deviation of the mean. The free-energy difference (2) is itself the *logarithm* of the average calculated, and can therefore be obtained with great precision, usually to within about $0.005NkT$ for $N = 108$. This high precision makes the technique ideally suited to the study of mixtures, where it is exactly the relatively small free-energy *differences* that are the physically important but computationally elusive property. In fact the techniques described here have now been exploited to determine the free energy and phase diagram of a model binary mixture with an upper critical solution temperature [19, 20].

4. Discussion

The basic idea of designing Monte Carlo experiments to sample the configuration space of the system according to an arbitrary distribution of states can be useful in several ways. First, such sampling procedures render estimates of free-energy differences of the form $\langle \exp(-\Delta(U/kT)) \rangle_0$ practical and efficient. This in turn eliminates the need of the more roundabout and expensive thermodynamic integrations, and is particularly advantageous when the system undergoes a phase transition, since it appears that by applying umbrella-sampling which embraces a stable reference system, information can be collected in or near the unstable regions of the phase diagram of the complete system without the usual associated convergence problems. Secondly, because umbrella distributions of the type described here span a much larger region of

Note on p. 159

configuration space than do Boltzmann-weighted Monte Carlo experiments on either the reference system or the system of interest, a correspondingly greater amount of information about the system can be obtained simply by collecting and later re-weighting the appropriate data. For example, Levesque and Verlet [12] used data from five Monte Carlo runs to establish a path of integration between a soft-sphere fluid and the Lennard–Jones system whereas a single umbrella-sampling suffices for 32 particles at high temperature. For a fluid of 32 Lennard–Jones particles at fixed density only two umbrella-sampling experiments are needed to sample the regions of configuration space appropriate to all temperatures between that of the triple point and twice that of the critical point, whereas numerous runs (of similar length) would be required for conventional $1/T$ integrations. The cost efficiency of the method is therefore high. Such sampling gains could likewise be realized for the other commonly used ensembles (isothermal–isobaric and grand canonical) with relatively straight-forward modifications of the techniques used here for the canonical case.

In fact, although the choice of ΔU as the single argument of the weighting function w defined in (4) was a natural one for the free-energy difference problem, the general case in which w is any arbitrary function of the coordinates can extend the usefulness of the biased sampling to a wider range of problems. For example, we have used such a sampling scheme to measure the communal entropy of a hard-sphere fluid [4], though the sampling distributions used did not turn out to be particularly efficient. The method has been quite successful, however, in Monte Carlo experiments designed to get particular microscopic information [21] (rather than merely macroscopic thermal averages) where the sampling problems are different in nature but similar in magnitude to those associated with measuring free-energy differences according to Eq. (1).

Superficially, the most serious limitation of the sampling techniques described here may appear to be the lack of a direct and straightforward way of determining the weighting function to use for a given problem. Instead, $w(\mathbf{q}^N)$ must be determined by a trial-and-error procedure for each case, often beginning with the information available from the distribution in a very short Boltzmann-weighted experiment which is then broadened in stages through subsequent short test runs with successively greater bias of the sampling. What this rather inelegant procedure lacks aesthetically is more than compensated by the efficiency of the ultimate umbrella-sampling experiment. The test runs require a small amount of time relative to the final production run and are necessary anyway, even in the absence of biasing weights, in order to age the system. One cannot expect to replace this trial-and-error procedure with a trustworthy a priori estimate of the correct value of $w(\mathbf{q}^N)$, not only because of the great variety of problems to which the technique might be applied, but because a guess sufficiently accurate to work would constitute prior knowledge of $f_0(\Delta U^*)$ and hence of ΔA itself. At present, the most efficient trial-and-error procedure involves an interaction between the trial computer results and human judgment. It is important to emphasize, however, that the necessary experience seems to be very readily obtained. A possible embellishment of the technique would be to program the computer to carry out itself the trial-and-error development of a good weighting function.

ACKNOWLEDGMENT

The authors acknowledge the financial assistance of the National Research Council of Canada, and G. Torrie is grateful for scholarship assistance from the same source.

REFERENCES

1. N. A. METROPOLIS, A. W. ROSENBLUTH, M. N. ROSENBLUTH, A. H. TELLER, AND E. TELLER, *J. Chem. Phys.* **21** (1953), 1087.
2. G. TORRIE AND J. P. VALLEAU, *Chem. Phys. Lett.* **28** (1974), 578.
3. J. P. VALLEAU AND D. N. CARD, *J. Chem. Phys.* **57** (1972), 5457.
4. G. TORRIE, J. P. VALLEAU, AND A. BAIN, *J. Chem. Phys.* **58** (1973), 5479.
5. G. N. PATEY AND J. P. VALLEAU, *Chem. Phys. Lett.* **21** (1973), 297.
6. C. H. BENNETT, private communication.
7. W. G. HOOVER, M. ROSS, K. W. JOHNSON, D. HENDERSON, J. A. BARKER, AND B. C. BROWN, *J. Chem. Phys.* **52** (1970) 4931.
8. I. R. MCDONALD AND K. SINGER, *Discuss. Faraday Soc.* **43** (1967), 40.
9. I. R. MCDONALD AND K. SINGER, *J. Chem. Phys.* **50** (1969), 2308.
10. J.-P. HANSEN, *Phys. Rev. A* **2** (1970) 221.
11. J.-P. HANSEN AND L. VERLET, *Phys. Rev.* **184** (1969), 151.
12. D. LEVESQUE AND L. VERLET, *Phys. Rev.* **182** (1969), 307.
13. E. M. GOSLING AND K. SINGER, *Pure Appl. Chem.* **22** (1970), 303; J. P. VALLEAU AND S. G. WHITTINGTON, *J. Chem. Soc. Faraday Trans.* **2** (1973), 1004; E. M. GOSLING AND K. SINGER, *J. Chem. Soc. Faraday Trans.* **2** (1973), 1009.
14. I. R. MCDONALD AND K. SINGER, *Mol. Phys.* **23** (1972), 29.
15. These data are presented in the form of an equation for free-energy *differences* between states in the dense fluid region, designed to fit the ensemble averages of pressure and energy obtained by Monte Carlo experiments at a large number of state points. The absolute free energies plotted in Fig. 3 result from forcing agreement with the present results at the single point $N\sigma^3/V = 0.80$, $kT/\epsilon = 1.00$.
16. The total entropy values actually reported have first been converted back into excess entropies in reduced units and then to free energies using the internal energies of McDonald and Singer [14] for the corresponding state points. This introduces an additional uncertainty in $A_c/N\epsilon$ of less than 0.01 and is therefore not important for the comparison being made here.
17. L. VERLET, *Phys. Rev.* **159** (1967), 98.
18. W. W. WOOD, Monte Carlo studies of simple liquid models, *in* "Physics of Simple Liquids" (H. N. V. Temperly, J. S. Rowlinson, and G. S. Rushbrooke, Eds.), North-Holland, Amsterdam, 1968.
19. G. TORRIE, Doctoral Thesis, University of Toronto, 1975.
20. G. TORRIE AND J. P. VALLEAU, *J. Chem. Phys.*, in press.
21. G. N. PATEY AND J. P. VALLEAU, *J. Chem. Phys.* **63** (1975), 2334.

Note to Reprint II.7

Applications of "umbrella sampling" or of closely related techniques are described in (Torrie et al., J. Chem. Phys. 58 (1973) 5479; Torrie and Valleau, Chem. Phys. Lett. 28 (1974) 578; Ng et al., Mol. Phys. 38 (1979) 781; Okazaki et al., J. Chem. Phys. 71 (1979) 2421; Lee and Scott, J. Chem. Phys. 73 (1980) 4591; Shing and Gubbins, Mol. Phys. 43 (1981) 717; Nakanishi et al., J. Chem. Phys. 76 (1982) 629; Fixman, J. Chem. Phys. 78 (1983) 4223).
Recent developments are discussed by Frenkel in the article referred to in Note II.5.6.

CHAPTER III

Transport and Non-Equilibrium Molecular Dynamics

In the study of transport processes, the equivalence between Monte Carlo and molecular dynamics methods is lost, since the order in which a succession of Monte Carlo configurations is generated is unrelated to the dynamical evolution of the system. This chapter, which is concerned with dynamic rather than static properties, is therefore restricted to a discussion of molecular dynamics techniques.

If values of the macroscopic transport coefficients are to be extracted from the results of a microcopic simulation, relations are needed that express the former in terms of the latter. The necessary theoretical framework is provided by the linear-response theory of Kubo [1], in which an explicit connection is established between the transport coefficients and correlation functions of spontaneous fluctuations in dynamical variables around their equilibrium values. In particular, transport coefficients in the linear regime are expressible as time integrals of the appropriate time-correlation functions [2]. Examples of these so-called Green–Kubo relations, or of the equivalent, generalized Einstein relations, can be found many times in the accompanying reprints. Linear-response theory may be regarded as the statistical-mechanical version of the fluctuation-regression hypothesis of Onsager [3].

Any of the relevant time-correlation functions can be written either as an ensemble average or as a time average along a phase-space trajectory, the equivalence of the two being ensured by the ergodic hypothesis. Written as a time average, the correlation function of two dynamical variables A and B, say, is

$$\langle A(t)B(0)\rangle = \lim_{T\to\infty}(1/T)\int_0^T \mathrm{d}\tau\, A(t+\tau)B(\tau),$$

which is in a form suitable for calculation by molecular dynamics. With the advent of computer "experiments", it therefore became possible to compute the transport coefficients of model fluids without the introduction of any simplifying assumption. This approach was pioneered by Alder, Gass and Wainwright (Reprint III.1) in their famous comparison of the results of molecular dynamics simulations of hard spheres with the predictions of Enskog's kinetic theory, and by Levesque, Verlet and Kürkijarvi (Reprint III.2), who were the first to undertake a systematic study of the collective dynamic properties and transport coefficients of a "realistic" model fluid.

Linear-response theory can be generalized to deal with non-linear phenomena, but the formalism becomes much heavier. The non-linear coefficients are expressed in terms of higher-order time-correlation functions. Such functions are not easily obtained by simulation, since the statistical errors tend to be large; this is the likely explanation of why conventional molecular dynamics methods have thus far not been used in the study of non-linear processes. The general question of the precision attainable in the calculation of time-correlation functions by computer simulation is discussed by Zwanzig and Ailawadi in Reprint III.3. It is shown that the statistical uncertainty caused by the fact that the time average is computed over a finite time interval T goes as $T^{-1/2}$, with a coefficient related to the characteristic relaxation time of the correlation function. A distinction is drawn between single-particle properties, such as the velocity autocorrelation function, and collective properties, such as the stress autocorrelation function, that depend explicitly on the phase-space coordiantes of all particles in the system. Zwanzig and Ailawadi show that the mean-square error in collective properties is typically a factor N (for an N-particle system) larger than for single-particle properties computed in the same simulation.

The large statistical uncertainties that are inherent in efforts to calculate the collective transport coefficients (shear viscosity, thermal conductivity, etc.) by conventional, equilibrium methods have led to a search for computationally more efficient, non-equilibrium schemes. There are, however, other reasons to attempt the simulation of systems out of equilibrium. Non-equilibrium simulations are of help in pinning down the range of validity of the assumptions that underlie most theoretical descriptions of transport phenomena. They can be used, in particular, to improve our understanding of the way in which a microscopic system approaches equilibrium, to confirm the microscopic validity of the linear laws that govern the diffusion of mass, momentum or energy, and to determine the limits of the linear regime and the corrections to the linear relations between the fluxes and the thermodynamic forces from which the fluxes arise. The remainder of this chapter is therefore largely devoted to what are called "non-equilibrium molecular dynamics" (NEMD) methods.

Approach to equilibrium

If an isolated system is in an unlikely initial condition as a result either of an external perturbation or of a spontaneous fluctuation, it will move towards a more probable dynamical state: put in simpler terms, the system will relax towards equilibrium. That such a process would occur in dilute gases was already predicted by Boltzmann on the basis of the H-theorem, and one of the first applications that Alder and Wainwright [4] made of molecular dynamics was to study the same phenomenon numerically for a fluid of hard spheres. Alder and Wainwright showed that a hard-sphere system prepared in an unlikely initial state possesses an H-function that decays almost monotonically with time towards its equilibrium value; this corresponds to what is usually called "kinetic behaviour". In a later paper (Reprint III.4), Orban and Bellemans consider the objection raised by Loschmidt to Boltzmann's arguments, namely that for every microscopic state through which a system passes on its approach to equilibrium there exists, by time reversal, another state in which the system is moving away from equilibrium ("anti-kinetic behaviour"). Orban and Bellemans analyze the anti-kinetic situations that are created in a fluid of hard disks by instantaneous reversal of the velocities of all particles. They find that the subsequent anti-kinetic behaviour of the system can be reduced, or even suppressed, by small, random errors in the velocity-inversion procedure, so that the system again evolves towards equilibrium. When equilibrium is reached, the system continues to show small fluctuations. The latter are not usually of interest in a macroscopic description, but their dynamic correlations play an important role in the Green–Kubo treatment of the transport coefficients.

NEMD simulations

All the methods that are loosely grouped together under the title of "non-equilibrium molecular dynamics" can be used to study both linear and non-linear transport. In what follows, we classify the methods as either inhomogeneous (boundary conditions not periodic) or homogeneous (consistent with some type of periodic boundary condition).

Inhomogeneous methods. One of the earliest NEMD schemes was a method devised by Ashurst and Hoover [5] to simulate steady shear flow. A pair of non-periodic boundary layers were used, each consisting of a small number of particles with the same spatial correlations as the fluid in the central cell, and shear flow was induced by moving the boundary layers in opposite directions along parallel planes. The boundary layers also had to act as thermostats in order to extract the heat generated by the viscous flow. A similar idea was used by Ashurst [6] to simulate the flow of heat between two reservoirs maintained at constant temperatures, and variants were later developed [7,8] in which stochastic methods were used both to shear the system and to control the temperature. Numerical results obtained in these ways were found to be in good agreement with calculations of transport coefficients via the Green–Kubo approach. This gave confidence in the use of NEMD methods and also stimulated the use of such techniques outside the linear regime. On the other hand, there is inevitably some difficulty in the interpretation of results obtained for small systems with large inhomogeneities. For example, systems under shear that are cooled at the boundaries tend to develop large temperature gradients. At present, therefore, these methods are rarely used to study transport processes, except in cases where interest is specifically in the effects associated with inhomogeneous boundary conditions. A good example is the recent study [9] of the long-range correlations between hydrodynamic fluctuations that exist in an inhomogeneous, non-equilibrium, steady state.

Homogeneous methods. As in the inhomogeneous case, the earliest homogeneous techniques were designed for the study of shear flow. In the work of Lees and Edwards [10], this was achieved by the use of moving (but periodic) boundary conditions, while Gosling et al. [11] produced a periodic shear-flow profile by application of a sinusoidal, transverse force field to the particles. Both these schemes in their original forms suffered from the fact that large perturbations were needed if a measurable response was to be obtained. The use of strong fields leads to substantial heating, and a stationary state cannot be reached unless a device is introduced to act as a thermostat.

The problems associated with the use of large fields are avoided if the method of "subtraction of trajectories" is adopted [12,13]. This is a technique that exploits the physical content of linear-response theory, namely that a time-correlation function measures the response of a system to an infinitesimal perturbation designed to create a certain flux. In a simulation, the measurement is made by following perturbed and unperturbed trajectories that start from the same phase point of the unperturbed system; the response to the perturbation is then obtained as the difference in the relevant dynamical variable along the two phase-space trajectories. This procedure is repeated as often as is necessary to obtain adequate statistics. The success of the subtraction method relies on a cancellation of the large thermal fluctuations, so that the difference contains only the systematic response. The principle of the method and a variety of applications are discussed in Reprint III.5. The techniques described there can be used to study the relaxation of any imposed fluctuation. However, in order to obtain a zero-frequency response, such as a hydrodynamic transport coefficient, it is necessary that the time during which the perturbed and unperturbed trajectories remain correlated be greater than the characteristic decay time of the fluctuations of interest. In practice, the fact that phase-space trajectories diverge exponentially in time [14] means that this situation is not always realized.

The use of small perturbations eliminates the necessity to introduce a thermostat into the system, but it leaves unresolved the question of what form the perturbation should take in a particular problem. The construction of the driving force appropriate to an arbitrary flow (steady or not) that is also compatible with the periodic boundary conditions poses a far from trivial problem. Singer, Singer and Fincham (Reprint III.6) have used the Lees–Edwards boundary conditions in combination with the method of subtraction to study shear viscosity, and Gillan [15] has employed a modified form of Newton's equations to study heat flow without the introduction of a thermal gradient. It is only recently, however, that a systematic approach to the problem has been developed. In the case of shear and bulk viscosity, Hoover et al. (Reprint III.7) have used a homogeneous scheme based on a hamiltonian perturbation ("Doll's tensor"), of the general form

$$\mathscr{H}_{\mathrm{p}} = \sum_{i=1}^{N} \boldsymbol{q}_i \boldsymbol{p}_i : \nabla \boldsymbol{u},$$

where the imposed flow field, possibly time-dependent, is described by the velocity-gradient tensor $\nabla \boldsymbol{u}$. The perturbation is applied, not only to the particles in the central cell, but also to the periodic images; in the case of planar Couette flow, the Doll's-tensor hamiltonian contains the moving boundary conditions of Lees and Edwards. A slightly different modification of the equations of motion was subsequently suggested by Ladd [17]. The result (the "SLLOD tensor" *) is not derivable from a hamiltonian, but contains the Lees–Edwards perturbation as a special case. The problem of heat flow has been studied by Evans [18], who has devised a perturbation technique similar to that of Gillan but which has the advantage, unlike Gillan's method, of conserving the net linear momentum of the system. Conservation of momentum is of particular importance when strong external fields are applied. The method of Evans has recently been generalized to the simulation of thermal transport in fluid mixtures [19].

Where large perturbations are used, a further device is required in order to prevent the system from heating up. One useful method is an isokinetic scheme [20,21] in which the equations of motion are altered in such a way as to keep the total kinetic energy of the particles constant; in equilibrium, the method gives rise to a variant of the canonical ensemble. Another possible thermostat is the one described by Hoover (Reprint IV.6), discussion of which we postpone until the next chapter. A general approach in which a "driving force" does mechanical work in maintaining a thermodynamic flux, while "constraint forces" extract the resulting heat, has been developed by Evans et al. (Reprint III.8) on the basis of Gauss' principle of least constraint. Gauss'

* The name "Doll's tensor" is the responsibility of B. Hoover while SLLOD, needless to say, is not an obscure mathematician... For illumination, see ref. [16].

principle is a generalization of the Lagrange equations of the first kind for holonomic constraints, and can be used to derive the equations of motion of particles subject to velocity-dependent constraints, such as that required to maintain a fixed total kinetic energy. In the latter case, the method leads back to the isokinetic scheme postulated earlier.

The final result of the efforts sketched above is a family of non-hamiltonian (sometimes, less properly, called non-newtonian) equations of motion that can be used to simulate non-equilibrium steady states. Care is needed, however, in the study of systems far from equilibrium [22]. It is not immediately obvious how these non-equilibrium techniques fit into the conventional framework of transport theory. For example, can linear-response theory be applied in standard form to the non-hamiltonian perturbations used to create the thermodynamic fluxes? Is the validity of linear-response theory destroyed by the incorporation of mechanical heat baths? A recent paper by Holian and Evans (Reprint III.9) shows that in both respects the modified equations of motion are compatible with linear-response theory and, in addition, that the equations provide a means of studying non-linear response in a systematic way.

An interesting alternative application of the new algorithms is in the simulation of equilibrium states in ensembles other than the microcanonical [23]. Until now, however, most studies of other ensembles have been based on methods that are described in the next chapter.

References

[1] R. Kubo, J. Phys. Soc. Jpn. 12 (1957) 570.

[2] L.P. Kadanoff and P.C. Martin, Ann. Phys. (USA) 24 (1963) 419.
J.M. Luttinger, Phys. Rev. 135 (1964) A1505.
J.L. Jackson and P. Mazur, Physica 30 (1964) 2295.

[3] L. Onsager, Phys. Rev. 37 (1931) 405.
L. Onsager, Phys. Rev. 38 (1931) 2256.

[4] B.J. Alder and T.E. Wainwright, in: Proceedings of the International Symposium on Statistical Mechanical Theory of Transport Processes, Brussels 1956, ed. I. Prigogine (Interscience, New York, 1958).

[5] W.T. Ashurst and W.G. Hoover, Phys. Rev. Lett. 31 (1973) 206.
W.G. Hoover and W.T. Ashurst, Adv. Theor. Chem. 1 (1975) 1.
W.T. Ashurst and W.G. Hoover, Phys. Rev. A 11 (1975) 658.

[6] W.T. Ashurst, in: Advances in Thermal Conductivity, eds R.L. Reisbig and H.J. Sauer Jr (University of Missouri Press, Rolla, 1976).

[7] A. Tenenbaum, G. Ciccotti and R. Gallico, Phys. Rev. A 25 (1982) 2778.

[8] C. Trozzi and G. Ciccotti, Phys. Rev. A 29 (1984) 916.

[9] M. Mareschal and E. Kestemont, Phys. Rev. A 30 (1984) 1158.

[10] A.W. Lees and S.F. Edwards, J. Phys. C 5 (1972) 1921.

[11] E.M. Gosling, I.R. McDonald and K. Singer, Mol. Phys. 26 (1973) 1475.

[12] G. Ciccotti and G. Jacucci, Phys. Rev. Lett. 35 (1975) 789.

[13] G. Ciccotti, G. Jacucci and I.R. McDonald, Phys. Rev. A 13 (1976) 426.

[14] S.D. Stoddard and J. Ford, Phys. Rev. A 8 (1974) 1504.
[15] M.J. Gillan, AERE Harwell Rep. R9332 (1978).
M.J. Gillan and M. Dixon, J. Phys. C 16 (1983) 869.
[16] C.H. Facett, Kewpie, the doll that built a villa in Capri, Hobbies 63 (1958) 41.
[17] A.J.C. Ladd, Mol. Phys. 53 (1984) 459.
[18] D.J. Evans, Phys. Lett. A 91 (1982) 457.
[19] D. MacGowan and D.J. Evans, Phys. Rev. A 34 (1986) 2133.
M.J. Gillan, AERE Harwell Rep. TP1189 (1986).
[20] W.G. Hoover, A.J.C. Ladd and B. Moran, Phys. Rev. Lett. 48 (1982) 1818.
[21] D.J. Evans and G.P. Morriss, Chem. Phys. 77 (1983) 63.
[22] D.J. Evans and G.P. Morriss, Phys. Rev. Lett. 56 (1986) 2172.
J.J. Erpenbeck, Phys. Rev. A 35 (1987) 218.
[23] D.J. Evans and G.P. Morriss, Comput. Phys. Rep. 1 (1984) 297.

THE JOURNAL OF CHEMICAL PHYSICS VOLUME 53, NUMBER 10 15 NOVEMBER 1970

Studies in Molecular Dynamics. VIII. The Transport Coefficients for a Hard-Sphere Fluid*

B. J. ALDER, D. M. GASS, AND T. E. WAINWRIGHT

Lawrence Radiation Laboratory, University of California, Livermore, California

(Received 22 June 1970)

The diffusion coefficient, the shear and bulk viscosity, and the thermal conductivity have been evaluated by means of their Einstein expressions by molecular dynamics computation over the entire fluid region. The autocorrelation functions for the different transport coefficients as well as their various components (such as kinetic and potential) have been obtained also. The results are compared to the predictions of the Enskog theory which involves a nearly exponential autocorrelation function. The observed deviations from an exponentially decaying autocorrelation function persist for many mean collision times, indicating that highly collective effects must be involved. The largest deviations for the transport coefficients occur near solid densities for the viscosity which is about twice as large as the Enskog prediction and for the diffusion coefficient which is about a factor of 2 smaller. In conformity with the Stokes relation, the product of diffusion and viscosity is found to be nearly constant over the entire fluid density range and in nearly quantitative agreement with the theoretically predicted constant using slipping boundary conditions. The deviations from the Enskog theory for the thermal conductivity are barely perceptible within the few percent accuracy of the data. The same is true for the bulk viscosity with its larger inaccuracies.

I. INTRODUCTION

One of the purposes of computer calculations of transport coefficients is to gain insight into the nature of many-body correlations. These correlations have heretofore been discussed for the diffusion coefficient of hard spheres at a series of densities, using the empirically determined difference between the autocorrelation function predicted by the uncorrelated theory of Enskog and that calculated by molecular dynamics.[1] By observing the amplitude and sign as well as the time scale over which the many-body contributions persist, it was possible to propose qualitative explanations of their origin which could be verified by further computer experiments.[2] In the long time limit a hydrodynamic vortex model could quantitatively account for these effects.[3] The primary aim of the present article is to present the many-body contributions for other transport coefficients of hard spheres at a series of densities. A quantitative discussion of the hydrodynamic model is postponed in the belief that further work will show that it explains the long-time, many-body effects for all the transport coefficients.

The question of whether significant correlations extend for long times (measured relative to the relaxation times for the uncorrelated system) has important theoretical implications. Such long persisting phenomena can obviously not be described as few-body effects, which are all that can be analyzed rigorously by analytical means. The method that is available to describe them is in terms of a continuum model, as was done for diffusion. A very important point which arises in connection with the long-time behavior of the correlations is the question of convergence of the transport coefficients. An attempt to calculate the initial density correction to the low-density (Boltzmann) transport coefficients[4] showed that the triplet collision contributions for disks (and quadruplet for spheres) persisted for long times and led, in fact, to divergent transport coefficients. Convergence was obtained[5] by summing all cyclical collision histories, whose physical analog might be considered to be the vortex motion proposed from the hydrodynamic model. However, these two approaches led to different conclusions; for example, the two-dimensional diffusion coefficient, on the basis of the vortex model, is divergent.[3] This hydrodynamic model, however, involves viscosity, and its long-time behavior affects that of the diffusion coefficient.

Unfortunately, the very long-time behavior of the correlation functions is hard to obtain numerically, particularly for spheres and for low density. In fact, for the diffusion coefficient it was necessary to study systems of about a thousand particles in two dimensions at intermediate and high densities before inter-

TABLE I. The ratio of the values of diffusion coefficients here calculated to those derived from the Enskog theory at a series of densities.

v/v_0	D/D_E[a]	D/D_E[b]	α_H[c]	F[d]	D/D_E[e]
100	1.00	1.02		0.306	1.02
20	1.01	1.03	0.019		1.04
10	1.03				
5	1.09		0.298	0.608	1.16
4	1.13			0.533	
3	1.16	1.22	0.525	0.761	1.34
2.5	1.17				
2	1.06	1.14	0.529	0.932	1.27
1.8	0.95		0.529	0.958	1.15
1.6	0.70	0.76	0.341	0.983	0.84
1.5	···	0.55	0.171	0.992	0.58

[a] For 108 particles with a typical uncertainty in the results of ±0.01.
[b] For 500 particles with a typical uncertainty in the results of ±0.01.
[c] $\rho(s)=\alpha/s^{3/2}$, $\alpha_H=(16/3)(6y\eta_0/5\pi\eta)^{3/2}(v_0\sqrt{2}/v)^{1/2}$.
[d] $\alpha=\alpha_H F$, where F is a correction factor of the hydrodynamic α_H due to the local diffusion of particles; $F^{2/3}=\eta/\eta_0(\eta/\eta_0+6/5\,D_E/D_0)^{-1}$.
[e] Extrapolated to infinite systems on basis of hydrodynamic theory.

ference effects, due to the presence of periodic boundaries, could be delayed sufficiently to clearly observe the long-time tail. In three dimensions, the systems would have to be much larger (by approximately the $\frac{3}{2}$ power) and the running time correspondingly longer, but even in two dimensions the calculations proved to be barely within the practical limitations of present day computers. Long runs are necessary to reduce sufficiently the statistical errors at long times. For the transport coefficients other than diffusion, there is the additional complication that the amount of statistics gathered per collision is reduced by a factor containing the number of particles in the system. This is because for each time step every particle diffuses individually, but any stress or energy fluctuation is an event involving the system as a whole. This is not quite as serious as it appears at first (though it prevented getting accurate results a few years ago with older computers) since the analysis required at each time-step is much smaller than in the case of diffusion and thus the program runs faster. Nevertheless, these transport coefficients have not been as accurately determined as the diffusion coefficient in this work, and therefore the discussion of the behavior of the correlation at long times is postponed till the study of large systems of disks can be completed with adequate statistics.

This paper presents, first, a method by which transport coefficients and their corresponding autocorrelation functions can be calculated. Because of the impulsive nature of the hard-sphere force, the standard expression[6] is slightly recast. Inasmuch as the results will be discussed in relation to the Enskog theory,[7] that theory is resummarized in terms of the autocorrelation function approach[8] rather than the original kinetic theory approach. This formulation allows expressions to be given for the initial values and slopes (with respect to time) of the autocorrelation functions which can be evaluated analytically in the low-density approximation and compared to machine results.

The results are primarily given for systems of 108 particles, but a few systems of 500 particles were studied as well. The purpose is to explore the number dependence of the properties in order to obtain results appropriate to macroscopic systems. The number dependence of some of the initial properties of the autocorrelation functions can be calculated analytically at all densities since they are independent of many-body correlations. At intermediate times the number dependence can be inferred from the molecular dynamics calculations. At long times the number dependence can be deduced from the hydrodynamic model. The results for the diffusion coefficient are then given for an infinite system based on these extrapolations. For the other transport coefficients no such extrapolation is attempted at present. Nevertheless, the inverse proportionality of the viscosity and the diffusion coefficient is tested in order to establish the validity of this experimental observation based on a macroscopic model[9] of a microscopic phenomenon. This comparison is limited by the few percent accuracy of the computed transport coefficients. Bulk viscosity is only briefly discussed because it is subject to much larger statistical error,

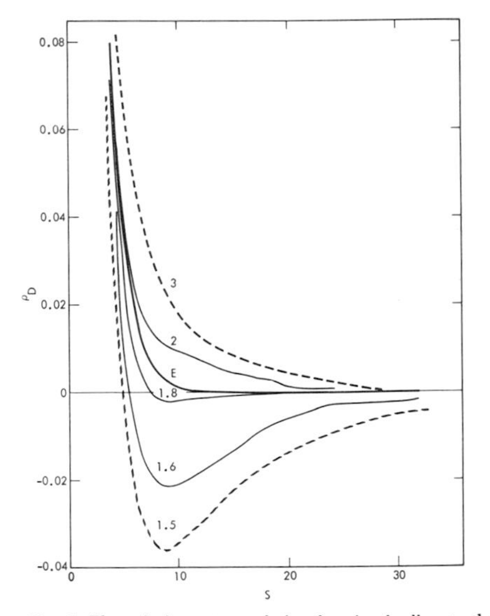

FIG. 1. The velocity autocorrelation function leading to the diffusion coefficient, ρ_D, as a function of time, s, measured in terms of mean collision times, at a series of densities indicated by the value of v/v_0. The curve labeled E is the pure exponential given by the Enskog theory. The solid curves represent results for 108 particles and the dashed curves for 500 particles.

caused by the fact that it must be calculated by the regression of fluctuations about a nonzero mean.

Tabulated transport coefficients can be compared at very low density to the predictions of the first-order density corrections to the Boltzmann theory.[10] As pointed out earlier,[1] such a comparison is of marginal significance because it is not correct to go below densities for which a particle traverses a small system in only a few mean free paths. Furthermore, the three-body correction to the Boltzmann equation is so small that highly accurate data are required before its quantitative effect can be observed numerically. A comparison of the calculated transport coefficients with high-temperature data for the rare gases[11] is more significant, since that can establish whether significant improvement is obtained by taking into account the more accurate hard-sphere results over the previously used, uncorrelated Enskog theory. Finally, some typical autocorrelation functions are shown graphically for the various transport coefficients and their components, primarily at densities where there are large deviations from the Enskog theory. The aim is to establish their qualitative nature; a quantitative tabulation is not justified in view of the accuracy of the calculated results.

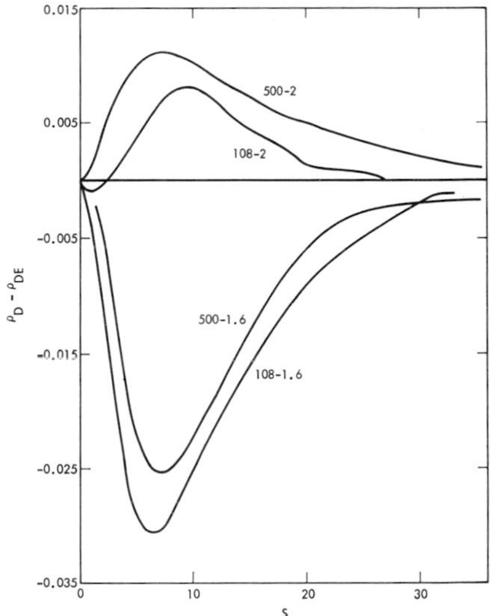

FIG. 3. As Fig. 2 except at two higher densities and for two different numbers of particles (108 and 500).

II. METHOD

The diffusion coefficient D has previously been evaluated by two equivalent formulas which lead to identical results. One utilizes the familiar Einstein expression[6]

$$D=(1/2s)\langle[x_i(t+s)-x_i(t)]^2\rangle, \qquad (1)$$

where it is understood that the infinite time limit is to be taken. The x component of the position of particle i at some arbitrary initial time t is designated by $x_i(t)$ and at some later time, $t+s$, by $x_i(t+s)$. The angular bracket indicates an average over an equilibrium ensemble or equivalently, as it is done on the computer, over a set of initial times, t, sampled from a long run of a single system after it has reached equilibrium. The calculation is carried out for each particle separately. The average of the square of the distance traveled by each particle in each of the three directions, after that quantity has reached a steady state increase with time, yields the diffusion coefficient.

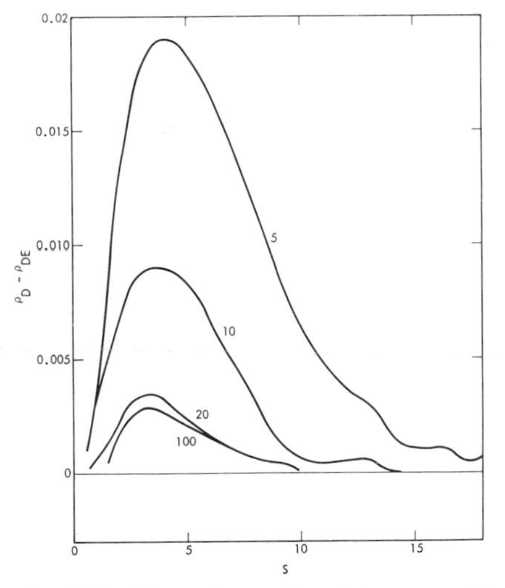

FIG. 2. The difference between the velocity autocorrelation function leading to the diffusion coefficient, ρ_D, and the pure exponential one predicted by the Enskog theory, ρ_{DE}, as a function of time, s, measured in terms of mean collision times at a series of densities indicated by the value of v/v_0.

That calculation actually turned out to be less convenient than the alternative one which involves only the velocities of the particles. This is because, in keeping track of the positions of particles, it is necessary to know whether any particle has left its original cell and crossed over into one of the neighboring cells which, because of the periodic boundary conditions, are displaced images of the original cell. By invoking the classical equations of motion, independence of the results on the choice of the initial time t, and by a change of variables, the diffusion coefficient can be rewritten as

$$D=\int_0^\infty \langle\dot{x}_i(t)\dot{x}_i(t+s)\rangle ds, \qquad (2)$$

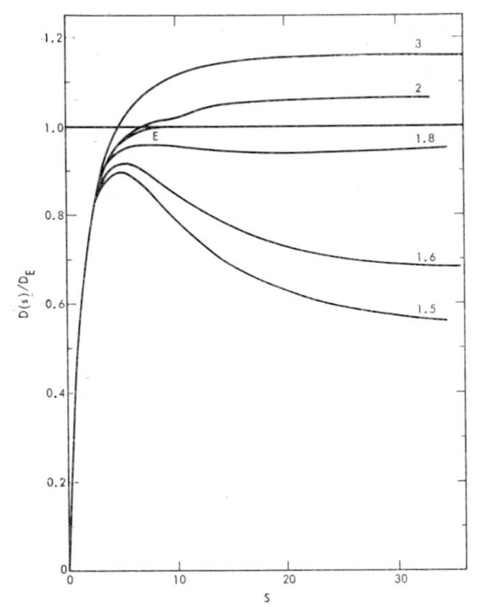

FIG. 4. The diffusion coefficient $D(s)$ as a function of time (or the integral of the autocorrelation function as a function of the upper limit of integration) relative to the prediction of the diffusion coefficient at infinite time by the Enskog theory, D_E, where s is measured in terms of mean collision times at a series of densities indicated by the value of v/v_0. The curve labeled E is the pure exponential result given by the Enskog theory.

where $\dot{x}_i$ is the x component of the velocity of particle i. Once the numerical equivalence of these two expressions was established, Eq. (2) was used exclusively from thereon because of its greater calculational ease.

The other transport coefficients can be similarly evaluated by two equivalent expressions.[6] For example, the shear viscosity η can be written either in Einstein form,

$$\eta=\frac{m^2}{2VkTs}\langle[\sum_{i=1}^{N}\dot{x}_i(t+s)y_i(t+s)-\dot{x}_i(t)y_i(t)]^2\rangle, \quad (3)$$

or in terms of an autocorrelation function,

$$\eta=(VkT)^{-1}\int_0^\infty\langle J_{xy}^\eta(t)J_{xy}^\eta(t+s)\rangle ds, \quad (4)$$

where

$$J_{xy}^\eta=m\sum_{i=1}^{N}\dot{x}_i\dot{y}_i+\tfrac{1}{2}\sum_{i\neq j}F_{ij}^x y_{ij} \quad (5)$$

is the xy component of the microscopic stress tensor. The proof of equivalence is analogous to that discussed for the diffusion coefficient, namely, that for any dynamic variable G corresponding to a particular transport coefficient, the expression

$$(1/2s)\langle[G(t+s)-G(t)]^2\rangle \quad (6)$$

is identical in the long-time limit to

$$\tfrac{1}{2}(d/ds)\langle[G(t+s)-G(t)]^2\rangle \quad (7)$$

and to

$$\int_0^s\langle\dot{G}(t)\dot{G}(t+s)\rangle ds. \quad (8)$$

Thus the shear viscosity dynamical variable is to be identified with

$$G=m\sum_{i=1}^{N}\dot{x}_i y_i$$

and hence $\dot{G}$ with J_{xy}^η. The stress tensor element J_{xy}^η has consequently two components, the first of which is called the kinetic term and the second the potential term. The potential term for hard spheres contributes only at a collision. At a collision time t_c between particles i and j, $\dot{G}$ contains two potential contributions, namely, $m\ddot{x}_iy_i+m\ddot{x}_jy_j$, which, since the change of momentum for particle i, $\Delta\dot{x}_i$, must be equal and opposite to that of particle j, can be rewritten as $m\ddot{x}_iy_{ij}$ or $F_{ij}^xy_{ij}$, where $F_{ij}^x=m\Delta\dot{x}_i\delta(t-t_c)$ is the x component of the force exerted by particle j on i and y_{ij} is the y component of the separation of the two particles at contact.

Since, for hard spheres, the force consists of a δ function at each collision, the autocorrelation form is more difficult to use directly, and hence the transport

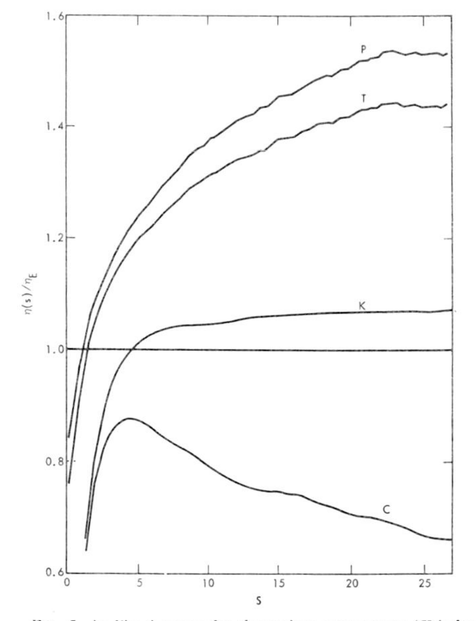

FIG. 5. As Fig. 4 except for the various components (K is for kinetic, P is for potential, and C is for cross between kinetic and potential) and for the total (T) shear viscosity η at a single density, $v/v_0=1.6$.

TABLE II. The ratio of the values of shear viscosity and its various components here calculated to those derived from the Enskog theory at a series of densities.

v/v_0	η/η_E	$\eta^K/\eta_E{}^K$	$\eta^C/\eta_E{}^C$	$\eta^P/\eta_E{}^P$	Run[b]
100	1.01±0.02	1.01±0.02	1.05±0.06	1.07±0.04	2
20	1.00±0.02	1.00±0.02	1.00±0.03	1.01±0.02	6
20[a]	1.00±0.04	1.01±0.04	0.99±0.04	1.04±0.03	10
10	0.99±0.04	1.00±0.04	0.95±0.04	0.97±0.02	1
5	0.99±0.05	1.05±0.05	0.90±0.07	0.98±0.07	1
3	1.02±0.01	1.06±0.01	1.01±0.03	1.01±0.01	10
2	1.11±0.06	1.05±0.03	0.87±0.07	1.17±0.06	2
2[a]	1.10±0.04	1.13±0.03	0.93±0.05	1.13±0.05	10
1.8	1.10±0.03	1.08±0.03	0.73±0.04	1.16±0.03	2
1.6	1.44±0.07	1.08±0.02	0.67±0.08	1.54±0.08	2
1.5[a]	2.16±0.09	1.13±0.05	0.63±0.13	2.31±0.09	10

[a] Results for 500 particles; the remainder are 108-particle results.

[b] Length of run in millions of collisions.

coefficients were evaluated by means of Eq. (7), without, however, having to be explicitly concerned whether a particle crossed into its periodic image. This was accomplished by considering the quantity

$$\Delta G=[G(t+s)-G(t)]=\int_t^{t+s}\dot{G}(\Upsilon)d\Upsilon$$

$$=\sum_{\text{coll}}^{s}\left(m\sum_{i=1}^{N}\dot{x}_i\dot{y}_i\right)s_c+\sum_{\text{coll}}^{s}m\Delta\dot{x}_iy_{ij}, \qquad (9)$$

which involves only relative positions and velocities. The first sum is evaluated at time t and subsequently changes only at a collision because one pair of molecules changes velocity. That term is multiplied by the time between successive collisions, s_c. The second sum represents the changes in G due to the succession of impulsive momentum changes because of collisions which have occurred within the interval s.

ΔG was evaluated for integer multiples of Δs, where Δs was about $\frac{1}{2}$ a mean collision time for about 100 such intervals. The appropriate 100 differences squared as given by Eq. (6) could then be formed and divided by the corresponding time interval. This allowed the transport coefficient to be evaluated for one initial condition, t, at 100 values of s over a sufficiently long time interval so that the difference of G squared over the time reached a constant value. The averaging over initial conditions, t, was usually performed every time the function G itself was tabulated, that is, about every $\frac{1}{2}$ mean collision time a new transport coefficient calculation was initiated, although sometimes an integer multiple thereof was chosen. The advantage of this procedure is that the 100 G values already available can also be used to form differences, appropriate for the transport coefficient with different initial t's. These differences are continuously formed and averaged in as the calculation advances each time step s, so that when, for example, the 101st time step is completed, the second initial condition calculation is completed, all the other (100) new differences squared are calculated and added to the appropriate time difference storage locations. This then allows the first G value to be erased since all possible differences with it have already been performed. Thus, the memory requirements of 100 G values at any point in the calculation are quite small.

These small memory requirements allow a number of transport coefficients to be evaluated simultaneously as the dynamic collision history is generated. This is all the more important inasmuch as the calculational time required for the determination of the transport coefficients is a small fraction (about 10%) of the time required to generate the collision history. Thus the thermal conductivity λ, for which

$$G=\sum_{i=1}^{N}x_iE_i,$$

and the bulk viscosity κ, for which

$$G=m\sum_{i=1}^{N}x_i\dot{x}_i,$$

TABLE III. The ratio of the values of thermal conductivity and its various components here calculated to those derived from the Enskog theory at a series of densities.

v/v_0	λ/λ_E	$\lambda^K/\lambda_E{}^K$	$\lambda^C/\lambda_E{}^C$	$\lambda^P/\lambda_E{}^P$
100	0.98±0.02	0.98±0.02	0.99±0.03	1.00±0.02
20	0.99±0.02	0.99±0.02	1.00±0.03	1.02±0.03
20[a]	0.99±0.02	1.00±0.02	0.95±0.03	1.03±0.03
5	0.97±0.03	0.94±0.02	0.96±0.04	1.03±0.04
3	1.00±0.01	1.00±0.01	0.96±0.01	1.02±0.01
2	1.02±0.03	1.00±0.03	0.97±0.04	1.04±0.03
2[a]	1.07±0.03	1.11±0.04	1.04±0.05	1.07±0.03
1.8	1.03±0.02	0.99±0.02	1.00±0.03	1.03±0.03
1.6	1.05±0.02	1.00±0.03	1.03±0.03	1.05±0.02
1.5[a]	1.05±0.03	1.01±0.03	1.02±0.04	1.05±0.03

[a] Results for 500 particles; the remainder are 108-particle results.

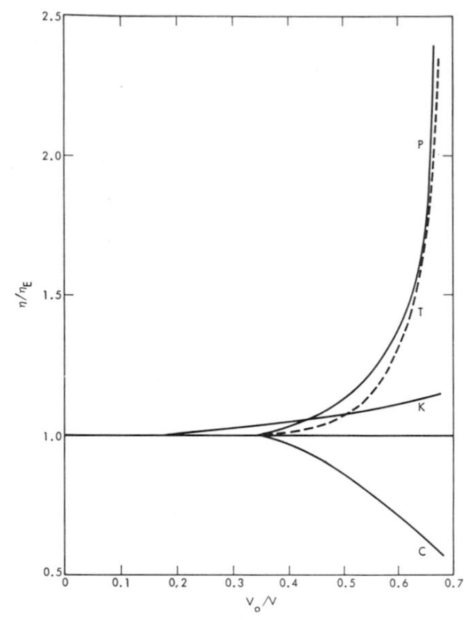

FIG. 6. The shear viscosity (T), and its kinetic (K), potential (P), and cross (C) components as a function of density, v_0/v, where v_0 is the close packed volume of hard spheres, relative to the predictions of the Enskog theory, η_E.

were also calculated concurrently with the viscosity, by

$$\lambda=(2VkT^2s)^{-1}\langle[\sum_{i=1}^{N} x_i(t+s)E_i(t+s)-x_i(t)E_i(t)]^2\rangle \tag{10}$$

and

$$\kappa+\tfrac{4}{3}\eta=(2VkTs)^{-1}$$
$$\times\langle[\sum_{i=1}^{N} mx_i(t+s)\dot{x}_i(t+s)-mx_i(t)\dot{x}_i(t)-pVs]^2\rangle, \tag{11}$$

where E_i is the sum of the potential and kinetic energy of particle i.

The extra pV term in the bulk viscosity arises from the fact that $\dot{G}$ averaged over an equilibrium ensemble is not equal to zero as it is for all the other dynamical variables treated here, but is equal to the external force, pV, by the virial theorem. Hence $\langle\Delta G\rangle=pVs$ must be subtracted since the bulk viscosity is related to the autocorrelation function of the fluctuation of the diagonal elements of the stress tensor about its mean. Since the mean is itself determined on the computer and shifts slightly as the calculation proceeds, the results for the bulk viscosity are much less accurately determined. Additionally, the subtraction of $(\frac{4}{3})\eta$, which itself also is determined with some uncertainty, causes further errors in the bulk viscosity values.

The computer calculations not only determine the x or xy components of the transport coefficients as explicitly written above, but the other two components (y, z or xz, yz) as well. The deviations of the three components from their mean can then be used as a measure of the statistical error of the results. A better measure was attained by dividing a run into about 10 batches, making sure that the 10 separate results formed approximately a Gaussian distribution, from which the mean and its statistical uncertainty could be determined.

The transport coefficients could be more accurately determined from Eq. (7) than from its equivalent form given by Eq. (6). This point can simply be illustrated by an example in which the transport coefficient approaches its long-time behavior exponentially, that is, as

$$\int_0^s \exp(-a\Upsilon)d\Upsilon=a^{-1}(1-e^{-as}).$$

Equation (6) is then equal to the integral of the above divided by s, namely, $1/a-1/a^2s+e^{-as}/a^2s$. Although both expressions in the long-time limit give the same result, in the second case it is necessary to proceed until $1/a^2s$ becomes small.

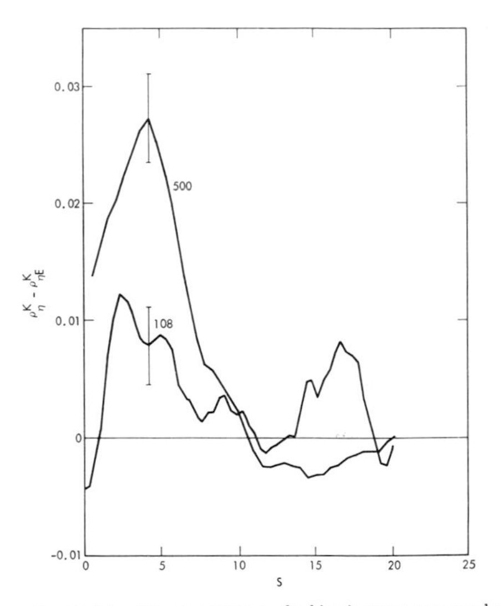

FIG. 7. The difference between the kinetic stress autocorrelation function for the viscosity ρ_η^K and the pure exponential one predicted by the Enskog theory, $\rho_{\eta E}{}^K$, as a function of time, s, measured in terms of mean collision times at a density of $v/v_0=2$ for 500 and 108 particles. The autocorrelation functions are normalized relative to the initial value of the pure exponential Enskog theory. The vertical bars indicate the typical statistical error in the molecular dynamic results.

Equation (7) was thus used to calculate the transport coefficient as a function of time, s, not only for the purpose of determining whether the calculation had been carried out far enough in s to reach a constant (plateau) value, but also to ascertain how that value had been reached. The manner in which the long-time limit in Eqs. (7) and (8) is reached reveals in much greater detail than the values of transport coefficients the physical processes that contribute to a given non-equilibrium property. That detail is most clearly demonstrated in the autocorrelation function itself rather than the integral of it as a function of the upper limit [see Eqs. (7) and Eq. (8)]. Hence the autocorrelation functions $\rho(s)$ were also calculated by numerically differentiating Eq. (7), which is identical to Eq. (8) and leads to

$$\rho(s)=\langle \dot{G}(t)\dot{G}(t+s)\rangle=\tfrac{1}{2}(d^2/ds^2)\langle[G(t+s)-G(t)]^2\rangle. \tag{12}$$

Still more detail can be obtained by calculating the kinetic and potential parts of the transport coefficients and their corresponding autocorrelation functions separately rather than only the total. The diffusion coefficient involves only a kinetic part, but the viscosity, as Eq. (5) shows, involves a kinetic and a potential part. When ΔG, as shown in Eq. (9), is substituted into Eq. (7), three separate terms result, a correlation of the kinetic term (K) with itself, a correlation of the potential term with itself (P), and finally a cross correlation (C) of the kinetic and potential terms. As a check, the kinetic autocorrelation function was also calculated directly by substituting the first term of Eq. (5) into Eq. (12). For the cross and potential terms this is not possible, because of the impulsive nature of the force and hence both terms are obtained only by using the second equality of Eq. (12).

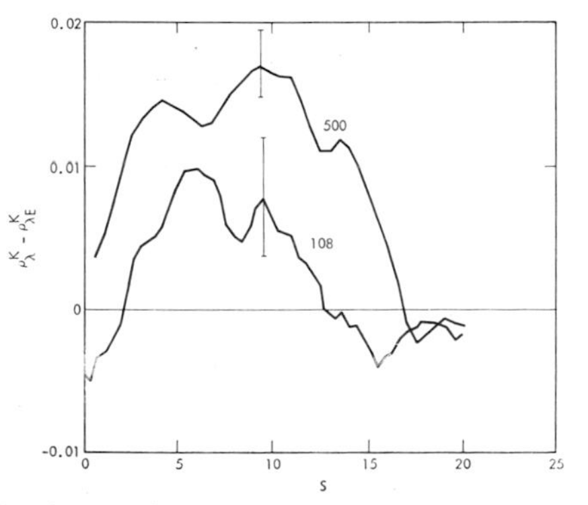

FIG. 8. As Fig. 7 except for kinetic thermal flux autocorrelation function for the thermal conductivity, $\rho_\lambda{}^K$.

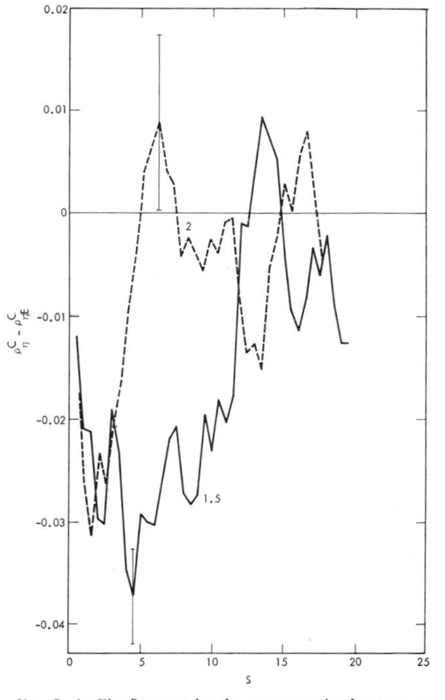

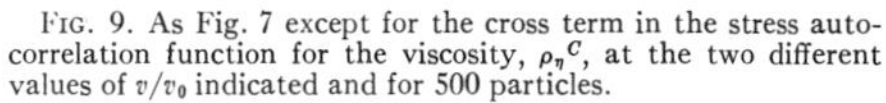

FIG. 9. As Fig. 7 except for the cross term in the stress autocorrelation function for the viscosity, $\rho_\eta{}^C$, at the two different values of v/v_0 indicated and for 500 particles.

III. ENSKOG THEORY

The Enskog theory[7] with which the molecular dynamics results will be compared makes predictions for the kinetic, cross, and potential terms based on the molecular chaos approximation. This approximation, that a given particle collides successively with other particles which are statistically uncorrelated at any density, implies that all rate processes can be scaled by the collision rate since per collision a mean of any quantity is independent of the density. The virial theorem leads to the prediction that this collision rate is rigorously proportional to $PV/NkT-1=y$. The proportionality constant is the average momentum exchanged per collision. It follows that the integral of any autocorrelation function

$$\int_0^\infty \rho(s)\,ds,$$

whose integrand is normalized initially to unity, i.e., at $s=0$, should first of all be scaled simply by y^{-1}, since it has the dimension of time. The normalization of the potential part of the autocorrelation function involves

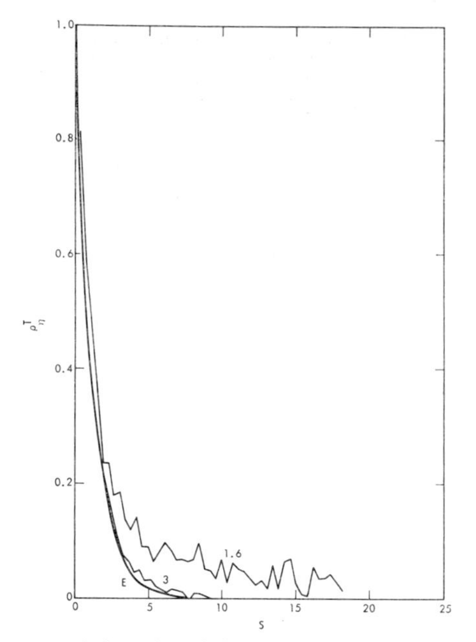

FIG. 10. Comparison of the stress autocorrelation function (total) for the viscosity, $\rho_\eta{}^T$, as a function of time, s, measured in terms of mean collision times, at the two different values of v/v_0 indicated with the pure exponential given by the Enskog theory (E).

[see Eq. (9), which must be squared] an equilibrium ($s=0$) average of $(\Delta\dot{x})^2$ or $(F_{ij}{}^x)^2$ which, by the virial theorem, as just stated, is itself proportional to y^2. The cross term normalization involves only the first power of the rate of momentum exchange, and hence the functional form of the Enskog theory is apparent on merely dimensional grounds,

$$\eta/\eta_0 = (B/V)(\eta_E{}^K/y + \eta_E{}^C + \eta_E{}^P y). \tag{13}$$

The viscosity is expressed in terms of its low-density (Boltzmann) value, η_0, in such a way that $\eta_E{}^K$ is unity, since y in the low-density limit approaches B/V, the second virial coefficient divided by the volume. The Enskog theory thus predicts that $\eta_E{}^K$, $\eta_E{}^C$, $\eta_E{}^P$ are pure numbers, since all density dependences have been scaled out. The objective of these molecular dynamic calculations is to find the density dependence of η^κ, η^C, and η^P and thus evaluate the effect of correlations.

The calculation of $\eta_E{}^K$, $\eta_E{}^C$, $\eta_E{}^P$ was done originally by Enskog using the Boltzmann equation approach. That equation was modified only to take into account that at higher densities the size of the molecules could no longer be neglected compared to their average distance of separation. In order to compare the computer results more directly with predictions of this theory, it is desirable to formulate the results in terms of the autocorrelation function approach. For that purpose the initial values and slopes of the autocorrelation functions are calculated in the low-density limit.

For the initial value of the kinetic term, only averages over a Boltzmann distribution of velocities have to be performed, and these can be done exactly at any density. The initial rate of decay similarly can be evaluated by calculating the change in the kinetic term caused by the collision of only two particles. If then the autocorrelation function is assumed to decay exponentially, the initial value and the initial rate of decay serve to define it completely. The initial rate of decay of the kinetic part of the autocorrelation function is correctly given by the above calculation at any density since it depends only on first collisions. Any correlation in successive collisions which would be expected to destroy the exponential character of the decay would affect only the higher derivatives of the autocorrelation function. It can thus be predicted that the kinetic autocorrelation functions are correctly calculated at all densities to about one mean collision time, since it takes on the average longer than that to have two successive collisions.

The result of such a calculation is that for diffusion

$$D_E{}^K = \frac{2}{3}\int_0^\infty e^{-2s/3} ds, \tag{14}$$

for viscosity

$$\eta_E{}^K = \frac{4}{5}\int_0^\infty e^{-4s/5} ds, \tag{15}$$

and for thermal conductivity

$$\lambda_E{}^K = \frac{8}{15}\int_0^\infty e^{-8s/15} ds, \tag{16}$$

where s is the time measured in relation to the mean collision time. For diffusion, only this kinetic term contributes. For all the transport coefficients the above re-

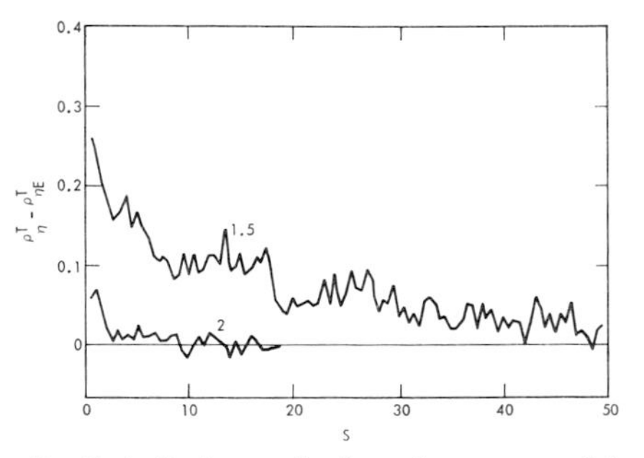

FIG. 11. As Fig. 7 except for the total stress autocorrelation function, $\rho_\eta{}^T$, at the two different values of v/v_0 indicated and for 500 particles.

sults are equivalent to only the zeroth-order Sonine polynomial solution of the Boltzmann equation. Inasmuch as the autocorrelation function is not quite exponential even in the limit of low densities, a small correction has to be applied to each transport coefficient, which is calculated by the higher-order Sonine polynomials. This correction can be ignored for present purposes. The Enskog theory is thus considered to give rise to exponential autocorrelation functions whose relaxation times are very short, of the order of one mean collision time.

The initial value of the cross autocorrelation function can also be calculated exactly at any density since it involves a correlation of particle velocities with the momentum exchanged in a single collision. The initial decay of this autocorrelation function, however, involves second collisions of these particles and therefore requires for its calculation the molecular chaos approximation. The calculated initial slopes can therefore be expected to be only accurate at low densities. The assumption of exponential decay leads to

$$\eta_E{}^C=\frac{16}{25}\int_0^\infty e^{-4s/5}ds=\tfrac{4}{5}, \tag{17}$$

$$\lambda_E{}^C=\frac{16}{25}\int_0^\infty e^{-8s/15}ds=\tfrac{6}{5}. \tag{18}$$

These results are identical to those of Enskog.[7] The decay times of the cross term are also exactly the same as those of the kinetic term.

The potential part of the autocorrelation function for the viscosity has previously been evaluated from the impulsive change in momentum which particles experience upon collision.[8] $\eta_E{}^P$ was thus considered as a sum over collisions rather than as an integral over time. The effect of the first collisions, $\eta_1{}^P=48/25\pi$, could be calculated exactly. The effect of the second collision,

$$\eta_2{}^P=128/125(3\sqrt{3}+\pi), \tag{19}$$

TABLE IV. The ratio of the values of bulk viscosity here calculated to those derived from the Enskog theory at a series of densities.

v/v_0	κ/κ_E[b]
100	0.9
20	1.1
20[a]	1.0
5	0.9
3	0.98±0.07
2	0.9
2[a]	1.2
1.8	1.1
1.6	1.1
1.5	0.6

[a] Results for 500 particles; the remainder are 108-particle results.
[b] The error in the result is generally estimated to be ±0.2 except for the long run at $v/v_0=3$, where the uncertainty is indicated.

could again only be calculated in the molecular chaos or low-density approximation. An analogous assumption to an exponentially decaying autocorrelation function for this sum is that

$$\eta_E{}^P=\eta_1{}^P/(1-\eta_2{}^P/\eta_1{}^P)=0.765. \tag{20}$$

The result is close to that of Enskog, 0.761. Alternatively, the potential autocorrelation function can be considered as consisting of an initial delta function ($\eta_1{}^P$) and a subsequently decaying function which may be assumed to be exponential with the same relaxation time as the kinetic and cross terms. Such a function can be adjusted to fit the Enskog results:

$$\eta_E{}^P=\frac{48}{25\pi}+0.120\int_0^\infty e^{-4s/5}ds=0.761. \tag{21}$$

The advantage of this representation is that it is more directly comparable to the machine calculation. It makes also more apparent that both the intercept and slope of the potential autocorrelation function can be expected to be density dependent; only $\eta_1{}^P$ can be calculated exactly.

For the bulk viscosity it is not difficult to show, by a procedure analogous to the shear viscosity, that only the potential term contributes. Calculation of the contribution of first collisions for which $\kappa_1{}^P=16/5\pi$ yields exactly the Enskog result. This would imply that in the low-density approximation c is zero in the expression

$$\kappa_E{}^P=16/5\pi+c\int_0^\infty e^{-4s/5}ds, \tag{22}$$

where it has again been assumed that the autocorrelation function decays with the same relaxation time as that of the kinetic part of the shear viscosity. This result contradicts the calculation of $\kappa_2{}^P$, a problem as yet unresolved.

For the thermal conductivity, $\lambda_1{}^P$ has been evaluated as $32/25\pi$. The value of $\lambda_2{}^P$ has been estimated by assuming an exponential decay, as in the kinetic and cross terms, and by assuming that the Enskog result is correct in the low-density limit. Thus,

$$\lambda_E{}^P=\frac{32}{25\pi}+0.186\int_0^\infty e^{-8s/15}ds=0.755. \tag{23}$$

The foregoing results for the Enskog theory are applicable to infinitely large systems, but it would be desirable to know the results for finite systems of N particles, as studied on the computer. For this purpose it is necessary to evaluate the initial values and slopes of the various correlation functions for a finite system. This can be done rigorously only for the kinetic terms and the initial value of the cross term; for the other terms, the N dependence can only be evaluated in the low-density limit as discussed before. Inasmuch as the present study concentrates on the long-time behavior of the autocorrelation functions, the initial values and slopes have not been very accurately evaluated except

TABLE V. Test of Stokes relation $D\eta\sigma/kT=1/3\pi=0.106$ (sticks) or $1/2\pi=0.159$ (slips).

v/v_0	y	$D\eta/D_E\eta_E$[a]	$D\eta/D_E\eta_E$[b]	$D\eta/D_E\eta_E$[c]	$D_E\eta_E\sigma/kT$	$D\eta\sigma/kT$
5	0.89	1.08		1.15	0.135	0.156
3	2.06	1.18		1.37	0.107	0.147
2	4.86	1.18	1.25	1.40	0.112	0.157
1.8	6.45	1.05		1.26	0.117	0.147
1.6	9.20	1.01		1.21	0.124	0.150
1.5	11.49	···	1.23	1.30	0.129	0.168

[a] For 108 particles.
[b] For 500 particles.
[c] For infinite systems.

for the initial value of the kinetic term. The kinetic autocorrelation function, furthermore, is the only one which shows a statistically significant N dependence. For these reasons the analysis of the N dependence is confined to a discussion of the initial value of the kinetic term.

There are two major N dependent effects. One arises from the constraints imposed in a finite system by the conservation conditions. For example, if the velocity of one particle is known to be $\dot{x}_i$, the sum of the velocities of the remaining particles

$$\sum_{j\neq i}^{N} \dot{x}_j$$

must be equal to $-\dot{x}_i$, since momentum is conserved. In other words, there is a counterflow in which particle j has on the average a velocity, $-\dot{x}_i/(N-1)$. For the shear viscosity, as an illustration,

$$\rho_{KE}^{\eta}(0)=\sum_{i=1}^{N}\sum_{j=1}^{N}\langle\dot{x}_i\dot{y}_i\dot{x}_j\dot{y}_j\rangle=\langle\sum_{i=1}^{N}\dot{x}_i^2\dot{y}_i^2\rangle$$

$$+\langle\sum_{i=1}^{N}\dot{x}_i\dot{y}_i\sum_{j\neq i}^{N}\dot{x}_j\dot{y}_j\rangle=\langle\sum_{i=1}^{N}\dot{x}_i^2\dot{y}_i^2\rangle[1+(N-1)^{-1}], \quad (24)$$

where, in the first equality, the i and j terms have been written separately, and in the last equality the sum over $j\neq i$ has been replaced by the single counterflow velocities for the $\dot{x}_j$ and $\dot{y}_j$ for each of the $N-1$ terms in the sum. Similarly, for the thermal conductivity an average over the velocities involving all but one particle must be carried out with a Maxwell–Boltzmann velocity distribution whose mean is $-\dot{x}_i/(N-1)$ and whose temperature T_i has to be determined from

$$\tfrac{1}{2}m\dot{x}_i^2+(N-1)\tfrac{3}{2}kT_i=\tfrac{3}{2}kTN. \quad (25)$$

This leads to the result that $\rho_{KE}^{\lambda}(0)$ for an infinite system has to be multiplied by $(1+10/3N)$ for a finite system.

The other N-dependent effect arises from the fact that the Maxwell–Boltzmann velocity distribution for finite systems does not have a long-range tail and hence has a cutoff at an upper limit. Furthermore, the exponential function must be represented by its finite approximation. This leads to a further correction of $(1-2/3N)$ for the viscosity and $(1-16/3N)$ for the thermal conductivity, and hence a total correction of $(1+1/3N)$ for the viscosity and $(1-2/N)$ for the thermal conductivity. The thermal conductivity has a much larger coefficient for its N dependence, because a higher moment (sixth power) of the velocity distribution is involved. The higher moments are, of course, much more affected by the cutoff of the distribution at high velocities.

The initial value of the diffusion coefficient is unaffected by the size of the system. This is because the second moment of the velocity distribution is independent of N to order $1/N^2$. There is, however, an N dependence of the slope of the autocorrelation function since it involves collisions with particles which on the average have a drift velocity of $-\dot{x}_i/(N-1)$. In fact, the hydrodynamic model leads to the result that a correction of $1/N$ must be added to the velocity autocorrelation function at long times for finite systems. This correction is valid till interference due to boundary effects appears, at which time the autocorrelation function can no longer represent an infinite system anyway.

IV. RESULTS

The results for the diffusion coefficient have already been published[3] in part and need here be only briefly summarized. In Table I the data for 108 and 500 particles are collected. They were obtained, as were all the other data presented below, on a CDC 6600 computer. Runs were typically a million collisions long, requiring on the order of 10 h of machine time for 108 particles. The 500-particle runs were a few million collisions long requiring a similar amount of machine time, but fewer averages of the diffusion coefficient were taken. As Fig. 1 shows, the runs were sufficiently long to determine a smooth autocorrelation function with a typical uncertainty of 0.001 in ρ_D independent of s. The graphs of the autocorrelation function make evident the structure that causes diffusion coefficients to be first larger than Enskog values, then smaller as the density increases.

It was the attempt to understand the enhanced diffusion coefficient at intermediate densities that led to the proposal of the hydrodynamic vortex model. This en-

TABLE VI. Initial values of the autocorrelation functions here calculated relative to the values derived from the Enskog theory.

A. Kinetic term

v/v_0	$\rho_\eta^K(0)/\rho_{\eta E}^K(0)$	$\rho_\lambda^K(0)/\rho_{\lambda E}^K(0)$
20	0.998	0.970
5	1.002	0.959
2	0.995	0.977
1.8	1.001	0.973
1.6	1.002	0.981
Theory[a]	1.003	0.982

B. Cross and potential term[c]

v/v_0	$\rho_\eta^C(0)/\rho_{\eta E}^C$	$\rho_\lambda^C(0)/\rho_{\lambda E}^C$	$\rho_\eta^P(0)/\rho_{\eta E}^P$	$\rho_\lambda^P(0)/\rho_{\lambda E}^P$
100	0.98	1.00	0.97	1.00
20	1.03	1.00	1.08	0.96
20[b]	1.01	0.98		
5	0.99	0.95	0.90	1.00
3	0.97	0.98	1.17	1.11
2	0.96	0.96	1.50	1.00
2[b]	0.96	1.02	1.37	1.07
1.8	0.98	0.99	1.65	1.04
1.6	0.98	1.00	1.65	1.08
1.5[b]	0.98	1.03	1.74	1.11

[a] For 108 particles; the uncertainty of the results is such that there is no disagreement with the theory.

[b] Results for 500 particles; the remainder are 108-particle results.

[c] Excludes initial δ function value; accuracy of the results are of the order of 2%.

hancement is more clearly evident from Fig. 2. This figure shows that the positive structure relative to the exponentially decaying autocorrelation function increases in magnitude with density in the low-density region. Furthermore, the peak of the positive structure progressively moves to longer times as the density increases, till at a density of $v/v_0=2$ it appears near 10 collision times (see Fig. 3). Some small positive structure in the autocorrelation function is present even in the low-density limit and has the same origin as the Sonine polynomial correction to the Boltzmann result. This effect can account for the structure at $v/v_0=100$ and 20 but not at higher density.

One of the characteristics of the positive structure also is that it depends strongly on the number of particles, as Fig. 3 shows. Furthermore, for larger systems the positive structure has a longer tail. The vortex model qualitatively accounts for all these facts.

The maximum positive effect is expected at a mean collision time roughly given by the number of nearest neighbors. Interference effects are correctly predicted at the time when a sound wave travels the length of the system. This interference causes the positive tail of the smaller system to be distorted at an earlier time, that is, of the order of 10 collision times for 108 particles and of the order of 20 collision times for 500 particles, depending slightly on the density. Finally, the hydrodynamic model justifies a $1/N$ correction to be added to the autocorrelation function by an adjustment of the coordinate system necessitated by the conservation of momentum requirement imposed on a finite system molecular dynamics calculation.

Even at the higher densities, where the autocorrelation function has a negative structure, the vortex arguments still apply. As Fig. 3 shows, the qualitative number dependence is the same as at lower density. The cause of the negative structure is ascribed to a reversal of the particle's motion at higher densities after a few mean collision times due to backscattering. Thus a typical particle still sets up initially a vortex pattern, but because of the higher density built up in front of it, this particle on the average changes direction. The long-time tail would still be expected to be positive, but because of the smallness of the system with its boundary interferences it has not been seen. Nevertheless, a $(1/N)$-dependent correction should be added to the long-time negative tail, observed for example in the $v/v_0=1.5$ curve in Fig. 1 or $v/v_0=1.6$ curves in Fig. 3, which would bring the curves closer to the zero base line.

The finite system results for the diffusion coefficient have been extrapolated to infinite systems, as Table I shows, on the basis of the observed N dependence and by the long-time behavior predicted by the vortex model. These corrections are rather large. The hydrodynamic model predicts that the autocorrelation function decays with the inverse $\frac{3}{2}$ power of the time and with a coefficient that depends upon the viscosity, as shown in Table I. The basis for calculating the coefficient is entirely analogous to the one for two di-

Notes on p. 181

TABLE VII. Initial slopes of the autocorrelation functions[a] here calculated relative to the values derived from the Enskog theory.

v/v_0	$a_\eta^K(0)/a_{\eta E}^K(0)$	$a_\lambda^K(0)/a_{\lambda E}^K(0)$	$a_\eta^C(0)/a_{\eta E}^C(0)$	$a_\lambda^C(0)/a_{\lambda E}^C(0)$
100	1.004	0.989	0.98	0.98
20	1.004	0.995	0.98	1.00
20[a]	0.981	1.000	0.97	1.04
5	0.981	1.014	1.08	1.04
3	0.982	0.992	1.09	0.98
2	0.998	1.000	1.16	0.97
2[a]	0.963	0.983	1.10	0.98
1.8	0.999	0.997	1.09	0.99
1.6	1.003	0.989	1.07	0.97
1.5[b]	0.989	0.981	1.08	0.97

[a] The estimated inaccuracy in the slopes for the potential term of the viscosity (40%) and thermal conductivity (10%) are too large to make any deviation from the theoretical slope or any trend with density apparent except for the more accurate results at $v/v_0=3$, where $a_\eta^P(0)/a_{\eta E}^P(0)=1.24\pm0.1$ and $a_\lambda^P(0)/a_{\lambda E}^P(0)=0.76\pm0.05$. The values given in the table are accurate within about 2% except for greater inaccuracies in the shorter $v/v_0=5$ run.

[b] Results for 500 particles; the remainder are 108-particle results.

mensions, except that $\frac{2}{3}$ of the initial momentum (instead of $\frac{1}{2}$) is involved in the vortex flow. In three dimensions as opposed to two dimensions, the diffusion coefficient does converge to a steady-state value. How this plateau value is reached is shown in Fig. 4. In all cases it takes at least 10–20 collision times. At the highest density ($v/v_0=1.5$), which represents a typical fluid near its solidification point, it takes longer, on the order of 30 mean collision times. This is evident from the autocorrelation function of Fig. 1. The important point to recognize is that the negative structure ascribed to backscattering has the same magnitude and time scale as the positive structure and hence is not readily describable by a model involving only a few particles. A continuum description involving a generalization of the vortex model to viscoelastic hydrodynamics might possibly be successful at this time scale.

The results for the viscosity are given in Table II together with the lengths of the various runs. The code requires about an hour for a quarter of a million collisions to calculate all the transport coefficients except diffusion. Figure 5, analogous to Fig. 4, shows how the various components of the viscosity at a given density reach their plateau values. At the higher densities it again requires on the order of 30 collision times. The rugged nature of the curves shows that the data are not as accurate as they are for diffusion. This figure, and even more clearly, Fig. 6, shows that the various components differ in their deviation from the Enskog theory, both in sign and in magnitude. The kinetic component deviates positively, but no more than about 10%. None of the terms deviate significantly up to densities of $v/v_0=3$, and in that region therefore the Enskog theory gives a very accurate description of the data. At higher densities, the cross term deviates negatively and the potential term positively. The total viscosity follows the potential part very closely, inasmuch as at high densities it contributes more than 90%. At the highest densities departures from the Enskog theory are substantial, by a factor of about 2.

Figure 7 shows that the general characteristics as well as the number dependence of the kinetic autocorrelation function for the viscosity, and indeed for the thermal conductivity also (see Fig. 8), are the same as those for diffusion. The initial small negative dip in the 108 results can be traced to the larger number-dependent correction of the initial slope of the autocorrelation function for the smaller system. The correlation functions for viscosity and thermal conductivity are not as accurately determined; their uncertainty of about 0.01 is nearly independent of s, and makes the observed positive structure in the viscosity at large s for the 108-particle system of doubtful significance. In fact, the similarity in the behavior of the kinetic terms suggests that the hydrodynamic vortex model is applicable to them all. Longer runs for larger two-dimensional systems are in progress in order to investigate the decay of the autocorrelation functions for the viscosity and thermal conductivity at long times.

The cause of negative deviation of the cross term from the Enskog theory manifests itself in a negative structure in the correlation function as shown in Fig. 9. This negative structure persists for longer times at higher density. Similarly, the positive structure shown in Fig. 10 in the total autocorrelation function for the viscosity becomes prominent at higher densities and lasts for many mean collision times. At $v/v_0=3$ the autocorrelation function is nearly exponential in agreement with the Enskog theory. Even at $v/v_0=2$, the deviations from exponential behavior, as shown in Fig. 11, are barely significant. At $v/v_0=1.5$, corresponding to a liquid near solidification, however, the viscosity correlation function shows a long positive structure, which causes the viscosity to be over twice as large as that predicted by the Enskog theory. The question of whether this also indicates that the potential part of the viscosity has a nonexponentially decaying autocorrelation function at long times awaits the analysis of the more accurate two-dimensional results. The deviations of the thermal conductivity from the Enskog theory, on the other hand, are much smaller (see Table III). In fact, they are so small that they are barely outside the statistical significance of the results. In view of this, the structure in the autocorrelation functions is almost lost within the numerical noise of the results and no graphs are given.

For the bulk viscosity also, the data are too imprecise to decipher any deviations from the Enskog theory, as Table IV shows, with the exception of the result at $v/v_0=1.5$. The bulk viscosity was obtained by subtracting from the computer results an empirical value of $-(pV)s$, which is nearly the same as that calculated from the virial theorem. The procedure used was to fit a least-squares line at long times to the computer results obtained from Eq. (11) without the $(pV)s$ term. The remaining constant plateau value is then identified with $\kappa+(\frac{4}{3})\eta$. The same procedure can be applied to the potential part only since the other components do not contribute. Somewhat more accurate results are obtained in that way.

The values of diffusion and shear viscosity can be used to test the validity of the Stokes model on a molecular scale.[9] The Stokes relation, that the product of the diffusion coefficient and the viscosity is a constant related to the diameter of a spherical object moving through a fluid, has been found to be approximately applicable to molecules in many pure fluids and mixtures. The conclusion from the model, that as the diffusion decreases the viscosity increases, is valid for hard spheres as shown in Table V. The correlations which made the viscosity increase over the Enskog theory by a factor of about 2 near solid density are compensated for by correlations which made the diffusion coefficient decrease by nearly that much. The quantitative application of the Stokes model that the diameter of the spherical ball can be extrapolated to microscopic dimensions also works rather well. Table V shows that the Enskog theory predicts a value of the constant nearly in agreement with the Stokes prediction. The density range covered is the entire fluid region, that is, densities for which $v/v_0\leq 5$, where the system would be above the critical density if there were attractive forces. The small rise observed at $v/v_0=5$ is an indication of the inapplicability of the model in the gas phase. The Enskog theory predicts that

$$\frac{D_E\eta_E\sigma}{kT}=\frac{5\sqrt{2}\pi}{96v/v_0}\left(\frac{1}{y^2}+\frac{0.8}{y}+0.761\right), \qquad (26)$$

which blows up at low density. Over a threefold range in volume, however, the Stokes relation is remarkably precise, and the use of more accurate transport coefficients from molecular dynamics improves that picture. The detailed molecular picture leads to the conclusion that the effective diameter to be used in the Stokes relation with slipping boundary conditions nearly quantitatively matches the actual one. This is true despite the fact that the velocity autocorrelation function is not exponential as required for the application of the Stokes theory.

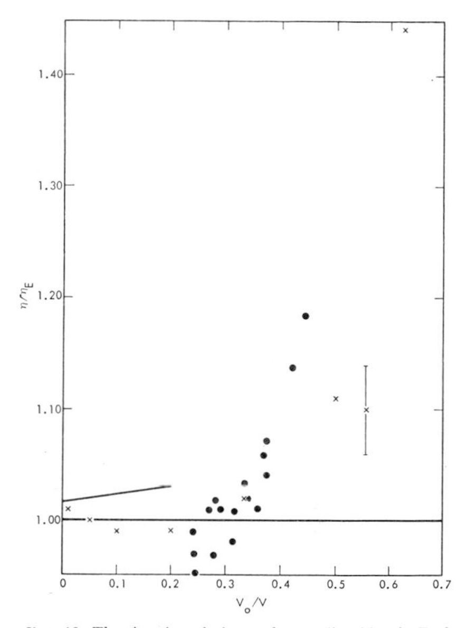

FIG. 12. The viscosity relative to that predicted by the Enskog theory as a function of density, v_0/v, where v_0 is the close packed volume of spheres. The crosses are the 108 results of molecular dynamics with the vertical bar at a value of v_0/v near 0.6 showing typical statistical errors in the machine computation. Note the cross in the upper right-hand corner for the highest density graphed. The circle represents the results from an analysis of the high-temperature viscosity of argon and krypton. The straight line drawn at low density has an intercept at $v_0/v=0$ given by the Sonine polynominal correction to the pure exponential autocorrelation results for the viscosity in the Boltzmann limit and a slope as calculated by the triplet collision correction to the Enskog theory.

It was thought worthwhile to list the initial values and slopes for the autocorrelation function separately, although, as pointed out before, they have not been evaluated very precisely except for the kinetic terms. Table VI lists the kinetic terms separately and compares them with the number-dependent theoretical results. As can be seen, the smaller dependence of the viscosity upon N and the larger dependence of the thermal conductivity is verified within the accuracy of the data. The cross terms agree with the theoretical results for the infinite systems within their uncertainty. The data are not accurate enough to establish any N dependence. The potential term, however, shows agreement with the theoretical results only at low density. At higher densities both the viscosity and thermal conductivity show significant deviations from the theoretical result, which was based on the molecular chaos approximation.

As for the slopes given in Table VII, the kinetic

Notes on p. 181

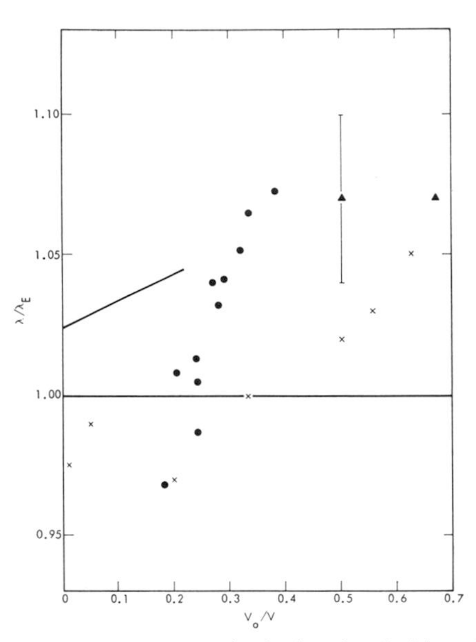

FIG. 13. As Fig. 12 except for the thermal conductivity and that in addition the 500 particle results are plotted as triangles. On the other hand, the circles represent data for argon only.

terms are predicted accurately by the theoretical results for infinite systems. The calculated slopes of the cross term by aid of the molecular chaos approximation show agreement only at low density in the case of viscosity. The slopes of the thermal conductivity cross term do not show large deviations from theory, but then neither did the total thermal conductivity. It seems that many-body correlations do not affect the thermal conductivity very much. The slopes of the potential terms are too inaccurately known to be tabulated, except at $v/v_0=3$ where deviations from the low-density theory are already apparent.

The results can be compared not only to the Enskog theory but also to predictions of the density corrections to the Boltzmann equation. The triple collision correction to the low-density Enskog theory has been evaluated for spheres to be[10]

$$\eta/\eta_E=1.016+0.071v_0/v, \qquad (27)$$

$$\lambda/\lambda_E=1.024+0.095v_0/v. \qquad (28)$$

The straight lines in the graphs in Figs. 12 and 13 show these small density corrections. There is no real inconsistency with the molecular dynamics data because of the statistical noise in the latter and also because of the finite size of the system.

Finally, comparison is made of the theoretical results with experimental determinations of the transport coefficients of the rare gases at high temperatures.[11] This is a refinement of the comparison made earlier when only results from the Enskog theory were available. The question is whether these high-temperature experiments conform more closely to the more accurate results by molecular dynamics and hence whether hard spheres can accurately describe the actual phenomena or whether the effect of the soft repulsive potential of real atoms and the effects of the attractive forces must be taken into account. The comparison given in Figs. 12 and 13 indicates that the many-body corrections of hard spheres certainly improve agreement with experiments. For both the viscosity and thermal conductivity deviations from the Enskog theory are even larger than those indicated by molecular dynamics results. However, the following things must be kept in mind: First, the disagreement is very small, only a few percent, so that the predictions of the hard-sphere theory are remarkably accurate. Second, the molecular dynamics results must be extrapolated to infinite numbers of particles. Such extrapolation appears to increase the values for the hard-sphere viscosity and also for the thermal conductivity, which would still further narrow the discrepancy. Final comparison hence must await more reliable means for making such extrapolations.

ACKNOWLEDGMENTS

We wish to thank Mary Ann Mansigh for her superb programming efforts and John Dymond for his analysis of the experimental data.

* Work performed under the auspices of the U.S. Atomic Energy Commission.

[1] B. J. Alder and T. E. Wainwright, Phys. Rev. Letters **18**, 988 (1967).
[2] B. J. Alder and T. E. Wainwright, J. Phys. Soc. Japan Suppl. **26**, 267 (1968).
[3] B. J. Alder and T. E. Wainwright, Phys. Rev. A **1**, 18 (1970).
[4] J. R. Dorfman and E. G. D. Cohen, Phys. Letters **16**, 124 (1965); J. Weinstock, Phys. Rev. **140**, A460 (1965).
[5] K. Kawasaki and I. Oppenheim, Phys. Rev. **139**, A1763 (1965).
[6] E. Helfand, Phys. Rev. **119**, 1 (1960); D. M. Gass, J. Chem. Phys. **51**, 4560 (1969).
[7] S. Chapman and T. G. Cowling, *The Mathematical Theory of Non-Uniform Gases* (Cambridge U. P., London, 1939).
[8] T. Wainwright, J. Chem. Phys. **40**, 2932 (1964).
[9] S. A. Rice and P. Gray, *The Statistical Mechanics of Simple Liquids* (Interscience, New York, 1965).
[10] J. V. Sengers, in *Lectures in Theoretical Physics Vol. IX C Kinetic Theory*, edited by W. Brittin (Gordon and Breach, New York, 1967), p. 335.
[11] J. H. Dymond and B. J. Alder, J. Chem. Phys. **48**, 343 (1968).

Notes to Reprint III.1

1. The wavelength and frequency-dependent linear transport coefficients, transverse-current correlation functions and dynamic structure factor of the hard-sphere fluid have been computed by Alley and Alder (Phys. Rev. A 27 (1983) 3158) and by Alley et al. (Phys. Rev. A 27 (1983) 3174)) who also compare their results with the predictions of kinetic theory.

2. More precise results for the shear viscosity of hard spheres have subsequently been obtained by Erpenbeck and Wood (J. Stat. Phys. 24 (1981) 455).

PHYSICAL REVIEW A VOLUME 7, NUMBER 5 MAY 1973

Computer "Experiments" on Classical Fluids. IV. Transport Properties and Time-Correlation Functions of the Lennard-Jones Liquid near Its Triple Point

D. Levesque and L. Verlet
*Laboratoire de Physique Théorique et Hautes Energies, Orsay, France**

Juhani Kürkijarvi
University of Helsinki, Finland
(Received 14 September 1972)

A molecular-dynamics "experiment" was performed for a system of 864 particles interacting through a Lennard-Jones potential. The state considered was in the immediate neighborhood of the triple point. The total duration of the "experiment" was quite large: It corresponds to 10^{-9} sec in the case of argon. Transport coefficients were calculated using the standard Kubo formulas. They are compared with the prediction of a simple hard-sphere model. It is shown that, as in the case of the hard-sphere fluid near solidification, the Kubo-correlation function relative to the shear viscosity presents a tail extending at large time. The inclusion of this tail turns out to be essential in explaining the transverse-correlation function and the dynamical-structure factor, which shows, for the lowest wave vectors accessible in this study, a characteristic Brillouin doublet structure. Using the hydrodynamical model of Zwanzig and Bixon, it is shown that the introduction of the long-time tail in the Kubo-correlation function for the viscosity explains the negative plateau of the velocity-autocorrelation function observed near the triple point by Rahman and others.

I. INTRODUCTION

In this paper, we shall report and analyze the results of a computer simulation of argon near its triple point. The aim of this "experiment" was to obtain information on the transport coefficients and the time-dependent correlation functions.

As in the preceding papers of this series[1,2] 864 atoms were considered: enclosed in a cubic box of side L (with periodic boundary conditions). They interact through a Lennard-Jones (LJ) potential

$$V(r)=4\epsilon\left[(\sigma/r)^{12}-(\sigma/r)^{6}\right]. \quad (1.1)$$

We choose σ, ϵ, and $\tau_0=(m\sigma/48\epsilon)^{1/2}$ as length, energy, and time units, respectively. When we make a comparison with real argon, we choose $\sigma=3.405$ Å, $\epsilon=119.8k_B$, $\tau_0=3.112\times10^{-13}$ sec.

Using the method described elsewhere,[1] the integration of the equation of motion was carried out for 100 800 integration steps of $0.032\tau_0$. This corresponds to a total time of 10^{-9} sec in argon. During the integration we calculated the quantities whose time correlation enters in the various Kubo formulas for the transport coefficients. These quantities as well as the coordinates and velocities of the particles are kept on tape.

Section II is devoted to a discussion of the thermodynamics of the state which is considered in this study.

In Sec. III, the results for the shear viscosity, the bulk viscosity, and the thermal conductivity are given. They are interpreted successfully with a hard-sphere model which can be built using the very complete results obtained by Alder, Gass, and Wainwright[3] for the transport coefficients of the hard-sphere gas. The more striking results of this section are the following: The Kubo correlation function for the shear viscosity η presents a tail extending to large times; due to the large incompressibility of the liquid near the triple point, the bulk viscosity ξ tends to be small; because of these two effects, the ratio ξ/η is not of the order of 1 as generally expected, but much smaller, of the order of $\frac{1}{4}$. All these results agree very well with similar properties of the hard-sphere gas near solidification.[3]

We then proceed (Sec. V) to study the correlation of transverse currents. Our results complement with an increased accuracy those obtained by Rahman.[4] They are concentrated in a region of relatively long wavelength, in order to study the generalization of linearized hydrodynamics.[5-8] As a first approximation, we analyze our results in terms of a viscoelastic theory with a k-dependent relaxation time. Shear waves appear as predicted by the theory for $k \gtrsim \sigma^{-1}$ (0.3 Å^{-1} in argon).

The simple viscoelastic theory appears to be inadequate at low wave vectors; it yields shear-wave peaks which are too broad and too low. The more solidlike behavior of the molecular-dynamics results can only be accounted for if one also introduces at finite k's the long-time tail observed at $k=0$. This tail is described by a second exponential with a large relaxation time. It tends to disappear when k increases and is seen to be related to a collective effect involving a small group of particles.

The dynamical-structure factor $S(k, \omega)$ is then computed and analyzed (Sec. V). It is shown that for $k \lesssim \sigma^{-1}$ this quantity still presents a secondary maximum which is the remainder of the Brillouin doublet as a function of ω. An analysis of the data can be made in terms of three parameters: a frequency-dependent longitudinal viscosity; a k-dependent thermal conductivity; and a k-dependent ratio $\gamma(k)$. Assuming a single-relaxation-time form for the longitudinal viscosity, an excellent fit can be obtained. A disturbing element of this fit is, however, that the limit of $\gamma(k)$ when k goes to zero turns out to be unreasonably large. This failure can be traced back to the neglect of the long-time tail in the generalized viscosity. The inclusion of this tail, as in the transverse case, leads to a completely satisfactory description of the data.

In Sec. V, we reexamine the model which was proposed by Zwanzig and Bixon[9] for the description of the velocity-autocorrelation function. In this model the motion of a LJ molecule was approximated by that of a hard sphere moving in a viscoelastic medium. The constants entering the model are all given by the molecular-dynamics computation. A single-relaxation-time viscoelastic theory leads to results similar to those obtained by Zwanzig. The velocity-autocorrelation function obtained from molecular dynamics is fairly well reproduced, but the model gives rise to oscillations when the "experimental" v.a.f. exhibits a negative plateau at large times. The inclusion of the large-time tail results in a frequency-dependent viscosity coefficient which provides a correct description of the long-time behavior of the velocity-autocorrelation function.

II. THERMODYNAMICAL CONSIDERATIONS

The molecular-dynamics computation reported in this paper has been made for the reduced density $\rho = 0.8442$ and the reduced temperature $T = 0.722$. This state is very near the triple point of the LJ potential[10] which is characterized by $\rho_t = 0.85 \pm 0.01$ and $T_t = 0.68 \pm 0.01$. It is also in the immediate neighborhood of the solidification line; the value of the maximum of the structure factor is equal to 2.76, whereas it reaches the value 2.85 on the solidification line.[10] Using the above-mentioned reduction constant, our state corresponds in argon to $\rho = 1.418$ g/cm^3, $T = 86.5$ °K. The triple point of argon is quite close: $\rho = 1.435$ g/cm^3, $T = 83.8$ °K. For our state we obtain for the compressibility factor $P/\rho kT = 0.25$ and for the configurational energy per particle $U_i/N = -6.08$.

We can get the specific heat at constant volume through the fluctuations of the kinetic and potential energy.[11] The first method yields the value 2.6, the other 2.8. We thus choose $c_V = 2.7 \pm 0.1$.

We remark here that, although we have made a computation 100 times longer than most of those usually made for continuous potentials, the error in the specific heat remains of the order of 5%.

The fluctuation of the product of the potential energy and of the virial yields[10]

$$\left(\frac{\partial P}{\partial T}\right)_V = 6.41 \pm 0.2 .$$

We obtain the inverse compressibility by extrapolating the structure factor $S(k)$ computed directly (Sec. V) to zero wave vector. We thus obtain

$$\beta\left(\frac{\partial P}{\partial \rho}\right)_T = 24.7 \pm 0.5 .$$

Using the prolongation procedure described by one of us,[12] with the help of which the complete $g(r)$ from the limited amount of information provided by the molecular-dynamics computation is obtained, we obtain

$$\beta\left(\frac{\partial P}{\partial \rho}\right)_T = 24.0 .$$

Combining these three thermodynamics derivatives, we obtain

$$\gamma = \frac{c_P}{c_V} = 1 + \left(\frac{\partial P}{\partial T}\right)_V^2 \frac{1}{\rho c_V} k_B T \left(\frac{\partial \rho}{\partial P}\right)_V$$
$$= 1.86 \pm 0.1 .$$

From this, we obtain for the speed of sound in reduced units, $c = 0.831$ or $= 906$ m/sec. This is compatible with the experimental value[13] $c = 876$ m/sec. The experimental value for γ is equal to 2.

It is interesting to take this opportunity to examine the predictions of the thermodynamic perturbation theory[14,15] concerning these various thermodynamic derivatives. Using the expressions given in Ref. 15, we obtain

$$c_V = 2.57 , \qquad \gamma = 1.99 ,$$

$$\left(\frac{\partial P}{\partial T}\right)_V = 6.42 , \quad \beta\left(\frac{\partial P}{\partial \rho}\right)_T = 22.6 .$$

III. TRANSPORT COEFFICIENTS

A. Shear Viscosity

The shear viscosity is obtained through the well-known Kubo-like formula

$$\eta = \int_0^\infty \eta(t)\, dt \tag{3.1}$$

with

$$\eta(t) = \frac{\rho}{3k_B T} \sum_G \frac{\langle \tau^{xy}(0)\, \tau^{xy}(t)\rangle}{N} , \tag{3.2}$$

where the sum is to be made on the circular permutation of the indices, and where τ^{xy} component of the microscopic stress tensor given by

$$\tau^{\alpha\beta}(t)=\sum_{i=1}^{i=N}\left(Mv_{\beta}{}^{\alpha}-\frac{1}{2}\sum_{j\neq i}\frac{r_{ij}{}^{\alpha}r_{ij}{}^{\beta}}{r_{ij}{}^{\beta}}\frac{\partial V}{\partial r_{ij}}\right). \tag{3.3}$$

The average over the initial time, indicated by the bracket in (3.2), is done over 27 000 values for the initial time separated by $0.128\tau_0$. The noise level on $\eta(t)$ can be estimated to be about 2–3% of its value for $t=0$, which is equal to the infinite-frequency shear modulus at zero wave vector $G_\infty(0)$. This shear modulus, which can be expressed in terms of the two-body radial-distribution function,[16,17] turns out to be especially simple in the case of the LJ potential, as Zwanzig and Mountain have shown[16]:

$$G_\infty(0)=3P/\rho-\tfrac{24}{5}(U_i/N)-2k_BT, \tag{3.4}$$

where P is the pressure and U_i/N is the configurational energy per particle. Using the thermodynamical results given in Sec. I we obtain

$$G_\infty(0)=23.9 .$$

The evaluation of $\eta(0)$ made using (3.2) yields $\eta(0)=24.7$. The shear viscosity obtained by integrating $\eta(t)$ up to a time of $12.8\tau_0$ is $\eta=27.9\pm2$.

McDonald and Singer[18] have recently calculated the shear viscosity by producing an actual shear on a system of 256 LJ particles.

This method succeeds in calculating the shear viscosity with much less computational effort than in the present paper. It should be noted, however, that this method unavoidably yields a shear viscosity determined for a finite value of the wave length, namely L.

Using the results obtained in Sec. IV, one can understand that these authors obtain a shear viscosity smaller than ours by around 30%.

We shall interpret the value obtained for the shear viscosity with the help of the hard-sphere model already used in the case of the self-diffusion constant.[1] This model was established using preliminary hard-sphere results.[19] Due to the appearance of the more complete and accurate hard-sphere results of Ref. 3, we give here again a detailed explanation of the model. We replace the LJ molecules by hard spheres of diameter d. There is obviously some arbitrariness in choosing this diameter.[20] We find that a good choice consists of taking for the diameter d the value which enables one to fit the equilibrium structure factor $S(k)$ with the analytical hard-sphere structure factor obtained by Wertheim and Thiele as the solution of the PY equation.[21]

For the state we consider, we have $d=1.02$. The packing fraction $\xi=\frac{1}{6}\pi\rho d^3$ is then equal to 0.47. Using the Carnahan–Starling[22] expression for the hard-sphere pressure, we obtain

$$y=(P/\rho k_BT)_{\rm HS}-1=9.7 .$$

The Enskog mean collision time $\tau_{\rm col}$ can then be obtained. This enables us to connect the scale of time of the hard-sphere gas with that of the LJ molecules:

$$\tau_{\rm col}=(2d/y)(\pi/3T)^{1/2}=0.226 . \tag{3.5}$$

We see, by the way, that the computation of the LJ transport coefficients which we have made includes 10^7 equivalent hard-sphere collisions. We thus expect the same kind of accuracy as in Alder's longest computations.[3]

The model yields for the diffusion constant

$$D=D_{\rm HS}^E f_D(\xi), \tag{3.6}$$

where $D_{\rm HS}^E$ is the Enskog value for the hard-sphere diffusion constant,

$$D_{\rm HS}^E=\tfrac{1}{32}\tau_{\rm col}T, \tag{3.7}$$

and $f_D(\xi)$ is the correction to the Enskog diffusion constant empirically determined by Alder *et al.*[3] from molecular-dynamics computations on the hard-sphere system. We now apply this model to obtain the diffusion constant of the LJ fluid.[2] We have used in Ref. 2 the preliminary hard-sphere results of Alder and Wainwright.[19] The model yielded diffusion constants which were too large by 20 or 30%. With the new results[3] for $f_D(\xi)$, the discrepancy is reduced to about 10%. In view of the basic roughness of this hard-sphere model it does not make sense to try to improve those results by choosing some other definition of the equivalent hard-sphere diameter.

For the shear viscosity, we found a similar expression

$$\eta=\eta_{\rm HS}^E f_\eta(\xi), \tag{3.8}$$

where $f_\eta(\xi)$ is the empirical correction[3] to the Enskog approximation for the hard-sphere viscosity $\eta_{\rm HS}^E$ which is given by

$$\eta_{\rm HS}^E=(10\,\xi/dy\tau_{\rm col})(1/y+0.8+0.761y). \tag{3.9}$$

For the state considered in this paper, we get $f_\eta(\xi)=1.54$ and $\eta=32.8$. The agreement with the "exact" result is good. The use of the hard-sphere model is seen to overestimate the shear viscosity and to underestimate the diffusion constant.

A consequence of the model is the following: Alder, Gass, and Wainwright[3] have shown that for the hard-sphere gas at high density, Stokes's law with slip boundary conditions holds within 10%. The hard-sphere model, therefore, implies that the Stokes relation also holds for the LJ molecules. It reads, with the present notations,

$$\eta=T/2\pi dD . \tag{3.10}$$

For the state under consideration, we have[2] $D=0.0047$. Equation (3.10) gives $\eta=24$, in good

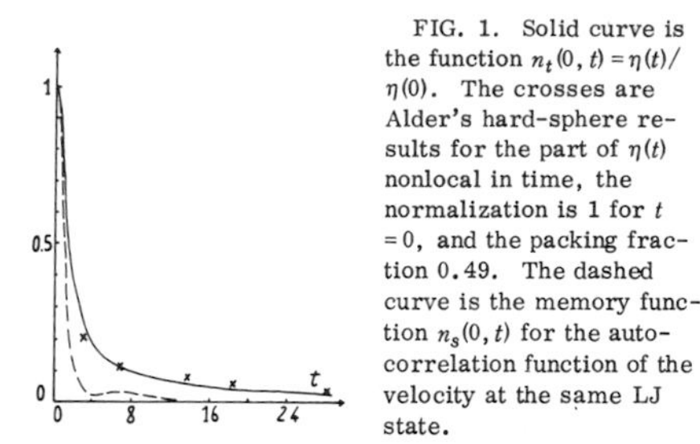

FIG. 1. Solid curve is the function $n_t(0, t) = \eta(t)/\eta(0)$. The crosses are Alder's hard-sphere results for the part of $\eta(t)$ nonlocal in time, the normalization is 1 for $t = 0$, and the packing fraction 0.49. The dashed curve is the memory function $n_s(0, t)$ for the autocorrelation function of the velocity at the same LJ state.

agreement with the molecular-dynamics result.

In Fig. 1, we have plotted the function

$$\eta_t(0, t) = \eta(t)/G_\infty(0) , \tag{3.11}$$

which is, as we shall see (Sec. IV), the memory function for the transverse-current correlation function at zero wave vector. We see that this function presents a very long tail slowly decaying at large times. We show in Fig. 1, for the sake of comparison, the memory function for the autocorrelation function $n_s(0, t)$ for the same state. As was shown in Ref. 2, this function has a fairly large extension in time which corresponds to the well-known negative-plateau region in the velocity-autocorrelation function near the triple point.[23] The tail of $n_t(0, t)$ is seen to have a different shape and a slower decay.

A long tail in $\eta(t)$ was also observed by Alder, Gass, and Wainwright for the hard-sphere gas near solidification. In Fig. 1 we have shown by crosses the hard-sphere results: they correspond to the part of $\eta(t)$ which is nonlocal in time with a normalization of one for $t = 0$. The value of the packing fraction is 0.49. Due to the difference in normalization and packing fraction, the quantitative agreement with the LJ data is coincidental. What we want to emphasize is the remarkable similarity of the slow decay in time. This leads to a large enhancement of the shear viscosity when one gets near solidification.

Using the above-defined hard-sphere model, the value of the packing fraction is a constant along the solidification line. Using (3.5) and (3.9), we thus have

$$\frac{\eta}{\eta_t} = \left(\frac{T}{T_t}\right)^{1/2} \left(\frac{d}{d_t}\right)^{-2} . \tag{3.12}$$

At the triple point of experimental argon we get for the hard-sphere viscosity

$$\eta_t = 27.8 = 3.64 \times 10^{-3} \text{ P} .$$

Taking into account the variation of d along the solidification line which can be derived using the LJ transition data of Ref. 10, we obtain

$$\eta \simeq \eta_t (T/T_t)^{0.63} . \tag{3.13}$$

If there is some truth in the hard-sphere model, the experimental situation appears rather puzzling. At the triple point of argon, Boon, Legros, and Thomas[24] obtain $\eta = 2.89 \times 10^{-3}$ P. On the other hand, at a slightly different temperature ($T = 88.5$ °K), very close to the transition line, de Bock *et al.*[25] obtain $\eta = 3.34 \times 10^{-3}$ P which, in view of (3.13), would yield a value of 3.25×10^{-3} P for the viscosity of the triple point, in apparent contradiction with the value obtained by Boon *et al.* For the state considered in the present study, these authors give the value $\eta = 2.71 \times 10^{-3}$ P, which differs significantly from the value we obtain for the LJ fluids, i.e., $\eta = 3.64 \times 10^{-3}$ P.

B. Bulk Viscosity

The Kubo formula for the bulk viscosity can be expressed as

$$\xi = \int_0^\infty \xi(t)\, dt , \tag{3.14}$$

with

$$\xi(t) = \frac{\rho}{9k_B T} \sum_{\alpha, \beta = 1, 2, 3} \langle [\tau^{\alpha\alpha}(t) - \langle \tau^{\alpha\alpha} \rangle] \times [\tau^{\beta\beta}(0) - \langle \tau^{\beta\beta} \rangle] \rangle . \tag{3.15}$$

The function $\xi(t)/\xi(0)$ is plotted in Fig. 2. It is seen that it is a rapidly decreasing function with no appreciable long-time tail. This behavior is in accordance with the hard-sphere results.[3] Zwanzig and Mountain[16] have shown that

$$\xi(0) = K_\infty - K_0 , \tag{3.16}$$

where K_∞ and K_0 are the infinite and zero-frequency bulk moduli, respectively. For K_∞, we have the relation[16]

$$K_\infty = \tfrac{5}{3} G_\infty + 2\rho k_B T (P/\rho k_B T - 1) = 38.7 . \tag{3.17}$$

For K_0, the adiabatic bulk modulus, we have

$$K_0 = \rho\gamma \left(\frac{\partial P}{\partial \rho}\right)_T = 28.3 \pm 2 . \tag{3.18}$$

We therefore find $K_\infty - K_0 = 10.4 \pm 2$, which

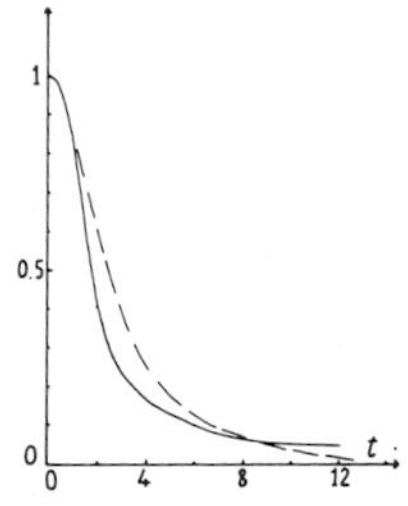

FIG. 2. Solid and dashed curves represent, respectively, $\chi(t)/\chi(0)$ and $K(t)/K(0)$, the normalized Kubo integrands for the bulk viscosity and the heat conductivity (the time unit is 10^{-13} sec).

Notes on p. 192

agrees, within the expected errors, with the result of the molecular-dynamics computation,

$$\xi(0)=12.3\pm 1 .$$

The computation also yields $\xi=7.3\pm 0.8$, and, therefore, the ratio

$$\xi/\eta=0.26\pm 0.05 .$$

For the hard-sphere fluid, Enskog theory gives for that ratio

$$\frac{\xi^E_{\rm HS}}{\eta^E_{\rm HS}}=\frac{1.002y}{0.761y+0.8+y^{-1}} . \tag{3.19}$$

We obtain $\xi^E_{\rm HS}/\eta^E_{\rm HS}=1.1$ for the state considered in this study. This Enskog ratio must be multiplied by the correction factor empirically determined by Alder's group,[3] which for the hard-sphere system near the transition line is of the order of $\frac{1}{4}$. Once again, the hard-sphere model works very well. The low value of ξ/η near the transition line appears to be due to the combination of two factors: the large value of the shear viscosity η owing to the appearance of a long tail in $\eta(t)$ near solidification; the low value of the bulk viscosity, which in the present computation can be related with the appearance of a very high adiabatic bulk modulus when one gets near to the transition line.

This low value of the ratio ξ/η near solidification predicted by the hard-sphere model and "observed" in our computer calculation seems to contradict the existing experiments[13] which yields $\xi/\eta \simeq 0.8$ near the triple point. It would be interesting to have newer, more precise experimental material related to this problem.

C. Thermal Conductivity

The thermal conductivity is given by

$$K=\int_0^\infty K(t)\,dt , \tag{3.20}$$

with

$$K(t)=\langle \mathcal{T}^\alpha_\epsilon(t)\,\mathcal{T}^\alpha_\epsilon(0)\rangle , \tag{3.21}$$

when the energy density flux is given by

$$\mathcal{T}^\alpha_\epsilon(t)=\left(\sum_{i=1}^{i=N}\frac{mv_i^2}{2}+\sum_{i=1}^{i=N}\sum_{j\neq i}\frac{V(r_{ij})}{2}\right)v_i^\alpha - \sum_{i=1}^{i=N}\sum_{j\neq i}\frac{\partial V(r_{ij})}{\partial r_{ij}{}^\alpha}\,\frac{\vec{r}_{ij}\cdot\vec{v}_i}{2} . \tag{3.22}$$

Due to our use of the expression of the thermal conductivity due to Luttinger,[26] we have made the computation with the last term replaced by

$$-\frac{1}{2}\sum_{i=1}^{i=N}\sum_{j\neq i}[\vec{\nabla} V(r_{ij})\cdot\vec{r}_{ij}]\,v_i{}^\alpha .$$

We believe, however, that our results are essentially correct: The contribution of the autocorrelation of this erroneous term to the total heat conductivity is of the order of 0.1%. Its cross terms with the kinetic- and potential-energy fluxes amount to less than 1%. A direct inspection of the corrected terms shows that they should be of the same order, and therefore negligible. Furthermore, we shall see in Sec. V that the extrapolation of the k-dependent heat conductivity which appears in the analysis of $S(k,\omega)$ is in good agreement with the value obtained directly.

$K(t)/K(0)$ is plotted in Fig. 2. As is the case for hard spheres, there appears to be no long tail as a function of time. For the value of the thermal conductivity, we obtain $K=2.14$.

Using again the hard-sphere model, we obtain in the Enskog approximation

$$K^E_{\rm HS}=\tfrac{25}{32}(\xi/\pi d^2)\,c_V^{\rm HS}(\tfrac{1}{3}\pi T)^{1/2}(1/y+1.2+0.755y) , \tag{3.23}$$

where $c_V^{\rm HS}$ is the hard-sphere specific heat, equal to 1.5.

Making the Alder correction,[3] we obtain

$$K=f_K(\xi)\,K^E_{\rm HS}=1.9 .$$

The agreement is quite satisfactory but may be coincidental: it may be argued that we should have tried to apply the hard-sphere model not to K, but rather to the quantity

$$a=K/\rho c_V ,$$

which appears naturally in the theory. In that case, the agreement is completely destroyed.

The comparison with experiment turns out to be disappointing. The molecular-dynamics computation gives

$$a=K/\rho c_V=0.94$$
$$=3.5\times 10^{-3}\ {\rm cm^2/sec}.$$

Using the experimental data quoted by Naugle *et al.*,[13] we get

$$a=1.68\times 10^{-3}\ {\rm cm^2/sec} ,$$

which differs by a factor 2. We may notice that Chung and Yip[6] use, in the analysis of Rahman's[4] computation, the value $a=3.4\times 10^{-3}$ cm^2/sec, which agrees very well with ours. The origin of this number is unfortunately not clear.

IV. TRANSVERSE-CURRENTS CORRELATION

The transverse-current correlation function is defined as

$$C_t(k,t)=k^2\left\langle \sum_{i=1}^{i=N} v_i{}^x(t)\,e^{-ikz_i(t)}\sum_{j=1}^{i=N} v_j{}^x(0)\,e^{ikz_j(0)}\right\rangle , \tag{4.1}$$

where $\vec{k}$ is along the z axis.

TABLE I. k is the mean value of the group of vectors $\vec{k}$ considered in the computation of the eight functions $S(k, \omega)$, with the corresponding multiplicity. Δk is the difference between the mean norm of the vector $\vec{k}$ and the norm of the largest or smallest vector of each group.

k	Multiplicity	Δk
0.6235	3	0
0.7526	6	0.13
0.8817	3	0
1.3667	13	0.28
1.9319	19	0.23
2.5348	39	0.28
3.1822	55	0.23
3.8124	72	0.23

Let us introduce its Fourier–Laplace transform:

$$\tilde{C}_t(k, \omega) = \int_0^\infty e^{i\omega t} C_t(k, t)\, dt . \qquad (4.2)$$

The behavior of $C_t(k, t)$ at small times and the hydrodynamic limit are included through the use of the memory-function formalism.[5–8]

Let us write

$$\tilde{C}_t(k, \omega) = \frac{\omega_0^2}{-i\omega + \omega_t^2 \tilde{n}_t(k, \omega)} , \qquad (4.3)$$

where

$$\omega_0^2 = C_t(k, 0) = k^2 (k_B T/m) \qquad (4.4)$$

and[27]

$$\omega_t^2 = \frac{1}{\omega_0^2} \frac{d^2 C_t}{dt^2}\bigg|_{t=0} = \omega_0^2 + \frac{\rho}{m} \int d\vec{r}\, g(r)\, \frac{\partial^2 V}{\partial x^2}\, (1 - e^{ikx})$$

$$= \frac{k^2}{\rho m} G_\infty(k) . \qquad (4.5)$$

This last equation defines $G_\infty(k)$, the k-dependent shear modulus. In order to obtain the correct hydrodynamic limit, we must have

$$\eta = G_\infty(0) \tilde{n}_t(0, 0). \qquad (4.6)$$

The average over initial states implied in (4.1) was made over 33 600 states with an interval of $0.128\tau_0$. $C_t(k, t)$ was calculated for wave vectors whose components were the first multiples of $2\pi/L$.

All the vectors whose length was comprised between k and $k + \Delta k$ were bunched together. The average of k as well as the multiplicity of each group is given in Table I.

We have plotted in Fig. 3 as a function of ω the results of the computation for $\tilde{C}_t(k, \omega)$. We see that, except for the lowest value of the wave vector, $\mathrm{Re}\tilde{C}_t(k, \omega)$ presents a maximum for a nonzero frequency. This is characteristic of the existence of shear waves. If we take for the memory function a simple relaxation form

$$n_t(k, t) = e^{-t/\tau_t(k)} , \qquad (4.7)$$

it is easy to see that $\mathrm{Re}\tilde{C}_t(k, \omega)$ presents a maximum for ω different from zero if $k > k_C$ where

$$k_c = \left(\frac{m\rho}{2G_\infty(k)}\right)^{1/2} \frac{1}{\tau_t(k)} . \qquad (4.8)$$

If we neglect the k dependence, we easily get this limiting wave vector. Using (4.6) and (4.7), we have

$$\tau_t(0) = \eta/G_\infty(0) = 1.17 . \qquad (4.9)$$

We see then from (4.8) that shear waves appear when k is larger than $0.79\sigma^{-1}$.

In the relaxation approximation, $\tilde{C}_t(k, \omega)$ depends only, for each value of k, on the parameter $\tau_t(k)$. It is determined from the computer data through a least-squares fit either in the t or the ω variable: The results turn out to be identical. The relaxation times thus obtained are given in Table II and shown in Fig. 4, as circles. We see that the extrapolation to the hydrodynamic limit (4.9) is quite smooth. We have also shown in Fig. 4 the results with crosses of Chung and Yip[6] who have analyzed Rahman's molecular-dynamics computation for the state $\rho = 0.83$, $T = 0.635$. The agreement is seen to be quite good.

In Fig. 3, we compare $\mathrm{Re}\tilde{C}_t(k, \omega)$ obtained in the relaxation approximation with the molecular-dynamics results. We see that the agreement is not very good. In particular, for long wavelengths, the shear-wave peaks are very much flattened out when the relaxation approximation is used. This discrepancy is owing to our neglect of the long-time tail of the memory function $n_t(k, t)$. In order to appreciate the effect of this tail, let us represent the transverse memory function through the two-time exponential formula

$$n_t(k, t) = [1 - \alpha(k)] e^{-t/\tau_<(k)} + \alpha(k) e^{-t/\tau_>} . \qquad (4.10)$$

We see, using the machine data for $n_t(0, t)$ shown in the Sec. III (Fig. 1), that an excellent fit is ob-

TABLE II. Values of $\tau_t(k)$, $\tau_<(k)$, and $\tau_t'(k)$ in reduced units. The values of $\tau_<(k)$ and $\tau_t'(k)$ given by least-squares fits of $C_t(k, \omega)$ and $S(k, \omega)$ are almost the same. Their difference might be due mostly to the statistical errors of the molecular-dynamics computations.

k	τ_t	$\tau_<$	τ_t'
0	1.17	0.64	0.64
0.7526	0.91	0.70	0.70
1.3667	0.74	0.65	0.58
1.9319	0.66	0.60	0.54
2.5348	0.58	0.54	0.50
3.1822	0.52	0.50	0.45
3.8124	0.46	0.44	0.45

Notes on p. 192

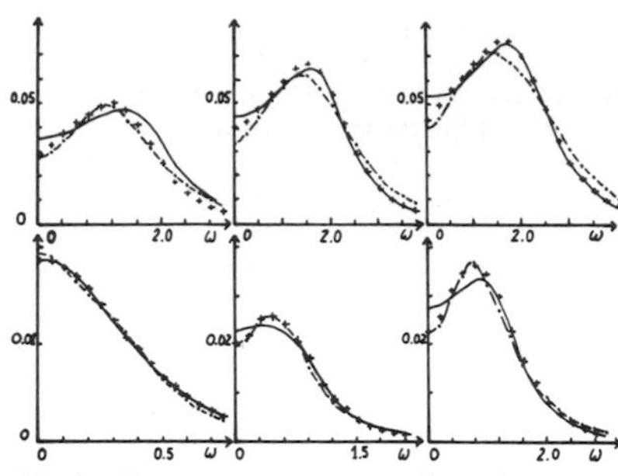

FIG. 3. Transverse-current correlation functions $\mathrm{Re}C_t(k,\omega)/2\pi$ as a function of ω (units τ_0^{-1}). From left to right, and from bottom to top, the six curves are for $k=0.752$, 1.366, 1.931, 2.534, 3.182, and 3.812 (units σ^{-1}). The dash-dot line is the molecular-dynamics results and the solid curve is the relaxation-approximation results. The crosses are the results of the approximation with the memory function (4.10).

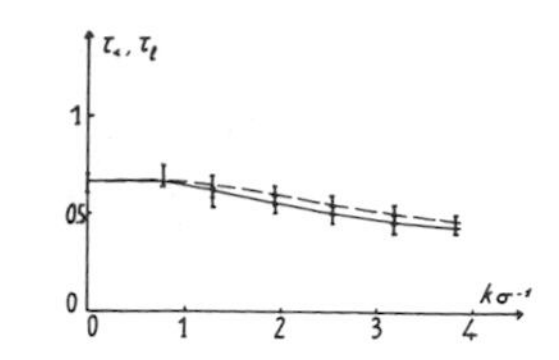

FIG. 5. Short relaxation time in the two-term approximations for the memory function of $n_t(k,t)$ and $n_l(k,t)$. The solid curve is $\tau_l'(k)$, and the dashed curve $\tau_<(k)$ (time in units of τ_0).

tained with $\tau_> = 4.72$, $\alpha(0)=0.128$, and

$$\tau_<(0) = [\eta/G_\infty(0) - \alpha\tau_>]/(1-\alpha) = 0.63 .$$

We see the appearance of two times: one $\tau_<$, which is of the order of the molecular-relaxation time, the other $\tau_>$, which is much larger.

We expect that the tail is due to some collective effect. The spatial extension of this tail can be measured through the decay of $\alpha(k)$ and is seen to be of the order of several interparticle distances.

Fitting the molecular dynamics $\tilde{C}_t(k,\omega)$ with the memory function given by (4.10), we obtain the curves shown in Figs. 5 and 6 for $\tau_<(k)$ and $\alpha(k)$. In the latter case, a smooth extrapolation to $k=0$ seems to lead to a value of $\alpha(0)$ which is smaller than 0.128 and may be of the order of 0.1. It can be seen that, given the errors on $C_t(k,t)$, and those, somewhat smaller, affecting $\eta(t)$, there is no real discrepancy there.

We see in Fig. 3 that an excellent fit is obtained for the wave vectors between 1 and 2 σ^{-1}. For the lowest value of k ($k=0.752\sigma^{-1}$), shear waves appear in the model and not in the molecular-dynamics results. This discrepancy is, however, within the computational uncertainties. The presence of the long-time tail lowers the value of the wave vector for which shear waves appear: It is easy to see that in the two-relaxation-time approximation this critical wave vector k_c is determined by solving for the smallest root of

$$\omega_t^4\,\alpha(1-\alpha)\,\tau_>\,\tau_<\,(\tau_>-\tau_<)^2 - 2\omega_t^2[\alpha\tau_>^2 + (1-\alpha)\,\tau_<^2] + 1 = 0 ,$$

where ω_t^2 is related to k by (4.5).

We thus obtain $k_c = 0.56\sigma^{-1}$ (with $\alpha=0.128$) instead of $k_c=0.78\sigma^{-1}$ in the single-relaxation-time approximation.

The lowering of the value of the critical wave vector is one consequence of the long-time tail. Another consequence, which is more important, is the large enhancement of the shear wave peaks in agreement with the molecular-dynamics experiments. This enhancement may be thought of as a precursory of the solidlike behavior in the neighborhood of the transition line.

For the largest value of k considered here, $\alpha(k)$ is negligible, and the single-relaxation-time approximation is practically recovered. This discrepancy of the fit with molecular dynamics is due to the well-known inadequacy of the exponential memory function which becomes more apparent when the small time region is weighed more, i.e., at high k's. In order to illustrate this statement, we plot in Fig. 7, for the case where $k=0.75\sigma^{-1}$ and $k=3.8\sigma^{-1}$, the memory function directly determined using (3.3). It is compared with the two-exponential approximation (4.10) (dotted curves).

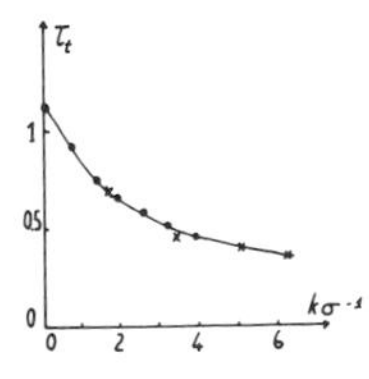

FIG. 4. Single relaxation time $\tau_t(k)$ for the transverse-current correlation function. The circles are our result and the crosses are those of Chung and Yip (Ref. 6) (time in units of τ_0).

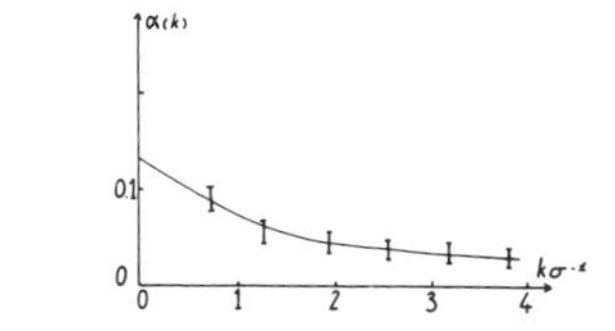

FIG. 6. Coefficient $\alpha(k)$ of the long-time tail of the transverse memory function (4.10).

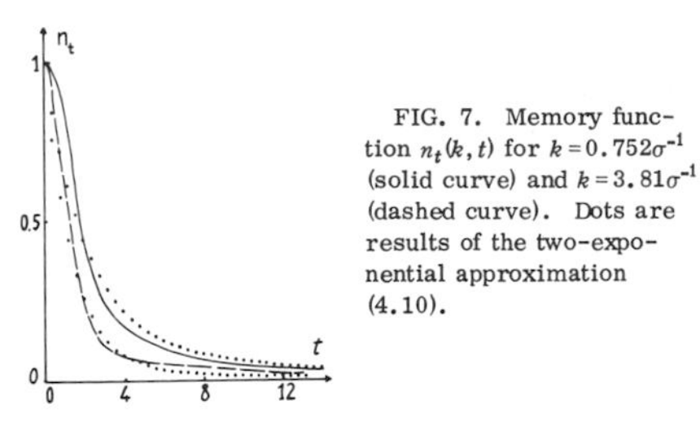

FIG. 7. Memory function $n_t(k,t)$ for $k=0.752\sigma^{-1}$ (solid curve) and $k=3.81\sigma^{-1}$ (dashed curve). Dots are results of the two-exponential approximation (4.10).

V. LONGITUDINAL-CURRENT CORRELATION AND DYNAMICAL-STRUCTURE FACTOR

The longitudinal-current correlation function is given by

$$C_l(k,t)=k^2\left\langle \sum_{i=1}^{i=N} v_i^z(t)\,e^{-ikz_i(t)} \sum_{j=1}^{j=N} v_j^z(0)\,e^{ikz_j(0)}\right\rangle . \tag{5.1}$$

The Fourier–Laplace transform of this function is

$$\tilde{C}_l(k,\omega)=\omega_0^2\bigg/\left(-i\omega+\frac{\omega_0^2}{-i\omega S(k)}+\tilde{N}_l(k,\omega)\right), \tag{5.2}$$

where $N_l(k,t)$ is the memory function for longitudinal currents. For small wave vectors and frequency, this must reduce to the hydrodynamical limit[5]

$$\tilde{C}_l(k,\omega)=\omega_0^2\bigg/\left(-i\omega+\frac{\omega_0^2}{-i\omega S(k)}+D_l k^2+\frac{\omega_0^2}{S(k)}\frac{\gamma-1}{-i\omega+ak^2}\right), \tag{5.3}$$

with

$$D_l=(1/m\rho)\left(\tfrac{4}{3}\eta+\xi\right). \tag{5.4}$$

The small-time expansion of $C_l(k,t)$ defines the value of $N_l(k,0)$. A form which includes both this information and the hydrodynamical limit is

$$\tilde{N}_l(k,\omega)=\left(\omega_l^2-\frac{\omega_0^2}{S(k)}\gamma(k)\right)\tilde{n}_l(k,\omega)+\frac{\omega_0^2}{S(k)}\frac{\gamma(k)-1}{-i\omega+a(k)k^2}, \tag{5.5}$$

with

$$n_l(k,0)=1. \tag{5.6}$$

For ω_l^2, we have the following expression,[27] in terms of the two-body potential:

$$\omega_l^2=\frac{1}{\omega_0^2}\left.\frac{d^2C_l}{dt^2}\right|_{t=0}=3\omega_0^2+\frac{\rho}{m}\int d\vec{r}\,g(r)\,\frac{\partial^2 V}{\partial z^2}\,(1-e^{ikz})$$

$$=\frac{k^2}{\rho m}\left[\tfrac{4}{3}G_\infty(k)+K_\infty(k)\right]. \tag{5.7}$$

This last relation defines $K_\infty(k)$. It reduces to (3.17) for $k=0$. Equation (5.5) includes the assumption that the ω dependence of the generalized thermal conductivity can be neglected.

In terms of the correlation function for longitudinal currents, the dynamical-structure factor is easily obtained as

$$S(k,\omega)=(1/\pi)\,\mathrm{Re}[\tilde{C}_l(k,\omega)/\omega^2]. \tag{5.8}$$

The molecular-dynamics experiments yield the Fourier transform of this function, the so-called intermediate scattering function,

$$F(k,t)=\langle\rho_k(0)\,\rho_{-k}(t)\rangle/N. \tag{5.9}$$

There, the time average includes 56.000 values for the initial time with a time step of $0.128\tau_0$.

The wave vectors have been grouped as in the transverse case, but with one exception: The lowest group for the transverse currents belonging to $k=0.75\sigma^{-1}$ is the average of two groups of wave vectors of equal lengths $k=0.63\sigma^{-1}$ and $k=0.88\sigma^{-1}$, which are calculated separately in the present case. The results obtained from the computer experiment for $S(k,\omega)$ are shown in Fig. 8. It is seen that, for the smallest wave vectors considered here, there is a secondary peak at finite frequency. Its maximum corresponds to a value of ω/k nearly equal to the macroscopic sound velocity. This structure could be observed experimentally using the long-wavelength neutrons available in high-flux reactors.

As in the transverse case, we first try a simple relaxation approximation by writing

$$\tilde{n}_l(k,\omega)=\frac{1}{-i\omega+1/\tau_l(k)}. \tag{5.10}$$

We can fit the curves of Fig. 8 very well and ob-

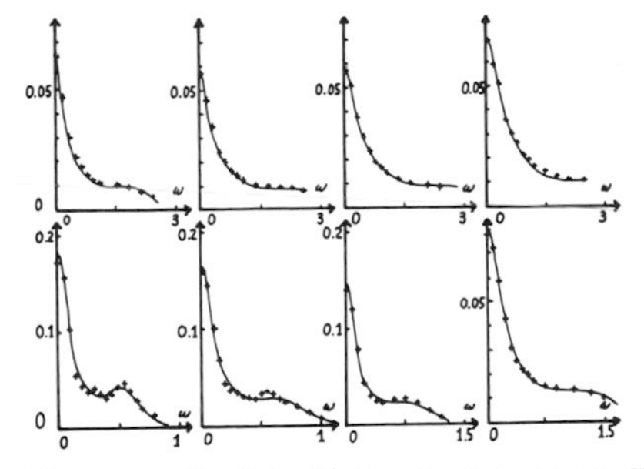

FIG. 8. Function $2S(k,\omega)/\pi$ for eight values of $k=0.623$, 0.752, 0.881, 1.366, 1.931, 2.534, 3.182, and 3.812, from left to right and from bottom to top (units of τ_0^{-1}). The crosses are the molecular-dynamical results and the solid curve is the representation with the memory function $N_l(k,\omega)$ (5.5).

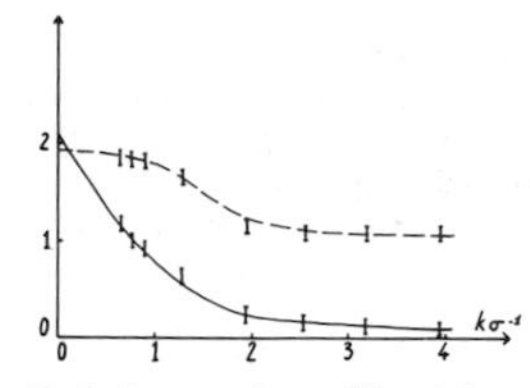

FIG. 9. Dashed curve gives $\gamma(k)$ and the solid curve $a(k)/\rho C_V$ which are two parameters of the memory function $N_l(k, \omega)$. [$a(k)/\rho C_V$ is in reduced units.]

tain the parameters $a(k)$, $\gamma(k)$, and $\tau_l(k)$.

This fit presents some satisfactory aspects: $\tau_l(k)$ tends for large wavelength to its hydrodynamic limit,

$$\tau_l(0) = (\tfrac{4}{3}\eta + \xi)/(\tfrac{4}{3}G_\infty + K_\infty - K_0) . \qquad (5.11)$$

The same is true for $a(k)$. Also, when k increases, $\gamma(k)$ tends to 1. This latter quantity, however, behaves quite badly when k is small. For $k = 0.63$, we obtain $\gamma(k) = 2.4$, and the extrapolation of the values obtained through the fit gives $\gamma(0) = 3.4$ instead of the "exact" value $\gamma = 1.86$. If we impose, for the lowest wave vector, a value of $\gamma(k)$ smaller than 2, the fit is completely spoiled.

We can trace this puzzling result to our neglect of the long-time tail which is related to the shear viscosity. Our simplified memory function depends on two times: a first time τ_l which is of the order of 0.6 and another time $1/ak^2$ which for the smaller wave vector is of the order of 3. This second time is substantially larger than the first one. It is of the same magnitude as the long relaxation time $\tau_>$ of the transverse memory function. It appears from these results that the inclusion of the long-time tail in the transverse memory function is necessary. When it is neglected, the fit can only be obtained by artificially raising the coefficient $\gamma(k) - 1$ of the thermal diffusion term: Then due to the similarity in the relaxation times, this term offers, for wave vectors of the order of 0.6, a fair simulation of the tail term which has been neglected.

We shall complete this analysis with the inclusion of the tail term in the longitudinal memory function. Due to the closeness of the bulk relaxation time with $\tau_<(0)$, we can use for the longitudinal memory function the expression

$$n_l(k, t) = [1 - \alpha_l(k)] e^{-t/\tau_l'(k)} + \alpha_l(k) e^{-t/\tau_>} \qquad (5.12)$$

with

$$\alpha_l(k) = \tfrac{4}{3}\eta(0)\, \alpha(k)/[\tfrac{4}{3}\eta(0) + \xi(0)] . \qquad (5.13)$$

The free parameters are now $a(k)$, $\gamma(k)$, and $\tau_l'(k)$. $a(k)/\rho G_V$ is plotted in Fig. 9; we see that it extrapolates correctly to the value of the heat conductivity given in Sec. III. $\gamma(k)$ now tends smoothly to the value $\gamma = 1.86$ which should be reached for $k = 0$. It should be noted that $\gamma(k)$ is practically equal to 1 as soon as k is of the order of 2. The coupling with the thermal modes is then negligible. $\tau_l'(k)$ is very close to $\tau_<(k)$ for all values of k's. These relaxation times have been obtained independently by fitting the longitudinal and transverse correlation function, respectively.

Given the computational uncertainties, the difference between those two relaxation times is not significant. The fit obtained for $S(k, \omega)$ is, on the whole, very good. For the highest values of k considered in this study (k around 3, i.e., of the order of 1 Å^{-1} in argon), there are clear discrepancies for large frequencies. For these relatively high wave vectors the tail term and the heat-coupling term are negligible. We are, therefore, in the rather uninteresting case where a simple memory function, short ranged in time, is sufficient. The discrepancies are due, as in the transverse case, to our use of a memory function of the exponential type, and should be removed by a more careful choice of the shape of that memory function. Much more interesting appears to be the region of lower wave vectors, now accessible with the help of high-flux reactors: it should be possible to observe there the collective effects which we have tried to describe in this paper.

VI. CONNECTION WITH VELOCITY-AUTOCORRELATION FUNCTION

Zwanzig and Bixon[9] have generalized the Stokes expression for the friction constant to finite frequencies. As in the derivation of the Stokes law, the particle whose self-motion is studied is coupled to the medium, described macroscopically, through hard-sphere boundary conditions. In order to calculate the frequency-dependent force acting on the particle, one has to solve the linearized Navier-Stokes equation with transport coefficients generalized to finite frequencies. This can be done analytically if the full coupling with thermal diffusion and the k dependence of the transport coefficients are neglected.

Zwanzig and Bixon have used a simple visco-

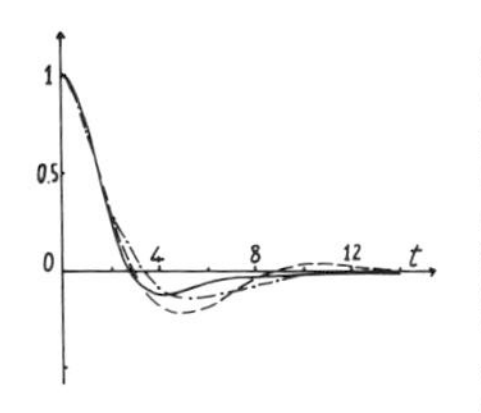

FIG. 10. Autocorrelation function of the velocity (time units 10^{-13} sec) obtained in the Zwanzig-Bixon model (dashed curve), and in the case of the representation (4.10) for $\eta(t)$ (dash-dot curve). The solid curve is the molecular-dynamics results.

elastic theory for the transport coefficient. We repeat their computation because we feel that it is interesting to see what becomes of their result when no parameter at all is adjusted: here, everything is available from the molecular-dynamics computation. As in Sec. III, we choose slip boundary conditions and a hard-sphere diameter chosen to fit the structure factor ($d=1.02$). The computation involves the isothermal velocity of sound given in Sec. II and the shear and bulk viscosities given in Sec. III. The shear and bulk relaxation times are obtained as $\eta/\eta(0)$ and $\xi/\xi(0)$, respectively. We then obtain the dotted curve of Fig. 10. We see that a surprisingly good agreement is obtained. An important feature is missing: The velocity-autocorrelation function of the LJ fluid near solidification presents an extended negative region, whereas the model leads to oscillations. Zwanzig and Bixon suggest that a more sophisticated frequency dependence of the transport coefficients might remove this discrepancy.

We shall show that this is indeed the case. We keep the same single-relaxation form for the bulk viscosity but use (4.10) to build a frequency-dependent shear viscosity with two relaxation times:

$$\tilde{\eta}(\omega)=\eta\left(\frac{1-\alpha}{1-i\omega\tau_<}+\frac{\alpha}{1-i\omega\tau_>}\right), \qquad (6.1)$$

where $\alpha_1 \tau_>$ and $\tau_<$ are given by (4.11).

Using this expression for the shear viscosity, we obtain the dash-dot curve in Fig. 10. This curve now presents a negative tail at the few percent level which coincides at large times with the molecular-dynamics results. The discrepancies at smaller times are evidently due to the basic roughness of the model. The general agreement appears more remarkable when one tries to vary the parameters of the model. For instance a variation of the shear viscosity by 20% leads to results which are clearly worse than before.

We cannot use a value of α smaller than 0.128, which is as we have seen a rather high value. The results would be improved by increasing that value to 0.15. It is probable that the thermal diffusion which was neglected acts as a supplementary long-wavelength long-time damping which can be simulated by increasing the role of the long-time tail of the shear viscosity a little.

A last remark is the following. For low frequencies, the friction can be represented by

$$\tilde{\gamma}(\omega)=\frac{2\pi d\eta}{m}-\frac{2\pi d^2 i}{3}\left(\frac{i\omega\rho\eta}{m}\right)^{1/2}+O(\omega). \qquad (6.2)$$

As pointed out by Zwanzig and Bixon, due to the square root (6.2), the frequency spectrum of the velocity-autocorrelation function will show a cusp for small ω's. The size of the domain of frequency where this root term predominates determines the extent of the time region when the $t^{-3/2}$ behavior due to this cusp should appear. It is to be noted that the role of the long-time tail in $\eta(t)$ is to reduce very substantially the size of the cusp (it appears only for $\omega \lesssim 0.05$).

The times where the velocity autocorrelation could become positive and behave like $t^{-3/2}$ are, in that case, so high as to lead to completely unobservable effects in a molecular-dynamics computation.

VII. CONCLUSIONS

The study of the LJ fluid near its triple point has demonstrated the existence of a tail extending at large times in the Kubo function which defines the shear viscosity. This tail has observable consequences. First, there should be an enhancement of the shear viscosity and a lowering of the ratio of the bulk to the shear viscosities when one approaches the solidification line. A further and more direct evidence for this tail could be obtained through the analysis of coherent-neutron-scattering experiments to be made with the long-wavelength neutron available with the high-flux reactors.

We are now undertaking a similar study at lower density.

ACKNOWLEDGMENT

The authors had many interesting and stimulating discussions with Professor B. J. Alder.

---

*Laboratoire associé au Centre National de la Recherche Scientifique. Postal address: Laboratoire de Physique Théorique et Hautes Energies Bâtiment 211, Université Paris Sud, Centre d'Orsay, 91, Orsay, France.

[1]L. Verlet, Phys. Rev. **159**, 98 (1967).

[2]D. Levesque and L. Verlet, Phys. Rev. A **2**, 2514 (1970).

[3]B. J. Alder, D. M. Gass, and T. E. Wainwright, J. Chem. Phys. **53**, 3813 (1970).

[4]A. Rahman, in *Neutron Inelastic Scattering* (International Atomic Energy Agency, Vienna, 1968), Vol. I, p. 561.

[5]L. P. Kadanoff and P. C. Martin, Ann. Phys. (N.Y.) **24**, 419 (1969).

[6]C. H. Chung and S. Yip, Phys. Rev. **182**, 323 (1969).

[7]A. Z. Akcasu and E. Daniels, Phys. Rev. A **2**, 926 (1970).

[8]N. K. Ailawadi, A. Rahman, and R. Zwanzig, Phys. Rev. A **4**, 1616 (1971).

[9]R. Z. Zwanzig and M. Bixon, Phys. Rev. A **2**, 2906 (1970).

[10]J. P. Hansen and L. Verlet, Phys. Rev. **184**, 151 (1969).

[11]J. L. Lebowitz, J. K. Percus, and L. Verlet, Phys. Rev. **153**, 250 (1967).

[12]L. Verlet, Phys. Rev. **165**, 201 (1968).

[13]D. G. Naugle, J. H. Lunsford, and J. R. Singer, J. Chem. Phys. **45**, 4669 (1966).

[14]J. D. Weeks, D. Chandler, and H. C. Andersen, J. Chem. Phys. **54**, 5237 (1971).

[15]L. Verlet and J. J. Weis, Phys. Rev. A **5**, 939 (1972).

[16]R. Zwanzig and R. D. Mountain, J. Chem. Phys. **43**, 4464 (1965).

Notes on p. 192

[17]P. Schofield, Proc. Phys. Soc. Lond. **88,** 149 (1966).
[18]J. Mc Donald and K. Singer (private communication).
[19]B. J. Alder and T. E. Wainwright, Phys. Rev. Lett. **18,** 988 (1967).
[20]M. Ross and P. Schofield, J. Phys. C **4,** L305 (1971).
[21]M. Wertheim, Phys. Rev. Lett. **10,** 321 (1963); E. J. Thiele , J. Chem. Phys. **38,** 1959 (1963).
[22]N. F. Carnahan and K. E. Starling, J. Chem. Phys. **51,** 635 (1969).
[23]A. Rahman, Phys. Rev. A **136,** 405 (1965).
[24]J. P. Boon, J. C. Legros, and G. Thomas, Physica (Utr.) **33,** 547 (1967).
[25]A. de Bock, W. Grevendonk, and W. Herreman, Physica (Utr.) **37,** 227 (1967).
[26]J. M. Luttinger, Phys. Rev. A **135,** 1505 (1964).
[27]P. G. de Gennes, Physica (Utr.) **25,** 825 (1959).

Notes to Reprint III.2

1. More recent equilibrium molecular dynamics calculations by Schoen and Hoheisel (Mol. Phys. 56 (1985) 653) and by Levesque and Verlet (Mol. Phys. 61 (1987) 143), together with NEMD results, have shown that the value given in this paper for the shear viscosity of the Lennard–Jones fluid is about 30% too large. Since the new results show no significant size dependence (see the table in the following note), and since there was no programming error, Levesque and Verlet suggest that the large value obtained in their earlier work was caused by a long-lived metastability in the system. Such an effect could give rise to an artificially enhanced tail in the shear-viscosity integrand (see fig. 1 in the paper), and hence to an overestimate of the shear viscosity itself.

2. The shear-viscosity integrand in fig. 1 has been inadvertently replotted in fig. 2.

3. The error in the formula for the heat current (see eq. (3.22) and the discussion that follows) has been rectified in later calculations by the same authors (Levesque and Verlet, Mol. Phys. 61 (1987) 143). The new results for the triple-point transport coefficients (at $\rho^* = 0.8442$ and in units of the paper) are given in the table below, where N is the number of particles, n the number of time steps, U denotes Univac and C Cray.

T^*	N	n	η^*	ξ^*	K^*	Computer
0.728	108	192000	20.6	7.8	1.02	U
0.715	256	64000	20.2	6.2	1.02	U
0.728	864	24000	18.2	8.3	0.72	C
0.733	864	36000	19.7	7.9	1.02	U

4. For more recent calculations of the intermediate scattering function and dynamic structure factor of the Lennard–Jones fluid, see Schoen and Hoheisel (Mol. Phys. 56 (1985) 653).

PHYSICAL REVIEW VOLUME 182, NUMBER 1 5 JUNE 1969

Statistical Error Due to Finite Time Averaging in Computer Experiments*

Robert Zwanzig and Narinder K. Ailawadi
University of Maryland, College Park, Maryland 20742
(Received 13 March 1969)

In using a computer experiment to calculate the time correlation function of some dynamical variable, the ensemble average over an equilibrium distribution in phase space often is replaced by a time average over a finite interval T. It is shown here that the statistical uncertainty due to this kind of average is of the order of $1/T^{1/2}$. The coefficient of the square root is related to a characteristic relaxation time of the correlation function. If *only* time averaging is performed, the resulting statistical uncertainty in typical computer experiments may be of the order of 20%.

Recently much attention has been given to the numerical evaluation of time correlation functions by means of computer experiments.[1-3] In this article we present a theoretical estimate of one particular source of error inherent in such calculations. The error is associated with replacement of an equilibrium ensemble average by a time average over a finite time interval.

In classical statistical mechanics, the time correlation function $C_\infty(t)$ of a dynamical quantity $A(t)$ is defined as

$$C_\infty(t) = \langle A(s)A(s+t)\rangle \ . \tag{1}$$

Here, $A(t)$ is the numerical value of some property A of a given system at time t; it is determined by the trajectory of the system point in phase space. The average denoted by $\langle\ \rangle$ is an ensemble average over an equilibrium distribution of initial system points in phase space. Note that $C_\infty(t)$ is independent of the arbitrary time s; this is a consequence of Liouville's theorem.

According to ergodic theory, the ensemble average can be replaced by an *infinite* time average for almost all initial system points in phase space,

$$C_\infty(t) = \lim_{T\to\infty} C_T(t) , \tag{2}$$

where

$$C_T(t) = \frac{1}{T}\int_0^T ds A(s)A(s+t) \tag{3}$$

is an average over the finite time interval T for a single initial point in phase space.

In typical computer experiments it is not easy to construct an accurate ensemble average. Each member of the ensemble corresponds to a repetition of an experiment with randomly chosen initial conditions. If the ensemble contains n members (i.e., the experiment is repeated n

times), the relative error of the estimated ensemble average is of order $1/n^{1/2}$. This is a consequence of the law of large numbers. For a relative error of 1%, about 10^4 repetitions are needed.

For this reason one usually performs a time average. If the time interval T is large enough, then we expect that only a single trajectory in phase space (a single computer experiment) will suffice. However, because of the nature of computer experiments, one can average over a finite interval only. So the question arises: How large an interval T is required in order to attain some specified accuracy?

In this paper we show that the relative error in a time correlation function obtained by averaging over a single trajectory for an interval T is of the order of $1/T^{1/2}$, and we present an explicit expression for the coefficient of the square root. We give numerical estimates of the error for a typical though hypothetical case.

Our argument is based on the assumption that the quantity $A(t)$ is a Gaussian random variable. Because this is not necessarily true when $A(t)$ is a dynamical quantity, our results are expected to be plausible estimates, but not rigorous. At present we do not know of any way to correct for non-Gaussian behavior.

We want to know how well an ensemble average is approximated by a finite time average. The difference between the two averages is denoted by $\Delta(t)$,

$$\Delta(t) = C(t) - C_\infty(t). \tag{4}$$

(For simplicity the subscript T on $C_T(t)$ is omitted.)

The deviation $\Delta(t)$ is a random quantity. It depends on the initial state of the system, or equivalently, on the trajectory in phase space followed by a single system. Its statistical properties may be characterized by its first and second moments.

Since the ensemble average is invariant to displacement of the time origin, Eq. (4) may be replaced by

$$\Delta(t) = \frac{1}{T}\int_0^T ds[A(s)A(s+t) - \langle A(s)A(s+t)\rangle]. \tag{5}$$

It is clear that the first moment (taken with respect to the ensemble) vanishes,

$$\langle \Delta(t)\rangle = 0. \tag{6}$$

The second moment of the deviation is

$$\langle \Delta(t_1)\,\Delta(t_2)\rangle = \frac{1}{T^2}\int_0^T ds_1 \int_0^T ds_2 \times [\langle A(s_1)A(s_1+t_1)A(s_2)A(s_2+t_2)\rangle - \langle A(s_1)A(s_1+t_1)\rangle\,\langle A(s_2)A(s_2+t_2)\rangle]. \tag{7}$$

Very little can be done with this expression in its general form. However, it can be reduced to a more manageable form by the assumption that $A(t)$ is a Gaussian random variable.

If $A(t)$ is Gaussian, then the average of a product of four A's can be expressed in terms of products of averages of two A's,

$$\langle A_1A_2A_3A_4\rangle = \langle A_1A_2\rangle\langle A_3A_4\rangle + \langle A_1A_3\rangle\langle A_2A_4\rangle + \langle A_1A_4\rangle\langle A_2A_3\rangle. \tag{8}$$

But the average of a pair of A's is just the time correlation function $C_\infty(t)$. Thus the second moment of the deviation is

$$\langle\Delta(t_1)\Delta(t_2)\rangle = \frac{1}{T^2}\int_0^T ds_1\int_0^T ds_2 \times [C_\infty(s_2-s_1)C_\infty(s_2-s_1+t_2-t_1) + C_\infty(s_2-s_1+t_2)C_\infty(s_2-s_1-t_1)]. \tag{9}$$

The integrals in Eq. (9) may be estimated in two ways. First, we can replace the ensemble averages $C_\infty(t)$ by the corresponding time averages, and use information about $C(t)$ obtained in computer experiments to calculate the integrals.

But a more useful and convenient estimate can be found by focussing attention on a typical limiting case. The correlation function $C_\infty(t)$ usually decays to zero within some characteristic time of order τ. In typical computer experiments, the averaging time T is much larger than τ; the ratio T/τ may be, for example, about 20. Also, we are concerned with deviations $\Delta(t)$ for times such that the correlation function is not negligibly small. This means that the times t_1 and t_2 in Eq. (9) are themselves of order τ, and are much smaller than T.

In the integrals, s_1 and s_2 both vary from 0 to T. But because the correlation functions vanish for times of order τ, the dominant contribution to the integrals comes from values of s_2 that are within a range of order τ from s_1. Further, the integrand is approximately independent of s_1 when T is large, so that one integration can be done immediately, leading to a factor T. The other integration, originally from 0 to T, may be replaced by an integration from $-\infty$ to $+\infty$ so that

$$\langle\Delta(t_1)\,\Delta(t_2)\rangle \simeq \frac{2}{T}\int_{-\infty}^{\infty} ds[C_\infty(s)]^2 = \frac{4}{T}\int_0^{\infty} ds[C_\infty(s)]^2. \tag{10}$$

Now we *define* a mean relaxation time τ by

$$\tau = 2\int_0^\infty ds[C_\infty(s)]^2/[C_\infty(0)]^2 . \qquad (11)$$

[Note that if $C_\infty(t)$ decays exponentially with the relaxation time τ, then the above equation is satisfied identically. This is the main reason for defining τ as in Eq. (11).]

In this way we find that the second moment of the deviation is approximately independent of t_1 and t_2, and has the value

$$\langle \Delta(t_1)\Delta(t_2)\rangle \simeq \frac{2\tau}{T}[C_\infty(0)]^2 . \qquad (12)$$

The preceding result will be used to discuss the statistical properties of two different quantities that are of interest in computer experiments,

$$\mathrm{I} = C(t)/C_\infty(0) \qquad (13)$$

and $\mathrm{II} = C(t)/C(0)$. (14)

The difference is in the denominator. In case I, the initial value of the ratio is not necessarily unity, because $C(t)$ is a random quantity while $C_\infty(t)$ is fixed. In case II, both numerator and denominator are random, but the initial value of the ratio is unity by definition. We will see that somewhat greater accuracy can be obtained from computer experiments in case II, as a result of cancellation of randomness for short times.

In case I, the deviation from the ensemble average is

$$\Delta_{\mathrm{I}}(t) = C(t)/C_\infty(0) - C_\infty(t)/C_\infty(0) . \qquad (15)$$

It is clear that the mean deviation vanishes,

$$\langle \Delta_{\mathrm{I}}(t)\rangle = 0 . \qquad (16)$$

The second moment of the deviation can be found from Eq. (12), and is

$$\langle[\Delta_{\mathrm{I}}(t)]^2\rangle \simeq 2\tau/T. \qquad (17)$$

This holds for values of the time t that are of order τ and much smaller than T.

It is convenient to introduce the normalized ensemble averaged correlation function $R(t)$,

$$R(t) = C_\infty(t)/C_\infty(0) . \qquad (18)$$

If we use one standard deviation as an indication of the error to be expected in determining $R(t)$ by computer experiments in case I, we obtain

$$R(t) = C(t)/C_\infty(0) \pm (2\tau/T)^{1/2} . \qquad (19)$$

It should be noted that the error is independent of t, so that this procedure will not give even the initial value $R(0) = 1$ correctly.

In a typical computer experiment where $T/\tau = 25$, the resulting accuracy of $R(t)$ determined this way is

$$R(t) = C(t)/C_\infty(0) \pm 0.28 . \qquad (19a)$$

Higher accuracy can be obtained, at least for short times, by using case II. Here we want to know how well $R(t)$ is approximated by $C(t)/C(0)$. The corresponding deviation is

$$\Delta_{\mathrm{II}}(t) = C(t)/C(0) - R(t) . \qquad (20)$$

By Eq. (4) this may be transformed to

$$\Delta_{\mathrm{II}}(t) = \frac{\Delta(t) - R(t)\Delta(0)}{C_\infty(0) + \Delta(0)} . \qquad (21)$$

On expansion of the denominator to first order, we obtain

$$\Delta_{\mathrm{II}}(t) = \frac{1}{C_\infty(0)}[\Delta(t) - R(t)\Delta(0)] - \frac{\Delta(0)}{[C_\infty(0)]^2}[\Delta(t) - R(t)\Delta(0)] + \cdots. \qquad (22)$$

The mean value of $\Delta_{\mathrm{II}}(t)$ does not vanish. If we use the estimate given in Eq. (12), we find

$$\langle \Delta_{\mathrm{II}}(t)\rangle \simeq \frac{2\tau}{T}[R(t) - 1] . \qquad (23)$$

It will turn out, however, that a standard deviation obtained from the second moment of $\Delta_{\mathrm{II}}(t)$ is of order $(\tau/T)^{1/2}$. So the mean value of $\Delta_{\mathrm{II}}(t)$ is of a smaller order of magnitude, and may be neglected.

The variance of $\Delta_{\mathrm{II}}(t)$ is easily found from Eq. (12), and is

$$\langle[\Delta_{\mathrm{II}}(t)]^2\rangle - [\langle \Delta_{\mathrm{II}}(t)\rangle]^2 \simeq \frac{2\tau}{T}\left(1 - \frac{2\tau}{T}\right)[1 - R(t)]^2 \simeq \frac{2\tau}{T}[1 - R(t)]^2. \qquad (24)$$

Evidently the correction from the square of the first moment is of a smaller order of magnitude, so we omit it.

Consequently, our estimate of $R(t)$ from a computer experiment in case II is

$$R(t) = C(t)/C(0) \pm (2\tau/T)^{1/2}[1 - R(t)] . \qquad (25)$$

In the error term we may replace $R(t)$ by the observed $C(t)/C(0)$ without appreciable effect. On subtracting unity from both sides, the result appears a bit simpler:

Note on p. 196

$$R(t) - 1 = \left[\frac{C(t)}{C(0)} - 1\right] \times \left[1 \pm \left(\frac{2\tau}{T}\right)^{\frac{1}{2}}\right] . \tag{26}$$

As an illustration, we take a hypothetical example where $T/\tau = 25$. The following table gives "measured" values of $C(t)/C(0)$ for several times, and the resulting estimate of $R(t)$:

t (psec)	$C(t)/C(0)$	$R(t)$
0	1.0	1.0 ± 0.0
0.1	0.9	0.9 ± 0.03
0.2	0.3	0.3 ± 0.20
0.3	− 0.1	− 0.1 ± 0.31
1	0	0 ± 0.28

[In some computer experiments, T/τ has been as large as 100. Then the error estimates given under $R(t)$ in this table should be halved.]

It appears that only the early stages of the decay of a time correlation function can be computed with any confidence by averaging over about 25 relaxation times.

If one also performs an average over an ensemble containing n members, one can obtain an improvement in the error by a factor $1/n^{1/2}$. The reason is that this extra ensemble average is roughly equivalent to considering a single system for a time interval nT.

The preceding discussion suggests that it may not be feasible economically to achieve much greater accuracy (say, 1%) by computer experiments using only time and ensemble averaging. It should be observed, however, that greater accuracy *has* been achieved in special cases, where another kind of average is possible. An example is the calculation of the velocity correlation function

$$\sigma(t) = \langle v_1(0) v_1(t) \rangle \tag{27}$$

of an individual particle in a liquid. If all N particles in the system are identical, then the velocity correlation function is independent of the particle number, so that

$$\sigma(t) = \frac{1}{N} \sum_{j=1}^{N} \langle v_j(0) v_j(t) \rangle . \tag{28}$$

This extra averaging over particle labels is roughly equivalent to averaging over an ensemble containing N systems. When a single system contains 400 particles, and the time average is taken over 25 relaxation times, an accuracy of about 1% is expected. Useful results have been obtained in this way.[1-3]

---

*This research was supported in part by the National Science Foundation under Grant GP 7652, and in part by the Office of Naval Research under Grant N00014-67-A-0239-0002.

[1] B. J. Alder and T. Wainwright, in Transport Processes in Statistical Mechanics, edited by I. Prigogine (Interscience Publishers, Inc., N. Y., 1958), pp. 97-131.

[2] A. Rahman, Phys. Rev. 136, A405 (1964).

[3] G. D. Harp and B. J. Berne, J. Chem. Phys. 49, 1249 (1968).

Note to Reprint III.3

The statistical errors involved in the calculation of spectroscopic properties are discussed by Frenkel (in: Intermolecular Spectroscopy and Dynamical Properties of Dense Systems, ed. J. van Kranendonk (North-Holland, Amsterdam, 1980)). An estimate of the mean-square noise in the power spectrum is given in Note 43 of (Frenkel and McTague, J. Chem. Phys. 72 (1980) 2801).

VELOCITY-INVERSION AND IRREVERSIBILITY IN A DILUTE GAS OF HARD DISKS

J. ORBAN * and A. BELLEMANS
University of Brussels, Belgium.

Received 22 April 1967

Anti-kinetic situations created in a dilute gas of hard disks by suddenly reversing all velocities, are shown to present some character of instability.

Consider a homogeneous dilute gas so that its entropy is directly related to the H-function of the velocities. The kinetic equation of Boltzmann predicts that this system always goes towards equilibrium, with H decreasing monotonically with time. This *kinetic* behavior, which agrees with macroscopic observations cannot be absolutely general however. A first objection (Zermelo) based on Poincaré's recurrence theorem, is easily bypassed for macroscopic systems, because the duration of an observation is always much shorter than the recurrence time in phase space. A second objection (Loschmidt) is more serious: from each microscopic state such that the system approaches equilibrium (*kinetic* behavior), another state can be generated by reversing all molecular velocities, so that the system goes away from equilibrium (*anti-kinetic* behavior). Recently, Prigogine and coworkers [1] derived an evolution equation describing both behaviors. Besides the usual collision term, it contains a new part depending on the initial correlations, which determines the kinetic or the anti-kinetic evolution; it can be proved that, for initially uncorrelated particles, the spontaneous correlations which appear, are of the kinetic type.

Irreversibility can then be understood as follows. The correlations existing at present in our terrestrial system have been created by the clustering of uncorrelated particles, during a time interval which, however long it may be, is still much shorter than the recurrence time of Poincaré. These correlations are therefore of the *kinetic* type. We can of course exert constraints on selected subsystems but our ability of altering their internal correlations is usually not strong enough as to generate an anti-kinetic evolution.

Although the creation of an anti-kinetic behavior in actual systems seems precluded, one could think of generating it mathematically. Consider a system of interacting molecules, starting from an uncorrelated state. One follows the evolution of the system during a certain time interval t_0, observing a decrease of H and the appearance of correlations. Then one suddenly reverses all molecular velocities, forcing the system to move backwards in positional space and to exhibit an *anti-kinetic* behavior: H increases with time, reaches its initial value at time $2\ t_0$,

* Boursier IRSIA.

620

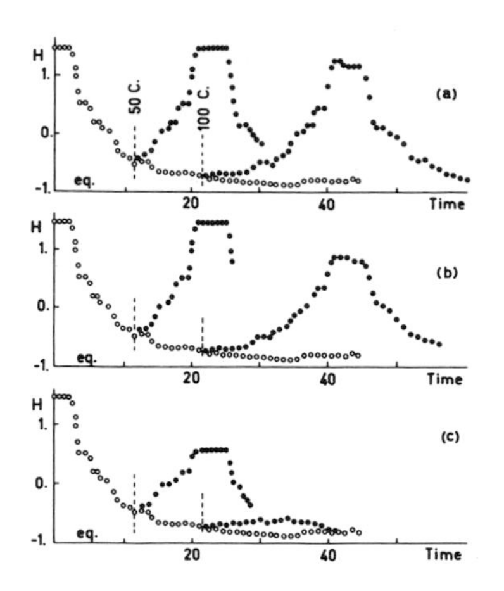

Fig. 1. Plot of H with time (in arbitrary units) showing the kinetic (o) and the anti-kinetic (•) evolutions for velocity-inversions taking place at 50 or 100 collisions, with random errors 10^{-8}, 10^{-5} and 10^{-2}respectively.

and then decreases again towards equilibrium.

Such an anti-kinetic situation has been considered by Balescu on a simple example [1]. More complicated ones can be worked out on computers by the method of *molecular dynamics*. We studied the evolution of a two-dimensional dilute gas of 100 hard disks in a periodic square box on a IBM 7040 (density: 0.04 with respect to close packing). The conditions at time 0 are: (a) all velocities have same absolute magnitude (unity) but are oriented at random (total momentum is zero) and (b) the molecules sit on the vertices of a square network in the box. As time proceeds, molecular collisions tend to realize a Maxwellian distribution and H decreases. The time interval for reaching equilibrium roughly corresponds to 200 collisions; however, after a fewer number of collisions (50 or 100) we invert all velocities. Part a of fig. 1 shows the behavior of H. When reverting all velocities after 50 collisions, H retraces its path exactly back to its initial value, and next starts to decrease again. When the velocity-inversion is carried out after 100 collisions, the system does not correctly find the way back to its initial state. It is misled by rounding-errors in the computations, although their magnitude is less than 10^{-8}, and H goes trough a maximum value lower than the initial one.

This shows that anti-kinetic situations present some kind of instability, in the sense predicted by Balescu: assume the velocity-inversion to be carried out with small random errors; then the anti-kinetic behavior of the system tends to be reduced or even suppressed [1]. Further illustrations of this instability are shown in parts b and c of fig. 1; here the velocity-inversions are carried out with random errors 10^{-5} and 10^{-2} respectively. As expected the anti-kinetic behavior of the system is always weakened and eventually nearly destroyed.

We thank Professors I. Prigogine and R. Balescu for helpful discussions.

References

1. I. Prigogine and P. Résibois, Estratto dagli Atti del Simposio Lagrangiano, (Accad. della Scienze di Torino, 1964);
R. Balescu, Physica, to be published.

* * * * *

Journal of Statistical Physics, Vol. 21, No. 1, 1979

"Thought-Experiments" by Molecular Dynamics

G. Ciccotti,[1] G. Jacucci,[2] and I. R. McDonald[3]

Received November 2, 1978

A method for studying the dynamical properties of liquids by molecular dynamics simulation is described. Its basis is the measurement of the response to a weak applied field of appropriate character. The explicit form of the mechanical perturbation is worked out in several cases, and details are given of the numerical techniques used in implementing the method.

KEY WORDS: Molecular dynamics; transport properties; linear response theory.

1. INTRODUCTION

The purpose of this paper is to present in general form a nonequilibrium method for studying the dynamical properties of liquids in the context of molecular dynamics calculations. The conventional approach is to compute the time correlation functions that describe the decay of spontaneous fluctuations in the variables of interest. The correlation functions in question are properties of the system in equilibrium, but from linear response theory we know that they can also be related to the response of the system to a weak external field of appropriate character. Our approach is more direct, since we choose to measure the response itself; to that extent our work may be regarded as a realization of the "thought-experiments" of Kubo and Luttinger.[1-4] A number of results obtained in this way have already been reported,[5-7] but until now we have progressed in a largely intuitive fashion. The more formal development given here is intended to give a firm theoretical basis to the method. Given this starting point, it should be easy to generalize the approach to cover a much wider range of phenomena than those we discuss explicitly. At the same time we take the opportunity of describing in

[1] Istituto di Fisica "G. Marconi," Università di Roma, Rome, Italy.
[2] Dipartimento di Matematica e Fisica, Libera Università di Trento, Povo, Trento, Italy.
[3] Department of Chemistry, Royal Holloway College, Egham, Surrey, England.

greater detail than hitherto the numerical techniques used in implementing the method.

In attempting to characterize the dynamical behavior of a system in terms of the response to a weak applied field we encounter at once two major problems. The first is the old question of how to represent in mechanical terms the generalized forces required to excite the responses that are of interest, including in particular the representation of the gradients necessary to induce a current of momentum or of energy. This has as a corollary the question of how to incorporate such forces into the molecular dynamics algorithm. The second difficulty concerns the measurement of the response, since in the linear regime this will be, at best, comparable in magnitude with the thermal fluctuations in the system. If these two problems can be overcome, we shall have at our disposal a method for probing systematically the dynamics of systems of interacting particles, including a variety of small cross effects which are impractical to study by the usual equilibrium techniques.

Let us first formulate the problem in very general terms. If $\langle\dot{\alpha}_\mu(\mathbf{r})\rangle_t$ is the flux induced in a microscopic variable $\alpha_\mu(\mathbf{r})$, then in a linear approximation we can write

$$\langle\dot{\alpha}_\mu(\mathbf{r})\rangle_t = \sum_\nu \int_{-\infty}^{t} dt' \int d\mathbf{r}\, L_{\mu\nu}(\mathbf{r}-\mathbf{r}', t-t') F_\nu(\mathbf{r}', t') \tag{1.1}$$

where $\langle\cdots\rangle_t$ denotes an average on a nonequilibrium ensemble and F_ν is an external field conjugate to the variable α_ν, i.e., we assume that the interaction with the system is described by the Hamiltonian

$$\mathscr{H}_I(t) = -\sum_\nu \int d\mathbf{r}\, \alpha_\nu(\mathbf{r}) F_\nu(\mathbf{r}, t) \tag{1.2}$$

Given the correct choice of α_i, the system of equations (1.1) reduces to the phenomenological laws of irreversible thermodynamics in the limit when the applied fields vary slowly in space and time. This means that with any *thermodynamic* force appearing in the phenomenological laws we may associate a *mechanical* force, which is the gist of the indirect Kubo method.[4] In the more general case ($k, \omega > 0$) we may rewrite Eq. (1.1) in the form

$$\langle\dot{\alpha}_\mu(\mathbf{k})\rangle_\omega = \sum_\nu L_{\mu\nu}(\mathbf{k}, \omega) \hat{F}_\nu(\mathbf{k}, \omega) \tag{1.3}$$

If we choose the set of variables $\alpha_\mu(\mathbf{r})$ to be the microscopic conserved variables associated with the independent fluxes appearing in the expression for entropy production, the quantities $L_{\mu\nu}(\mathbf{k}, \omega)$ form the tensor of kinetic coefficients which in the limit $k, \omega \to 0$ (taken in the right order) yields the transport coefficients of hydrodynamics.

Given the correspondence between thermodynamic and mechanical forces, it is clear that once we have identified the explicit form of the mechanical forces the method of molecular dynamics can be used to study irreversible processes in a direct way by exploitation of Eq. (1.1). By computing the fluxes induced by weak external fields of appropriate character we could, in particular, evaluate the various elements in the matrix of kinetic coefficients. Before this program can be carried through successfully, a solution must be found to the second problem mentioned above. The device we use is a "subtraction" technique, the justification for which is the fact that a nonequilibrium average obtained by applying a perturbation to the system can be transformed into an average over an equilibrium ensemble by the operation

$$\langle A \rangle_t \equiv (A, \rho(t)) = (A, \exp(i\mathscr{L}t)\, \rho_0) = (\exp(-i\mathscr{L}t)\, A, \rho_0) \equiv \langle \exp(-i\mathscr{L}t)\, A \rangle_0 \tag{1.4}$$

where $\mathscr{L}$ is the Liouville operator of the perturbed system, ρ_0 is the equilibrium probability distribution, and $\rho(t)$ is the perturbed probability distribution, with $\rho(t = 0) = \rho_0$; the symbol $(\cdots, \cdots)$ is used to denote an integral over phase space and $\langle \cdots \rangle_0$ denotes an equilibrium average. Thus, in order to improve the signal-to-noise ratio in the calculation of $\langle A \rangle_t$, we subtract from $\exp(-i\mathscr{L}t)\, A$ the quantity $\exp(-i\mathscr{L}_0 t)\, A$, where $\mathscr{L}_0$ is the unperturbed Liouville operator, and compute the difference

$$\Delta A(t) = \exp(i\mathscr{L}t)\, A - \exp(-i\mathscr{L}_0 t)\, A \tag{1.5}$$

the equilibrium average of which is

$$\langle \Delta A(t) \rangle_0 = \langle A \rangle_t - \langle A \rangle_0 \tag{1.6}$$

This procedure yields the dynamical response we wish to measure with an accuracy which is high if t is not too large. The reason for this is simply the fact that *random* fluctuations in the two terms in Eq. (1.5) are highly correlated and therefore largely cancel, leaving only the systematic part, i.e., the response to the perturbation.

At this point it might be helpful to describe briefly what we do in practice. We carry out a molecular dynamics run in the normal way, but in addition, at regular intervals in time, we apply a small perturbation to the system, which in the general case involves adjusting the coordinates and momenta of all particles in a prescribed way. Thereafter the paths of the particles in perturbed and unperturbed trajectories are followed simultaneously and the time variation of the response is calculated as the difference in the relevant dynamical variable. This yields the *mechanical* response defined by Eq. (1.5), and the *statistical* response represented by Eq. (1.6) is obtained via the ergodic

Notes on p. 220

theorem by averaging the mechanical response over a number (typically 50–100) of such pairs of trajectories. In each case the maximum useful information is obtained if the perturbed trajectory is followed for a time which is somewhat larger than the longest relaxation time characterizing the dynamical process under investigation.

The outline of the remainder of the paper is as follows. In Section 2 we give a general account of the method in the framework of linear response theory; in Section 3 we discuss the choice of mechanical perturbation and the character of the different responses; in Section 4 we give some details concerning the numerical solution of the equations of motion; and in Section 5 we summarize the advantages and limitations of method and consider some possible generalizations. For simplicity we restrict the detailed discussion to the case of monatomic systems.

2. GENERAL DEVELOPMENT

Let $\alpha(\mathbf{r})$ be a dynamical variable of the form

$$\alpha(\mathbf{r}) = \sum_{i=1}^{N} \alpha_i \, \delta(\mathbf{r} - \mathbf{r}_i) \tag{2.1}$$

with

$$\alpha(\mathbf{r}, t) = \exp(-i\mathscr{L}_0 t) \, \alpha(\mathbf{r}) \tag{2.2}$$

where α_i is a property of particle i, and may, however, be a function of the phase space coordinates $\mathbf{r}_j$, $\mathbf{p}_j$ of *all* particles j ($j = 1$ to N). Now suppose that the system is subjected to a weak external field $\phi(\mathbf{r}, t)$ which couples to the variable α. The total Hamiltonian in the presence of the perturbation is

$$\mathscr{H} = \mathscr{H}_0 + \mathscr{H}_I(t) \tag{2.3}$$

where $\mathscr{H}_0$ is the equilibrium Hamiltonian and $\mathscr{H}_I(t)$ represents the interaction between the system and the field

$$\mathscr{H}_I(t) = -\int d\mathbf{r} \, \alpha(\mathbf{r})\phi(\mathbf{r}, t) = -\sum_{i=1}^{N} \alpha_i \phi(\mathbf{r}_i, t) \tag{2.4}$$

If the perturbation is applied at $t = 0$, then for any observable property of the system, O say, the mean change $\langle O(\mathbf{r})\rangle_t$ induced by the perturbation after a time t is given by[1,8]

$$\langle O(\mathbf{r})\rangle_t = \beta \int d\mathbf{r}' \int_{-\infty}^{t} dt' \, \langle O(\mathbf{r}, t)\dot{\alpha}(\mathbf{r}', t')\rangle_0 \phi(\mathbf{r}', t') \tag{2.5}$$

where $\beta = 1/k_B T$ and, for simplicity, we have assumed that $\langle O(\mathbf{r})\rangle_0 = 0$; clearly Eq. (2.5) corresponds to the statistical response of Eq. (1.6). In writing

Eqs. (2.4) and (2.5) we have omitted the appropriate contraction of vectorial indices which must appear if α is not a scalar.

We now consider the special case when α is a conserved quantity, obeying a conservation law of the general form

$$\dot{\alpha}(\mathbf{r}, t) + \nabla \cdot \mathcal{J}^{\alpha}(\mathbf{r}, t) = 0 \tag{2.6}$$

where $\mathcal{J}^{\alpha}$ is the corresponding current; $\mathcal{J}^{\alpha}$ may be either a vector (if α is a scalar) or a second-rank tensor (if α is a vector). Equation (2.5) can then be rewritten as

$$\langle O(\mathbf{r})\rangle_t = \beta \int d\mathbf{r}' \int_{-\infty}^{t} dt' \, \langle O(\mathbf{r}, t)\mathcal{J}^{\alpha}(\mathbf{r}', t')\rangle_0 \, \nabla\phi(\mathbf{r}', t') \tag{2.7}$$

or, since the time correlation function appearing in Eq. (2.7) is a function only of the differences $\mathbf{r} - \mathbf{r}'$ and $t - t'$, as

$$\langle O(\mathbf{r})\rangle_t = \beta \int d\mathbf{r}' \int_{-\infty}^{t} dt' \, \langle O(\mathbf{r} - \mathbf{r}', t - t')\mathcal{J}^{\alpha}(0, 0)\rangle_0 \, \nabla\phi(\mathbf{r}', t') \tag{2.8}$$

Taking the Fourier transform in space, we obtain

$$\langle \hat{O}(\mathbf{k})\rangle_t = \beta V \int_{-\infty}^{t} dt' \, \langle \hat{O}(\mathbf{k}, t)\hat{\mathcal{J}}^{\alpha}(-\mathbf{k}, t')\rangle_0 i\mathbf{k}\phi(\mathbf{k}, t') \tag{2.9}$$

where for the transform of a function $f(\mathbf{r})$ we use the definition

$$\hat{f}(\mathbf{k}) = (1/V) \int d\mathbf{r} \, f(\mathbf{r}) \exp(-i\mathbf{k}\cdot\mathbf{r}) \tag{2.10}$$

In interpreting Eq. (2.8) or Eq. (2.9) it is again necessary to incorporate the appropriate contraction between $\mathcal{J}^{\alpha}$ and $\nabla\phi$ or between $\hat{\mathcal{J}}^{\alpha}$ and $\mathbf{k}\hat{\phi}$.

Given an interaction of the type displayed in Eq. (2.4), the equations of motion in Hamiltonian form are

$$\dot{\mathbf{r}}_i = \frac{\partial \mathcal{H}}{\partial \mathbf{p}_i} = \frac{\mathbf{p}_i}{m} - \phi(\mathbf{r}_i, t)\frac{\partial \alpha_i}{\partial \mathbf{p}_i} \tag{2.11}$$

$$\dot{\mathbf{p}}_i = -\frac{\partial \mathcal{H}}{\partial \mathbf{r}_i} = -\frac{\partial \mathcal{H}_0}{\partial \mathbf{r}_i} + \sum_{j=1}^{N} \phi(\mathbf{r}_j, t)\frac{\partial \alpha_j}{\partial \mathbf{r}_i} + \alpha_i \frac{\partial \phi(\mathbf{r}_i, t)}{\partial \mathbf{r}_i} \tag{2.12}$$

and in Newtonian form are

$$m\ddot{\mathbf{r}}_i = -\frac{\partial \mathcal{H}_0}{\partial \mathbf{r}_i} + \sum_{j=1}^{N} \phi(\mathbf{r}_j, t)\frac{\partial \alpha_j}{\partial \mathbf{r}_i} + \alpha_i \frac{\partial \phi(\mathbf{r}_i, t)}{\partial \mathbf{r}_i} - m\frac{d}{dt}\left(\phi(\mathbf{r}_i, t)\frac{\partial \alpha_i}{\partial \mathbf{p}_i}\right) \tag{2.13}$$

The final term on the right-hand side of (2.13) appears only when the perturbation is dependent on velocity, as is true in the cases discussed in Sections 3.4 and 3.5. If the perturbation is an impulsive force, care is needed in taking the derivative with respect to time; note that d/dt denotes a total derivative.

In writing Eq. (2.4) we have assumed that the nature of the applied field is such that the interaction with the system can be represented by an additional term in the Hamiltonian. However, we may also wish to study induced transverse currents (of number, mass, charge, etc.). In such cases the perturbation is a transverse field and cannot be described in this simple way. Nonetheless, if the applied field is a function only of coordinates, an equation similar to Eq. (2.5) can still be derived, the quantity $\dot{\alpha}$ taking the form

$$\dot{\alpha}(\mathbf{r}) = \sum_{i=1}^{N} \gamma_i \dot{\mathbf{r}}_i \, \delta(\mathbf{r} - \mathbf{r}_i) \tag{2.14}$$

where γ_i is the particle property which couples to the external field. Thus Jackson and Mazur[3] have shown that if the system is subjected at time $t = 0$ to a weak vectorial field $\boldsymbol{\xi}(\mathbf{r}, t)$ such that

$$\nabla \times \boldsymbol{\xi} \neq 0, \qquad \nabla \cdot \boldsymbol{\xi} = 0 \tag{2.15}$$

and the force acting on particle i is given by

$$\mathbf{F}(\mathbf{r}_i, t) = \gamma_i \boldsymbol{\xi}(\mathbf{r}_i, t) \tag{2.16}$$

then the mean change in a variable O is

$$\langle O(\mathbf{r}) \rangle_t = \beta \int d\mathbf{r}' \int_{-\infty}^{t} dt' \, \langle O(\mathbf{r}, t) \sum_{i=1}^{N} \gamma_i \dot{\mathbf{r}}_i(t') \times \delta\{\mathbf{r}' - \mathbf{r}_i(t')\} \rangle_0 \cdot \boldsymbol{\xi}(\mathbf{r}', t') \tag{2.17}$$

or

$$\langle \hat{O}(\mathbf{k}) \rangle_t = \beta V \int_{-\infty}^{t} dt' \, \langle \hat{O}(\mathbf{k}, t) \sum_{i=1}^{N} \gamma_i \dot{\mathbf{r}}_i(t') \times \exp\{i\mathbf{k} \cdot \mathbf{r}_i(t')\} \rangle_0 \cdot \hat{\boldsymbol{\xi}}(\mathbf{k}, t') \tag{2.18}$$

The total Hamiltonian of the system is now unchanged by the applied field, which appears instead as an additional term in the equations of motion. The latter now take the simpler form given by

$$\dot{\mathbf{r}}_i = \partial \mathscr{H}_0 / \partial \mathbf{p}_i = \mathbf{p}_i / m \tag{2.19}$$

$$\dot{\mathbf{p}}_i = -\partial \mathscr{H}_0 / \partial \mathbf{r}_i + \gamma_i \boldsymbol{\xi}(\mathbf{r}_i, t) \tag{2.20}$$

We have so far said nothing about the explicit time dependence of the applied field. This can be separated from the spatial dependence of the perturbation by writing

$$\phi(\mathbf{r}, t) = \phi(\mathbf{r}) f(t) \tag{2.21}$$

or, in the case of a transverse field,

$$\boldsymbol{\xi}(\mathbf{r}, t) = \boldsymbol{\xi}(\mathbf{r}) f(t) \tag{2.22}$$

which in either event represents a spatial field modulated by a scalar function $f(t)$. The form of $f(t)$ is arbitrary, but the choice of greatest practical interest is

$$f(t) = \delta(t - t_0) \tag{2.23}$$

where $\delta(t)$ is the Dirac delta function. This represents an impulsive force applied at $t = t_0$; to simplify the formulas we shall always assume that $t_0 = 0$. From Eqs. (2.9) and (2.17) it is easy to see that the response to a delta-function perturbation is the time correlation function itself. In some earlier calculations we used a step-function perturbation of the form

$$\begin{aligned} f(t) &= 0 \qquad \text{if} \quad t < t_0 \\ &= 1 \qquad \text{if} \quad t > t_0 \end{aligned} \tag{2.24}$$

representing a steady field switched on at $t = t_0$. In this case the response is the time integral of the correlation function. It follows that with a proper choice of the variables α and O the limiting drift current, namely

$$\lim_{t \to \infty} \langle \hat{O}(\mathbf{k}) \rangle_t$$

is closely related to a certain transport coefficient. In most circumstances, however, the structure of the correlation function is also of interest and for that reason alone use of a delta-function perturbation has generally more to recommend it. In addition, when the applied field is velocity dependent, integration of the equations of motion is more cumbersome when a step function is used. The subsequent discussion is therefore limited to the case of delta-function perturbations.

3. FORM OF THE MECHANICAL PERTURBATION

We want now to show how the general formalism of Section 2 can be adapted to specific cases. To avoid widening the discussion too far, we shall focus most of our attention on the important case of dynamical variables satisfying a conservation law. It is clear on general grounds that the most straightforward way of exciting a current $\mathscr{J}^\alpha$, say, is by applying a field which couples directly to the corresponding conserved variable α. This allows a study of the autocorrelation function of the quantity $\hat{\mathscr{J}}^\alpha(\mathbf{k}, t)$ from which, in the limit $k, \omega \to 0$, a transport coefficient can be extracted. It is equally clear, however, that any such perturbation will simultaneously give rise to the full set of cross effects described by the matrix of kinetic coefficients. In multi-component systems the number of cross effects is much larger than in pure

Notes on p. 220

fluids; this makes it natural to attempt the measurement of cross transport coefficients such as thermal diffusivity, quantities which are extremely difficult to determine by the standard molecular dynamics method.

The detailed way in which the mechanical equations of motion are modified by the perturbation depends on the character of the variable α. Thus the mechanical perturbation has a different form according to whether α_i is (i) independent of the phase space variables (as when $\mathcal{J}^\alpha$ is the longitudinal particle current); (ii) a function of the momentum $\mathbf{p}_i$ (as when $\mathcal{J}^\alpha$ is a component of the stress tensor); or (iii) a function of $\mathbf{p}_i$ and of all coordinates $\mathbf{r}_j$, $j = 1$ to N (as when $\mathcal{J}^\alpha$ is the energy current). We shall consider each of these in turn, together with the extension of (i) to the case of transverse currents. Other possibilities can be envisaged, but are of less direct physical interest. Apart from the brief discussion of mixtures given in Section 3.3, we limit the discussion to the case of one-component systems.

3.1. Longitudinal Particle Current

We consider first the case of an external field which couples to the particle density $n(\mathbf{r})$, the Fourier components of which are defined by

$$\hat{n}(\mathbf{k}) = (1/V) \sum_{i=1}^{N} \exp(-i\mathbf{k}\cdot\mathbf{r}_i) \tag{3.1}$$

The corresponding conservation law is given by

$$\dot{\hat{n}}(\mathbf{k}, t) + i\mathbf{k}\cdot\hat{\mathbf{J}}^n(\mathbf{k}, t) = 0 \tag{3.2}$$

where

$$\hat{\mathbf{J}}^n(\mathbf{k}) = (1/V) \sum_{i=1}^{N} \dot{\mathbf{r}}_i \exp(-i\mathbf{k}\cdot\mathbf{r}_i) \tag{3.3}$$

is the particle current; from Eq. (3.2) we see that density fluctuations are linked only to the longitudinal component of the current.

The perturbation in this example is of the form

$$\mathscr{H}_I(t) = -\int d\mathbf{r}\, n(\mathbf{r})\phi(\mathbf{r}, t) = -V \sum_k \hat{n}(-\mathbf{k})\phi(\mathbf{k}, t) \tag{3.4}$$

where $\phi(\mathbf{r}, t)$ is a scalar field (with dimensions of energy) and the sum on $\mathbf{k}$ runs over all wave vectors allowed by the periodic boundary conditions used in the molecular dynamics calculations. In practice, given the assumption of linearity, we are concerned only with the response to a single Fourier component of the external field, $\mathbf{k}$ say, which we shall assume to be of the form $\mathbf{k} = (k, 0, 0)$. The external field can therefore be written as

$$\phi(\mathbf{r}, t) = \Phi \exp(ikx)\, \delta(t) \tag{3.5}$$

and Eq. (3.4) simplifies to

$$\mathscr{H}_I(t) = -V\hat{n}(-\mathbf{k})\phi(\mathbf{k}, t) = -V\hat{n}(-\mathbf{k})\Phi\,\delta(t) \tag{3.6}$$

By identifying α with n we see from Eq. (2.9) that

$$\begin{aligned}\langle\hat{O}(\mathbf{k})\rangle_t &= \beta V\sum_\nu \langle\hat{O}(\mathbf{k}, t)\hat{J}_\nu{}^n(-\mathbf{k}, 0)\rangle_0 ik_\nu\Phi \\ &= \beta V\langle\hat{O}(\mathbf{k}, t)\hat{J}_x{}^n(-\mathbf{k}, 0)\rangle_0 ik\Phi\end{aligned} \tag{3.7}$$

In particular, the longitudinal particle current induced by a field of unit strength is given by

$$\frac{1}{\Phi}\langle\hat{J}_x{}^n(\mathbf{k})\rangle_t = \beta V\langle\hat{J}_x{}^n(\mathbf{k}, t)\hat{J}_x{}^n(-\mathbf{k}, 0)\rangle_0 ik = \frac{i\beta V}{k}C_l(k, t) \tag{3.8}$$

where

$$C_l(k, t) = k^2\langle\hat{J}_x{}^n(\mathbf{k}, t)\hat{J}_x{}^n(-\mathbf{k}, 0)\rangle_0 \tag{3.9}$$

is the equilibrium longitudinal current autocorrelation function.

Equation (3.8) shows that the response in $\mathbf{k}$ space to the field (3.5) is purely imaginary. This means only that the induced current is out of phase with the driving field by exactly $\frac{1}{2}\pi$ and it is easy to show that the response in $\mathbf{r}$ space to a real applied field is purely real. Taking the inverse transform of Eq. (3.8) for the case when only one Fourier component is excited, we see that

$$(1/\Phi)\langle J_x{}^n(\mathbf{r})\rangle_t = (i\beta V/k)C_l(\mathbf{k}, t)\exp(ikx) \tag{3.10}$$

so the response to a real field of the form

$$\phi(\mathbf{r}, t) = \tfrac{1}{2}\phi\{\exp(ikx) + \exp(-ikx)\}\,\delta(t) = \Phi\,\delta(t)\cos kx \tag{3.11}$$

is given by

$$\begin{aligned}\frac{1}{\Phi}\langle J_x{}^n(\mathbf{r})\rangle_t &= \frac{i\beta V}{2k}C_l(\mathbf{k}, t)\{\exp(ikx) + \exp(-ikx)\} \\ &= \frac{-\beta V}{k}C_l(k, t)\sin kx\end{aligned} \tag{3.12}$$

But from Eq. (3.8) we see that the real part of the corresponding Fourier component of the induced current is zero. Thus

$$\frac{1}{\Phi}\left\langle\sum_{i=1}^N \dot{x}_i\sin kx_i\right\rangle_t = -\frac{V^2}{k}C_l(k, t) \tag{3.13}$$

where the left-hand side gives the response to a field varying as cos kx. If, on the other hand, we choose to perturb the system with a field varying as sin kx, we must look for the response in the real part of the current.

From the general result given by Eq. (3.7) we see that a perturbation of the form of (3.5) will also induce change in other dynamical variables, the stress tensor (momentum current) **T** and the energy current $\mathbf{J}^e$, for example. However, the stress tensor is linked to the particle current by the conservation law

$$m\dot{\hat{\mathbf{J}}}^n(\mathbf{k}, t) + i\mathbf{k}\cdot\hat{\mathbf{T}}(\mathbf{k}, t) = 0 \tag{3.14}$$

Thus observation of the response in the diagonal element $\hat{T}_{xx}$ yields only the time derivative of $C_l(k, t)$. This particular type of cross effect is therefore rather uninteresting. Observations of the thermal response, given by

$$(1/\Phi)\langle \hat{J}_x{}^e(\mathbf{k})\rangle_t = -\beta V\langle \hat{J}_x{}^e(\mathbf{k}, t)\hat{J}_x{}^n(-\mathbf{k}, 0)\rangle_0 ik \tag{3.15}$$

provides a measure of the coupling between the particle current and the flow of energy.

3.2. Transverse Particle Current

In the case of the transverse component of the particle current the required perturbation is of the general form of Eq. (2.22) with the restrictions imposed by (2.15). Typically, therefore, choosing again the vector $\mathbf{k} = (k, 0, 0)$ and setting $\gamma_i = 1$ for all i, the applied force field takes the form

$$\xi_y(\mathbf{r}, t) = \Xi \exp(ikx)\, \delta(t), \qquad \xi_x(\mathbf{r}, t) = \xi_z(\mathbf{r}, t) = 0 \tag{3.16}$$

On substituting in Eq. (2.17), wc obtain

$$\langle \hat{O}(\mathbf{k})\rangle_t = \beta V\langle \hat{O}(\mathbf{k}, t)\hat{J}_y{}^n(-\mathbf{k}, 0)\rangle_0 \Xi \tag{3.17}$$

which is the transverse analog of Eq. (3.7). In particular, the response seen in the transverse current itself is

$$\frac{1}{\Xi}\langle \hat{J}_y{}^n(\mathbf{k})\rangle_t = \beta V\langle \hat{J}_y{}^n(\mathbf{k}, t)\hat{J}_y{}^n(-\mathbf{k}, 0)\rangle_0 = \frac{\beta V}{k^2} C_t(k, t) \tag{3.18}$$

where

$$C_t(k, t) = k^2\langle \hat{J}_y{}^n(\mathbf{k}, t)\hat{J}_y{}^n(-\mathbf{k}, 0)\rangle_0 \tag{3.19}$$

is the transverse current autocorrelation function. The response is in this case purely real, showing that the induced current is now in phase with the driving force. The response to a real force

$$\boldsymbol{\xi}(r) = (0, \Xi \cos kx, 0) \tag{3.20}$$

is therefore

$$\frac{1}{\Xi}\left\langle \sum_{i=1}^{N} \dot{y}_i \cos kx_i \right\rangle_t = \frac{\beta V^2}{k^2} C_t(k, t) \tag{3.21}$$

The possibility of studying cross-correlation effects again exists, but these are likely to be small, since in the hydrodynamic limit the transverse current is decoupled from all other fluctuating variables.

3.3. Binary Mixtures

The arguments of the two preceding sections are easily generalized to multicomponent systems. Consider, for example, the case of a binary mixture. If N_s is the number of particles of species s, with $N = N_1 + N_2$, then the Fourier components of the partial densities may be defined as

$$\hat{n}_s(\mathbf{k}) = \frac{1}{V} \sum_{i=1}^{N_s} \exp(-i\mathbf{k}\cdot\mathbf{r}_{is}), \qquad s = 1, 2 \tag{3.22}$$

where $\mathbf{r}_{is}$ denotes the coordinates of particle i of species s. The total number density is then given by

$$\hat{n}(\mathbf{k}) = \hat{n}_1(\mathbf{k}) + \hat{n}_2(\mathbf{k}) \tag{3.23}$$

but in addition we may introduce the mass density

$$\hat{m}(\mathbf{k}) = m_1\hat{n}_1(\mathbf{k}) + m_2\hat{n}_2(\mathbf{k}) \tag{3.24}$$

and, for uncharged systems, the concentration

$$\hat{c}(\mathbf{k}) = c_2\hat{n}_1(\mathbf{k}) - c_1\hat{n}_2(\mathbf{k}) \tag{3.25}$$

where m_s is the mass of a particle of species s and $c_s = N_s/N$. In the case of charged fluids the analog of (3.25) is the charge density

$$\hat{q}(\mathbf{k}) = q_1\hat{n}_1(\mathbf{k}) + q_2\hat{n}_2(\mathbf{k}) \tag{3.26}$$

where q_s is the charge carried by particles of species s.

Each of the densities (3.24)–(3.26) obeys a conservation law similar to Eq. (3.2) and both the longitudinal and transverse components of the corresponding currents can be induced by the methods already discussed. The only extra step required is to identify α_i (or γ_i) with the appropriate scalar quantity—mass, concentration, or charge—appearing in the definitions of the densities. This makes it straightforward to study a wide variety of cross effects: the coupling of concentration currents with the flow of energy, the mixing of modes of optical and acoustic character, and so on. Of greater importance, perhaps, is the fact that transverse currents of concentration or charge can be studied even in the limit $k = 0$. This offers a practical means of measuring electrical conductivity. By way of illustration, consider the case of a monovalent molten salt, for which we may use the notation $s = +, -$, with $q_+ = e$ and $q_- = -e$. If a homogeneous force field of the form

$$\boldsymbol{\xi} = (0, \Xi, 0) \tag{3.27}$$

is applied and if γ_i is set equal to the charge on particle i, we find as a special case of Eq. (2.16) that the response in the y component of the microscopic electric current $\mathbf{I}$ defined by

$$\mathbf{I} = e \sum_{i=1}^{N_+} \dot{\mathbf{r}}_{i+} - e \sum_{j=1}^{N_-} \dot{\mathbf{r}}_{j-} \tag{3.28}$$

is

$$(e/\Xi)\langle I_y \rangle_t = \beta \langle I_y(t) I_y(0) \rangle_0 \tag{3.29}$$

It follows immediately from the well-known formula of Kubo[1] that the electrical conductivity σ is given in terms of the response by

$$\sigma = (e/V\Xi) \int_0^\infty dt \, \langle I_y \rangle_t \tag{3.30}$$

A similar argument applied in the case of uncharged fluids leads to an expression for the interdiffusion coefficient.

3.4. Stress Tensor Fluctuations

The coefficient of shear viscosity η_s is related through a Kubo formula to the long-wavelength limit of the autocorrelation function of an off-diagonal element of the stress tensor and the longitudinal viscosity $\eta_l = \frac{4}{3}\eta_s + \xi$ (ξ being the bulk viscosity) is similarly linked to the autocorrelation function of a diagonal element of T. Specifically[2,9]

$$\eta_s = \beta V \int_0^\infty dt \, \langle \hat{T}_{xy}(0, t) \hat{T}_{xy}(0, 0) \rangle_0 \tag{3.31}$$

$$\eta_l = \beta V \int_0^\infty dt \, \langle \{\hat{T}_{xx}(0, t) - P\}\{\hat{T}_{xx}(0, 0) - P\} \rangle_0 \tag{3.32}$$

In writing Eq. (3.32) we have taken account of the fact that $\langle \hat{T}_{xx}(0) \rangle_0$ is equal to the pressure P. In the case of pairwise additive forces the microscopic stress tensor is given by

$$\mathbf{T}(\mathbf{k}) = \frac{1}{V} \sum_{i=1}^{N} m \dot{\mathbf{r}}_i \dot{\mathbf{r}}_i \exp(-i\mathbf{k} \cdot \mathbf{r}_i) + \frac{1}{2V} \sum_{i \neq j}^{N} \sum^{N} \frac{\mathbf{r}_{ij} \mathbf{r}_{ij}}{r_{ij}} v'(r_{ij}) Q_{ij}(\mathbf{k}) \tag{3.33}$$

where $v(r)$ is the pair potential, $v'(r) = dv(r)/dr$, $\mathbf{r}_{ij} = \mathbf{r}_j - \mathbf{r}_i$, and

$$Q_{ij}(\mathbf{k}) = [\exp(-i\mathbf{k} \cdot \mathbf{r}_j) - \exp(-i\mathbf{k} \cdot \mathbf{r}_i)]/i\mathbf{k} \cdot \mathbf{r}_{ij} \tag{3.34}$$

In studying the viscous modes of the system it is clear from the earlier discussion that the most suitable form of applied field is one which couples to the momentum density and consequently must be a *vector* field. In this case the perturbation is velocity dependent, having the form

$$\mathscr{H}_I = -\int d\mathbf{r}\, \mathbf{J}^n(\mathbf{r})\cdot\boldsymbol{\phi}(\mathbf{r}, t) = -V\sum_{\mathbf{k}} \hat{\mathbf{J}}^n(\mathbf{k})\cdot\hat{\boldsymbol{\phi}}(\mathbf{k}, t) \tag{3.35}$$

which corresponds to identifying α_i with $\dot{\mathbf{r}}_i$. If we write $\boldsymbol{\phi}(\mathbf{r}, t)$ (which has the dimensions of momentum) as

$$\boldsymbol{\phi}(\mathbf{r}, t) = \boldsymbol{\phi}(\mathbf{r})\,\delta(t) \tag{3.36}$$

we see from the conservation law (3.14) that the response in the general case is given by

$$\langle\hat{O}(\mathbf{k})\rangle_t = \frac{\beta V}{m}\sum_{\mu}\sum_{\nu}\langle\hat{O}(\mathbf{k}, t)\hat{T}_{\mu\nu}(-\mathbf{k}, 0)\rangle_0 ik_\mu\hat{\phi}_\nu(k) \tag{3.37}$$

It is clear from Eq. (3.37) that by varying the form of $\boldsymbol{\phi}(\mathbf{r})$ we can couple the field to a diagonal element of $\hat{\mathbf{T}}$, an off-diagonal element, or some combination of the two. Choosing

$$\boldsymbol{\phi}(\mathbf{r}) = (0, \phi\exp(ikx), 0) \tag{3.38}$$

and assuming, as always, that $\mathbf{k}$ is parallel to the x axis, we find that Eq. (3.37) reduces to

$$\langle\hat{O}(\mathbf{k})\rangle_t = (\beta V/m)\langle\hat{O}(\mathbf{k}, t)\hat{T}_{xy}(-\mathbf{k}, 0)\rangle_0 ik\Phi \tag{3.39}$$

In particular, choosing for $\hat{O}$ the xy component of $\hat{T}(\mathbf{k})$, we see that the response is given by

$$\langle\hat{T}_{xy}(\mathbf{k})\rangle_t = (\beta V/m)\langle\hat{T}_{xy}(\mathbf{k}, t)\hat{T}_{xy}(-\mathbf{k}, 0)\rangle_0 ik\Phi \tag{3.40}$$

As in the case of the longitudinal particle currents, it follows from Eq. (3.40) that the response to a field varying as $\cos kx$ is in the imaginary part of $\hat{T}_{xy}(\mathbf{k}, t)$. Thus in the limit $k \to 0$ we find that

$$\eta_s = (m/k\Phi)\int_0^\infty dt\, \lim_{k\to 0}\langle \mathrm{Im}\, \hat{T}_{xy}(\mathbf{k})\rangle_t \tag{3.41}$$

In practice, in contrast to the case of the electrical conductivity, we are unable to take the $k \to 0$ limit, and it is well known that we cannot interchange the limiting operation and the integration, since the integral then vanishes. The second of these difficulties can be avoided by replacing the upper limit of integration by a time τ which is finite but sufficiently large for the response to be essentially zero. However, the result thereby obtained for η_s will be useful only if the smallest value of k consistent with the periodic boundary conditions is sufficiently small to give an adequate estimate of the integrand in Eq. (3.41). For $k > 0$ we can study both the stress–stress autocorrelation itself and cross-correlations such as that between the stress tensor and the energy

Notes on p. 220

current. The latter quantity plays a role in generalized hydrodynamic descriptions of density fluctuations in liquids.[10,11]

In Fig. 1 we show some results obtained for the wavelength dependence of the function $C(t) = \langle \mathrm{Im}\, \hat{T}_{xy}(k)\rangle_t$ for the case of a Lennard-Jones fluid at a reduced number density $n\sigma^3 = 0.75$ and mean reduced temperature $k_B T/\epsilon = 1.15$; the calculations were made on a system of 256 particles in a cubic box. It is clear from the figure that the response is a rapidly varying function of k. In particular, the marked growth with increasing k of the region of negative autocorrelation means that the apparent shear viscosity obtained from Eq. (3.31) decreases rapidly with k. Choosing values of σ and ϵ appropriate to argon,[12] we find that for the smallest accessible wave vector the apparent shear viscosity is $\eta_s = 0.44 \times 10^{-3}$ g cm^{-1} sec^{-1}. This is approximately three times smaller than the experimental value[13] and it is clear that the major part of the discrepancy is due to the fact that the wavelength associated with the perturbation is too small for Eq. (3.41) to be useful. Improvement could be sought by, for example, extending the length of the molecular dynamics box in one direction. On the other hand, the shear viscosity can also be related to the response in transverse current to a shearing field, which was discussed in Section 3.2. A simple hydrodynamic[3,13] argument shows that the shear viscosity can be expressed as

$$\eta_s^{-1} = \int_0^\infty dt \lim_{k\to 0} (k^2/n^2\Xi)\langle \hat{J}_y{}^n(\mathbf{k})\rangle_t \tag{3.42}$$

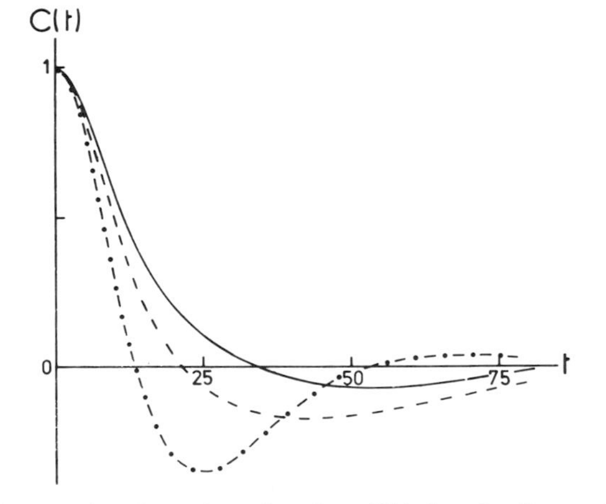

Fig. 1. The wave number-dependent function $C(t)$ for the Lennard-Jones fluid at $n\sigma^3 = 0.75$, $k_B T/\epsilon = 1.15$, normalized to unity at $t = 0$. Full curve: $k\sigma = 0.899$; dashed curve: $k\sigma = 1.798$; dash-dotted curve: $k\sigma = 4.495$. The unit of time is $h = 0.032(m\sigma^2/48\epsilon)^{1/2}$, equal to 10^{-14} sec in the case of argon.

where $\langle \hat{J}_y{}^n(\mathbf{k})\rangle_t$ is the response defined by Eq. (3.18). We are now much more favourably placed for taking the $k = 0$ limit, since the work of Levesque *et al.*[14] on the Lennard-Jones fluid and our own calculations for molten salts[6,15] show that the integral in (3.42) is only weakly k dependent at small k. Typically the extrapolation from the smallest accessible wavenumber to the limit $k = 0$ involves a correction of approximately 20–30%.

If, in place of (3.38), we choose

$$\boldsymbol{\phi}(\mathbf{r}) = (\Phi \exp(ikx), 0, 0) \tag{3.43}$$

Eq. (3.39) becomes

$$\langle \hat{O}(\mathbf{k})\rangle_t = (\beta V/m)\langle \hat{O}(\mathbf{k}, t)\hat{T}_{xx}(-\mathbf{k}, 0)\rangle_0 ik\Phi \tag{3.44}$$

and substitution of $\hat{T}_{xx}$ for $\hat{O}$ leads to an equation analogous to (3.40). This can be treated in the same way to yield an expression for the longitudinal viscosity. The constant term appearing in Eq. (3.32) does not enter the equation analogous to (3.41), since $\langle \hat{T}_{xx}(\mathbf{k})\rangle_0 = 0$ for $k \neq 0$.

3.5. Energy Fluctuations

The formula analogous to (3.31) [or (3.32)] for the thermal conductivity is

$$\lambda T = \beta V \int_0^\infty dt\, \langle \hat{J}_x{}^e(0, t)\hat{J}_x{}^e(0, 0)\rangle_0 \tag{3.45}$$

where $\hat{\mathbf{J}}^e$ is the energy current, defined as

$$\hat{\mathbf{J}}^e(\mathbf{k}) = \frac{1}{V}\sum_{i=1}^{N} e_i\dot{\mathbf{r}}_i \exp(-i\mathbf{k}\cdot\mathbf{r}_i) + \frac{1}{2V}\sum_{i\neq j}^{N}\sum^{N} \mathbf{r}_{ij}\cdot\dot{\mathbf{r}}_{ij}\frac{v'(r_{ij})}{r_{ij}}\mathbf{r}_{ij}Q_{ij}(\mathbf{k}) \tag{3.46}$$

e_i is the energy of particle i

$$e_i = (|\mathbf{p}_i|^2/2m) + \sum_{j>i}^{N} v(r_{ij}) \tag{3.47}$$

and the local energy density

$$\hat{e}(\mathbf{k}) = (1/V)\sum_{i=1}^{N} e_i \exp(-i\mathbf{k}\cdot\mathbf{r}_i) \tag{3.48}$$

is linked to the energy current by the conservation law

$$\dot{\hat{e}}(\mathbf{k}, t) + i\mathbf{k}\cdot\hat{\mathbf{J}}^e(\mathbf{k}, t) = 0 \tag{3.49}$$

To study thermal fluctuations it is clearly appropriate to apply a field which couples to the energy density. In this case the interaction with the system is represented by the Hamiltonian

$$\mathscr{H}_I(t) = -\int d\mathbf{r}\, e(\mathbf{r})\phi(\mathbf{r}, t) = -V\sum_{\mathbf{k}} \hat{e}(-\mathbf{k})\hat{\phi}(\mathbf{k}, t) \tag{3.50}$$

where ϕ is a dimensionless scalar field. This corresponds to setting $\alpha_i = e_i$. Proceeding along the now familiar lines, we see that the response in a variable $\hat{O}$ is

$$\langle\hat{O}(\mathbf{k})\rangle_t = \beta V \sum_{\nu} \langle\hat{O}(\mathbf{k}, t)\hat{J}_\nu{}^e(-\mathbf{k}, 0)\rangle_0 ik_\nu\phi(\mathbf{k}) \tag{3.51}$$

In particular, taking $\mathbf{k} = (k, 0, 0)$ and $\hat{O} = \hat{J}_x{}^e$, we find that

$$\langle\hat{J}_x{}^e(\mathbf{k})\rangle_t = \beta V\langle\hat{J}_x{}^e(\mathbf{k}, t)\hat{J}_x{}^e(-\mathbf{k}, 0)\rangle_0 ik\Phi \tag{3.52}$$

Thus

$$\lambda T = \int_0^\infty dt \lim_{k\to 0}[\langle \mathrm{Im}\, \hat{J}_x{}^e(\mathbf{k})\rangle_t/k\Phi] \tag{3.53}$$

if the applied field varies as $\cos kx$. The remarks following Eq. (3.41) are also relevant to the calculation of the thermal conductivity. As we have recently shown,[7] the apparent thermal conductivity is a rapidly varying function of k, but extrapolation to $k = 0$ yields a result for the Lennard-Jones fluid in fair agreement with experimental data on argon. As in the case of the shear viscosity, the calculation could again be improved by increasing the size of the molecular dynamics box in one dimension. It is also possible that the expression for the thermal conductivity in terms of the heat current (see Appendix B of Ref. 2) would yield a result which converges more rapidly to the $k = 0$ limit. By choosing $\hat{O} = \hat{J}_x{}^n$ or $\hat{O} = \hat{T}_{xy}$ in Eq. (3.51) we can again study the coupling between energy current and particle current or energy current and the stress tensor. On grounds of symmetry the cross-correlation functions determining the response must be identical to those introduced in earlier sections.

4. SOLUTION OF THE EQUATIONS OF MOTION AND MEASUREMENT OF THE RESPONSE

In solving the equations of motion of the particles we use for the most part the central-difference or "leap-frog" algorithm described by Verlet.[16] At $t = 0$, however, the perturbation appears as an impulsive force and some modification of the algorithm is needed.

Let h be the time step in the numerical integration. Making a Taylor expansion forward and backward in time about $t = 0$, we find

$$\mathbf{r}_i(h) = \mathbf{r}_i(0+) + \dot{\mathbf{r}}_i(0+)h + \tfrac{1}{2}\ddot{\mathbf{r}}_i(0+)h^2 + O(h^3) \tag{4.1}$$

$$\mathbf{r}_i(-h) = \mathbf{r}_i(0-) - \dot{\mathbf{r}}_i(0-)h + \tfrac{1}{2}\ddot{\mathbf{r}}_i(0-)h^2 + O(h^3) \tag{4.2}$$

Thus, to terms of order h^3, the predicted coordinates at $t = h$ are given by

$$\mathbf{r}_i(h) = 2\mathbf{r}_i(0-) - \mathbf{r}_i(-h) + \ddot{\mathbf{r}}_i(0-)h^2 + \{\mathbf{r}_i(0+) - \mathbf{r}_i(0-)\} + \{\dot{\mathbf{r}}_i(0+) - \dot{\mathbf{r}}_i(0-)\}h + \tfrac{1}{2}\{\ddot{\mathbf{r}}_i(0+) - \ddot{\mathbf{r}}_i(0-)\}h^2 \quad (4.3)$$

The first three terms on the right-hand side are the customary ones: the three succeeding terms arise from discontinuities in (i) position, (ii) velocity, and (iii) acceleration. Not all the additional terms will contribute in every case. In particular, when the perturbing field acting on particle i is a function solely of its position $\mathbf{r}_i$, the only discontinuity which appears is that in $\dot{\mathbf{r}}_i$.

To see in detail what happens at $t = 0$ it is helpful to rewrite Eq. (2.13) in a form in which the singularity in time appears explicitly. This we achieve by writing

$$\ddot{\mathbf{r}}_i(t) = \mathbf{P}(t) + \mathbf{Q}(t)\,\delta(t) + (d/dt)\{\mathbf{R}(t)\,\delta(t)\} \quad (4.4)$$

where $\mathbf{P}$, $\mathbf{Q}$, and $\mathbf{R}$ are the regular functions of t defined by

$$m\mathbf{P}(t) = -\partial\mathscr{H}_0/\partial\mathbf{r}_i \quad (4.5)$$

$$m\mathbf{Q}(t) = \alpha_i\frac{\partial\phi(\mathbf{r}_i)}{\partial\mathbf{r}_i} + \sum_{j=1}^{N}\phi(\mathbf{r}_j)\frac{\partial\alpha_j}{\partial(\mathbf{r}_i)} \quad (4.6)$$

$$\mathbf{R}(t) = -\phi(\mathbf{r}_i)\,\partial\alpha_i/\partial\mathbf{p}_i \quad (4.7)$$

In the transverse case the function $\mathbf{P}(t)$ retains the same form, $\mathbf{R}(t)$ is zero, and

$$\mathbf{Q}(t) = \gamma_i\xi(\mathbf{r}_i) \quad (4.8)$$

By integrating Eq. (4.4) we find that

$$\dot{\mathbf{r}}_i(0+) - \dot{\mathbf{r}}_i(0-) = \lim_{t\to 0+}\int_{0-}^{t}\ddot{\mathbf{r}}_i(t')\,dt' = \mathbf{Q}(0) + \mathbf{R}(0)\,\delta(t) \quad (4.9)$$

The first term on the right-hand side of (4.9) is just the discontinuity in velocity appearing in Eq. (4.3) and the second term represents an impulsive change in velocity at $t = 0$. The latter gives rise to a discontinuity in coordinates of the form

$$\mathbf{r}_i(0+) - \mathbf{r}_i(0-) = \mathbf{R}(0) \quad (4.10)$$

This last result follows immediately from integration of the delta-function term in Eq. (4.9).

From the discussion just given we see that if the perturbation is velocity independent, the effect of the applied field appears only as a discontinuity in $\dot{\mathbf{r}}_i$. In the more general case the function $\mathbf{R}(t)$ is nonzero [cf. Eq. (2.13)] and in consequence there is additionally a discontinuity in $\mathbf{r}_i$. This in turn implies a discontinuity in $\ddot{\mathbf{r}}_i$, since the total intermolecular force acting on particle i is a function of the coordinates of all interacting particles. Formulas giving the discontinuities in $\mathbf{r}_i$ and $\dot{\mathbf{r}}_i$ in particular cases can now be obtained by inserting in Eq. (4.6) and (4.7) the corresponding choices for α_i and the applied field, making the appropriate contraction of indices when the latter are vectorial quantities.

We now turn to the question of the computation of the response. The basic procedure we use is illustrated schematically in Fig. 2. The thick curve represents the trajectory in phase space which the system follows in the absence of any applied field. Branching off from this are the trajectories resulting from switching on a perturbation at times 1, 2, 3, 4,.... The response to the perturbation for $t > 0$ is given by

$$\begin{aligned}\langle \hat{O}(\mathbf{k}) \rangle_t &= \langle \exp(-i\mathscr{L}t)\, \hat{O}(\mathbf{k}) \rangle_0 \\ &= \lim_{\tau \to \infty} \tau^{-1} \int_0^{\tau} d\tau' \exp(i\mathscr{L}_0 \tau') \exp(-i\mathscr{L}t)\, \hat{O}(\mathbf{k}) \qquad (4.11)\end{aligned}$$

It is important to note that the average appearing on the right-hand side of (4.11) is taken over the unperturbed trajectory; the Liouville operators $\mathscr{L}$ and $\mathscr{L}_0$ have the same meaning as in Eq. (1.5). In principle, Eq. (4.11) could be used to calculate directly the mean change in $\hat{O}$ due to the perturbation, since we can always formulate the problem in such a way that $\langle \hat{O}(\mathbf{k}) \rangle_0 = 0$. In practice, because the unperturbed trajectory is not of infinite length, the mean value of $\hat{O}$ along that trajectory will invariably have some small but nonzero value. Furthermore, the perturbation used is very small; typically the parameter Φ (or Ξ) is chosen such that the changes in coordinates and momenta are of order one part in 10^6. Thus the systematic response is in general much smaller than the statistical fluctuations. We therefore choose to measure the response in the manner described by Eqs. (1.5) and (1.6), that is to say, by averaging the *difference* in the variable of interest in perturbed and

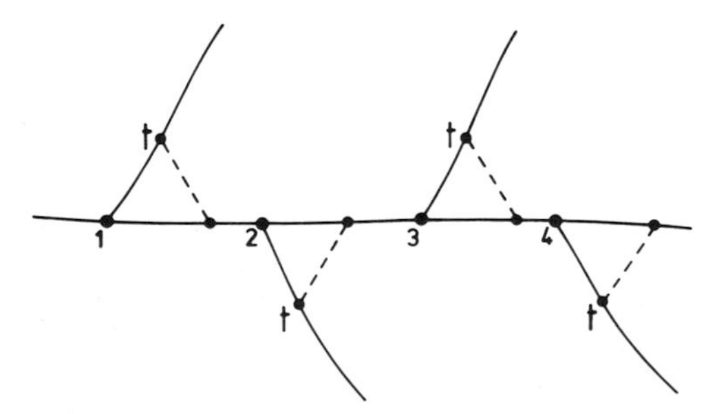

Fig. 2. Schematic illustration of the procedure used in calculating the response. The perturbation is assumed to be switched on at times 1, 2, 3, 4,..., and the broken lines link corresponding points on the perturbed and unperturbed trajectories. Calculation of the difference in the value of a given dynamical variable at two such points yields the mechanical response discussed in the text.

unperturbed trajectories at a time t after the field is switched on. For times which are not too long the improvement achieved is dramatic.

We come finally to a question of purely computational significance. Let us write the coordinates at time $t = h$ in the form

$$\mathbf{r}_i(h) = \mathbf{r}_{i0}(h) + \delta\mathbf{r}_i(h) \tag{4.12}$$

where $\mathbf{r}_{i0}(h)$ are the coordinates at time h in the absence of the perturbation and $\delta\mathbf{r}_i(h)$ is determined by the discontinuities on the right-hand side of (4.3). In principle, we must integrate $\mathbf{r}_i$ and $\mathbf{r}_{i0}$ separately. From the practical point of view this is the simplest way to proceed, but on the other hand it may be computationally more economical to solve instead an approximate differential equation for $\delta\mathbf{r}_i$ itself. The exact equation of motion for $\delta\mathbf{r}_i(t)$ is

$$\begin{aligned} m\,\delta\ddot{\mathbf{r}}_i(t) &= m\ddot{\mathbf{r}}_i(t) - m\ddot{\mathbf{r}}_{i0}(0) \\ &= \mathbf{F}_i(\{\mathbf{r}_j(t)\}) - \mathbf{F}_i(\{\mathbf{r}_{j0}(t)\}) \\ &\quad + \mathbf{Q}(t)\,\delta(t) + (d/dt)\{\mathbf{R}(t)\,\delta(t)\} \end{aligned} \tag{4.13}$$

where $\mathbf{F}_i$ is the force acting on particle i. We can now approximate the difference in internal forces by the first term in the Taylor expansion

$$\begin{aligned} \delta\mathbf{F}_i(t) &= \mathbf{F}_i(\{\mathbf{r}_j(t)\}) - \mathbf{F}_i(\{\mathbf{r}_{j0}(t)\}) \\ &\simeq \sum_{k=1}^{N} \frac{\partial}{\partial\mathbf{r}_{k0}(t)}\mathbf{F}_i(\{\mathbf{r}_{j0}(t)\})\cdot\delta\mathbf{r}_k(t) \end{aligned} \tag{4.14}$$

If the potential energy is pairwise additive, Eq. (4.14) may be rewritten as

$$\delta\mathbf{F}_i(t) = \sum_{j\neq i}^{N}\left\{\frac{\delta\mathbf{r}_{ij}}{\mathbf{r}_{ij}}\,v'(r_{ij}) - \frac{\mathbf{r}_{ij}}{r_{ij}}\frac{d}{dr_{ij}}\left(\frac{v'(r_{ij})}{r_{ij}}\right)\mathbf{r}_{ij}\cdot\delta\mathbf{r}_{ij}\right\} \tag{4.15}$$

where

$$\delta\mathbf{r}_{ij} = \delta\mathbf{r}_j(t) - \delta\mathbf{r}_i(t) \tag{4.16}$$

and $\mathbf{r}_{ij}$ is to be evaluated at time t. The justification for this linearization is the observation that the mechanical response is linear with respect to the applied force over an extremely wide range. The equation of motion for $\delta\mathbf{r}_i(t)$ for $t > 0$ is now solved in the usual way by writing

$$\delta\mathbf{r}_i(t+h) = -\delta\mathbf{r}_i(t-h) + 2\delta\mathbf{r}_i(t) + \delta\mathbf{F}_i(t)h^2 \tag{4.17}$$

where $\delta\mathbf{F}_i(t)$ is given by Eq. (4.14). Equation (4.17) is to be solved subject to the initial condition that

$$\delta\mathbf{r}_i(0) = \mathbf{r}_i(0+) - \mathbf{r}_i(0-) \tag{4.18}$$

Notes on p. 220

with $\delta\mathbf{r}_i(h)$ given by Eq. (4.3). The advantage of proceeding in this fashion lies in the fact that $\delta\mathbf{F}_i(t)$ can be computed at rather small cost in the same loop as the calculation of $\mathbf{F}_i(t)$ along the equilibrium trajectory.

If this linearization is used, it is clearly appropriate to measure a response which is also linearized with respect to the phase space variables. This may be achieved by approximating the mechanical response in the form

$$\exp(-i\mathscr{L}t)\,\hat{O}(\mathbf{k}) - \exp(-i\mathscr{L}_0 t)\,\hat{O}(\mathbf{k}) \simeq \sum_{i=1}^{N} \frac{\partial}{\partial \mathbf{r}_i(t)}\,\hat{O}(\mathbf{k}, t)\cdot\delta\mathbf{r}_i(t) + \frac{\partial}{\partial \mathbf{p}_i(t)}\,\hat{O}(\mathbf{k}, t)\cdot\delta\mathbf{p}_i(t) \tag{4.19}$$

from which the statistical response can be obtained in the manner already described. Proceeding in this way we find, for example, that the induced particle current [cf. Eq. (1.5)] is given in linearized form by

$$\Delta\hat{J}_x{}^n(\mathbf{k}, t) \simeq \frac{1}{V}\sum_{i=1}^{N} \{\delta\dot{x}_i(t) - ik\dot{x}_i(t)\,\delta x_i(t)\}\exp\{-ikx_i(t)\} \tag{4.20}$$

This linearization is consistent with the linearization of the equations of motion, and furthermore it is much the most sensible method available for computing the response when the equations of motion are solved in linearized form, since in that case the quantities known are not the perturbed coordinates and velocities themselves, but only the differences $\delta\mathbf{r}_i(t)$ and $\delta\dot{\mathbf{r}}_i(t)$. Reconstruction of the perturbed trajectory would obviously be very wasteful. Complete linearization also allows the computations to be carried out in single precision on, for example, IBM computers. Double-precision arithmetic is essential when the full equations of motion are solved along the perturbed trajectory. It must be said, however, that for the stress tensor and the energy current the calculation of the linearized response adds substantially to the length of the computation. When complete linearization is adopted the strength of the perturbing field enters only as a multiplicative factor.

5. DISCUSSION

The method we have described differs from other nonequilibrium molecular dynamics techniques[13,17–19] primarily insofar as the perturbation used is very small. This has two important consequences. First, the results obtained may legitimately be interpreted in the framework of linear response theory. Second, we avoid the systematic heating up of the system, energy drift, and other undesirable effects associated with the use of strong external fields. The method therefore represents a direct and economical means of studying the dynamical properties of liquids. Furthermore, again in contrast to other work, the method yields the complete, frequency-dependent response of the system as described by the corresponding time correlation function. The saving in cost which can be achieved is very considerable. In the calculation of electrical conductivity, for example, we find[6] a reduction in computing time by a factor of approximately five compared with that required with the conventional equilibrium technique.[20] Its main disadvantage is the fact that, except in special cases, it is limited to the study of the response to a disturbance having a finite wavelength. This creates difficulties when the quantity sought is the value of a transport coefficient. A less serious problem is the fact that it cannot be used for the study of very long-time (low-frequency) behavior, since in that case the correlation between perturbed and unperturbed trajectories is lost and the subtraction technique is no longer useful. In practical terms this limits its application to times which, for monatomic systems, are typically of order 200 integration steps. The accuracy of the algorithm which is used is here an important factor. In particular there is a strong case for adopting a more accurate scheme for calculating the velocities of the particles than that usually adopted with the leap-frog scheme. We should point out finally that though we have limited the discussion to the case of collective dynamical properties, the method is also very well suited to the study of single-particle motion. In general, of course, such phenomena are easily studied by the usual methods. However, in the case of molecules in solution, particularly very dilute solutions, there is obvious scope for application of a suitably modified form of the method we describe. By proceeding in this way it should be straightforward to compute quantities such as the mobility of an ion in a polar solvent, the reorientational correlation times of, say, a single molecule in an inert gas medium, the intrinsic viscosity of polyatomic molecules, and so on. We hope to return to some of these questions in a later publication.

REFERENCES

1. R. Kubo, *J. Phys. Soc. Jpn.* **12**:570 (1957).
2. J. M. Luttinger, *Phys. Rev.* **135**:A1505 (1964).
3. J. L. Jackson and P. Mazur, *Physica* **30**:2295 (1964).
4. R. Zwanzig, *Ann. Rev. Phys. Chem.* **16**:67 (1965).
5. G. Ciccotti and G. Jacucci, *Phys. Rev. Lett.* **35**:789 (1975).
6. G. Ciccotti, G. Jacucci, and I. R. McDonald, *Phys. Rev. A* **13**:426 (1976).
7. G. Ciccotti, G. Jacucci, and I. R. McDonald, *J. Phys. C: Solid State* **11**:L509 (1978).
8. R. Kubo, *Rep. Prog. Phys.* **29**:255 (1966).
9. E. Helfand, *Phys. Rev.* **119**:1 (1960).
10. J. R. D. Copley and S. W. Lovesey, *Rep. Prog. Phys.* **38**:461 (1975).
11. J. P. Hansen and I. R. McDonald, *Theory of Simple Liquids* (Academic Press, London, 1976).
12. A. Michels, H. Wijker, and Hk. Wijker, *Physica* **15**:627 (1949).
13. E. M. Gosling, I. R. McDonald, and K. Singer, *Mol. Phys.* **26**:1475 (1973).

Notes on p. 220

14. D. Levesque, L. Verlet, and J. Kürkijarvi, *Phys. Rev. A* **7**:1690 (1973).
15. G. Jacucci, I. R. McDonald, and A. Rahman, *Phys. Rev. A* **13**:1581 (1976).
16. L. Verlet, *Phys. Rev.* **163**:201 (1968).
17. W. T. Ashurst and W. G. Hoover, *Phys. Rev. Lett.* **31**:206 (1973).
18. W. T. Ashurst and W. G. Hoover, in *Theoretical Chemistry: Advances and Perspectives*, H. Eyring and D. Henderson, eds. (Academic Press, New York, 1975), Vol. 1.
19. D. J. Evans, *Mol. Phys.*, to appear.
20. J. P. Hansen and I. R. McDonald, *Phys. Rev. A* **11**:2111 (1975).

Notes to Reprint III.5

1. At long times, the statistical errors in the correlation functions computed by the subtraction technique become very large. For a discussion of the numerical accuracy of the method, see (Paolini et al., Phys. Rev. A 34 (1986) 1355).

2. It has recently been suggested (Hoover and Posch, Phys. Lett. A 113 (1985) 82) that the exponential divergence of trajectories, which limits the applicability of the subtraction technique, could be eliminated by imposition of an appropriate, non-holonomic constraint.

Molecular Physics, 1980, Vol. 40, No. 2, 515–519

Determination of the shear viscosity of atomic liquids by non-equilibrium molecular dynamics

by K. SINGER, J. V. L. SINGER and D. FINCHAM
Department of Chemistry, Royal Holloway College,
Egham TW20 0EX, Surrey

(*Received* 26 *November* 1979 ; *revision received* 4 *January* 1980)

The reliable determination of the shear viscosity η_s by the evaluation of the appropriate Green–Kubo integral in molecular dynamics (MD) calculations with realistic interaction potentials is difficult, if not impracticable (Levesque *et al.* [1], Evans and Street [2]).

Several non-equilibrium methods have been devised to circumvent this problem. Ashurt and Hoover [3] calculate the stress produced in a liquid between parallel walls which move in opposite directions. Gosling *et al.* [4] obtain a wavelength dependent $\eta_s(\lambda)$ as the ratio between a sinusoidal external force and a similar velocity field. The boundary conditions proposed by Lees and Edwards (LE) [5] produce a constant velocity gradient extending from $z=-\infty$ to ∞. The LE boundary conditions have been applied in different MD methods to the calculation of η_s of the hard sphere fluid (Naitoh and Ono [6]) and of a molecular liquid simulated by Lennard-Jones atom–atom potentials by Evans [7].

Although these methods have yielded reasonable results they have a common undesirable feature : in order to attain a satisfactory signal/noise ratio it is necessary to apply rates of shear which are by several orders of magnitude larger than those occurring in real systems. Concomitant effects are : an increase of temperature which can only be prevented by a rather drastic interference with the equations of motion, distortion of the radial distribution function, and values of η_s which may depend on the rate of shear [3].

The differential non-equilibrium method of Ciccotti *et al.* [8 (a, b), 9] avoids drastic departures from equilibrium. Starting from a given phase point one generates both an unperturbed and a slightly perturbed trajectory. The response to the perturbation is calculated as the difference of the appropriate property along the two trajectories after the onset of steady state conditions ; the procedure is repeated as often as may be necessary, starting from different phase points (method A). The success of the method rests on the cancellation of large thermal fluctuations as a result of the differential technique : the advantage of the method lies in the fact that it is possible to use perturbations—in this case rates of shear—which are of the same magnitude as those which occur in real experiments. An alternative to the repetition of ' segments ' of trajectory with and without perturbation, is the computation of first order perturbed trajectories by straightforward Taylor expansion of the trajectory with respect to a virtual perturbation (Hubbard and Beeby [10]). The truncation after the linear terms in this expansion is justified because the

0026-8976/80/4002 0515 $02·00

perturbations corresponding to macroscopic laminar flow are extremely small on a molecular scale. In this case more than one perturbed trajectory may be generated concurrently with the unperturbed trajectory (method B). Ciccotti *et al.* [9] have computed $\eta_s(\lambda)$ for molten salts by adapting the technique of Gosling *et al.* [4] to their differential method. The drawback lies in the need to extrapolate to $\lambda=\infty$ if the hydrodynamic η_s is required.

The present work combines the differential method of Ciccotti *et al.* with the LE boundary conditions to avoid, on the one hand, drastic departures from equilibrium, and λ-extrapolations on the other.

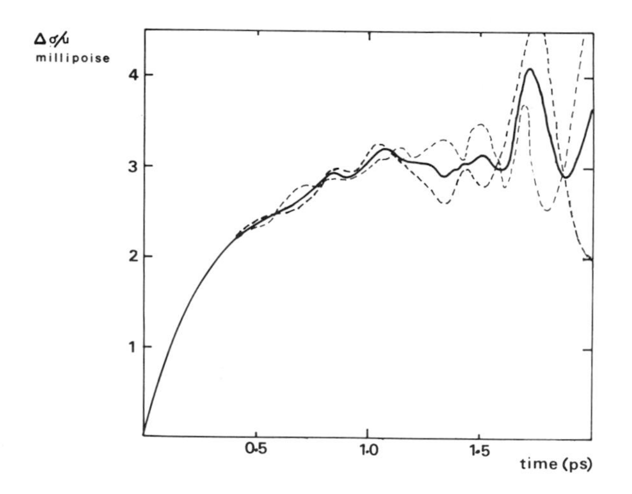

The variation with time of $\Delta\sigma/u$ (for run No. 3 averaged over the 30 segments of the run) showing the rise to the steady state plateau. The broken lines show the results for σ_{xz}/u and σ_{yz}/u, the solid line is the mean.

The LE boundary condition used to produce the perturbed trajectories is implemented as follows :

(1) a small constant velocity gradient $u=dv_x/dz$ (of the order 1 s^{-1}) is set up initially by substitution of $\dot{x}_i+z_iu$ for $\dot{x}_i$ for all centre of mass velocities ;

(2) when a particle at time t leaves the MD cube at $z_i=\pm L/2$ it re-enters at $z_i=\mp L/2$ with $\dot{x}_i$ changed to $\dot{x}_i\mp Lu$ and with x_i changed to $x_i\mp Lut$;

(3) a similar displacement $x'_j(t)=x_j(t)\pm Lut$ occurs when in the formation of the nearest image distance $z_i(t)-z_j(t)$ is changed to $z_i(t)-z'_j(t)$ $(z'_j(t)=z_j(t)\mp L)$.

The conditions suffice to maintain the velocity gradient constant within ± 2 per cent during the ' segment '. This is monitored by calculation of

$$\overline{dv_x/dz}=\sum_1^N \dot{x}_j z_j \Big/ \sum_1^N z_j^2. \tag{1}$$

The approximate constancy of the gradient within the box was monitored by determination of the mean value of $\dot{x}$ within ten slabs of thickness $L/10$.

In the course of the segment the difference of the xz-component of the stress tensors

$$\sigma_{xz} = \sum_{1}^{N} \left(m_i \dot{x}_i z_i + \sum_{j>i}^{N} \frac{x_{ij} z_{ij}}{r_{ij}} \frac{d\phi(r_{ij})}{dr_{ij}} \right) \tag{2}$$

for perturbed and unperturbed trajectories is calculated. Method (B) was used and most of the results are averages obtained from dv_x/dz and dv_y/dz gradients generated simultaneously. The shear viscosity is obtained from (1) and (2)

$$\eta_s = \sigma_{xz} / \overline{dv_x/dz}$$

as the average over 25–100 segments.

Computations were carried out for several state points of the Lennard-Jones liquid simulating argon ($\epsilon/k_B = 119{\cdot}8$, $\sigma = 0{\cdot}3405$ nm). The results should, by virtue of the corresponding states principle [11], also be valid for other rare gases (after appropriate scaling).

The accuracy of the method depends critically on the determination of

$$\Delta\sigma_{\alpha\beta} = \sigma_{\alpha\beta}{}^{\text{pert}} - \sigma_{\alpha\beta}{}^{\text{unpert}}$$

in the steady state. At densities and temperatures not too close to the triple point, $\Delta\sigma_{\alpha\beta}$ rises rapidly to a recognizable plateau which is somewhat blurred by statistical noise of $\sim \pm 5$ per cent. At the end of the plateau oscillations in $\Delta\sigma_{\alpha\beta}$ rapidly become dominant: the perturbed and unperturbed trajectories drift apart in the manner of the spreading ink filaments in the well-known simile.

In the neighbourhood of the triple point the plateau in $\Delta\sigma_{\alpha\beta}$ is so short as to be barely perceptible. An example is shown in the figure. Reproducible results are obtained if the ' plateau ' is taken to be defined by the criteria

(1) $|\Delta\sigma_{xz} - \Delta\sigma_{yz}| < \sim 0{\cdot}1 \times \frac{1}{2}(\Delta\sigma_{xz} + \Delta\sigma_{yz}) = 0{\cdot}1 \overline{\Delta\sigma_{\alpha\beta}}$;
(2) the time gradients of $\Delta\sigma_{xz}$ and $\Delta\sigma_{yz}$ are approximately zero;
(3) the relevant $\Delta\sigma_{\alpha\beta}$ is the highest value of $\Delta\bar{\sigma}_{\alpha\beta}$ for which both criteria (1) and (2) apply.

The estimates of the plateau obtained in this manner were in several cases confirmed by fitting the data to the form $y = A[(1 - \exp(-Bt))]$ in such a manner that the statistical weight of the data point varies inversely to the statistical noise (defined by $|\Delta\sigma_{xz} - \Delta\sigma_{yz}|$). The A values agreed within the margin of error with the values resulting from the criteria (1)–(3). Table 2 illustrates the increase in noise ($[\Delta\sigma_{xz} - \Delta\sigma_{yz}]$) with time; the points on the curve lying close to the assumed plateau are shown in italic type.

To reduce statistical noise Ciccotti *et al.* recommend that the discontinuous change of the forces at the cut-off distance $r = r_c$ be avoided. This can be done (*a*) by making the forces $\mathbf{F}(\mathbf{r}_{ij})$ go to zero in a short interval $r_c - \Delta\epsilon \leqslant r_{ij} \leqslant r_c$, or, (*b*) by the use of ' shifted ' forces: $\mathbf{F}'(\mathbf{r}_{ij}) = \mathbf{F}(\mathbf{r}_{ij}) - \mathbf{F}(\mathbf{r}_c)$. Another source of noise, the inaccuracy of the integration algorithm, can be reduced by the use of a shorter time step. The precaution (*a*) was applied in most runs listed in table 1 and simulations were carried out in cubes containing 108 particles or in double cubes ($L_x = L_y = L$, $L_z = 2L$) with $N = 216$. In runs 3, 5 and 12, cubes with $N = 256$ were used, the precaution (*a*) was dropped, but the time-step was reduced from 2×10^{-14} to 1×10^{-14} s. In addition, in order to eliminate any possible artifacts arising from vortex motions and to improve the statistics, the

Note on p. 226

Table 1. Comparison of η_s obtained by MD for the Lennard-Jones liquid with the experimental η_s for argon.

Run No.	V (cm³)	T (K)	P_{MD} (atm)	P_{exp} (atm)	N	Number of MD steps (×10³)	η_s (10^{-3} p)	η_{exp} B [11]	η_{exp} D [12]	η_{exp} N [13]	η_{exp} H [14]
(1)	27·8	83·5	105	104	216	11·6	3·25	2·95			
(2)	27·8	85·6	100	136	216	11·6	3·05	2·80	(3·70)	2·95	
(3)	27·8	86·0	122	150	256	6·0	3·10	2·80		2·95	
(4)	28·16	84·6	6	52	216	17·4	2·84	2·75			
(5)	28·16	86·5	53	113	256	4·6	2·68	2·60	3·20	2·70	
(6)	28·20	86·0	11	69		14·0	2·60	2·60	3·30	2·75	
(7)	28·50	87·3	−6	31	216	14·0	2·40	2·45		2·50	2·75
(8)	29·70	93·7	−44	23	108	14·0	2·00		2·30		
(9)	31·2	106·9	21	46	216	17·4	1·70		1·70	1·50	1·60
(10)	31·2	109·7	58	80	108	14·0	1·60		1·70	1·50	1·40
(11)	33·0	118·1	50	60	108	14·0	1·34		1·30		1·30
(12)	33·0	119·2	43	73	256	5·0	1·28		1·30		1·20

1 atm = $1{\cdot}013 \times 10^5$ N m^{-2}.

Table 2. Signal and noise.

	$\Delta t/10^{-14}$ s	60	70	80	90	100	110	120	130	140	150	160	170
Run No. (1)													
$V=27{\cdot}8$ cm³	$\frac{1}{2}\lvert\eta_{xz}+\eta_{yz}\rvert$			2·61	2·63	2·74	3·03	*3·21*	3·24	3·34	3·74	3·89	3·72
$T=83{\cdot}5$ K	$\lvert\eta_{xz}-\eta_{yz}\rvert$			0·10	0·12	0·18	0·38	0·06	0·44	0·16	0·10	0·85	1·15
Run No. (3)													
$V=27{\cdot}8$ cm³	$\frac{1}{2}\lvert\eta_{xz}+\eta_{yz}\rvert$			2·88	2·88	*3·16*	*3·16*	*3·04*	2·96	3·00	3·12	2·96	4·04
$T=86{\cdot}0$ K	$\lvert\eta_{xz}-\eta_{yz}\rvert$			0·12	0·12	0·16	0·12	0·12	0·66	0·20	0·72	0·32	0·70
Run No. (10)													
$V=31{\cdot}2$ cm³	$\frac{1}{2}\lvert\eta_{xz}+\eta_{yz}\rvert$	1·48	1·52	*1·59*	*1·58*	1·63	1·63	1·87	0·17	0·157			
$T=109{\cdot}7$ K	$\lvert\eta_{xz}-\eta_{yz}\rvert$	0·02	0·03	0·02	0·02	0·20	0·22	0·06	0·25	0·28			

xz and yz components of the stress tensors in these runs were symmetrized : $\Delta\sigma_{\alpha\beta}{}^{\mathrm{sym}} = \frac{1}{2}(\Delta\sigma_{\alpha\beta} + \Delta\sigma_{\beta\alpha})$. Table 1 shows that neither this nor changes in N had any systematic effect on the results.

The kinetic part of the stress tensor (not listed in table 1) is small : it rises from 2 to 3 per cent at $\eta_s \sim 3 \times 10^{-3}$ p to 8 to 10 per cent at $\eta_s \sim 1{\cdot}3 \times 10^{-3}$ p (1 p $= 0{\cdot}1$ N m^{-2} s).

On the basis of the first of the three criteria given above, we estimate the margin of statistical error to be of the order of $+5$ per cent.

The entries in table 1 show that at high η_s the agreement with the results of Naugle *et al.* [13] and to a lesser extent of Boon *et al.* [11] is fairly satisfactory. At lower η_s there is reasonable agreement with the data of de Bock *et al.* [12], Naugle *et al.* [13] and Hellemans *et al.* [14]. The quoted experimental data have been obtained by inter- or extrapolation. De Bock *et al.* [12] quote η_s as function of pressure ; to obtain the corresponding molar volumes, the data of van Itterbeek *et al.* [15] were used. The highest viscosity is put in brackets because it results from a large extrapolation in temperature.

The computed values agree well with those of Ashurst and Hoover [3 (a)] at high densities but are somewhat higher than theirs (~ 10 per cent) at lower densities. In particular, we do not find the very high value of 4·0 p. obtained by Levesque *et al.* [1] near the triple point. (Ashurst and Hoover attribute this to a ' width effect '.) The method should be applicable to molecular liquids if the coordinates, velocities and accelerations in (1) and (2) are those of the centres of mass. Work on this is in progress.

We are indebted to Mrs. E. M. Adams and to Dr. R. Renaud who carried out the fitting of $y = A[1 - \exp(-Bt)]$ referred to in the text to the MD data ; and we are grateful to Dr. I. R. McDonald for useful discussions. We also wish to acknowledge the generous allocation of computing time by the University of London Computer Centre.

References

[1] Levesque, D., Verlet, L., and Kürkijarvi, J., 1973, *Phys. Rev.* A, **7**, 1690.
[2] Evans, D. J., and Streett, W. B., 1978, *Molec. Phys.*, **36**, 161.
[3] (a) Ashurst, W. T., and Hoover, W. G., 1973, *Phys. Rev. Lett.*, **31**, 206. (b) Hoover, W. G., and Ashurst, W. T., 1975, *Theoretical Chemistry, Advances and Perspectives*, Vol. 1, edited by H. Eyring and D. Henderson (Academic Press), p. 1. (c) Ashurst, W. T., and Hoover, W. G., 1975, Sandia Report SAND 77–8614.
[4] Gosling, E. M., McDonald, I. R., and Singer, K., 1973, *Molec. Phys.*, **26**, 1475.
[5] Lees, A. W., and Edwards, S. F., 1972, *J. Phys.*, **15**, 1921.
[6] Naitoh, T., and Ono, S., 1976, *Phys. Lett.* A, **57**, 448 ; 1979, *J. chem. Phys.*, **70**, 4515.
[7] Evans, D. J., 1979, *Molec. Phys.*, **37**, 1745.
[8] (a) Ciccotti, G., and Jacucci, G., 1975, *Phys. Rev. Lett.*, **35**, 789. (b) Ciccotti, G., Jacucci, G., and McDonald, I. R., 1979, *J. statist. Phys.*, **21**, 1.
[9] Ciccotti, G., Jacucci, G., and McDonald, I. R., 1976, *Phys. Rev.* A, **13**, 426.
[10] Hubbard, J., and Beeby, J. L., 1969, *J. Phys.* C, **2**, 556.
[11] Boon, J. P., Legros, J. C., and Thomaes, G., 1967, *Physica*, **33**, 547.
[12] De Bock, A., Grevendonk, W., and Herreman, W., 1967, *Physica*, **37**, 227.
[13] Naugle, D. G., Lunsford, J. H., and Singer, J. R., 1966, *J. chem. Phys.*, **45**, 4669.
[14] Hellemans, J., Zink, H., and van Paemel, O., 1970, *Physica*, **46**, 395.
[15] Van Itterbeek, A., and Verbeke, O., 1960, *Physica*, **26**, 931 ; 1963, *Ibid.*, **29**, 742.

Note on p. 226

Note to Reprint III.6

The method described in this paper has been applied successfully to the calculation of the shear viscosity of a molecular liquid (n-butane) by Maréchal et al. (Mol. Phys. 61 (1987) 33).

PHYSICAL REVIEW A VOLUME 22, NUMBER 4 OCTOBER 1980

Lennard-Jones triple-point bulk and shear viscosities. Green-Kubo theory, Hamiltonian mechanics, and nonequilibrium molecular dynamics

William G. Hoover
Department of Applied Science, University of California at Davis/Livermore, Livermore, California 94550 and Lawrence Livermore National Laboratory, Livermore, California 94550

Denis J. Evans
Research School of Physical Sciences, Australian National University, Canberra, ACT 2600, Australia and National Bureau of Standards, Boulder, Colorado 80303

Richard B. Hickman
Lawrence Livermore National Laboratory, Livermore, California 94550

Anthony J. C. Ladd
Department of Applied Science, University of California at Davis/Livermore, Davis, California 95616

William T. Ashurst
Sandia Livermore Laboratory, Livermore, California 94550

Bill Moran
Department of Applied Science, University of California at Davis/Livermore, Livermore, California 94550 and Lawrence Livermore National Laboratory, Livermore, California 94550

(Received 24 March 1980)

A new Hamiltonian method for deformation simulations is related to the Green-Kubo fluctuation theory through perturbation theory and linear-response theory. Numerical results for the bulk and shear viscosity coefficients are compared to corresponding Green-Kubo calculations. Both viscosity coefficients depend similarly on frequency, in a way consistent with enhanced "long-time tails."

I. INTRODUCTION

Fluid flow can be treated at a variety of levels[1] by including only some of the difficulties associated with compressibility, viscosity, heat conduction, entropy production, gravity, and turbulence. In discussing shockwaves, only the last two of these complications can be ignored. The unique feature of shock compression[2] is the abrupt transformation of a fluid or solid from one equilibrium state to another. In a dense fluid this transformation can take place in approximately one atomic vibration time.[3] The details of this irreversible transformation process depend upon the transient transport of momentum and energy within the shockwave front. With pressure jumps of tens of kilobars occurring in distances of only a few atomic diameters it is not clear *a priori* that a continuum point of view is appropriate to shockwaves at all. Nevertheless, it is known that the predicted shockwave profiles from the simplest reasonable continuum model—the Navier-Stokes equations, with compressibility, viscosity, conduction, and entropy production included—agree fairly well with profiles from atomistic computer simulations.[3,4]

Viscosity is the physical property which dominates shockwave structure. Viscosity describes the extra work required when deformation takes place rapidly, rather than slowly and reversibly. General deformations include both changes in shape and in size so that two different viscosity coefficients, shear (for shape) and bulk (for size) are required to describe the dependence of work on deformation rate.

In many flow problems changes in shape, involving only shear viscosity, are much more important than changes in size. In shockwave problems the bulk viscosity is equally important. Because the fluid density may change by a factor of two, the viscous irreversibility associated with rapid compression must be included. Because the only operational theory for bulk viscosity in dense fluids, the Enskog theory,[5] is inadequate,[6] we have undertaken a study of dense-fluid bulk viscosity. We have developed a new method for simulating fluid deformation, and here compare it with previous bulk-viscosity calculations. The new method can also be applied to shear flows, and has some interesting connections with the more usual Green-Kubo methods for calculating transport coefficients.

Green and Kubo showed that the transport coefficients describing nonequilibrium flows of mass, momentum, and energy can be expressed in terms of the decay of equilibrium fluctuations of velocity, stress, and heat flux.[7] Numerical decay calculations have been carried out for both hard[8] and soft[9] interparticle potential functions. Motivated both by the long times required for accurate equilibrium fluctuation calculations, and by the desire to develop an independent computational method as a check, several groups[10-14] have carried out direct measurements of the transport coefficients using nonequilibrium molecular dynamics. In these direct approaches steady or oscillatory hydrodynamic states are maintained by performing external work. Phenomenological hydrodynamics is then used to relate the resulting mass, momentum, and energy fluxes to transport coefficients. The direct calculations are often more efficient than is the indirect approach of fluctuation theory, but suffer from the drawback that the results of the calculations must be extrapolated to the small-gradient limit of macroscopic hydrodynamics. The direct calculations make it possible to generate nonequilibrium distribution functions and to study nonlinear effects.

Shear flow and heat flow have both been simulated by carrying out molecular-dynamics calculations with "reservoirs" which maintain constant velocities and temperatures at the boundaries.[12,15] The reservoirs themselves were kept at fixed velocities and temperature by external forces. Such forces do work on both the reservoirs' centers of mass and on the half-width of the reservoirs' velocity distributions relative to the center-of-mass velocities. These reservoir calculations provided estimates for the thermal conductivity and for the shear viscosity η which appears in Newton's phenomenological model[16] for stress in a flowing fluid:

$$\vec{\sigma} = [\sigma_{eq} + \lambda \vec{\nabla}\cdot\vec{u}]\vec{I} + \eta[\vec{\nabla}\vec{u} + \vec{\nabla}\vec{u}^{t}]. \quad (1)$$

In (1) the stream velocity $\vec{u}$ varies in space and time. The viscous contributions to the stress tensor $\vec{\sigma}$, over and above the equilibrium stress, are proportional to the symmetric tensors $\vec{\nabla}\cdot\vec{u}\vec{I}$, where $\vec{I}$ is the unit tensor, and $\vec{\nabla}\vec{u} + \nabla\vec{u}^{t}$, where t indicates transpose. The "second viscosity coefficient[17]" λ can be expressed in terms of the bulk viscosity $\eta_v = \lambda + \frac{2}{3}\eta$.

The bulk viscosity describes the extra stress due to dilation in the absence of shear. Because this effect depends upon volume change—a change of thermodynamic state—bulk viscosity cannot be measured in a steady-state-reservoir experiment. If the volume is cycled over a small range, with a frequency ω, the constitutive relation (1) implies that in addition to the elastic stress, proportional to the strain and the adiabatic (frequency-dependent) bulk modulus B, there will also be a viscous stress proportional to the strain rate and the bulk viscosity. The bulk viscosity can be obtained either by determining the strain-rate component of stress, or by averaging the work done, by the forces cycling the volume, over a complete cycle of dilation and compression:

$$\int dW = \int \sigma dV = \int_0^{2\pi} d\omega t (3\xi V_0 \cos\omega t)(\sigma_{eq} + 3B\xi \sin\omega t + 3\eta_v \xi\omega \cos\omega t) = 9\pi\xi^2 V_0 \omega\eta_v . \quad (2)$$

In (2) ξ is the maximum one-dimensional strain amplitude. In a complete cycle, the elastic part of the stress does no work.

We have developed a method for simulating such a cyclic process by using nonequilibrium molecular dynamics.[6] The viscosities resulting from such nonequilibrium simulations can be compared with Green-Kubo results and used to interpret the shockwave profiles from computer experiments.[3] The new data should also stimulate improvements in two areas of dense-fluid transport theory—the data show that the Enskog theory of transport is inadequate for soft potentials and that the mode-coupling[18] estimates of transport-coefficient frequency dependence are much too small[14] near the triple point.

In the present work we first indicate the relation of our numerical method to the Green-Kubo theory. We then apply the method to a dense-fluid state near the triple point. For this numerical work we have chosen to study the Lennard-Jones potential

$$\phi(r) = 4\epsilon[(\sigma/r)^{12} - (\sigma/r)^{6}], \quad (3)$$

at the same reduced density $N\sigma^3/V = 0.8442$, and reduced temperature $kT/\epsilon = 0.722$ studied by Levesque *et al.*[9] Notice that in (3), in Sec. IV of the text, in the tables, and in the figures, σ and ϵ represent potential parameters rather than stress and strain.

II. HAMILTONIAN FOR ADIABATIC DEFORMATION

Any mechanical flow can be described by specifying the space and time dependence of the "strain-rate tensor" $\vec{\nabla}\vec{u}$. The tensor describes the rate at which any macroscopic coordinate q changes with time:

$$\dot{\vec{q}} = \vec{q}\cdot\vec{\nabla}\vec{u}. \quad (4)$$

The two simplest such flows are homogeneous plane Couette flow (with du_x/dy the only nonzero element of $\vec{\nabla}\vec{u}$, for instance) and homogeneous dilation (with $\vec{\nabla}\vec{u}$ proportional to the unit tensor $\overline{I}$). If we consider a microscopic collection of N particles with coordinates $\vec{q}$ (now, and in what follows, using $\vec{q}$ to indicate a *set* of coordinates) and potential energy Φ, the application of the purely *mechanical* deformation (4) for a short time changes the potential energy: $\dot{\Phi} = -V\overline{P}_\phi : \vec{\nabla}\vec{u}$, where $\overline{P}_\phi$ is that part of the pressure tensor which depends upon the interparticle forces. In a *thermodynamic* deformation we expect to do work against the kinetic part $\overline{P}_k$, of the pressure too: $\dot{K} = -V\overline{P}_k : \vec{\nabla}\vec{u}$. This work is done *exactly* if we choose to vary the momenta in a way parallel to (4):

$$\dot{\vec{p}} = -\vec{\nabla}\vec{u}\cdot\vec{p}. \tag{5}$$

A microscopic Hamiltonian which incorporates not only the coordinate and momentum changes from (4) and (5), but also the usual changes from inertia and interparticle forces, is

$$\mathcal{H} = \Phi(\vec{q}) + K(\vec{p}) + \vec{q}\vec{p} : \vec{\nabla}\vec{u}. \tag{6}$$

The microscopic equations of motion derived from (6) are

$$\begin{aligned} \dot{\vec{q}} &= \partial\mathcal{H}/\partial\vec{p} = (\vec{p}/m) + \vec{q}\cdot\vec{\nabla}\vec{u}, \\ \dot{\vec{p}} &= -\partial\mathcal{H}/\partial\vec{q} = \vec{F} - \vec{\nabla}\vec{u}\cdot\vec{p}. \end{aligned} \tag{7}$$

The microscopic representation of the pressure tensor for a fluid with pairwise additive forces:

$$V\overline{P} = \vec{q}_{ij}\vec{F}_{ij} + (\vec{p}\vec{p}/m), \tag{8}$$

can then be used to establish that the coupled equations of motion (7) satisfy exactly the first law of thermodynamics for adiabatic flow:

$$\dot{E} = -V\overline{P} : \vec{\nabla}\vec{u}, \tag{9}$$

where E is the internal energy $\Phi + (p^2/2m)$.

Thus the Hamiltonian (6) has the desirable feature of providing equations of motion consistent with thermodynamics. This Hamiltonian has other applications too. Anderson[19] has just arrived independently at the same equations of motion (7). In his work, the pressure is treated as an independent variable to which the strain-rate responds. In our work the roles of pressure and strain-rate are reversed. Before proceeding to numerical applications of the equations of motion, we discuss the connection of the Hamiltonian and Green-Kubo fluctuation theory.

III. CONNECTIONS WITH GREEN-KUBO THEORY

There are two different ways to relate our perturbed Hamiltonian to conventional Green-Kubo theory. Let us consider first a treatment resembling one of the several sketched by Zwanzig.[7] For simplicity we choose a particular strain-rate tensor $\vec{\nabla}\vec{u} = \dot{\epsilon}\overline{I}$; this choice describes a homogeneous isotropic dilation. An analogous treatment applies for shear flows. Because the perturbed Hamiltonian describing this system includes a velocity gradient proportional to ϵ, we expect that in the limit of small strain rates $\dot{\epsilon} \sim 0$, a thermodynamic system described by the Hamiltonian

$$\mathcal{H} = \Phi + K + \dot{\epsilon}\vec{q}\vec{p} : \overline{I}, \tag{10}$$

could also be correctly described by Newton's phenomenological model (1):

$$\begin{aligned} \langle PV\rangle_{\text{noneq}} &= \langle PV\rangle_{\text{eq}} - 3\dot{\epsilon}\eta_v V \\ &= \tfrac{1}{3}\langle \vec{q}_{ij}\cdot\vec{F}_{ij} + (p^2/m)\rangle_{\text{noneq}}. \end{aligned} \tag{11}$$

If we introduce the Hamiltonian (10) into the ordinary canonical probability distribution

$$f_{\text{noneq}}/f_{\text{eq}} = \exp[-\dot{\epsilon}\vec{q}\vec{p} : \overline{I}/kT], \tag{12}$$

we find a simple expression for the bulk viscosity η_v:

$$-3\dot{\epsilon}\eta_v V = \tfrac{1}{3}\langle[-\dot{\epsilon}\vec{q}\cdot\vec{p}/kT][\vec{q}_{ij}\cdot\vec{F}_{ij} + (p^2/m)]\rangle_{\text{eq}}. \tag{13}$$

The virial theorem[5] can then be used to express the instantaneous pressure in (13) in terms of the dot product $\vec{q}\cdot\vec{p}$, giving

$$3PV = \vec{q}_{ij}\cdot\vec{F}_{ij} + (p^2/m) = 3\overline{P}V + (d/dt)(\vec{q}\cdot\vec{p}), \tag{14}$$

where $\overline{P}$ is the long-time-average pressure. Because the average value of $\vec{q}\cdot\vec{p}$ vanishes at equilibrium, (13) and (14) can be combined to give

$$\eta_v = \tfrac{1}{18} VkT\langle(d/dt)(\vec{q}\cdot\vec{p})^2\rangle_{\text{eq}}. \tag{15}$$

This last relation can then be converted into the usual Green-Kubo autocorrelation form by writing the $\vec{q}\cdot\vec{p}$ as integrals of pressure fluctuations:

$$\begin{aligned} \eta_v &= \lim_{\tau\to\infty} \frac{V}{2\tau kT}\int_0^\tau ds \int_0^\tau dt\,\langle \delta P(s)\delta P(t)\rangle_{\text{eq}} \\ &= \frac{V}{kT}\int_0^\infty dt\,\langle \delta P(0)\delta P(t)\rangle_{\text{eq}}, \end{aligned} \tag{16}$$

where δP is $P - \overline{P}$. An essential step in this heuristic derivation is the smoothed, or coarse grained, evaluation of the time derivative in (15). The derivative approaches the value given by Newton's phenomenological model only at times exceeding microscopic relaxation times.

We next consider a more convincing derivation of (16) from (10), based on linear response theory.[20] This treatment resembles Kubo's calculation of the electrical conductivity.[7] Linear response theory considers the effect of adding a perturbation $-A(\vec{q},\vec{p})a(t)$ to the Hamiltonian, where A is a function of the coordinates and momenta and $a(t>0)$ is a function of time. The theory expresses

Note on p. 234

the time behavior of the (arbitrary) response function R in terms of the correlation of dA/dt and R at different times:

$$\langle R(\vec{q},\vec{p})\rangle_{\text{noneq}} = \frac{1}{kT}\int_0^t ds\, a(s)\langle \dot{A}(0)R(t-s)\rangle_{\text{eq}}. \quad (17)$$

If we select for the response function the time derivative of the trace of Doll's tensor,

$$R \equiv \frac{d}{dt}(\vec{q}\cdot\vec{p}) = 3PV(t) - 3\bar{P}V(t), \quad (18)$$

where the long-time-average pressure $\bar{P}[=\bar{P}(0) - 3B^s\epsilon(t)]$ is evaluated at the volume $V(t)$ and internal energy $E(t)$,[21] and if we use the phenomenological viscoelastic equation of state,[22,23] valid for small strains and strain rates, we have

$$\langle R(t)\rangle_{\text{noneq}} = 9[B^s - B(\omega)]V\epsilon(t) - 9\eta_v(\omega)V\dot{\epsilon}(t)$$

$$= -\frac{V^2}{kT}\int_0^t ds\, 9\dot{\epsilon}(t-s)\langle \delta P(0)\delta P(s)\rangle_{\text{eq}}, \quad (19)$$

where ϵ is the strain, $\epsilon = \xi \sin\omega t$, and $\delta P = P - \bar{P}$. The short-time limit of (19) reproduces Zwanzig and Mountain's relation[23] between the infinite-frequency bulk modulus and equilibrium pressure fluctuations. For long times the upper limit in the integral can be replaced by infinity; (19) can then be separated into two independent equations, one for $\sin\omega t$ and one for $\cos\omega t$. These establish the well-known results for the frequency-dependent bulk modulus $B(\omega)$ and bulk viscosity $\eta_v(\omega)$:

$$B(\omega) - B^s = \frac{V}{kT}\int_0^\infty d\omega\, t\sin\omega t\,\langle \delta P(0)\delta P(t)\rangle_{\text{eq}}, \quad (20)$$

$$\eta_v(\omega) = \frac{V}{kT}\int_0^\infty dt\,\cos\omega t\langle \delta P(0)\delta P(t)\rangle_{\text{eq}}. \quad (21)$$

Thus we obtain the Newtonian liquid model from the microscopic equations as a direct long-time limit of linear-response theory. In the case that $\vec{\nabla}\vec{u}$ is chosen to correspond to a shear flow, a similar calculation provides the Green-Kubo formulas for the shear modulus $G(\omega)$ and the shear viscosity $\eta(\omega)$. Although the bulk and shear relations are "well-known," the methods used here to derive them are remarkably direct. In the next section we consider numerical applications of the Hamiltonian for adiabatic deformation.

IV. LENNARD-JONES TRIPLE-POINT CALCULATIONS

The Lennard-Jones thermodynamic state $N\sigma^3/V = 0.8442$, $kT/\epsilon = 0.722$ has been studied exhaustively.[9,12] This state corresponds to liquid argon near the triple point if σ and ϵ/k are given the values 3.405 Å and 119.8 K. The published Green-Kubo shear viscosity[9] has recently been supplemented by unpublished calculations carried out by Levesque in France and Pollock in America. We have also extended the earlier steady homogeneous-shear calculations,[12] which treated successively wider systems of 108, 2×108, and 3×108 particles, by carrying out a calculation with a width eight times that of a 108-particle cube. These results are all summarized in Table I. The unpublished results of Levesque and Pollock for 108 to 500 particles agree fairly well with each other and with the experimental shear viscosity for liquid argon, expressed in terms of the atomic mass m, σ, and ϵ. The French 864-particle data, both published and unpublished, deviate from the rest. The directly calculated reservoir calculations are also summarized in the table, and agree with all but the 864-particle results. The two homogeneous-shear calculations for the shear viscosity use slightly different (steady *versus* oscillatory) algorithms—Denis Evans will publish details of his calculations (Table II) separately.

We have verified that the present perturbed-Hamiltonian method reproduces correctly the

TABLE I. Green-Kubo, reservoir, and homogeneous-shear values for the Lennard-Jones shear viscosity in the vicinity of the triple point. These calculations were carried out at a reduced density $N\sigma^3/V$ of 0.8442 and typically include 10^5 time steps.

N	kT/ϵ	$\eta\sigma^2/(m\epsilon)^{1/2}$	Type	Source
108	0.728	2.97	GK	Levesque
256	0.715	2.92	GK	Levesque
256	0.722	2.6 ± 0.1	GK	Pollock
500	0.722	3.2 ± 0.2	GK	Pollock
864	0.722	3.85	GK	Levesque
864	0.722	4.03	GK	Ref. 9
108–324	0.722	2.95 ± 0.2	R, H	Ref. 12
108×8	0.715	3.0 ± 0.15	H	Present work (steady shear)
108	0.722	3.18 ± 0.1	H	(Table II) (oscillatory)
Experimental estimate:		3.0		Ref. 25

TABLE II. Lennard-Jones shear viscosity near the triple point obtained by applying homogeneous oscillatory isothermal shear. These results were all obtained with Lennard-Jones's potential truncated at 2.5 σ and with a timestep of 0.007 $\sigma(m/\epsilon)^{1/2}$; $N=108$, $N\sigma^3/V=0.8442$, and $kT/\epsilon=0.722$. Amplitude times frequency $\xi\omega$, frequency ω, and number of shearing cycles are listed.

$\xi\omega\sigma(m/\epsilon)^{1/2}$	$\omega\sigma(m/\epsilon)^{1/2}$	$\eta\sigma^2/(m\epsilon)^{1/2}$	Cycles
0	0[a]	3.18±0.1	
0.10	8.98	1.27	770
0.20	1.12	2.70	25
0.20	2.24	2.18	100
0.20	4.49	1.74	100
0.20	8.98	1.22	200
0.20	11.98	1.10	533
0.20	14.96	0.90	333
0.30	1.12	2.41	32
0.30	2.24	2.33	80
0.30	4.49	1.72	100
0.30	8.98	1.25	200
0.30	11.97	1.14	267

[a] Extrapolation from Ref. 14.

shear viscosities already obtained using external reservoirs. To enhance the importance of the kinetic contribution to the shear flow (almost negligible at the triple point) we carried out a shear-flow simulation at a reduced density of 0.45 with a reduced temperature of 2.16. Both this perturbed-Hamiltonian calculation and the external-reservoir calculation give 108-particle viscosities of $(0.45\pm0.02)(m\epsilon)^{1/2}/\sigma^2$, with nearly equal contributions from the kinetic and potential parts of the momentum flux. A trial calculation was carried out to assess the importance of the perturbation force $F_y=-(du_x/dy)p_x$; when this essential term was omitted the kinetic contribution to the shear viscosity was reduced from $0.23\ (m\epsilon)^{1/2}/\sigma^2$ to nearly zero.

For bulk viscosity the only previous calculations used the Green-Kubo method—there is no bulk-viscosity analog for the reservoir calculations used to simulate shear flow. The Green-Kubo results are summarized in Table III. Again we have included recent unpublished calculations carried out by Levesque. We have carried out a series of lengthy calculations using the equations of motion

$$\dot{\vec{q}}=(\vec{p}/m)+\dot{\epsilon}\vec{q}\,,\quad \dot{\vec{p}}=\vec{F}-\dot{\epsilon}\vec{p}\,. \tag{7'}$$

It is convenient to solve these first-order equations using a standard packaged routine.[24] We also add to the set of $6N$ equations the adiabatic equation for conservation of energy:

$$\dot{E}=-3\dot{\epsilon}VP\,. \tag{9'}$$

The integrated energy change over a cycle of dilation and compression can then be compared with the change in the internal energy over the cycle, calculated from $\Phi+K$ with the initial and the final coordinates and momenta. We chose a timestep such that these two independent estimates of the hysteresis agreed to about one part in ten thousand. At the end of every compressional cycle the particle momenta p were rescaled so that the next cycle would begin with the desired initial internal energy. After completing most of the calculations we found that the computation could be made considerably faster by adding a small term, proportional to r^{+6}, to the pair potential to make the forces vanish continuously at the potential cutoff.

Each calculation began in a body-centered-cubic initial state with a Maxwell-Boltzmann velocity distribution chosen to give the same thermal (i.e., relative to a perfect crystal) energy per particle as that found for 864 particles by Levesque *et al.*[9] Melting was enhanced by the bcc structure and a check of the temperature indicated that there was no difficulty in melting to form a liquid state. Although only the first cycle appeared obviously anomalous we took the precaution of discarding the first ten cycles.

The adiabatic external work for each cycle can be separated into potential and kinetic components, but these have no particularly simple significance

TABLE III. Comparison of Green-Kubo bulk viscosities for the Lennard-Jones potential with the present calculations. The densities and temperatures for Levesque's unpublished calculations correspond to those given in Table I.

N	$\eta_v\sigma^2/(m\epsilon)^{1/2}$	Type	Source
108	1.13	GK	Levesque
256	0.89	GK	Levesque
864	1.04	GK	Levesque
864	1.05	GK	Ref. 9
54	1.55		Present work
Experimental estimate:	2.0		Ref. 25

Note on p. 234

for the Lennard-Jones potential. It is interesting to point out that for the simpler inverse-power potentials, the virial-theorem relation between the pressure and the energy,

$$3PV = 2K + n\Phi , \tag{22}$$

allows us to calculate separately the strain-rate dependence of the potential and kinetic energies and to relate these two terms to the strain-rate dependence of the pressure, which is still given by (22). In general the potential contribution to the bulk viscosity is $-\frac{1}{2}n$ times the kinetic contribution for an inverse-nth-power pair potential.

The link between the potential and kinetic parts of the (constant-energy) pressure fluctuations leads to interesting conclusions. For the inverse-nth-power potential, the ratio of the "potential" to "cross" to "kinetic" terms in the Green-Kubo bulk viscosity integrand is *exactly* $\frac{1}{4}n^2$ to $-n$ to 1.

For a general force law, linear-response theory, applied to the many-body Hamiltonian (10), can be used to show directly that *exactly* half the cross term contributes to the kinetic part of the bulk (or shear) viscosity; the remaining half contributes to the potential part. Thus, in the inverse-12th-power "soft-sphere" case, the potential "long-time tail" for bulk viscosity is 36 times larger than the kinetic one. The simple relationships between the potential and kinetic parts of the pressure fluctuations have been verified numerically for the inverse-12th-power soft-sphere potential in a series of bulk-viscosity calculations.[6] A numerical analysis for the Lennard-Jones potential should be carried out.

The numerical results of our Lennard-Jones triple-point calculations are given in Table IV

TABLE IV. Perturbed-Hamiltonian bulk viscosity for the nearest-image Lennard-Jones potential at $N\sigma^3/V = 0.8442$ and $kT/\epsilon = 0.722$. The amplitude ξ, frequency ω, and number of dilation-compression cycles are listed.

N	ξ	$\omega\sigma(m/\epsilon)^{1/2}$	$\eta_b\sigma^2/(m\epsilon)^{1/2}$	Cycles
54	0.02	1	1.10 ± 0.06	800
54	0.02	2	0.82 ± 0.04	500
54	0.02	3	0.71 ± 0.02	1000
54	0.02	4	0.61 ± 0.02	1000
54	0.02	5	0.52 ± 0.02	200
54	0.02	6	0.51 ± 0.01	1000
54	0.02	7	0.50 ± 0.02	500
54	0.02	8	0.45 ± 0.02	500
54	0.02	9	0.48 ± 0.02	500
54	0.02	10	0.45 ± 0.02	200
54	0.01	10	0.48 ± 0.03	1000
128	0.02	10	0.45 ± 0.02	200
250	0.02	10	0.45 ± 0.01	200

(see also Fig. 1). Each calculation depends upon three separate parameters; the number of particles N, the strain amplitude ξ, and the frequency ω. The data show that the number dependence is small, at least at high frequency. The dependence on strain amplitude is harder to assess. Small amplitudes give large fluctuations in the hysteresis per cycle, while large strain amplitudes include a wider range of densities. For the most part our calculations were limited to a single maximum amplitude $\xi = 0.02$.

The bulk and shear viscosities vary similarly with frequency. The homogeneous-shear data (see Fig. 1) were calculated with both the temperature and the frequency held constant. These shear-viscosity results, for the range of frequencies and strain rates corresponding to our own bulk-viscosity calculations, can be described by the empirical relation

$$\eta\sigma^2/(m\epsilon)^{1/2} \sim 3.18 - 0.65(m/\epsilon)^{1/4}(\sigma\omega)^{1/2} . \tag{23}$$

If this dependence *actually* holds in the MHz to GHz range of laboratory experiments, it should be

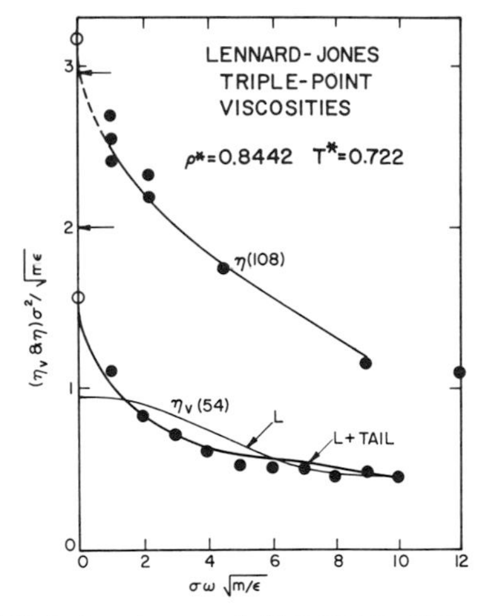

FIG. 1. Computer-generated homogeneous-shear isothermal shear viscosities and perturbed-Hamiltonian bulk viscosities are shown as filled circles. Experimental estimates of the low-frequency viscosities for liquid argon are indicated by the horizontal arrows. The phenomenological fit (23) is shown for the shear viscosity. In the bulk-viscosity case the L line gives the frequency-dependent viscosity from a numerical integration of Levesque's data as shown in Fig. 2. The L+tail line indicates the effect of an enhanced long-time tail corresponding to Eq. (24) of the text. The intercepts for both the shear and bulk fits are shown as open circles.

possible to observe noticeable frequency-dependent effects.

The bulk-viscosity results have a form similar to the shear data. If we use the same square-root dependence of viscosity on frequency, the extrapolated hydrodynamic bulk viscosity lies close to current experimental estimates.[25] At the same time, this extrapolation

$$\eta_v \sigma^2/(m\epsilon)^{1/2} \sim 1.55 - 0.55(m/\epsilon)^{1/4}(\sigma\omega)^{1/2}, \quad (24)$$

lies considerably above the Green-Kubo (zero-frequency) estimates. It is difficult to settle the question of the long-time or low-frequency dependence of the viscosities by numerical calculation. Levesque's and Pollock's Green-Kubo data indicate considerable number dependence at long times and our own results cannot be pushed to lower frequencies without substantial improvements in the efficiency of numerical simulations.

Nevertheless, a self-consistent picture of the long-time and low-frequency behavior does emerge if we combine Levesque's bulk-viscosity integrand—his data are shown for 108 and 864 particles in Fig. 2—with the long-time tail consistent with the low-frequency relation (24). The coefficient required, $0.55/(2\pi)^{1/2} = 0.22$ in the units of Fig. 2, is only slightly less than the $0.65/(2\pi)^{1/2} = 0.26$ required by Evans's shear-viscosity data.

The result of adding the tail correction to the Green-Kubo data is shown in Fig. 1. The tail changes the overall curvature of the plot from negative to positive and brings about excellent agreement between the equilibrium and nonequilibrium data. The calculated bulk viscosity, 1.55 in the units of Fig. 1, is not too far below the experimental estimate[25] for liquid argon 2.0. The good agreement linking the Green-Kubo correlation function to our nonequilibrium simulations and to experiment is gratifying. It suggests that the present methods can be used with confidence for other thermodynamic states and for other force laws.

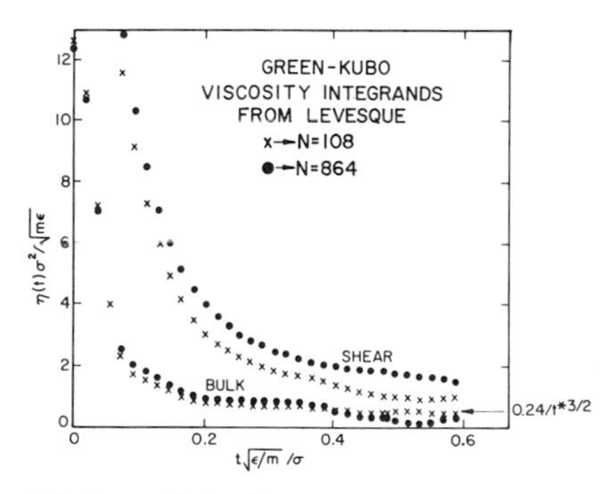

FIG. 2. Equilibrium fluctuation correlation functions calculated by Levesque. The nonequilibrium calculations described in the text suggest a long-time tail at about the level indicated by the arrow (0.26 for shear, 0.22 for bulk).

The shear-viscosity results are less consistent. Integration of the large-system ($N = 864$) Green-Kubo integrand, with or without an appended long-time tail, gives viscosities substantially higher than either the small-system or experimental estimates.

V. DISCUSSION

The perturbed-Hamiltonian approach to nonequilibrium deformation is aesthetically pleasing because it is so closely related to thermodynamics and equilibrium fluctuation theory. This same Hamiltonian should prove to be useful in attempts to understand theoretically the frequency and amplitude dependence of the viscosities.

It would be useful to find an analogous formulation for diffusion and heat conduction, but our attempts to do this for conduction have failed. It is easy to use an extra force proportional to each particle's energy fluctuation to drive a homogeneous isothermal heat current with nonequilibrium molecular dynamics. It is not so easy to find a simple isochoric (as opposed to adiabatic) analog of the first law of thermodynamics. Nevertheless, we expect that the heat current resulting from the perturbation just described, will provide a perfectly useful approach to thermal conductivity. We expect to carry out such calculations for comparison with the earlier reservoir and Green-Kubo work.[9,15]

The present bulk-viscosity results show once again[6] that the Enskog theory is a poor approximation for potentials as soft as r^{-12}. The hard-sphere prediction, underlying that theory, that the frequency changes in the bulk and shear moduli are similar at high density fails for soft potentials. For soft forces the high- and low-frequency bulk moduli are similar, so that the bulk viscosity is relatively small. The present data underscore the need for theoretical understanding of dense-fluid transport. In particular, the mode-coupling predictions,[18] even if they turn out to be correct for frequencies below those which can be studied in computer simulations, are grossly in error for the frequencies studied here. Because even a relatively crude theory would be welcome, it seems possible that models based on cell theories incorporating perturbed equations of motion will turn out to be useful.

Note on p. 234

Notes added in proof. (i) The Doll's-tensor Hamiltonian can also be used to obtain equilibrium fluctuation expressions for nonequilibrium distribution functions. See D. J. Evans, W. G. Hoover, and A. J. C. Ladd, Phys. Rev. Lett. 45, 124 (1980). (ii) Bill Wood and Bob Dorfman kindly pointed out to us that in Ernst, Hauge, and van Leeuwen's work [J. Stat. Phys. 15, 7 (1975)], the kinetic part of the pressure fluctuation is *defined* to be zero. We alert the reader that this peculiar choice is different from ours, as described following Eq. (22).

ACKNOWLEDGMENTS

Gary Doolan made useful suggestions for implementing a vectorized version of our calculations on the CRAY computer. We thank Roy Pollock for carrying out the two shear-viscosity Green-Kubo calculations in Table I. We are particularly grateful to Dr. Levesque for making his unpublished data available. In view of the confusion surrounding Fig. 2 of Ref. 9, the new data were indispensible in correlating the equilibrium and nonequilibrium viscosities. This work was performed under the auspices of the U.S. Department of Energy at the Lawrence Livermore National Laboratory under contract No. W-7405-ENG and supported, in part, at UCDavis, by the Army Research Office, Research Triangle Park, North Carolina. Partial accounts of this work were presented by W. G. Hoover at the M. S. Green Statistical Physics Symposium, 24–25 April 1980, Gaithersburg, Maryland and at the Sitges (Spain) Symposium, 9–13 June 1980.

[1]B. Lé Mehauté, *An Introduction to Hydrodynamics and Water Waves* (Springer, Berlin, 1976).
[2]Y. B. Zeldovich and Y. P. Raizer, *Physics of Shock Waves and High-Temperature Hydrodynamic Phenomena* (Academic, New York, 1967).
[3]V. Y. Klimenko and A. N. Dremin, in *Detonatsiya, Chernogolovka*, edited by O. N. Breusov *et al.* (Akad. Nauk, Moscow, SSSR, 1978), p. 79; B. L. Holian, W. G. Hoover, B. Moran, and G. K. Straub, Phys. Rev. A (to be published).
[4]W. G. Hoover, Phys. Rev. Lett. 42, 1531 (1979).
[5]J. O. Hirschfelder, C. F. Curtiss, and R. B. Bird, *Molecular Theory of Gases and Liquids* (Wiley, New York, 1954).
[6]W. G. Hoover, A. J. C. Ladd, R. B. Hickman, and B. L. Holian, Phys. Rev. A 21, 1756 (1980); W. G. Hoover and R. B. Hickman, Bull. Am. Phys. Soc. 25, 550 (1980).
[7]R. W. Zwanzig, Ann. Rev. Phys. Chem. 16, 67 (1965).
[8]B. J. Alder, D. M. Gass, and T. E. Wainwright, J. Chem. Phys. 53, 3813 (1970).
[9]D. Levesque, L. Verlet, and J. Kürkijarvi, Phys. Rev. A 7, 1690 (1973). Note that the "bulk"-viscosity integrand in Fig. 2 of LVK is actually the *shear*-viscosity integrand and that the quoted thermal conductivity is in error by about a factor of two. (D. Levesque, private communications.)
[10]E. M. Gosling, I. R. McDonald, and K. Singer, Mol. Phys. 26, 1475 (1973).
[11]A. W. Lees and S. F. Edwards, J. Phys. C 5, 1921 (1972).
[12]W. T. Ashurst and W. G. Hoover, Phys. Rev. A 11, 658 (1975).
[13]T. Naitoh and S. Ono, J. Chem. Phys. 70, 4515 (1979).
[14]D. J. Evans, Phys. Lett. 74A, 229 (1979) and J. Stat. Phys. 22, 81 (1980).
[15]W. G. Hoover and W. T. Ashurst, in *Theoretical Chemistry, Advances and Perspectives*, edited by H. Eyring and D. Henderson (Academic, New York, 1975), Vol. 1, p. 1.
[16]S. G. Brush, *Kinetic Theory* (Pergamon, New York, 1972), Vol. 3.
[17]L. Rosenhead, Proc. R. Soc. London Ser. A 226, 1 (1954).
[18]Y. Pomeau and P. Résibois, Phys. Rep. 19, 63 (1975).
[19]H. C. Andersen, J. Chem. Phys. 72, 2384 (1980).
[20]D. A. McQuarrie, *Statistical Mechanics* (Harper and Row, New York, 1976), p. 507.
[21]J. A. McLennan, Prog. Theor. Phys. Jpn. 30, 408 (1963).
[22]J. Frenkel, *Kinetic Theory of Liquids* (Dover, New York, 1955).
[23]R. W. Zwanzig and R. D. Mountain, J. Chem. Phys. 43, 4464 (1965).
[24]L. F. Shampine and M. K. Gordon, *Computer Solution of Ordinary Differential Equations* (Freeman, San Francisco, 1975).
[25]S. A. Mikhailenko, B. G. Dudar, and V. A. Schmidt, Fiz. Nizk. Temp. 1, 224 (1975) [Sov. J. Low Temp. Phys. 1, 109 (1975)].

Note to Reprint III.7

It is curious that this paper is best known for its use of the "Doll's tensor" Hamiltonian, a form of perturbation that had been proposed much earlier (Luttinger, Phys. Rev. 135 (1964) A1505), whereas the main innovation – the combination of the perturbation with periodic boundary conditions – is nowhere emphasized in the article. The necessary combination is achieved by applying the perturbation both to the particles in the central cell and to their periodic images; for a good discussion, see the paper by Maréchal et al. referred to in the Note to Reprint III.6.

PHYSICAL REVIEW A VOLUME 28, NUMBER 2 AUGUST 1983

Nonequilibrium molecular dynamics via Gauss's principle of least constraint

Denis J. Evans
Research School of Chemistry, Australian National University, Canberra, Australian Capital Territory 2600, Australia

William G. Hoover, Bruce H. Failor, Bill Moran, and Anthony J. C. Ladd
Department of Applied Science, University of California at Davis–Livermore and Lawrence Livermore National Laboratory, Livermore, California 94550
(Received 14 February 1983)

Gauss's principle of least constraint is used to develop nonequilibrium molecular-dynamics algorithms for systems subject to constraints. The treatment not only includes "nonholonomic" constraints—those involving velocities—but it also provides a basis for simulating nonequilibrium steady states. We describe two applications of this new use of Gauss's principle. The first of these examples, the isothermal molecular dynamics of a three-particle chain, can be treated analytically. The second, the steady-state diffusion of a Lennard-Jones liquid, near its triple point, is studied numerically. The measured diffusion coefficient agrees with independent estimates from equilibrium fluctuation theory and from Hamiltonian external fields.

I. INTRODUCTION

Many-body systems have been simulated by solving Newton's equations of motion, ever since the development of computers. Early constant-energy Newtonian dynamic studies were concerned with the nature of equilibrium, and the approach to it. But, to drive realistic nonequilibrium systems undergoing shear or compressional flows or heat flows, a new kind of *non*equilibrium molecular dynamics was needed.

About ten years ago, nonequilibrium molecular-dynamics (NEMD) simulations using heat reservoirs and moving boundaries began to be developed, but on a case-by-case basis, without an underlying basic formalism. Very recently, Hoover and Evans discovered that Karl Friedrich Gauss, in 1829, had enunciated a fundamental dynamical principle (Gauss called it *the* most fundamental principle) which can fruitfully be applied to a wide range of irreversible phenomena and used to generate NEMD algorithms. We describe and illustrate our new treatment of Gauss's principle of least constraint.

In dynamical problems "holonomic" constraints are used to restrict *coordinates* only. Such constraints are used to fix bond lengths or angles at average values, thereby avoiding small-amplitude high-frequency motions. The equations of motion for a system with these geometric constraints were written by Ryckaert, Ciccotti, and Berendsen,[1] using Lagrange multipliers. Their numerical method satisfies the constraint equations continuously. Here we point out that this same technique[1] can be generalized to the *non*holonomic (*velocity-dependent*) case in which the constraint forces can do work on the system. These constrained equations can be found directly from Gauss's principle of least constraint.[2]

This simple extension of the kinds of constraints considered permits us to describe homogeneous nonequilibrium steady states. In characterizing such a state at least two constraints are used. The first "driving" constraint sets the value of the thermodynamic force or flux and the second "stabilizing" constraint fixes a thermodynamic variable in order to maintain a steady nonequilibrium state. Typically the driving constraint is a strain rate or heat flux, and the stabilizing constraint corresponds to constant energy or constant temperature.[3–5]

A steady irreversible process produces entropy, through irreversible heating or mixing, and must therefore incorporate a compensating mechanism for extracting heat or separating molecular species. Heat is ordinarily removed, by conduction, at isothermal boundaries. However, by using stabilizing constraints to maintain a steady state we can avoid the need for physical boundaries and substantially reduce the dependence of our results on system size. The techniques described here are remarkably flexible, permitting steady states to be maintained either by thermodynamic forces or by the conjugate thermodynamic fluxes.

Because these constraints are novel, we illustrate them first with a familiar system, simple enough for complete analysis: the three-particle Hooke's-law chain. We then apply the same techniques to a complicated many-body problem, the simulation of steady diffusive flow in a model liquid.

II. GAUSS'S PRINCIPLE OF LEAST CONSTRAINT (REF. 6)

It is not widely appreciated that just over 150 years ago Gauss formulated a mechanics more general than Newton's. Gauss's formulation applies to systems which are subject to constraints, either holonomic or nonholonomic. Gauss stated that the trajectories actually followed would deviate as little as possible, in a least-squares sense, from the unconstrained Newtonian trajectories.[6]

If the constraints do no work on the system then it is possible to prove that Newton's and Gauss's formulations are equivalent.[2] The equivalence holds also in the nonholonomic case with linear homogeneous work-performing constraints.[7–9] But in the general case the "proofs" of Gauss's principle of least constraint require the addition of postulates or assumptions to Newton's equations of motion. Here we follow Appell[10] and Ray,[8] simply accepting without proof the validity of Gauss's principle.

To reduce Gauss's principle to a form suitable for numerical work, we first introduce constraints into the equations of motion. It is necessary to treat holonomic constraints $g(r,t)=0$ and nonholonomic constraints $g(r,\dot{r},t)=0$ separately. In the holonomic case, two differentiations with respect to time give the relation restricting the acceleration $\ddot{r}$

$$\ddot{r}\frac{\partial g}{\partial r}+2\dot{r}\frac{\partial^2 g}{\partial r\partial t}+\dot{r}^2\frac{\partial^2 g}{\partial r^2}+\frac{\partial^2 g}{\partial t^2}=0 . \tag{1a}$$

In the nonholonomic case only a single time differentiation is required

$$\ddot{r}\frac{\partial g}{\partial \dot{r}}+\dot{r}\frac{\partial g}{\partial r}+\frac{\partial g}{\partial t}=0 . \tag{1b}$$

In *either* case the acceleration $\ddot{r}$ is constrained to lie on the constant-g hypersurface by a restriction of the general form

$$n(r,\dot{r},t)\cdot\ddot{r}+w(r,\dot{r},t)=0 , \tag{2}$$

where the explicit functional forms of n and w can be obtained from the imposed constraint g according to Eqs. (1) above.

The explicit forms of $n(r,\dot{r},t)$ and $w(r,\dot{r},t)$ are not essential in the general treatment outlined in Eqs. (3)–(6) below. At all times the dynamical state of the system $r,\dot{r}$ is confined to the hypersurface which satisfies Eq. (2). If the constraints (1) were absent, then the unconstrained motion of the system calculated from

$$m\ddot{r}_u=F , \tag{3}$$

could leave the constraint hypersurface. Gauss's formulation prevents such a violation by adding an acceleration normal to the surface

$$\ddot{r}_c=\ddot{r}_u-\lambda[n(r,\dot{r},t)/m] \tag{4}$$

with λ chosen to satisfy the restriction (2), and thereby (1)

$$\lambda=(n\cdot\ddot{r}_u+w)/[n\cdot(n/m)] . \tag{5}$$

This added acceleration can be expressed in terms of a constraint force F_c:

$$\ddot{r}_c=(F+F_c)/m=F/m-\lambda(n/m) , \tag{6}$$

where the instantaneous value of the Lagrange multiplier λ is chosen to satisfy the constraint (1) according to Eq. (5).

There are *many* different ways to project unconstrained accelerations back onto the constrained hypersurface. Gauss's principle of least constraint states that the actual constrained motion should be obtained by the normal projection technique just described. An alternative description of this simple principle is the statement that the mean-square value of the constraint force divided by the particle mass $\langle F_c^2/m\rangle$ should be minimized.

III. APPLICATION TO AN ISOTHERMAL LINEAR CHAIN (REF. 11)

The harmonic one-dimensional chain is a familiar prototype for ordered crystalline solids. Each particle in such a chain interacts with its neighbors through a Hooke's-law potential. The equations of motion have the form

$$m\ddot{x}_i=\kappa(x_{i+1}-2x_i+x_{i-1}) , \tag{7}$$

where x_{i+1} and x_{i-1} are the coordinates of the particles adjacent to that at x_i. Because this problem is linear, the various "normal-mode" solutions can be superposed to match any initial conditions (the set of x_i and $\dot{x}_i$), and to follow the constant-energy dynamical development of the chain in time.

In *non*equilibrium problems it is convenient to specify isothermal conditions to extract irreversibly generated heat. For that reason we consider here *isothermal,* as opposed to equilibrium isoenergetic, dynamics for the linear chain. The isothermal restriction complicates the microscopic dynamics by imposing a collective constraint, coupling together the previously independent normal modes. But on the other hand, the macroscopic, thermodynamic behavior of the chain is simplified by preventing temperature fluctuations. Our definition of temperature is based on the ideal-gas thermometer; that is, temperature is proportional to the kinetic energy. Thus the constant-temperature constraint has the form

$$g(r,\dot{r},t)=\sum(m\dot{x}^2/2)-E_{\rm kin}=0 . \tag{8}$$

The constraint is *nonholonomic,* because it includes velocities, and is also nonlinear. The functions n and w corresponding to the isothermal constraint (8) are, respectively, $m\dot{x}$ and 0. Gauss's principle leads to the constrained accelerations

$$\ddot{x}_c=F/m-\zeta\dot{x} ,$$
$$\zeta=\sum(F\dot{x})\Big/\sum m\dot{x}^2 . \tag{9}$$

The collective variable ζ plays the role of a friction coefficient, but it takes on both positive and negative values as time goes on, as required to keep temperature constant. The set of Eqs. (9) cannot be solved analytically, but numerical solutions show that the system is stable and well behaved and that the time-averaged potential energy approaches the equipartition value as the number of masses in the chain increases.

The two-particle chain is uninteresting, because the fixed-kinetic-energy constraint allows no accelerations—the two particles move to infinity at constant speeds. The constrained three-particle chain with fixed center of mass is the simplest interesting problem, because all velocities change with time and cover a broad range of dynamical states (see Fig. 1). We consider the three-particle problem here in detail. We use displacement coordinates (measured relative to a minimum-energy configuration) and periodic boundaries. We have the restrictions

$$x_1+x_2+x_3=0 , \tag{10a}$$

$$\dot{x}_1+\dot{x}_2+\dot{x}_3=0 , \tag{10b}$$

$$\dot{x}_1^2+\dot{x}_2^2+\dot{x}_3^2=2E_{\rm kin}/m . \tag{10c}$$

The dynamical state of the three-mass system $\{x_1,x_2,x_3,\dot{x}_1,\dot{x}_2,\dot{x}_3\}$ can then be described by three independent variables: x_1, x_2, and $\dot{x}_1$, for instance. The remaining coordinates and velocities follow from the constants of the motion (10).

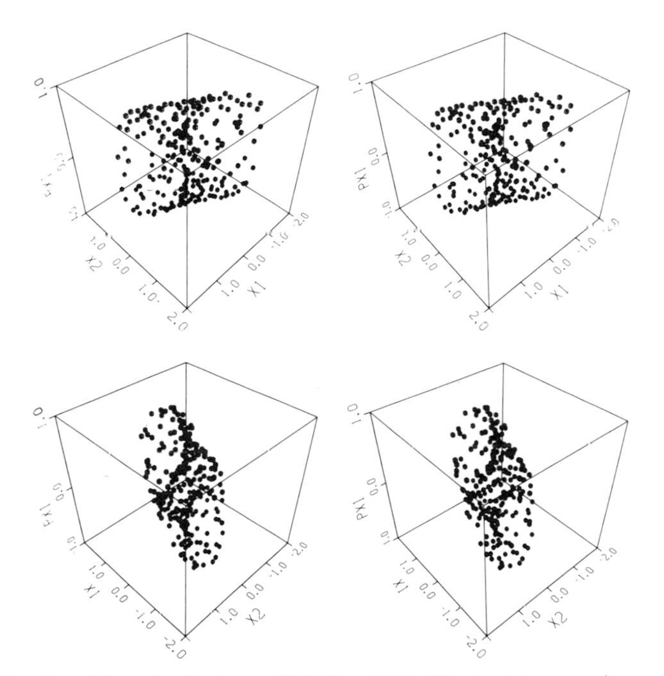

FIG. 1. Two stereo views of the motion of a three-particle isothermal chain. Discrete points x_1, x_2, $m\dot{x}_1$ corresponding to a numerical solution of Eqs. (11) of the text are plotted. The corresponding Newtonian chain would describe an ellipse in this space.

The isothermal equations of motion then become

$$m\ddot{x}_1 = -3\kappa x_1 - \zeta m\dot{x}_1 \ ,$$
$$m\ddot{x}_2 = -3\kappa x_2 - \zeta m\dot{x}_2 \ , \qquad (11)$$
$$\zeta = -3\kappa[x_1\dot{x}_1 + x_2\dot{x}_2 + (x_1+x_2)(\dot{x}_1+\dot{x}_2)]/2E_{\text{kin}} \ .$$

The problem can be simplified, analytically, by introducing plane polar coordinates r and θ in place of the particle coordinates x_1 and x_2:

$$r\cos\theta = \sqrt{3}(x_1+x_2) \ ,$$
$$r\sin\theta = x_1 - x_2 \ . \qquad (12)$$

In polar form the problem reduces to one-dimensional (radial) motion in an effective potential which includes the angular momentum. This transformed description of the problem establishes that the isothermal chain dynamics consists of nonlinear oscillations between the two turning points of the effective potential

$$v_{\text{eff}}(r) = A\exp(Br^2)/r^2 \ , \qquad (13)$$

where the constants A and B depend upon the initial conditions.

A typical series of representative points (x_1, x_2, $m\dot{x}_1$), from a numerical solution of equations (11), is displayed in Fig. 1. The motion is relatively complex, compared to the Newtonian solution, which is an ellipse in the same space.

Gauss's principle provides a unique solution to the isothermal initial-value problem. As noted above there are *many* other motions which preserve the kinetic energy. For instance, consider the Lagrangian equations of motion involving the nonphysical momentum p_x

$$p_x = m\dot{x}(1+\lambda) \ ,$$
$$\dot{p}_x = -3\kappa x \ . \qquad (14)$$

These equations, with the constraint (10) governing the time development of λ, are identical to the constrained Gaussian set (11) if the Lagrange multiplier vanishes, so that p_x is equal to $m\dot{x}$. Then the Lagrangian and Gaussian values of $\ddot{x}$ differ. In any other case, Eqs. (14) lead to different values of $\ddot{x}$ from the same initial conditions. Because p_x is not a physical momentum the initial conditions $\{x,\dot{x}\}$ are insufficient to specify a well-posed problem. Once a particular initial choice of λ has been made, the motion can again be studied in polar coordinates, with an effective potential which is the reciprocal of a sixth-order polynomial in r. The resulting Lagrangian dynamics, Eq. (14), resembles qualitatively the Gaussian dynamics of Fig. 1. The two approaches can be made to coincide by

Notes on p. 240

continuously forcing the Lagrange multiplier in Eq. (14) to vanish.

IV. HAMILTONIAN ALGORITHM FOR SELF-DIFFUSION

To complement the periodic and homogeneous treatments of momentum[12] and heat flows[13] we develop here a Hamiltonian method for determining the self-diffusion coefficient D. D can alternatively be calculated from the equilibrium Green-Kubo expression

$$D=\int_0^\infty \langle \dot{x}(0)\dot{x}(t)\rangle_{\text{eq}}dt \ . \tag{15}$$

Many calculations have been based on this method, starting with the hard-sphere calculations of Alder and Wainwright.[14] An alternative approach introduces an external field into the Hamiltonian which couples to a particle property q_i analogous to electric charge. We call this property "color" rather than "charge" to emphasize that it does not enter into interparticle interactions. The many-body Hamiltonian

$$H_0=\sum(m\dot{r}^2/2)+\sum\sum\phi \tag{16}$$

has added to it a perturbing external field at time 0

$$H=H_0-\sum qxE, \quad t>0 \ . \tag{17}$$

The external field E stimulates a current density J

$$J_x=(1/V)\sum q\dot{x} \ , \tag{18}$$

analogous to an electric current density. For simplicity we choose q_i equal to -1 for $i\leq n/2$ and q_i equal to $+1$ for the remaining particles. The linear-response theory[15] establishes that the limiting small-field nonequilibrium current density can be written in terms of an equilibrium color conductivity memory function σ:

$$\langle J(t)\rangle_{\text{ne}}=\int_0^t \sigma(t-\Delta t)E(\Delta t)d\Delta t \ ,$$
$$\sigma(t)=(V/kT)\langle J(0)J(t)\rangle_{\text{eq}} \ . \tag{19}$$

The current-density autocorrelation function is simply related to the velocity autocorrelation function in Eq. (15). This is because $\langle x_j\rangle$ is $-x_1/(N-1)$ for $j=2,3,\ldots,N$, which can be used to show that the steady color conductivity is proportional to the self-diffusion coefficient D

$$\sigma=\int_0^\infty \sigma(t)dt=N^2D/[(N-1)VkT] \ . \tag{20}$$

Thus the self-diffusion coefficient can be determined by carrying out a series of constant-field simulations and extrapolating the resulting conductivities to the zero-field limit. See Fig. 2.

The work done by the external field would normally cause the system to heat up at a rate proportional to E^2 in the small-field limit. This heating can be eliminated by carrying out the calculation at fixed temperature. In the numerical work described in Sec. VI we do this by rescaling the y and z components of the velocity distribution to maintain the corresponding second moments at fixed values.[16,17]

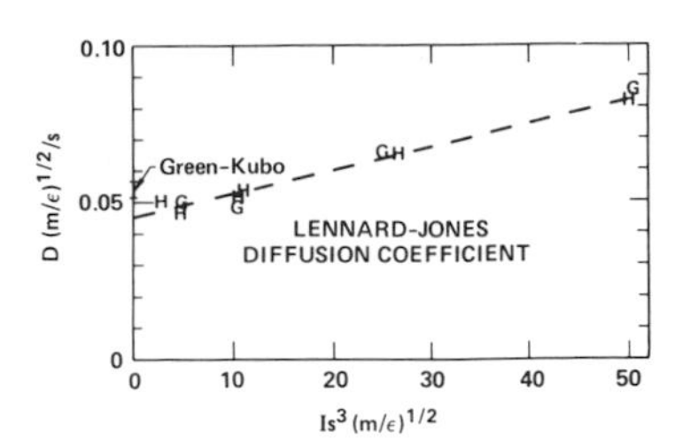

FIG. 2. Self-diffusion coefficient for a Lennard-Jones liquid at a reduced density $Ns^3/V=0.85$ and reduced temperature $kT/\epsilon=1.08$. Calculations according to Gaussian dynamics and using Hamiltonian external fields are indicated by G and H, respectively. Green-Kubo calculation, from Ref. 18, is probably an overestimate, as explained in Sec. VI of the text.

V. GAUSSIAN ALGORITHM FOR SELF-DIFFUSION

The diffusion problem can be treated directly using Gauss's principle of least constraint. A driving constraint g_d provides a constant current, and a stabilizing constraint g_s provides constant y and z temperatures

$$g_d=\sum(q\dot{x})-I=0 \ ,$$
$$g_s=\sum m(\dot{y}^2+\dot{z}^2)/2-NkT=0 \ . \tag{21}$$

The equations of motion include contributions from the two Lagrange parameters used to satisfy these constraints

$$m\ddot{x}=F_x-\lambda_d q \ ,$$
$$m\ddot{y}=F_y-\lambda_s m\dot{y} \ ,$$
$$m\ddot{z}=F_z-\lambda_s m\dot{z} \ . \tag{22}$$

The multipliers can be identified explicitly by multiplying the equations of motion (22) by (q/m), $\dot{y}$, and $\dot{z}$, and summing over all particles

$$\sum(q\ddot{x})=\frac{dI}{dt}=0=\sum(qF_x/m)-(\lambda_d/m)\sum q^2 \ ,$$
$$\sum(m\dot{y}\ddot{y}+m\dot{z}\ddot{z})=0=\sum(\dot{y}F_y+\dot{z}F_z)-\lambda_s\sum m(\dot{y}^2+\dot{z}^2) \ . \tag{23}$$

Thus the Lagrange multipliers are functions of time which depend only upon the particle colors, velocities, and forces

$$\lambda_d=\sum(qF_x)\Big/\sum(q^2) \ ,$$
$$\lambda_s=\sum(\dot{y}F_y+\dot{z}F_z)\Big/\sum m(\dot{y}^2+\dot{z}^2) \ . \tag{24}$$

The forces of constraint associated with the current density and the temperature do work on the system at the rates

$$P_d=\sum(qF_x)\sum(q\dot{x})\Big/\sum(q^2)$$

and

$$P_s=-\sum(\dot{y}F_y+\dot{z}F_z) \ , \tag{25}$$

respectively.

The color conductivity can be determined either from the driving power P_d associated with maintaining the current or with the stabilizing power P_s used to keep the temperature constant:

$$\sigma = I_0^2/V\langle P_d\rangle_{ne} = -I_0^2/V\langle P_s\rangle_{ne} \ . \tag{26}$$

The angular brackets in Eq. (26) indicate long-time steady-state averages. For short times P_d tends to exceed $-P_s$ by the rate at which energy is being stored in the system through its developing nonequilibrium distribution function.

Although this nonequilibrium diffusion problem is a relatively simple one, it is clear that Gauss's principle makes it possible to formulate many problems in a convenient way. The principle is particularly valuable in establishing nonequilibrium steady states suitable for theoretical analysis.

VI. SIMULATION RESULTS

Levesque and Verlet[18] determined the self-diffusion coefficient for a dense liquid composed of 864 particles interacting with the Lennard-Jones potential

$$\phi = 4\epsilon[(s/r)^{12} - (s/r)^6] \ . \tag{27}$$

They used the Green-Kubo formula [Eq. (15)] and considered a temperature $1.08\epsilon/k$ and number density $(N/V) = 0.85/s^3$ close to the liquid triple point. In our 108-particle calculations, at this same thermodynamic-state point, we truncate the potential at $r = 2.5s$.

Calculations using the Hamiltonian algorithm described in Sec. IV have been carried out by Snook, Evans, and Isbister.[19] For comparison with those results we have solved the Gaussian equations of motion (using a fifth-order predictor-corrector "Gear" integration) with a time step of $0.002(m/\epsilon)^{1/2}s$. The constraints maintained constant current and temperature to an accuracy of plus or minus 0.05% in 10 000 time steps. Typical run lengths were 15 000 steps.

Figure 2 shows the diffusion coefficient [related to the color conductivity by Eq. (20)] as a function of current. It is noteworthy that the two nonequilibrium methods are self-consistent within the statistical uncertainties. The equilibrium Green-Kubo conductivity, which appears to exceed the nonequilibrium conductivities, is probably slightly overestimated. The intermediate-time negative velocity correlation function was truncated in carrying out the Green-Kubo calculation according to Eq. (15).

The Gaussian method is more efficient than the Hamiltonian one because the transition time required to reach steady-state conditions is substantially reduced. In the Hamiltonian simulation approximately 10 000 time steps are necessary to attain the steady state from an initial equilibrium state. The accuracy with which the constraints are satisfied, and the balance between the work done and heat rejected, are convenient checks on the numerical work.

VII. CONCLUSION

In classical mechanics, constraints are used to simplify the analysis of dynamical systems. These constraints can be removed entirely if a full analysis is made of the interaction of the system and its surroundings. Constraints are used to replace the (possibly very complex) dynamics of the surroundings by their net effect (through the forces of constraint) on the system of interest. Gauss's principle of least constraint provides us with a systematic means for correctly formulating the equations of motion in such constrained systems.

Gauss's principle is ideal for developing nonequilibrium computer simulations. It allows us to project out of the motion the degrees of freedom corresponding to external reservoirs, replacing these thermodynamic baths by forces of constraint. We believe that the fundamental properties of nonequilibrium steady states can best be determined in this way.

There are several indications that Gauss's principle, as an extension of Newtonian mechanics, is consistent with statistical mechanics and kinetic theory. First, the isothermal dynamics is consistent with the Gibbs canonical ensemble.[3] A set of systems initially distributed canonically in the phase space remains distributed canonically, despite the changes in the energies of the systems making up the ensemble. Second, the isothermal dynamics, applied to a low-density shear-flow problem, predicts exactly the same shear and normal stresses as those derived from the Burnett-level solution of the Boltzmann equation.[20] Finally, equilibrium configurational properties, calculated using the isothermal equations of motion, can be shown to agree with the same properties calculated using the canonical ensemble of Gibbs.[5]

Here we have stressed the application of Gaussian dynamics to nonequilibrium steady states. It is evident that similar calculations can be applied to nonsteady problems and to systems in other ensembles, such as constant pressure.[21]

ACKNOWLEDGMENTS

We thank Jim Olness and Andy Combs for their numerical work on the constant-temperature linear-chain dynamics. We thank Jim Dufty and Jim Haile for useful conversations and correspondence, and Donna Clifford for technical support. This work was supported at the University of California at Davis, by the Academy of Applied Science, the U. S. Air Force Office of Scientific Research, the U. S. Army Research Office, and the Electric Power Research Institute. This work was carried out, at the Lawrence Livermore National Laboratory, under the auspices of the U. S. Department of Energy, under Contract No. W-7405-ENG-48 with the University of California.

Notes on p. 240

[1]J.-P. Ryckaert, G. Ciccotti, and H. J. C. Berendsen, J. Comp. Phys. 23, 327 (1977).
[2]L. A. Pars, *A Treatise on Analytic Dynamics* (Ox Bow, Woodbridge, Conn., 1979); E. T. Whittaker, *A Treatise on the Analytical Dynamics of Particles and Rigid Bodies* (Cambridge University Press, Cambridge, 1927).
[3]W. G. Hoover, A. J. C. Ladd, and B. Moran, Phys. Rev. Lett. 48, 1818 (1982).
[4]D. J. Evans, J. Chem. Phys. 78, 3297 (1983).
[5]J. M. Haile and S. Gupta (unpublished).
[6]K. F. Gauss, J. Reine Angew. Math. IV, 232 (1829).
[7]S. P. Liu, Am. J. Phys. 49, 750 (1981).
[8]J. R. Ray, Am. J. Phys. 34, 406 (1966); 34, 1202 (1966); 40, 179 (1972).
[9]E. J. Saletan and A. H. Cromer, Am. J. Phys. 38, 892 (1970).
[10]P. Appell, Acad. Sci., Compt. Rend. 152, 1197 (1911).
[11]W. G. Hoover, in Proceedings of the National Bureau of Standards Conference on Nonlinear Fluid Phenomena, Boulder, Colorado, June, 1982, Physica (Utrecht) 118A, 111 (1983).
[12]W. G. Hoover, D. J. Evans, R. B. Hickman, A. J. C. Ladd, W. T. Ashurst, and B. Moran, Phys. Rev. A 22, 1690 (1980).
[13]D. J. Evans, Phys. Lett. 91A, 457 (1982).
[14]B. J. Alder and T. Wainwright, in *Proceedings of the International Symposium on Transport Processes in Statistical Mechanics, Brussels, 1956*, edited by I. Prigogine (Interscience, New York, 1958), pp. 97–131.
[15]D. A. McQuarrie, *Statistical Mechanics* (Harper and Row, New York, 1976), pp. 507–509.
[16]W. T. Ashurst, Ph.D. dissertation, University of California at Davis—Livermore, 1974 (unpublished).
[17]D. J. Evans, Mol. Phys. 37, 1745 (1979).
[18]D. Levesque and L. Verlet, Phys. Rev. A 2, 2514 (1970).
[19]I. K. Snook, D. J. Evans, and D. J. Isbister (unpublished).
[20]W. G. Hoover, Annual Meeting of the Society of Rheology, Evanston, Illinois, October, 1982 (unpublished).
[21]D. J. Evans and G. P. Morriss, J. Chem. Phys. (in press).

Notes to Reprint III.8

1. Gauss' principle can be used only as a guide to writing down equations of motion that may be of value in statistical-mechanical applications. It appears to give useful results for non-holonomic constraints that are homogeneous functions of velocities (e.g. temperature and pressure), but it fails in the heat-flow problem, where the heat current is an inhomogeneous function (Hoover et al., J. Stat. Phys. 37 (1984) 109).

2. The equilibrium distribution function corresponding to gaussian, isokinetic equations of motion has been derived by Evans and Morriss (Phys. Lett. A 109 (1983) 433) and by Nosé (Reprint IV.5). The fact that use of a gaussian thermostat does not change the form of the Green–Kubo relations is discussed by Evans and Morriss (Chem. Phys. 87 (1984) 451) and by Evans and Holian (J. Chem. Phys. 83 (1985) 4069).

3. Gauss' principle can be used to write down equations of motion corresponding to a constant thermodynamic flux. This makes possible the simulation of a "Norton ensemble", for which the flux rather than the force is the independent variable. The fluctuation formula for the resistance in the linear regime was worked out and tested by Evans and Morriss (Phys. Rev. A 31 (1985) 3817).

Classical response theory in the Heisenberg picture

B. L. Holian[a] and D. J. Evans
Research School of Chemistry, Australian National University, Canberra ACT 2601, Australia

(Received 2 April 1985; accepted 27 June 1985)

A new derivation of classical response theory is presented from the Heisenberg point of view, where observable properties are followed along a classical many-body trajectory. The exact equivalence of the Heisenberg (observable) and Schrödinger (distribution function) pictures is shown and the nature of their relationship is clarified. The formal expression is derived for the response to time-dependent external fields, to all orders.

I. INTRODUCTION

In this paper, we present a new derivation of classical response theory, both linear and nonlinear, from the point of view of the so-called "Heisenberg" picture, where one follows the trajectory of an observable in a system perturbed from equilibrium by an externally imposed force. The usual derivation[1] takes the conjugate point of view, in that the distribution function is followed in time—the so-called "Schrödinger" picture. A previous attempt[2] at a derivation in the Heisenberg picture glossed over some important conceptual points and gave incorrect results, which we will rectify. In particular, the previous Heisenberg result differed from the Schrödinger, in that the claim was made that a one-term nonperturbative response was obtainable from a linear perturbation to the equations of motion. More importantly, our approach leads to a new result in that the expression for the nonlinear response to a time-dependent external force can be obtained. We show, too, how the Schrödinger and Heisenberg pictures are related, with new insight into the meaning of each one. (Obviously, we are using the terms Schrödinger and Heisenberg somewhat liberally, since our aim is to describe classical statistical mechanical systems. We hope that the intended analogy will aid those who are more used to thinking about quantum mechanical systems. Indeed, the present work may provide useful clues for those who wish to prove the analogous results for the quantum case.[3])

Our objective, then, is to calculate the long-time, hydrodynamic response of an observable B in a macroscopic system subjected at time $t = 0$ to an external, time-dependent driving force $\dot{\epsilon}(t)$. Our notation will be a compromise between that of Kubo's original Schrödinger picture derivation and that of linear (Navier–Stokes) hydrodynamics keeping in mind the problem of computing the appropriate linear transport coefficient η, the shear viscosity, for planar Couette flow in a fluid. In that case, the field $\dot{\epsilon}$ is the shear rate, $(\partial/\partial y)u_x$, of the fluid velocity $\mathbf{u}$. The observable of interest B is $J = P_{yx} V$, the shear momentum flux, with $P_{yx} = P_{xy}$ the shear component of the pressure tensor and V the volume in d spatial dimensions of an N-particle system. In the thermodynamic limit (N, $V \to \infty$, $N/V = \text{const}$), at long times ($t \to \infty$), and for small steady shear rates [$\dot{\epsilon}(t) = \dot{\epsilon} \to 0$], all of which we will abbreviate by the symbol "lim", the hydrodynamic response is given by

$$\lim \langle J(t)\rangle = \langle J\rangle_0 - \eta V\dot{\epsilon} + O(\dot{\epsilon}^2),$$

where $\langle J(t)\rangle$ is the time-dependent nonequilibrium ensemble average of J, and $\langle J\rangle_0$ is the equilibrium value, which, for shear flow in fluids, is zero.

The nonequilibrium ensemble average of an obervable B can be thought of in two equivalent ways. First, in the Heisenberg, or observable-based picture we imagine performing a series of experiments differing only in the intitial values of the dN coordinates q and momenta p. This is the point of view we would take in performing nonequilibrium molecular dynamics (NEMD) computer experiments, for example. We denote phase space by $\Gamma = (q, p, \xi)$, a $(2dN + 1)$-dimensional vector; notice that we have included an extra dynamical variable— a thermostatting coefficient ξ. The equations of motion presented in the next section and described in more detail in the Appendix, include this extra degree of freedom to simulate by homogeneous feedback the effect of a large thermal reservoir. By homogeneous, we mean that the contact with the reservoir occurs throughout the volume V of our system, affecting all N particles in some small way all of the time. By feedback, we mean that ξ changes with time in such a way that the mean kinetic energy of the N particles is fixed (as well as certain other equilibrium properties and their distribution). We will refer to the special case $\xi = 0$ as "adiabatic," since in that case, dissipative heating due to the external field is not extracted from the N-particle system. Each initial phase Γ is weighted by its equilibrium phase-space distribution function $f_0(\Gamma)$. Each such $t = 0$ starting point behaves like a Lagrangian "mass" point, following a trajectory $\Gamma(t)$ perturbed from the ordinary one by the external force. (In fact, this perturbation in the trajectory will not be small, even after only a few mean collision times, but may rather diverge exponentially in time from the unperturbed one.) Just as in the Lagrangian formulation of fluid mechanics, we can imagine that the phase-space mass point has a differential box $d\Gamma$ surrounding it which changes shape with time as the phase point follows its trajectory. (See Fig. 1.) The differential probability, or mass $f_0(\Gamma)d\Gamma$ for the mass point remains constant, but the value of the observable changes implicitly in time according to $B(t) = B[\Gamma(t)]$, hence, $\partial B/\partial t = 0$. The fundamental postulate of nonequilibrium statistical mechanics therefore states that

[a] Permanent address: Los Alamos National Laboratory, Los Alamos, NM 87545.

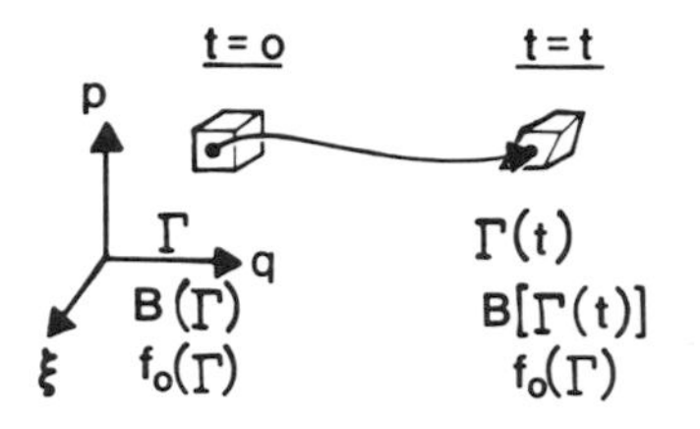

FIG. 1. The observable-based (Heisenberg), or Lagrangian picture of phase-space trajectories: the observable B changes with time, the distribution f_0 is fixed.

$$\langle B(t)\rangle = \int d\Gamma\, f_0(\Gamma)\, B\,[\Gamma(t)];$$

i.e., we average the experimental measurements of the observable at time t over all possible initial conditions (all of phase space), weighted by the equilibrium distribution function.

The second view is the Schrödinger, or distribution-based picture, where now the Γ refers *not* to the initial value of a point in phase space, but to a stationary point (for all times) inside a stationary differential box $d\Gamma$. Just as in the Eulerian formulation of fluid mechanics, the observable takes on a fixed value for all time $B(\Gamma)$, while mass points with different probabilities flow through the box. (See Fig. 2.) Now, the distribution function $f(\Gamma,t)$ changes in time from its $t=0$ initial equilibrium value $f_0(\Gamma)$. The nonequilibrium ensemble average can be written in the equivalent conjugate Schrödinger form:

$$\langle B(t)\rangle = \int d\Gamma\, B(\Gamma) f(\Gamma,t),$$

i.e., we weight each (stationary) value of Γ and $B(\Gamma)$ by the distribution function at time t and add them all up. Let us emphasize that these two pictures of response theory are conjugate and equivalent to each other, with quite different meanings for what is meant by Γ, for example. Because it is easier to expand $f(t)$ about the equilibrium distribution f_0, whose mathematical properties are well known, than it is to expand an arbitrary observable, the Schrödinger picture has

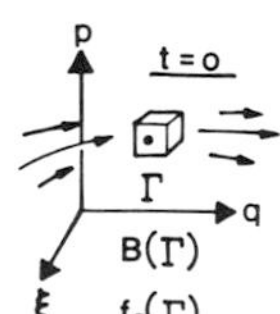

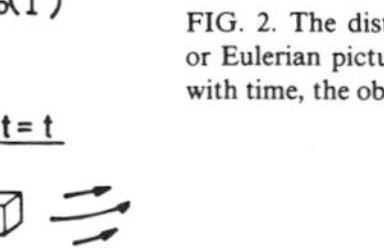

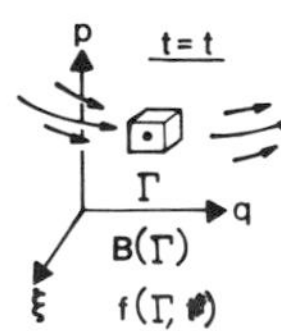

FIG. 2. The distribution-based (Schrödinger), or Eulerian picture: the distribution f changes with time, the observable B is fixed.

advantages over the more physically intuitive Heisenberg formulation. We will show how the Schrödinger picture can be obtained from the Heisenberg, and vice versa.

II. HEISENBERG–SCHRÖDINGER EQUIVALENCE

In the Heisenberg picture we first write down the equations of motion in what may seem a rather arbitrary fashion:

$$\dot{\Gamma}(\Gamma,t) = \begin{cases} \dot{q}(\Gamma,t) = \dfrac{p-p_0}{m} + \dot{\epsilon}(t)\mathscr{Q}(q,p) \\ \dot{p}(\Gamma,t) = F(q) - \dot{\epsilon}(t)\mathscr{P}(q,p) - \xi\cdot(p-p_0) \\ \dot{\xi}(\Gamma,t) = \dot{\xi}(p), \end{cases} \tag{1}$$

where $Np_0 = \Sigma p$ is the total linear momentum, $\mathscr{Q}$ (in the case of shear flow) is a coordinate-like phase function, and $\mathscr{P}$ is momentum-like ($\mathscr{Q}$, $\mathscr{P}$, and $\dot{\epsilon}$ can be tensorial quantities, in general). For shear flow, we can use the so-called "Sllod" tensor dynamics[4] (transposed "Dolls" or "$q\,p$" tensor) where $\mathscr{Q}_x = y$ and $\mathscr{P}_x = p_y$, otherwise zero. (For comparison with notation elsewhere, our flux $J = \dot{A}$ and external field $\dot{\epsilon} = -F_{\rm ext}$ in Kubo,[1] while $\mathscr{Q} = -A_p$ and $\mathscr{P} = -A_q$ in Evans and Morriss.[5]) Note that Γ is the initial $t=0$ phase, while $\dot{\epsilon}(t)$ is the external field at current time t. This apparent inconsistency allows us to invent a time-translating phase-space propagator (sometimes called p propagator) $U(t)$ that advances the implicitly time-dependent phase variables from the initial condition Γ, along the trajectory, to the current phase $\Gamma(t)$:

$$\Gamma(t) = U(t)\Gamma.$$

$U(t)$ acts on all functions of Γ that appear to the right of it, as distinct from $\dot{\epsilon}(t)$ which is left alone. The equations of motion at time t are then given by $\dot{\Gamma}[\Gamma(t),t] = U(t)\,\dot{\Gamma}(\Gamma,t)$. The observable $B(t)$ at time t is $B[\Gamma(t)] = U(t)\; B(\Gamma)$. Since $\partial B/\partial t = 0$, the full-time derivative $\dot{B}(t)$ is

$$\begin{aligned} \frac{d}{dt} B\,[\Gamma(t)] &= \dot{\Gamma}[\Gamma(t),t]\cdot\frac{\partial}{\partial\Gamma} B\,[\Gamma(t)] \\ &= U(t)\dot{\Gamma}(\Gamma,t)\cdot\frac{\partial}{\partial\Gamma} B(\Gamma) \\ &= U(t)\, iL(t)\, B(\Gamma) \\ &= \frac{\partial}{\partial t} U(t)\, B(\Gamma), \end{aligned}$$

where we have introduced the Liouville operator (sometimes called p Liouvillean)

$$iL(t) = \dot{\Gamma}(\Gamma,t)\cdot\frac{\partial}{\partial\Gamma},$$

which acts on functions of *initial* phase Γ and contains the *current* field $\dot{\epsilon}(t)$ through the time-displaced equations of motion $\dot{\Gamma}(\Gamma,t)$ we first wrote down. It is crucial to appreciate that the Liouville operator and associated propagator act on the initial phase, not on the current phase. If this mathematical legality is glossed over, incorrect conclusions can be drawn.[2]

The equation of motion for the propagator is

$$\frac{\partial}{\partial t} U(t) = U(t)\, iL(t),$$

whose formal solution is the right-hand-sided time exponential, $\exp_R$,

$$U(t) = \exp_R \int_0^t ds\, iL(s)$$
$$= \sum_{n=0}^{\infty} \int_0^t ds_1 \ldots \int_0^{s_{n-1}} ds_n\, iL(s_n)\ldots iL(s_2) iL(s_1),$$

which has the property that time derivaties bring down the Liouville operator on the right-hand side of the exponential.

We shall establish two important properties of the Liouville operator. First

$$\int d\Gamma\, f_0(\Gamma) iL(t) B(\Gamma) = \int d\Gamma\, f_0(\Gamma)\dot{\Gamma}(\Gamma,t)\cdot\frac{\partial}{\partial\Gamma} B(\Gamma)$$
$$= f_0 \dot{\Gamma} B \,|_{-S_\Gamma}^{+S_\Gamma} - \int d\Gamma\, B(\Gamma)\frac{\partial}{\partial\Gamma}\cdot [f_0(\Gamma)\dot{\Gamma}(\Gamma,t)]$$
$$= -\int d\Gamma\, B(\Gamma) i\mathscr{L}(t) f_0(\Gamma),$$

where we have integrated by parts and made use of the fact that the surface, or boundary term vanishes either because (1) f_0 vanishes at the $\pm S_\Gamma$ boundaries of phase space, or (2) f_0, $\dot{\Gamma}$, and B are periodic functions that take on the same values at the opposite periodic boundaries. With the defintion of the phase-space compressibility,

$$\Lambda = \frac{\partial}{\partial\Gamma}\cdot\dot{\Gamma}(\Gamma,t),$$

which is simply a multiplicative, rather than linear partial differential operator, and which vanishes in the special case that the equations of motion can be derived from a Hamiltonian, we may write the distribution function Liouville operator (sometimes called f Liouvillean) as

$$i\mathscr{L}(t) = \dot{\Gamma}(\Gamma,t)\cdot\frac{\partial}{\partial\Gamma} + \frac{\partial}{\partial\Gamma}\cdot\dot{\Gamma}(\Gamma,t) = iL(t) + \Lambda.$$

The second and most important property is the derivation of the Schrödinger expression for $\langle B(t)\rangle$ from the Heisenberg picture, using the first result in repeated integrations by parts:

$$\int d\Gamma\, f_0(\Gamma) B[\Gamma(t)] = \int d\Gamma\, f_0(\Gamma) U(t) B(\Gamma)$$
$$= \sum_{n=0}^{\infty}\int_0^t ds_1\ldots\int_0^{s_{n-1}} ds_n \times \int d\Gamma\, f_0(\Gamma) iL(s_n)[\ldots iL(s_1) B(\Gamma)]$$
$$= \sum_{n=0}^{\infty}\int_0^t ds_1\ldots\int_0^{s_{n-1}} ds_n \times \int d\Gamma\, [-i\mathscr{L}(s_n)]\, f_0(\Gamma) iL(s_{n-1})[\ldots iL(s_1) B(\Gamma)]$$
$$= \sum_{n=0}^{\infty}\int_0^t ds_1\ldots\int_0^{s_{n-1}} ds_n \times \int d\Gamma\, \{[-i\mathscr{L}(s_1)]\ldots[-i\mathscr{L}(s_n)]\, f_0(\Gamma)\}\cdot B(\Gamma)$$
$$= \int d\Gamma\, B(\Gamma) U^+(t) f_0(\Gamma)$$
$$= \int d\Gamma\, B(\Gamma) f(\Gamma,t),$$

where we have introduced the distribution function propagator (sometimes called f propagator)

$$U^+(t) = \exp_L - \int_0^t ds\, i\mathscr{L}(s)$$
$$= \sum_{n=0}^{\infty}(-1)^n \int_0^t ds_1 \ldots \int_0^{s_{n-1}} ds_n\, i\mathscr{L}(s_1) i\mathscr{L}(s_2)\ldots i\mathscr{L}(s_n).$$

This is the usual left-hand sided time exponential whose time derivatives bring down the f Liouvillean on the left-hand side of the exponential, so that the equation of motion for $U^+(t)$ is

$$\frac{\partial}{\partial t} U^+(t) = -i\mathscr{L}(t) U^+(t).$$

The f propagator U^+ moves $f_0(\Gamma)$ forward in time according to

$$f(\Gamma,t) = U^+(t) f_0(\Gamma),$$

and f obeys the equation of motion

$$\frac{\partial f}{\partial t} = \frac{\partial}{\partial t} U^+(t) f_0(\Gamma)$$
$$= -i\mathscr{L}(t) U^+(t) f_0(\Gamma)$$
$$= -\left[\dot{\Gamma}(\Gamma,t)\cdot\frac{\partial}{\partial\Gamma} + \frac{\partial}{\partial\Gamma}\cdot\dot{\Gamma}(\Gamma,t)\right] f(\Gamma,t)$$
$$= -\frac{\partial}{\partial\Gamma}\cdot [f(\Gamma,t)\dot{\Gamma}(\Gamma,t)],$$

which we immediately recognize as the complete Liouville equation for the time evolution of the phase-space distribution function. This equation (sometimes referred to as the "non-Liouville" equation[5]) can be derived in a manner perfectly analogous to the derivation of the Eulerian mass conservation equation of hydrodynamics, with f playing the role of the mass density ρ and $\dot{\Gamma}$ the role of the fluid velocity $\mathbf{u} = \dot{\mathbf{r}}$:

$$\frac{\partial}{\partial t}\rho(\mathbf{r},t) = -\frac{\partial}{\partial\mathbf{r}}\cdot[\rho(\mathbf{r},t)\mathbf{u}(\mathbf{r},t)].$$

We emphasize at this point that in deriving the Schrödinger picture from the Heisenberg, we have also transformed the meaning of the phase variables Γ from the Lagrangian to the Eulerian meaning. That is, in the Heisenberg picture Γ was the initial phase in a trajectory; in the Schrödinger picture Γ is some particular fixed point in phase space. When we integrate over all Γ, we cover all of phase space in either picture, since both sets are complete. Note that the time-displaced Lagrangian equations of motion $\dot{\Gamma}(\Gamma,t)$ have the *identical* form in the Eulerian picture: given a point in phase space Γ, regardless of how a mass point arrived there, the equations of motion at time t are obtained by plugging in Γ and the current value of the field $\dot{\epsilon}(t)$. The careful derivation we have presented here is absolutely essential for establishing the equivalence of the Heisenberg and Schrödinger pictures. It should also now be clear, given the properties of left- and right-exponential propagators, how the mathematical steps in arriving at the Schrödinger (Eulerian) picture from the Heisenberg (Lagrangian) picture are completely reversible.

Notes on p. 247

III. RESPONSE THEORY

Having arrived at the Schrödinger picture of nonequilibrium statistical mechanics from the Heisenberg picture, we can proceed to look at the consequences for linear and nonlinear response theory. Let us first separate the adjoint Liouville operator into an unperturbed, but thermostatted part $i\mathscr{L}_0$, and a part due to the perturbation from the time-dependent field L_1:

$$i\mathscr{L}(t) = iL(t) + \Lambda$$
$$= i\mathscr{L}_0 + iL_1(t).$$

The distribution-function propagator then becomes

$$U^+(t) = \exp_L\left[-it\mathscr{L}_0 - \int_0^t ds\, iL_1(s)\right].$$

The unperturbed propagator is

$$U_0^+(t) = \exp_L(-it\mathscr{L}_0).$$

In terms of U_0^+ and L_1, we may express the full propagator using the so-called "Dyson decomposition"[6]:

$$U^+(t) = U_0^+(t) - \int_0^t ds\, U_0^+(t-s) iL_1(s) U^+(s),$$

which is consistent with the equation of motion for the propagator

$$\frac{\partial}{\partial t} U^+(t) = -i\mathscr{L}_0 U_0^+(t) - U_0^+(0) iL_1(t) U^+(t)$$
$$- \int_0^t ds(-i\mathscr{L}_0) U_0^+(t-s) iL_1(s) U^+(s)$$
$$= -i\mathscr{L}_0 U_0^+(t) - iL_1(t) U^+(t)$$
$$- i\mathscr{L}_0[U^+(t) - U_0^+(t)]$$
$$= [-i\mathscr{L}_0 - iL_1(t)] U^+(t)$$
$$= -i\mathscr{L}(t) U^+(t).$$

This result, namely the Dyson decomposition for time-dependent fields, is new. In the Appendix, we show that $i\mathscr{L}_0 f_0 = 0$, so that the unperturbed propagator acting on the equilibrium distribution function leaves it unchanged:

$$U_0^+(t) f_0(\Gamma) = f_0(\Gamma),$$

hence, the nonequilibrium distribution function in the Schrödinger (Eulerian) picture is

$$f(\Gamma,t) = U^+(t) f_0(\Gamma)$$
$$= \left[1 - \int_0^t ds\, U_0^+(t-s) iL_1(s) U^+(s)\right] f_0(\Gamma)$$
$$= \Big\{1 - \int_0^t ds_1\, U_0^+(t-s_1) iL_1(s_1)\Big[U_0^+(s_1)$$
$$- \int_0^{s_1} ds_2\, U_0^+(s_1-s_2) iL_1(s_2) U^+(s_2)\Big]\Big\} f_0(\Gamma)$$
$$= \Big[1 + \sum_{n=1}^{\infty} (-1)^n \int_0^t ds_1 \int_0^{s_1} ds_2$$
$$\ldots \int_0^{s_{n-1}} ds_n$$
$$\times U_0^+(t-s_1) iL_1(s_1) U_0^+(s_1-s_2) iL(s_2)$$
$$\ldots U_0^+(s_{n-1}-s_n) iL_1(s_n)\Big] f_0(\Gamma)$$

by repeated application of the Dyson decomposition. The first two terms form the basis of Kubo's derivation of linear response theory. The response of the observable B is given by the expression for the fundamental postulate of nonequilibrium statistical mechanics, first in the Heisenberg picture, then in the Schrödinger:

$$\langle B(t)\rangle = \int d\Gamma\, f_0(\Gamma) B[\Gamma(t)]$$
$$= \int d\Gamma\, B(\Gamma) f(\Gamma,t)$$
$$= \int d\Gamma\, B(\Gamma) U^+(t) f_0(\Gamma)$$
$$= \int d\Gamma\, B(\Gamma) f_0(\Gamma) + \sum_{n=1}^{\infty} (-1)^n$$
$$\times \int_0^t ds_1 \ldots \int_0^{s_{n-1}} ds_n \int d\Gamma\, B(\Gamma)$$
$$\cdot U_0^+(t-s_1) iL_1(s_1) U_0^+(s_1-s_2) iL_1(s_2)\ldots$$
$$\cdot U_0^+(s_{n-1}-s_n) iL_1(s_n) f_0(\Gamma).$$

Since $iL_1(t) f_0 = \beta J \dot\epsilon(t) f_0$ for either the adiabatic case or Nosé–Hoover[7] thermostatting, and since $U_0^+(t) J f_0 = f_0 U_0(-t) J = f_0 J_\xi(-t)$ (see the Appendix for details), we may write

$$\langle B(t)\rangle = \langle B\rangle_0 - \beta \int_0^t ds\, \dot\epsilon(s) \int d\Gamma\, f_0(\Gamma) B(\Gamma) J_\xi(s-t)$$
$$+ \beta \int_0^t ds_1 \int_0^{s_1} ds_2\, \dot\epsilon(s_2) \int d\Gamma\, B(\Gamma) e^{-i(t-s_1)\mathscr{L}_0}$$
$$\times iL_1(s_1) f_0(\Gamma) J_\xi(s_2-s_1) + \mathscr{O}(\epsilon^3).$$

For convenience, we have written the maximum possible strain as $\epsilon = |\max(\dot\epsilon)| t$. It is clear that the response to linear perturbations to the equation of motion give rise to nonlinear terms.

The linear susceptibility is defined as

$$\chi_B(t) = \beta \int d\Gamma\, f_0(\Gamma) B(\Gamma) J_\xi(-t)$$
$$= \beta \langle B e^{-itL_\xi} J\rangle_0$$
$$= \beta \langle B(0) J(-t)\rangle_0$$
$$= \beta \langle J e^{itL_\xi} B\rangle_0$$
$$= \beta \langle J(0) B(t)\rangle_0.$$

The last step demonstrates the time-translational invariance of equilibrium covariance functions, as obtained from the properties of the (thermostatted) equilibrium propagator. The linear response is then

$$\langle B(t)\rangle = \langle B\rangle_0 - \int_0^t ds\, \dot\epsilon(s) \chi_B(t-s) + \mathscr{O}(\epsilon^2)$$
$$= \langle B\rangle_0 - \int_0^t ds\, \dot\epsilon(t-s) \chi_B(s) + \mathscr{O}(\epsilon^2)$$
$$= \langle B\rangle_0 - \beta \int_0^t ds\, \dot\epsilon(t-s) \langle J(0) B(s)\rangle_0 + \mathscr{O}(\epsilon^2).$$

The linear transport coefficient η is obtained by setting $B=J$, $\dot\epsilon(t)=\dot\epsilon$ (steady field), and taking the limits $N,V\to\infty(N,V=\text{const})$, $t\to\infty$, $\dot\epsilon\to 0$,

$$\lim\langle J(t)\rangle = \langle J\rangle_0 - \eta V\dot\epsilon + \mathcal{O}(\dot\epsilon^2)$$
$$= \langle J\rangle_0 - \beta\dot\epsilon\int_0^\infty dt\,\langle J(0)J(t)\rangle_0 + \mathcal{O}(\dot\epsilon^2),$$

whence we obtain the Kubo[1] expression for η:

$$\eta = \frac{1}{VkT}\int_0^\infty dt\,\langle J(0)J(t)\rangle_0,$$

the time integral of the autocovariance function of the flux J.

We now make some comments regarding the order of limits that we take on $\langle J(t)\rangle$. The Taylor series we have written down for the response as products of time integrals over L_1 can, conservatively speaking, be said to be conditionally convergent for a steady field $\dot\epsilon(t)=\dot\epsilon$ if the product of the strain rate $\dot\epsilon$ and the total time t, i.e., the total strain, $\epsilon=\dot\epsilon t$, is small: $|\epsilon|\ll 1$. For a given length of time t, which is at least the one or two mean free times for a steady state to be reached (perhaps longer if the thermostatting relaxation time is greater than one or two mean free times),[8] the maximum strain rate is limited by $\dot\epsilon=\epsilon/t$. Is this mathematical constraint on $\dot\epsilon$ consistent with physical constraints on the experiments to be performed and averaged over?

First of all, the maximum strain for a problem such as the uniaxial compression in the measurement of longitudinal viscosity should be small, so that the volume of the system does not change appreciably. Obviously, the Taylor series convergence criterion is consistent with this physical requirement. However, for shear viscosity, where volume does not change, no such physical constraint is necessary, so that the mathematical constraint is unnecessarily stringent.

Secondly, for an adiabatic experiment, we desire that the temperature not rise too far from its initial value:

$$\frac{\dot H_0 t}{2K} = -\frac{J\dot\epsilon t}{d(N-1)kT} \cong -\frac{(-\eta V\dot\epsilon)\dot\epsilon t}{d(N-1)kT} \cong \frac{\eta\dot\epsilon^2 t}{dnkT} \ll 1$$

or $\dot\epsilon \ll (dnkT/\eta t)^{1/2}$, which is easily satisfied by the more stringent Taylor series restriction ($dnkT/\eta$ is typically of order unity for dense fluids; $n=N/V$ is the number density). However, for the thermostatted case, no such physical constraint need be applied.

The third consideration is the relationship between the maximum time allowed and the size of the system. A suggestion of this relationship is afforded by the linear response term, which is a time integral of an equilibrium convariance function. If the upper limit on that integral exceeds the time for a sound-wave disturbance to cross the periodic system (side length of order $V^{1/d}$), then it can be argued that the time correlation no longer approximates that of the infinite system. Thus the maximum value of t consistent with an N-particle system is $t<V^{1/d}/c\sim N^{1/d}$, where c is the adiabatic sound speed in the equilibrium system. For a dense fluid where $N\sim 100$ (a typical size for molecular dynamics computer simulations), t is then of the order of, say, ten mean free times, which is sufficient time for the nonequilibrium system to achieve a steady state.

We can therefore conclude that the limits, when taken in the following manner for the macroscopic, long-time, small-field hydrodynamic response—$N\to\infty(N/V=\text{const})$, $t\sim N^{1/d}$, $\dot\epsilon\sim N^{-1/d}$—will satisfy the mathematical convergence criterion as well as physical constraints for the worst case of adiabatic compression. For thermostatted shearing, it is not clear what the Taylor series convergence implies, physically

IV. CONCLUSIONS

We have shown in this paper the correct interpretation of response theory (linear and nonlinear) in both Heisenberg and Schrödinger pictures. The Heisenberg picture requires a careful Lagrangian formulation of the time evolution of an observable along a trajectory. The resulting right-hand-sided propagator for the observable in a time-dependent field can be expressed in terms of the equilibrium (possibly thermostatted) propagator using the Dyson decomposition

$$U(t) = U_0(t) + \int_0^t ds\, U(s)iL_1(s)U_0(t-s).$$

The relationship to the Schrödinger picture—that is, the equivalence—is then obtained by an Eulerian formulation of the rate of change of the distribution function. The resulting linear response of the system can be obtained in terms of the usual Kubo expression for the linear transport coefficient. The nonlinear response for time-varying fields can also be written down, at least formally, to all orders in the strain $\epsilon=\dot\epsilon t$ (moreover, all orders exist). Finally, we show that the most conservative of interpretations, on both mathematical and physical grounds, indicate that the response theory derived herein is applicable to times long enough for each particle to have suffered many collisions (Maxwell relaxation times)—even in systems as small as those usually studied by NEMD. Van Kampen's[9] objection to the theory is that a Taylor series in a particle's trajectory, taken to linear order in the perturbation, gets one no farther than a collision or two, whereupon the theory falls apart. We see that his objection is perhaps too stringent.

ACKNOWLEDGMENTS

It is a pleasure to thank William Hoover, Gary Morriss, and Michael Collins, whose helpful comments and sharp-eyed criticisms spurred us on down the right track. One of us (BLH) would like to thank the Research School of Chemistry, Australian National University, for providing a stimulating environment for this work, which was performed under appointment as Visiting Research Fellow; additional support was obtained from the U. S. Department of Energy.

APPENDIX

Here, we present details about Liouville operators and time propagators for the equations of motion [Eq. (1)]. The Liouville operator is given by

Notes on p. 247

$$iL(t) = \dot{\Gamma}(\Gamma,t)\cdot\frac{\partial}{\partial\Gamma} = \Sigma\left(\dot{q}\cdot\frac{\partial}{\partial q} + \dot{p}\cdot\frac{\partial}{\partial p}\right) + \dot{\xi}\frac{\partial}{\partial\xi}$$
$$= iL_0 - \xi\Sigma(p-p_0)\cdot\frac{\partial}{\partial p} + \dot{\xi}\frac{\partial}{\partial\xi} + iL_1(t),$$

where the adiabatic ($\xi\equiv 0$), unperturbed ($\dot{\epsilon}\equiv 0$), Newtonian dynamics Liouvillean is

$$iL_0 = \Sigma\left(\frac{p-p_0}{m}\cdot\frac{\partial}{\partial q} + F\cdot\frac{\partial}{\partial p}\right)$$

the thermostatted, unperturbed p Liouvillean is

$$iL_\xi = iL_0 - \xi\Sigma(p-p_0)\cdot\frac{\partial}{\partial p} + \dot{\xi}\frac{\partial}{\partial\xi},$$

and the time-dependent perturbation is

$$iL_1(t) = \dot{\epsilon}(t)\Sigma\left(\mathscr{Q}\cdot\frac{\partial}{\partial q} - \mathscr{P}\cdot\frac{\partial}{\partial p}\right).$$

With this, we can compute the rate of change of the internal energy in the system $\dot{H}_0$. The equilibrium Hamiltonian internal energy function is

$$H_0(q,p) = K(p) + \Phi(q),$$

where $\Phi(q)$ is the potential energy, and the kinetic energy relative to the center-of-mass motion is

$$K(p) = \Sigma\frac{(p-p_0)^2}{2m}.$$

Internal velocities and forces in a Newtonian system, i.e., adiabatic (unthermostatted) and unperturbed, are given by

$$\frac{\partial H_0}{\partial p} = \frac{\partial K}{\partial p} = \frac{p-p_0}{m},$$
$$-\frac{\partial H_0}{\partial q} = -\frac{\partial\Phi}{\partial q} = F(q).$$

For a function that depends on (q,p) only through its dependence on the equilibrium Hamiltonian H_0 (such as H_0 itself or f_0, the equilibrium distribution function), we may use the chain rule to write

$$iL(t) = \dot{H}_0\frac{\partial}{\partial H_0} + \dot{\xi}\frac{\partial}{\partial\xi}$$
$$= [-2K\xi - J\dot{\epsilon}(t)]\frac{\partial}{\partial H_0} + \dot{\xi}\frac{\partial}{\partial\xi},$$

where the dissipative flux J is

$$J = \Sigma\left(\frac{p-p_0}{m}\cdot\mathscr{P} + F\cdot\mathscr{Q}\right).$$

The rate of change of energy in the system is

$$\dot{H}_0 = \frac{d}{dt}H_0 = iL(t)H_0 = -2K\xi - J\dot{\epsilon}(t).$$

(Note that $\partial H_0/\partial t = 0 = \partial H_0/\partial\xi$.)

The phase-space compressibility, which is a multiplicative operator, is given by

$$\Lambda = \frac{\partial}{\partial\Gamma}\cdot\dot{\Gamma}(\Gamma,t) = \Sigma\left(\frac{\partial}{\partial q}\cdot\dot{q} + \frac{\partial}{\partial p}\cdot\dot{p}\right) + \frac{\partial\dot{\xi}}{\partial\xi}$$
$$= -\xi\Sigma\frac{\partial}{\partial p}\cdot(p-p_0)$$
$$+\dot{\epsilon}(t)\Sigma\left(\frac{\partial}{\partial q}\cdot\mathscr{Q} - \frac{\partial}{\partial p}\cdot\mathscr{P}\right),$$

since $\dot{\xi}$ was assumed not to depend explicitly on ξ. We restrict ourselves to equations of motion where

$$\Sigma\left(\frac{\partial}{\partial q}\cdot\mathscr{Q} - \frac{\partial}{\partial p}\cdot\mathscr{P}\right) = 0,$$

so that in the adiabatic case ($\xi\equiv 0$), phase space is incompressible (i.e., $\partial/\partial\Gamma\cdot\dot{\Gamma} = 0$, by analogy with fluid mechanics). This is the so-called[5] assumption of "adiabatic incompressibility of phase space" or "AIΓ," for short. In this case, Λ is no longer an explicit function of t. Note that

$$\Sigma\frac{\partial}{\partial p}\cdot(p-p_0) = \Sigma\frac{\partial}{\partial p}\cdot\left(p - \frac{1}{N}\Sigma p'\right)$$
$$= N\cdot d\cdot\left(1-\frac{1}{N}\right) = d(N-1) = g,$$

the number of degrees of freedom in momentum space. (The center-of-mass coordinate $Mq_0 = \Sigma\, mq$, where $M = \Sigma m$ is the total mass, and the center-of-mass momentum $Np_0 = \Sigma p$ are conserved by our equations of motion, i.e., $N\dot{p}_0 = 0 = M\dot{q}_0$, as long as $\Sigma\mathscr{P} = 0 = \Sigma m\mathscr{Q}$. The total internal force on the system is $\Sigma F = 0$.) Thus, we find that the phase-space compressibility is

$$\Lambda = -g\xi.$$

We can now separate the f Liouvillean $i\mathscr{L}(t) = iL(t) + \Lambda$ into an unperturbed part, either adiabatic or thermostatted, $i\mathscr{L}_0$, which does not explicitly depend on time, and a perturbed part that is linear in the external field $\dot{\epsilon}(t)$, and is therefore explicitly time dependent, $iL_1(t)$:

$$i\mathscr{L}(t) = i\mathscr{L}_0 + iL_1(t),$$
$$i\mathscr{L}_0 = iL_0 - \xi\Sigma(p-p_0)\cdot\frac{\partial}{\partial p} + \dot{\xi}\frac{\partial}{\partial\xi} - g\xi$$
$$= iL_\xi - g\xi.$$

If $\xi\equiv 0$, then $i\mathscr{L}_0 = iL_\xi = iL_0$, the equilibrium (Newtonian) Liouvillean. If the function operated on by either of these Liouvilleans depends on (q,p) through H_0, then

$$i\mathscr{L}_0 = -2\xi K\frac{\partial}{\partial H_0} + \dot{\xi}\frac{\partial}{\partial\xi} - g\xi,$$
$$iL_1(t) = -J\dot{\epsilon}(t)\frac{\partial}{\partial H_0}.$$

One of many forms of thermostatting which have been proposed, at least for equilibrium systems, is that due to Nosé[7(a)] and reformulated along more physically meaningful lines by Hoover,[7(b)] where one starts with a quasicanonical distribution function and derives the $\dot{\xi}$ equation of motion from the requirement $i\mathscr{L}_0 f_0 = 0$. The equilibrium distribution function is postulated to be

$$f_0(\Gamma) = \frac{1}{Z(\beta,\tau)}\exp[-\beta H_0(q,p) - \tfrac{1}{2}g\tau^2\xi^2],$$

where $\beta = 1/kT$ (T is the temperature, k is the Boltzmann's constant), τ is a parameter (a thermostatic relaxation time), and

$$Z(\beta,\tau) = \int dq \int dp \int d\xi \times \exp[-\beta H_0(q,p) - \tfrac{1}{2} g\tau^2\xi^2].$$

Since $\partial f_0/\partial H_0 = -\beta f_0$ and $\partial f_0/\partial \xi = -g\tau^2\xi f_0$,

$$i\mathscr{L}_0 f_0 = 0 = 2\xi K\beta f_0 - g\tau^2\xi\dot{\xi} f_0 - g\xi f_0 = g\xi f_0\left(\frac{2K}{gkT} - 1 - \tau^2\dot{\xi}\right),$$

so that for ξ not always zero and $K_0 = \tfrac{1}{2} gkT$,

$$\dot{\xi} = \frac{1}{\tau^2}\left(\frac{K}{K_0} - 1\right)$$

is the Nosé–Hoover equation of motion for the thermostatting coefficient. Note that

$$i\mathscr{L}_0 J f_0 = \frac{\partial}{\partial \Gamma}\cdot(J f_0 \dot{\Gamma}) = J\frac{\partial}{\partial \Gamma}\cdot(f_0\dot{\Gamma}) + f_0\dot{\Gamma}\cdot\frac{\partial}{\partial \Gamma} J = J i\mathscr{L}_0 f_0 + f_0 iL_\xi J = f_0 iL_\xi J,$$

which implies that

$$U_0^+(t) J f_0 = e^{-it\mathscr{L}_0} J f_0 = [1 - t(i\mathscr{L}_0) + \tfrac{1}{2} t^2 (i\mathscr{L}_0)^2 - + \ldots] J f_0 = f_0[1 - t(iL_\xi) + \tfrac{1}{2} t^2 (iL_\xi)^2 - + \ldots] J = f_0 e^{-itL_\xi} J = f_0 U_0(-t) J = f_0 J_\xi(-t),$$

where $J_\xi(-t)$ is the result of propagating J backwards in time under equilibrium thermostatting. Finally, we have that

$$iL_1(t) f_0 = \beta J\dot{\epsilon}(t) f_0.$$

This result (as well as $i\mathscr{L}_0 f_0 = 0$) is identical to the adiabatic case, provided that f_0 is canonical in the adiabatic case. However, the equilibrium Newtonian equations of motion in the adiabatic case generate, via time averages, the microcanonical (isoenergetic) ensemble. This ergodic inconsistency, pointed out by Kubo,[10] is not a problem in the Nosé–Hoover thermostatting, where the equilibrium equations of motion generate, via time averages, the canonical ensemble.

Other forms of thermostatting can be included within our general framework, including Gaussian isokinetic (velocity scaling in the small time-step Δt limit) and isoenergetic ("enostatting"). Results are similar, in that ergodic consistency is obtained. Also, to relative order $1/N$ in both isokonetic and isoenergetic cases, the formal expression for the linear response $iL_1 f_0$ is the same as for Nosé–Hoover thermostatting.

[1]R. Kubo, J. Phys. Soc. Jpn. **12**, 570 (1957).
[2]W. M. Visscher, Phys. Rev. A **10**, 2461 (1974); see also R. C. Mjolsness and W. M. Visscher, Phys. Fluids **15**, 1854 (1972).
[3]See, for example, L. P. Kadanoff and P. C. Martin, Ann. Phys. **24**, 419 (1963).
[4]D. J. Evans and G. P. Morriss, Phys. Rev. A **30**, 1528 (1984), citing private communication from W. G. Hoover and A. J. C. Ladd; Dolls tensor dynamics is discussed in W. G. Hoover, D. J. Evans, R. B. Hickman, A. J. C. Ladd, W. T. Ashurst, and B. Moran, *ibid.* **22**, 1690 (1980).
[5]For a recent review of nonequilibrium molecular dynamics methods, see D. J. Evans and G. P. Morriss, Comp. Phys. Rep. **1**, 297 (1984).
[6]D. J. Evans and G. P. Morriss, Chem. Phys. **87**, 451 (1984).
[7]a)S. Nosé, J. Chem. Phys. **81**, 511 (1984); b) W. G. Hoover, Phys. Rev. A **31**, 1695 (1985).
[8]D. J. Evans and B. L. Holian, J. Chem. Phys. (to be published). (Details on the Nosé–Hoover thermostatting in nonequilibrium molecular dynamics are discussed therein.)
[9]N. G. Van Kampen, Phys. Norv. **5**, 279 (1971).
[10]R. Kubo, Int. J. Quantum. Chem. **16**, 25 (1982).

Notes to Reprint III.9

1. The methods developed in this paper have been used by Evans and Holian (J. Chem. Phys. 83 (1985) 4069) in the derivation of linear-response theory for the case of a system coupled to a Nosé–Hoover thermostat (Reprint IV.6).

2. The mathematics of time-ordered exponential operators is illustrated for the simple case of the driven harmonic oscillator by Holian (J. Chem. Phys. 84 (1986) 1762).

CHAPTER IV

Other Ensembles

The original, Metropolis version of the Monte Carlo method (Reprint I.1) was restricted to the simulation of systems at constant temperature, volume and particle number. In other words, it represented a scheme for sampling from a canonical (NVT) ensemble. By contrast, the natural ensemble for molecular dynamics calculations is the microcanonical one, in which the total energy, volume and particle number are fixed parameters. The choice of ensemble becomes irrelevant in the thermodynamic limit, but this is not necessarily true for the small system sizes used in computer simulations. Furthermore, there are many physical phenomena that are more conveniently discussed in frameworks other than those provided by either the canonical or microcanonical ensemble. In the problem of adsorption of a gas on a solid surface, for example, contact with the real world is most easily established by choosing the surface area and chemical potential as fixed quantities [1], while there are many physical processes that are commonly studied under conditions of constant pressure and temperature. In order to treat such situations, it would be useful to have available techniques of simulation appropriate to a wider range of ensembles than those discussed so far.

The extension of the Monte Carlo method to other ensembles is comparatively straightforward, since the importance-sampling technique can be used in any problem for which it is possible to construct the appropriate Markov chain. Wood [2], for example, has shown that the Metropolis scheme is easily adapted to the isobaric-isothermal (NPT) ensemble, and has used the method in a study of hard disks in two dimensions. The procedure described by Wood is computationally very efficient, but suitable only for hard-core systems. A scheme that is convenient for use with continuous potentials is described in Reprint IV.1, where it is applied in a calculation of excess thermodynamic properties of mixtures of Lennard–Jones fluids. The key idea in these and other extensions of the basic method is to sample not only the particle coordinates but also the variable that is conjugate to the new fixed parameter; in the isobaric case, the new variable to be sampled is the system volume V.

In the grand canonical ensemble, the chemical potential is a fixed quantity, and the Monte Carlo method involves sampling from the conjugate variable N. This approach was pioneered by Norman and Filinov [3], Adams [4], and

Rowley et al. [5]. A general description of simulations in the grand canonical ensemble and a discussion of the subtleties involved in constructing the correct Markov chain are given in Reprint IV.2, together with results obtained for a number of model electrolyte solutions. A limitation of the grand canonical method is the fact that it becomes very inefficient at high densities, because the probability of acceptance of a trial move in which the number of particles is changed is then very small. The difficulty here is similar to that encountered in the particle-insertion method discussed in Chapter II; other sampling techniques [6,7] have been devised in attempts to circumvent the problem, but progress in this direction has been slow. The great advantage of the grand canonical approach is the fact that the chemical potentials of all species in the system are known. The calculation of free energies is therefore a straightforward task.

In a conventional molecular dynamics simulation, the newtonian equations of motion of a system of N particles in a (periodic) volume V are integrated numerically. In the absence of external perturbations, the total energy E and total linear momentum $\boldsymbol{p}$ are conserved. Hence, if the system is ergodic, time averages along a phase-space trajectory are equivalent to averages in a microcanonical, constant-momentum (NVE; $\boldsymbol{p}$) ensemble. As we have already pointed out, such an ensemble is often not the most convenient one for the problem in hand. However, unlike the analogous developments of the Monte Carlo method, it is only recently that generalizations of molecular dynamics to other ensembles have emerged. The reason for this is simple. Since Newton's equations of motion lead naturally to the microcanonical ensemble, any extension to a different ensemble involves a degree of artificiality, and there is no well-defined way of deriving the modified equations of motion. Most of the progress that has occurred has its inspiration in a paper by Andersen (Reprint IV.3). Andersen's approach is based on the introduction of one or more degrees of freedom that are additional to the degrees of freedom of the particles of the physical system. In the specific case considered by Andersen, the extra "coordinate"is the volume V. This new dynamical variable is coupled to the particle coordinates because a change in V corresponds to a homogeneous scaling of all centre-of-mass positions. Associated with the extra "coordinate" is a new "momentum" and a new "mass". Newton's equations are then solved for the "extended system" consisting of the particles and the additional dynamical variable; the conserved quantity is no longer the sum of kinetic and potential energies of the particles but, instead, a quantity closely related to the total enthalpy H of the system. The resulting trajectories represent a simulation of a system at a constant pressure that appears as a parameter in the equations of motion. In the same article, a method is described for the dynamical simulation of a system at constant temperature as well as constant pressure (NPT ensemble). The method used to control the temperature is stochastic in nature, and the resulting constant-*NPT* dynamics is not strictly deterministic.

Soon after the appearance of Andersen's paper, Parrinello and Rahman [8] proposed a scheme that allows for fluctuations in shape as well as volume of the periodic cell. The Parrinello–Rahman method is ideally suited to the study of phase transitions in crystalline solids. Such transitions are virtually excluded in the usual approach because a periodic boundary condition tailored to fit one phase is in general incompatible with the crystal structure of the other. In a later article (Reprint IV.4), Parrinello and Rahman showed how their method could be further generalized to include the simulation of solids subject to an external stress.

The method used by Parrinello and Rahman is an extension of the *NPH*-ensemble approach of Andersen. As in the original work, there is a "mass" – in fact a "mass" tensor – associated with the dynamics of the unit cell. In general, of course, it would be more convenient to carry out the simulations at constant temperature rather than constant enthalpy. As an alternative to the stochastic scheme proposed by Andersen, Nosé [9] has developed a dynamical method for controlling the temperature of the system. A further dynamical variable is introduced, having the meaning of a time-scaling factor, through which the momenta of the particles are coupled to an external heat bath. A unified treatment of this and other extended-system methods is given in Reprint IV.5, which shows how the constant-pressure and constant-temperature approaches can be combined to give a purely dynamical scheme for carrying out simulations in the *NPT* ensemble. Hoover (Reprint IV.6) was later able to show that Nosé's equations of motion can be written in an equivalent form in which the new variables play the role of dynamical friction coefficients. The equations of motion remain reversible in time, but no time scaling is required and the interpretation of the equations is simpler than in Nosé's original formulation of the problem. Berendsen et al. [10] have proposed another method in which the equations of motion are similar to those used by Nosé and Hoover, without being reversible in time, but the connection between time averages and ensemble averages in this case has yet to be established.

The uses to which the new molecular dynamics methods can be put are well illustrated by the work described in Reprint IV.7, where an attempt is made to throw light on the nature of a recently discovered phase of solid helium. The simulations described in the paper show that a transformation occurs at high pressure from a face-centred cubic to a body-centred cubic structure. This suggests that the new phase is body centred cubic. However, a firm conclusion concerning the stability of the body-centred cubic phase observed in the simulations requires a computation of the free energies of all the phases concerned. The results of such a calculation are contained in Reprint IV.8.

They show that for the model system studied in Reprint IV.7, the body-centred cubic phase is indeed stable with respect to the face-centred cubic structure, but it is unstable with respect to the liquid. The message that emerges from these two papers is clear. When two or more phases are (meta)stable over the same range of temperature and pressure, the Parrinello–Rahman method must be supplemented by free-energy calculations of the type described in Chapter II before a thermodynamic phase transition can be unambiguously located.

Since it is now possible to carry out both Monte Carlo and molecular dynamics simulations in a variety of ensembles, the question arises as to which of the two approaches is best suited to a particular problem. If interest is focussed solely on static properties, there is no simple answer; if information on dynamic properties is also wanted, molecular dynamics is the only option. It must be emphasized, however, that the extended-system methods do not yield the true dynamics of the physical, many-particle system. A feature of all such methods is the appearance of an inertial factor for each additional degree of freedom that is introduced. If the "masses" are large, there is little change in the dynamics of the particles, but the rate of equilibration is slow; if they are small, there is inevitably a strong perturbation of the dynamics. Nonetheless, there is evidence to suggest that reliable results can be obtained for simple dynamic properties. For example, it has been found [11] that the reorientational motion of molecules in disordered crystals is not strongly dependent on the values of the extra "masses". On the other hand, it is widely believed that the values chosen for the "masses" can have an effect on the kinetics of any phase transition that may occur. For the computation of static properties, Monte Carlo and molecular dynamics are, in principle, equivalent, but little is known about their relative efficiencies. Sometimes it may even be advantageous to equilibrate by one method and to produce the results of interest by the others [12].

No account of computer simulation in different ensembles would be complete wihout some reference to the ensemble dependence of fluctuations in bulk properties. The paper by Lebowitz, Percus and Verlet (Reprint IV.9) discusses this important question and also examines the relevance of the theoretical expressions to computer "experiments" on small systems.

References

[1] D. Nicholson and N.G. Parsonage, Computer Simulation and the Statistical Mechanics of Adsorption (Academic Press, London, 1982).

[2] W.W. Wood, J. Chem. Phys. 48 (1968) 415.
W.W. Wood, in: Physics of Simple Liquids, eds H.N.V. Temperley, G.S. Rushbrooke and J.S. Rowlinson (North-Holland, Amsterdam, 1968).
W.W. Wood, J. Chem. Phys. 52 (1970) 729.

[3] G.E. Norman and V.S. Filinov, High Temp. Res. USSR 7 (1969) 216.

[4] D.J. Adams, Mol. Phys. 28 (1974) 1241.
D.J. Adams, Mol. Phys. 29 (1975) 307.
[5] L.A. Rowley, D. Nicholson and N.G. Parsonage, J. Comput. Phys. 17 (1975) 401.
L.A. Rowley, D. Nicholson and N.G. Parsonage, Mol. Phys. 31 (1976) 365; 389.
L.A. Rowley, D. Nicholson and N.G. Parsonage, J. Comput. Phys. 26 (1978) 66.
[6] M. Mezei, Mol. Phys. 40 (1980) 901.
[7] V. Vlachy and A.D.J. Haymet, J. Chem. Phys. 84 (1986) 5874.
[8] M. Parrinello and A. Rahman, Phys. Rev. Lett. 45 (1980) 1196.
[9] S. Nosé, Mol. Phys. 52 (1984) 255.
[10] H.J.C. Berendsen, J.P.M. Postma, W.F. van Gunsteren, A. Di Nola and J.R. Haan, J. Chem. Phys. 81 (1984) 3684.
[11] S. Nosé and M.L. Klein, J. Chem. Phys. 78 (1983) 6928.
[12] J.J. Erpenbeck and W.W. Wood, J. Stat. Phys. 24 (1981) 455.

MOLECULAR PHYSICS, 1972, VOL. 23, NO. 1, 41–58

NpT-ensemble Monte Carlo calculations for binary liquid mixtures

by I. R. McDONALD
Department of Chemistry, Royal Holloway College (University of London), Englefield Green, Surrey

(*Received* 8 *November* 1971)

A Monte Carlo method for the calculation of thermodynamic properties in the isothermal–isobaric ensemble is described. Application is made to the calculation of excess thermodynamic properties (enthalpy, volume and Gibbs free energy) of binary mixtures of Lennard–Jones 12 6 liquids. Comparison is made with the predictions of a number of theories of liquid mixtures ; the so-called van der Waals one-fluid model and the variational theory of Mansoori and Leland are both found to give excellent results. The accuracy attainable in estimates of the excess properties is discussed in terms of statistical fluctuations in various calculated quantities and the advantages and disadvantages of the method are examined in relation to calculations by the more familiar constant-volume method.

1. INTRODUCTION

Computer experiments on model systems [1] have in recent years provided much valuable information on the thermodynamic, structural and transport properties of classical dense fluids. The success of these methods rests primarily on the fact that a model containing a relatively small number of particles (usually several hundred) is in general found to be sufficient to simulate the behaviour of a macroscopic system. Two distinct techniques of computer simulation have been developed ; these are known as the method of molecular dynamics and the Monte Carlo method. In molecular dynamics the equations of motion of a system of interacting particles are solved and equilibrium properties are determined from time-averages taken over a sufficiently long time interval. The Monte Carlo procedure requires the generation of a series of configurations of the particles of the model in a way which ensures that the configurations are distributed in phase space according to some prescribed probability density. The mean value of any configurational property determined from a sufficiently large number of configurations provides an estimate of the ensemble-average value of that quantity ; the nature of the ensemble average depends upon the chosen probability density. These machine calculations provide what is essentially exact information on the consequences of a given intermolecular force law. Application has been made to hard spheres and hard disks, to particles interacting through a Lennard–Jones 12–6 potential function and other continuous potentials of interest in the study of simple fluids, and to systems of charged particles.

The major advantage of molecular dynamics over the Monte Carlo method is that it allows the study of time-dependent phenomena. On the other hand the

Monte Carlo method has a flexibility which gives it a special value in certain applications. In particular, as Wood [2] has pointed out, the method may in principle be adapted to the calculation of average quantities in any of the standard statistical mechanical ensembles. However, this possibility has not been exploited to any great extent and applications of the Monte Carlo method which have been described in the literature have for the most part been confined to calculations in the usual Gibbs petit-canonical, constant volume or NvT-ensemble. Some calculations for hard disks and hard spheres in the isothermal–isobaric or NpT-ensemble [2–4] and for a lattice gas in the grand canonical ensemble [5] have also been published. All other calculations, including those for smooth potentials, have been made in the NvT-ensemble, excepting only the preliminary report of the present investigation which was given some time ago [6] and some very recent work on the phase transitions of the 12–6 and coulombic systems [7].

The purpose of the work described here is to extend the Monte Carlo method to the calculation in the NpT-ensemble of equilibrium properties of systems of molecules interacting through the Lennard–Jones 12–6 potential function. Application is made to binary mixtures of 12–6 fluids in the liquid range of density and temperature. The equilibrium properties of the one-component 12–6 fluid have been very extensively studied both by molecular dynamics [8, 9] and by the conventional Monte Carlo NvT-method [10–13] and further calculations by the NpT-method would seem to be superfluous. In any case the NpT-method does not appear to have any marked advantages when applied to pure fluids. The problem of mixtures, however, is quite different. A preliminary account has appeared [14] of calculations for mixtures of 12–6 fluids by the NvT-method and results for mixtures of hard spheres were reported some years ago [15, 16]. Otherwise there has been little published work on the application of computer simulation to such systems. Furthermore, the NpT-ensemble is a natural choice for the study of liquid mixtures, particularly of the excess properties, because experimental data are recorded at effectively constant (usually near-zero) pressure and theories of mixtures are commonly formulated under the assumption of constant-pressure mixing. Data obtained by the NvT-method may be processed in such a way as to provide information on changes in thermodynamic properties on mixing at constant pressure, a possibility which has been well exploited by Singer and Singer [14], but the calculations are lengthy and represent an additional source of error. The NpT-method has disadvantages of its own but it does have the merit of yielding the required results in an appealingly direct manner.

Recent advances in the statistical thermodynamics of liquid mixtures [17–23] make this an appropriate time to report on Monte Carlo investigations of such systems. A number of theories for the calculation of excess thermodynamic properties have been proposed and it is desirable to test these not only against experimental data on real systems, in which case the comparison is confused by uncertainties in the intermolecular potentials, but also against the exact results obtained by computer simulation. A small number of the results tabulated below have already been used for this purpose by several authors [19–23]. It should be noted here that the method of molecular dynamics is less useful than the Monte Carlo method in the calculation of excess properties because the molecular dynamics ' experiment ' does not proceed under isothermal conditions and the time-averaged temperature cannot be specified in advance except within rather wide limits.

The purpose of the present paper is to describe the application of the NpT-method to the study of systems of particles interacting through the 12–6 potential. The computational problems which are involved and the accuracy which may be attained in the calculation of various thermodynamic properties are discussed and the advantages and disadvantages of the method are assessed in comparison with Monte Carlo calculations at constant volume. In order to illustrate the use of the NpT-method the excess thermodynamic properties of binary liquid mixtures of Ar, Kr, CH_4, N_2, O_2 and CO are computed and compared with the results of various theories. A systematic study of the effect on the thermodynamic properties and structure which results from changes in pressure, temperature and composition and in intermolecular potential parameters is now in progress and the results will be reported in a later publication.

2. The NpT-ensemble

Some important relations for the NpT-ensemble are recalled here for the sake of easy reference. Detailed accounts may be found in the book by Hill [24] and the review article by Wood [2]. The latter contains a discussion of the NvT- and NpT-ensembles with special reference to Monte Carlo calculations and the development given below is largely based on Wood's formulation of the problem.

The configurational Gibbs free energy of a system of N particles at a temperature T and pressure p may be written in the form

$$G(N, p, T) = -\beta^{-1} \ln \Delta \tag{1}$$

with

$$\Delta = \Lambda(2\pi m\beta/h^2)^{3N/2}(1/N!) \int_0^\infty dv \exp(-\beta p v) \int_v d\mathbf{r}^N \exp(-\beta\Phi(\mathbf{r}^N)), \tag{2}$$

where Λ is a multiplicative factor the form of which is of no concern here [2]; Φ is the total potential energy of a configuration denoted symbolically by $\mathbf{r}^N$; and the integral over the variable v is to be evaluated for a constant shape of the volume enclosing the particles.

The NpT-ensemble average of a function $f(\mathbf{r}^N, v)$ is given by

$$\langle f(\mathbf{r}^N, v)\rangle = \frac{\int_0^\infty dv \exp(-\beta p v) \int_v d\mathbf{r}^N f(\mathbf{r}^N, v) \exp(-\beta\Phi(\mathbf{r}^N))}{\int_0^\infty dv \exp(-\beta p v) \int_v d\mathbf{r}^N \exp(-\beta\Phi(\mathbf{r}^N))}. \tag{3}$$

In the Monte Carlo calculation the particles are confined to a cube of fluctuating edge L. This makes it convenient to introduce the scaled coordinates

$$\boldsymbol{\alpha}_i = L^{-1}\mathbf{r}_i, \tag{4}$$

so that the integrals over the particle coordinates in equation (3) become integrals over the unit cube ω. Equation (3) may then be written as

$$\langle f([L\boldsymbol{\alpha}]^N, v)\rangle = \frac{\int_0^\infty dv \exp(-\beta p v)v^N \int_\omega d\boldsymbol{\alpha}^N f([L\boldsymbol{\alpha}]^N, v) \exp(-\beta\Phi([L\boldsymbol{\alpha}]^N, L))}{\int_0^\infty dv \exp(-\beta p v)v^N \int_\omega d\boldsymbol{\alpha}^N \exp(-\beta\Phi([L\boldsymbol{\alpha}]^N, L))} \tag{5}$$

Note on p. 271

which represents an average in the $(3N+1)$-dimensional space of the variables $\{v, \boldsymbol{\alpha}_1, \ldots, \boldsymbol{\alpha}_N\}$ with a probability density proportional to the pseudo-Boltzmann weight factor

$$\exp(-\beta p v - \beta\Phi([L\boldsymbol{\alpha}]^N, L) + N \ln v). \tag{6}$$

The details of the Monte Carlo procedure designed to calculate averages such as (5) are as follows. Let the total potential energy of a given configuration of N particles within a cube of volume v' be Φ'. A trial configuration is generated according to the rules

$$\boldsymbol{\alpha}_i \rightarrow \boldsymbol{\alpha}_i + \lambda \mathbf{R}^\alpha, \tag{7}$$

$$L \rightarrow L + \mu R^L \tag{8}$$

where the particle i is chosen either cyclically or at random, the quantities R_x^α, R_y^α, R_z^α and R^L are chosen randomly and uniformly within the interval $(-1, +1)$, λ is a displacement parameter and μ is a volume change parameter. Let the total potential energy of the new configuration be Φ'' and let the new volume of the cube be v''. The quantity

$$W = (\Phi'' - \Phi') + p(v'' - v') - N\beta^{-1} \ln (v''/v') \tag{9}$$

is calculated and the new configuration is chosen to replace the old one with a probability P given by

$$\left.\begin{aligned} P &= 1, \quad \text{if} \quad W \leqslant 0; \\ P &= \exp(-\beta W), \quad \text{if} \quad W > 0. \end{aligned}\right\} \tag{10}$$

Repetition of this procedure gives rise to a chain of configurations which are distributed in phase space with a probability density proportional to the pseudo-Boltzmann weight factor (6). (In forming the chain a configuration is counted again if the trial configuration generated from it is rejected.) Estimates of the molar configurational internal energy U and molar volume V may therefore be obtained from the mean values of Φ and v calculated for a sufficiently long chain. The specific heat and compressibility may be obtained from the mean-square fluctuations in, respectively, $(\Phi + pv)$ and v by the application of well-known fluctuation theorems for the NpT-ensemble [24] but the results are subject to large errors. The mean value of the total intermolecular virial function Ψ may be used to determine the equilibrium pressure $\langle p \rangle$ from the virial theorem. If the calculations are to be consistent the quantity $\langle p \rangle$ should be equal to the value of the chain parameter p_{MC}; the requirement that $\langle p \rangle \approx p_{\mathrm{MC}}$ provides a useful check on the reliability of the computations.

3. Computational details and results

The calculations reported here have been made for samples of 108 particles. The usual boundary conditions are used in which the basic cube is surrounded by periodic images of itself, each containing the same number of particles in the same relative positions. All contributions to Φ and Ψ which arise from interactions between pairs of particles separated by a distance less than $\frac{1}{2}L$ (one-half of the current cube-length) are calculated explicitly and contributions from particles separated by a greater distance are obtained by integration over a uniform particle density. The maximum number of pair interactions which have to be evaluated explicitly in the calculation of Φ (or Ψ) is therefore $\frac{1}{2}N(N-1)$. If the total potential energy Φ' and virial Ψ' of a configuration are known, together with all individual pair terms, then the calculation of Φ'' and Ψ'' for a trial configuration may be made in two stages. Firstly, the interactions of the displaced particle with all other particles must be recalculated. This involves a maximum of $(N-1)$ interactions because the remaining $\frac{1}{2}(N-1)(N-2)$ terms are unaltered. Secondly, the changes resulting from the alteration in volume from v' to v'' must be determined. In general this can be done only by recalculating all $\frac{1}{2}N(N-1)$ interactions but for potentials of the Lennard–Jones type the same results may be obtained by a straightforward scaling procedure. In the particular case of the 12–6 potential the changes in Φ and Ψ which result solely from the change in volume are given by

$$\Phi''-\Phi' = \Phi'(2y-y^2-1)+\Psi'(y/6-y^2/6), \tag{11}$$

$$\Psi''-\Psi' = \Phi'(12y^2-12y)+\Psi'(2y^2-y-1), \tag{12}$$

where $y=(v'/v'')^2$. An equivalent simplification of the problem is found for the coulomb potential but calculations based, say, on a Kihara potential would be considerably more lengthy. Use of the scaled coordinates $\boldsymbol{\alpha}_i$ means that the numerical values of the coordinates do not change when the volume is altered.

Liquid	ϵ/k(K)	σ(A)
Ar	119·8	3·405
Kr	167·0	3·633
CH_4	152·0	3·74
N_2	101·3	3·612
O_2	119·8	3·36
CO	104·2	3·62

Table 1. Intermolecular potential parameters for pure liquids.

Results have been obtained for models chosen to simulate a number of simple liquid mixtures: Ar + Kr at 115·8 K, Ar + CH_4 and CO + CH_4 at 91·0 K, and Ar + N_2, Ar + CO and O_2 + N_2 at 83·8 K. These are systems for which a great deal of theoretical and experimental work has been reported [23, 25]. All calculations have been carried out at zero pressure, i.e. with $p_{MC}=0$. The system Ar + Kr has been studied over a range of composition but for the other systems only the case of the equimolar mixture has been considered. The interaction parameters used for the pure components, shown in table 1, are those deduced by Streett and Staveley [26] from experimental data on liquid densities. The cross-interaction parameters are calculated from the combining rules

$$\sigma_{12}=\tfrac{1}{2}(\sigma_{11}+\sigma_{22}) \quad \text{(Berthelot rule)}, \tag{13}$$

$$\epsilon_{12}=\xi(\epsilon_{11}\epsilon_{22})^{1/2}, \tag{14}$$

where the choice $\xi=1$ (the Lorentz rule) is made.

Note on p. 271

(*a*) Total properties

System	x_1	$10^{-6}t$	H (J mol^{-1})	V (cm^3 mol^{-1})	$\langle p \rangle$ (bar)
Ar + Kr (115·8 K)	0·25	2·5	−7733 ± 17	33·22 ± 0·06	−0·0 ± 0·6
	0·398	2·0	−7206 ± 23	32·96 ± 0·08	+1·6 ± 0·7
	0·5	2·0	−6841 ± 17	32·82 ± 0·08	+1·1 ± 0·7
	0·602	2·0	−6473 ± 20	32·76 ± 0·08	+0·2 ± 0·6
	0·75	2·0	−5929 ± 17	32·77 + 0·07	−0·0 ± 0·7
Ar + CH_4 (91·0 K)	0·5	3·0	−7180 ± 12	31·81 ± 0·04	+0·3 ± 0·4
CO + CH_4 (91·0 K)	0·5	2·0	−6508 ± 16	35·47 ± 0·06	−0·3 ± 0·6
Ar + N_2 (83·8 K)	0·5	4·0	−5419 ± 9	31·66 ± 0·05	−0·1 ± 0·4
Ar + CO (83·8 K)	0·5	2·0	−5530 ± 15	31·61 ± 0·05	+0·0 ± 0·6
O_2 + N_2 (83·8 K)	0·5	2·0	−5417 ± 15	31·08 ± 0·06	+0·4 ± 0·8

(*b*) Component properties

System	x_1		$\langle \beta\Phi_{ij}/N \rangle$			$\langle \beta\Psi_{ij}/N \rangle$	
		1–1	1–2	2–2	1–1	1–2	2–2
Ar + Kr (115·8 K)	0·25	−0·33 ± 0·01	−2·58 ± 0·02	−5·13 ± 0·02	−0·09 ± 0·03	+0·41 ± 0·07	+2·67 ± 0·06
	0·398	−0·82 ± 0·01	−3·35 ± 0·02	−3·31 ± 0·01	−0·06 ± 0·03	+0·98 ± 0·08	+2·07 ± 0·06
	0·5	−1·32 ± 0·01	−3·48 ± 0·02	−2·31 ± 0·01	−0·06 ± 0·04	+1·50 ± 0·07	+1·55 ± 0·04
	0·602	−1·90 ± 0·01	−3·37 ± 0·02	−1·45 ± 0·01	+0·18 ± 0·04	+1·71 ± 0·05	+1·11 ± 0·04
	0·75	−2·96 ± 0·01	−2·63 ± 0·02	−0·57 ± 0·01	+0·80 ± 0·05	+1·69 ± 0·06	+0·51 ± 0·03
Ar + CH (91·0 K)	0·5	−1·77 ± 0·01	−4·72 ± 0·02	−3·00 ± 0·01	+0·07 ± 0·07	+1·22 ± 0·08	+1·70 ± 0·07
CO + CH_4 (91·0 K)	0·5	−1·63 ± 0·01	−4·23 ± 0·02	−2·74 ± 0·02	−0·27 ± 0·07	+1·39 ± 0·10	+1·88 ± 0·07
Ar + N_2 (83·8 K)	0·5	−1·97 ± 0·01	−3·92 ± 0·02	−1·89 ± 0·01	+1·18 ± 0·03	+1·50 ± 0·05	+0·33 ± 0·04
Ar + CO (83·8 K)	0·5	−1·93 ± 0·01	−4·06 ± 0·02	−1·95 ± 0·01	+1·06 ± 0·06	+1·42 ± 0·08	+0·52 ± 0·06
O_2 + N_2 (83·8 K)	0·5	−1·91 ± 0·01	−3·95 ± 0·02	−1·92 ± 0·01	+1·06 ± 0·04	+1·58 ± 0·06	+0·36 ± 0·06

Table 2. Calculated thermodynamic properties of binary liquid mixtures.

The values obtained for the enthalpy H (which in this work is equal to U) and V and for the contributions made to Φ and Ψ by 1–1, 1–2 and 2–2 pair interactions are shown in table 2. Also listed there are the total numbers of configurations, or chain lengths, t, on which the estimates of the NpT-ensemble averages are based. The quoted values of t do not include those configurations which were first generated in order to bring the model system near to equilibrium; a minimum of 5×10^5 such equilibration steps were taken for each system studied. The quoted statistical errors are the standard errors in the mean determined from sub-averages over groups of 10^5 configurations, an accepted procedure in Monte Carlo calculations [2]; these sub-averages will be referred to in the discussion as 'local' averages. The statistical errors display a moderately consistent pattern, amounting to approximately 0·3 per cent or less in H and 0·2 per cent or less in V. The relative errors in the individual $\langle \Phi_{ij} \rangle$ $(i, j = 1, 2)$ are significantly greater than those in the total potential energy and very large errors are found in the $\langle \Psi_{ij} \rangle$. The error in $\langle \Psi \rangle$, however, is small in every case because the constraint that $\langle p \rangle \approx p_{\mathrm{MC}}$ is imposed; for $p_{\mathrm{MC}} = 0$ it follows that $\beta \langle \Psi \rangle / N \approx 3$. The agreement between $\langle p \rangle$ and p_{MC} is good and the same is found to be true for other (unpublished) calculations at higher pressures.

The statistical errors in H (or U) shown in table 2 are substantially greater than those which would arise in NvT-calculations of U based on chains of comparable length because internal energy is sensitive to small changes in volume; the same comment applies to the errors in the terms $\langle \Phi_{ij} \rangle$ which are important in the calculation of the excess Gibbs free energy (see below). If the variation of U and its component terms along an isotherm were required it would clearly be preferable to use the NvT-method. In the study of liquid mixtures, however, the data usually required are the thermodynamic properties of the system at a specified pressure. This information may be obtained from NvT-calculations by generating chains at more than one density along an isotherm and making the appropriate interpolations. Such a procedure is wasteful of computer time and it is more economical and almost equally accurate to use a generalization to the case of mixtures of a method of parameter extrapolation which has been successfully applied to the case of one-component fluids [13]. Application of this method, which is the basis of the work of Singer and Singer [14], allows the use of data obtained at a particular v, T point to determine thermodynamic properties at a neighbouring state point provided that the changes in volume and temperature are not too great. If parameter extrapolation is used, however, the accuracy of the values obtained for thermodynamic properties at the pressure of interest is limited by the magnitude of the error in the determination of pressure in the original NvT-calculation. Previous experience [13] has shown that for $t = 5 \times 10^5$ and densities and temperatures in the liquid range the statistical error in pressure in NvT-calculations is approximately 15–20 bar in the case of argon; the error is expected to decrease roughly as $t^{1/2}$ to reach a value of 5–10 bar at chain lengths typical of those employed in the present work. Taking for the compressibility a value of $2{\cdot}5 \times 10^{-4}$ bar^{-1} it follows that such an error in pressure gives rise to an error of approximately 0·1–0·3 per cent in an estimate obtained for the volume at a given pressure. Thus the statistical errors associated with the NpT- and NvT-estimates of the volume of the mixture are likely to be of similar magnitude; the same may be shown to be true of the error in internal energy.

A careful examination of the fluctuations in various average values as a function

Note on p. 271

of chain length is always a useful exercise in any Monte Carlo calculation and is particularly important in the present work because the quantities of interest, namely the excess thermodynamic properties, are themselves small and, in some cases, not much greater than the statistical error in the corresponding property of the mixture. Control charts showing the extent to which the 'local' averages vary during the length of a complete run prove to be especially valuable sources of information. In particular it is found that fluctuations in certain quantities are significantly correlated with each other. There is, for example, a strong positive correlation between Φ_{11} and Φ_{22} and between each of these quantities and v. On the other hand there is a negative (and weaker) correlation between Φ_{12} and v. In other words the fluctuations in v which characterize the NpT-chain are accompanied by an interchange of energy between like and unlike pair interactions. This has the effect of damping the fluctuations in Φ, thereby reducing the statistical error in the estimate obtained for H. There is also a strong serial correlation in the fluctuations in the component properties ; this suggests that the statistical errors in these quantities may not be very reliably estimated. The rate of convergence of the 'overall' averages as a function of the parameters λ and μ has not been studied systematically. In applications of the NvT-method a rule-of-thumb used by workers in the field is to assign to the displacement parameter a value such that approximately one-half of all trial configurations are rejected. This rule is less useful in the present work because different choices for λ and μ may give rise to the same rate of rejection. In the calculations reported here the values used were in the ranges $\lambda/L=0{\cdot}02$–$0{\cdot}03$ and $\mu/L=0{\cdot}01$–$0{\cdot}04$, leading to rejection rates of between 50 and 65 per cent. The convergence of the 'overall' averages does not appear to be sensitive to the choice of λ and μ within these ranges but the matter has not been investigated in sufficient detail to justify any more precise statement.

4. Excess thermodynamic properties

The molar excess enthalpy H^{E} and excess volume V^{E} may be obtained directly from the data given in table 2 if the properties of the pure components are known. In the early stages of this work it was proposed to carry out separate NpT-calculations for both pure components and for the mixture but it is clear that more accurate values of the excess properties may be obtained by calculating the properties of the pure liquids from curves fitted to the large quantity of Monte Carlo data now available for the one-component 12–6 fluid and this approach is the one adopted here. The equations used for the zero-pressure internal energy $U(p=0)$ and volume $V(p=0)$ are [27] :

$$U(p=0)/N_{\mathrm{A}}\epsilon=-8{\cdot}69614+3{\cdot}04195(kT/\epsilon)+0{\cdot}785383(kT/\epsilon)^2, \tag{15}$$

$$N_{\mathrm{A}}\sigma^3/V(p=0)=1{\cdot}06804-0{\cdot}164783(kT/\epsilon)-0{\cdot}206539(kT/\epsilon)^2, \tag{16}$$

where N_{A} is the Avogadro number.

The calculation of the molar excess Gibbs free energy G^{E} is made in a less direct manner. Adapting the procedure used by Singer and Singer [14] to the case of the NpT-ensemble G^{E} is calculated as the sum

$$G^{\mathrm{E}}=\Delta G^{\mathrm{I}}+\Delta G^{\mathrm{II}}+\Delta G^{\mathrm{III}}. \tag{17}$$

The superscripts I, II and III in equation (17) refer to the successive steps in the following process: firstly, the values ϵ_{ii}, σ_{ii} $(i=1, 2)$ for appropriate amounts of the two components are changed to the common values ϵ_{ref}, σ_{ref} which are conveniently set equal to ϵ_{12}, σ_{12}; secondly, the two identical liquids resulting from step I are mixed to form a reference liquid; and, thirdly, the 1–1, 1–2 and 2–2 interaction parameters are changed from ϵ_{ref}, σ_{ref} to the required values ϵ_{11}, ϵ_{12}, ϵ_{22}, σ_{11}, σ_{12}, σ_{22}. The quantity ΔG^{I} may be calculated from a knowledge of the Gibbs free energy of the pure components and the reference liquid; step II is an ideal mixing process and therefore ΔG^{II} is zero; and ΔG^{III} may be evaluated from the equation

$$\Delta G^{III}=\int_{\mathbf{X}_{ref}}^{\mathbf{X}_{mix}} d\mathbf{X}\,(\partial G(N, p, T; \mathbf{X})/\partial \mathbf{X}) \tag{18}$$

where $\mathbf{X}_{mix}$, $\mathbf{X}_{ref}$ are used to denote the arrays of interaction parameters in the final mixture and the reference liquid, respectively, the latter being regarded as a mixture of identical components. Thus

$$\mathbf{X}_{ref}\equiv(\epsilon_{ref}, \sigma_{ref};\ \epsilon_{ref}, \sigma_{ref};\ \epsilon_{ref}, \sigma_{ref}), \tag{19}$$

$$\mathbf{X}_{mix}\equiv(\epsilon_{11}, \sigma_{11};\ \epsilon_{12}, \sigma_{12};\ \epsilon_{22}, \sigma_{22}). \tag{20}$$

From equations (1), (2) and (3) it follows that

$$\left.\begin{aligned}\partial G(N, p, T; \mathbf{X})/\partial \mathbf{X} &= -(\beta\Delta)^{-1}(\partial\Delta/\partial\mathbf{X})\\ &= \langle\partial\Phi(\mathbf{r}^N; \mathbf{X})/\partial\mathbf{X}\rangle\end{aligned}\right\} \tag{21}$$

and therefore

$$\partial G(N, p, T;\ \epsilon_{ij})/\partial\epsilon_{ij}=\langle\Phi_{ij}\rangle/\epsilon_{ij}, \tag{22}$$

$$\partial G(N, p, T;\ \sigma_{ij})/\partial\sigma_{ij}=-\langle\Psi_{ij}\rangle/\sigma_{ij}. \tag{23}$$

The quantities $\langle\Phi_{ij}\rangle/\epsilon_{ij}$ and $\langle\Psi_{ij}\rangle/\sigma_{ij}$ are expected to change only slowly with changes in ϵ_{ij} and σ_{ij}; over small ranges of the interaction parameters they may be assumed to vary linearly with, respectively, ϵ_{ij} and σ_{ij}. If this approximation is used over the entire range between $\mathbf{X}_{ref}$ and $\mathbf{X}_{mix}$ then, in the special case when $\epsilon_{ref}=\epsilon_{12}$ and $\sigma_{ref}=\sigma_{12}$, equation (18) becomes

$$\begin{aligned}\Delta G^{III}=(N_A/2N)\{&[\langle\Phi_{11}\rangle/\epsilon_{11}+x_1{}^2\langle\Phi_{ref}\rangle/\epsilon_{12}](\epsilon_{11}-\epsilon_{12})\\ &+[\langle\Phi_{22}\rangle/\epsilon_{22}+x_2{}^2\langle\Phi_{ref}\rangle/\epsilon_{12}](\epsilon_{22}-\epsilon_{12})\\ &-[\langle\Psi_{11}\rangle/\sigma_{11}+x_1{}^2\langle\Psi_{ref}\rangle/\sigma_{12}](\sigma_{11}-\sigma_{12})\\ &-[\langle\Psi_{22}\rangle/\sigma_{22}+x_2{}^2\langle\Psi_{ref}\rangle/\sigma_{12}](\sigma_{22}-\sigma_{12})\},\end{aligned} \tag{24}$$

where $\langle\Phi_{ref}\rangle$, $\langle\Psi_{ref}\rangle$ are the equilibrium potential energy and virial of the N molecules of the reference liquid at the temperature and pressure of interest. Trial calculations for the equimolar mixture Ar+Kr based on separate Monte Carlo chains for parameter values intermediate between $\mathbf{X}_{ref}$ and $\mathbf{X}_{mix}$ show that the error introduced by the use of the linear approximation over the whole range is negligible. The evaluation of G^E also requires a knowledge of the zero-pressure

Note on p. 271

Gibbs free energy $G(p=0)$ of the pure liquids (including the reference system) and for this the following equation based on Monte Carlo data [27] is used :

$$G(p=0)/N_A kT = -8{\cdot}69614(\epsilon/kT) - 0{\cdot}304195 \ln (kT/\epsilon) - 0{\cdot}785383(kT/\epsilon) - 3 \ln \sigma + C, \quad (25)$$

where C is a constant.

(a) H^E (J mol^{-1})

System	MC	Expt	APM	vdW1	vdW2	Pert	Var	PY
$Ar+Kr$ (115·8 K)	-29 ± 17		$+162$	-30	$+18$	-49	-31	-49
$Ar+CH_4$ (91·0 K)	-60 ± 12	$+103$	$+223$	-55	-11	-76	-34	
$CO+CH_4$ (91·0 K)	$+15\pm12$	$+105$	$+96$	$+26$	$+52$	$+18$	$+30$	
$Ar+N_2$ (83·8 K)	$+16\pm9$	$+51$	$+78$	$+43$	$+28$	$+25$	$+42$	
$Ar+CO$ (83·8 K)	$+37\pm15$		$+80$	$+35$	$+22$	$+18$	$+37$	
O_2+N_2 (83·8 K)	$+39\pm15$		$+118$	$+52$	$+33$	$+25$		

(b) V^E (cm^3 mol^{-1})

System	MC	Expt	APM	vdW1	vdW2	Pert	Var	PY
$Ar+Kr$ (115·8 K)	$-0{\cdot}69\pm0{\cdot}06$	$-0{\cdot}52$	$+0{\cdot}09$	$-0{\cdot}68$	$-0{\cdot}47$	$-0{\cdot}73$	$-0{\cdot}73$	$-0{\cdot}78$
$Ar+CH_4$ (91·0 K)	$-0{\cdot}22\pm0{\cdot}04$	$+0{\cdot}17$	$+0{\cdot}61$	$-0{\cdot}23$	$-0{\cdot}17$	$-0{\cdot}36$	$-0{\cdot}14$	
$CO+CH_4$ (91·0 K)	$-0{\cdot}76\pm0{\cdot}06$	$-0{\cdot}32$	$-0{\cdot}30$	$-0{\cdot}71$	$-0{\cdot}50$	$-0{\cdot}69$	$-0{\cdot}75$	
$Ar+N_2$ (83·8 K)	$-0{\cdot}25\pm0{\cdot}05$	$-0{\cdot}18$	$-0{\cdot}02$	$-0{\cdot}25$	$-0{\cdot}20$	$-0{\cdot}30$	$-0{\cdot}26$	
$Ar+CO$ (83·8 K)	$-0{\cdot}17+0{\cdot}05$	$+0{\cdot}10$	$+0{\cdot}05$	$-0{\cdot}19$	$-0{\cdot}16$	$-0{\cdot}25$	$-0{\cdot}17$	
O_2+N_2 (83·8 K)	$-0{\cdot}28\pm0{\cdot}06$	$-0{\cdot}31$	$+0{\cdot}08$	$-0{\cdot}28$	$-0{\cdot}21$	$-0{\cdot}35$		

(c) G^E (J mol^{-1})

System	MC	Expt	APM	vdW1	vdW2	Pert	Var	PY
$Ar+Kr$ (115·8 K)	$+46\pm7$	$+84$	$+139$	$+46$	$+61$	$+33$	$+47$	$+37$
$Ar+CH_4$ (91·0 K)	-14 ± 6	$+74$	$+159$	-17	$+9$	-28	-12	
$CO+CH_4$ (91·0 K)	$+77\pm7$	$+115$	$+111$	$+83$	$+84$	$+67$	$+76$	
$Ar+N_2$ (83·8 K)	$+35\pm5$	$+34$	$+57$	$+39$	$+27$	$+29$	$+42$	
$Ar+CO$ (83·8 K)	$+26\pm5$	$+57$	$+55$	$+29$	$+20$	$+21$	$+28$	
O_2+N_2 (83·8 K)	$+38\pm5$	$+39$	$+80$	$+43$	$+29$	$+30$		

Table 3. Excess thermodynamic properties of binary liquid mixtures ($x_1=x_2=\frac{1}{2}$).

The calculated values of H^E, V^E and G^E for each equimolar system studied are given in table 3 ; the composition-dependence of the excess properties of the system $Ar+Kr$ is shown in figures 1 to 3. The quoted statistical errors are those arising from statistical fluctuations in the calculations for the mixtures and do not include possible errors resulting from the use of equations (15), (16) and (25). The errors in H^E and V^E are therefore the same as those in H and V which are listed in table 2. (The comparison with the predictions of various analytical theories which is made in table 3 and, in particular, in figure 1 suggests that the tabulated errors in H^E may be too large.) The small values listed for the errors in G^E call for some comment because G^E is calculated, via equation (24), from the individual contributions to $\langle\Phi\rangle$ and $\langle\Psi\rangle$ for which the statistical errors, particularly in the $\langle\Psi_{ij}\rangle$, are known to be large. Numerically the situation is most easily understood by considering a specific (but typical) example. For the case of $CO+CH_4$ the contributions to ΔG^{III} made by successive terms on the right-hand side of equation (24) are, respectively, $+265$, -344, $+3$ and -16 J mol^{-1}. (ΔG^I for this system is $+169$ J mol^{-1}.) Two facts should be noted. Firstly, the terms involving $\langle\Psi_{ij}\rangle$ are small and of opposite sign ; thus the large errors associated with these quantities do not have a serious effect on the accuracy of the estimate obtained for G^E. As these terms are also the ones involving the 12–6 parameters σ_{11} and σ_{22} it is clearly implied that mixtures of molecules differing only in size will have very small values of G^E. Secondly, the terms involving $\langle\Phi_{ij}\rangle$, though of considerably greater magnitude than those involving $\langle\Psi_{ij}\rangle$, are also of opposite sign and the statistical error in the sum is greatly reduced by the strong positive correlation between fluctuations in Φ_{11} and Φ_{22}. Thus the relative errors in the individual terms in equation (24) are large but the absolute error in ΔG^{III} itself is very much smaller.

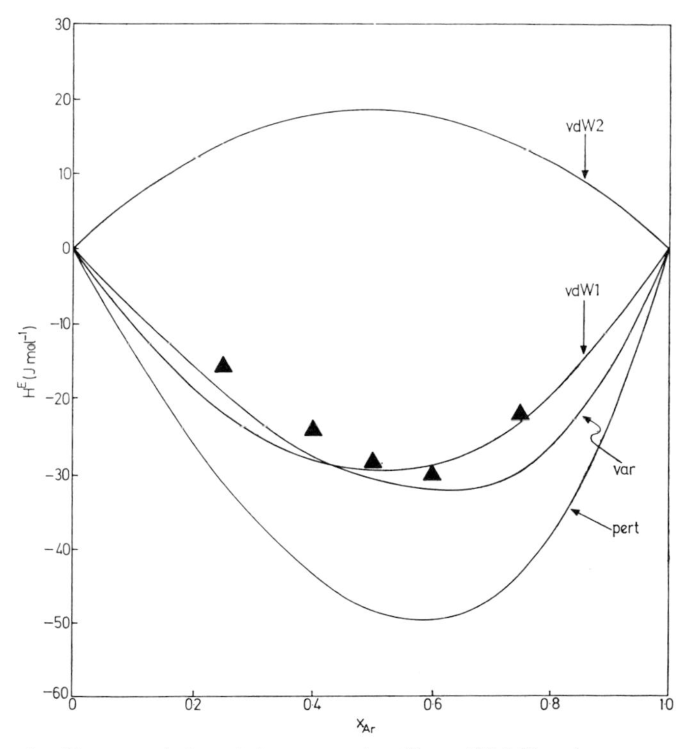

Figure 1. Excess enthalpy of the system Ar + Kr at 115·8 K and zero pressure. The points represent the Monte Carlo results and the curves give the predictions of various theories.

Note on p. 271

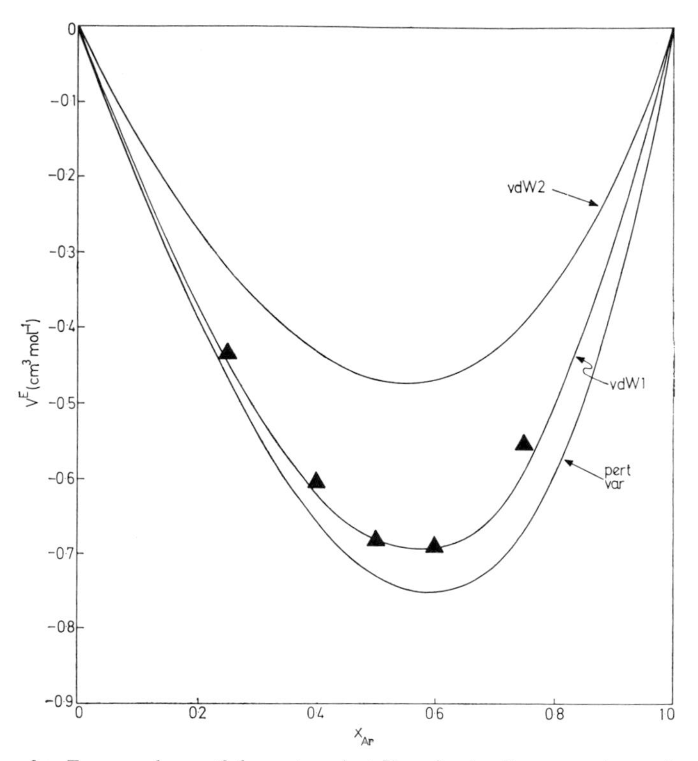

Figure 2. Excess volume of the system Ar + Kr ; for details see caption to figure 1.

Also shown in table 3 and figures 1 to 3 are values of the excess properties predicted on the basis of a number of theories of mixtures. (Note that in certain cases the results of different theories are indistinguishable on the scale to which the graphs are drawn.) The theories considered here fall into two groups. The first comprises the so-called *n*-fluid (corresponding-states) theories in which the properties of a mixture of 12–6 components are taken to be those of either a single hypothetical 12–6 substance with suitably averaged parameters $\tilde{\epsilon}$ and $\tilde{\sigma}$ (one-fluid theories) or of an ideal mixture of such substances. Application of theories of this type requires a knowledge of the reduced thermodynamic properties of the 12–6 system as a function of reduced temperature and pressure. The examples considered in this paper are the two-fluid version of the Average Potential Model (APM) of Prigogine and his collaborators [25, 28] (i.e. the ‘ refined version II ’ [25]) and the one- and two-fluid versions of the more recently developed van der Waals model (vdW1, vdW2) of Leland *et al.* [17, 18]. Numerical results are obtained with the help of equations (15), (16) and (25). In theories of the second type, which for convenience are referred to here as *ab initio* theories, separate computations must be made of the thermodynamic properties of each of the pure components and of the mixture. The *ab initio* theories discussed here are the perturbation approach (pert) of Leonard *et al.* [19], the variational calculation (var) of Mansoori and Leland [20] and the Percus–Yevick theory (PY) [22]. Results obtained by means of the variational theory are taken from the work of Mansoori [21] with corrections made to take account of small differences in the 12–6 parameters.

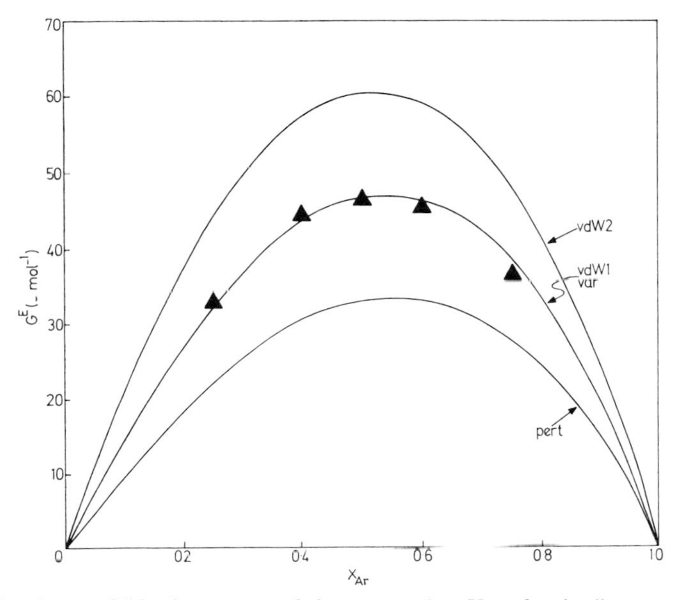

Figure 3. Excess Gibbs free energy of the system Ar + Kr; for details see caption to figure 1.

The comparison made in table 3 and figures 1 to 3 shows that there is excellent agreement between the predictions of both the one-fluid version of the van der Waals model and the variational theory and the results of the Monte Carlo calculations. The excess properties obtained from these two theories are everywhere in agreement with the Monte Carlo results within, or close to, the limits of statistical error in the latter. The results given by perturbation theory are slightly inferior; the values obtained for both H^E and G^E appear to be systematically too low but V^E is accurately predicted. The Percus–Yevick theory also gives good results in the one case studied but both the Average Potential Model and, perhaps surprisingly, the two-fluid version of the van der Waals model are clearly unsatisfactory as theories of mixtures.

The three *ab initio* theories are found to give very much better results for the properties of *mixing* than for the properties of the mixtures themselves. The variations with composition of the enthalpy and volume of the mixture Ar + Kr as given by the perturbation and variational theories are compared with the corresponding Monte Carlo data in figures 4 and 5. Inspection of these graphs makes it clear that the theories give rise to systematic errors, particularly in the calculation of volume, but these errors largely cancel in the case of the excess properties. It is interesting that the slight superiority of the variational theory in

Note on p. 271

the calculation of excess properties is achieved in spite of the fact that the perturbation results for the total properties of the mixtures contain significantly smaller errors. In view of the discrepancies between the Monte Carlo calculations and the perturbation results the good agreement between the latter and the experimental values of the volume of the mixture (figure 5) must be regarded as fortuitous. Improved agreement between the machine calculations and the experimental data could be obtained by making different choices for the 12–6 parameters for Ar and Kr.

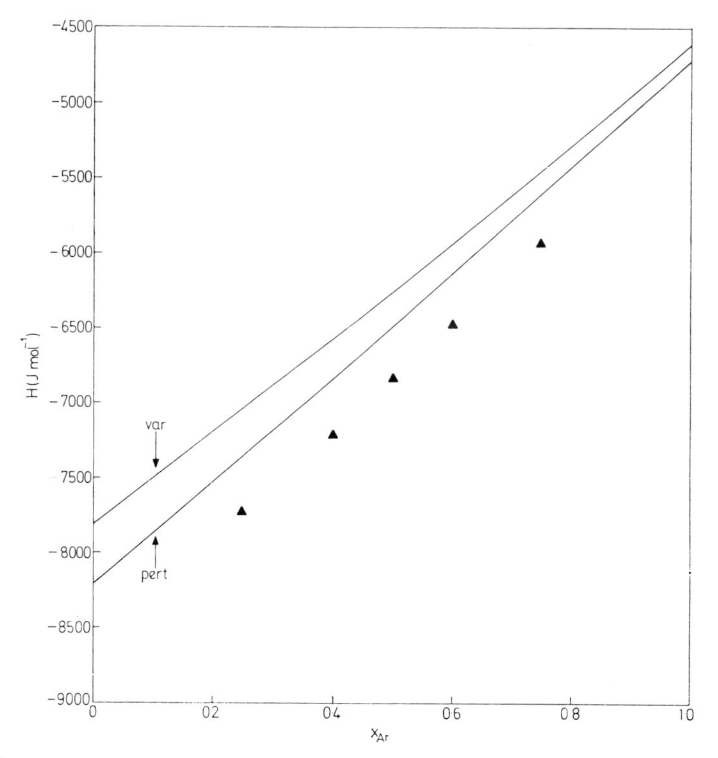

Figure 4. Enthalpy of the system Ar + Kr ; for details see caption to figure 1.

The results of the Monte Carlo calculations are compared with the available experimental data [23, 29] on excess properties in table 3 (for equimolar mixtures) and figures 6 and 7 (for the system Ar + Kr). Agreement is poor but may be substantially improved by treating the quantity ξ in equation (14) as an adjustable parameter. The values which must be assigned to ξ in order to bring the calculated G^E into agreement with the experimental results are easily determined from the Monte Carlo data on $\langle\Phi_{12}\rangle$ by making use of equation (22). The adjusted values of ξ (denoted by the symbol ξ_{expt}) which are obtained in this way are listed in table 4. The resulting changes in H^E and V^E are most easily determined by applying the van der Waals one-fluid theory ; the calculated changes may then

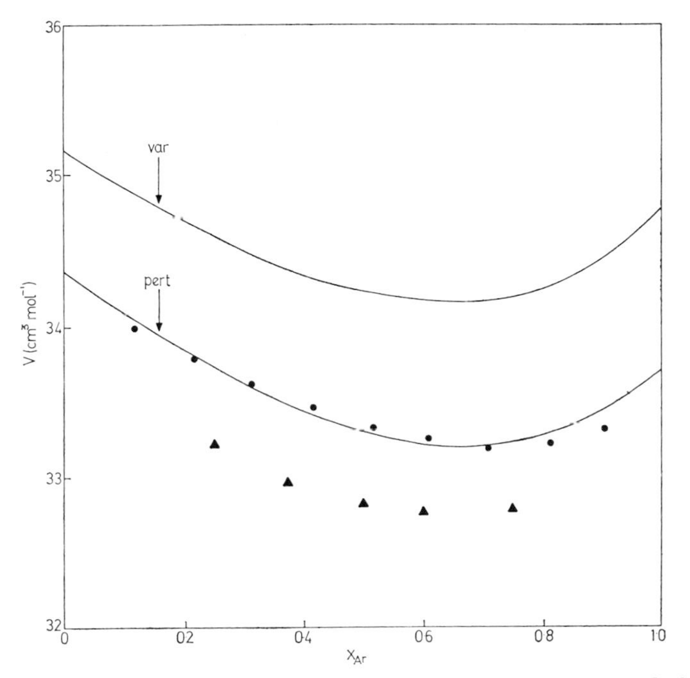

Figure 5. Volume of the system Ar + Kr ; the triangles represent the Monte Carlo results and the circles give the experimental values. For other details see caption to figure 1.

System	ξ_{expt}	H^E (J mol^{-1}) MC	H^E (J mol^{-1}) Expt	V^E (cm^3 mol^{-1}) MC	V^E (cm^3 mol^{-1}) Expt
Ar + Kr (115·8 K)	0·989	+29		−0·60	−0·52
Ar + CH_4 (91·0 K)	0·975	+57	+103	−0·11	+0·17
CO + CH_4 (91·0 K)	0·988	+70	+105	−0·68	−0·32
Ar + N_2 (83·8 K)	1·001	+34	+51	−0·25	−0·18
Ar + CO (83·8 K)	0·989	+79		−0·10	+0·10
O_2 + N_2 (83·8 K)	0·999	+42		−0·28	−0·31

Table 4. Comparison of experimental values of excess thermodynamic properties with values calculated for the case $\xi = \xi_{expt}$ $(x_1 = x_2 = \frac{1}{2})$.

be added to the Monte Carlo estimates of the excess properties calculated for the case $\xi = 1$ in order to obtain values of H^E and V^E corresponding to $\xi = \xi_{expt}$. The results are displayed in table 4 and figures 6 and 7. (Note that for the system Ar + Kr the value obtained for ξ_{expt} for the equimolar mixture is used throughout the entire range of composition.) In general there is a significant improvement

Note on p. 271

between calculated and experimental values. Further improvement could be obtained by varying the arithmetic-mean rule for σ_{12} but it is questionable whether any significance could be attached to the results of such manipulations. However, the conclusions reached here about the merits of the geometric-mean rule for ϵ_{12} are in general agreement with the results of other recent work on this topic [17–19]. In particular it is found that ξ_{expt} is less than unity for all systems except $Ar+N_2$; even in this case it is found that ξ_{expt} exceeds unity only by a very small amount which could be accounted for by the combined errors in the calculations and the experimental results. However, the values of ξ_{expt} obtained here are in all cases greater than those recently deduced on the basis of perturbation theory [19]; this arises from the fact that there are systematic differences between the Monte Carlo calculations and the perturbation results for G^{E}.

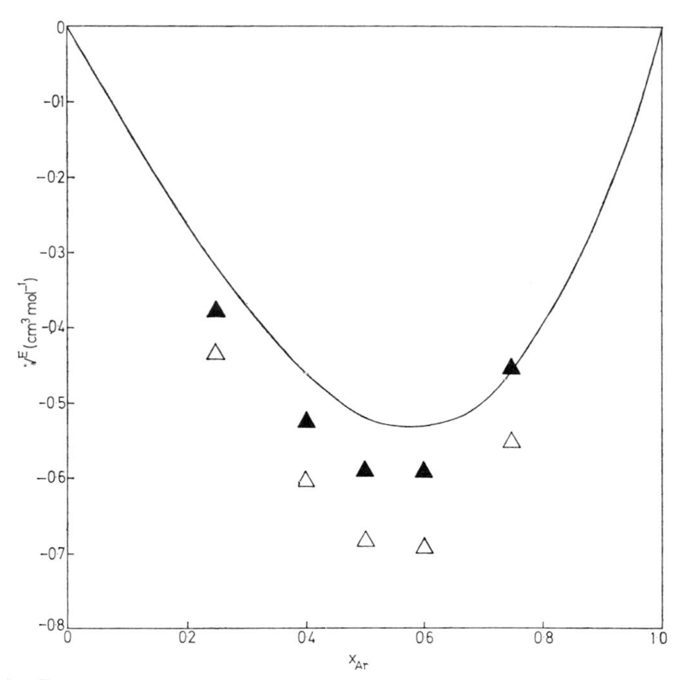

Figure 6. Excess volume of the system Ar + Kr. The open triangles give the results of Monte Carlo calculations for the case $\xi = 1$; the filled triangles are results for the case $\xi = \xi_{\text{expt}}$ calculated in the manner described in the text. The curve shows the experimental values. For other details see caption to figure 1.

5. Conclusions

The work described in this paper provides some indication of both the scope and the limitations of the NpT-method in its application to liquid mixtures. The most serious drawback to its use is one likely to be common to any study of mixtures by computer simulation, namely that the method makes a very heavy demand upon computing time. The NpT- and NvT-methods yield data of approximately equal statistical reliability for chains of equal length but the NpT-calculations are slightly more complicated and the total running time is therefore some 10 per cent greater. However the advantage held by the NvT-method cannot always be fully realized. Parameter extrapolation leads to large errors if the change in volume exceeds approximately 2 per cent and therefore the density used in the NvT-calculation must be chosen so as to ensure that the equilibrium pressure is not too far from the required value ; this requires some prior knowledge of the properties of the mixture which, in general, can be obtained only from additional Monte Carlo runs.

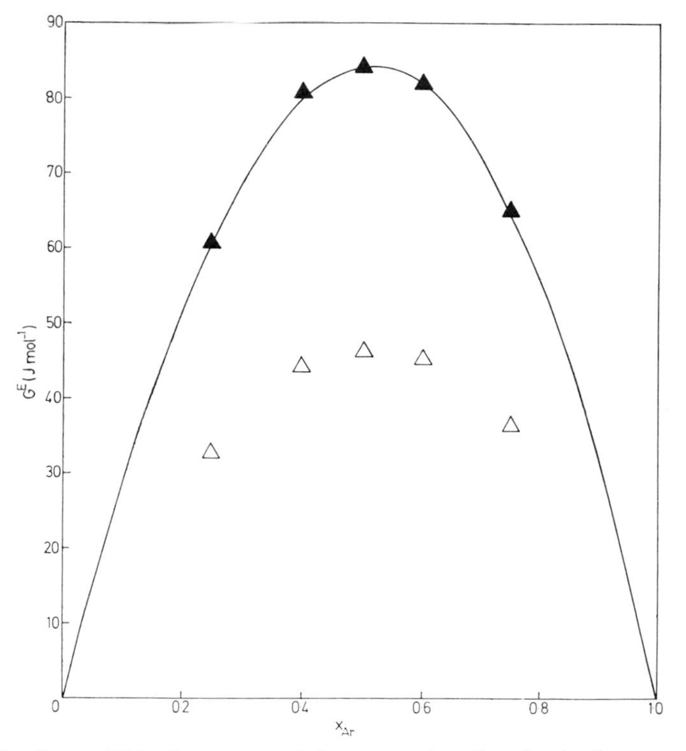

Figure 7. Excess Gibbs free energy of the system Ar + Kr ; for details see caption to figure 6.

For the specific systems studied here the main feature which emerges is the excellence of the predictions made by both the variational theory and the van der Waals one-fluid model. The magnitude of the statistical errors in the Monte Carlo calculations are such as to preclude at present any firm statement concerning the relative merits of these two theories ; it is hoped that this will become possible when the more systematic investigation referred to in the introduction is complete. Perturbation theory (in the particular form considered here) gives less accurate values for the excess properties but is more successful than the variational approach

Note on p. 271

in reproducing the thermodynamic properties of the mixture itself. Comparison with experimental data leads to the now familiar conclusion that in general the geometric-mean rule overestimates the energy cross-interaction parameter by some 1–2 per cent.

I am grateful to Dr. D. Henderson and Dr. G. Ali Mansoori for providing me with their theoretical results prior to publication, to Dr. K. Singer for helpful discussions about results obtained by the NvT-method, to the Institute of Computer Science of the University of London for a very generous allocation of computer time and to the Science Research Council for financial support.

References

[1] McDonald, I. R., and Singer, K., 1970, *Q. Rev. chem. Soc.*, **24**, 38.
[2] Wood, W. W., 1968, *Physics of Simple Liquids* edited by H. N. V. Temperley, J. S. Rowlinson, and G. S. Rushbrooke (North-Holland Publishing Co.), Chap. 5.
[3] Wood, W. W., 1968, *J. chem. Phys.*, **48**, 415.
[4] Wood, W. W., 1970, *J. chem. Phys.*, **52**, 729.
[5] Chesnut, D. A., and Salsburg, Z. W., 1963, *J. chem. Phys.*, **38**, 2861.
[6] McDonald, I. R., 1969, *Chem. Phys. Lett.*, **3**, 241.
[7] Vorontsov, P. N., Elyashevich, A. M., Morgenshtern, L. A., and Chasovskikh, V. P., 1970, *High Temp. Rs.*, **8**, 275.
[8] Verlet, L., 1967, *Phys. Rev.*, **159**, 98.
[9] Verlet, L., 1968, *Phys. Rev.*, **165**, 201.
[10] Wood, W. W., and Parker, F. R., 1957, *J. chem. Phys.*, **27**, 720.
[11] McDonald, I. R., and Singer, K., 1967, *Disc. Faraday Soc.*, **43**, 40.
[12] McDonald, I. R., and Singer, K., 1967, *J. chem. Phys.*, **47**, 4766.
[13] McDonald, I. R., and Singer, K., 1969, *J. chem. Phys.*, **50**, 2308.
[14] Singer, J. V. L., and Singer, K., 1970, *Molec. Phys.*, **19**, 279.
[15] Smith, E. B., and Lea, K. R., 1963, *Trans. Faraday Soc.*, **59**, 1535.
[16] Rotenberg, A., 1965, *J. chem. Phys.*, **43**, 4377.
[17] Leland, T. W., Rowlinson, J. S., and Sather, G. A., 1968, *Trans. Faraday Soc.*, **64**, 1447.
[18] Leland, T. W., Rowlinson, J. S., Sather, G. A., and Watson, I. D., 1969, *Trans. Faraday Soc.*, **65**, 2034.
[19] Leonard, P. J., Henderson, D., and Barker, J. A., 1970, *Trans. Faraday Soc.*, **66**, 2439.
[20] Mansoori, G. A., and Leland, T. W., 1970, *J. chem. Phys.*, **53**, 1931.
[21] Mansoori, G. A. (to be published).
[22] Grundke, E. W., Henderson, D., and Murphy, R. D., 1971, *Can. J. Phys.*, **49**, 1593.
[23] Henderson, D., and Leonard, P. J., 1971, *Physical Chemistry—An Advanced Treatise* edited by H. Eyring, D. Henderson, and W. Jost, (Academic Press), Vol. 8, Chap. 7.
[24] Hill, T. L., 1956, *Statistical Mechanics* (McGraw-Hill Book Co., Inc.), Chaps. 3 and 4.
[25] Bellemans, A., Mathot, V., and Simon, M., 1967, *Adv. chem. Phys.*, **11**, 117.
[26] Streett, W. B., and Staveley, L. A. K., 1967, *J. chem. Phys.*, **47**, 2449.
[27] McDonald, I. R., and Singer, K., 1972, *Molec. Phys.*, **23**, 29.
[28] Prigogine, I., 1957, *The Molecular Theory of Solutions* (North-Holland Publishing Co.).
[29] Davies, R. H., Duncan, A. G., Saville, G., and Staveley, L. A. K., 1967, *Trans. Faraday Soc.*, **63**, 855.

Note to Reprint IV.1

For more recent tests of theories of simple liquid mixtures, see (Hoheisel et al., Mol. Phys. 49 (1983) 159, Shing and Gubbins, Mol. Phys. 49 (1983) 1121, and Gubbins et al., J. Phys. Chem. 87 (1983) 4573). Note that the coefficient of the second term on the right-hand side of eq. (25) should be -3.04195 (cf. eq. (15)).

Primitive model electrolytes. I. Grand canonical Monte Carlo computations

John P. Valleau and L. Kenneth Cohen

Lash Miller Chemical Laboratories, University of Toronto, Toronto, Ontario, Canada M5S 1A1

(Received 20 December 1979; accepted 21 February 1980)

Monte Carlo calculations in the grand canonical ensemble are described for coulombic systems, and carried out for 1:1, 2:2, 2:1, and 3:1 aqueous electrolytes in the primitive model with equal ion sizes. Energies and activity coefficients are obtained, and the scope and reliability of the method is discussed.

I. INTRODUCTION

Once upon a time this laboratory published[1,2] Monte Carlo results for the restricted primitive model of 1 : 1 aqueous electrolytes. The results were useful in evaluating various theoretical treatments of the electrolyte problem. At the same time similar Monte Carlo (MC) investigations were carried out for the 3 : 1 and 2 : 2 cases.[3] These data are of still greater interest, but they were not published: many of them are reported in later papers of this series.

We were unwilling to publish those results earlier because we were not confident of the accuracy of the osmotic coefficients, and hence of the free energies, that were obtained. The experiments were carried out in the canonical ensemble, as is most Monte Carlo work. Like the pressure in uncharged systems, the osmotic coefficient requires the values of the pair correlation functions at contact of the particles, and these are obtained by extrapolation of the pair function data. In the ionic case, however, the pair functions for unlike ions vary extremely rapidly near contact, and the extrapolation is correspondingly dubious even with quite precise pair function data. Using conventional techniques the free energy and activity coefficients would be obtained by integration of the osmotic coefficient (with respect to ln$c$) and so our values of those quantities were also in doubt, although the internal energy and the structural data were both good in themselves.

In the case of the 1 : 1 electrolytes[1,2] the problem is not so severe, and the extrapolations seemed adequate. Furthermore, those free energies were subsequently obtained by a second and independent method[4] which confirmed the conventional results. Application of that technique (multistage sampling) to the other systems would have been somewhat tedious, but a later development of the idea using non-Boltzmann sampling techniques[5,6]—so-called "umbrella sampling"—would be quite efficient.

A still better approach is available, however, for the low ionic densities in which we are interested: the problem is a natural one for the grand canonical Monte Carlo methods[6,7] (GCMC). In GCMC work one fixes the chemical potential, along with the temperature and the volume. The Markov chain allows fluctuation of the concentration, and one finds eventually the mean concentration (and energy, etc.) corresponding to the particular values chosen for the fixed parameters. The possibility of such calculations has been pointed out often,[7] but it is only recently, with the work of Norman and Filinov,[8] Adams,[9,10] and Nicholson, Rowley, and Parsonage,[11,12] that attempts have been made to exploit the method.

The GCMC method appears not to be useful at high densities (e.g., for liquids). This is because steps in the Markov chain which lead to changes of concentration are then exceedingly rare: one is unlikely to find room to insert particles at random positions in a dense system. As a consequence the concentration does not fluctuate adequately and convergence is prohibitively slow. There are no such problems at low densities, however, and this makes GCMC ideal for studying dilute electrolyte solutions.

The present article (I) describes the application of GCMC methods to electrolyte solutions, and reports results for 1 : 1, 2 : 2, 2 : 1, and 3 : 1 aqueous electrolytes in the primitive model with equal ion sizes. The following paper (II) reports some of the earlier canonical Monte Carlo (CMC) data for 2 : 2 electrolytes, and uses both sets of data to examine the (by now many) theoretical approximations for the problem and to comment on the structure of such ionic solutions. The third paper (III) does the same thing for the unsymmetrical 2 : 1 and 3 : 1 cases. A subsequent paper (IV) will report GCMC results for the 1 : 1 aqueous primitive model with *unequal* ion sizes.

This article (I) is therefore primarily methodological, and begins (Sec. II) by discussing the theory of GCMC for the electrolytic case and choosing an appropriate transition matrix for the MC Markov chain. After other details of the computations are described (Sec. III), the results are reported (Sec. IV). Most of the *scientific* discussion of these results is left for later papers; here we are interested in the efficiency and reliability of the GCMC results.

II. GRAND CANONICAL MONTE CARLO THEORY

To specify a configuration in the grand canonical ensemble requires the number and the locations of the particles of each species present. Suppose π_i is the probability of a configuration i in the grand canonical ensemble. We require a convenient Markov chain among such states having a limiting distribution proportional to the distribution $\{\pi_i\}$. The Markov chain is defined by the stochastic matrix $||p_{ij}||$ of the probabilities p_{ij} of transitions from state i to state j. A sufficient (though not a necessary[13,14]) condition for the irreducible chain to have the correct limiting distribution is

$$\frac{p_{ij}}{p_{ji}} = \frac{\pi_j}{\pi_i} \tag{1}$$

and we seek such transition probabilities. As usual we attempt, at each step of the chain, a "trial" change of state, and this trial move is then accepted or rejected in such a way as to lead to (1). Suppose the probability of a *trial* step i to j is q_{ij}, and the probability of *acceptance* of that trial step is f_{ij} for $i \neq j$. Then

$$p_{ij} = q_{ij} f_{ij}, \quad (i \neq j) \ , \tag{2}$$

$$p_{ii} = 1 - \sum_{j \neq i} p_{ij} \ .$$

There is of course wide latitude in the choice of $\| q_{ij} \|$, and then also of $\| f_{ij} \|$, satisfying (1). We examine simple choices of $\| q_{ij} \|$ and $\| f_{ij} \|$, of the sort proposed by Adams,[9] but elaborated to deal with the multicomponent ionic solutions.

Monte Carlo experiments can be regarded as describing an ensemble either of labeled particles or of unlabeled particles. It turns out to be most convenient to regard the particles as unlabeled, and this is done in what follows. (An alternative formulation in terms of labeled particles is of course possible.[15])

We now describe a scheme for carrying out a grand canonical Monte Carlo computation. Consider an electrolyte which on solution dissociates into $\nu_+ L_0$ cations of charge Z_+ and $\nu_- L_0$ anions of charge Z_- per mole (where L_0 is Avogadro's number). Evidently $\nu_+ Z_+ + \nu_- Z_- = 0$, and $\nu \equiv \nu_+ + \nu_-$ ions can form an electrically neutral combination. At each step of the chain either (i) one tries to add or to delete an electrically neutral combination of ν ions, or (ii) one tries to move one ion to a new location. The steps are done as follows:

(i) *Additions or deletions* are attempted with equal probability P; ν_+ cations and ν_- anions are added or deleted in a single step. In a trial addition each of the ν ions is inserted anywhere in the box with equal probability. It is convenient (and valid) to regard the box as consisting of V discrete sites (V extremely large). Then if state j is obtained from state i by *addition*, $N_j^+ = N_i^+ + \nu_+$ and $N_j^- = N_i^- + \nu_-$ in an obvious notation, and evidently

$$q_{ij} = \frac{P}{V^\nu \nu_+! \nu_-!} \ . \tag{3}$$

In a trial deletion any set of ν_+ cations and ν_- anions is removed, with equal probability, so (with i obtained from j by deletion)

$$q_{ji} = \frac{P N_i^+! N_i^-!}{N_j^+! N_j^-! \nu_+! \nu_-!} \ . \tag{4}$$

It is interesting that the "underlying matrix" $\| q_{ij} \|$ is not symmetric. The probability of a configuration i in an ensemble of unlabeled particles will be

$$\pi_i = \frac{1}{\Pi} \frac{1}{\Lambda_+^{3N_i^+} \Lambda_-^{3N_i^-}} \exp[\beta(\mu_+ N_i^+ + \mu_- N_i^-) - \beta U_i] \ , \tag{5}$$

where μ_+ and μ_- are the chemical potentials of the ions, Π is the grand canonical partition function, $\Lambda_+ \equiv h/(2\pi m_+ kT)^{1/2}$, and U_i is the configurational energy of the state. Combining Eqs. (1)–(5), we obtain (still with $N_j^+ = N_i^+ + \nu_+$, $N_j^- = N_i^- + \nu_-$)

$$\frac{f_{ij}}{f_{ji}} = \frac{q_{ji}}{q_{ij}} \frac{\pi_j}{\pi_i} = \frac{V^\nu}{\Lambda_+^{3\nu_+} \Lambda_-^{3\nu_-}} \frac{N_i^+! N_i^-!}{N_j^+! N_j^-!} \exp[\beta\mu - \beta(U_j - U_i)], \tag{6}$$

where $\mu \equiv \nu_+ \mu_+ + \nu_- \mu_-$ is the chemical potential of the electrolyte. Extending a definition due to Adams,[2] we introduce

$$B \equiv \beta(\mu - \mu_{\text{ideal}}) + \ln N^{+\nu_+} N^{-\nu_-}$$
$$= \beta\mu + \ln \frac{V^\nu}{\Lambda_+^{3\nu_+} \Lambda_-^{3\nu_-}} \ , \tag{7}$$

which leads to

$$\frac{f_{ij}}{f_{ji}} = \frac{N_i^+! N_i^-!}{N_j^+! N_j^-!} \exp[B - \beta(U_j - U_i)] \ . \tag{8}$$

Evidently [cf. Eq. (7)] fixing B fixes μ at fixed temperature and volume. If the acceptance probabilities f_{ij}, f_{ji} are made to conform to Eq. (8) with some particular value of B, the corresponding deviation from ideality, $\mu - \mu_{\text{ideal}}$, may be obtained from it once the expectation values of the concentrations N^+ and N^- are known. Of course

$$\beta(\mu - \mu_{\text{ideal}}) = \nu \ln \gamma_\pm \ , \tag{9}$$

where $\gamma_\pm$ is the mean ionic activity coefficient in the McMillan–Mayer system.

A simple way of realizing the result (8) is to set

$$f_{ij} = \min\{1, f_{ij}/f_{ji}\} \quad \text{for addition,}$$
$$f_{ji} = \min\{1, f_{ji}/f_{ij}\} \quad \text{for deletion} \tag{10}$$

of particles, where on the right hand sides of the expressions the ratio f_{ij}/f_{ji} is given by Eq. (8) after finding the value of $(U_j - U_i)$.

(ii) *Particle moves* are attempted with a probability $(1 - 2P)$ at each step, and are carried out exactly as in the canonical case. That is, a particle is chosen at random and moved to a random position in a volume element surrounding its original position. The volume element is chosen so that the part of the transition matrix describing these trial moves is symmetric: $q_{ij} = q_{ji}$ for moves. It is usual to choose a position within a cube or sphere centered on the original position of the particle; a cube of side δ was used in the computations described here. The move is accepted with a probability f_{ij} given by

$$f_{ij} = \min\{1, \exp[-\beta(U_j - U_i)]\} \ . \tag{11}$$

III. COMPUTATIONAL DETAILS

The calculations are for the primitive model with ions of equal size. Thus the potential energy of interaction u_{ab} between two ions a, b separated by a distance r_{ab} is given by

$$u_{ab} = \frac{Z_a Z_b e^2}{\epsilon r_{ab}} \quad \text{if } r_{ab} > R$$
$$= \infty \quad \text{if } r_{ab} < R \ , \tag{12}$$

where $Z_a e$, $Z_b e$ are the charges on the ions (with $-e$ the electronic charge), ϵ is the dielectric constant chosen to represent the screening effect of the solvent, and R is

Notes on p. 278

TABLE I. Results for 1 : 1 electrolyte with $R=4.25$ Å and $\epsilon T=2.345\times10^4$. The columns are as follows: L is the length of the cubical periodic box; B the chemical potential input parameter [Eq. (7)]; No. Steps refers to the length of the MC run, while No. Disc. shows how much of the run was regarded as aging and ignored in the averaging; $2P$ is the proportion of steps in which additions or deletions were attempted, and the next column gives the acceptance rate for such steps; the average number of particles N and the resulting molarity M are followed by the data for the activity coefficients and the reduced configurational energies.

L (Å)	B	$10^{-3}\times$ No. Steps	$10^{-3}\times$ No. Disc.	$2P$	Acc. rates Add/Del	$\langle N\rangle$	$\langle M\rangle$	$-\langle\ln\gamma_\pm\rangle$	$-\langle U\rangle/\langle N\rangle kT$
165.21	6.722	40	16	0.20	0.76	62.9 ± 0.8	0.01158 ± 0.00015	0.087 ± 0.013	0.116 ± 0.004
80.99	6.471	100	10	0.20	0.64	64.1 ± 0.4	0.1002 ± 0.0006	0.232 ± 0.006	0.277 ± 0.004
47.36	6.415	100	10	0.20	0.47	63.4 ± 0.6	0.495 ± 0.004	0.249 ± 0.009	0.460 ± 0.007
37.59	6.679	100	20	0.20	···	63.4 ± 1.0	0.991 ± 0.016	0.117 ± 0.016	0.556 ± 0.014
37.59	6.683	312	26	0.20	0.33	64.2 ± 0.4	1.003 ± 0.006	0.127 ± 0.006	0.558 ± 0.006
29.84	7.468	100	20	0.20	0.14	63.8 ± 1.1	1.994 ± 0.034	−0.271 ± 0.017	0.672 ± 0.021
26.06	8.660	720	120	0.20	0.04	65.1 ± 0.4	3.052 ± 0.019	−0.847 ± 0.006	0.740 ± 0.008
32.84	10.046	720	60	0.20	0.04	129.9 ± 0.7	3.045 ± 0.016	−0.849 ± 0.005	0.731 ± 0.007
23.68	10.300	3024	504	0.20	0.008	66.1 ± 0.3	4.13 ± 0.02	−1.652 ± 0.005	0.800 ± 0.006
29.84	11.686	1680	336	0.20	0.008	131.0 ± 0.6	4.09 ± 0.02	−1.661 ± 0.005	0.786 ± 0.006
22.238	11.366	2400	200	0.67	0.002	63.05 ± 0.24	4.76 ± 0.02	−2.232 ± 0.004	0.830 ± 0.005
28.02	12.752	1560	156	0.67	0.002	125.75 ± 0.57	4.75 ± 0.02	−2.235 ± 0.005	0.813 ± 0.008

the diameter of each ion. The properties of the system depend on R and on the product ϵT, as well as on the ionic charge types. R and ϵT were chosen to match closely those used earlier in the CMC calculations[3]: for the 1 : 1 case $R=4.25$ Å and $\epsilon T=2.345\times10^4$ deg (cf. $R=4.25$ Å and $\epsilon T=2.339\times10^4$ for CMC), and for the 2 : 2, 2 : 1, and 3 : 1 cases $R=4.2$ Å and $\epsilon T=2.342\times10^4$ (cf. $R=4.2$ Å and $\epsilon T=2.336\times10^4$). At room temperature the parameter ϵT corresponds to the dielectric constant of water. Of course the results can equally well be interpreted as corresponding to a lower ϵ at a higher temperature, for example, a plasma at about 2.34×10^5 °K. In a similar way the results we describe as those of 2 : 2 electrolyte may be interpreted as a 1 : 1 electrolyte at a value of ϵT reduced by a factor of 4 (e.g., an ϵ of 19.6 at 25 °C).

Periodic boundary conditions were used, and the repeating box was cubical. The energy calculations were done in the minimum image approximation (MI): that is the potential energy of a particle is evaluated by summing the pair interactions with only one image of each of the other particles, the image closest to the particle in question. This corresponds to a cubical cutoff having the dimensions and orientation of the repeating box. It has been shown[1,16] that a spherical cutoff is totally unacceptable for coulombic systems, and this is readily understood.[14] Many calculations on Coulombic systems have approximated the truly periodic boundary conditions by doing Ewald summations, although it seems that this will introduce very nonphysical forces and correlations.[14] Fortunately at low ionic densities the Ewald and MI approximations are expected to agree with each other and to be physically meaningful. Brush, Sahlin, and Teller[17] compared the two approximations for the one-component plasma and found that they agreed as long as the dimensionless parameter

$$\Gamma=\left(\frac{4\pi N}{6V}\right)^{1/3}\frac{Z^2e^2}{\epsilon kT} \qquad (13)$$

was below about 10. (Here N/V is the number density and Ze the ionic charge.) Our own tests[1,14] and those of Hoskins and Smith[16] on two-component plasmas confirm that for low Γ the methods agree well while for higher Γ they give totally different results. Up to $\Gamma\sim10$ the discrepancy is comparable to the statistical uncertainty. For all of the calculations reported here Γ is substantially less than 10 (e.g., for the 2 : 2 system at 2 M, $\Gamma=4.9$), so the energy approximation should be trustworthy. For the more concentrated solutions this was tested, however, by carrying out parallel runs with systems of double the usual size. One expects the results to vary with the system's size if the energy approximation is becoming inadequate; the results are discussed below.

Two "experimental" parameters will affect the convergence rate of the Markov chain. These are the "step size" δ allowed when a particle move is attempted, and the ratio $2P/(1-2P)$ of attempted additions and deletions to attempted particle moves. The former was chosen to make the acceptance rate for moves well behaved; for the denser systems it was adjusted to give an acceptance rate close to 50%. Little is known about the best way to choose P, and we made no systematic study of this. At low densities the acceptance rates for addition and deletion are substantial, and we chose $P=0.1$. A high densities the acceptance rate becomes small, however, and it was necessary to increase P in order to get adequate variation of concentration during the runs. Data on P and the addition/deletion acceptance rates are reported. Since pair correlation function data for most of these systems had been obtained in the earlier CMC calculations,[3] it was not collected in the runs reported here. (For systems with unequal ion sizes, reported in Paper IV, CMC results were not available and pair correlations were studied in the GCMC runs.)

IV. RESULTS AND DISCUSSIONS

Tables I–IV refer to the 1 : 1, 2 : 2, 2 : 1, and 3 : 1 systems, in that order. In each table the last three columns report the physical results: activity coefficients $\gamma_\pm$ and configurational energies U at various concentrations.

TABLE II. Results for 2 : 2 electrolyte with $R = 4.2$ Å and $\epsilon T = 2.342 \times 10^4$. The columns are described in the legend of Table I.

L (Å)	B	$10^{-3}\times$ No. Steps	$10^{-3}\times$ No. Disc.	$2P$	Acc. rates Add/Del	$\langle N\rangle$	$\langle M\rangle$	$-\langle \ln\gamma_\pm\rangle$	$-\langle U\rangle/\langle N\rangle kT$
174.48	5.190	1248	104	0.20	0.36	65.63 ± 0.50	0.01025 ± 0.00008	0.896 ± 0.008	1.208 ± 0.019
102.04	3.872	630	60	0.20	0.25	58.36 ± 0.56	0.0456 ± 0.0004	1.437 ± 0.010	1.774 ± 0.031
59.675	2.632	540	60	0.20	0.16	65.07 ± 0.52	0.254 ± 0.002	2.166 ± 0.008	2.465 ± 0.032
75.186	4.018	630	60	0.20	0.15	127.47 ± 1.43	0.249 ± 0.003	2.146 ± 0.011	2.450 ± 0.044
44.571	1.952	540	60	0.20	0.12	63.11 ± 0.97	0.592 ± 0.009	2.476 ± 0.015	2.841 ± 0.043
37.590	1.603	630	60	0.20	0.099	62.16 ± 0.82	0.971 ± 0.013	2.635 ± 0.013	3.102 ± 0.065
32.139	1.570	1456	208	0.20	0.065	63.89 ± 1.12	1.597 ± 0.028	2.679 ± 0.017	3.383 ± 0.095
29.837	1.592	630	60	0.20	0.051	63.41 ± 0.54	1.982 ± 0.017	2.660 ± 0.009	3.528 ± 0.046
37.592	2.986	630	60	0.20	0.045	123.48 ± 1.33	1.929 ± 0.021	2.630 ± 0.011	3.484 ± 0.059
26.065	1.946	2520	280	0.20	0.024	61.01 ± 0.37	2.860 ± 0.017	2.445 ± 0.006	3.771 ± 0.038
26.065	1.946	2464	154	0.20	0.024	61.96 ± 0.39	2.904 ± 0.018	2.460 ± 0.006	3.785 ± 0.038
32.840	3.332	1512	126	0.20	0.018	127.30 ± 1.31	2.984 ± 0.031	2.487 ± 0.010	3.878 ± 0.070
23.682	1.946	2464	308	0.40	0.019	51.40 ± 0.31	3.212 ± 0.019	2.273 ± 0.006	3.815 ± 0.037
29.837	3.332	1464	244	0.40	0.008	118.13 ± 2.38	3.691 ± 0.074	2.413 ± 0.020	4.265 ± 0.171
23.682	3.060	2400	200	0.20	0.008	62.84 ± 0.42	3.927 ± 0.026	1.917 ± 0.007	3.984 ± 0.044

TABLE III. Results for 2 : 1 electrolyte with $R = 4.2$ Å and $\epsilon T = 2.342 \times 10^4$. The columns are described in the legend of Table I.

L (Å)	B	$10^{-3}\times$ No. Steps	$10^{-3}\times$ No. Disc.	$2P$	Acc. rates Add/Del	$\langle N\rangle$	$\langle M\rangle$	$-\langle \ln\gamma_\pm\rangle$	$-\langle U\rangle/\langle N\rangle kT$
328.4	10.111	280	40	0.20	0.71	62.65 ± 0.26	0.000979 ± 0.000004	0.131 ± 0.004	0.129 ± 0.002
151.6	9.518	280	40	0.20	0.54	63.60 ± 0.38	0.01009 ± 0.00006	0.343 ± 0.006	0.362 ± 0.005
88.68	8.630	380	40	0.20	0.40	60.07 ± 0.50	0.0477 ± 0.0004	0.582 ± 0.008	0.637 ± 0.010
51.86	7.992	280	40	0.20	0.24	65.30 ± 0.42	0.259 ± 0.002	0.878 ± 0.006	1.035 ± 0.011
38.94	7.720	280	80	0.20	0.15	61.71 ± 0.68	0.578 ± 0.006	0.913 ± 0.011	1.255 ± 0.022
32.84	8.120	280	40	0.20	0.078	62.68 ± 0.47	0.979 ± 0.007	0.795 ± 0.007	1.416 ± 0.018
39.37	10.359	672	96	0.20	0.057	124.39 ± 1.08	1.128 ± 0.010	0.734 ± 0.009	1.451 ± 0.020
31.24	8.280	672	96	0.20	0.059	62.93 ± 0.41	1.141 ± 0.007	0.745 ± 0.007	1.463 ± 0.016
29.20	8.720	720	60	0.20	0.036	64.04 ± 0.40	1.423 ± 0.009	0.616 ± 0.006	1.540 ± 0.017
29.20	8.720	840	60	0.20	0.036	64.23 ± 0.40	1.427 ± 0.009	0.619 ± 0.006	1.542 ± 0.016
36.79	10.799	720	60	0.20	0.035	126.27 ± 0.96	1.403 ± 0.011	0.602 ± 0.008	1.523 ± 0.019
27.38	9.180	1152	96	0.20	0.020	62.83 ± 0.35	1.694 ± 0.009	0.444 ± 0.006	1.600 ± 0.014
34.49	11.259	960	96	0.20	0.019	125.53 ± 0.55	1.692 ± 0.007	0.443 ± 0.004	1.589 ± 0.013
25.93	9.849	1584	132	0.32	0.010	62.54 ± 0.30	1.984 ± 0.010	0.216 ± 0.005	1.654 ± 0.013
32.67	11.928	1200	240	0.32	0.010	123.80 ± 0.73	1.964 ± 0.012	0.206 ± 0.006	1.640 ± 0.015
24.75	10.906	1584	132	0.60	0.0036	64.68 ± 0.26	2.360 ± 0.009	−0.102 ± 0.004	1.718 ± 0.012
31.18	12.985	1200	240	0.60	0.0033	131.35 ± 1.27	2.396 ± 0.023	−0.087 ± 0.010	1.726 ± 0.029
23.76	12.226	2000	200	0.80	0.00089	66.91 ± 0.39	2.760 ± 0.016	−0.508 ± 0.006	1.779 ± 0.018
29.94	14.305	1440	180	0.80	0.00093	134.8 ± 1.4	2.780 ± 0.029	−0.501 ± 0.010	1.777 ± 0.026

TABLE IV. Results for 3 : 1 electrolytes with $R = 4.2$ Å and $\epsilon T = 2.342 \times 10^4$. The columns are described in the legend of Table I.

L (Å)	B	$10^{-3}\times$ No. Steps	$10^{-3}\times$ No. Disc.	$2P$	Acc. rates Add/Del	$\langle N\rangle$	$\langle M\rangle$	$-\langle \ln\gamma_\pm\rangle$	$\langle U\rangle/\langle N\rangle kT$
298.4	13.410	420	60	0.20	0.53	64.54 ± 0.31	0.001008 ± 0.000005	0.252 ± 0.005	0.298 ± 0.006
375.9	16.182	480	60	0.20	0.53	129.0 ± 1.3	0.001008 ± 0.000010	0.252 ± 0.010	0.282 ± 0.005
138.5	11.770	420	60	0.20	0.30	64.00 ± 0.48	0.01000 ± 0.00007	0.654 ± 0.008	0.756 ± 0.012
80.99	9.890	420	60	0.20	0.18	59.88 ± 0.34	0.0468 ± 0.0003	1.058 ± 0.006	1.202 ± 0.013
80.99	10.110	420	60	0.20	0.18	65.23 ± 0.48	0.0510 ± 0.0004	1.088 ± 0.007	1.236 ± 0.016
47.36	8.046	420	60	0.20	0.087	61.17 ± 0.64	0.2390 ± 0.0025	1.540 ± 0.010	1.759 ± 0.030
35.376	7.560	420	60	0.20	0.036	60.80 ± 0.53	0.570 ± 0.005	1.655 ± 0.009	2.101 ± 0.030
29.835	8.120	1440	120	0.20	0.011	62.38 ± 0.46	0.975 ± 0.007	1.541 ± 0.007	2.344 ± 0.028
34.62	10.466	1200	60	0.50	0.0079	106.32 ± 0.83	1.064 ± 0.008	1.488 ± 0.008	2.378 ± 0.031
27.48	9.080	1680	120	0.60	0.0035	65.04 ± 0.49	1.301 ± 0.010	1.343 ± 0.008	2.489 ± 0.031
34.62	11.852	960	60	0.50	0.0029	131.7 ± 1.1	1.317 ± 0.011	1.355 ± 0.009	2.489 ± 0.034
25.51	10.390	2400	240	0.80	0.00091	64.65 ± 0.46	1.617 ± 0.012	1.009 ± 0.007	2.605 ± 0.028
32.14	13.162	1800	200	0.80	0.00078	130.68 ± 0.94	1.634 ± 0.012	1.020 ± 0.007	2.610 ± 0.032

Notes on p. 278

The other columns report technical information on the GCMC runs. The physical results are also shown in Figs. 1 and 2.

The scientific implications of these results are discussed in Papers II and III, in conjunction with the CMC results reported in those papers. Here we examine methodological matters relating to the GCMC method.

At very low densities the acceptance rates for addition and deletion steps are of course high, and convergence of the chains is rapid. However, as the density increases that acceptance rate falls off rather quickly. In order to sample adequately one then needs to go to very long runs, and at a quite modest density GCMC calculations become impractical. This can be examined for each of our systems in the tables. Of course in this electrolyte work we are adding or deleting more than one particle at a time, so the acceptance rate is lower than it would be for a simpler system. The runs reported for the higher concentrations pushed the GCMC technique pretty hard: they managed to use addition/deletion acceptance rates as low as ~ 0.001. In order to do this it was necessary to increase the proportion $2P/(1-2P)$ of attempted additions or deletions drastically, and also

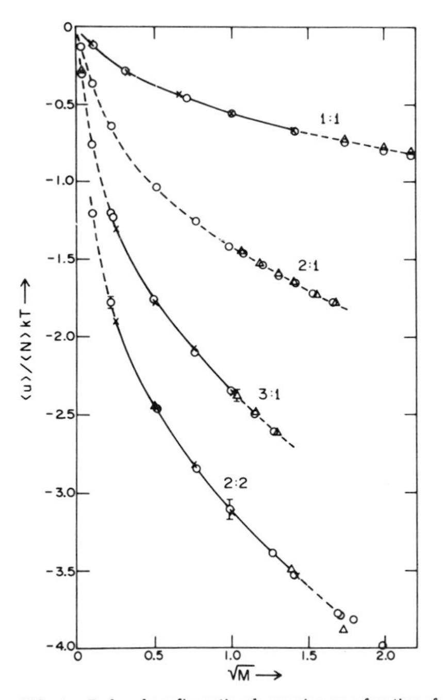

FIG. 1. Reduced configurational energies as a function of concentration. Open circles refer to GCMC results with $N \sim 64$, open triangles to systems about twice as large. The crosses and solid curves show the very precise results from CMC work. A few typical error bars are shown.

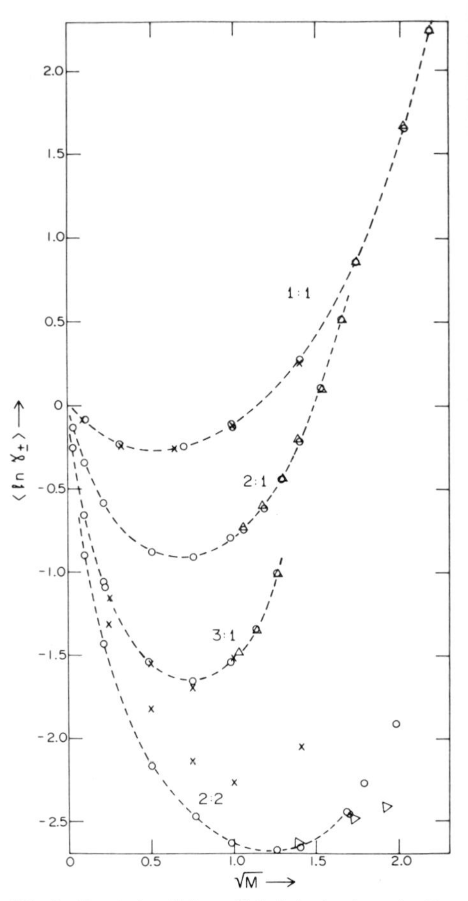

FIG. 2. Mean ionic activity coefficients (as $\ln\gamma_{\pm}$) as a function of concentration. Open circles refer to GCMC results with $N \sim 64$, open triangles to systems with $N \sim 128$. Crosses give some apparent values obtained from CMC osmotic coefficients: except for the 1 : 1 system they are dubious.

to do very long runs, often of a few million configurations. (CMC runs for these systems require only ~ 100 000 configurations.)

These techniques enabled one to go to 4.75 M, 2.78 M, and 1.63 M for the 1 : 1, 2 : 1, and 3 : 1 systems, respectively (or $\rho/\rho_0 = 0.31$, 0.26, and 0.21, where ρ_0 is the close-packed density). The upper concentration depends strongly on the valence type, of course, since one is adding 2, 3, or 4 particles, respectively, in the three types. The upper concentration limit for the 2 : 2 case seems to depend on other factors which are discussed below.

It is well known that the nonideality $\nu \ln\gamma_\pm = \beta(\mu - \mu_{\text{ideal}})$ of electrolytes goes through a minimum in the concentration ranges studied here. It follows (if the number of particles N is to be kept roughly constant) that the parameter B will also go through a minimum as the concentration is increased for approximately constant N, and this can be seen in the tables. One might imagine that this would lead to ambiguous MC results, two concentrations corresponding to a single value of B. This is not the case, however, since for *fixed volume* the two concentrations would have to be such that [cf. Eqs. (7) and (9)] the difference in $\ln\gamma_\pm$ was given by

$$\Delta \ln\gamma_\pm = -\frac{1}{\nu} \Delta \ln(N^{+\nu^+}N^{-\nu^-}) = -\Delta \ln M,$$

but $\ln\gamma_\pm$ always decreases much less rapidly than $-\ln M$; hence the results are single valued.

For most of the runs the volumes and the values of B were chosen to give roughly $N \sim 64$ particles. The number dependence of the thermodynamic properties for dilute solutions was studied previously,[1,3] and it is known that a system of this size gives adequate scientific results. For the denser systems studied here, where one might worry about the adequacy of the energy approximation, parallel runs were performed with double the volume and an appropriately adjusted B [cf. Eq. (7)]. These results appear in the figures as open triangles to distinguish them from the smaller systems. It is evident that, except for the 2 : 2 systems, the agreement is entirely satisfactory, in most cases well within the statistical uncertainty of the results. In going to a larger system one expects to find a very slightly less negative energy (since the cancelling charge surrounding each ion is allowed to spread over a larger volume); in the CMC work this was found to raise the energy by ~ 1% on doubling the size from $N = 64$ (and by ~ 2% in extrapolating to an infinite system). The N dependence of the GCMC results is not inconsistent with that, although it may be even weaker for the 2 : 1 and 3 : 1 systems. The very large 2 : 2 discrepancies are discussed separately.

Comparison with the CMC results[1-3] on the same systems provides a useful check on the GCMC results. One expects the *precision* of the GCMC *energy* estimates to be less than those of the CMC, due to the density fluctuations. This is borne out by the results, in which the standard deviations of the energies are several times larger than in the CMC case, in spite of the longer runs. The *accuracy* is all one could hope for, however: in Fig. 1 the solid lines show the very precise CMC results for $N = 64$, and the agreement of the GCMC results is everywhere complete.

In the case of the activity coefficients, on the other hand, it is the GCMC results which are expected to be more precise, as discussed above. Figure 2 shows (as crosses) some CMC results obtained by Gibbs–Duhem integration of the osmotic coefficients.[3] For the 1 : 1 case the extrapolations of $g_{+-}(r)$ to $r = R$ were not difficult, and the results were checked by using multistage sampling,[4] so it is not surprising to find the total agreement with the new GCMC results. The free energy results for the 1 : 1 system have now been obtained in three different ways, in fact, with perfect consistency. Most of the 3 : 1 CMC results for $\ln\gamma_\pm$ turn out to be fairly good, in spite of our misgivings about them. For the 2 : 2 case, however, the CMC results are quite wrong. This is undoubtedly due mostly to lack of success in extrapolating $g_{+-}(r)$ back to $r = R$. The extrapolation was done by using a quadratic fit to the small-r values of $g_{+-}(r)$; the errors correspond to too-high values of $g_{+-}(R)$. It is possible to get lower (or higher) values of $g_{+-}(R)$ by extrapolating differently, but there is no *a priori* way to choose the best extrapolation. These results justify our reluctance to publish CMC results for the osmotic coefficients of the higher-valence electrolytes.

The GCMC results appear totally satisfactory except for the 2 : 2 system at concentrations > 2 M. There the number dependence appears to be large and erratic. In particular the result at 3.69 M (with $N = 118$) should coincide with that at 3.21 M (with $N = 51$), instead of which they are totally discrepant in energy and density and thus activity. The apparent number dependencies at 1.9 M and 2.9 M are just believable as random error, but also look really to be discrepant. Although the small-system results ($N \sim 64$) fall on a credible curve for both energy and activity, the peculiar behaviour of the large-system ($N \sim 128$) results puts all of the results into question.

It is natural to seek an explanation in terms of slow convergence due to low addition/deletion acceptance rates. In fact, however, these 2 : 2 runs show acceptance rates very much *higher* than those which were easily handled for the other systems. Furthermore, control charts of N and U show neither long correlation times nor any tendency to drift, so simple slow convergence seems to be ruled out. The other obvious explanation is an enhanced number dependence due to breakdown of the minimum-image energy approximation. All the evidence[14,16] suggests that in this case the larger system should have a *less* negative energy, however, while these results show a *lower* energy. Furthermore, the fact that the density discrepancies are in different directions seems puzzling on this hypothesis, while previous tests[3,16,17] indicate that the minimum image approximation remains quite accurate at the values of Γ in question here.

So (always supposing that no error crept into the programme!) the problem looks to be one of quasiergodicity. In this connection it is of interest that there is believed to be a phase change in Coulombic systems with a critical point at a temperature somewhat lower than ours and at a somewhat uncertain density.[18,19] [Some results of Vorontsov-Vel'yaminov and Chasovskikh,[18] based on MC computations, suggest that there might be a critical point in a 2–2 "aqueous" electrolyte (i.e., with $\epsilon = 78.4$) near 200 Å and 3.6 M.] It is natural to ask whether the apparent quasiergodicity of the 2–2 results is associated with such a phase transition; that is the subject of a separate study now underway. Meanwhile it seems prudent not to use the results for the 2–2 system at high concentrations.

The scientific implications of the results are discussed in the following two papers[20,21] of the series.

Notes on p. 278

ACKNOWLEDGMENTS

The authors acknowledge with thanks the financial support of the National Research Council of Canada. They have enjoyed useful discussions with Glenn Torrie and Stuart Whittington, and the helping hand of Mike McNeal. Jay Rasaiah pointed out long ago the problems of estimating osmotic coefficients from CMC data.

[1]D. N. Card and J. P. Valleau, J. Chem. Phys. **52**, 6232 (1970).

[2]J. C. Rasaiah, D. N. Card, and J. P. Valleau, J. Chem. Phys. **56**, 248 (1972).

[3]D. N. Card, thesis, University of Toronto, 1972.

[4]J. P. Valleau and D. N. Card, J. Chem. Phys. **57**, 5457 (1972).

[5]G. M. Torrie and J. P. Valleau, J. Comput. Phys. **23**, 187 (1977).

[6]J. P. Valleau and G. M. Torrie, "A Guide to Monte Carlo for Statistical Mechanics. 2. Byways," in *Modern Theoretical Chemistry, Vol. 5A, Statistical Mechanics: Equilibrium Techniques*, edited by B. Berne (Plenum, New York, 1977), p. 169.

[7]W. W. Wood, in *Physics of Simple Liquids*, edited by H. N. V. Temperley, G. S. Rushbrooke, and J. S. Rowlinson (North-Holland, Amsterdam, 1968), Chap. 5.

[8]G. E. Norman and V. S. Filinov, High Temp. (USSR) **7**, 216 (1969).

[9]D. J. Adams, Mol. Phys. **28**, 1241 (1974).

[10]D. J. Adams, Mol. Phys. **29**, 307 (1975).

[11]L. A. Rowley, D. Nicholson, and N. G. Parsonage, J. Comput. Phys. **17**, 401 (1975).

[12]L. A. Rowley, D. Nicholson, and N. G. Parsonage, Mol. Phys. **31**, 365, 389 (1976); see also J. Comput. Phys. **26**, 66 (1978).

[13]J. M. Hammersley and D. C. Handscomb, *Monte Carlo Methods* (Methuen, London, 1964).

[14]J. P. Valleau and S. G. Whittington, "A Guide to Monte Carlo for Statistical Mechanics. 1. Highways," in *Modern Theoretical Chemistry, Vol. 5A: Statistical Mechanics: Techniques*, edited by B. Berne (Plenum, New York, 1977), p. 137.

[15]If one wishes to regard the particles as labeled, one must imagine a process of fixing the labels to the particles in a consistent way in spite of the additions and deletions: every system with N^+ cations and N^- cations should involve a particular set of labels, e.g., integers from 1 to N^+ for the cations and from 1 to N^- for the anions. One way of achieving this would be to assign those labels to the particles, at random, after each attempted addition or deletion. We could then impute, for the labeled states [cf. Eqs. (3) and (4)]

$$q_{ij}=\frac{P}{V^{\nu}\nu_+!\,\nu_-!}\,\frac{1}{N_j^+!\,N_j^-!}$$

and

$$q_{ji}=\frac{P}{N_j^+N_j^-\nu_+!\,\nu_-!}\,\frac{1}{N_i^+!\,N_i^-!}\ .$$

However, the probability π_i of a configuration i for labeled particles will be [cf. Eq. (5)]

$$\pi_i=\frac{1}{\Pi}\,\frac{1}{N_i^+!\,N_i^-!}\,\frac{1}{\Lambda_+^{3N_i^+}\Lambda_-^{3N_i^-}}\exp[\beta(\mu_+N_i^+ + \mu_-N_i^-)-\beta U_i]\ ,$$

and thus the ratio of acceptance probabilities (6) would be unchanged. This point of view seems, however, cumbersome and artificial, since no such labeling in fact takes place.

[16]C. S. Hoskins and E. R. Smith (private communication).

[17]S. G. Brush, H. L. Sahlin, and E. Teller, J. Chem. Phys. **45**, 2102 (1966).

[18]B. P. Chasovskikh and P. N. Vorontsov-Vel'iaminov, High Temp. (USSR) **14**, 174 (1976).

[19]G. Stell, K. C. Wu, and B. Larsen, Phys. Rev. **37**, 1369 (1976).

Notes to Reprint IV.2

1. One other paper in the series has so far appeared (Valleau et al., J. Chem. Phys. 72 (1980) 5942). The methods described in the present paper can also be applied to the inhomogeneous case; see, for example (van Megen and Snook, J. Chem. Phys. 73 (1980) 4656; Torrie and Valleau, J. Chem. Phys. 73 (1980) 5807; Torrie and Valleau, J. Phys. Chem. 86 (1982) 3251).

2. The quantities ν'_+ and ν'_- in eqs. (3) and (4) and in Note 15 should appear in the numerator and not, as printed, in the denominator. Because of cancellation in the final expressions, the error does not affect any of the results.

3. It is important that a random choice be made at each step between particle moves, additions and deletions. If, for example, moves were to be strictly alternated with additions/delections, a single step in the Markov chain would consist of an attempted move followed by an attempted addition/deletion. For a Markov chain constructed in this way, it is easy to think of pairs of states for which the principle of microscopic reversibility is violated. Microscopic reversibility is only a sufficient, not a necessary condition, but it is not obvious that such a procedure would satisfy the more general requirement that $\sum_i u_i p_{ij} = u_j$ (see Reprint I.2, eq. (8)).

Molecular dynamics simulations at constant pressure and/or temperature[a)]

Hans C. Andersen

Department of Chemistry, Stanford University, Stanford, California 94305
(Received 10 July 1979; accepted 31 October 1979)

In the molecular dynamics simulation method for fluids, the equations of motion for a collection of particles in a fixed volume are solved numerically. The energy, volume, and number of particles are constant for a particular simulation, and it is assumed that time averages of properties of the simulated fluid are equal to microcanonical ensemble averages of the same properties. In some situations, it is desirable to perform simulations of a fluid for particular values of temperature and/or pressure or under conditions in which the energy and volume of the fluid can fluctuate. This paper proposes and discusses three methods for performing molecular dynamics simulations under conditions of constant temperature and/or pressure, rather than constant energy and volume. For these three methods, it is shown that time averages of properties of the simulated fluid are equal to averages over the isoenthalpic-isobaric, canonical, and isothermal-isobaric ensembles. Each method is a way of describing the dynamics of a certain number of particles in a volume element of a fluid while taking into account the influence of surrounding particles in changing the energy and/or density of the simulated volume element. The influence of the surroundings is taken into account without introducing unwanted surface effects. Examples of situations where these methods may be useful are discussed.

I. INTRODUCTION

Molecular dynamics and Monte Carlo methods have become important tools for the study of fluids.[1,2] They have been used to study the equilibrium and transport properties of model atomic liquids, such as the hard-sphere fluid,[3] the Lennard-Jones fluid,[4] and models for molecular liquids.[5]

The Monte Carlo method, as developed by Metropolis *et al.*[6] and extended in various ways,[1(a),(b)] is a procedure for evaluating configuration space equilibrium averages for constant temperature ensembles, such as the canonical ensemble and the isothermal-isobaric ensemble. In the canonical ensemble, the temperature T, volume V; and number of particles N are specified in advance, and an algorithm is used to generate a sequence of configurations. The average of any property over this sequence is an approximation to the measured value of that property for the thermodynamic state with the specified values of N, V, and T. Similarly, for the isothermal-isobaric ensemble, N, the pressure P, and T are specified in advance, and properties are averaged over a sequence of generated configurations.

In the molecular dynamics method, the Newtonian equations of motion of a set of N particles in volume V are solved numerically. The total energy E of the system is conserved as the system moves along its trajectory. The average of any property over the trajectory is an approximation to the measured value of that property for the thermodynamic state with the specified values of N, V, and E. Such an average is equivalent to an average over a microcanonical ensemble if the trajectory passes through all parts of phase space that have the specified energy.

[a)]This work was supported by the NSF-MRL Program through the Center for Materials Research at Stanford University and by the National Science Foundation through grant CHE78-09317.

In some situations it is desirable to perform simulations at constant temperature and/or pressure. For example, in studying dilute solutions, it is worthwhile to simulate both the pure solvent and the dilute solution at the same temperature and pressure. This corresponds to the usual experimental situation in which partial molar quantities of the solute are measured, and the comparison of calculated results with experiment is facilitated.[5(b)] Also, in studies of the glass transition in atomic fluids, it is helpful to be able to manipulate the pressure and temperature of the surroundings of the fluid being simulated.[4(d)] To achieve isothermal and/or isobaric conditions, with the appropriate energy and/or volume fluctuations, it has been necessary to use Monte Carlo methods, rather than molecular dynamics. An advantage of the molecular dynamics method, over the Monte Carlo method, is that molecular dynamics gives information about the time dependence and magnitude of fluctuations of position and momentum variables away from their equilibrium values, while Monte Carlo deals only with position variables and gives no information about the time dependence of fluctuations. Thus, in order to be able to specify the temperature and/or pressure of a simulation, it has been necessary to use the Monte Carlo method and thereby forgo the possibility of obtaining dynamical information from the same simulation.

The object of this paper is to present and discuss molecular dynamics methods for simulating a fluid subject to a constant pressure, constant temperature, or constant temperature and pressure. The trajectory averages for these three types of simulations correspond to averages over the isoenthalpic-isobaric, canonical, and isothermal-isobaric ensembles, respectively. These methods have the advantages of isothermal and/or isobaric simulation without sacrificing a dynamical description of the fluid.

A molecular dynamics calculation can simulate the motion of only a small number of particles (typically,

 0021-9606/80/042384-10$01.00

between 50 and 1000). A physical system with this number of particles is more like a droplet than a bulk fluid, and its properties would be strongly affected by its surface. In order to eliminate the surface and obtain results for a bulk liquid, periodic boundary conditions[6] are ordinarily used. These conditions make all points in the simulated fluid equivalent to all other points. It is generally assumed that a molecular dynamics calculation with periodic boundary conditions gives results equivalent to the properties of a small volume embedded in a bulk sample of the material, provided that the length of the sample being simulated in several multiples of the correlation length of the bulk fluid.

However, a small group of 50 to 1000 particles in a volume element in a fluid is constantly subjected to energy and volume fluctuations due to the surrounding fluid. These fluctuations are responsible for equilibrating the volume element to the temperature and pressure of the surrounding fluid. The usual molecular dynamics method in effect suppresses these fluctuations and keeps the volume and energy for the N particles fixed. The methods discussed in this paper allow these fluctuations to be simulated, without introduction of an unwanted surface, by distributing the effect of the fluctuations throughout the volume of the simulated fluid.

These methods will be discussed for the special case of an atomic fluid. In Sec. II, the fluid of interest is discussed. Sections III–V discuss molecular dynamics simulations at constant pressure, constant temperature, and constant temperature and pressure, respectively. Section VI discusses potential applications of these techniques.

II. SYSTEM OF INTEREST

We imagine that the system of interest to be simulated is an atomic fluid. (All the methods and results are easily generalized to the case of molecular fluids.) In this section we will discuss the classical equations of motion for a fluid and define several types of ensemble averages.

The fluid of interest is N atoms, with coordinates $\mathbf{r}_1, \mathbf{r}_2, \ldots, \mathbf{r}_N$ in a cubic volume V with periodic boundary conditions. Each component of each coordinate is a number between 0 and $V^{1/3}$. In using periodic boundary conditions, we imagine that if particle i is at $\mathbf{r}_i$, there is a set of image particles at positions $\mathbf{r}_i + \mathbf{n}V^{1/3}$, where $\mathbf{n}$ is a vector with integer components. The potential energy of the atoms is

$$U(\mathbf{r}^N) = \sum_{i<j} u(r_{ij}) . \tag{2.1}$$

In this sum, r_{ij} is to be interpreted as the distance between $\mathbf{r}_i$ and either $\mathbf{r}_j$ or the nearest image of particle j, whichever is closer. Thus,

$$r_{ij} = |\mathbf{r}_{ij}| \equiv \min_{\mathbf{n}} |\mathbf{r}_i - \mathbf{r}_j + \mathbf{n}V^{1/3}| . \tag{2.2}$$

This is the minimum image convention. In all summations over particles, the lower and upper limits are 1 and N, respectively.

The Lagrangian for the fluid is

$$\mathcal{L}_1(\mathbf{r}^N, \dot{\mathbf{r}}^N) = \frac{m}{2}\sum_i \dot{\mathbf{r}}_i \cdot \dot{\mathbf{r}}_i - \sum_{i<j} u(r_{ij}) , \tag{2.3}$$

where m is the mass of an atom. The momenta are defined as

$$\mathbf{p}_i \equiv \frac{\partial \mathcal{L}_1(\mathbf{r}^N, \dot{\mathbf{r}}^N)}{\partial \dot{\mathbf{r}}_i} = m\dot{\mathbf{r}}_i . \tag{2.4}$$

The Hamiltonian is

$$\mathcal{H}_1(\mathbf{r}^N, \mathbf{p}^N; V) = \sum_i \dot{\mathbf{r}}_i \cdot \mathbf{p}_i - \mathcal{L}_1 = (2m)^{-1}\sum_i \mathbf{p}_i \cdot \mathbf{p}_i + \sum_{i<j} u(r_{ij}) . \tag{2.5}$$

The Hamiltonian equations of motion are

$$\frac{d\mathbf{r}_i}{dt} = \frac{\partial \mathcal{H}_1}{\partial \mathbf{p}_i} = \frac{\mathbf{p}_i}{m} , \tag{2.6a}$$

$$\frac{d\mathbf{p}_i}{dt} = -\frac{\partial \mathcal{H}_1}{\partial \mathbf{r}_i} = -\sum_{j(\neq i)} \hat{\mathbf{r}}_{ij} u'(r_{ij}) , \tag{2.6b}$$

where u' denotes the derivative of u and $\hat{\mathbf{r}}_{ij}$ denotes a unit vector in the direction of $\mathbf{r}_i - \mathbf{r}_j$, using the minimum image convention.

A measurable structural or thermodynamic property F of the system is associated with a function $F(\mathbf{r}^N, \mathbf{p}^N; V)$ of the mechanical state of the system. The usual assumption of statistical thermodynamics is that the measured F is equal to the ensemble average of the function F over a suitably chosen ensemble of states. In this paper we are concerned with four ensembles: the microcanonical (NVE) ensemble, the canonical (NVT) ensemble, the isothermal–isobaric (NPT) ensemble, and the isoenthalpic–isobaric (NPH) ensemble.

The microcanonical ensemble average of the function F will be denoted $F_{NVE}(N, V, E)$, where the subscript denotes the nature of the ensemble and the arguments denote the numerical values of N, V, and E. It is defined as

$$F_{NVE}(N, V, E) \equiv [N!\,\Omega(N, V, E)]^{-1} \int_V d\mathbf{r}^N \int d\mathbf{p}^N \times \delta[\mathcal{H}_1(\mathbf{r}^N, \mathbf{p}^N; V) - E] F(\mathbf{r}^N, \mathbf{p}^N; V) , \tag{2.7}$$

where

$$\Omega(N, V, E) \equiv (N!)^{-1} \int_V d\mathbf{r}^N \int d\mathbf{p}^N \delta[\mathcal{H}_1(\mathbf{r}^N, \mathbf{p}^N; V) - E] \tag{2.8}$$

is the microcanonical ensemble partition function. Here $\delta[x]$ denotes the Dirac δ function. The $\mathbf{r}_i$ integrations extend over the volume V and the $\mathbf{p}_i$ integrations extend over all values from $-\infty$ to $+\infty$ for all components. The canonical average is

$$F_{NVT}(N, V, T) \equiv [N!\,Q(N, V, T)]^{-1} \int_V d\mathbf{r}^N \int d\mathbf{p}^N \times \exp\left(-\frac{\mathcal{H}_1(\mathbf{r}^N, \mathbf{p}^N; V)}{kT}\right) F(\mathbf{r}^N, \mathbf{p}^N; V) , \tag{2.9}$$

where

$$Q(N, V, T) \equiv (N!)^{-1} \int_V d\mathbf{r}^N \int d\mathbf{p}^N \exp\left(-\frac{\mathcal{H}_1(\mathbf{r}^N, \mathbf{p}^N; V)}{kT}\right) \tag{2.10}$$

and k is Boltzmann's constant. The isothermal-isobaric ensemble average is

$$F_{NPT}(N,P,T)=[N!\,\Delta(N,P,T)]^{-1}\int_0^\infty dV\int_V d\mathbf{r}^N\int d\mathbf{p}^N \times \exp\left(-\frac{[PV+\mathcal{H}_1(\mathbf{r}^N,\mathbf{p}^N;V)]}{kT}\right)F(\mathbf{r}^N,\mathbf{p}^N;V)\,, \qquad (2.11)$$

where

$$\Delta(N,P,T)=(N!)^{-1}\int_0^\infty dV\int_V d\mathbf{r}^N\int d\mathbf{p}^N \times \exp\left(-\frac{[PV+\mathcal{H}_1(\mathbf{r}^N,\mathbf{p}^N;V)]}{kT}\right). \qquad (2.12)$$

Finally, the isoenthalpic-isobaric ensemble average is

$$F_{NPH}(N,P,H)=[N!\,\Gamma(N,P,H)]^{-1}\int_0^\infty dV\int_V d\mathbf{r}^N\int d\mathbf{p}^N \times \delta[\mathcal{H}_1(\mathbf{r}^N,\mathbf{p}^N;V)+PV-H]F(\mathbf{r}^N,\mathbf{p}^N;V)\,, \qquad (2.13)$$

where

$$\Gamma(N,P,H)=(N!)^{-1}\int_0^\infty dV\int_V d\mathbf{r}^N\int d\mathbf{p}^N \times \delta[\mathcal{H}_1(\mathbf{r}^N,\mathbf{p}^N;V)+PV-H]\,. \qquad (2.14)$$

This ensemble is not as commonly used as the others. It is related to the NPT ensemble in the same way as the microcanonical ensemble is related to the canonical ensemble. Thus $\Delta(N,P,T)$ is the Laplace transform of $\Gamma(N,P,H)$ with respect to H, just as $Q(N;V,T)$ is the Laplace transform of $\Omega(N,V,E)$ with respect to E, with $(kT)^{-1}$ being the Laplace transform variable in both cases.

These ensembles are equivalent for the calculation of thermodynamic quantities.[7,8] Thus, for example, if T in the NVT ensemble is chosen so that the average value of the energy is E, then the NVE ensemble with these same values for N, V, and E will give the same value for any thermodynamic property for large values of N. More precisely, if F is an intensive property, then F_{NVE}, F_{NVT}, F_{NPT}, and F_{NPH} are equal except for differences of order N^{-1}, if the parameters of each ensemble are chosen so that all ensembles have the same average value of N, V, and E. (If F is extensive, the differences are of order N^0.) Because of these differences, the fluctuations of thermodynamic quantities about their average values are different. In this sense, the various ensembles are not equivalent.

Next we define a trajectory average or time average. Suppose $\mathbf{r}^N(t)$, $\mathbf{p}^N(t)$, and $V(t)$ are specified in some way for $t\geq 0$. Then the trajectory average of a function $F(\mathbf{r}^N,\mathbf{p}^N;V)$ is defined as

$$\overline{F}=\lim_{T\to\infty}T^{-1}\int_0^T dt\,F(\mathbf{r}^N(t),\mathbf{p}^N(t);V(t))\,, \qquad (2.15)$$

provided the limit exists.

In the usual molecular dynamics method, N and V are fixed, an initial choice of $\mathbf{r}^N(0)$ and $\mathbf{p}^N(0)$ is made, and Hamilton's equations of the form of (2.6) or their equivalent are solved numerically. Then trajectory averages are calculated for thermodynamic properties. The energy $\mathcal{H}_1$ is conserved along the trajectory, and the hypothesis is made that the trajectory spends equal times in all equal volumes with that same value of energy. It follows that

$$\overline{F}=F_{NVE}(N,V,E)\,, \qquad (2.16)$$

provided $\overline{F}$ is calculated from (2.15) using the solution of (2.6) and the values of N, V, and E on the right are those that correspond to the calculated trajectory. In other words, the trajectory average is equal to the microcanonical ensemble average, and hence the latter can be calculated from the molecular dynamics trajectory.[9]

In Secs. III–V, we show that there are ways of generating trajectories so that the trajectory average of any function is equal to the ensemble average of that function over the NPH, NVT, or NPT ensembles. These ways are modifications of the usual molecular dynamics, and thus they permit the use of molecular dynamics for the calculation of averages over these other ensembles.

III. MOLECULAR DYNAMICS AT CONSTANT PRESSURE

At constant pressure, the volume of a system of N particles fluctuates. To describe such fluctuations, we devise a molecular dynamics method in which the volume is a dynamical variable rather than a fixed quantity. The result will be a way of calculating trajectories so that the trajectory average of any property is equal to the NPH ensemble average of that property.

In the constant-pressure molecular dynamics method we replace the coordinates $\mathbf{r}_i$, $i=1,\ldots,N$, of the atoms by scaled coordinates, ρ_i, $i=1,\ldots,N$, defined in the following way

$$\rho_i=\mathbf{r}_i/V^{1/3},\quad i=1,2,\ldots,N\,. \qquad (3.1)$$

For $\mathbf{r}_i$ in the box of volume V, each component of ρ_i is a dimensionless number between zero and one. Consider the following Lagrangian, in which a new variable Q appears.

$$\mathcal{L}_2(\rho^N,\dot\rho^N,Q,\dot Q)=\tfrac{1}{2}mQ^{2/3}\sum_{i=1}^N\dot\rho_i\cdot\dot\rho_i-\sum_{i<j=1}^N u(Q^{1/3}\rho_{ij})+\tfrac{1}{2}M\dot Q^2-\alpha Q\,. \qquad (3.2)$$

(In evaluating the potential energy term, we use the minimum image convention for the ρ vectors in the cube of unit volume.) If we interpret Q as the volume V, then the first two terms on the right are just the Lagrangian of Eq. (2.3) expressed in terms of the new variables. The third term is a kinetic energy for the motion of Q, and the fourth represents a potential energy, $+\alpha Q$, associated with Q. Here α and M are constants.

This Lagrangian can be given a physical interpretation. Suppose the fluid to be simulated is in a container of variable volume. The fluid can be compressed by a piston. Thus, Q, whose value is the volume, is the coordinate of the piston, αV is a pV potential derived from an external pressure α acting on the piston, and

M is the mass of the piston. The piston is not of the usual cylindrical type that expands or contracts the system along only one direction; instead, a change in Q causes an isotropic expansion or contraction. This interpretation is not entirely consistent with Eq. (3.2). If Eq. (3.1) holds with $V=Q$, then

$$\dot{\mathbf{r}}_i = Q^{1/3}\dot{\rho}_i + \tfrac{1}{3}Q^{-2/3}\dot{Q}\rho_i \,, \qquad (3.3)$$

and the kinetic energy of the atoms should contain terms with factors of $\dot{Q}$ arising from the second term on the right. Such terms do not appear in Eq. (3.2).

Despite the absence of a consistent physical interpretation, Eq. (3.2) gives a well-defined Lagrangian, and we now analyze the dynamics it generates. We will call this system "the scaled system" to distinguish it from "the original system" whose Lagrangian is $\mathcal{L}_1$. The momentum conjugate to ρ_i will be denoted π_i.

$$\pi_i = \frac{\partial \mathcal{L}_2}{\partial \dot{\rho}_i} = mQ^{2/3}\dot{\rho}_i \,. \qquad (3.4)$$

The momentum conjugate to Q will be denoted Π.

$$\Pi = \frac{\partial \mathcal{L}_2}{\partial \dot{Q}} = M\dot{Q} \,. \qquad (3.5)$$

The Hamiltonian is

$$\mathcal{H}_2(\rho^N, \pi^N, Q, \Pi) = \sum_{i=1}^{N} \dot{\rho}_i \cdot \pi_i + \dot{Q}\Pi - \mathcal{L}_2(\rho^N, \dot{\rho}^N, Q, \dot{Q})$$

$$= (2mQ^{2/3})^{-1}\sum_{i=1}^{N} \pi_i \cdot \pi_i + \sum_{i<j=1}^{N} u(Q^{1/3}\rho_{ij}) + (2M)^{-1}\Pi^2 + \alpha Q \,. \qquad (3.6)$$

The Hamiltonian equations of motion are:

$$\frac{d\rho_i}{dt} = \frac{\partial \mathcal{H}_2}{\partial \pi_i} = \frac{\pi_i}{mQ^{2/3}} \qquad (3.7a)$$

$$\frac{d\pi_i}{dt} = -\frac{\partial \mathcal{H}_2}{\partial \rho_i} = -Q^{1/3} \sum_{j(\neq i)=1}^{N} \frac{\rho_{ij} u'(Q^{1/3}|\rho_{ij}|)}{|\rho_{ij}|} \qquad (3.7b)$$

$$\frac{dQ}{dt} = \frac{\partial \mathcal{H}_2}{\partial \Pi} = \frac{\Pi}{M} \qquad (3.7c)$$

$$\frac{d\Pi}{dt} = -\frac{\partial \mathcal{H}_2}{\partial Q} = -(3Q)^{-1}\Big(-2(2mQ^{2/3})^{-1}\sum_{i=1}^{N} \pi_i \cdot \pi_i + Q^{1/3}\sum_{i<j} \rho_{ij} u'(Q^{1/3}\rho_{ij}) + 3\alpha Q\Big) \,. \qquad (3.7d)$$

These equations of motion for the scaled system can be solved numerically to give the coordinates and momenta as a function of time. Such molecular dynamics calculations give trajectory for the scaled system: $\rho^N(t)$, $\pi^N(t)$, $Q(t)$, and $\Pi(t)$.

The trajectory average of any function, $G(\rho^N, \pi^N, Q, \Pi)$, of the coordinates and momenta of the scaled system is defined as in Eq. (2.15).

$$\overline{G} \equiv \lim_{T\to\infty} T^{-1}\int_0^T dt\, G(\rho^N(t), \pi^N(t), Q(t), \Pi(t)) \,. \qquad (3.8)$$

We assume that this time average is equal to an ensemble average of G over an NE ensemble, i.e., an ensemble with fixed energy and fixed number of particles. (This can be regarded as a microcanonical, NVE, ensemble for the scaled system, in which V is unity since the coordinates ρ_i are constrained to lie within a dimensionless unit volume.) It follows that

$$\overline{G} = G_{NE}(N, E) \qquad (3.9)$$

where

$$G_{NE}(N,E) = [N!\,\Omega_2(N,E)]^{-1}\int_1 d\rho^N \int d\pi^N \int_0^\infty dQ \int_{-\infty}^{\infty} d\Pi \times \delta[\mathcal{H}_2(\rho^N, \pi^N, Q, \Pi) - E]\,G(\rho^N, \pi^N, Q, \Pi) \qquad (3.10)$$

and

$$\Omega_2(N,E) \equiv (N!)^{-1}\int_1 d\rho^N \int d\pi^N \int_0^\infty dQ \int_{-\infty}^{\infty} d\Pi \times \delta[\mathcal{H}_2(\rho^N, \pi^N, Q, \Pi) - E] \,. \qquad (3.11)$$

In these two integrals, each ρ_i is integrated over the unit cube. The value of E to choose on the right-hand side of (3.9) is the constant energy of the trajectory used to calculate the left-hand side.

We now define a correspondence between the scaled system and the original system. This correspondence is given by

$$V = Q \,, \qquad (3.12a)$$

$$\mathbf{r}_i = Q^{1/3}\rho_i \,, \qquad (3.12b)$$

$$\mathbf{p}_i = \pi_i / Q^{1/3} \,. \qquad (3.12c)$$

Every state of the scaled system corresponds to a unique value of V and a unique point in the phase space of the original system for that volume V. (Note that Π does not appear in these equations, so each V and phase-space point in the original system corresponds to a manifold of states of the scaled system.)

Using this correspondence, the calculated trajectory for the scaled system can be used to generate a trajectory for the original system. Along this latter trajectory, the volume varies with time.

$$V(t) = Q(t) \,, \qquad (3.13a)$$

$$\mathbf{r}_i(t) = Q(t)^{1/3}\rho_i(t) \,, \qquad (3.13b)$$

$$\mathbf{p}_i(t) = \pi_i(t)/Q(t)^{1/3} \,. \qquad (3.13c)$$

The equations of motion for this trajectory can be derived from (3.13) and (3.7).

$$\frac{d\mathbf{r}_i}{dt} = \frac{\mathbf{p}_i}{m} + \tfrac{1}{3}\mathbf{r}_i \frac{d\ln V}{dt} \,, \qquad (3.14a)$$

$$\frac{d\mathbf{p}_i}{dt} = -\sum_{j(\neq i)=1}^{N} \hat{r}_{ij} u'(\mathbf{r}_{ij}) - \tfrac{1}{3}\mathbf{p}_i \frac{d\ln V}{dt} \,, \qquad (3.14b)$$

$$\frac{Md^2V}{dt^2} = -\alpha + \Big(\frac{2}{3}\sum_{i=1}^{N} \frac{\mathbf{p}_i \cdot \mathbf{p}_i}{2m} - \frac{1}{3}\sum_{i<j=1}^{N} r_{ij} u'(r_{ij})\Big)\Big/ V \,. \qquad (3.14c)$$

These equations are not the same as Hamilton's equations for the original system. Compare Eqs. (2.6). In the limit, however, that the mass of the piston, M, becomes infinitely large and $dV/dt=0$ initially, these equations become equivalent to the dynamical equations for the original system.

The trajectory defined by Eq. (3.14) can be used to calculate time averages of any function $F(\mathbf{r}^N, \mathbf{p}^N; V)$ according to Eq. (2.15). The most important result of this section is that *the time average of any F calculated from this trajectory is equal to the ensemble average of F for an isoenthalpic-isobaric ensemble in which the pressure is* α; i.e.,

$$\bar{F} = F_{NPH}(N, \alpha, H) , \tag{3.15}$$

for some appropriate value of H, except for negligible errors.

The proof is straightforward. For any function $F(\mathbf{r}^N, \mathbf{p}^N; V)$, we can define a corresponding $G(\rho^N, \pi^N, Q, \Pi)$ by

$$G(\rho^N, \pi^N, Q, \Pi) = F(Q^{1/3}\rho^N, \pi^N/Q^{1/3}; Q) . \tag{3.16}$$

Then it is clear that

$$\bar{G} = \bar{F} , \tag{3.17}$$

where the left-hand side is defined in Eq. (3.8) and the right-hand side in Eq. (2.15). Combining (3.17), (3.9), and (3.10), we find

$$\bar{F} = [N!\,\Omega_2(N, E)]^{-1} \int_1 d\rho^N \int d\pi^N \int_0^\infty dQ \int_{-\infty}^{\infty} d\Pi \times \delta[\mathcal{H}_2(\rho^N, \pi^N, Q, \Pi) - E] F(Q^{1/3}\rho^N, \pi^N/Q^{1/3}; Q). \tag{3.18}$$

The variables of integration are converted to V, $\mathbf{r}^N$, and ρ^N to obtain

$$\bar{F} = \frac{\int_{-\infty}^{\infty} d\Pi \int_0^\infty dV \int_V d\mathbf{r}^N \int d\mathbf{p}^N \,\delta[\mathcal{H}_1(\mathbf{r}^N, \mathbf{p}^N; V) + \alpha V + (2M)^{-2}\Pi^2 - E] F(\mathbf{r}^N, \mathbf{p}^N; V)}{\int_{-\infty}^{\infty} d\Pi \int_0^\infty dV \int_V d\mathbf{r}^N \int d\mathbf{p}^N \,\delta[\mathcal{H}_1(\mathbf{r}^N, \mathbf{p}^N; V) + \alpha V + (2M)^{-1}\Pi^2 - E]} , \tag{3.19}$$

which looks very much like an ensemble average of F, except for the Π integration, which plays no role in any ensemble for the original system. In both the numerator and denominator, however, the integrand for fixed Π is closely related to the isoenthalpic–isobaric ensemble. Using (2.13) and (2.14), we find

$$\bar{F} = \frac{\int_{-\infty}^{\infty} d\Pi\, \Gamma(N, \alpha, E - \Pi^2/2M) F_{NPH}(N, \alpha, E - \Pi^2/2M)}{\int_{-\infty}^{\infty} d\Pi\, \Gamma(N, \alpha, E - \Pi^2/2M)} \tag{3.20}$$

The ensemble average in the numerator can be expanded in a power series in $\Pi^2/2M$.

$$F_{NPH}(N, \alpha, E - \Pi^2/2M) = F_{NPH}(N, \alpha, E) - \frac{\Pi^2}{2M}\frac{\partial F_{NPH}(N, \alpha, E)}{\partial H} + O(N^{-2}) . \tag{3.21}$$

The correction term is proportional to $\partial^2 F_{NPH}/\partial H^2$. We have assumed that F represents an intensive property in estimating the order of magnitude of the correction term. (If F were extensive, the correction would be of order N^{-1}.) Thus

$$\bar{F} = F_{NPH}(N, \alpha, E) - \frac{\partial F_{NPH}(N, \alpha, E)}{\partial H} \times \frac{\int_{-\infty}^{\infty} d\Pi\, \Gamma(N, \alpha, E - \Pi^2/2M)(\Pi^2/2M)}{\int_{-\infty}^{\infty} d\Pi\, \Gamma(N, \alpha, E - \Pi^2/2M)} + O(N^{-2}) . \tag{3.22}$$

The ratio of the two integrals can easily be shown to be equal to $\overline{\Pi^2/2M}$. [To show this, use Eqs. (3.9) and (3.10) for $G = \Pi^2/2M$, and transform the resulting integrals to integrals over $\mathbf{r}^N$, $\mathbf{p}^N$, and V, as we did in going from (3.18) to (3.19).] Therefore,

$$\bar{F} = F_{NPH}(N, \alpha, E) - \overline{\Pi^2/2M}\, \partial F_{NPH}(N, \alpha, E)/\partial H + O(N^{-2})$$
$$= F_{NPH}(N, \alpha, E - \overline{\Pi^2/2M}) + O(N^{-2}) . \tag{3.23}$$

Q.E.D. Compare Eq. (3.15). (If F had represented an extensive quantity, the correction term would be of order N^{-1}.) In deriving this result, we have assumed that $\overline{\Pi^2/2M}$ is of order N^0. The variable Π is a momentum conjugate to a coordinate that is coupled to $6N$ other position and momentum variables. The momentum Π appears in the Hamiltonian only in a quadratic energy term, $\Pi^2/2M$. The average, $\overline{\Pi^2/2M}$, will therefore be $\frac{1}{2}kT$, where T is the temperature corresponding to the fixed N and E of the scaled system. It follows that $\overline{\Pi^2/2M}$ is intensive.

In other words, when the trajectory used to calculate time averages of the original system is obtained from a trajectory of the scaled system, the time average of any F is the *NPH* ensemble average of that F. The value of P for the ensemble is the value of α in the Lagrangian of the scaled system. The value of H for the ensemble is the energy of the trajectory of the scaled system minus the time average kinetic energy associated with the motion of Q.

Using the equivalence of the various ensembles, we can conclude from (3.23) that

$$\bar{F} = F_{NVE}(N, \bar{V}, E - \overline{\Pi^2/2M} - \alpha\bar{V}) + O(N^{-1})$$
$$= F_{NVT}(N, \bar{V}, T) + O(N^{-1})$$
$$= F_{NPT}(N, \alpha, T) + O(N^{-1}) , \tag{3.24}$$

where $T = k^{-1}\overline{\Pi^2/2M}$. Here the N^{-1} errors arise from the slight lack of equivalence of the ensembles. Note, however, that the error term in (3.23) is of order N^{-2}. Thus, the mean-square fluctuations in the *NPH* are equal to those that occur along the trajectory.

The results of this section provide a basis for a molecular dynamics simulation method for constant pressure. To simulate a fluid with Lagrangian $\mathcal{L}_1$, we construct the analogous scaled system with Lagrangian $\mathcal{L}_2$. The trajectory of the scaled system is converted into a trajectory for the fluid with volume fluctuations. Time averages along this trajectory can then be calculated. These time averages are equal to ensemble averages corresponding to the thermodynamic state with the desired pressure. In constructing the scaled system, we have to decide on the values of the constants α, E, and M.

The quantity α, which is a parameter appearing in $\mathcal{L}_2$ and in the equations of motion, is chosen to be the value of the pressure of the fluid thermodynamic state to be simulated.

The quantity E, which is the energy of the trajectory of the scaled system, is equal to the enthalpy of the fluid thermodynamic state to be simulated, except for a small correction of $\frac{1}{2}kT$. Thus, if desired, the value of this enthalpy can be precisely chosen in advance.

The quantity M is a parameter appearing in $\mathfrak{L}_2$ and in the equations of motion. It can be interpreted as the mass of a piston whose motion expands or compresses the fluid. The trajectory averages calculated from a simulation are independent of the value of M, as long as M is finite and positive. [This follows from Eq. (3.23) by the following argument. A particular value of E for the scaled system implies a particular temperature of the scaled system and, hence, a particular value of $\overline{\Pi^2/2M}$ that is independent of M. Hence, despite the appearance of M on the right-hand side of Eq. (3.23), the right-hand side is independent of M.] Hence, any finite positive value can be chosen, if the only goal of the simulation is to calculate equilibrium averages.

If the goal is also to simulate the dynamics of atoms in a small volume under constant pressure conditions, then it is important to choose an appropriate value of M. Equation (3.14c) can be interpreted as Newton's third law for the coordinate of a piston on which two forces act. The first is $-\alpha$ and the second is

$$\left(\frac{2}{3}\sum_{i=1}^{N}\frac{\mathbf{p}_i\cdot\mathbf{p}_i}{2M}-\frac{1}{3}\sum_{i<j=1}^{N}r_{ij}u'(r_{ij})\right)\Big/V .$$

The first is the negative of the external applied pressure and the second is the internal pressure of the fluid. An imbalance between these two forces causes an acceleration of the piston. The coordinate of the piston will fluctuate as the motion of the atoms causes the internal pressure to fluctuate. The time scale for this volume fluctuation will be determined by the mass of the piston, M. For a small sample of fluid imbedded in a much larger sample of fluid, the volume of the small sample will also fluctuate in response to an imbalance between the internal and external pressure. The time scale for the fluctuation of the volume of a sample of fluid is approximately equal to the length of the sample divided by the speed of sound in the sample. Thus, it is desirable to choose the mass M so that the time scale for the fluctuations of Q in the scaled system is approximately equal to $Q^{1/3}$ divided by the speed of sound in the fluid.

IV. MOLECULAR DYNAMICS AT CONSTANT TEMPERATURE

At constant temperature, the energy of a system of N particles fluctuates. In order to simulate such a system, we need some mechanism for introducing energy fluctuations. It might be possible to do this by inventing one or more additional degrees of freedom, as we did in the constant pressure case. We have not been able to do this in a practical way. Instead, we resort to the use of stochastic forces that act on the atoms of the sample and change their kinetic energy. The result will be a way of calculating trajectories so that the trajectory average of any property is equal to the NVT ensemble average of this quantity.

In the constant temperature molecular dynamics method, the equations of motion of the N particles in volume V are the Hamiltonian equations, Eq. (2.6), supplemented by a stochastic collision term in the equation for $d\mathbf{p}_i/dt$. Each stochastic collision is an instantaneous event that affects the momentum of one particle. The collisions suffered by a particle occur in accord with a Poisson process,[10,11] and the times at which different particles suffer collisions are statistically uncorrelated. Between stochastic collisions, the state of the system evolves in accordance with Eq. (2.6).

To perform the simulation we must first choose the numerical values of two parameters: T and ν. The first, T, is the desired temperature of the sample. The second, ν, is the mean rate at which each particle suffers stochastic collisions. The probability that a particular particle suffers a stochastic collision in any small time interval Δt is $\nu\Delta t$.

The times at which each particle suffers a collision is decided before beginning the simulation. This can be done by using random numbers to generate the values for the time intervals between successive collisions of a particle, such intervals being distributed according to

$$P(t)=\nu e^{-\nu t} , \tag{4.1}$$

where $P(t)\Delta t$ is the probability that an interval between collisions is between t and $t+\Delta t$. (Alternatively, as the calculation proceeds, random numbers can be used to decide which particles are to suffer collisions in any small time interval.)

We pick an initial set of positions and momenta $\mathbf{r}^N(0)$ and $\mathbf{p}^N(0)$, and integrate the Hamiltonian equations of motion until the time of the first stochastic collision. Suppose the particle suffering the collision is i. The value of the momentum of particle i after the collision is chosen at random from a Boltzmann distribution at temperature T. The change in momentum takes place instantaneously. All other particles are unaffected by the collision. Then the Hamiltonian equations for the entire collection of particles are integrated until the time of the next stochastic collision. This process is then repeated.

The result of this constant temperature molecular dynamics procedure is a trajectory, specified by $\mathbf{r}^N(t)$ and $\mathbf{p}^N(t)$, for N particles in a volume V with periodic boundary conditions. This trajectory can be used to calculate time averages of any function $F(\mathbf{r}^N, \mathbf{p}^N; V)$ according to Eq. (2.15). The central result of this section is that *under certain conditions* (see below) *the time average of any F calculated from this trajectory is equal to the ensemble average of F for the canonical ensemble in which the temperature is* T, i.e.,

$$\overline{F}=F_{NVT}(N,V,T) . \tag{4.2}$$

The proof of this theorem is very similar to the proof of the theorem that is basic to the use of Monte Carlo simulations to perform canonical ensemble averages.[11] First we note that the constant temperature molecular dynamics procedure generates a Markov chain in phase space.[12] The states in phase space for a finite number of particles are not countable. However, in practice the calculations will be performed using a finite number of significant figures, and so we assume that the number of states is countable. Thus we can apply the many powerful theorems about Markov chains with a countable number of states.

For certain Markov chains, the probability that the simulated system is in each state at time t approaches a limit as $t \to \infty$, and this limiting probability distribution is independent of the initial state of the system. Then the time average of F along the trajectory is equal to an ensemble average calculated with the unique limiting distribution. Sufficient conditions for this to be true are that the chain must have stationary transition probabilities, must be irreducible and aperiodic, and must have an invariant probability distribution.[13] The invariant probability distribution is the unique limiting distribution.

Let $a_j^{(n)}$ be the probability that the state of the system is j at the nth time. Then for a Markov chain with stationary transition probabilities,

$$a_j^{(n+1)} = \Sigma_k a_k^{(n)} p_{kj} , \tag{4.3}$$

where p_{kj}, the probability of making a transition from state k to state j is one time step, is independent of n. For the constant-temperature molecular dynamics procedure, the transitions are caused by Hamiltonian motion and stochastic collisions. This procedure is consistent with these equations and satisfies the definition of a Markov chain.[14]

An invariant probability distribution u_i is defined as one that satisfies the equations

$$u_j = \Sigma_k u_k p_{kj} \tag{4.4}$$

$$\Sigma_j u_j = 1 . \tag{4.5}$$

If

$$a_j^{(n)} = u_j , \tag{4.6}$$

it follows that

$$a_j^{(n+1)} = u_j , \tag{4.7}$$

and

$$a_j^{(m)} = u_j \tag{4.8}$$

for all $m \geq n$. An invariant distribution, hence, is one that, if it were the actual distribution at one time, would remain the distribution for all times. For the constant-temperature molecular dynamics procedure, the canonical distribution, which is

$$[N!Q(N, V, T)]^{-1} \exp[-\mathcal{H}_1(\mathbf{r}^N, \mathbf{p}^N; V)/kT]$$

is an invariant probability distribution if T in this formula is the same as the T that governs the stochastic collisions. This follows because, as a consequence of Liouville's theorem, the Hamiltonian motion leaves this distribution unchanged[15] and because stochastic collisions of the type discussed above obviously leave the distribution unchanged.

To prove that an irreducible chain is aperiodic,[16] it is sufficient[17] to prove that there is at least one state i with $p_{ii} > 0$. Consider a state in which the potential energy of the particles is a local or global minimum and in which all the momenta are zero. (Such a state must exist if the energy of the system has a lower bound.) This state has a nonzero value of p_{ii}.

The remaining condition is that the chain be irreducible, i.e., that every state can be reached from every other state in a finite amount of time. Since this is probably not true under all circumstances, we prefer to keep this as a condition in the statement of the theorem. Hence the theorem should be restated as: *if the Markov chain generated by the constant temperature molecular dynamics procedure is irreducible in phase space, the time average of any F calculated from a trajectory is equal to the ensemble average of F for the canonical ensemble in which the temperature is T*, i.e.,

$$\bar{F} = F_{NVT}(N, V, T) . \tag{4.2}$$

Q.E.D.

Next we must consider the conditions under which the Markov chain is irreducible. First let us consider the Markov chain that is generated by Hamiltonian motion without stochastic collisions. This is obviously not irreducible in phase space because Hamiltonian motion conserves energy and momentum. If the system is started in one state, only states on the same manifold of constant energy and momentum can be reached. Each stochastic collision in the modified molecular dynamics procedure can change both the energy and momentum of the system. One can easily devise a sequence of pairs of stochastic collisions that will change both the energy and momentum of the system in any desired way. Thus by stochastic collisions it is possible to make transitions from any energy–momentum manifold to any other such manifold. Suppose that the chain generated by Hamiltonian motion is irreducible on the manifold of states of constant energy and momentum; that is, suppose that a Hamiltonian trajectory starting on any state eventually passes through every state of the same energy and momentum. It follows that the chain for the modified molecular dynamics procedure is irreducible in all of phase space, and the theorem applies.

One of the ways of justifying the use of molecular dynamics trajectories to calculate microcanonical ensemble averages is to assume that the Hamiltonian motion is irreducible on a manifold of constant energy and momentum. The paragraph above shows that this same assumption for all energies and momenta justifies the use of the constant-temperature molecular dynamics method to calculate canonical ensemble averages.

Irreducibility of Hamiltonian motion on the constant energy–momentum manifolds is a sufficient but not necessary condition for irreducibility of the constant-temperature molecular dynamics motion in all phase space. It is easy to imagine that the stochastic collisions increase the freedom of motion in phase space to such an extent that the latter motion is irreducible, even though the former is not.

A major cause for concern about lack of irreducibility of the motion generated by the constant-temperature molecular dynamics procedure arises from the possibility that at high density the system may be trapped in certain regions of configuration space from which it cannot depart either in an infinitely long time or in the finite time of an actual simulation. Stochastic collisions may be of little help in eliminating this possibility because they do not directly affect the coordinates of the particles. If this situation actually exists for a particular fluid of interest and if the region of configuration space in which the system is trapped is atypical of most of the available configuration space, then trajectory averages calculated by the constant-temperature molecular dynamics will not be accurate approximations to canonical ensemble averages. These are conditions under which the usual molecular dynamics method and Monte Carlo method should also be expected to fail.

To perform a constant-temperature molecular dynamics calculation, it is necessary to choose a value of ν, the stochastic collision frequency for a particle. Equation (4.2) assures us, however, that the calculated trajectory averages are independent of the choice of ν if the irreducibility condition holds. In particular, the mean-square fluctuation of the total energy from its average value is independent of ν. The time dependence of the fluctuations will be very sensitive to ν, however. It is reasonable to choose ν so that the time for the decay of energy fluctuations along the trajectory will be the same as the time for decay of energy fluctuations of a small volume of real liquid surrounded by a much larger volume. This optimum value of ν can be estimated by the following procedure.

Consider a small sample of matter with volume V surrounded by a much larger heat bath of similar matter at temperature T. Suppose there is a temperature fluctuation in the small sample so that its average temperature is $T+\Delta T$. The small sample will gain or lose energy at a rate proportional to the temperature difference ΔT and to the thermal conductivity κ. By dimensional analysis, the rate of heat gain (in energy per unit time) is easily shown to be $-a\kappa\Delta T V^{1/3}$, where a is a dimensionless constant that depends on the shape of the sample and upon the temperature distribution within the small sample. The dynamics of this sample can be simulated by the constant-temperature molecular dynamics technique. Each stochastic collision changes the energy of the system by $-\frac{3}{2}k\Delta T$, since the average kinetic energy of a particle before collision is $\frac{3}{2}k(T+\Delta T)$ and after collision is $\frac{3}{2}kT$. The total rate of occurrence of stochastic collisions is $N\nu$. Hence the rate of energy gain is $-\frac{3}{2}N\nu k\Delta T$. If we equate these two expressions for the rate of energy gain, we find

$$\nu = \tfrac{2}{3}a\,\kappa V^{1/3}/kN = \tfrac{2}{3}a\,\kappa/k\rho^{1/3}N^{2/3}\,, \tag{4.9}$$

where

$$\rho = N/V \tag{4.10}$$

is the number density of particles. If the stochastic collisions are to simulate the effects of the surroundings of a collection of N atoms in a fluid, the collision frequency should be that given in this formula. Note that ν is of order $N^{-2/3}$, and the total collision rate for a sample is of order $N^{1/3}$. If some estimate of κ is available, then the proper choice of ν can be estimated using (4.10).

A crude but instructive estimate of ν can be obtained by imagining that N is unity, i.e., that the small system being simulated contains only one particle. Then all the other molecules comprise the heat bath, and it is reasonable to suppose that the required stochastic collision frequency should be the actual collision frequency ν_c for a particle. Applying Eq. (4.9) then, we find

$$\nu_c = (2a/3)\kappa/k\rho^{1/3}\,. \tag{4.11}$$

Hence (4.9) can be rewritten as

$$\nu = \nu_c/N^{2/3}\,. \tag{4.12}$$

For large enough N, the stochastic collision frequency will be much smaller than the intermolecular collision frequency. Therefore, for most of the time, most of the molecules will be moving according to the conservative equations of motion for a closed system. The stochastic interruptions will be infrequent, but they will cause the energy of the system to relax to a value appropriate to the temperature T at a rate appropriate for a system of N particles and will cause the energy to fluctuate about its equilibrium value with the magnitude appropriate to a canonical ensemble.

V. MOLECULAR DYNAMICS AT CONSTANT TEMPERATURE AND PRESSURE

At constant temperature and pressure, the energy, pressure, and enthalpy of a system of N particles fluctuate. In order to simulate such a system, we need some method for simulating these fluctuations. This can be done by introducing stochastic collisions into the constant-pressure molecular dynamics method. The result is a way to calculate trajectories so that the time average of any function F is equal to the NPT ensemble average of this quantity.

In the constant-temperature constant pressure molecular dynamics method we start with the Hamiltonian equations (3.7) for the scaled system. In addition, we imagine instantaneous stochastic collisions that affect the momentum of one particle at a time. As in Sec. IV, the collisions suffered by a particle occur in accord with a Poisson process, and the times at which the different particles suffer collisions are statistically uncorrelated. The mean frequency for these collisions is chosen according to the considerations discussed at the end of Sec. IV. (We could also have stochastic collisions that change the momentum Π of the piston.) Between stochastic collisions, the state of the system evolves in accordance with Eqs. (3.7).

Each stochastic collision is instantaneous, and so it occurs at a particular value of Q, the volume coordinate. The Boltzmann distribution for particle momentum π_i for a particular value of Q is proportional to

$$\exp[-\pi_i\cdot\pi_i/2mQ^{2/3}kT]\,.$$

The effect of each stochastic collision is to replace the momentum of the affected particle by a new value chosen at random from this distribution. (Note that the mean-square value of π_i depends on the value of Q.)

The calculated trajectory of the scaled system can

then be converted into a trajectory of the original system using the correspondence in Eq. (3.13). This latter trajectory can be used to calculate time averages of any function $F(\mathbf{r}^N, \mathbf{p}^N; V)$ according to Eq. (2.15). The most important result of this section is that *the time average of any F calculated from this trajectory is equal to the ensemble average of F for an isothermal–isobaric ensemble in which the pressure is α and the temperature is T*; i.e.,

$$\overline{F} = F_{NPT}(N, \alpha, T) , \tag{5.1}$$

where α is the parameter in the scaled Lagrangian and T is the temperature governing the effect of the stochastic collisions.

The proof of this theorem will merely be outlined, since it is similar to those in Secs. III and IV. For the function $F(\mathbf{r}^N, \mathbf{p}^N; V)$, we define the function $G(\rho^N, \pi^N, Q, \Pi)$ by Eq. (3.16). It follows that

$$\overline{F} = \overline{G} , \tag{5.2}$$

where the left-hand side is defined in Eq. (2.15) and the right-hand side in (3.8). Then we assume that the Hamiltonian motion of the scaled system is irreducible on a manifold of constant energy and constant total particle momentum in the phase space of the scaled system. Stochastic collisions will change the energy and momentum. It follows, as in Sec. IV, that the Markov chain generated by the combination of Hamiltonian motion and stochastic collisions is irreducible in the entire phase space of the scaled system. If $\alpha > 0$, an aperiodic state can be found by minimizing the $\mathcal{H}_2$. The stationary distribution for this Markov chain is a Boltzmann distribution at temperature T. It follows that

$$\overline{G} = G_{NT}(N, T) \equiv [N!\, Q_2(N, T)]^{-1} \int_1 d\rho^N \int d\pi^N \int_0^\infty dQ \int_{-\infty}^\infty d\Pi \times \exp\left(-\frac{\mathcal{H}_2(\rho^N, \pi^N, Q, \Pi)}{kT}\right) G(\rho^N, \pi^N, Q, \Pi) \tag{5.3}$$

where

$$Q_2(N, T) \equiv (N!)^{-1} \int_1 d\rho^N \int d\pi^N \int_0^\infty dQ \int_{-\infty}^\infty d\Pi \times \exp\left(-\frac{\mathcal{H}_2(\rho^N, \pi^N, Q, \Pi)}{kT}\right) . \tag{5.4}$$

Because Π appears in $\mathcal{H}_2$ only in a quadratic term that contains no other coordinates or momenta and because G is independent of Π [see Eq. (3.16)], the Π integrals in the numerator and denominator of Eq. (5.3) cancel. When the integrations over ρ^N, π^N, and Q are converted to integrations over $\mathbf{r}^N, \mathbf{p}^N$, and V using Eq. (3.12), the result is

$$\overline{G} = F_{NPT}(N, \alpha, T) . \tag{5.5}$$

Eq. (5.1) follows from (5.2) and (5.5).

Q.E.D.

VI. COMMENTS AND POTENTIAL APPLICATIONS

The methods discussed in this paper allow the dynamics of a system of a small number of particles at constant temperature and/or pressure to be simulated by calculation of trajectories in various ways. They simulate the effect of surrounding particles without creating undesirable surfaces. Moreover, the methods simulate not only the forces that drive the system to equilibrium at a given value of the temperature and/or pressure, but also the forces that cause the energy and/or volume of the system to fluctuate about their equilibrium values. The main conclusions of this paper are a set of theorems relating time averages along the calculated trajectory to ensemble averages in various ensembles.

The emphasis of the discussion has been on the calculation of ensemble averages of functions $F(\mathbf{r}^N, \mathbf{p}^N; V)$. An interesting, but still open, question is whether time correlation functions calculated along these trajectories are related to transport coefficients. For example, consider the momentum autocorrelation function, $C_{pp}(t)$. As a time integral along a trajectory it would be defined as

$$C_{pp}^{(t)}(t) \equiv N^{-1} \Sigma_i \lim_{T\to\infty} T^{-1} \int_0^T d\tau\, \mathbf{p}_i(t+\tau) \cdot \mathbf{p}_i(t) .$$

As an ensemble average, it would be defined as

$$C_{pp}^{(e)}(t) \equiv \langle N^{-1} \Sigma_i \mathbf{p}_i \cdot \mathbf{P}_i(t; \mathbf{r}^N, \mathbf{p}^N, V) \rangle$$

where $\mathbf{P}_i(t; \mathbf{r}^N, \mathbf{p}^N, V)$ is the momentum at time t of particle i, given that the state of the system at time 0 is $(\mathbf{r}^N, \mathbf{p}^N)$ in volume V, and the angular brackets denote an average over the ensemble of interest. In the calculation of $\mathbf{P}_i(t; \mathbf{r}^N, \mathbf{p}^N, V)$, the volume V is fixed and the motion should follow purely Hamiltonian dynamics. $C_{pp}^{(e)}(t)$ is related to the self-diffusion coefficient. Is it true that $C_{pp}^{(e)}(t) = C_{pp}^{(t)}(t)$, when the trajectory used on the right corresponds to the ensemble on the left in the sense of Secs. III–V of this paper?

More generally, it is true that the time evolution of any property of an N particle system simulated using the methods of Secs. III–V is the same as that of a real system of N particles surrounded by a much larger amount of similar matter at constant temperature and/or pressure? It is plausible that this is true for the constant-temperature–constant-pressure simulation if the parameters M and ν are chosen to make the time scales for the decay of volume and energy fluctuations in the simulation equal to values appropriate to the real system. It is even more plausible that properties calculated in simulations at constant pressure and temperature for small N are more similar to real properties than the usual constant energy constant volume simulations for the same small N. However, we have not been able to prove these conjectures.

These simulation methods are likely to be useful in a number of situations. (1) When a system is to be simulated for only a small number of thermodynamic states, these methods can be used to insure that desired values of temperature and/or pressure are achieved in each simulation. (2) The methods allow energy and density fluctuations to take place, thus perhaps facilitating the study of such processes as nucleation and phase separation which involve large fluctuations. (3) The methods allow the investigation of how a system responds to finite rates of heating, cooling, compression, and expansion. This may be useful for the study of the glass transition that takes place when a liquid is cooled and for the study of bubble formation that takes place when a liquid is heated and decompressed. (4) The methods allow the simulation of nonequilibrium processes such as chemical reactions and phase changes that release a large amount of energy. If such a process is simulated with the usual molecular dynamics method using a small system that has a small heat capacity, the temperature will rise to an unrealistic extent. In the case of chemical reactions, the rise in temperature may increase the rate of the reverse reaction in an unrealistic way. In the case of a phase transition to a phase of lower energy, the rise in temperature may destroy the newly formed nuclei of the emerging phase. The constant temperature simulations would eliminate these problems by allowing the extra energy to be released to the surroundings at a physically reasonable rate.

[1]The following references are review articles on the molecular dynamics and Monte Carlo methods. (a) J. P. Valleau and S. G. Whittington, in *Statistical Mechanics, Part A: Equilibrium Techniques*, edited by B. J. Berne (Plenum, New York, 1977), p. 137; (b) J. P. Valleau and G. M. Torrie, *ibid.*, p. 169; (c) J. J. Erpenbeck and W. W. Wood, in *Statistical Mechanics, Part B: Time-Dependent Processes*, edited by B. J. Berne (Plenum, New York, 1977), p. 1; (d) J. Kushick and B. J. Berne, *ibid.*, p. 41; (e) W. W. Wood, in *Physics of Simple Liquids*, edited by H. N. V. Temperley, G. S. Rushbrooke, and J. S. Rowlinson (North-Holland, Amsterdam, 1968), p. 116; (f) F. H. Ree, in *Physical Chemistry, An Advanced Treatise*, edited by D. Henderson (Academic, New York, 1971), Vol. VIIIA, p. 157; (g) B. J. Berne and D. Forster, Annu. Rev. Phys. Chem. 22, 563 (1971).

[2]The following references are review articles on the theory of liquids. (a) P. A. Egelstaff, Ann. Rev. Phys. Chem. 24, 159 (1973); (b) H. C. Andersen, *ibid.* 26, 145 (1975); (c) J. A. Barker and D. Henderson, Rev. Mod. Phys. 48, 587 (1976); (d) W. B. Streett and K. E. Gubbins, Annu. Rev. Phys. Chem. 28, 373 (1977); (e) D. Chandler, Annu. Rev. Phys. Chem. 29, 441 (1978).

[3]See, for example: B. J. Alder and T. E. Wainwright, Phys. Rev. Lett. 18, 988 (1967); W. G. Hoover and F. H. Ree, J. Chem. Phys. 49, 3609 (1968); L. V. Woodcock, J. Chem. Soc. Faraday II 72, 1667 (1976).

[4]See, for example: (a) A. Rahman, Phys. Rev. 136, A405 (1964); (b) L. Verlet, Phys. Rev. 159, 98 (1967); *ibid.* 165, 201 (1968); (c) A. Rahman, M. Mandell, and J. McTague, J. Chem. Phys. 64, 1564 (1976); (d) H. R. Wendt and F. F. Abraham, Phys. Rev. Lett. 41, 1244 (1978).

[5]See, for example: (a) F. H. Stillinger and A. Rahman, J. Chem. Phys. 60, 1545 (1974); (b) J. Owicki and H. A. Scheraga, J. Am. Chem. Soc. 99, 7413 (1977); (c) H. Popkie, H. Kistenmacher, and E. Clementi, J. Chem. Phys. 59, 1323 (1973); (d) P. J. Rossky and M. Karplus, J. Am. Chem. Soc. 101, 1913 (1979); (e) W. B. Streett and D. J. Tildesley, Proc. R. Soc. London A 348, 485 (1976).

[6]N. Metropolis, A. W. Metropolis, M. N. Rosenbluth, A. H. Teller, and E. Teller, J. Chem. Phys. 21, 1087 (1953).

[7]T. L. Hill, *An Introduction to Statistical Thermodynamics* (Addison-Wesley, Reading, 1960), p. 38.

[8]T. L. Hill, *Statistical Mechanics* (McGraw-Hill, New York, 1956), p. 110.

[9]In the dynamics calculation, the momentum of the system is also conserved. Thus, it would be more accurate to say that the trajectory average is equal to an ensemble average in which the total momentum is specified in addition to N, V, and E.

[10]W. Feller, *An Introduction to Probability Theory and Its Applications*, 3rd ed. (Wiley, New York, 1950), Vol. I, pp. 446–448.

[11]E. Parzen, *Stochastic Processes* (Holden-Day, San Francisco, 1962), pp. 117–123.

[12]Strictly speaking, it generates a Markov process rather than a Markov chain, since the time is a continuous rather than discrete variable. The dynamical equations describing the process will in practice be solved by using a discrete grid of times, and we assume that it is correct to regard the process as a chain and apply theorems that have been proven for chains.

[13]Reference 10, p. 394.

[14]Reference 10, p. 374.

[15]Any distribution function that expresses the probability density at a point in phase space as a function of only the value of the Hamiltonian at that point is invariant with respect to the motion generated by that Hamiltonian.

[16]An aperiodic chain is a chain all of whose states are aperiodic. See Ref. 10, p. 387, for the definition of an aperiodic state.

[17]If such a state exists, it is aperiodic (Ref. 10, p. 387), and hence the chain is aperiodic if it is irreducible (Ref. 10, p. 391).

Polymorphic transitions in single crystals: A new molecular dynamics method

M. Parrinello
University of Trieste, Trieste, Italy

A. Rahman
Argonne National Laboratory, Argonne, Illinois 60439

(Received 1 July 1981; accepted for publication 14 August 1981)

A new Lagrangian formulation is introduced; it can be used to make molecular dynamics (MD) calculations on systems under the most general, externally applied, conditions of stress. In this formulation the MD cell shape and size can change according to dynamical equations given by this Lagrangian. This new MD technique is well suited to the study of structural transformations in solids under external stress and at finite temperature. As an example of the use of this technique we show how a single crystal of Ni behaves under uniform uniaxial compressive and tensile loads. This work confirms some of the results of static (i.e., zero temperature) calculations reported in the literature. We also show that some results regarding the stress-strain relation obtained by static calculations are invalid at finite temperature. We find that, under compressive loading, our model of Ni shows a bifurcation in its stress-strain relation; this bifurcation provides a link in configuration space between cubic and hexagonal close packing. It is suggested that such a transformation could perhaps be observed experimentally under extreme conditions of shock.

PACS numbers: 64.70.Kb, 61.50.Ks

I. INTRODUCTION

The behavior of solids under the combined effects of external stress and of temperature has considerable practical relevance. Yet even in the idealized case of a perfect crystal, a detailed microscopic picture of such effects is still lacking. Most of the theoretical studies have been confined to conditions at zero temperature; in addition a perfect and prefixed crystalline arrangement of the atoms has been assumed. These two assumptions may lead to useful insights for relatively small values of the stress and temperature. However, it is obviously desirable to be able to study the behavior of solids at normal temperatures and high levels of external stress. In particular at high values of the stress spontaneous defect generation and/or crystal structure transformation become possible; this makes the assumption of a perfect, even if elastically distorted, crystalline arrangement untenable. Furthermore, these processes are sensitive to temperature variations as well.

The experience of the past two decades has shown that molecular dynamics (MD) calculations can provide a valuable tool to investigate nonharmonic effects in solids.[1] The authors have recently developed[2] a new MD method which allows a crystalline system to modify its structure if the temperature and external stress conditions make such a modification favorable. This method is therefore pertinent to the discussion of stress and temperature effects just mentioned above.

It is our purpose to present work based on this new MD method; the model system which we have studied has been described in a series of papers by Milstein and collaborators.[3–5] It is a system of classical particles interacting via a pairwise additive potential of the Morse type. The parameters of the potential have been adjusted to reproduce the elastic constants and the lattice parameter of Ni.[3] We have used this model of Ni for our study. The significance of the results for laboratory experiments will be dealt with in the last section.

Two reasons have prompted our interest in this model. First, Milstein and collaborators[3–5] have made extensive calculations of stress-strain relations for this model of Ni. Their calculations are a convenient check for our method of calculation at least for those values of temperature and stress where their theoretical approach is expected to be valid.

Second, in a recent paper an interesting possibility has been suggested by Milstein and Farber.[5] They have considered an fcc crystal of Ni under a uniform tensile [100] load. As the load is increased the crystal stretches in the direction of the load and contracts in the lateral directions. Beyond a certain value of the load and hence of the extension in the [100] direction, a new path in the stress-strain relation becomes possible along which the tetragonal symmetry is broken. Along this new branch, with *decreasing* load, the system starts to *expand* in one of the two lateral directions and eventually its extension in that lateral direction "catches up" with the [100] extension, the system ending up, at zero load, again in a tetragonal face-centered structure.

We show here that at finite temperature the above-mentioned "bifurcation" does not occur. Instead very close to the bifurcation point the system actually fails. Moreover the zero load tetragonal face-centered states (except the fcc one) mentioned in Ref. 5 as possible stable states spontaneously evolve into an hcp structure even at very low temperature.

A new result we have found is that, for this model of Ni, upon uniform uniaxial compression, the fcc structure transforms into an hcp arrangement.

II. MOLECULAR DYNAMICS AT CONSTANT EXTERNAL STRESS

Molecular dynamics (MD) methods have been extensively used in the past to study a variety of physical systems; for a detailed description the reader is referred to the book by Hansen and McDonald.[6] However, for the sake of clarity we shall briefly review the main features of MD.

It is a method for studying classical statistical mechanics of well-defined systems through a numerical solution of Newton's equations. A set of N classical particles have coordinates $\mathbf{r}_i$, velocities $\dot{\mathbf{r}}_i$ and masses $m_i, i = 1,...,N$. The particles interact through a potential $V_N(\mathbf{r}_1,...,\mathbf{r}_N)$ which, in most investigations is taken to be:

$$V_N = \frac{1}{2}\sum_i\sum_j \phi(r_{ij}), \tag{2.1}$$

where $r_{ij} = |\mathbf{r}_{ij}| = |\mathbf{r}_i - \mathbf{r}_j|$. More general forms of V_N can also be used, perhaps at the cost of more computational labor. Newton's equations are then

$$m_i\ddot{\mathbf{r}}_i = \sum_{j\neq i}\frac{1}{r_{ij}}\frac{d\phi}{dr_{ij}}\mathbf{r}_{ij}, \quad i = 1,...,N, \tag{2.2}$$

and are solved numerically. As the system evolves in time it eventually reaches equilibrium conditions in its dynamical and structural properties; the statistical averages of interest are calculated from $\mathbf{r}_i(t)$ and $\dot{\mathbf{r}}_i(t), i = 1,...,N$, as temporal averages over the trajectory of the system in its phase space. For practical reasons N is restricted to at most a few thousand and for small systems surface effects are obviously very important. However, to simulate a bulk system the common practice is to use periodic boundary conditions. These are obtained by periodically repeating a unit cell of volume Ω containing the N particles by suitable translations. Periodic boundary conditions obviously give a system in which the N particles are always contained in a cell of volume Ω, and without loss of generality every particle can be thought of as being at the "center." In other words, the summation over j in Eq. (2.2) extends over the infinite system generated by the periodic boundary conditions.

As a consequence of V_N being a function of $\mathbf{r}_i$ only [Eq. (2.1)], the solution of Eq. (2.2) conserves the total energy E of the system; thus the statistical ensemble generated in a conventional MD calculation is a (Ω,E,N) ensemble or a microcanonical ensemble. We shall use (.....) to indicate quantities whose constancy characterizes a given statistical ensemble.

The restriction that the MD cell be kept constant in volume and in shape severely restricts the applicability of the method to problems involving crystal structure transformations; in such transformations changes in the shape of the cell most obviously play an essential role. (For example, in a plane four points at the vertices of a square together with one at the center can obviously be used to generate a square lattice of points; this can become a lattice of equilateral triangles only if the square is allowed to become a $1:\sqrt{3}$ rectangle.)

In order to overcome this difficulty we[2] have modified a method due to Andersen[7] so as to allow for changes in volume *and* shape of the MD cell containing a system of particles under constant external hydrostatic pressure. Note that in the method of Andersen[7] only changes in the volume of the MD cell were possible but not in its shape. Thus crystal structure transformations are inhibited in Andersen's method because of the suppression of the essential fluctuations, namely those in the shape of the MD cell.

This extra degree of flexibility was introduced into the MD method as follows[2]: As before the system consists of N particles in a cell that is periodically repeated to fill all space. However, the cell can have arbitrary shape and volume being completely described by three vectors **a**, **b**, and **c** that span the edges of the MD cell. The vectors **a**, **b**, and **c** can have different lengths and arbitrary mutual orientations. An alternative description is obtained by arranging the vectors as $\{\mathbf{a}, \mathbf{b}, \mathbf{c}\}$ to form a 3×3 matrix $\mathbf{h}$ whose columns are, in order, the components of **a**, **b**, and **c**. The volume is given by

$$\Omega = \|\mathbf{h}\| = \mathbf{a}\cdot(\mathbf{b}\wedge\mathbf{c}); \tag{2.3}$$

a, **b**, and **c**, in that order, are assumed to be a right-handed triad.

The position $\mathbf{r}_i$ of a particle i can be written in terms of $\mathbf{h}$ and of a column vector $\mathbf{s}_i$, with components ξ_i,η_i, and ζ_i, as

$$\mathbf{r}_i = \mathbf{h}\mathbf{s}_i = \xi_i\mathbf{a} + \eta_i\mathbf{b} + \zeta_i\mathbf{c}. \tag{2.4}$$

Obviously $0 \leqslant \xi_i,\eta_i,\zeta_i \leqslant 1$ is the range of variation of the numbers $\xi_i,\eta_i,\zeta_i, i = 1,...,N$. The images of $\mathbf{s}_i$ are at $\mathbf{s}_i + (\lambda,\mu,\nu)$ where λ, μ, and ν are integers from $-\infty$ to $+\infty$.

Let a prime $'$ denote a transpose of a vector or a tensor in the usual way. Then the square of the distance between i and j is given by

$$r_{ij}^2 = (\mathbf{s}_i - \mathbf{s}_j)'\mathbf{G}(\mathbf{s}_i - \mathbf{s}_j), \tag{2.5}$$

where the metric tensor $\mathbf{G}$ is

$$\mathbf{G} = \mathbf{h}'\mathbf{h}. \tag{2.6}$$

To complete the notation used here we note finally that the reciprocal space is spanned by the vectors

$$\frac{2\pi}{\Omega}\{\mathbf{b}\wedge\mathbf{c},\mathbf{c}\wedge\mathbf{a},\mathbf{a}\wedge\mathbf{b}\} \equiv \frac{2\pi}{\Omega}\boldsymbol{\sigma}. \tag{2.7}$$

The matrix $\boldsymbol{\sigma} \equiv \Omega\mathbf{h}'^{-1}$, carries information concerning the size and orientation of the MD cell.

A. The case when only hydrostatic pressure is applied

In Ref. (2) variability in the shape and size of the MD cell was obtained as follows: the usual set of $3N$ dynamical variables, that describe the positions of the N particles, was augmented by the nine components of $\mathbf{h}$. The time evolution of the $3N + 9$ variables was then obtained from the Lagrangian

$$\mathscr{L} = 1/2\sum_{i=1}^{N} m_i\dot{\mathbf{s}}_i'\mathbf{G}\dot{\mathbf{s}}_i - \sum_{i=1}^{N}\sum_{j>i}^{N}\phi(r_{ij}) + 1/2 W \,\mathrm{Tr}\, \dot{\mathbf{h}}'\dot{\mathbf{h}} - p\Omega, \tag{2.8}$$

where p is the hydrostatic pressure that we intended to impose on the system. We shall comment later on W, which has dimensions of mass. Whether such a Lagrangian is derivable from first principles is a question for further study; its validity can be judged, as of now, by the equations of motion and the statistical ensembles that it generates. From Eq. (2.8) the equations of motion are easily found. We get

$$\ddot{\mathbf{s}}_i = -\sum_{j\neq i} m_i^{-1}(\phi'/r_{ij})(\mathbf{s}_i - \mathbf{s}_j) - \mathbf{G}^{-1}\dot{\mathbf{G}}\dot{\mathbf{s}}_i, i = 1,...,N, \tag{2.9}$$

$$W\ddot{\mathbf{h}} = (\boldsymbol{\pi} - p)\boldsymbol{\sigma}, \tag{2.10}$$

where, using the usual dyadic notation, and writing $\mathbf{v}_i = \mathbf{h}\dot{\mathbf{s}}_i$,

$$\Omega\boldsymbol{\pi} = \sum_i m_i \mathbf{v}_i \mathbf{v}_i - \sum_i \sum_{j>i} (\phi'/r_{ij}) \mathbf{r}_{ij} \mathbf{r}_{ij}. \tag{2.11}$$

When $\mathbf{h}$ = constant, i.e., when the MD cell is time independent, $\dot{\mathbf{G}} = 0$, and due to Eq. (2.4), Eq. (2.9) becomes identical to Eq. (2.2). Of course the pressure in the system cannot be controlled; its value can be obtained from 1/3 of the trace of the average of $\boldsymbol{\pi}$ in the usual way.

The equations derived by Hoover *et al.*[8] have a close affinity with Eq. (2.9) above. Their two first-order equations of motion are equivalent to Eq. (2.9) on identifying $\dot{\mathbf{h}}\mathbf{h}^{-1}$ with their strain rate tensor transpose. Thus $\dot{\mathbf{h}}$ = (strain-rate tensor)$'\mathbf{h}$ and hence, as desired by them, $\mathbf{h}$ is driven by the strain-rate tensor. Their equations are thus suitable for the study of externally driven nonequilibrium phenomena.

Equation (2.10), however, allows the system to be driven by the dynamic imbalance between the externally applied stress and the internally generated stress tensor [the more general case is given in Eq. (2.25) below]; thus Eq. (2.10) allows one to study nonequilibrium phenomena driven by the above mentioned imbalance. In a state of equilibrium, making the external stress have an oscillatory time dependance will also allow one to study frequency dependent response of the system to external stimuli of various kinds. From Eq. (2.10) it also follows that the mass W determines the relaxation time for recovery from an imbalance between the external pressure and the internal stress. As discussed by Andersen[7], an appropriate choice for the value of W can make this relaxation time of the same order of magnitude as that of the relaxation of a small portion of a much larger sample. His suggestion is for a choice of W such that the above-mentioned relaxation time is of the same order of magnitude as the time L/c, where L is the MD cell size and c is the sound velocity. This obviously eliminates the arbitrariness in the choice of W and makes the calculation more realistic. However, if one is interested only in static averages, W can be chosen on the basis of computational convenience. In fact, in classical statistical mechanics, the equilibrium properties of a system are independent of the masses of its constituent parts.

From Eq. (2.8) one can construct the corresponding Hamiltonian following the usual rules of mechanics. Since the system is not subject to time dependent external forces this is a constant of motion. We get

$$\mathscr{H} = \sum_i 1/2 m_i \mathbf{v}_i^2 + \sum_i \sum_{j>i} \phi(r_{ij}) + 1/2 W \,\mathrm{Tr}\, \dot{\mathbf{h}}'\dot{\mathbf{h}} + p\Omega. \tag{2.12}$$

In equilibrium, at temperature T, $9/2 k_B T$ is contributed by the term with W and $3N/2\, k_B T$ by the other kinetic terms. Therefore to an accuracy of $3{:}N$ one finds that the constant of motion $\mathscr{H}$ is nothing but the enthalpy

$$H = E + p\Omega, \tag{2.13}$$

where

$$E = \sum_i 1/2 m_i \mathbf{v}_i^2 + \sum_i \sum_{j>i} \phi(r_{ij}). \tag{2.14}$$

Hence the Lagrangian in Eq. (2.8) generates a (p, H, N) ensemble.[7]

B. The case when a general stress is applied

The above formulation of MD, briefly presented before,[2] lends itself quite naturally to the introduction of *nonisotropic* external stress. This is not difficult to realize since in the classical theory of elasticity the notion of strain is intimately connected to the variations in the metric tensor and, as has surely been noticed, $\mathbf{G}$ is a natural constituent of our MD scheme.

In order to make the above remark practicable, we need to introduce, as is usually done in elasticity theory,[9] a reference state. Using the notions already introduced, this reference state of the system can be defined by its matrix $\mathbf{h}_0$ and volume $\Omega_0 = \|\mathbf{h}_0\|$. In this reference state a point in space given by the coordinate vector $\mathbf{s}$ is at the position

$$\mathbf{r}_0 = \mathbf{h}_0 \mathbf{s}. \tag{2.15}$$

A homogeneous distortion of the system changes $\mathbf{h}_0$ to $\mathbf{h}$, moving $\mathbf{r}_0$ to $\mathbf{r}$ where

$$\mathbf{r} = \mathbf{h}\mathbf{s} = \mathbf{h}\mathbf{h}_0^{-1}\mathbf{r}_0, \tag{2.16}$$

giving the displacement $\mathbf{u}$ due to the distortion:

$$\mathbf{u} = \mathbf{r} - \mathbf{r}_0 = (\mathbf{h}\mathbf{h}_0^{-1} - 1)\mathbf{r}_0. \tag{2.17}$$

Following Landau and Lifshitz[9] to define the strain tensor $\boldsymbol{\epsilon}$, and using x_μ to denote the components of $\mathbf{r}_0$,

$$\epsilon_{\lambda\mu} = 1/2\left(\frac{\partial u_\lambda}{\partial x_\mu} + \frac{\partial u_\mu}{\partial x_\lambda} + \sum \frac{\partial u_\nu}{\partial x_\mu}\frac{\partial u_\nu}{\partial x_\lambda}\right), \tag{2.18}$$

we find, using Eq. (2.6) which defines $\mathbf{G}$ and Eq. (2.17) which defines $\mathbf{u}$, that

$$\boldsymbol{\epsilon} = 1/2(\mathbf{h}_0'^{-1}\mathbf{G}\mathbf{h}_0^{-1} - 1) \tag{2.19}$$

(We give in an appendix the connection between this formal definition of $\boldsymbol{\epsilon}$ in Eq. (2.19) and that given in elementary text books.)

Having identified the strain $\boldsymbol{\epsilon}$, an expression for the elastic energy, V_{el}, can now be written. If $\mathbf{S}$ is the external stress[10] and p the hydrostatic pressure,

$$V_{el} = p(\Omega - \Omega_0) + \Omega_0 \,\mathrm{Tr}\,(\mathbf{S} - p)\boldsymbol{\epsilon}. \tag{2.20}$$

In the limit of small strain,

$$\mathrm{Tr}\boldsymbol{\epsilon} \simeq \Delta\Omega/\Omega_0 = (\Omega - \Omega_0)/\Omega_0. \tag{2.21}$$

Hence, when Eq. (2.21) is a valid approximation, we get the more familiar expression,

$$V_{el} \simeq \Omega_0 \,\mathrm{Tr}\, \mathbf{S}\boldsymbol{\epsilon}. \tag{2.22}$$

Otherwise, i.e. when Eq. (2.21) is not valid, we need Eq. (2.20) to get the correct description of the effects of hydrostatic pressure.

To generalize the Lagrangian of Eq. (2.8) we need to substitute V_{el} of Eq. (2.20) in place of $p\Omega$ in Eq. (2.8) for $\mathscr{L}$. This gives us the new Lagrangian $\mathscr{L}_s$,

$$\mathscr{L}_s = \mathscr{L} - 1/2\,\mathrm{Tr}\,\boldsymbol{\Sigma}\mathbf{G}, \tag{2.23}$$

where the symmetric tensor $\boldsymbol{\Sigma}$ is related to the stress $\mathbf{S}$:

$$\boldsymbol{\Sigma} = \mathbf{h}_0^{-1}(\mathbf{S} - p)\mathbf{h}_0'^{-1}\Omega_0. \tag{2.24}$$

In deriving Eq. (2.23) we have dropped inconsequential con-

Notes on p. 297

stant terms in the energy viz. $p\Omega_0$ and Ω_0 Tr$(\mathbf{S} - p)$. We have also used the identity Tr(**AB**) = Tr(**BA**).

Using Eq. (2.23) to write the Lagrangian equations of motion we get Eq. (2.9) as before but Eq. (2.10) is now replaced by

$$W\ddot{\mathbf{h}} = (\pi - p)\sigma - \mathbf{h}\Sigma. \tag{2.25}$$

It is easy to see that, analogous to Eq. (2.13), the Lagrangian $\mathscr{L}_s$ gives rise to a $(\mathbf{S}, H_s, N)$ ensemble where the generalized enthalpy is

$$H_s = E + V_{\rm el}, \tag{2.26}$$

where E is given by Eq. (2.14) and $V_{\rm el}$ by Eq. (2.20).

The equations of motion Eq. (2.25) imply that a state of equilibrium will necessarily give zero for the average value of the right side of Eq. (2.25). This makes, using the definition of Σ, and writing σ_0 for the equivalent of Eq. (2.7),

$$\langle(\pi - p)\sigma\rangle = \langle\mathbf{h}\rangle\mathbf{h}_0^{-1}(\mathbf{S} - p)\sigma_0. \tag{2.27}$$

This suggests, as is otherwise obvious intuitively, that for a system in equilibrium, to relate the constant matrix Σ [which controls the trajectories via Eq. (2.25)] to the external stress matrix **S** through Eq. (2.24), the most reasonable choice for the reference state appears to be $\mathbf{h}_0 = \langle\mathbf{h}\rangle$.

III. SUMMARY OF PREVIOUS STATIC CALCULATIONS

Before discussing our MD calculations in the next section and to put them in proper context, it is appropriate to recapitulate some of the results obtained by Milstein and collaborators.[3–5] In their model for Ni they assume a pairwise additive potential for the system, the pair potential being

$$\phi(r) = D\{\exp[-2\alpha(r - r_0)] - 2\exp[-\alpha(r - r_0)]\}. \tag{3.1}$$

The constants D, α, and r_0 are fixed from a fit to the elastic constants c_{11} and c_{12} and to the lattice constant a_0 of fcc nickel.[3] Using this potential, all the properties are calculated at zero temperature, starting with a perfect fcc arrangement. The details of the procedure to calculate the stress-strain relation are given in Ref. 5. We only recall that in these calculations the lengths a_1, a_2, and a_3 of the initially cubic cell are allowed to adjust to changing stress conditions, but the angles between the cell edges are constrained to remain right angles. Moreover, for each set of a_1, a_2, and a_3 values, the atoms are given appropriately modified lattice positions but of course without any thermal disorder. The results that are obtained by Milstein and Farber[5] are as follows, when the system is subjected to a homogeneous [100] load.

Along a "primary path" in the stress-strain relation (as indicated in Fig. 1) one has $a_1 \neq a_2 = a_3$ and along this path three structures can be identified at zero load (Fig. 1). One of the three obviously is the original undistorted fcc state for which $a_1 = a_2 = a_3 = a_0$. The second, $B^{(1)}$, is a bcc structure obtained in compression (i.e. $a_1 < a_0$) at $a_1 \simeq 0.80\, a_0$, $a_2 = a_3 = \sqrt{2}a_1$. The third, $T^{(1)}$, is a tetragonal state, also obtained in compression with $a_1 \simeq 0.75a_0, a_2 = a_3 \simeq 1.59a_1$. The fcc state is at the absolute minimum of energy, $B^{(1)}$ at a local maximum and $T^{(1)}$ at a local minimum.[5] From a consideration of the Born stability criteria one concludes[4] that $B^{(1)}$

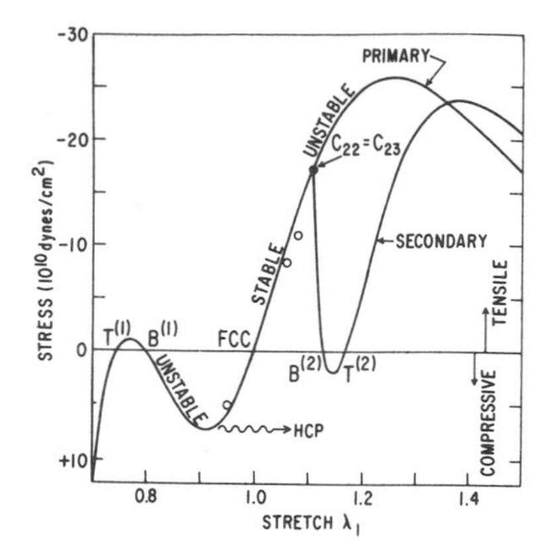

FIG. 1. Stress-strain relation under uniaxial load. O denote our results, lines are the static calculations of Milstein and Farber.[5] We found: (i) System failure at point (marked $c_{22} = c_{23}$) of intersection of the primary and secondary paths of Ref. 5; (ii) B and T tetragonal states on the zero load line spontaneously evolve into hcp structures; (iii) Under extreme compressive loading (marked by ⇝) the system changes to an hcp structure.

and $T^{(1)}$ should be unstable states.

Along the same "primary path" but when a_1 is increased under the action of a tensile load the system is predicted[4] to fail at a value of stress $= 16 \times 10^{10}$ dyne/cm^2 because at that point the Born stability condition $c_{22} - c_{23} > 0$ is violated. At this point $a_1 = 1.107\, a_0$.

However, it was discovered by Milstein and Farber[5] that at the point where $c_{22} = c_{23}$, a "secondary path" branches out of the primary path of extension. Along the secondary path, with decreasing load, the tetragonal symmetry is broken, i.e. $a_2 \neq a_3$ along this secondary path. This point, at which $c_{22} = c_{23}$, is hence a bifurcation point. Along the secondary path the zero load condition is encountered at two points where tetragonal symmetry is reestablished. One, a bcc state, $B^{(2)}$ at $a_1 = a_2 = \sqrt{2}a_3$ $= 1.1293\, a_0$ and the other a tetragonal state, $T^{(2)}$, with $a_1 = a_2 = 1.5701\, a_3 = 1.1696\, a_0$; $B^{(2)}$ is at a local energy maximum and $T^{(2)}$ at a local minimum.

The existence of this secondary path was envisaged in Ref. 5 as a possible mechanism for an fcc to bcc transition under conditions of a strictly uniaxial [100] tensile load.

IV. MOLECULAR DYNAMICS RESULTS

An MD calculation on Ni using the Lagrangian $\mathscr{L}_s$ of Eq. (2.23) makes it possible to check the validity of the results summarized above. This is because in MD with the Lagrangian $\mathscr{L}_s$ the restrictions of zero temperature and preassigned crystalline arrangement can both be removed.

As is customary we shall use reduced (i.e. dimensionless) quantities to specify various physical parameters. These reduced quantities will be denoted by an asterisk.

We shall use $D = 0.35059 \times 10^{-12}$ erg to be the unit of energy; $L = r_0/2^{1/6} = 2.2518$ Å that of length; m the mass of the Ni atom, that of mass. Then $\tau = (mL^2/D)^{1/2} = 0.375 \times 10^{-12}$ sec will be the unit of time. D and r_0 are the quantities occurring in the potential[3] Eq. (3.1). All calculations were made with $p^* = 0$, $W^* = 20$. (This choice of W^* is suggested by our previous experience[2] in using the Lagrangian $\mathscr{L}$). Since we were interested in static averages no attempt was made to calibrate W^* so as to obtain a realistic value for the relaxation times related with the behavior of Eq. (2.25). As discussed in Sec. 2 the choice of W^* only affects the calculation of dynamical correlations.

Throughout, we have monitored the structure of the system by calculating the pair correlation function $g(r)$ as was done previously.[2] In certain situations it was found useful to plot out all the particle coordinates in suitable two-dimensional "slices" of the three-dimensional system so as to get a visual impression of the structure.

Since the various MD calculations fall into distinct categories, these will now be identified through subsections and suitable subheadings.

A. Preparation of a system under conditions of zero stress

The genesis of this and all subsequent calculations was a 500-particle system of Ni atoms on a perfect fcc lattice. The initial value of $\mathbf{h}^*$ was $h^*_{ij} = 1^*_0\delta_{ij}, 1^*_0 = 7.8244$. Hence the MD cell at the start was a perfect cube. A small random displacement of the particles from the lattice sites and zero velocities provided the initial conditions for the ensuing dynamics. The equations of motion, Eqs. (2.9) and (2.25), were solved, with $\Delta t^* = 0.01$, using the predictor-corrector algorithm.[11] For this calculation Σ was put equal to zero in Eq. (2.25). Initially the temperature of the system was controlled with the standard procedures of MD. After a long period of "aging" to allow for the establishment of equilibrium, an MD run was made in which the temperature fluctuated around the mean value $T^* = 0.14$ (i.e.350° K). The MD cell remained a cube to high accuracy (i.e., the nondiagonal elements of $\mathbf{h}$ fluctuated around essentially zero values). The mean value of the three cell edges was the same, being $1^*_{0.14} = 7.88 \pm 0.02$. The pair correlation showed an unmodified fcc structure.

B Compressive uniaxial loading

A configuration (i.e., the values and the derivatives of all the dynamical variables) of the equilibrium run just described was used as the initial condition for a calculation in which a uniaxial [100] compressive load was applied. In other words Σ^*_{11} was nonzero positive, all other Σ^*_{ij} being 0. Σ^*_{11} was raised to a value $\Sigma^*_{11} = 15$ using two short intermediate runs at $\Sigma^*_{11} = 4$ and 8. Under the action of such a load the matrix $\mathbf{h}$ starts to change in a very well defined manner, i.e., the MD cell starts to distort away from its initial cubic shape. As expected there is a contraction in the [100] direction and an expansion in [010] and [001] directions while preserving the tetragonal symmetry to high accuracy. At $\Sigma^*_{11} = 15$ a long run of 5000 Δt^* was made. All averages were calculated with the last 2290 Δt^* of this run. We found $\langle h^*_{11}\rangle = 7.45 \pm 0.05, \langle h^*_{22}\rangle = \langle h^*_{33}\rangle = 8.08 \pm 0.05$, giving a "stretch" $\lambda_1 = \langle h^*_{11}\rangle / 1^*_{0.14} = 0.95$. The nondiagonal $\langle h^*_{ij}\rangle$ were zero to within ± 0.05. Using $\langle \mathbf{h}^*\rangle$ for $\mathbf{h}^*_0$ one gets, from Eq. (2.24), $S_{11} = 5.25 \times 10^{10}$ dynes/cm^2 for Ni, as seen in Fig. 1. This is in good agreement with the stress-strain relation given by Milstein and Farber.[5] (We note here that the nondiagonal elements of $\langle \mathbf{h}^*\rangle$ were so small as to be negligible; their inclusion or otherwise has no effect on the value quoted above for S_{11}).

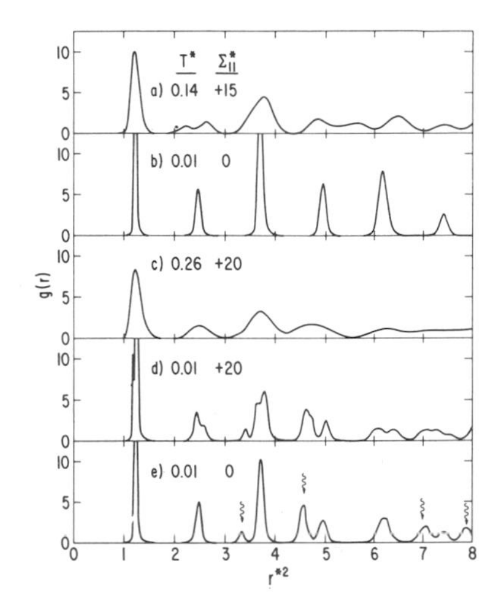

FIG. 2. Plot of pair correlation $g(r)$ to show structural transformation produced during compressive [100] loading. Note that abscissa denotes r^{*2} (not r^*). If r^*_i is the distance of the ith shell, r^{*2}_i/r^{*2}_1 have the values 1, 2, 3, 4, 5, etc. for fcc and 1, 2, 8/3, 3, 11/3, 4, 5, etc. for hcp ordering. (a) Shows fc tetragonal structure at $\Sigma^*_{11} = 15$ (compressive [100] load) and $T^* = 0.14$. Note the splitting of the second peak corresponding to a "stretch" $\lambda_1 = 0.95$ (Sec. B). (b) System reverts back to fcc structure on unloading. Removal of thermal effects by quenching[12] to $T^* = 0.01$ shows perfect fcc shell structure (Sec. B). (c) At $\Sigma^*_{11} = 20$ large changes occur in the MD cell parameters (shown in Fig. 3). $g(r)$ shows some differences at large r; compare with (a) above (Sec. C). (d) At $\Sigma^*_{11} = 20$, after completion of structural changes, quenching reveals shells of hcp ordering in addition to peak splitting as in (a) above (Sec. C). (e) Unloading the system shown in (d) to zero load, letting it equilibrate at $T^* = 0.21$ then quenching the system leads to unambiguous shell structure of hcp ordering (Sec. C).

The pair correlation of the system in equilibrium at $\Sigma^*_{11} = 15, T^* = 0.14$, is shown in Fig. 2(a). Note that $g(r)$, the pair correlation, is monitored as a function of $\mathbf{r}^2$; the second peak in Fig. 2(a) is split because the six original next nearest equidistant neighbors of the fcc structure break up into sets of two and four neighbors at slightly different distances. This splitting is not visible in the first peak because of thermal motion. The behavior of $\langle \mathbf{h}^*\rangle$ already made it clear that we had obtained a face-centered tetragonal structure. The $g(r)$ simply was a confirmation.

The fact that this equilibrium state under a uniaxial compressive load is reversibly connected to the original fcc

Notes on p. 297

state was easy to demonstrate. Using the end of the above $\Sigma^*_{11} = 15$, $T^* = 0.14$ equilibrium run as the initial condition, the load was reduced to $\Sigma^*_{11} = 0$ in several short runs with successively smaller values of Σ^*_{11}, the temperature in the final $\Sigma^*_{11} = 0$ run being $T^* = 0.14$. Finally, on reaching a no load condition, the system was quenched to a very low temperature to observe the pair correlation with the effects of thermal motion removed.[12] This is shown in Fig. 2(b) leaving no doubt that one has regained the original perfect fcc structure. Of course, all other indicators, namely the components of $\mathbf{h}^*$, pointed to the same conclusion.

C. Structure transformation under further compression

After completing the $\Sigma^*_{11} = 15$ study the load was raised to $\Sigma^*_{11} = 20$ and the dynamics was allowed to take its course according to the dictates of the equations of motion [Eqs. (2.9) and (2.25)].

The behavior of the system was remarkably different as might have been guessed by the title given to this subsection. Figure 3 shows the details of the changes with the passage of time. The MD cell, i.e., the $\mathbf{h}$ matrix, undergoes large and swift changes which cannot possibly be described as elastic deformations. In fact, as Fig. 3 shows, when equilibrium was reached the average values of the components of $\mathbf{h}^*$ were $\langle h^*_{11} \rangle = 5.54 \pm 0.03$, $\langle h^*_{22} \rangle = 9.76 \pm 0.06$,

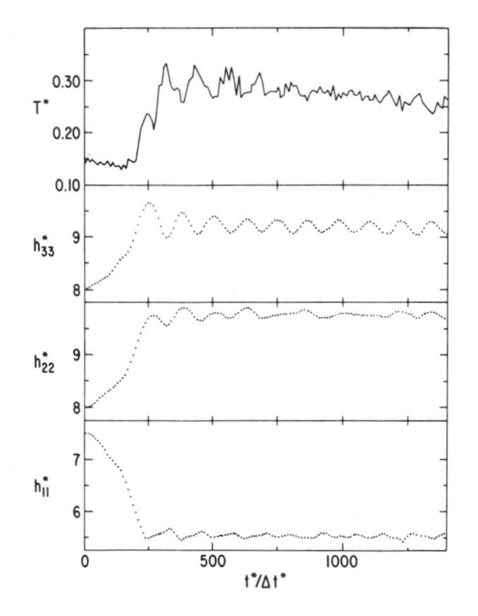

FIG. 3. Behavior of the MD cell parameters and the temperature of the system as it evolves in time when the compressive [100] load is increased from $\Sigma^*_{11} = +15$ to $+20$ (the latter value shown by ⤳ in Fig. 1). After a rapid change h^*_{ii} settle down to values at which an hcp structure can be accommodated in the MD cell (Fig. 2(c) shows the $g(r)$ at the end of the above time elapse). The rise in temperature is due to the release of elastic energy as the transformation occurs. See Sec. C for details.

$\langle h^*_{33} \rangle = 9.18 \pm 0.09$. The average of the nondiagonal elements was essentially zero. We note here that the tetragonal symmetry of the initially stressed state is destroyed as a result of this transformation; one gets instead an orthorhombic system, still under [100] compression. (See below for the description of this orthorhombic system under zero load.)

As seen in Fig. 3, contemporaneously with the rapid changes in $\mathbf{h}^*$, the temperature T^* increased from ~ 0.15 to ~ 0.30, finally settling down to ~ 0.25 or perhaps somewhat less. This is obviously a manifestation of an abrupt release of elastic energy in the relatively short time interval of $\sim 100\Delta t^*$ (or ~ 0.4 ps). Of course we recall that the $\Sigma^*_{11} = 15$ to 20 change was made in one Δt^*.

The pair correlation at the end of the $\Sigma^*_{11} = 20$ run is shown in Fig. 2(c). In spite of the large deformations in $\mathbf{h}$ mentioned above, the $g(r)$ in Fig. 2(c) is very similar to the one in Fig. 2(a) and does not show clear evidence of a new arrangement of particles. However, the quenching technique does indicate that a new structure has been formed. The $g(r)$ after quenching is shown in Fig. 2(d). This figure displays not only the shell structure of a hexagonal close packed system but it also shows that under this loaded state otherwise single shells show up as split into two. Visual examination of particle positions suitably displayed showed that stacking faults were present in the hcp arrangement. It is interesting to note here that the direction of the compressive stress is normal to the c axis of the new, close-packed structure.

Having obtained the above-described transformation under a compressive load we reduced the load from $\Sigma^*_{11} = 20$ to $\Sigma^*_{11} = 0$ using several intermediate steps. The temperature in the final $\Sigma^*_{11} = 0$ run was $T^* = 0.21$. There was no structural change evident during this process. To reveal the structure clearly at the end of this process (of reducing the load from a high value to zero) we used the usual quench technique. The $g(r)$ is shown in Fig. 2(e). The shell

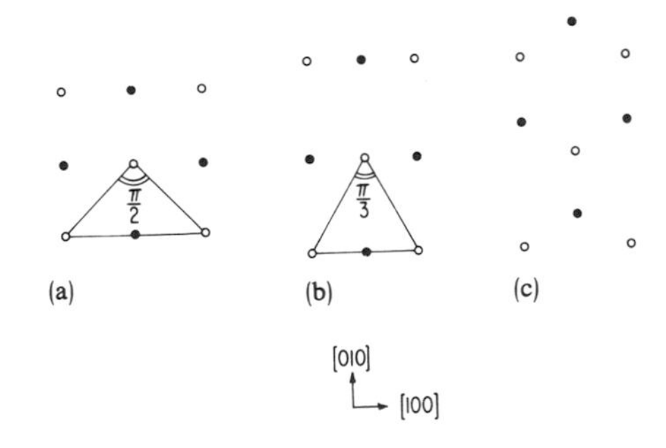

FIG. 4. Two planes of an fcc structure perpendicular to [001] are shown by ○ and ● respectively. (a)-(b) shows how the face-centered square structure changes to a triangular lattice on suitable compression in the [100] direction. (b)-(c) shows the necessary translation of the ● planes to achieve hcp ordering. At the same time spacing between ○ and ● planes has to correspond to the "c/a" value of an hcp arrangement, namely $\sqrt{8/3}$.

structure in Fig. 2(e) shows unambiguously an hcp arrangement. We recall that on reducing the load from $\Sigma^*_{11} = +15$ to $\Sigma^*_{11} = 0$ the system had gone back to its original fcc state (Sec.B 1).

The structure of the system which when quenched gave the $g(r)$ shown in Fig. 2(e) is of considerable interest. The distortions in the original cubic MD cell which allow an hcp structure to be accommodated were as follows: We found $\langle h^*_{22}\rangle/\langle h^*_{11}\rangle = \sqrt{3} \pm 0.01$, $\langle h^*_{33}\rangle/\langle h^*_{11}\rangle = \sqrt{8/3} \pm 0.01$, and an almost precise ($\langle h^*_{ij}\rangle \simeq 0.01, i \neq j$) rectangular parallelopiped. The ratio $\sqrt{3}$ is necessary to transform the [001] square-centered plains of the fcc structure into hexagonal plains [see Figs. 4(a)–4(b)]. The ratio $\sqrt{8/3}$ is the "c/a" value of the hcp structure. We note also that these distortions alone do not bring an fcc into an hcp structure; they have to be accompanied by the slip of alternate [001] planes as illustrated in Figs. 4(b)–4(c).

Given the highly coordinate nature of this transformation it is not surprising that in a rapid decompression defects such as stacking faults are produced.

D. System under tensile uniaxial loading

A study similar to the one just described was made by applying a tensile load; starting from the same configuration as was used to initiate the compressive runs, we raised the tensile load from 0 to $\Sigma^*_{11} = -20$ through a series of small intermediate steps at $\Sigma^*_{11} = -4, -8, -12, -16$. During these runs the MD cell remained tetragonal, elongated in the [100] direction and shortened equally in the [010] and [001] directions. At $\Sigma^*_{11} = -20$ the system was perfectly stable for the duration of a 4610 Δt^* MD run. During this run we obtained $T^* = 0.13$, $\langle h^*_{11}\rangle = 8.35 \pm 0.01$, $\langle h^*_{22}\rangle = \langle h^*_{33}\rangle = 7.75 \pm 0.02$, $\langle h^*_{ij}\rangle \simeq 0$ for $i \neq j$. The stress S evaluated from Eq. (2.24), using $\mathbf{h}^*_0 = \langle \mathbf{h}^* \rangle$, gave $S_{11} = -8.6 \times 10^{10}$ dyne/cm^2 for Ni, the strain, expressed as a stretch, being $\lambda_1 = \langle h^*_{11}\rangle/1^*_{0.14} = 1.06$. These values are in complete accord with the results of Milstein and Farber[5] as seen in Fig. 1.

Increasing the tensile load to $\Sigma^*_{11} = -25$ gave a system like the one just mentioned. The final values were $T^* = 0.13$, $\langle h^*_{11}\rangle = 8.49 \pm 0.01$, $\langle h^*_{22}\rangle = \langle h^*_{33}\rangle = 7.72 \pm 0.04$, $\langle h^*_{ij}\rangle \simeq 0$ for $i \neq j$. This gave $\lambda_1 = 1.08$ and $S_{11} = -10.9 \times 10^{10}$ dyne/cm^2 again in accord with Ref. 5 as shown in Fig. 1. Note the proximity to the $c_{22} = c_{23}$ bifurcation point of Milstein and Farber.[5]

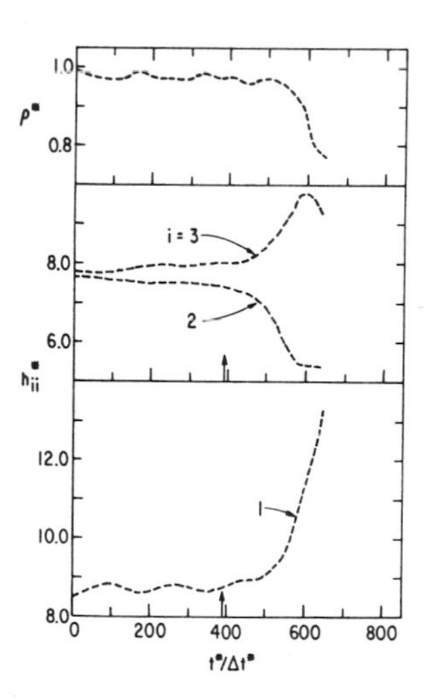

FIG. 5. At a tensile load of $\Sigma^*_{11} = -30$ the system fails after about 400 Δt. Initial breaking of tetragonal symmetry is clear ($h^*_{22} \neq h^*_{33}$). Arrow shows the critical moment after which even rapid unloading cannot prevent failure. Before the critical moment even slow unloading leads to the recovery of the original fcc state (see Fig. 1).

E. System failure under tensile uniaxial loading

On going from $\Sigma^*_{11} = -25$ to -30 the system "failed" as is described below. Figure 5 shows the values of h^*_{ii} as a function of time. The arrow indicates a region of time before which it has been found that it is possible to recuperate by rapidly reducing the load to more normal values. Beyond this time recovery may be difficult or impossible to achieve. This will be discussed in the following subsection.

As is clear from Fig. 5 the system does show a nontetragonal behavior at this increased load. This is in accord with the prediction of Ref. 5 regarding tetragonality in this region of strain, i.e., near the bifurcation point, made on the basis of static calculations. However, the fact of system failure really substantiates the conjectural remark of Milstein and Farber[5] that the very presence of the bifurcation point may lead to failure.

F. Recovery from failure

The configurations 100Δt^* before and after the time indicated by the arrow in Fig. 5 were used as the initial conditions in a series of calculations at tensile loads smaller than the $\Sigma^*_{11} = -30$ value which leads to failure.[13] This was done in an attempt to capture, but without success, a stable point on the secondary stress-strain path of Milstein and Farber[5] i.e., a system without tetragonal symmetry.

In the first attempt, with a configuration occurring earlier than the position of the arrow in Fig. 5, Σ^*_{11} was brought to zero in a series of intermediate runs of about 300 Δt^* each (i.e., with slow unloading), with Σ^*_{11} reduced by five units every time. The final $\Sigma^*_{11} = 0$ state was found to be a system in which a cubic MD cell was reestablished; the structure had fallen back on to the undistorted fcc state of the primary path (see Fig. 1).

In the second attempt, with a configuration occurring later than the position of the arrow in Fig. 5, the length of the intermediate runs was 100 Δt^* (i.e., rapid unloading). Already at $\Sigma^*_{11} = -25$, i.e., during the very first intermediate run, the differences between h^*_{22} and h^*_{33} tended to increase monotonically. This tendency continued during the $\Sigma^*_{11} = -20$ run and the system failed.

Notes on p. 297

We conclude that along the secondary path the system is unstable at high values of stress. As is shown in the next subsection *this is so even at zero load.*

G. Stable structures at zero load

As indicated in Sec. 3, the $B^{(2)}$ and $T^{(2)}$ states of Ref. 5 occur on the secondary path and at zero load. A calculation completely analogous to the one we reported[2] on a Lennard-Jones system was made on the Ni system now under consideration. Starting at $B^{(2)}$ i.e., with $h^*_{11} = h^*_{22} = \sqrt{2}\ h^*_{33} = 8.7826$, $h^*_{ij} = 0$ for $i \neq j, T^* = 0.11$ and $\Sigma^* = 0$, an MD calculation was initiated with the equations of motion given by $\mathscr{L}_s$ (which is the same as $\mathscr{L}$, because $\Sigma = 0$) of Eq. (2.23). The system spontaneously started to produce distortions in the MD cell and after only 890 Δt^* the angles between the otherwise orthogonal axes became $\sim 89°$, $89°$, and $102°$ with an uncertainty of $\pm 3°$ in each case), the lengths of the edges of the MD cell became $l^*_1 \simeq l^*_2 = 8.88 \pm 0.12$, $l^*_3 = 6.42 \pm 0.08$, while the temperature of the system rose to $T^* = 0.20$. These changes were accompanied by changes in the pair correlation function $g(r)$. As usual we quenched the system and this revealed a $g(r)$ corresponding to hcp ordering with stacking faults.

Instead of starting at the $B^{(2)}$ point, we started a calculation like the one described above from a configuration near the $T^{(2)}$ point on the secondary path of Ref. 5. This also transformed very fast to a system with hcp ordering with stacking faults. Note that the $T^{(2)}$ point is very little different from the $T^{(1)}$ state of the primary path. Since $T^{(1)}$ was found by Milstein[4] to be unstable it is not surprising that the same happens for $T^{(2)}$. Same remark applies a fortiori to the zero load bcc state since obviously it is the same state whether it is on one path or the other.

We did start a calculation from the $T^{(1)}$ state as the initial condition and, quite as we expected, the system evolved and gave an hcp state. In this case, the final state had no stacking faults.

We thus conclude that at zero load and finite (and rather low!) temperatures for Ni only the fcc structure is stable among the ones considered by Milstein and Farber.[5] However, as our calculations have clearly shown, a new, locally stable state is possible and this is an hcp structure.

The relative stability of the fcc and hcp structures is of course a difficult question to answer. An answer to this question requires a detailed and accurate computation of the free energy difference. From our calculations we can only infer that the local free energy minima in configuration space seem to be locally very stable at the low temperatures we have investigated.

V. CONCLUDING REMARKS

In this paper we have illustrated some of the possibilities opened up by our new MD method for the investigation of the elastic behavior of solids. Our calculations have reproduced results obtained by others using static methods.

In Fig. 1 we have displayed how the stress-strain relation in the system is in good agreement with the results of Milstein and Farber.[5] We can go even further to state that the departure of our results away from their static results is itself in the right direction since the system will be "softer" in the presence of thermal agitation; Fig. 1 shows this indeed to be so. However, because of its greater flexibility, the new method has allowed us to predict new results. A clear example of this is a genuine bifurcation point: the *fcc→hcp transition under compression*. In principle such a transition could have been predicted by static calculations[5] of Milstein and Farber if they had searched for *all* possibilities instead of restricting themselves to the case of face-centered tetragonal states. In an MD calculation the system follows a dynamical trajectory determined by well defined dynamical equations and is free to assume the crystal structure most suited to the interaction potential and the ambient conditions of temperature and stress. Being a fully dynamical method it even allows the study of the kinetics of temperature and stress induced transformations.

A few words of caution are in order especially if one wants to make comparisons with laboratory experiments. One limitation comes from the small system size and periodic boundary conditions used for the dynamical simulation. This probably reduces but certainly does not eliminate the occurrence of extended defects such as grain boundaries or dislocations; these defects play an important role in the plastic behavior of solids. This limitation is, however, not intrinsic and can be much reduced with the use of much larger MD systems. The possibility of creating extended defects probably accelerates the breakdown of the system under applied stress. Thus the tensile strength we have obtained ($S_{11} \sim 11 \times 10^{10}$ dyne/cm^2) is probably only an upper bound to the true value.

The second weakness of model calculations is the use of potentials like the one in Eq. (3.1). This potential is purely empirical without much justification from a microscopic point of view. Moreover, the normal state properties of crystalline systems are determined by values of the potential and its derivatives only at distances where the various neighbors are situated. The detail of the short-range repulsion probably plays a crucial role in the phenomena discussed in this paper and is not so well determined when one uses normal state properties of the crystal to determine the parameters of the empirical potential. Especially in Ni a proper account of the role of d-electrons may call for the use of many body forces in a more realistic calculation. This again is not a intrinsic problem with simulation methods.

Some of the conclusions reached in this paper are expected to be consequences only of the symmetry; as such they have a greater range of validity than the potential used in arriving at those conclusions. For instance it is possible that the fcc-hcp transition in a [100] compression is observable. Since the many-body terms in the "true" potential function are expected to be of short range their presence or absence cannot play a determining role in this particular structural transformation.

From the point of view of structural transformation in monotonic systems it will be most useful to have a systematic study of model systems with pair interactions of the type $1/r^m - 1/r^n$ when these systems are put under various forms of external stress. Such calculations and also the ones we have presented are an exact consequence (apart from problems related with the numerical solutions of differential equations) of the potential function used. As such they serve the function of data for testing approximate theoretical models of the phenomena under investigation.

There is the intriguing possibility that the fcc-hcp transition might actually occur in Ni single crystals under extreme conditions of shock.

ACKNOWLEDGMENT

This work was supported by the U.S. Department of Energy.

APPENDIX A

The connection between the formal definition of the strains, Eq. (2.19), and that given in elementary textbooks[14] is as follows. Consider a reference state in which the MD cell is a rectangular parallelopiped with edges parallel to the Cartesian reference frame. In this case $\mathbf{h}_0$ is diagonal, the diagonal elements being the lengths of the three edges. In a distorted state the MD cell is given by $\mathbf{h} = \mathbf{h}_0 + \Delta\mathbf{h}$ say. Up to linear terms in $\Delta\mathbf{h}$, we find from Eq. (2.6),

$$\mathbf{G} = \Delta\mathbf{h}'\mathbf{h}_0 + \mathbf{h}_0'\Delta\mathbf{h} + \mathbf{h}_0'\mathbf{h}_0. \quad \text{(A.1)}$$

Using this in Eq. (2.19), due to the diagonal form of $\mathbf{h}_0$ we get

$$\epsilon_{ii} = (h_{ii} - h_{0,ii})/h_{0,ii}, \quad \text{(A.2)}$$

and for $i \neq j$

$$\epsilon_{ij} = 1/2(h_{ij}/h_{0,jj} + h_{ji}/h_{0,ii}). \quad \text{(A.3)}$$

Equation (A2) connects the diagonal terms of the strain to the length variation of the edges of the cell while Eq. (A3) gives the expected relation between the off-diagonal terms and the changes in the angles between the edges.

Up to order Δh, the volume change is given, as in Eq. (2.21), by $\Delta\Omega = \Omega_0 \operatorname{Tr} \epsilon$.

[1] E. R. Cowley, G. Jacucci, M. L. Klein, and I. R. McDonald, Phys. Rev. B **14**, 1758 (1976).

[2] M. Parrinello and A. Rahman, Phys. Rev. Lett. **45**, 1196 (1980).

[3] F. Milstein, J. Appl. Phys. **44**, 3825 (1973).

[4] F. Milstein, J. Appl. Phys. **44**, 3833 (1973).

[5] F. Milstein and B. Farber, Phys. Rev. Lett. **44**, 277 (1980).

[6] J. P. Hansen and I. R. McDonald, *Theory of Simple Liquids* (Academic, London, 1976).

[7] H. C. Andersen, J. Chem. Phys. **72**, 2384 (1980).

[8] W. G. Hoover, D. J. Evans, R. B. Hickman, A. J. Ladd, W. T. Ashurst, and B. Moran, Phys. Rev. A **22**, 1690 (1980).

[9] L. D. Landau and E. M. Lifshitz, *Theory of Elasticity* (Pergamon, Oxford, 1959).

[10] Note that in the definition of $\mathbf{S}$ through Eq. (2.20) we have adopted a certain sign convention: $\mathbf{S} = +p$ corresponds to a system under hydrostatic pressure p.

[11] A. Rahman, in NATO Advanced Study Institute Series: *Correlation Functions and Quasiparticle Interactions in Condensed Matter*, edited by J. W. Halley (Plenum, New York, 1978).

[12] This procedure, used systematically in Ref. 2, is extremely useful for the purpose of resolving peaks in $g(r)$ which otherwise would merge in neighboring peaks especially if the corresponding coordination numbers are very different. For example, in an hcp structure the third and the fouth shells are very close to each other with a population, respectively, of 2 and 18 neighbors.

[13] In an MD calculation the "failure" of a system is manifested by a monotonic change in the h_{ii} and in an obvious disruption of the shell-like structure of $g(r)$. A monotonic drop in the value of the density is simply a consequence of the behavior of the h_{ii}.

[14] C. Kittel, *Introduction to Solid State Physics* (Wiley, New York, 1968).

Notes to Reprint IV.4

1. The extension of the Parrinello–Rahman method to molecular systems is discussed in Chapter V.

2. The Parrinello–Rahman method appears to be very well suited to the calculation of the elastic constants of a crystal. In principle, all that is required is to apply a stress to the system and measure the resulting strain, but Sprik et al. (Phys. Rev. B 29 (1984) 4368) have shown that this direct approach is computationally less efficient than the use of equilibrium fluctuation formulae. For other, recent calculations of elastic constants by computer simulation, see (Ray et al., Phys. Rev. B 32 (1985) 733; 33 (1986) 895; Impey et al., J. Chem. Phys. 83 (1985) 3638; Kluge et al., J. Chem. Phys. 85 (1986) 4028).

A unified formulation of the constant temperature molecular dynamics methods

Shuichi Nosé[a)]
Division of Chemistry, National Research Council of Canada, Ottawa, Ontario, Canada, K1A OR6

(Received 17 November 1983; accepted 15 March 1984)

Three recently proposed constant temperature molecular dynamics methods by: (i) Nosé (Mol. Phys., to be published); (ii) Hoover *et al.* [Phys. Rev. Lett. **48**, 1818 (1982)], and Evans and Morriss [Chem. Phys. **77**, 63 (1983)]; and (iii) Haile and Gupta [J. Chem. Phys. **79**, 3067 (1983)] are examined analytically via calculating the equilibrium distribution functions and comparing them with that of the canonical ensemble. Except for effects due to momentum and angular momentum conservation, method (i) yields the rigorous canonical distribution in both momentum and coordinate space. Method (ii) can be made rigorous in coordinate space, and can be derived from method (i) by imposing a specific constraint. Method (iii) is not rigorous and gives a deviation of order $N^{-1/2}$ from the canonical distribution (N the number of particles). The results for the constant temperature–constant pressure ensemble are similar to the canonical ensemble case.

I. INTRODUCTION

Recently, the extension of molecular dynamics (MD) methods to treat ensembles other than the traditional microcanonical ensemble has attracted considerable attention.

The constant pressure MD method, first introduced by Andersen[1] and subsequently extended by Parrinello and Rahman[2–4] to allow for changes of the MD cell shape, has demonstrated its usefulness in applications to structural changes in the solid state.[2–9] Recently, Heyes[10] employed a similar approach to that of Anderson but used real variables instead of the scaled variables.[1–4] A constant pressure MD method based on nonequilibrium MD technique was also proposed by Hoover *et al.*[11]

Several constant temperature MD methods have been proposed. The purpose of the present article is to examine and compare these methods and to establish a unified formalism for their derivation. Emphasis is placed on the static properties and on the equilibrium distribution function.

The earliest method for the constant temperature MD is a momentum scaling procedure, in which the velocities of the particles are scaled at each time step to maintain the total kinetic energy at a constant value.[12] This method has been used without demonstrated justification. Haile and Gupta[13] discussed how to add the constraint of constant kinetic energy to the equations of motion. As a special case, they proposed a constraint method based on a momentum scaling procedure. This is a refinement of the earlier method. It will be shown in Sec. III C that the equilibrium distribution function in the momentum scaling method deviates from the canonical distribution by order $N^{-1/2}$ (N the number of particles).

Anderson[1] proposed a hybrid of MD and Monte Carlo methods. In his approach, the particles change their velocities by stochastic collisions. The distribution of the velocities of the particles that collided is chosen to reproduce the canonical ensemble. Because of the sudden change of the velocities by collisions, the trajectory in the phase space is discontinuous.

Hoover *et al.*[14,15] and Evans[16] proposed a constraint MD method which was derived from a nonequilibrium MD formulation.[17,18] This method will be called the HLME method hereafter.

In this method, an additional term $-\alpha \mathbf{p}_i$ is added to the force term in Eq. (1.2) ($\mathbf{q}_i$, coordinate; $\mathbf{p}_i$, momentum of particle i),

$$\frac{d\mathbf{q}_i}{dt} = \mathbf{p}_i/m_i, \tag{1.1}$$

$$\frac{d\mathbf{p}_i}{dt} = -\frac{\partial \phi}{\partial \mathbf{q}_i} - \alpha \mathbf{p}_i. \tag{1.2}$$

Consequently, the equations are no longer in a canonical form. The parameter α is determined from the requirement that the total kinetic energy is constant,

$$\sum_i \mathbf{p}_i^2/2m_i = gkT/2 \tag{1.3}$$

or

$$\sum_i \mathbf{p}_i \frac{d\mathbf{p}_i}{dt}/m_i = 0. \tag{1.4}$$

Thus, we get

$$\alpha = -\Big(\sum_i \frac{\partial \phi}{\partial \mathbf{q}_i}\mathbf{p}_i/m_i\Big)\Big/\Big(\sum_i \mathbf{p}_i^2/m_i\Big). \tag{1.5}$$

This method can produce the canonical distribution in coordinate space if we set $g = 3N - 1$, where N is the number of particles (in the original papers[14,16] $g = 3N$). Further, in Sec. III B, it will be shown that the HLME equations are derived from the extended system (ES) method[19] by imposing a particular constraint.

The extended system method by Nosé[19] introduced an additional degree of freedom s, which acts as an external system for the physical system of N particles. If we choose an appropriate potential $gkT \ln s$, for the variable s, the equilibrium distribution function, projected onto the physical system from the extended system of the particles and the variable s, is exactly that of the canonical ensemble. The parameter g is an integer, essentially equal to the number of degrees of freedom in the physical system, but the exact value depends on the particular procedure.

[a)] Present address: Department of Physics, Faculty of Science and Technology, Keio University, 3-14-1 Hiyoshi, Kohoku-ku, Yokohama 223, Japan.

In the present article, the basic formalism is that of the extended system method and other methods are analyzed in this formulation. The details of the ES method are given in Sec. II. The equations of motion of the ES method with a constraint and the derivation of the HLME method are given in Sec. III. Extension to the constant temperature-constant pressure (TP) ensemble is discussed in Sec. IV. The method by Evans and Morriss[18] is derived from the ES method[19] in a similar fashion as for the canonical ensemble case. Most of the discussion is exact only if we ignore the momentum and the angular momentum conservation laws. The correction for the momentum conservation as well as other comments are given in Sec. V.

II. THE EXTENDED SYSTEM METHOD

A. A virtual variable formulation

We start from a method which seems to be most convenient to obtain the equilibrium distribution function. In the ES method,[19] an additional degree of freedom s is introduced which acts as an external system on the physical system of N particles, with coordinates $\mathbf{q}_i'$, masses m_i and potential energy $\phi(\mathbf{q}')$. We also introduce *virtual* variables (coordinate $\mathbf{q}_i$, momentum $\mathbf{p}_i$, and time t) which are related to the *real* variables $(\mathbf{q}_i', \mathbf{p}_i', t')$ by

$$\mathbf{q}_i' = \mathbf{q}_i, \tag{2.1}$$

$$\mathbf{p}_i' = \mathbf{p}_i/s, \tag{2.2}$$

$$t' = \int^t \frac{dt}{s}. \tag{2.3}$$

The real velocity $(d\mathbf{q}_i'/dt_i')$ is also expressed via a scaled form in the virtual variable formulation

$$\frac{d\mathbf{q}_i'}{dt'} = s\frac{d\mathbf{q}_i'}{dt} = s\frac{d\mathbf{q}_i}{dt}. \tag{2.4}$$

Thus, a simple interpretation of these transformations is scaling the time by $dt' = dt/s$. This is similar to scaling the coordinates in the constant pressure MD method.[1–4]

The Hamiltonian of the extended system of the particles and the variable s in terms of the virtual variables is postulated as

$$H = \sum_i \mathbf{p}_i^2/2m_i s^2 + \phi(\mathbf{q}) + p_s^2/2Q + gkT \ln s, \tag{2.5}$$

p_s is the conjugate momentum of s; Q is a parameter of dimension energy·(time)^2 and behaves as a *mass* for the motion of s; k is Boltzmann's constant; T the externally set temperature; the parameter g is essentially equal to the number of degrees of freedom of the physical system. However, its exact value will be chosen to satisfy the canonical distribution exactly at equilibrium. As we will discuss later, a logarithmic dependence of the potential on the variable s, $gkT \ln s$, is essential for producing the canonical ensemble.

We assume the Hamiltonian formalism can be applied to Eq. (2.5) with the virtual variables. The equations of motion are

$$\frac{d\mathbf{q}_i}{dt} = \frac{\partial H}{\partial \mathbf{p}_i} = \mathbf{p}_i/m_i s^2, \tag{2.6}$$

$$\frac{d\mathbf{p}_i}{dt} = -\frac{\partial H}{\partial \mathbf{q}_i} = -\frac{\partial \phi}{\partial \mathbf{q}_i}, \tag{2.7}$$

$$\frac{ds}{dt} = \frac{\partial H}{\partial p_s} = p_s/Q, \tag{2.8}$$

$$\frac{dp_s}{dt} = -\frac{\partial H}{\partial s} = \left(\sum_i \mathbf{p}_i^2/m_i s^2 - gkT\right)/s. \tag{2.9}$$

In Lagrangian form, these are

$$\frac{d}{dt}\left(m_i s^2 \frac{d\mathbf{q}_i}{dt}\right) = -\frac{\partial \phi}{\partial \mathbf{q}_i} \tag{2.10}$$

or

$$\frac{d^2\mathbf{q}_i}{dt^2} = -\frac{1}{m_i s^2}\frac{\partial \phi}{\partial \mathbf{q}_i} - \frac{2}{s}\frac{ds}{dt}\frac{d\mathbf{q}_i}{dt} \tag{2.11}$$

and

$$\frac{d}{dt}\left(Q\frac{ds}{dt}\right) = \left[\sum_i s^2\left(\frac{d\mathbf{q}_i}{dt}\right)^2/m_i - gkT\right]/s. \tag{2.12}$$

The conserved quantities are the Hamiltonian H, the total momentum $\Sigma_i \mathbf{p}_i$, and the angular momentum $\Sigma_i \mathbf{q}_i \times \mathbf{p}_i$.

$$\frac{dH}{dt} = \sum_i \left(\frac{\partial H}{\partial \mathbf{p}_i}\frac{d\mathbf{p}_i}{dt} + \frac{\partial H}{\partial \mathbf{q}_i}\frac{d\mathbf{q}_i}{dt}\right) + \frac{\partial H}{\partial p_s}\frac{dp_s}{dt} + \frac{\partial H}{\partial s}\frac{ds}{dt} = 0.$$

The conservation laws for the last two quantities are derived from Eq. (2.7) and the properties satisfied by the potential

$$\sum_i \frac{\partial \phi}{\partial \mathbf{q}_i} = 0$$

and

$$\sum_i \mathbf{q}_i \times \frac{\partial \phi}{\partial \mathbf{q}_i} = 0.$$

However, it should be noted here that during the ordinary type of simulations with periodic boundary condition the angular momentum is not conserved.

Because of the momentum and angular momentum conservation, the ensembles produced by the MD method are slightly different from the usual statistical mechanical ensembles.[20–22] These small deviations are ignored in the discussion in Secs. II–IV.

The partition function Z for N identical particles is obtained by integration of the equilibrium distribution function $\rho(x_1, x_2, \ldots)$ over the whole phase space.

$$Z = \frac{1}{N! h^{3N}} \int dx_1 \int dx_2 \ldots \rho(x_1, x_2, \ldots),$$

where h is Planck's constant and x_i is a generalized coordinate (the constant factors for ρ and Z are ignored hereafter). The projection of the equilibrium distribution function from the space (x_1, x_2) onto the space (x_1) is carried out by integrating with respect to the variable x_2,

$$\rho(x_1) = \int dx_2\, \rho(x_1, x_2).$$

In particular, we need a distribution function $\rho(\mathbf{p}', \mathbf{q}')$ that is projected from the extended system onto the physical system.

Notes on p. 306

In the extended system, the total Hamiltonian of Eq. (2.5) is conserved. Therefore, this method produces a microcanonical ensemble and the distribution function $\rho(\mathbf{p},\mathbf{q},p_s,s)$ is expressed as $\delta(H-E)$; $\delta(x)$ is the Dirac delta function. The shortened forms $d\mathbf{p} = d\mathbf{p}_1 d\mathbf{p}_2 ... d\mathbf{p}_N$, $d\mathbf{q} = d\mathbf{q}_1 d\mathbf{q}_2 ... d\mathbf{q}_N$, and $H_0(\mathbf{p},\mathbf{q}) = \Sigma_i \mathbf{p}_i^2/2m_i + \phi(\mathbf{q})$ are used. The partition function is

$$Z = \int dp_s \int ds \int d\mathbf{p} \int d\mathbf{q} \times \delta[H_0(\mathbf{p}/s,\mathbf{q}) + p_s^2/2Q + gkT \ln s - E]. \tag{2.13}$$

The virtual momenta $\mathbf{p}_i$ and coordinates $\mathbf{q}_i$ are transformed to the real variables $\mathbf{p}_i' = \mathbf{p}_i/s$, $\mathbf{q}_i' = \mathbf{q}_i$. The volume element is $d\mathbf{p}\, d\mathbf{q} = s^{3N} d\mathbf{p}' d\mathbf{q}'$. Hence

$$Z = \int dp_s \int d\mathbf{p}' \int d\mathbf{q}' \int ds \cdot s^{3N} \delta[H_0(\mathbf{p}',\mathbf{q}') + p_s^2/2Q + gkT \ln s - E]. \tag{2.14}$$

Because the argument of the δ function in the above equation has only one zero as a function of the variable s, we can employ the equivalence relation $\delta[f(s)] = \delta(s - s_0)/f'(s_0)$; s_0 is the zero of $f(s)$.

$$Z = \frac{1}{gkT} \int dp_s \int d\mathbf{p}' \int d\mathbf{q}' \int ds \cdot s^{3N+1} \times \delta(s - \exp\{-[H_0(\mathbf{p}',\mathbf{q}') + p_s^2/2Q - E]/gkT\}) \tag{2.15}$$

$$= \frac{1}{gkT} \exp\left[\left(\frac{3N+1}{g}\right) E/kT\right] \int dp_s \times \exp\left[-\left(\frac{3N+1}{g}\right) p_s^2/2QkT\right] \int d\mathbf{p}' \int d\mathbf{q}' \times \exp\left[-\left(\frac{3N+1}{g}\right) H_0(\mathbf{p}',\mathbf{q}')/kT\right].$$

If we choose $g = 3N + 1$, the partition function of the extended system is equivalent to that of the physical system in the canonical ensemble except for a constant factor:

$$Z = C \int d\mathbf{p}' \int d\mathbf{q}' \exp[-H_0(\mathbf{p}',\mathbf{q}')/kT],$$

and the equilibrium distribution function is

$$\rho(\mathbf{p}',\mathbf{q}') = \exp[-H_0(\mathbf{p}',\mathbf{q}')/kT]. \tag{2.16}$$

With the quasiergodic hypothesis which relates the time average along the trajectory to the ensemble average, the averages of any static quantities expressed as functions of $\mathbf{p}_i/s,\mathbf{q}_i$ along the trajectory determined by Eqs. (2.6)–(2.9), are exactly those in the canonical ensemble:

$$\lim_{t_0 \to \infty} \frac{1}{t_0} \int_0^{t_0} A(\mathbf{p}/s,\mathbf{q}) dt = \langle A(\mathbf{p}/s,\mathbf{q}) \rangle = \langle A(\mathbf{p}',\mathbf{q}',) \rangle_c, \tag{2.17}$$

$\langle ... \rangle$ and $\langle ... \rangle_c$ denote the ensemble average in the extended system and in the canonical ensemble, respectively. The first equivalence in Eq. (2.17) is achieved by sampling data points at integer multiples of the virtual time unit Δt. We call this *virtual time sampling*. In this sampling, the real time interval of each time step is unequal. If we sample using equal intervals in real time t' with $t_1 = \int_0^{t_0} dt/s$ (we can use an interpolation or the method in Sec. II B for this purpose), the result is a weighted average

$$\lim_{t_1 \to \infty} \frac{1}{t_1} \int_0^{t_1} A(\mathbf{p}/s,\mathbf{q}) dt' = \lim_{t_1 \to \infty} \frac{t_0}{t_1} \frac{1}{t_0} \int_0^{t_0} A(\mathbf{p}/s,\mathbf{q}) \frac{dt}{s}$$

$$= \left[\lim_{t_0 \to \infty} \frac{1}{t_0} \int_0^{t_0} A(\mathbf{p}/s,\mathbf{q}) \frac{dt}{s}\right] \Big/ \left(\lim_{t_0 \to \infty} \frac{1}{t_0} \int_0^{t_0} \frac{dt}{s}\right)$$

$$= \langle A(\mathbf{p}/s,\mathbf{q})/s \rangle \Big/ \left\langle \frac{1}{s} \right\rangle. \tag{2.18}$$

From comparison with Eq. (2.15), we find that if we choose g as $g = 3N$, the weighted average in the extended system [Eq. (2.18)] is identical to $\langle A(\mathbf{p}',\mathbf{q}') \rangle_c$. In virtual time sampling, g should be $3N + 1$, and in real time sampling, g must be $3N$.

B. Equations in real variables

The equations of motion [Eqs. (2.6)–(2.9)] can be transformed into the equations for the real variables $\mathbf{q}_i' = \mathbf{q}_i$, $\mathbf{p}_i' = \mathbf{p}_i/s$, $t' = \int^t dt/s$, $s' = s$, and $p_s' = p_s/s$:

$$\frac{d\mathbf{q}_i'}{dt'} = s\frac{d\mathbf{q}_i'}{dt} = s\frac{d\mathbf{q}_i}{dt} = \mathbf{p}_i/m_i s = \mathbf{p}_i'/m_i, \tag{2.19}$$

$$\frac{d\mathbf{p}_i'}{dt'} = s\frac{d\mathbf{p}_i'}{dt} = s\frac{d}{dt}(\mathbf{p}_i/s) = \frac{d\mathbf{p}_i}{dt} - \frac{1}{s}\frac{ds}{dt}\mathbf{p}_i = -\frac{\partial \phi}{\partial \mathbf{q}_i'} - s' p_s' \mathbf{p}_i'/Q, \tag{2.20}$$

$$\frac{ds'}{dt'} = s\frac{ds'}{dt} = s\frac{ds}{dt} = s'^2 p_s'/Q, \tag{2.21}$$

$$\frac{dp_s'}{dt'} = s\frac{dp_s'}{dt} = s\frac{d}{dt}(p_s/s) = \frac{dp_s}{dt} - \frac{1}{s}\frac{ds}{dt}p_s = \left(\sum_i p_s'^2/m_i - gkT\right)/s' - s' p_s'^2/Q. \tag{2.22}$$

Equations (2.19)–(2.22) are no longer canonical, since Eqs. (2.20) and (2.22) have additional force terms. H of Eq. (2.5), in terms of the real variables,

$$H' = \sum_i p_s'^2/2m_i + \phi(\mathbf{q}') + s'^2 p_s'^2/2Q + gkT \ln s \tag{2.23}$$

is not a Hamiltonian. This is a disadvantage of the real variable method. H' is still conserved

$$\frac{dH'}{dt'} = \sum_i \left(\frac{\partial H'}{\partial \mathbf{p}_i'} \frac{d\mathbf{p}_i'}{dt'} + \frac{\partial H'}{\partial \mathbf{q}_i'} \frac{d\mathbf{q}_i'}{dt'}\right) + \frac{\partial H'}{\partial p_s'} \frac{dp_s'}{dt'} + \frac{\partial H'}{\partial s'} \frac{ds'}{dt'} = 0.$$

The Lagrangian forms are

$$\frac{d}{dt'}\left(m_i s' \frac{d\mathbf{q}_i'}{dt'}\right) = -s' \frac{\partial \phi}{\partial \mathbf{q}_i'} \tag{2.24}$$

and

$$\frac{d}{dt'}\left(\frac{Q}{s'} \frac{ds'}{dt'}\right) = \sum_i m_i \left(\frac{d\mathbf{q}_i'}{dt'}\right)^2 - gkT. \tag{2.25}$$

The Lagrangian of the real time formulation is related to the original one by relation

$$L' = s(L + E), \tag{2.26}$$

E is the conserved value of H of Eq. (2.5). These equations are to be solved in real time, so g must be $g = 3N$.

C. Other potentials for the variable s

We can construct another class of constant temperature MD method by replacing the potential for s by (for example) $gkTs^n$ [$n > 0$, an integer and $g = (3N - n + 1)/n$]. The equilibrium distribution function can be readily obtained in a similar way.

$$\rho(\mathbf{p}',\mathbf{q}',p_s) = \{ [E - p_s^2/2Q - H_0(\mathbf{p}',\mathbf{q}')]/gkT \}^g$$
$$\times h\,[E - p_s^2/2Q - H_0(\mathbf{p}',\mathbf{q}')], \tag{2.27}$$

where $h(x)$ is the Heaviside function, $h(x) = 1$ for $x > 0$, and $h(x) = 0$ for $x < 0$. The Heaviside function is necessary to limit the range of the intergration in phase space. This constraint arises because the potential $gkTs^n$ is positive semidefinite; for some region in phase space ($\mathbf{p}',\mathbf{q}',p_s$), the equation $H = E$ does not have any real solution for s.

With the most favorable assumption $\langle s^n \rangle = 1$, the value of total energy $E = \langle H_0(\mathbf{p}/s,\mathbf{q})\rangle + \langle p_s^2/2Q \rangle + gkT$. Using the definition $H_1 = \langle H_0(\mathbf{p}/s,\mathbf{q})\rangle - H_0(\mathbf{p}',\mathbf{q}') + \langle p_s^2/2Q \rangle - p_s^2/2Q$, Eq. (2.27) is

$$\rho(\mathbf{p}',\mathbf{q}',p_s) = (1 + H_1/gkT)^g h\,(gkT + H_1).$$

Note that the distribution for the variable p_s cannot be separated from those of $\mathbf{p}'$ and $\mathbf{q}'$. From the expansion of $g\ln(1 + a/g)$ with respect to $1/g$; $g\ln(1 + a/g) = a - a^2/2g + a^3/3g^2...$, ρ is approximated as

$$\rho = \exp[H_1/kT - \tfrac{1}{2}(H_1/kT)^2/g]. \tag{2.28}$$

Singe H_1 is a quantity of order $N^{1/2}$, the leading term of the deviation of ρ from the canonical distribution is of order $N^{-1/2}$.

As we can see from the derivation of the equilibrium distribution function [see Eqs. (2.14) and (2.15)], it is related to the inverse function of the potential for s.

$$\rho(p',q',p_s) = \int ds\cdot s^{3N}\delta[H'' + f(s) - E]$$
$$= \int ds\cdot s^{3N}\delta(s - s_0)/f'(s_0) = s_0^{3N}/f'(s_0);$$

s_0 satisfies the relation $f(s_0) = E - H''$, thus $s_0 = f^{-1}(E - H'')$. H'' is that part of the Hamiltonian H [Eq. (2.5)] which is independent of s. Therefore, a logarithmic form is essential to produce the canonical distribution.

III. THE CONSTRAINT METHOD

A. Virtual variable formulation

The distribution function in momentum space is usually simple and the contribution of this term can be easily calculated in the canonical ensemble. Therefore, any method that produces the canonical distribution, even if only in coordinate space, can be useful in some situations. The standard way for this approach is to constrain the total kinetic energy term

$$\sum_i \frac{m_i}{2}\left(\frac{d\mathbf{q}_i'}{dt'}\right)^2 = \frac{g}{2}kT. \tag{3.1}$$

However, fluctuations of the total kinetic energy are suppressed by imposing this constraint. The Hamiltonian [Eq. (2.5)]

$$H = \sum_i \mathbf{p}_i^2/2m_i s^2 + \phi(\mathbf{q}) + p_s^2/2Q + gkT\ln s$$

is contrained by the conditions[13]

$$\frac{\partial H}{\partial s} = -\left[\sum_i \mathbf{p}_i^2/m_i s^2 - gkT\right]/s \equiv 0 \tag{3.2}$$

and

$$\frac{\partial H}{\partial p_s} = p_s/Q \equiv 0. \tag{3.3}$$

Equation (3.3) is trivial, and we ignore the $p_s^2/2Q$ term hereafter. The equations of motion for $\mathbf{q}_i$ and $\mathbf{p}_i$ maintain the same form as Eqs. (2.6) and (2.7):

$$\frac{d\mathbf{q}_i}{dt} = \frac{\partial H}{\partial \mathbf{p}_i} + \frac{\partial H}{\partial s}\frac{\partial s}{\partial \mathbf{p}_i} = \frac{\partial H}{\partial \mathbf{p}_i} = \mathbf{p}_i/m_i s^2, \tag{3.4}$$

$$\frac{d\mathbf{p}_i}{dt} = -\frac{\partial H}{\partial \mathbf{q}_i} - \frac{\partial H}{\partial s}\frac{\partial s}{\partial \mathbf{q}_i} = -\frac{\partial H}{\partial \mathbf{q}_i} = -\frac{\partial \phi}{\partial \mathbf{q}_i}, \tag{3.5}$$

but the value of s must be determined from Eq. (3.2), i.e.,

$$s = \left[\left(\sum_i \mathbf{p}_i^2/m_i\right)/gkT\right]^{\frac{1}{2}}. \tag{3.6}$$

The Hamiltonian [Eq. (2.5)] is still conserved:

$$\frac{dH}{dt} = \sum_i\left(\frac{\partial H}{\partial \mathbf{p}_i}\frac{d\mathbf{p}_i}{dt} + \frac{\partial H}{\partial \mathbf{q}_i}\frac{d\mathbf{q}_i}{dt}\right)$$
$$+ \frac{\partial H}{\partial p_s}\frac{dp_s}{dt} + \frac{\partial H}{\partial s}\frac{ds}{dt} = 0.$$

The partition function in this case is [recall that $H_0(\mathbf{p},\mathbf{q}) = \sum_i \mathbf{p}_i^2/2m_i + \phi(\mathbf{q})$]

$$Z = \int ds \int d\mathbf{p} \int d\mathbf{q}\, \delta[H_0(\mathbf{p}/s,\mathbf{q}) + gkT\ln s - E]$$
$$\times \delta\left\{s - \left[\left(\sum_i \mathbf{p}_i^2/m_i\right)/gkT\right]^{\frac{1}{2}}\right\}$$
$$= \int ds \int d\mathbf{p} \int d\mathbf{q}\, \delta\,[H_0(\mathbf{p}/s,\mathbf{q}) + \mathrm{gkT}\ln s - E\,]sgkT$$
$$\times \delta\left(s^2 gkT/2 - \sum_i \mathbf{p}_i^2/2m_i\right).$$

With the transformation $\mathbf{p}_i' = \mathbf{p}_i/s$, $\mathbf{q}_i' = \mathbf{q}_i$, we get

$$Z = \left[\int d\mathbf{p}'\delta\left(\sum_i \mathbf{p}_i'^2/2m_i - gkT/2\right)\right]\int d\mathbf{q}' \int ds\cdot s^{3N-1} gkT$$
$$\times \delta\,[gkT/2 + \phi(\mathbf{q}') + gkT\ln s - E\,]$$
$$= \left[\int d\mathbf{p}'\delta\left(\sum_i \mathbf{p}_i'^2/2m_i - gkT/2\right)\right]\int d\mathbf{q}'$$
$$\times \exp\left\{-\frac{3N}{g}[gkT/2 + \phi(\mathbf{q}') - E\,]/kT\right\}.$$

With $g = 3N$, we obtain the equilibrium distribution function

$$\rho(\mathbf{p}',\mathbf{q}') = \delta\left(\sum_i \mathbf{p}_i'^2/2m_i - gkT/2\right)\exp[-\phi(\mathbf{q}')/kT\,]. \tag{3.7}$$

Equations (3.4)–(3.6) produce the canonical distribution in

Notes on p. 306

coordinate space. Note that if we sample in real time, g must be $3N-1$.

B. Equations in real variables

The equations for the real variables $\mathbf{q}_i'$, $\mathbf{p}_i'$, t' are

$$\frac{d\mathbf{q}_i'}{dt'} = s\frac{d\mathbf{q}_i}{dt} = \mathbf{p}_i/m_i s = \mathbf{p}_i'/m_i, \tag{3.8}$$

$$\frac{d\mathbf{p}_i'}{dt'} = s\frac{d}{dt}(\mathbf{p}_i/s) = -\frac{\partial\phi}{\partial\mathbf{q}_i'} - \frac{ds}{dt}\mathbf{p}_i'. \tag{3.9}$$

These are formally equivalent to Eqs. (2.19) and (2.20). The derivative ds/dt is obtained via the differentiation of Eq. (3.2),

$$\sum_i \mathbf{p}_i \frac{d\mathbf{p}_i}{dt}/m_i = gkTs\frac{ds}{dt}$$

or

$$\frac{ds}{dt} = -\left(\sum_i \frac{\partial\phi}{\partial\mathbf{q}_i'}\mathbf{p}_i'/m_i\right)/gkT. \tag{3.10}$$

Equation (3.10) can be changed to

$$\frac{ds}{dt} = -\left(\sum_i \frac{\partial\phi}{\partial\mathbf{q}'_i}\mathbf{p}_i'/m_i\right)/gkT$$

$$= -\left(\sum_i \frac{\partial\phi}{\partial\mathbf{q}_i'}\frac{d\mathbf{q}_i'}{dt'}\right)/gkT = -\frac{d\phi}{dt'}/gkT. \tag{3.11}$$

If one sets the parameter $\alpha = ds/dt$, Eqs. (3.8)–(3.10) are identical to Eqs. (1.1), (1.2), and (1.5) of the HLME method.[14–18] These equations are based on real time sampling, so that if g is set equal to $3N-1$ we recover the canonical distribution. In the original papers,[14,16] $g = 3N$ was used.

As shown above, the HLME method is equivalent to the ES method with the additional constraint of Eq. (3.2).

C. Another constraint method

Any other choice but the logarithmic form for the potential function of s used in Eq. (2.5) leads to an ensemble different from the canonical one. As an example, we derive the equilibrium distribution function of the refined form of the momentum scaling (HG) method by Haile and Gupta.[13]

The starting Hamiltonian is (s corresponds to 1-ξ of Ref. 13)

$$H = \sum_i \mathbf{p}_i{}^2/2m_i s + \phi(q) + \frac{g}{2}kTs. \tag{3.12}$$

With the constraint

$$\frac{\partial H}{\partial s} = -\sum_i \mathbf{p}_i^2/2m_i s^2 + \frac{g}{2}kT \equiv 0 \tag{3.13}$$

or

$$s = \left[\left(\sum_i \mathbf{p}_i^2/m_i\right)/gkT\right]^{\frac{1}{2}},$$

the equations of motion are

$$\frac{d\mathbf{q}_i}{dt} = \mathbf{p}_i/m_i s = \mathbf{p}_i'/m_i, \tag{3.14}$$

$$\frac{d\mathbf{p}_i}{dt} = -\frac{\partial\phi}{\partial\mathbf{q}_i}. \tag{3.15}$$

The momenta $\mathbf{p}_i' = \mathbf{p}_i/s$ are considered to be the real momenta. These equations are identical to the traditional momentum scaling equations.[12,23] The partition function is

$$Z = \int ds \int d\mathbf{p} \int d\mathbf{q}\, \delta\left[\sum_i \mathbf{p}_i^2/2m_i s + \phi(\mathbf{q}) + \frac{g}{2}kTs - E\right]$$

$$\times \delta\left\{s - \left[\sum_i (\mathbf{p}_i^2/m_i)/gkT\right]^{\frac{1}{2}}\right\}.$$

With the transformation $\mathbf{p}_i' = \mathbf{p}_i/s$, we get

$$Z = \left[\int d\mathbf{p}'\, \delta\left(\sum_i \mathbf{p}_i'^2/2m_i - gkT/2\right)\right]\int d\mathbf{q}\int ds\cdot s^{3N-1}$$

$$\times \delta\,[gkTs + \phi(\mathbf{q}) - E\,]$$

$$= \left[\int d\mathbf{p}'\, \delta\left(\sum_i \mathbf{p}_i'^2/2m_i - gkT/2\right)\right]\int d\mathbf{q}$$

$$\times\{[E - \phi(\mathbf{q})]/gkT\}^{3N-1} h\,[E - \phi(\mathbf{q})]. \tag{3.16}$$

Assuming the most favorable case: $\langle s\rangle = 1$, $g = 3N-1$, and $E = gkT + \langle\phi\rangle$, the equilibrium distribution function of this method is

$$\rho(\mathbf{p}',\mathbf{q}) = \delta\left(\sum_i \mathbf{p}_i'^2/2m_i - gkT/2\right)$$

$$\times\{1 + [\langle\phi\rangle - \phi(\mathbf{q})]/gkT\}^g$$

$$\times h[gkT + \langle\phi\rangle - \phi(\mathbf{q})]. \tag{3.17}$$

In the same way as in Sec. II C, Eq. (3.17) deviates from $\exp[-\phi(\mathbf{q})/kT]$ by order $N^{-\frac{1}{2}}$. The coordinate part of $\rho(\mathbf{p}',\mathbf{q})$ in Eq. (3.17) is similar to the equilibrium distribution function of the microcanonical ensemble projected onto coordinate space,

$$\rho_{mc}(\mathbf{q}) = \int d\mathbf{p}\, \delta\left(\sum_i \mathbf{p}_i^2/2m_i + \phi - E\right)$$

$$= \int dP\, \delta(P^2 + \phi - E)\int d\mathbf{p}\, \delta\left[P - \left(\sum_i \mathbf{p}_i^2/2m_i\right)^{\frac{1}{2}}\right]$$

$$= C\int dP\cdot P^{3N-1}\delta(P^2 + \phi - E)$$

$$= C'\int dP\cdot P^{3N-2}\delta[P - (E-\phi)^{\frac{1}{2}}]$$

$$= C'(E-\phi)^{\frac{3}{2}N-1} = C''\left(1 + \frac{\langle\phi\rangle - \phi}{3NkT/2}\right)^{\frac{3}{2}N-1}, \tag{3.18}$$

where P is a radius of a $3N$ dimension sphere and $E = \frac{3}{2}NkT + \langle\phi\rangle$. Therefore, the difference between the canonical ensemble and the ensemble generated by the HG method[13] is the same order as we expect between the microcanonical and the canonical ensembles. The ensemble corrections of order N^{-1} for first order quantities (energy, virial, ...) and those of order one for quantities relating to fluctuation formulas (heat capacity, compressibility, ...) are generally expected.[21,22] For example, the fluctuation of the potential energy in the HG method is

$$\langle(\delta\phi)^2\rangle = N(kT)^2 3c_v^*/(3 + c_v^*) + O(1), \tag{3.19}$$

where $c_v^* = (1/Nk)(d\langle\phi\rangle/dT)$ is a heat capacity of the coordinate part.

IV. THE CONSTANT TEMPERATURE-CONSTANT PRESSURE (TP) ENSEMBLE

A. The extended system method

Combined with the constant pressure MD method of Anderson,[1] the canonical ensemble MD method can be readily extended to the TP ensemble.[19] Here we use the formulation for uniform dilation given by Anderson,[1] but the extension to the generalized form of the constant pressure simulation method by Parrinello and Rahman can be derived in a similar way.[2–4]

In the TP ensemble, the virtual variables $(\mathbf{q}_i, \mathbf{p}_i, s, V, t)$ are related to the real variables $(\mathbf{q}_i', \mathbf{p}_i', s, V, t')$ via scaling of the coordinates by $V^{1/3}$ and scaling of the time by s (V, the volume of a MD cell),

$$\mathbf{q}_i' = V^{\frac{1}{3}}\mathbf{q}_i, \tag{4.1}$$

$$\mathbf{p}_i' = \mathbf{p}_i/V^{\frac{1}{3}}s, \tag{4.2}$$

$$t' = \int^t \frac{dt}{s}. \tag{4.3}$$

The values of the components of the scaled coordinates $\mathbf{q}_i$ are limited to the range of 0 to 1. The Hamiltonian is

$$H = \sum_i \mathbf{p}_i^2/2m_i V^{\frac{2}{3}}s^2 + \phi(V^{\frac{1}{3}}\mathbf{q}) + p_s^2/2Q + gkT\ln s + p_V^2/2W + P_{ex}V, \tag{4.4}$$

where p_V is the conjugate momentum of V, W is a *mass* for the volume motion, and P_{ex} is the externally set pressure.

The equations of motion are

$$\frac{d\mathbf{q}_i}{dt} = \frac{\partial H}{\partial \mathbf{p}_i} = \mathbf{p}_i/m_i V^{\frac{2}{3}}s^2, \tag{4.5}$$

$$\frac{d\mathbf{p}_i}{dt} = -\frac{\partial H}{\partial \mathbf{q}_i} = -\frac{\partial \phi}{\partial \mathbf{q}_i} = -\frac{\partial \phi}{\partial \mathbf{q}_i'}V^{\frac{1}{3}}, \tag{4.6}$$

$$\frac{ds}{dt} = \frac{\partial H}{\partial p_s} = p_s/Q, \tag{4.7}$$

$$\frac{dp_s}{dt} = -\frac{\partial H}{\partial s} = \left[\sum_i \mathbf{p}_i^2/m_i V^{\frac{2}{3}}s^2 - gkT\right]\Big/s, \tag{4.8}$$

$$\frac{dV}{dt} = \frac{\partial H}{\partial p_V} = p_V/W, \tag{4.9}$$

$$\frac{dp_V}{dt} = -\frac{\partial H}{\partial V} = \left[\sum_i(\mathbf{p}_i^2/m_i V^{\frac{2}{3}}s^2 - \frac{\partial \phi}{\partial \mathbf{q}_i'}\mathbf{q}_i')\right]\Big/3V - P_{ex}. \tag{4.10}$$

In Lagrangian form, these are

$$\frac{d}{dt}\left(m_i V^{\frac{2}{3}}s^2\frac{d\mathbf{q}_i}{dt}\right) = -\frac{\partial \phi}{\partial \mathbf{q}_i} = -\frac{\partial \phi}{\partial \mathbf{q}_i'}V^{\frac{1}{3}}, \tag{4.11}$$

$$\frac{d}{dt}\left(Q\frac{ds}{dt}\right) = \left[\sum_i V^{\frac{2}{3}}s^2\left(\frac{d\mathbf{q}_i}{dt}\right)^2/m_i - gkT\right]\Big/s, \tag{4.12}$$

$$\frac{d}{dt}\left(W\frac{dV}{dt}\right) = \left\{\sum_i\left[V^{\frac{2}{3}}s^2\left(\frac{d\mathbf{q}_i}{dt}\right)^2/m_i - \frac{\partial \phi}{\partial \mathbf{q}_i'}q_i'\right]\right\}\Big/3V - P_{ex}. \tag{4.13}$$

The equilibrium distribution function is obtained in the same way as in Secs. II and III. We define $H_0(\mathbf{p},\mathbf{q}) = \Sigma_i \mathbf{p}_i^2/2m_i + \phi(\mathbf{q})$ as before. Then

$$Z = \int dp_V \int dV \int dp_s \int ds \int d\mathbf{p} \int d\mathbf{q}\,\delta[H_0(\mathbf{p}/V^{\frac{1}{3}}s, V^{\frac{1}{3}}\mathbf{q}) + p_s^2/2Q + gkT\ln s + p_V^2/2W + P_{ex}V - E].$$

The transformations equations (4.1) and (4.2) lead to

$$\begin{aligned} Z &= \int dp_V \int dp_s \int dV \int d\mathbf{p}' \int d\mathbf{q}' \int ds\, s^{3N}\delta[H_0(\mathbf{p}',\mathbf{q}') + p_s^2/2Q + gkT\ln s + p_V^2/2W + P_{ex}V - E] \\ &= C\int dp_V \int dp_s \int dV \int d\mathbf{p}' \int d\mathbf{q}' \exp\left[-\left(\frac{3N+1}{g}\right)\cdot(H_0(\mathbf{p}',\mathbf{q}') + p_s^2/2Q + P_V^2/2W - P_{exp}V - E)/kT\right] \\ &= C\int dV \int d\mathbf{p}' \int d\mathbf{q}' \int \exp\left[-\left(\frac{3N+1}{g}\right)\cdot(H_0(\mathbf{p}',\mathbf{q}') + P_{ex}V)/kT\right]. \end{aligned}$$

For virtual time sampling, with $g = 3N + 1$, the equilibrium distribution function is

$$\rho(\mathbf{p}',\mathbf{q}',V) = \exp[-(H_0(\mathbf{p}',\mathbf{q}') + P_{ex}V)/kT] \tag{4.14}$$

and the averages of any function of $\mathbf{p}',\mathbf{q}',V$ are identical with those in the TP ensemble.

$$\begin{aligned} \lim_{t_0\to\infty}\frac{1}{t_0}\int_0^{t_0} A(\mathbf{p}/V^{\frac{1}{3}}s, V^{\frac{1}{3}}\mathbf{q}, V)dt &= \langle A(\mathbf{p}/V^{\frac{1}{3}}s, V^{\frac{1}{3}}\mathbf{q}, V)\rangle \\ &= \langle A(\mathbf{p}',\mathbf{q}',V)\rangle_{TP}, \end{aligned} \tag{4.15}$$

$\langle\ldots\rangle_{TP}$ denotes the average in the TP ensemble. For real time sampling, g must be $3N$ in order to produce the TP ensemble.

In the same fashion as in Sec. II B, we can also get the equations for real variables.

B. The constraint method

The equations for the contraint method are Eqs. (4.5) and (4.6) and the constraints

$$\frac{\partial H}{\partial s} = -\left(\sum_i \mathbf{p}_i^2/m_i V^{\frac{2}{3}}s^2 - gkT\right)\Big/s \equiv 0, \tag{4.16}$$

$$\frac{\partial H}{\partial V} = -\left[\sum_i\left(\mathbf{p}_i^2/m_i V^{\frac{2}{3}}s^2 - \frac{\partial \phi}{\partial \mathbf{q}_i'}\mathbf{q}_i'\right) - 3P_{ex}V\right]\Big/3V \equiv 0, \tag{4.17}$$

$$\frac{\partial H}{\partial p_s} = p_s/Q \equiv 0, \quad \text{and} \quad \frac{\partial H}{\partial p_V} = p_V/W \equiv 0.$$

Notes on p. 306

The partition function is

$$Z = C\int d\mathbf{p}'\,\delta\Big(\sum_i p_i'^2/2m_i - gkT/2\Big)\int dV\int d\mathbf{q}'$$
$$\times \exp\Big\{-\Big(\frac{3N}{g}\Big)[gkT/2+\phi(\mathbf{q}')+P_{ex}V]/kT\Big\}\delta \quad (4.18)$$
$$\times\Big[(gkT-\sum_i\frac{\partial\phi}{\partial\mathbf{q}_i'}\mathbf{q}_i')/3-P_{ex}V\Big],$$

g is $3N$ for virtual time sampling and $3N-1$ for real time sampling.

The equations for real variables ($\mathbf{p}_i'=\mathbf{p}_i/V^{1/2}s$, $\mathbf{q}_i'=V^{1/3}\mathbf{q}_i$, and $t'=\int^t dt/s$) are

$$\frac{d\mathbf{q}_i'}{dt'}=\mathbf{p}_i'/m_i+\Big(\frac{1}{3V}\frac{dV}{dt'}\Big)\mathbf{q}_i' \quad (4.19)$$

and

$$\frac{d\mathbf{p}_i'}{dt'}=-\frac{\partial\phi}{\partial\mathbf{q}_i'}-\Big(\frac{1}{3V}\frac{dV}{dt'}\Big)\mathbf{p}_i'-\Big(\frac{ds}{dt}\Big)\mathbf{p}_i'. \quad (4.20)$$

The derivatives ds/dt, $1/3V\,dV/dt'$, are obtained via differentiation of Eqs. (4.16) and (4.17),

$$\frac{ds}{dt}+\frac{1}{3V}\frac{dV}{dt'}=-\Big(\sum_i\frac{\partial\phi}{\partial\mathbf{q}_i'}\frac{\mathbf{p}_i'}{m_i}\Big)/gkT \quad (4.21)$$

and

$$\frac{1}{3V}\frac{dV}{dt'}=-\Big(\sum_i\frac{\partial\phi}{\partial\mathbf{q}_i'}\frac{\mathbf{p}_i'}{m_i}+\sum_i\sum_j\frac{\partial^2\phi}{\partial\mathbf{q}_i'\partial\mathbf{q}_j'}\mathbf{q}_j'\frac{\mathbf{p}_i'}{m_i}\Big)\Big/ \quad (4.22)$$
$$\Big(9P_{ex}V+\sum_i\frac{\partial\phi}{\partial\mathbf{q}_i'}\mathbf{q}_i'+\sum_i\sum_j\frac{\partial^2\phi}{\partial\mathbf{q}_i'\partial\mathbf{q}_j'}\mathbf{q}_i'\mathbf{q}_j'\Big).$$

If we define parameters $\alpha=ds/dt$ and $\dot\epsilon=1/3V\,dV/dt'$, Eqs. (4.19)–(4.22) are identical to the equations given by Evans and Morriss[18] except that in this derivation the total kinetic energy $\Sigma^i\mathbf{p}_i'^2/2m_i$ has to be set equal to $(3N-1)kT/2$ and not to $3NkT/2$.

V. DISCUSSION

A. Comparison of the extended system method and the constraint method

The constant temperature MD methods for the canonical and the TP ensembles are reviewed in Secs. II–IV. The relation of these equations is listed in Table I. The HLME method[14–18] and Evans–Morriss method[18] were derived as a special case of the ES method.[19] Other methods except those mentioned here and that by Anderson[1] do not seem to give the rigorous canonical distribution.

The methods were first presented in the virtual variable formulation, then transformed to the real variable case. The virtual variable formulation is the backbone of the constant temperature method. In this form, the equations remain canonical and the proof of equivalence with the canonical ensemble is straightforward. However, the unequal time intervals are not convenient for simulations. The real time formulation is recommended for applications. It should be remembered that in this case the equations are no longer canonical.

Both constant temperature methods (the ES and the HLME) are still not ideal. The number of independent variables are listed in Table II. The ES method has more independent variables than the equivalent statistical mechanical ensemble. This is the reason why the ES method gives correct results for the static quantities, but the time evolution of s and/or V are dependent on the adjustable parameters Q and/or W. This arbitrariness is both a disadvantage and an advantage of this method. The calculation of the velocity autocorrelation functions in the constant pressure MD method[7] and in the constant temperature method[13,24,25] show no significant difference from that of the ordinary MD method. If the effects on the dynamics of the physical system are negligible, we can select the parameters Q and W to optimize the efficiency of the calculations.

In the constraint method, the number of independent variables is less than those of the statistical mechanical formulation, due to the addition of the constraints. Some of the static quantities in the constraint method are not exactly those appropriate to the canonical distribution. In the canonical ensemble, only the quantities dependent on the momentum are affected. In the TP ensemble, the pressure constaint depends both on the volume and on coordinate space.

If we define the instantaneous temperature T_i and pressure P_i by

$$\sum_j\mathbf{p}_j^2/2m_j=\frac{3N}{2}kT_i, \quad (5.1)$$

and

TABLE I. Relation between the various constant temperature methods.

	Extended system method		Constraint method
Virtual variable equations	Sec. II A, Nosé[a] Eqs. (2.5)–(2.12)	→ constraint Eq. (3.2) →	Sec. III A Eqs. (3.2)–(3.6)
	↓ transformation Eqs. (2.1)–(2.3)		↓ transformation [Eqs. (2.1)–(2.3)]
Real variable equations	Sec. II B Eqs. (2.19)–(2.25)		Sec. III B, HLME[b] Eqs. (3.8)–(3.10)

[a] Reference 19.
[b] References 14–18.

TABLE II. The number of independent variables in the various constant temperature MD methods. The numbers in brackets explicitly consider the momentum and angular momentum conservation.

	Extended system method	Statistical mechanics	Constraint method
Canonical Ensemble	$6N+1$ $(6N-5)$	$6N$	$6N-1$ $(6N-7)$
TP Ensemble	$6N+3$ $(6N-3)$	$6N+1$	$6N-1$ $(6N-7)$

$$\Big[\sum_j (\mathbf{p}_j^2/2m_j - \frac{\partial\phi}{\partial \mathbf{q}_j}\mathbf{q}_j)\Big]/3V = P_i, \tag{5.2}$$

the averages and the fluctuations of these quantities in the canonical and the TP ensemble are

$$\langle T_i \rangle = T, \tag{5.3}$$

$$\langle (T_i - T)^2 \rangle = \frac{2}{3N} T^2, \tag{5.4}$$

$$\langle P_i \rangle = P_{\rm ex}, \tag{5.5}$$

and

$$\langle (P_i - P_{\rm ex})^2 \rangle = -kT \Big\langle \frac{\partial P_i}{\partial V} \Big\rangle = \frac{kT}{9}\Big\langle \Big(5\sum_j \mathbf{p}_j^2/m_j - 2\sum_j \frac{\partial \phi}{\partial \mathbf{q}_j}\mathbf{q}_j + \sum_j\sum_k \frac{\partial^2\phi}{\partial \mathbf{q}_j \partial \mathbf{q}_k}\mathbf{q}_j\mathbf{q}_k \Big)/V^2 \Big\rangle. \tag{5.6}$$

The fluctuations of T_i and P_i [Eqs. (5.4) and (5.6)] are suppressed in the constraint method. In the formulation of statistical mechanics, T and $P_{\rm ex}$ are the temperature and pressure of the external system, and the values of the temperature and pressure in the physical system are defined only in an averaged sense by Eqs. (5.3) and (5.5).

B. Some comments

The proper choices for the values of the parameter g in the potential energy function for the variable s, $gkT \ln s$, are listed in Table III. These values depend on the nature of the method as well as on the type of sampling.

In the rigid molecule case, the kinetic energy term for the molecular rotation $\sum_i \frac{1}{2} s^2 \omega_i I_i \omega_i$ or $\sum_i \frac{1}{2} \mathbf{p}_{\omega_i} I_i^{-1} \mathbf{p}_{\omega_i}/s^3$ (ω_i, angular velocity; I_i, moment of inertia tensor; $\mathbf{p}_{\omega_i}$, the conjugate momentum of ω_i) are added to the Hamiltonian [Eq. (2.5)] and the number of degrees of freedom of the rotation must be added to g value.

TABLE III. Proper values of the parameter g in the potential function for s, $gkT \ln s$.

	Extended system method	Constraint method
Sampling in virtual time	$3N+1$	$3N$
Sampling in real time	$3N$	$3N-1$

The constant pressure MD method deviates by order N^{-1} from the constant enthalpy-constant pressure ensemble in any case, due to the kinetic energy term for the volume motion $p_V^2/2W$. Therefore, the effects of using different functional forms for this term [e.g., $A(V)p_V^2$] are much less significant.[4] However, in the canonical and the TP ensemble, methods, there are no intrinsic deviation of order N^{-1} to start with, so different forms for these kinetic terms (e.g., $p_s^2/2Q$ or $p_s^2/2Qs^2$, $p_V^2 V^2/2W$) can give rise to deviations of order N^{-1}. A careful choice of the functional form for the additional kinetic energy terms is important in these types of calculations.

Due to the conservation of the total momentum and angular momentum, MD methods produce ensembles that deviate slightly from the statistical mechanical ensembles.[20-22] The true number of independent variables (when the above conservation laws are taken into account) are bracketed in Table II. In practice, the angular momentum is not conserved in MD simulations if we employ periodic boundary conditions. Here, we only discuss the effect of the momentum conservation law.

The momentum conservation law holds when it is expressed in terms of virtual variables

$$\sum_i p_{ix} = p_x, \quad \sum_i p_{iy} = p_y, \quad \text{and} \quad \sum_i p_{iz} = p_z, \tag{5.7}$$

where p_x, p_y, p_z are constants. The partition function in the extended system is modified

$$Z = \int dp_s \int ds \int d\mathbf{p} \int d\mathbf{q}\, \delta(H-E)\, \delta(\sum_i p_{ix} - p_x)\delta(\sum_i p_{iy} - p_y)\delta(\sum_i p_{iz} - p_z).$$

With the transformation to real variables, it becomes

$$Z = \int dp_s \int d\mathbf{p}' \int d\mathbf{q}' \int ds \cdot s^{3N} \delta(H' - E)\delta(s\sum_i p'_{ix} - p_x) \times \delta(s\sum_i p'_{iy} - p_y)\delta(s\sum_i p'_{iz} - p_z)$$

$$= \int dp_s \int d\mathbf{p}' \int d\mathbf{q}' \int ds \cdot s^{3N-3} \delta(H' - E)\delta(\sum_i p'_{ix} - p_x/s) \times \delta(\sum_i p'_{iy} - p_y/s)\delta(\sum_i p'_{iz} - p_z/s).$$

If the total momentum $p_x = p_y = p_z = 0$, momentum space and coordinate space can be separated. Hence,

$$Z = \int dp_s \int d\mathbf{p}' \delta(\sum_i p'_{ix})\delta(\sum_i p'_{iy})\delta(\sum_i p'_{iz}) \times \int d\mathbf{q}' \int ds \cdot s^{3N-3}\delta(H' - E). \tag{5.8}$$

From above equation, if $g = 3N+3$, the equilibrium distribution function is

$$\rho(\mathbf{p}',\mathbf{q}') = \exp[-H_0(\mathbf{p}',\mathbf{q}')/kT]\delta(\sum_i p'_{ix})\delta(\sum_i p'_{iy})\delta(\sum_i p'_{iz}). \tag{5.9}$$

In momentum space, there is restriction due to momentum

conservation, but that has no effect in coordinate space. The instantaneous temperature should be defined as

$$\sum_i p_i'^2/2m_i = (3N-3)kT/2. \quad (5.10)$$

This follows from a decrease of the number of independent variables. In the case of the TP ensemble, the weight factor for the volume also changes.

Theoretically, static quantities are independent of the value chosen for the parameter Q. However, in practice because of the finite number of time steps, the equivalence equation (2.17) is not always satisfied. With small Q values, the degree of freedom associated with s tends to decouple from the physical system. On the other hand, large Q values lead to inefficient sampling of phase space. The most efficient calculation will be done by choosing same order of time scales for the physical system and the variable s. If we only consider the fluctuation of s around the averaged value $\langle s\rangle$, $s = \langle s\rangle + \delta s$,[19] Eq. (2.12) can be simplified as

$$Q\frac{d^2}{dt^2}\delta s = -\frac{2gkT}{\langle s\rangle^2}\delta s. \quad (5.11)$$

The frequency of this harmonic equation is

$$\omega^2 = \left(\frac{2gkT}{Q\langle s\rangle^2}\right). \quad (5.12)$$

We can choose a Q value such that ω^2 in Eq. (5.12) gives the same order of magnitude as the second moment of the frequency spectrum of the velocity autocorrelation function of the physical system.

This time scale approximately corresponds to the time taken for a sound wave to travel the nearest neighbor distance.

VI. SUMMARY

Three constant temperature MD methods are examined analytically. Except for effects due to momentum and angular momentum conservation, the ES method[19] gives rigorous equilibrium distribution functions in the canonical and in the TP ensembles. The HLME constraint method[14–18] give the canonical distribution only in coordinate space. Both the HLME and the Evans–Morriss method have been derived from the formulation of the ES method by imposing constraints. The Haile–Gupta method[13] on the other hand does not give the rigorous canonical distribution.

The virtual variable formulation is best suited for proof of the equivalence with the statistical mechanical ensembles. The equations based on the real variable formulations [Eqs. (2.19)–(2.25) or the HLME method] are recommended for applications.

The extension of the MD method to ensembles other than the microcanonical ensemble is formulated in a unified fashion. By introducing real variables and virtual variables the constant pressure MD method is generated from a scaling of the coordinates. The constant temperature MD method is obtained from a scaling of the time.

ACKNOWLEDGMENTS

The author thanks Mike Klein and Ray Somorjai for their interest and helpful discussions.

[1] H. C. Anderson, J. Chem. Phys. **72**, 2384 (1980).
[2] M. Parrinello and A. Rahman, Phys. Rev. Lett. **45**, 1196 (1980).
[3] M. Parrinello and A. Rahman, J. Appl. Phys. **52**, 7182 (1981).
[4] S. Nosé and M. L. Klein, Mol. Phys. **50**, 1055 (1983).
[5] M. Parrinello, A. Rahman, and P. Vashishta, Phys. Rev. Lett. **50**, 1073 (1983).
[6] S. Nosé and M. L. Klein, Phys. Rev. Lett. **50**, 1207 (1983).
[7] S. Nosé and M. L. Klein, J. Chem. Phys. **78**, 6928 (1983).
[8] R. W. Impey, S. Nosé, and M. L. Klein, Mol. Phys. **50**, 243 (1983); D. Lévesque, J.-J. Weis, and M. L. Klein, Phys. Rev. Lett. **51**, 670 (1983).
[9] R. G. Munro and R. D. Mountain, Phys. Rev. B **28**, 2261 (1983).
[10] D. M. Heyes, Chem. Phys. **82**, 285 (1983).
[11] W. G. Hoover, D. J. Evans, R. B. Hickman, A. J. C. Ladd, W. T. Ashurst, and B. Moran, Phys. Rev. A **22**, 1690 (1980).
[12] L. V. Woodcock, Chem. Phys. Lett. **10**, 257 (1971).
[13] J. M. Haile and S. Gupta, J. Chem. Phys. **79**, 3067 (1983).
[14] W. G. Hoover, A. J. C. Ladd, and B. Moran, Phys. Rev. Lett. **48**, 1818 (1982).
[15] A. J. C. Ladd and W. G. Hoover, Phys. Rev. B **28**, 1756 (1983).
[16] D. J. Evans, J. Chem. Phys. **78**, 3297 (1983).
[17] D. J. Evans, W. G. Hoover, B. H. Failor, B. Moran, and A. J. C. Ladd, Phys. Rev. A **28**, 1016 (1983).
[18] D. J. Evans and G. P. Morriss, Chem. Phys. **77**, 63 (1983); Phys. Lett. A **98**, 433 (1983).
[19] S. Nosé, Mol. Phys. (to be published).
[20] W. G. Hoover and B. J. Alder, J. Chem. Phys. **46**, 686 (1967).
[21] J. L. Lebowitz, J. K. Percus, and L. Verlet, Phys. Rev. **153**, 250 (1967).
[22] D. C. Wallace and G. K. Straub, Phys. Rev. A **27**, 2201 (1983).
[23] W. G. Hoover, Physica A **118**, 111 (1983), Eq. (1).
[24] D. Brown and J. H. R. Clarke, Mol. Phys. **51**, 1243 (1984).
[25] S. Nosé (unpublished work). The velocity autocorrelation functions in solid and fluid states with different Q values are identical to those of the ordinary MD method within statistical errors.

Notes to Reprint IV.5

1. Two misprints have been pointed out to us by S. Nosé. (a) Eq. (2.23) should read

$$H' = \sum_i p_i'^2/2m_i + \phi(\mathbf{q}') + s'^2 p_s'^2/2Q + gkT \log s.$$

(b) On p. 516, in the transformation of the expression for the partition function, "$-P_{\text{ext}}V$" should be replaced by "$+P_{\text{ext}}V$".

2. The paper referred to in the Abstract and in Note 19 was later published as (Nosé, Mol. Phys. 52 (1984) 255). It is pointed out there that the equations of motion in the real-variable formulation are invariant under the transformation $s_{\text{new}} = s_{\text{old}}/a$ (a is a constant). Thus, if s deviates greatly from unity, its value can be rescaled in order to maintain the precision with which the integration is performed.

3. The interpretation in terms of time scaling is not unique. Scaling of mass is another possibility.

4. Branka and Parrinello (Mol. Phys. 58 (1986) 989) have exploited a relation that exists between the statistical moments of s and the free energy in order to compute the thermal properties of a simple system (a cluster of Lennard–Jones particles).

PHYSICAL REVIEW A VOLUME 31, NUMBER 3 MARCH 1985

Canonical dynamics: Equilibrium phase-space distributions

William G. Hoover
Department of Applied Science, University of California at Davis—Livermore, Livermore, California 94550
(Received 18 September 1984)

Nosé has modified Newtonian dynamics so as to reproduce both the canonical and the isothermal-isobaric probability densities in the phase space of an N-body system. He did this by scaling time (with s) and distance (with $V^{1/D}$ in D dimensions) through Lagrangian equations of motion. The dynamical equations describe the evolution of these two scaling variables and their two conjugate momenta p_s and p_v. Here we develop a slightly different set of equations, free of time scaling. We find the dynamical steady-state probability density in an extended phase space with variables x, p_x, V, $\dot{\epsilon}$, and ζ, where the x are reduced distances and the two variables $\dot{\epsilon}$ and ζ act as thermodynamic friction coefficients. We find that these friction coefficients have Gaussian distributions. From the distributions the extent of small-system non-Newtonian behavior can be estimated. We illustrate the dynamical equations by considering their application to the simplest possible case, a one-dimensional classical harmonic oscillator.

I. INTRODUCTION

Classical "constant-temperature" calculations have been pursued for over a decade.[1,2] In this sense, "temperature" is a measure of the instantaneous kinetic energy in a system. Thus the corresponding dynamical equations include non-Newtonian accelerations designed to keep the kinetic energy $\sum p^2/2m$ constant. The non-Newtonian isothermal accelerations are useful in dissipative systems involving viscous flow, or heat flow, far from equilibrium. Such systems would heat rapidly in the absence of constraints. By now, many[3–6] distinct sets of differential equations of motion have been devised to keep the kinetic energy constant.

A somewhat different kind of constant-temperature calculation strives to reproduce the canonical phase-space distribution, so that the kinetic energy can fluctuate, with a distribution proportional to $\exp(-\sum p^2/2mkT)$. Obtaining the canonical distribution is desirable, at least in equilibrium work, in order to correlate the results of many-body simulations with Gibbs's and Jaynes's statistical mechanics. Andersen[7] has used occasional discontinuous "stochastic" collisions to induce the canonical distribution in many-body simulations.

Nosé achieved a major advance by showing that the canonical distribution can be generated with smooth, deterministic, and time-reversible trajectories. To do this he introduced a time-scale variable s, its conjugate momentum p_s, and a parameter Q. Nosé's augmented Hamiltonian[8]

$$H_{\text{Nosé}}=\Phi(q)+\sum p^2/2ms^2 +(X+1)kT\ln s+p_s^2/2Q\ , \tag{1}$$

contains a nonlinear collective potential in which the time-scale variable s oscillates. Thus the system, with X degrees of freedom, is coupled to a heat bath (described by the variables s and p_s). Nosé proved that the microcanonical distribution in the augmented set of variables is equivalent to a canonical distribution of the variables q,p', where the p' are the scaled momenta p/s. Thus the Hamiltonian (1) generates the canonical probability distribution *independent of the values chosen for $H_{\text{Nosé}}$ and Q.*

During the canonical-ensemble calculations just described, the volume V and temperature T are held fixed. Nosé demonstrated the usefulness of these ideas by carrying out several dense-fluid simulations using the Hamiltonian $H_{\text{Nosé}}$.

By allowing *length* to vary,[7] as well as time, Nosé generalized this work to include the isothermal-isobaric ensemble. These methods and ideas forge a remarkable link between the ensembles of statistical theory and atomistic dynamics. They suggest promising approaches for the investigation of nonequilibrium systems.

Here we exhibit steady-state (equilibrium) distributions for the new variables which play the role of thermodynamic friction coefficients. Our equations of motion are very much like Nosé's, but differ in that scaling of the time is not required. The new results for distributions make it possible to estimate finite-size effects on dynamical averages. In Sec. II we review Nosé's canonical equations of motion and introduce a version of them free of time scaling. In Sec. III we formulate the phase-space evolution of the many-body probability density $f_{NVT}(q,p,\zeta,Q)$ and exhibit a steady-state solution. We indicate the straightforward extension to include the isobaric case. With some additional effort, it seems likely that a stress-tensor version of this ensemble could be constructed along the lines pioneered by Rahman and Parrinello.[9] In the final section we illustrate the equations of motion with some representative trajectories for a single classical oscillator.

II. CANONICAL DISTRIBUTION FROM NON-NEWTONIAN DYNAMICS (REF. 10)

The equations of motion from Nosé's Hamiltonian (1) are

$$\dot{q}=p/ms^2,\ \dot{p}=F(q),\ \dot{s}=p_s/Q,$$
$$\dot{p}_s=\sum p^2/ms^3-(X+1)kT/s\ . \tag{2}$$

These coupled first-order equations take a simpler form if the time scale is reduced by s, so that $dt_{\rm old} \equiv s\, dt_{\rm new}$. All of the rates given in (2) can then be expressed as derivatives with respect to $t_{\rm new}$ (for which we will still use the superior dot notation)

$$\dot{q}=p/ms,\ \dot{p}=sF,\ \dot{s}=sp_s/Q,$$
$$\dot{p}_s=\sum p^2/ms^2-(X+1)kT\ . \tag{3}$$

The somewhat inconvenient variable s can then be eliminated from the equations (3) by rewriting the coordinate-evolution equations in terms of q, $\dot{q}$, and $\ddot{q}$:

$$\ddot{q}=\dot{p}/ms-(p/ms)\dot{s}/s=F/m-\dot{q}p_s/Q\equiv F(q)/m-\zeta\dot{q}\ . \tag{4}$$

The thermodynamic friction coefficient $\zeta\equiv p_s/Q$ which appears in the second-order equations (4) evolves in time according to a first-order equation

$$\dot{\zeta}=\left[\sum m\dot{q}^2-(X+1)kT\right]\Big/Q\ . \tag{5}$$

Nosé showed that the phase-space distribution resulting from the equations (2) is canonical in the variables $q,p/s$. In the next section we show that the distribution resulting from equations (4) and (5) can be made canonical too, and in such a way as to avoid time scaling. To do this we redefine $p\equiv m\dot{q}$ and replace Nosé's $X+1$ by X obtaining[10]

$$\dot{q}=p/m,\ \dot{p}=F(q)-\zeta p,\ \dot{\zeta}=\left[\sum p^2/\ m-XkT\right]\Big/Q\ . \tag{6}$$

Berendsen[6] has just suggested a close relative of (6) in which ζ rather than $\dot{\zeta}$ is proportional to $\Delta E_{\rm kin}\equiv\sum p^2/2m-XkT/2$. Notice that Berendsen's equations are not reversible in time. The equations (6) are much less severely damped than Berendsen's. An extreme opposite limiting case, in which $\Delta E_{\rm kin}$ is identically zero and time reversibility is retained, has been achieved by setting the friction coefficient equal to $(\sum Fp/m)/(\sum p^2/m)$ or, equivalently, by "velocity scaling."[1-4]

III. PHASE-SPACE EVOLUTION OF $f_{NVT}(q,p,\zeta)$

Because the variables q, p, and ζ used in (6) are *independent*, we can easily calculate the components of the flow of probability density $f(q,p,\zeta)$ in $(2X+1)$-dimensional space. The equations governing the motion in this space are *not* Hamiltonian. Therefore the derivatives $\partial\dot{q}/\partial q$ and $\partial\dot{p}/\partial p$ do not generally sum to zero. Thus the analog of Liouville's equation, expressing the conservative flow of probability with time, including flow in the ζ direction, is

$$\partial f/\partial t+\dot{q}\,\partial f/\partial q+\dot{p}\,\partial f/\partial p+\dot{\zeta}\,\partial f/\partial\zeta$$
$$+f[\partial\dot{q}/\partial q+\partial\dot{p}/\partial p+\partial\dot{\zeta}/\partial\zeta]=0\ . \tag{7}$$

Consider a density function f_{NVT} proportional to the following exponential:

$$f_{NVT}\propto\exp\left[-\left\{\Phi(q)+\sum p^2/2m+Q\zeta^2/2\right\}\Big/kT\right]\ . \tag{8}$$

The nonvanishing terms in (7) obtained from this density function are as follows:

$$\dot{q}\,\partial f/\partial q=(f/kT)\sum Fp/m,$$
$$\dot{p}\partial f/\partial p=(f/kT)\sum(-F+\zeta p)p/m,$$
$$\dot{\zeta}\,\partial f/\partial\zeta=(f/kT)\left[\left[-\sum p^2/m+XkT\right]\Big/Q\right]\zeta Q, \tag{9}$$
$$f\,\partial\dot{p}/\partial p=(f/kT)(-XkT\zeta)\ .$$

Inspection shows that these terms sum to zero, provided that the coefficient of kT in the dynamical equation (6) for the friction coefficient is chosen equal to the number of independent degrees of freedom in the set q,p. In the usual molecular dynamics simulation, with periodic boundaries, the center of mass and its velocity are fixed so that this number of degrees of freedom is $D(N-1)$ for a D-dimensional N-body system. Thus the canonical distribution (8) is a steady equilibrium solution of the flow equation (7) and satisfies the equations of motion (6).

In commenting on an earlier draft of this manuscript, Brad Holian pointed out that the phase-space distribution (8) can be used to *derive* the equation of motion for the friction coefficient ζ. To see this, note that the canonical distribution (8) satisfies (7) if, *and only if*, ζ follows the relaxation equation (6) of Nosé. *Thus Nosé's canonical equations of motion are unique.* Other relaxation equations, such as Berendsen's, cannot lead to the canonical distribution (8).

To extend these ideas to the isothermal-isobaric case is straightforward. Reduced coordinates $x\equiv q/V^{1/D}$ are introduced, as is also a fixed "external pressure" $P_{\rm ext}$ and relaxation time τ. The equations of motion

$$\dot{x}=p/mV^{1/D},\ \dot{p}=F-(\dot{\epsilon}+\zeta)p,\ \dot{\zeta}Q=\sum p^2/m-XkT\ , \tag{10}$$
$$\dot{\epsilon}=\dot{V}/DV,\ \ddot{\epsilon}=(P-P_{\rm ext})V/\tau^2kT\ ,$$

have the steady equilibrium solution $f_{NPT}\propto V^{N-1}\exp(-\Psi/kT)$, where

$$\Psi\equiv\Phi(xV^{1/D})+\sum p^2/2m+Q\zeta^2/2$$
$$+D\dot{\epsilon}^2\tau^2kT/2+P_{\rm ext}V\ . \tag{11}$$

IV. CANONICAL HARMONIC OSCILLATOR

To illustrate the changes in viewpoint discovered by Nosé we consider a one-dimensional harmonic oscillator with the mass, force constant, and initial values of q and p all taken to be unity. We consider equations for which the values of q^2 and p^2 have averaged values of unity. The microcanonical equations of motion

$$\dot{q}=p,\ \dot{p}=-q \tag{12}$$

generate closed elliptical trajectories in the two-dimensional qp phase space. See Fig. 1(a). For this same oscillator Nosé's canonical equations [with X in (2) taken to be zero and s initially unity] take the form

$$\dot{q}=p/s^2,\ \dot{p}=-q,\ \dot{s}=p_s/Q,\ \dot{p}_s=p^2/s^3-1/s\ . \tag{13}$$

Note on p. 310

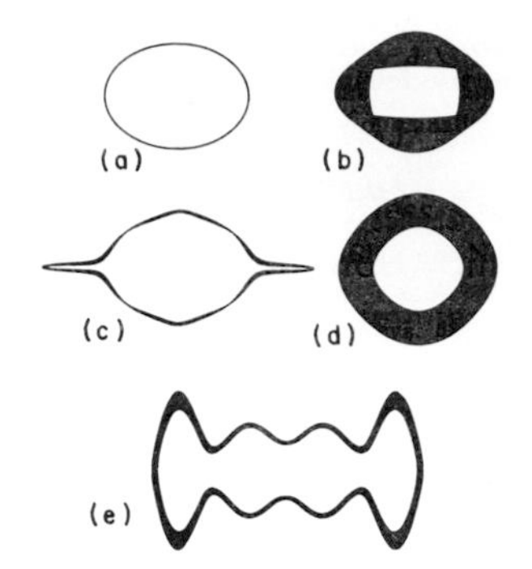

FIG. 1. (a) Elliptical orbit for an oscillator described by Eqs. (12). The abscissa is q, the ordinate is p. The major-to-minor axis ratio of unity has been increased in plotting to fit the Tektronix hard-copy screen area symmetrically. This same increase applies to each figure. All data were obtained on the Digital Equipment Corporation VAX 11/780 computer at the Physics Department (Lausanne) using a fourth-order Runge-Kutta integration in double precision with time steps in the range 0.01 down to 0.001. (b) Long-time qp trajectory for Eqs. (13) or (14) with initial values $q=1$, $p=1$, $s=1$, $p_s=0$, and $Q=1$. (c) Same as (b) with $Q=0.1$. (d) Long-time qp trajectory for Eqs. (15) with initial values $q=1$, $p=1$, $\zeta=0$, and $Q=1$. (e) Same as (d) with $Q=0.1$.

For large Q these equations simply reproduce the microcanonical behavior shown in Fig. 1(a). In Figs. 1(b) and 1(c) we show trajectories for $Q=1.0$ and 0.1 using the same initial conditions. For the larger Q, the trajectories in qp space gradually fill in a region between two limiting curves. For the smaller Q the trajectories develop more nearly singular turning points and the size of the filled region diminishes. When a new time is introduced, with $dt_{\rm old} \equiv s dt_{\rm new}$ and

$$\dot{q}=p/s,\ \dot{p}=-qs,\ \dot{s}=sp_s/Q,\ \dot{p}_s=(p/s)^2-1\ , \tag{14}$$

exactly the same trajectories are produced, but at different rates. This is a good check of the numerical integration.

Finally, if we abandon time scaling and redefine $p \equiv \dot{q}$ we have

$$\dot{q}=p,\ \dot{p}=-q-\zeta p,\ \dot{\zeta}=(p^2-1)/Q\ . \tag{15}$$

Solutions for these equations appear in Figs. 1(d) and 1(e). The small-Q limit of (15) can be inferred from these figures. The oscillator moves between widely-separated turning points at velocity ± 1.

These examples illustrate that a single oscillator is not sufficiently chaotic to reproduce the canonical distribution from a single initial condition. The trajectories are, however, stable and cover a relatively large part of the oscillator phase space for reasonable values of the parameter Q. For unreasonable values of Q (either very small or very large) it is not at all clear that even large systems will behave in a canonical (as opposed to microcanonical) way. A study of the number dependence and Q dependence of the phase-space density for a series of small systems might help to clarify this point.

ACKNOWLEDGMENTS

It is a pleasure to thank Shūichi Nosé and Carl Moser for stimulating conversations at the 1984 Centre Européen de Calcul Atomique et Moléculaire (Orsay, France) Workshop on Constrained Dynamics. Professor Nosé kindly made several comments, correcting and clarifying a previous version of this manuscript. Professor Philippe Choquard kindly provided local support and facilities for this work at Laboratoire de Physique Théorique, École Polytechnique Fédérale de Lausanne, Switzerland. The Academy of Applied Science supported related work at the University of California at Davis—Livermore as well as the cost of travel between California and Europe. This work was partially supported by the Lawrence Livermore National Laboratory under the auspices of the U.S. Department of Energy under Contract No. W-7405-ENG-48.

[1]W. T. Ashurst and W. G. Hoover, Phys. Rev. Lett. **31**, 206 (1973).

[2]L. V. Woodcock, Chem. Phys. Lett. **10**, 257 (1971).

[3]D. J. Evans, J. Chem. Phys. **78**, 3297 (1983).

[4]W. G. Hoover, A. J. C. Ladd, and B. Moran, Phys. Rev. Lett. **48**, 1818 (1982).

[5]J. M. Haile and S. Gupta, J. Chem. Phys. **79**, 3067 (1983).

[6]See H. J. C. Berendsen, J. P. M. Postma, W. F. van Gunsteren, A. DiNola, and J. R. Haak, J. Chem. Phys. **81**, 3684 (1984). For an oscillator this "new" approach gives (Lord) Rayleigh's equation [Philos. Mag. **15**, 229 (1883)].

[7]H. C. Andersen, J. Chem. Phys. **72**, 2384 (1980) and references quoted therein.

[8]S. Nosé, Mol. Phys. **52**, 255 (1984).

[9]M. Parrinello, A. Rahman, and P. Vashishta, Phys. Rev. Lett. **50**, 1073 (1983) and references quoted therein.

[10]S. Nosé, J. Chem. Phys. **81**, 511 (1984). See Sec. II B for equations equivalent to (2)—(6).

Note to Reprint IV.6

For further discussion of the ideas contained in this paper, see (Nosé, Mol. Phys. 57 (1986) 187; Hoover, Phys. Rev. A 34 (1986) 2499). Nosé also shows how it is possible to introduce more than one temperature-control variable, corresponding to different degrees of freedom (e.g. the translational and rotational motion of molecules).

New High-Pressure Phase of Solid ^{4}He Is bcc

Dominique Lévesque and Jean-Jacques Weis
Laboratoire de Physique Théorique et Hautes Energies, Université Paris-Sud, F-91405 Orsay, France

and

Michael L. Klein
Chemistry Division, National Research Council of Canada, Ottawa, Ontario K1A 0R6, Canada

(Received 6 June 1983)

The effect of isobarically heating solid ^{4}He at high density is investigated with use of constant-pressure molecular-dynamics calculations and a realistic interatomic pair potential. At $P \sim 16$ GPa the observed sequence of stable phases with increasing temperature is fcc → bcc → liquid. The presence of a new, thermally stabilized, bcc phase for high-density solid ^{4}He is thus confirmed.

PACS numbers: 67.80.Gb, 62.50.+p, 64.70.Dv, 64.70.Kb

The behavior of solid helium at high densities continues to attract attention.[1-6] The recent observation of a cusp point on the melting curve of solid ^{4}He around room temperature suggested a triple point and hence the presence of a new high-pressure solid phase.[6] While no direct evidence was obtained for a solid → solid phase transition, the characteristics of melting changed above the cusp point; the enhanced premelting and smaller volume change were interpreted as evidence for the presence of a bcc solid.[6]

The interatomic pair potential for He is well established.[7] Moreover, computer simulation techniques have now advanced to the point that it is possible to explore directly the relationship between an interatomic potential and its preferred crystal structure at finite temperature.[8,9] Hence, the question of a possible solid → solid phase transition occurring in solid ^{4}He at high pressures and the nature of the phases involved are amenable to direct investigation. Accordingly, we have carried out a series of isobaric molecular-dynamics simulations.[8] Under an external pressure corresponding to about 16 GPa, we observed the following sequence of stable phases with increasing temperature: fcc → bcc → liquid. The presence of a stable bcc phase preceding the melting of high-density ^{4}He thus confirms the speculations based upon the anomalous behavior of the melting curve.[6] If corresponding-states arguments are invoked, other insulating crystals will likely also exhibit a bcc phase when subjected to analogous conditions of temperature and pressure.

The constant-pressure molecular-dynamics (MD) technique that we employed is documented in the literature[8] and so we omit most of the details. Essentially, the equations of motion are integrated by standard techniques for a system of 432 ^{4}He atoms initially arranged on a body-centered tetragonal lattice with periodic boundary conditions. If the lengths of the basis vectors of the MD cell a, b, and c are in the ratios $1{:}1{:}\sqrt{2}$ the system is equivalent to an fcc lattice, whereas if the ratios of $a{:}b{:}c$ are 1:1:1 and the angles (a,b), (b,c), (a,c) are still 90°, the system is bcc. The new MD method allows the system to change both its volume and shape in response to any instantaneous imbalance between an externally applied pressure and spontaneously generated thermal stresses. By this technique several solid → solid phase transitions have been successfully investigated.[8,9] In particular, a previous study of the relative stability of fcc and bcc lattices employing the Lennard-Jones 12-6 potential established that at low temperature and pressures the close-packed structure is preferred.[8] Although still widely used as an effective pair potential for the rare gases, the 12-6 potential is unfortunately a poor approximation to their true potentials.[10]

Accordingly, in our simulations that are designed specifically to model He, we have employed a realistic pair potential.[7] Before describing our results we mention two questions, not specifically addressed here, that perhaps will deserve further study: quantum effects and three-body forces. Our defense for their omission from our calculations is based largely on pragmatic and intuitive arguments. First, it is known that quantum effects have only a modest influence on the location of the He melting line around room temperature.[4] Second, if solid He can be approximated by an oscillator model obeying the Grüneisen equation of state, then the leading quantum contribution to the pressure is given by $\Delta P_Q = (3\gamma RT/20V)(\theta/T)^2$. With use of extrapolations of lower-pressure data[3] to estimate the Grüneisen parameter γ and the Debye temperature θ, we obtain $\Delta P_Q V/RT \sim 1$ which is small compared with the pressure of interest to us here ($PV/RT \sim 20$).

Finally, we note that of all the rare gases, three-body forces are the smallest in helium[10]; we estimate $\Delta P_3 V/RT \sim 0.4$. Moreover, we are interested in a possible fcc → bcc phase transition and since it is known that at constant density three-body forces are essentially identical in these two structures[11] they are not likely to greatly influence such a transition. We now describe our results.

All of the calculations were carried out under a nominal pressure of 15.2 GPa but when due allowance is made for the contribution from ΔP_Q and ΔP_3 the effective external pressure is approximately 16 GPa. Figure 1(a) shows how at 310 K an initial ideal bcc structure, whose MD cell started with $a = b = c = 14.44$ Å and $(a,b) = (b,c) = (a,c) = 90°$, evolved spontaneously to a fcc-like structure whose MD cell had $\vec{a} = (12.52$ Å, -1.28 Å, $0)$, $\vec{b} = (1.40$ Å, 13.39 Å, 1.00 Å$)$, $\vec{c} = (0.03$ Å, -0.18 Å, 17.53 Å$)$, and $(a,b) = 90°$, $(b,c) = 89.2°$, $(a,c) = 89.6°$. The transformation was completed over a period of 1500 time steps of 1.23 fs and the resulting fcc structure was stable when followed for a further 6500 steps. We conclude therefore that under the conditions $T = 310$ K, $P = 16$ PGa, solid ^{4}He is fcc. However, at higher temperatures, for example at 360 K [see Fig. 1(b)], the reverse behavior was exhibited, namely starting from an fcc MD cell with $a = b = 12.85$ Å, $c = 18.17$ Å, $(a,b) = (b,c) = (a,c) = 90°$, the system evolved spontaneously to become predominantly bcc with an MD cell $\vec{a} = (12.04$ Å, -3.50 Å, 0.56 Å$)$, $\vec{b} = (-0.66$ Å, 12.52 Å, -0.10 Å$)$, $\vec{c} = (0.36$ Å, -0.97 Å, 20.35 Å$)$, $(a,b) = 109°$, $(b,c) = 95°$, and $(a,c) = 87°$. Between these two temperatures the system is observed to oscillate from fcc to bcc as a function of time but with a definite preference for bcc as the temperature increases. We have attempted to classify the character of these "mixed" crystals by monitoring the distribution of neighbors around a given atom and also the distribution of angles specified by the interatomic bond vectors. In this way individual atoms are classified as either bcc-like or fcc-like and we have an efficient means to follow the time evolution of the crystal structure. Typical plots are shown inset in Fig. 2(a), which also shows the variation of the enthalpy H as a function of temperature for the three phases. Figure 2(b) gives an analogous plot for the density variations.

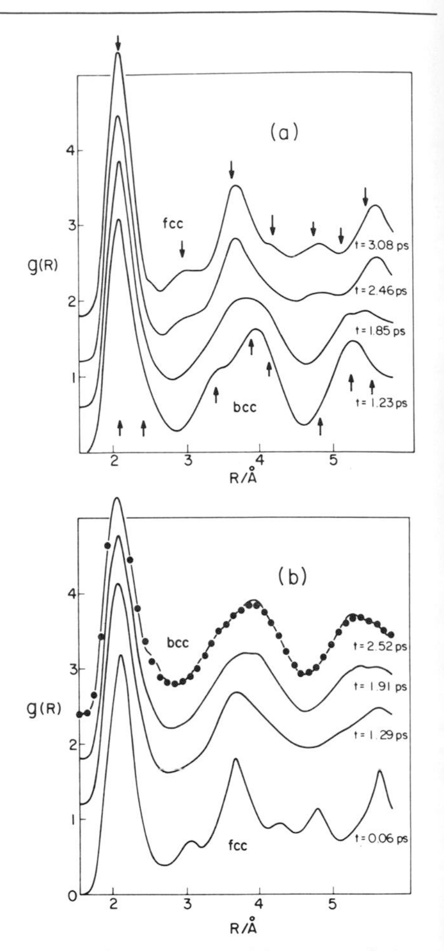

FIG. 1. (a) Time evolution of the pair distribution function, $g(r)$, for solid ^{4}He at $P = 16$ GPa and $T = 310$ K starting from a bcc lattice at $t = 1.23$ ps and finishing at $t = 3.08$ ps with an fcc lattice. The arrows indicate peak positions in the ideal bcc and fcc lattices. (b) Time evolution of $g(r)$ at $P = 16$ GPa and $T = 360$ K starting from an fcc lattice at $t = 0.06$ ps and finishing with a bcc lattice at $t = 2.52$ ps; the dots are results for $t = 6.47$ ps.

The arrow in Fig. 2(a) indicates the melting temperature predicted for the Aziz ^{4}He potential[7]

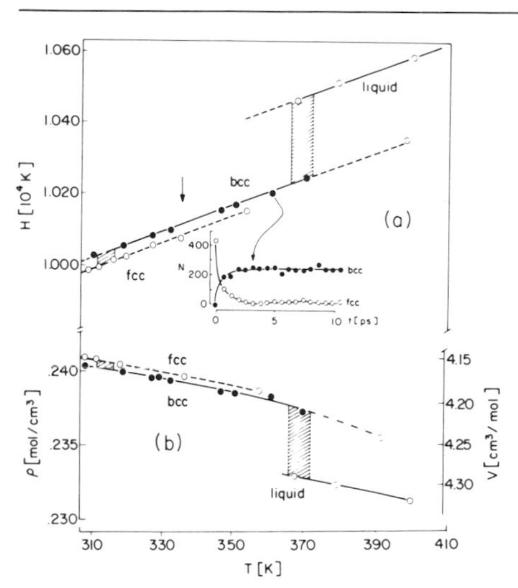

FIG. 2. (a) Temperature dependence of the enthalpy H for three phases of ^{4}He. The inset indicates the evolution of local structure (circles, fcc; dots, bcc). The arrow indicates the melting temperature of the fcc phase calculated in Ref. 6 (see text) while the shaded regions indicate our estimated transitions. (b) Temperature dependence of the density ρ for solid ^{4}He at $P = 16$ GPa.

based upon a perturbation theory that used as a reference system Monte Carlo results for an inverse twelfth-power potential.[4] In our constant-pressure molecular-dynamics calculations, the system is still solid under these conditions, and hence our results point to shortcomings in the perturbation theory, under these extreme conditions. We note from Fig. 2(b) that the volume change on melting is about 2.1% at $T \sim 370$ K. The experimental estimate[6] is certainly greater than 1%.

If we use corresponding-states scaling, an fcc-bcc-liquid triple point would occur in Xe at about $P = 1.2$ Mbar and $T = 10^4$ K. Unfortunately, under these conditions Xe may well no longer be an insulator[12-16] and in any case it may already be bcc for other reasons.[17] The rare gases Ne and Ar would appear to be more serious candidates to exhibit this phenomenon since their metallization occurs under more extreme conditions. However, the most promising candidate for study may well be solid hydrogen, which at room temperature and high pressure appears to be hcp.[18-20] We estimate a possible hcp-bcc-liquid triple point at $P = 26.5$ GPa and $T = 950$ K. In studies of solid H_2 the requisite pressure (density) has already been far exceeded, but studies have yet to be reported above room temperature.

In summary we have observed that at $P \sim 16$ GPa, solid fcc ^{4}He transforms to bcc before melting. Further work is needed to establish the range of existence of this new phase and to explore more deeply the reasons for its stability. These questions will be taken up elsewhere.

One of us (M.L.K.) thanks the Centre National de la Recherche Scientifique–National Research Council of Canada exchange agreement for sponsoring a visit to Orsay during which this work was completed, and Professor J. M. Besson for his comments. The Laboratoire de Physique Théorique et Hautes Energies is a Laboratoire associé au Centre National de Recherche Scientifique.

[1] J. M. Besson and J. P. Pinceaux, Science 206, 1073 (1979).

[2] J. P. Franck and W. B. Daniels, Phys. Rev. Lett. 44, 259 (1980).

[3] B. L. Mills, D. H. Liebenberg, and J. C. Bronson, Phys. Rev. B 21, 5137 (1980).

[4] P. Loubeyre and J. P. Hansen, Phys. Lett. 80A, 181 (1980).

[5] D. A. Young, A. K. McMahan, and M. Ross, Phys. Rev. B 24, 5119 (1981).

[6] P. Loubeyre, J. M. Besson, J. P. Pinceaux, and J. P. Hansen, Phys. Rev. Lett. 49, 1172 (1982); J. M. Besson, private communication.

[7] R. Aziz, V. P. S. Nain, J. S. Carley, W. L. Taylor, and G. T. McConville, J. Chem. Phys. 70, 4430 (1979).

[8] M. Parrinello and A. Rahman, Phys. Rev. Lett. 45, 1196 (1980), and 50, 1073 (1983), and J. Appl. Phys. 52, 7182 (1981).

[9] S. Nosé and M. L. Klein, Phys. Rev. Lett. 50, 1207 (1983).

[10] J. A. Barker, in *Rare Gas Solids*, edited by M. L. Klein and J. A. Venables (Academic, London, 1976).

[11] R. D. Murphy and J. A. Barker, Phys. Rev. A 3, 1037 (1971).

[12] K. S. Chan, T. L. Huang, T. A. Grybowski, T. J. Whetton, and A. L. Ruoff, Phys. Rev. B 26, 7116 (1982); D. A. Nelson and A. L. Ruoff, Phys. Rev. Lett. 42, 383 (1979).

[13] M. Ross and A. K. McMahan, Phys. Rev. B 21, 1658 (1980).

[14] K. Asaumi, T. Mori, and Y. Kondo, Phys. Rev. Lett. 49, 837 (1982).

[15] I. Makarenko, G. Weill, J. P. Itie, and J. M. Besson, Phys. Rev. B 26, 7113 (1982); K. Syassen, Phys. Rev. B 25, 6548 (1982); D. Schiferl, R. L. Mills, and L. E.

Trimmer, Solid State Commun. 46, 783 (1983).
[16]H. Niki, H. Nagara, H. Miyagi, and T. Nakamura, Phys. Lett. 79A, 428 (1980); J. Hama and S. Matsui, Solid State Commun. 37, 889 (1981).
[17]A. K. Ray, S. B. Trickey, and A. B. Kunz, Solid State Commun. 41, 351 (1982).
[18]I. F. Silvera and R. J. Wijngaarden, Phys. Rev. Lett. 47, 39 (1981).
[19]H. Shimizu, E. M. Brody, H. K. Mao, and P. M. Bell, Phys. Rev. Lett. 47, 128 (1981).
[20]J. van Straaten, R. J. Wijngaarden, and I. F. Silvera, Phys. Rev. Lett. 48, 97 (1982).

Stability of the High-Pressure Body-Centered-Cubic Phase of Helium

D. Frenkel

Department of Physics, Rijksuniversiteit Utrecht, 3508 TA Utrecht, The Netherlands

(Received 28 October 1985)

This paper reports absolute free-energy calculations of the fluid, body-centered-cubic, and face-centered-cubic phases of helium at $T = 327.04$ K. We find that at and around this temperature the model potential proposed by Aziz *et al.* does not yield a stable bcc phase. Quantum corrections do not alter this conclusion.

PACS numbers: 61.20.Ja, 64.60.Cn, 67.80.Gb

In 1982 Loubeyre *et al.*[1] reported the observation of a cusp in the melting curve of ^{4}He around room temperature. In Ref. 1 and in subsequent publications[2-4] it is suggested that the cusp in the melting curve is due to the appearance of a thermodynamically stable body-centered cubic (bcc) phase of helium between the fluid and the face-centered cubic (fcc) solid phase, for temperatures above 299 K. Levesque *et al.*[5] carried out constant-stress molecular-dynamics simulations[6] to study the nature of the phase transformations in dense helium. In these simulations the pair potential proposed by Aziz *et al.*[7] was used to model the intermolecular interactions in dense helium. Constant-stress Monte Carlo simulations on the same model system were carried out by Loubeyre, Levesque, and Weis.[8] Both sets of simulations indicate that above room temperature the fcc phase of ^{4}He transforms upon expansion (and/or heating) into a bcc solid. The latter phase is observed to melt upon further lowering of the density.

In the present paper I report absolute free-energy calculations of the fcc, bcc, and fluid phases of "Aziz" helium. Our simulation results are fully consistent with the findings of Refs. 6 and 8 on the *mechanical* stability of the different phases involved. However, I find that the Aziz potential cannot explain the *thermodynamical* stability of the bcc phase of ^{4}He around 300 K. I have also studied the effect of the lowest-order quantum corrections but these have little effect on the relative stability of the fcc and bcc phases.

The excess Helmholtz free energy $F^{\text{fl}}_{\text{ex}}(\rho)$ of the fluid phase at density ρ is most easily calculated by thermodynamic integration:

$$F^{\text{fl}}_{\text{ex}}(\rho)/NkT = \int_0^\rho [P(\rho')/\rho' kT - 1]/\rho' \, d\rho'. \tag{1}$$

In Eq. (1) P is the pressure at density ρ', T is the absolute temperature, and k is the Boltzmann constant. To carry out the integration in Eq. (1) I performed Monte Carlo simulations for a system of 256 particles at a number of densities along the $T = 327.04$ isotherm (see Fig. 1). This particular temperature was chosen because the simulations of Loubeyre, Levesque, and Weis[8] were carried out for the same value of T. In Fig. 1 I have included eight state points from Ref. 8. No systematic differences between my data and those of Loubeyre, Levesque, and Weis are observable. In addition, I evaluated the second virial coefficient of Aziz helium at 327.04 K. With use of this information the Monte Carlo data were fitted by a fifth-order polynomial in the density. The free energies of the fcc and bcc solids were evaluated by construction of a reversible path from the solid under consideration to an Einstein crystal with the same structure.[9] The method used in the present simulations differs slightly from the one described in Ref. 9. Let us consider a poten-

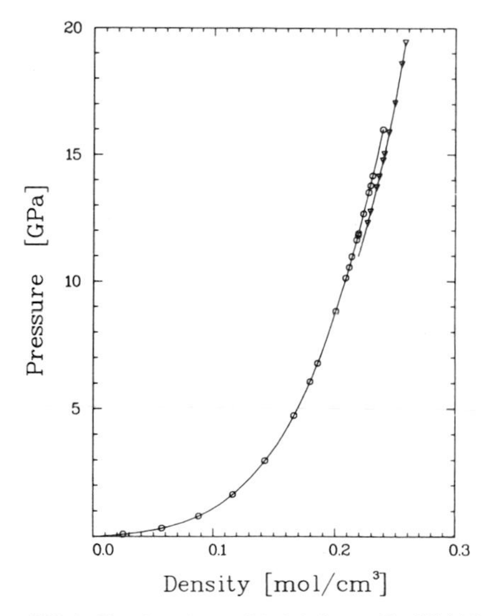

FIG. 1. Equation of state of Aziz helium at $T = 327.04$ K. Open circles, fluid phase; inverted triangles, fcc phase. Drawn curves, polynomial fits to the Monte Carlo data (see text). The bcc branch has been omitted from this figure.

tial energy function

$$U(\lambda) = V_0(\mathbf{r}_0^N) + \lambda[V(\mathbf{r}^N) - V_0] + (1-\lambda)K\sum_{i=1}^{N}(\mathbf{r}^i - \mathbf{r}_0^i)^2. \quad (2)$$

Here V_0 is the potential energy of the solid with all atoms at their lattice sites, $V(\mathbf{r})$ is the potential energy of an assembly of N particles interacting through the Aziz potential, and $(1-\lambda)K$ is a variable spring constant binding all atoms to their lattice sites. For $\lambda = 1$, $U(\lambda)$ describes the interaction of the crystal under consideration; for $\lambda = 0$, $U(\lambda)$ is the potential energy function of an Einstein crystal with the same structure. In practice, K was chosen such that the mean square displacement of the atoms in the Einstein crystal was approximately the same as in the unperturbed solid. As the absolute free energy of the Einstein crystal is known, the free energy of the helium crystal can be determined either by thermodynamic integration, using

$$\frac{\partial F}{\partial \lambda} = \langle V(\mathbf{r}^N) - V_0 - K\sum_{i=1}^{N}(\mathbf{r}^i - \mathbf{r}_0^i)^2 \rangle, \quad (3)$$

or by some other method (see, e.g., Frenkel and Ladd[10]). I computed F_{ex} for both the fcc and bcc phases of Aziz helium at $\rho_1 \equiv 0.239\,66$ mol/cm^3. Two methods were employed, namely thermodynamic integration, using a ten-point Gauss-Legendre quadrature, and Bennett's overlapping-distribution method[11] using twelve values for the coupling constant λ. Both methods gave identical results to within their respective estimated errors. In what follows I shall be using the thermodynamic integration results for F_{ex}^{fcc} and F_{ex}^{bcc}. All conclusions I shall arrive at below hold also for the overlapping-distribution results. At ρ_1 I obtain the following excess free energies: $F_{ex}^{fcc}/NkT = 10.4907 \pm 0.0012$ and $F_{ex}^{bcc}/NkT = 10.4994 \pm 0.0014$. Note that at ρ_1 the fcc phase is slightly, but significantly, more stable than the bcc phase. I carried out MC simulations on a 256-particle fcc crystal at seven densities and on a 250-particle bcc crystal at five densities. Combining my data with those of Ref. 8 I could fit the fcc and bcc isotherms by a three-term polynomial in the density (see Fig. 2). The finite-size corrections to the excess free energies were estimated to be smaller than the error in the thermodynamic integration results. Combining all the available data we are now in a position to compute the coexistence points of fluid, fcc, and bcc Aziz helium (see Table I). The coexistence lines have been indicated in Fig. 2. Note that we find that at $T = 327.04$ K, Aziz helium does not have a stable bcc phase. In order for a stable bcc phase to be at all possible at this temperature our estimate of $F_{ex}^{fcc} - F_{ex}^{bcc}$ has to be off by at least 5 standard deviations (much more, if we use the overlapping-distribution results). I also computed the dependence of the coexistence lines on the temperature using $dP/dT = \Delta H/T\Delta V$. As can be seen from Table I the resulting slopes are very nearly equal. This implies that there is a fairly wide temperature range around $T = 327$ K where the bcc phase is not stable.

At first sight the present findings may seem to be at odds with the constant-stress molecular-dynamics and MC simulations of Refs. 5 and 8. Both sets of simulations report a spontaneous transition from fcc to bcc as the pressure is decreased (or the temperature is increased) and back to fcc as the pressure is increased (temperature is decreased). There is, however, a considerable amount of hysteresis associated with the transformation fcc → bcc → fcc. In the constant-stress simulations a phase transformation occurs when the initial phase becomes mechanically unstable with respect to the other phase. The phase transformation takes place with an irreversible decrease of the Gibbs free energy of the system. The final state must be

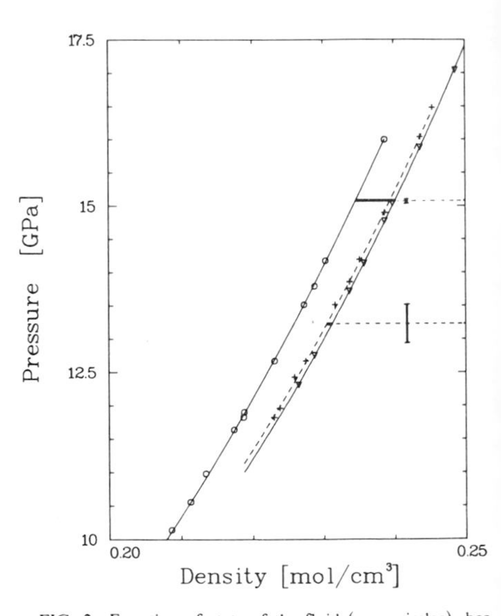

FIG. 2. Equation of state of the fluid (open circles), bcc solid (pluses), and fcc solid (inverted triangles) phases of Aziz helium around its melting point at $T = 327.04$ K. The drawn and dashed curves are polynomial fits to the Monte Carlo data. The horizontal dashed lines are the classical estimates for the coexistence pressures (see Table I). Estimated errors in these transition pressure are indicated by vertical bars. Note that this error estimate is almost ten times larger for the fcc-bcc transition than for the fcc-fluid transition.

TABLE I. Coexistence properties of Aziz helium. The left-hand column shows the classical transition pressure, coexistence densities, and slope of the coexistence line. The right-hand column was obtained from inclusion of quantum corrections to lowest order. Note that the estimated error in the transition pressure is about 10 times larger for the fcc-bcc transition (lower half) than for the fcc-liquid transition (upper half). The estimated error in the densities is almost exclusively due to the error in the coexistence pressure and has therefore not been indicated separately. Also indicated in the table is the excess Helmholtz free energy of the fcc phase at coexistence. The excess free energy is expressed in units of NkT. Knowledge of F^{ex} of one coexisting phase suffices to compute F^{ex} of the other. All data in this table were calculated at $T = 327.04$ K.

	Classical	Quantum $[O(\hbar^2)]$
$P_{\text{fcc-liquid}}$ (GPa)	15.07(4)	14.73(4)
ρ_{fcc} (mol/cm^3)	0.2401	0.2355
ρ_{liq} (mol/cm^3)	0.2345	0.2300
F^{ex}_{fcc}/NkT	10.528(2)	10.501(2)
dP/dT (GPa/K)	0.727(5)	0.737(5)
$P_{\text{fcc-bcc}}$ (GPa)	13.2(3)	12.9(3)
ρ_{fcc} (mol/cm^3)	0.2312	0.2265
ρ_{bcc} (mol/cm^3)	0.2305	0.2258
F^{ex}_{fcc}/NkT	9.771(2)	9.807(2)
dP/dT (GPa/K)	0.072(7)	0.072(7)

thermodynamically more stable than the initial state, but it need not be the true equilibrium state at that particular temperature and pressure. In the helium case the constant-stress simulations show a transition from fcc to bcc. The present calculations indicate that at the point where this transition occurs, the Gibbs free energy of the bcc phase is indeed lower than that of the fcc phase. But both phases are thermodynamically unstable with respect to the liquid. At the thermodynamic melting point of the fcc phase, the bcc phase has a higher Gibbs free energy than the fcc phase. At no point is the bcc phase more stable than both the liquid and the fcc phase. As the bcc phase is compressed in the constant-stress simulations, a transition takes place to the fcc phase. The present results show that this occurs at a point where the fcc phase is indeed thermodynamically stable. Hence the observations made in constant-stress simulations are not in contradiction with the present results. Our absolute free-energy calculations only show that the bcc solid observed in the earlier simulations is not a thermodynamically stable phase.

As ^{4}He is very light, quantum effects might be important even at room temperature. I have therefore computed the lowest-order $[O(\hbar^2)]$ quantum corrections to the free energy, using the method described by Hansen and Weis.[12] These quantum corrections were evaluated for the dense liquid and the fcc and bcc solids. From the quantum corrections to the free energy, we obtain the corresponding corrections to the pressure by numerical differentiation. The improved estimates for the coexistence points have been collected in Table I. Although quantum corrections do shift the transitions somewhat, they do not change the qualitative picture. Nor do quantum corrections affect dP/dT much (see Table I).

Hence, we must conclude that the Aziz potential for helium does not predict a stable bcc phase around room temperature. I stress once more that my results are compatible with the simulations of Refs. 6 and 8 because the observed limits of mechanical stability for the fcc and bcc phases are never in conflict with the relative thermodynamical stability that we compute.

I gratefully acknowledge discussions with M. L. Klein, R. leSar, and, in particular, P. Loubeyre, who kindly made the MC data of Ref. 8 available prior to publication.

[1]P. Loubeyre, J. M. Besson, J. P. Pinceaux, and J. P. Hansen, Phys. Rev. Lett. **49**, 1172 (1982).

[2]J. M. Besson, R. le Toullec, P. Loubeyre, J. P. Pinceaux, and J. P. Hansen, in *High Pressure in Science and Technology*, edited by C. Homan, R. K. MacCrone, and E. Whalley, Materials Research Society Symposia Proceedings No. 22 (North-Holland, New York, 1984), Vol. 2, p. 13.

[3]P. Loubeyre and J. P. Hansen, Phys. Rev. B **31**, 634 (1985).

[4]P. Loubeyre, D. Levesque, and J. J. Weis, to be published.

[5]D. Levesque, J. J. Weis, and M. L. Klein, Phys. Rev. Lett. **51**, 670 (1983).

[6]M. Parrinello and A. Rahman, Phys. Rev. Lett. **45**, 1196 (1980).

[7]R. Aziz, V. P. S. Nain, J. S. Carley, W. L. Taylor, and G. T. McConville, J. Chem. Phys. **70**, 4330 (1979).

[8]P. Loubeyre, D. Levesque, and J. J. Weis, to be published.

[9]D. Frenkel and A. J. C. Ladd, J. Chem. Phys. **81**, 3188 (1984).

[10]D. Frenkel, in Proceedings of the International School of Physics "Enrico Fermi," Course 97 (North-Holland, Amsterdam, to be published).

[11]C. H. Bennett, J. Comput. Phys. **22**, 245 (1976).

[12]J. P. Hansen and J. J. Weis, Phys. Rev. **188**, 314 (1969).

Note on p. 318

Note to Reprint IV.8

In table I, incorrect values are given for the excess free energy of the face-centred-cubic phase of Aziz helium at coexistence with the body-centred-cubic phase. The correct values are $F_{ex}^{FCC}/NkT = 9.736(2)$ (classical) and 9.685(2) (quantum mechanical). The latter are the values used in the calculations reported in the paper and the conclusions are therefore unaffected. These errors were brought to light by F.H. Ree.

PHYSICAL REVIEW VOLUME 153, NUMBER 1 5 JANUARY 1967

Ensemble Dependence of Fluctuations with Application to Machine Computations*

J. L. LEBOWITZ
Belfer Graduate School of Science, Yeshiva University, New York, New York

AND

J. K. PERCUS
Courant Institute of Mathematical Sciences, New York University, New York, New York

AND

L. VERLET†
Belfer Graduate School of Science, Yeshiva University, New York, New York

(Received 18 August 1966)

The standard theory of fluctuations in thermodynamic variables in various ensembles is generalized to nonthermodynamic variables: e.g., the mean-square fluctuations of the kinetic energy K in a classical microcanonical ensemble at fixed energy E is given, for large systems, by $\langle(\delta K)^2\rangle/\langle K\rangle = T[1-3/2C)$, where T is the temperature (corresponding to the energy E) and C is the specific heat per particle (in units of Boltzmann's constant). The general results may be expressed in terms of the asymptotic behavior of the Ursell functions in various ensembles. Applications are made to molecular dynamic computations where time averages correspond (via ergodicity) to phase averages in an ensemble with fixed energy and momentum. The results are also useful for time-dependent correlations.

I. INTRODUCTION

IT is generally believed (and partially proven[1]) that all thermodynamic properties of a physical system may be computed from any of the various Γ-space ensembles, e.g., canonical, grand-canonical, microcanonical, constant-total-momentum, isobaric, etc., commonly used in statistical mechanics. The reason for this is that in the thermodynamic limit (size of system $\to\infty$) appropriate to the various ensembles, the expected values of phase functions corresponding to "intensive" or per-unit-volume (per particle) properties of the system are independent of the ensemble used. Care must be exercised, of course, in the region of thermodynamic singularities, i.e., at phase transitions, and we shall not discuss the relevant extensions here.

The kinetic energy per particle

$$k=\frac{K}{N}=\frac{1}{N}\sum_{i=1}^{N} p_i^2/2m \tag{1.1}$$

is an example of such an intensive property. Its value may be computed in a canonical ensemble (c.e.) with fixed reciprocal temperature β, particle number N, and volume Ω, it may also be computed in a microcanonical ensemble (m.e.) at fixed N, Ω, and energy per particle

$$\epsilon=H/N=K/N+\Phi/N,$$

where

$$\Phi=\tfrac{1}{2}\sum{}' \varphi(q_i-q_j) \tag{1.2}$$

is the potential energy. The results differ only by terms $o(N)$ as $N\to\infty$, with $\rho=N/\Omega$ kept fixed.[2] That is, separating explicitly the fixed parameters,

$$\langle k(Q^N,P^N)\,|\,\beta,\rho,N\rangle=\langle k(Q^N,P^N)\,|\,\bar\epsilon,\rho,N\rangle+o(N), \tag{1.3}$$

where Q^N, P^N denote the full system phase space of coordinates and momenta $(q_1,\cdots q_N,\,p_1,\cdots p_N)$, and

$$\bar\epsilon=\bar\epsilon(\beta,\rho,N)\equiv\langle H(Q^N,P^N)/N\,|\,\beta,\rho,N\rangle \tag{1.4}$$

is given the value obtained from the canonical ensemble. Here the volume Ω is fixed by rigid or periodic boundary conditions, the precise nature of which is unimportant to our considerations.

An equation similar to (1.3) holds for the virial

$$\langle \mathcal{W}(Q^N,P^N)\rangle\equiv\left\langle\sum_i\left[\frac{1}{3m}p_i^2+\tfrac{1}{6}\sum_j(\mathbf{q}_i-\mathbf{q}_j)\cdot\frac{\partial\varphi(\mathbf{q}_i-\mathbf{q}_j)}{\partial\mathbf{q}_i}\right]\right\rangle$$

$$\equiv\langle\mathcal{W}_K+\mathcal{W}_\Phi\rangle=NP/\rho, \tag{1.5}$$

where P is the pressure. In general, we expect an ensemble-independent average in the thermodynamic limit for the value per particle of a function A of the form

$$A(\mathbf{Q}^N,\mathbf{P}^N)=\sum_{i=1}^{N}a(\mathbf{q}_i,\mathbf{p}_i;\,\mathbf{Q}^N,\mathbf{P}^N),$$

---

* Supported in part by the U. S. Air Force Office of Scientific Research and the U. S. Atomic Energy Commission.

† Permanent address: Faculté des Sciences, Laboratoire de Physique, Théorique et Hautes Energies, Bâtiment 211, 91 Orsay, France.

[1] See, e.g., M. E. Fisher, Arch. Ratl. Mech. Anal. 17, 377 (1964); R. Griffith, J. Math. Phys. 6, 1447 (1965). J. Van der Linden, Physica 32, 642 (1966).

[2] Explicit forms for the $o(N)$ terms are given in J. L. Lebowitz and J. K. Percus, Phys. Rev. 124, 1673 (1961) and references quoted therein. Some of these have been used and verified explicitly in machine computations with 1000 particles by J. L. Anderson, J. K. Percus, and J. Steadman (to be published).

where

$$a(\mathbf{q}_i,\mathbf{p}_i;\mathbf{Q}^N,\mathbf{P}^N)=\sum_{j_1\cdots j_n} a(\mathbf{q}_i,\mathbf{p}_i;\mathbf{q}_{j_1},\mathbf{p}_{j_1},\cdots\mathbf{q}_{j_n},\mathbf{p}_{j_n}), \quad (1.6)$$

and

$$a(\mathbf{q}_i,\mathbf{p}_i;\mathbf{q}_{j_1}\cdots\mathbf{p}_{j_n})\to 0 \quad\text{when}\quad |\mathbf{q}_i-\mathbf{q}_{j_s}|\to\infty.$$

Here n is independent of N, and a vanishes sufficiently strongly at ∞ so that the integral over the $\mathbf{q}_{j_s}$ exists. For large enough systems, we can then use any convenient ensemble to compute expectation values of quantities like A.

The situation is quite different, however, when we consider fluctuations. Let us define

$$L_{\alpha\gamma}=\langle A_\alpha A_\gamma\rangle-\langle A_\alpha\rangle\langle A_\gamma\rangle \quad (1.7)$$

in any ensemble, for quantities of the form (1.6). Then in the thermodynamic limit (in a single phase system), we expect

$$\lim_{\langle N\rangle\to\infty}\left[\frac{1}{\langle N\rangle}L_{\alpha\gamma}\right]=l_{\alpha\gamma} \quad (1.8)$$

to exist, where we have used $\langle N\rangle$ instead of N to include the possibility of a grand ensemble. The quantity $l_{\alpha\gamma}$ will however *not* be independent of the ensemble. For example, if $A_\alpha=A_\gamma=H$, then $l_{\alpha\gamma}=0$ in the m.e. but is proportional to the specific heat per particle in the c.e. In comparing classical-canonical with grand-canonical ensemble expectations, the momentum distributions are identical. The relation between fluctuations in functions of the coordinates is then obtainable from the asymptotic properties of the spatial distribution $n_{s+t}(\mathbf{q}_1,\cdots\mathbf{q}_{s+t})$ when the set of s particles is far removed from the set of t particles. Indeed, it has been shown that[3]

$$\text{g.c.e.:}\quad n_{s+t}(z)\to n_s(z)n_t(z), \quad (1.9)$$

$$\text{c.e.:}\quad n_{s+t}(N)\to n_s(N)n_t(N)-\frac{\rho T}{N}\chi\left(\rho\frac{\partial n_s(N)}{\partial\rho}\right)\times\left(\rho\frac{\partial n_t(N)}{\partial\rho}\right), \quad (1.10)$$

in the grand-canonical and canonical ensemble, respectively, where χ is the isothermal compressibility. It is our purpose here to obtain a general and useful relation between fluctuations in different ensembles. (Our analysis here will be restricted to classical systems, the generalization to quantum systems may involve problems of commutation relations.)

II. EXPECTATIONS UNDER TRANSFORMATIONS OF ENSEMBLE

Suppose that we know the expected value of a quantity $A(\mathbf{R})$, where $\mathbf{R}$ denotes a point of the full phase space, in an ensemble specified by extensive variables V_1, V_2, $\cdots$, as well as by parameters which will not be specified since they will not be altered. Hence

$$\langle A|V_1,V_2,\cdots\rangle=\int W(\mathbf{R}|V_1,V_2,\cdots)A(\mathbf{R})d\mathbf{R}/W(V_1,V_2\cdots), \quad (2.1)$$

where $W(\mathbf{R}|V_1,V_2\cdots)$ is the appropriate statistical weight and

$$W(V_1,V_2,\cdots)=\int W(\mathbf{R}|V_1,V_2\cdots)d\mathbf{R}=e^{-\Psi(V_1,V_2\cdots)} \quad (2.2)$$

the associated partition function, with the property that in the thermodynamic limit,

$$\lim_{\langle N\rangle\to\infty}\frac{1}{\langle N\rangle}\Psi(V_1,V_2\cdots)=\psi(v_1,v_2\cdots) \quad (2.3)$$

exists, with $v_i=V_i/\langle N\rangle$. A Legendre transformation to intensive variables X_1, $X_2\cdots$ now results from the definition

$$W(\mathbf{R}|X_1,X_2,\cdots)=\int\cdots\int W(\mathbf{R}|V_1,V_2\cdots)\times\exp[-\sum X_iV_i]d\mathbf{V}, \quad (2.4)$$

so that

$$e^{-\Psi(X_1,X_2\cdots)}=\int\cdots\int e^{-[\Psi(V_1,V_2\cdots)+\sum X_iV_i]}d\mathbf{V}, \quad (2.5)$$

and

$$\langle A|X_1,X_2\cdots\rangle=e^{\Psi(X_1,X_2\cdots)}\int\cdots\int\langle A|V_1,V_2,\cdots\rangle\times\exp-[\Psi(V_1,V_2\cdots)+\sum X_iV_i]d\mathbf{V}. \quad (2.6)$$

In particular, we have from (2.5) and (2.6) the basic expectations and fluctuations[4] ($\bar{V}_i\equiv\langle V_i\rangle$, $\delta V_i\equiv V_i-\bar{V}_i$)

$$\bar{V}_i(X_1,X_2,\cdots)=-\partial\Psi(X_1,X_2\cdots)/\partial X_i, \quad (2.7)$$

$$\langle\delta V_i\delta V_j|X_1,X_2,\cdots\rangle=\partial^2\Psi(X_1,X_2\cdots)/\partial X_i\partial X_j=-\partial\bar{V}_i/\partial X_j=-\partial\bar{V}_j/\partial X_i. \quad (2.8)$$

In the thermodynamic limit, the exponent $\Psi(\mathbf{V})+\sum X_iV_i$ in (2.5) and (2.6) goes to infinity as $\langle N\rangle$, so that the distribution in V space, in units of $\langle N\rangle$, becomes infinitely sharp. Hence by a steepest descent expansion, or any equivalent technique,[3,5] $\langle A|\mathbf{X}\rangle$ will be given at finite $\langle N\rangle$ by $\langle A|\langle\mathbf{V}(\mathbf{X})\rangle\rangle$ plus a correction series in ascending powers of $\langle N\rangle^{-1}$. The series is most directly obtained by a Taylor expansion about the

[3] J. L. Lebowitz and J. K. Percus, Phys. Rev. **122**, 1673 (1961); see also A. Meeron and Siegart, J. Math. Phys. **7**, 741 (1966); Z. Salsburg, J. Chem. Phys. **44**, 3090 (1966), where extensive use is made of these relations.

[4] See, e.g., L. D. Landau and E. M. Lifshitz, *Statistical Physics* (Addison-Wesley Publishing Company, Reading, Massachusetts, 1958).

[5] G. Horwitz, J. Math. Phys. (to be published).

limiting form:

$$\langle A|\mathbf{V}\rangle=\langle A|\langle\mathbf{V}(\mathbf{X})\rangle\rangle+\sum\delta V_i(\partial/\partial\bar{V}_i)\langle A|\langle\mathbf{V}(\mathbf{X})\rangle\rangle+\tfrac{1}{2}\sum\delta V_i\delta V_j(\partial^2/\partial\bar{V}_i\partial\bar{V}_j)\langle A|\langle\mathbf{V}(\mathbf{X})\rangle\rangle+\cdots,$$

followed by an average defined by (2.6):

$$\langle A|\mathbf{X}\rangle=\langle A|\langle\mathbf{V}\rangle\rangle+\tfrac{1}{2}\sum\langle\delta V_i\delta V_j|\mathbf{X}\rangle\frac{\partial^2\langle A|\langle\mathbf{V}\rangle\rangle}{\partial\bar{V}_i\partial\bar{V}_j}+O(\langle A\rangle/\langle N\rangle). \quad (2.9)$$

Since numerical computations are simplest in a $\mathbf{V}$ ensemble while analytical computations are simpler in an $\mathbf{X}$ ensemble, (2.9) is more useful in its inverse form, now reading

$$\begin{aligned}\langle A|\langle\mathbf{V}\rangle\rangle&=\langle A|X\rangle-\tfrac{1}{2}\sum\langle\delta V_i\delta V_j|\mathbf{X}\rangle\frac{\partial^2\langle A|\mathbf{X}\rangle}{\partial V_i\partial V_j}+O(\langle A\rangle/\langle N\rangle)\\&=\langle A|\mathbf{X}\rangle+\tfrac{1}{2}\sum\frac{\partial V_i}{\partial X_j}\frac{\partial^2\langle A|\mathbf{X}\rangle}{\partial V_i\partial V_j}\\&=\langle A|\mathbf{X}\rangle+\tfrac{1}{2}\sum\frac{\partial}{\partial X_j}\frac{\partial}{\partial V_j}\langle A|\mathbf{X}\rangle\\&=\langle A|\mathbf{X}\rangle+\tfrac{1}{2}\sum\frac{\partial}{\partial X_j}\frac{\partial X_i}{\partial V_j}\frac{\partial}{\partial X_i}\langle A|\mathbf{X}\rangle, \quad (2.10)\end{aligned}$$

where $\mathbf{X}=\mathbf{X}(\mathbf{V})$ is the inverse function to $\langle\mathbf{V}(\mathbf{X})\rangle$. Applying (2.10) to AB, we have as well the transformation formula for fluctuations:

$$\begin{aligned}\langle\delta A\delta B|\mathbf{V}\rangle&=\langle AB|\mathbf{V}\rangle-\langle A|V\rangle\langle B|\mathbf{V}\rangle\\&=\langle\delta A\delta B|\mathbf{X}\rangle-\sum\langle\delta V_i\delta V_j|\mathbf{X}\rangle\times\left(\frac{\partial\langle A|\mathbf{X}\rangle}{\partial V_i}\right)\left(\frac{\partial\langle B|\mathbf{X}\rangle}{\partial V_j}\right) \quad (2.11)\\&=\langle\delta A\delta B|\mathbf{X}\rangle+\sum\frac{\partial X_i}{\partial V_j}\frac{\partial\langle A|\mathbf{X}\rangle}{\partial X_i}\frac{\partial\langle B|\mathbf{X}\rangle}{\partial X_j}, \quad (2.12)\end{aligned}$$

now to relative order $O(1)$. The generalization of these results to transformations between ensembles with mixed intensive and extensive variables, e.g., isobaric and grand canonical, is straightforward. [Equation (2.11) can also be derived, formally at least, by considering generalized ensembles in which the functions A and B play the role of additional $\mathbf{V}_i$'s with respect to which the ensembles are always canonical, using thermodynamic fluctuation theory and then setting the corresponding $\mathbf{X}_i$'s equal to zero.]

We note here that the functions $A(R)$ and $B(R)$ could also depend on the time t. In particular, transport coefficients can be expressed, via the Kubo relations, as expectations of fluctuations in quantities like $A(R_t)$ and $B(R)$, where R_t is the point in phase space at which R arrives after a time t. Since $\langle A(R_t)\rangle=\langle A(R)\rangle$, the relation between $\langle\delta A(R_t)\delta B(R)\rangle$ in different ensembles is independent of t and (2.11) then yields a general relation between expressions for transport coefficients in different ensembles.[6]

Asymptotic Form of the Correlation Functions

A particularly useful application of the general formalism is to the distribution and Ursell functions in the various ensembles. This leads to a generalization of (1.10) from which the relation between fluctuations in different ensembles may be found. Letting $y_i=(q_i,p_i)$, we have for any ensemble[7]

$$n_s(z_1,\cdots,z_s)=\langle\sum_{i_1\neq\cdots\neq i_s}\delta(y_{i_1}-z_1)\cdots\delta(y_{i_s}-z_s)\rangle, \quad (2.13)$$

which is of the form (1.6) for a given set of z_i's, the n_s now being, however, $O(1)$ rather than $O(\langle N\rangle)$. Application of (2.10) then shows at once that, with obvious notation,

$$n_s(\mathbf{V})-n_s(\mathbf{X})=\tfrac{1}{2}(\partial/\partial\mathbf{X})(\partial/\partial\mathbf{V})n_s(\mathbf{X}). \quad (2.14)$$

Equation (2.14) directly implies (2.10) as $\langle A\rangle$ may be generally written as an integral over the n_s.

Further, if we introduce the generating functional

$$n[f]\equiv\sum\frac{1}{s'}\int n_s(y_1,\cdots,y_s)f(y_1)\cdots f(y_s)dy_1\cdots dy_s, \quad (2.15)$$

then

$$n[f|V]=n[f|\mathbf{X}]+\tfrac{1}{2}(\partial/\partial\mathbf{X})(\partial/\partial\mathbf{V})n[f|\mathbf{X}].$$

But then to the same order in $\langle N\rangle$, the generating functional for the Ursell distribution[7] $F_s(y_1,\cdots y_s)$, where

$$F_1(y_1)=n(y_1),\ F_2(y_1,y_2)=n_2(y_1,y_2)-n_1(y_1)n_1(y_2),\ \cdots \quad (2.16)$$

has the form

$$F[f|\mathbf{V}]=\ln n[f|\mathbf{V}]=\ln n[f|\mathbf{X}]+\frac{1}{2}\frac{1}{n[f|\mathbf{X}]}\frac{\partial}{\partial\mathbf{X}}\cdot\frac{\partial}{\partial\mathbf{V}}n[f|\mathbf{X}] \quad (2.17)$$

or

$$F[f|\mathbf{V}]-F[f|\mathbf{X}]=\frac{1}{2}\frac{\partial}{\partial\mathbf{X}}\cdot\frac{\partial}{\partial\mathbf{V}}F[f|\mathbf{X}]+\frac{1}{2}\frac{\partial}{\partial\mathbf{X}}F[f|\mathbf{X}]\cdot\frac{\partial}{\partial\mathbf{V}}F[f|\mathbf{X}]. \quad (2.18)$$

If the intensive ensemble is a grand ensemble and we are considering the limit as a set of s particles diverge

[6] See, e.g., M. S. Green, Phys. Rev. 119, 829 (1960); R. Zwanzig, Ann. Rev. Phys. Chem. 16, 67 (1965) for review and references.

[7] J. L. Lebowitz and J. K. Percus, J. Math. Phys. 4, 1495 (1963).

Notes on p. 324

from a set of t particles, then it is known that[3]

$$F_{s+t} \to 0. \tag{2.19}$$

It follows then from (2.18) that

$$F_{s+t}(\mathbf{V}) \to -\frac{1}{2}\frac{\partial}{\partial \mathbf{V}} F_s(\mathbf{X}) \cdot \frac{\partial \mathbf{V}}{\partial \mathbf{X}} \cdot \frac{\partial}{\partial \mathbf{V}} F_t(\mathbf{X}), \tag{2.20}$$

which is a generalization of (1.10).

It should be noted that n_s and F_s are quantities $O(1)$, so that the right sides of (2.14) and (2.18) are $O(\langle N\rangle^{-1})$. These corrections are therefore important only when used in finding the expectations of quantities like fluctuations which involve integrations over the volume of the system without any "cutoff." When that happens, the first term on the right of (2.18) will give a contribution of $O(1)$, while the second term will be of $O(\langle N\rangle)$ recovering the general form (2.11).

III. FLUCTUATIONS IN MOLECULAR DYNAMIC COMPUTATIONS

In molecular dynamics calculations on high-speed digital computers,[8–10] the nature of the system parameters is quite rigidly fixed. The calculations are done by setting initial conditions for several hundred particles interacting via some pair potential and restricted to a box with periodic boundaries, and then solving the classical equations of motion. Waiting a certain amount of time for the system to "thermalize," time averages of quantities like A_α or fluctuations are then computed:

$$\langle A_\alpha\rangle_t = \frac{1}{T}\int_{t_0}^{t_0+T} A_\alpha(\mathbf{Q}^N(t),\mathbf{P}^N(t))dt, \tag{3.1}$$

$$I_{\alpha\gamma} = \langle A_\alpha A_\gamma\rangle_t - \langle A_\alpha\rangle_t \langle A_\gamma\rangle_t.$$

Assuming the system to be ergodic, these time averages will coincide with ensemble averages at fixed values of any of the uniform constants of the motion which exist. Consequently, the system is specified by extensive parameters: total energy $H=E$, total momentum $\mathbf{M}$, total particle number N, total volume Ω.

Let us now compare this microcanonical ensemble of molecular dynamics computations with the corresponding canonical ensemble at fixed reciprocal temperature β, center-of-mass velocity $\mathbf{v}$. Total volume Ω and particle number N are fixed in both. We shall suppose that the system is maintained at $\mathbf{M}=0$, so that $\mathbf{v}=0$ as well. Now for the canonical ensemble,

$$\langle(\delta H)^2\rangle = -\partial\langle H\rangle/\partial\beta = N\beta^{-2}C = N\beta^{-2}(C_K + C^i), \tag{3.2}$$

where $C_K=\frac{3}{2}$ is the kinetic component of the specific heat per particle, in units of Boltzmann's constant k, and C^i is the potential component. Further, at $\mathbf{v}=0$,

$$\langle \delta H \delta M_i\rangle = -\partial\langle \mathbf{M}\rangle_i/\partial\beta = 0,$$
$$\langle(\delta M)^2\rangle = \langle \mathbf{M}\rangle^2 = 3N/\beta = -\partial\langle\mathbf{M}\rangle/\partial \mathbf{v}\beta. \tag{3.3}$$

Equations (2.10) and (2.11) now become

$$\langle A\,|\,E,\mathbf{M}=0\rangle = \langle A\,|\,\beta,\mathbf{v}=0\rangle - \frac{1}{2}\frac{\partial}{\partial\beta}\frac{\beta^2}{NC}\frac{\partial}{\partial\beta} \times\langle A\,|\,\beta,\mathbf{v}=0\rangle - \frac{1}{2}\frac{\partial}{\partial\mathbf{v}}\frac{1}{3N\beta}\cdot\frac{\partial}{\partial \mathbf{v}} \times\langle A\,|\,\beta,\mathbf{v}\rangle|_{\mathbf{v}=0}, \tag{3.4}$$

$$\langle \delta A\delta B\,|\,E,\mathbf{M}=0\rangle = \langle\delta A\delta B\,|\,\beta,\mathbf{v}=0\rangle - \frac{\beta^2}{NC}\frac{\langle A\rangle_t}{\partial\beta}\frac{\langle B\rangle_t}{\partial\beta} - \frac{1}{3N\beta}\frac{\langle A\rangle_t}{\partial\mathbf{v}}\cdot\frac{\langle B\rangle_t}{\partial\mathbf{v}}\bigg|_{\mathbf{v}=0}. \tag{3.5}$$

We shall now use (3.5) to obtain theoretical values for the fluctuations $\langle(\delta K)^2\rangle$ and $\langle\delta W\delta\Phi\rangle$ defined in (1.1) and (1.5), which were "experimentally" measured by Verlet[10] and Rahman.[9]

$$\langle(\delta K)^2\,|\,\beta,\mathbf{v}=0\rangle = -(\partial/\partial\beta)\langle K\rangle = 3N/2\beta^2, \tag{3.6}$$

so that from (3.4),

$$\frac{1}{N}\langle(\delta K)^2\,|\,E,\mathbf{M}=0\rangle = \frac{3}{2\beta^2}\left(1-\frac{3}{2C}\right) = \frac{1}{N}\langle(\delta\Phi)^2\,|\,E,\mathbf{M}=0\rangle. \tag{3.7}$$

In other words, the kinetic-energy fluctuation is a direct measure of the specific heat, and of course, since $K+\Phi$ is constant, this is identical with the potential-energy fluctuation. Similarly, $\partial\mathcal{W}/\partial\mathbf{v}=0$ at $\mathbf{v}=0$ by parity and we find

$$N^{-1}\langle\delta\mathcal{W}_\Phi\delta\Phi\,|\,E,\mathbf{M}=0\rangle$$
$$= -N^{-1}\langle\delta\mathcal{W}_\Phi\delta K\,|\,E,\mathbf{M}=0\rangle$$
$$= N^{-1}\langle\delta\mathcal{W}_\Phi\delta\Phi\,|\,\beta,\mathbf{v}=0\rangle + N^{-1}\left(\frac{\partial\beta}{\partial E}\right)\frac{\partial\langle\mathcal{W}_\Phi\rangle}{\partial\beta}\frac{\partial\langle\Phi\rangle}{\partial\beta}$$
$$= (3T^2/2C)(\partial/\partial T)[(P/\rho)-T], \tag{3.8}$$

where use has been made of the relation $(\partial/\partial\beta)\langle\mathcal{W}\,|\,\beta\rangle = -\langle\delta\mathcal{W}\delta H\,|\,\beta\rangle$.

Application to Computer "Experiments"

The specific heat can be computed, using (3.7), from either the kinetic-energy or the potential-energy fluctuations, which are theoretically equal in a machine computation, where the equations of motion are inte-

[8] B. J. Alder and T. E. Wainwright, J. Chem. Phys. 33, 1439 (1960).
[9] A. Rahman, Phys. Rev. 136, A405 (1964).
[10] L. Verlet, Phys. Rev. (to be published).

TABLE I. Interaction part C^i of the specific heat per atom, in units of the Boltzmann constant. (1): numerical differentiation of the internal energy. (2): calculated value using (3.9) with the results (1) in the right-hand side. (3): direct calculation of C^i using (3.9).

$\rho\sigma^3$	T/ϵ	(1) C^i	(2) C^i	(3) C^i
0.85	2.89	0.73	0.63	0.59
	2.20	0.79	0.78	0.78
	1.21	0.95	0.89	0.84
	1.13	0.99	0.86	0.78
	0.88	1.11	1.19	1.24
0.75	2.84	0.56	0.495	0.47
	0.827	0.88	0.885	0.89
0.45	4.62	0.20	0.23	0.25
	2.93	0.26	0.23	0.22
	1.71	0.28	0.42	0.46
	1.51	0.28	0.30	0.31

TABLE II. $\rho^{-1}\partial P/\partial T-1$. (1) numerical differentiation of the equation of state. (2) calculated value using (3.8).

$\rho\sigma^3$	T/ϵ	(1) $\left(\frac{1}{\rho}\frac{\partial P}{\partial T}-1\right)$	(2) $\left(\frac{1}{\rho}\frac{\partial P}{\partial T}-1\right)$
0.85	2.89	4.3	3.3
	2.20	4.4	4.0
	1.21	4.9	4.6
	1.13	5.0	4.8
	0.88	6.0	6.4
0.75	2.84	3.4	3.0
	0.827	4.9	4.7

grated at constant density. This is only approximately so in the calculations performed by one of us,[10] some of the results of which will be given now as examples of the preceding considerations; there, the integration algorithm determines the positions at various times. The velocities are then calculated by a numerical differentiation which introduces a small apparent fluctuation of the total energy. Numerically, however, this effect is very small: The fluctuations in potential and kinetic energy differ by less than 1% when the time averages are performed on 1200 time steps of 10^{-14} sec each, using a system of 864 particles simulating argon molecules with the help of a Lennard-Jones potential $\varphi(r)=4\epsilon[(\sigma/r)^{12}-(\sigma/r)^6]$. From (3.6), we can write for the part C^i of the specific heat due to the interaction

$$C^i=C-\tfrac{3}{2}$$
$$=(\beta^2/N)\tfrac{2}{3}C(\langle\Phi^2\rangle-\langle\Phi\rangle^2). \quad (3.9)$$

In Table I [column labeled (1)] are given some values of C^i, deduced from numerical differentiation of the interaction part of the internal energy with respect to the temperature along the three isochores, for which the particle density is of 0.85, 0.75, and 0.4 in units σ^{-3}. The precision is of the order of 5%. The error on the fluctuation of the potential energy is much larger, as may be seen from Table I. In the column labeled (2) are given the values obtained when, in the right-hand side of (3.8), C has been replaced by the value (1), obtained by numerical differentiation of the internal energy. In the last column are given the values of C^i derived entirely from (3.9). Although the values obtained from the first procedure—which was used to display a more direct test of (3.9)—are somewhat better, it is seen that the statistical errors are rather large and may reach 20%. On the other hand, the results are sufficiently precise to illustrate the use of (3.9) to calculate the specific heat in the microcanonical ensemble. The agreement with experiment is certainly good for the isochore at $\rho\sigma^3=0.45$, where the experimental data of Levelt[11] and the internal energy calculated from molecular dynamics agree very well. For the two high-density isotherms, we know that the equation of state $(P/\rho kT)$ derived from the molecular dynamics computations agree very well with experiment. No direct comparison is possible for the internal energies. Some experimental data are available at temperatures and densities rather near some of the points of Table I:

For $\rho\sigma^3=0.84$, $(T/\epsilon)=0.836$, the experimental value is $C^i=1.07$, which should be compared with the value 1.11 at $\rho\sigma^3=0.85$, $T/\epsilon=0.88$; for $\rho\sigma^3=0.798$, $T/\epsilon=0.92$, $C^i=0.96$ experimentally; at $\rho\sigma^3=0.75$, $T/\epsilon=0.827$, $C^i=0.88$, from the molecular dynamics computation. The results are in reasonable agreement.

Also, a comparison can be made with the data which can be extracted from Rahman's work[9]; for about 800 steps of 10^{-14} sec, at $T/\epsilon=0.79$, $\rho\sigma^3=0.82$, the value $C^i=0.81$ can be derived from the temperature fluctuation given by Rahman.

The same kind of analysis can be made using the fluctuation of the product of the virial and the potential energy [Eq. (3.8)]. As above, we compare the results derived, using the formula where C has been obtained from the computed internal energy, with the results we get by differentiating the equation of state [column labeled (2) in Table II, precision around 5%]. Again, the statistical error on the fluctuation is larger, but the over-all agreement is satisfactory. The agreement with experiment is also good in view of the fact that the equation of state obtained from computation fits very well with the argon data.

ACKNOWLEDGMENT

One of us (J.L.) would like to acknowledge useful discussions with R. Griffith.

[11] J. M. H. Levelt, Physica 26, 361 (1960).

Notes on p. 324

Notes to Reprint IV.9

1. J.K. Percus has provided some clarification of the mathematics of Section 3. (a) The quantity $\mathbf{v}$ is the negative centre-of-mass velocity. (b) Application of eq. (2.10) proceeds via the insertion of

$$\begin{bmatrix} \partial\beta/\partial E & \partial\beta\mathbf{v}/\partial E \\ \partial\beta/\partial\mathbf{M} & \partial\beta\mathbf{v}/\partial\mathbf{M} \end{bmatrix} = -\begin{bmatrix} \beta^2/NC & \beta\mathbf{v}/NC \\ \beta\mathbf{v}/NC & (\mathbf{vv}+\beta\mathbf{I})/NC \end{bmatrix}$$

(c) The effect of setting $\mathbf{v}=\mathbf{0}$ after carrying out the differentiation with respect to $\mathbf{v}$ in eq. (2.10) is to obtain, in place of the printed eqs. (3.4) and (3.5), the expressions,

$$\langle A \mid E, \mathbf{M}=0\rangle = \langle A\rangle - \frac{1}{2}\frac{\partial}{\partial\beta}\frac{\beta^2}{NC}\frac{\partial}{\partial\beta}\langle A\rangle - \frac{1}{2}\frac{3\beta}{NC}\frac{\partial}{\partial\beta}\langle A\rangle - \frac{1}{2}\frac{\beta}{Nm}\frac{\partial}{\partial\mathbf{v}}\cdot\frac{\partial}{\partial\mathbf{v}}\langle A\rangle, \qquad (3.4)$$

$$\langle \delta A\delta B \mid E, \mathbf{M}=0\rangle = \langle \delta A\delta B\rangle - \frac{\beta^2}{NC}\frac{\partial\langle A\rangle}{\partial\beta}\frac{\partial\langle B\rangle}{\partial\beta} - \frac{1}{Nm}\frac{\partial\langle A\rangle}{\partial\mathbf{v}}\cdot\frac{\partial\langle B\rangle}{\partial\mathbf{v}} \qquad (3.5)$$

The results in eqs. (3.7) and (3.8) are unaffected by these changes.

2. Fluctuation formulae for the elastic constants were first derived in the canonical ensemble by Squire et al. (Physica 42 (1969) 388). For a discussion of the formulae in other ensembles, see Ray et al. (Phys. Rev. B 33 (1986) 895).

3. Fluctuation formulae for molecular systems can be found in (Cheung, Mol. Phys. 33 (1977) 519).

4. Another simple and elegant method for the derivation of fluctuation formulae for the microcanonical ensemble has recently been given by Pearson et al. (Phys. Rev. A 32 (1985) 3030) and its extension to the isoenthalpic–isobaric ensemble has been described by Ray and Graben (Phys. Rev. A 34 (1986) 2517).

CHAPTER V

Molecular and Ionic Systems

Up till now, our attention has been concentrated almost exclusively on systems for which the pair potential is both spherically symmetric and short ranged. When one or other of these simplifications is absent, new problems arise both in the techniques that are required for the simulations and in the interpretation of the results.

Molecular systems

Papers devoted to molecular liquids and solids now form a large fraction of the published output of work on the computer simulation of condensed phases. Liquids that have been studied in this way range from highly idealized models such as the Stockmayer fluid, through nitrogen, the halogens and simple organic substances, to hydrogen-bonded systems and assemblies of chain molecules and other large molecules of biological interest. Much attention has also been paid to a variety of molecular solids, particularly to those in which there is a high degree of orientational disorder – the so-called plastic crystals. The article on liquid water by Rahman and Stillinger that we reprint here (Reprint V.1) was not the first to describe the simulation of a molecular system. The Monte Carlo study of water by Barker and Watts (Reprint I.8) had appeared two years earlier, while Harp and Berne [1] had previously published some molecular dynamics results for a simple model of liquid carbon monoxide. The Rahman–Stillinger paper is nonetheless a landmark in the history of computer simulation, not only for its content, but also for the fact that it alerted a wide scientific community to the potential value of such methods for the study of complex systems of chemical, biological or technological importance. The paper itself is remarkably comprehensive. Nearly all the main concerns of later papers on molecular liquids are already addressed here in some form, including possible ways of characterizing the static order, the pattern of association in hydrogen-bonded systems, the decay of orientational correlations of various types, the nature of the dielectric response in systems of polar molecules, the role of boundary conditions, and so on.

Any attempt to carry out molecular dynamics calculations on an assembly of molecules rather than atoms brings with it the question of how best to integrate the equations of motion of particles having vibrational and rotational degrees of freedom. In the case of a rigid body, the classic approach to the problem involves a separation of internal and centre-of-mass coordinates. The centre-of-mass motion can be handled by any of the methods used for atomic systems, including the predictor–corrector method of Reprint I.6 and the central-difference algorithm popularized by Verlet (Reprint I.7), and the orientation may be expressed in terms of Euler angles – or of polar angles, if the molecule is linear. However, the coupled equations of motion that determine the time evolution of the Euler angles α, β and γ have a singularity at $\sin \beta = 0$. (The notation used here is that of Reprint V.1; for a linear molecule, β becomes the polar angle θ.) One way of overcoming this difficulty is to redefine the coordinate frame associated with the molecule in question whenever $\sin \beta$ becomes dangerously small, but the more usual procedure is to adopt an algorithm that is free of singularities from the outset. The most commonly used schemes are described in two of the papers reprinted here: the method of constraints (Reprint V.2), in which the equations of motion are solved in cartesian form, and a method that involves a transformation from Euler angles to a set of quaternions (Reprint V.3). The two methods are of similar computational efficiency, but the method of constraints has the advantage of being easily applicable to systems of flexible molecules. The extensions to constant-pressure molecular dynamics are described in Reprints V.4 (for the method of constraints) and V.5 (for the method of quaternions). Reprint V.5 also contains a detailed discussion of the adaptation of the Parrinello–Rahman method (Reprint IV.4) to the study of structural phase transitions in molecular crystals.

Monte Carlo methods have been comparatively little used for molecular systems. They have found application, however, in the study of certain potential models of theoretical interest, including particles with non-spherical hard cores and dipolar hard spheres, where molecular dynamics techniques meet with difficulties. When applied to rigid molecules, the basic Monte Carlo scheme involves no new point of principle relative to that used for atoms, but care is needed in devising methods for the sampling of internal angles in the non-rigid case [2].

Ionic systems

When the interparticle potentials are short ranged, pair energies and forces are almost always computed with the help of the "spherical-cutoff" convention described in Reprints I.2 and I.7. In other words, the potential is set equal to zero beyond a cutoff distance that is typically equal to three or four particle diameters. Allowance can later be made for the effect of the neglected interactions on certain calculated properties of the system. The use of such a procedure is clearly much less appropriate for long-range potentials, i.e. potentials that decay (in three dimensions) no faster than r^{-3} at large r. The problem is seen at its most acute in the case of Coulomb systems. For example, use of the spherical-cutoff convention in the simulation of a molten salt could lead to the unacceptable situation in which the truncation sphere around a given ion was not electrically neutral. Electroneutrality is guaranteed if the potential is truncated at the surface of the periodic cell centred on the ion of interest rather than the surface of an inscribed sphere, but other problems then arise [3,4] and the method is reliable only at low charge densities, as in the case of the work described in Reprint IV.2. We shall see below that use of a truncated potential also leads to difficulties in the study of polar systems.

The key pioneering papers on the simulation of ionic fluids are those of Brush et al. [3] on the classical, one-component plasma, and of Woodcock and Singer [5] on molten potassium chloride modelled by pair potentials of the Born–Mayer–Huggins type. Both these groups of workers used the Monte Carlo method for their calculations. The paper reprinted here (Reprint V.6) describes an early molecular dynamics study of a simple model of a monovalent molten salt in which the cations and anions differ only in the signs of their charges; the special object of interest are those collective modes of the melt that represent the liquid-state analogues of the longitudinal acoustic and optic phonons of an ionic crystal. Monte Carlo and molecular dynamics methods have subsequently been applied to a wide variety of Coulomb systems, including molten salts of several types, one and two-component plasmas, fast-ion conductors and other disordered ionic solids, electrolyte solutions (see Reprint IV.2) and charged interfaces. For a recent review, see ref. [6]. The work of Turq et al. [7] on electrolyte solutions is of particular technical interest for its elaboration of the method of "brownian dynamics", the special feature of which is the fact that the forces acting on the ions are partly stochastic in character. Extensions of this approach have been exploited in other fields, notably in the study of protein dynamics [8]; an example of such an application is given in Chapter VI (Reprint VI.5).

In work on ionic systems, the coulombic energies and forces are generally calculated by the Ewald method, a numerical technique best known for its use in the evaluation of the Madelung constants of ionic crystals. Many derivations of the basic formulae can be found in the literature; one that is particularly well suited to the needs of computer simulation is that of Hansen [6], while the implementation of the method is discussed in Appendix A of Reprint V.5. The effect of the Ewald transformation is to express the total Coulomb energy per cell of the infinite, periodic array as the sum of two rapidly convergent series. One of the series is a sum in real space of a short-range pair potential; the other is a sum over wavevectors that are commensurate with the periodic boundary condition. The objection can be made that use of the Ewald method overemphasizes the essentially artificial periodicity of the system, but no generally acceptable alternative is at present available, at least in the three-dimensional case. Hansen et al. [9] have described a different approach, but in practice this is useful only for two-dimensional problems.

Polar molecules

Although the dipole–dipole interaction is long ranged, decaying as r^{-3}, the spherical-cutoff convention has been used successfully in many simulations of polar systems, including the work on liquid water described in Reprint V.1. The inadequacies of the convention become apparent, however, when dielectric properties are the main object of study. The dielectric constant of a polar fluid is related to the magnitude of long-wavelength fluctuations in the dipole-moment density of the sample; these fluctuations are strongly dependent on the boundary conditions or, in the context of a simulation, on the way in which the dipolar term in the pair potential is treated. Truncation of the potential has the effect of greatly suppressing the long-wavelength fluctuations, which in turn implies that certain angular components of the pair distribution function are very different from those appropriate to the infinite-system limit.

Two methods have been widely adopted in attempts to circumvent the problems sketched above. The first involves the use of the Ewald method either in its original form, which is applicable in cases where the electrostatic interaction between molecules is described in terms of point charges, or in the modification due to Kornfeld, which is used in situations where interactions between point dipoles are involved [10]. The second approach is more physical in nature; this is the reaction-field method of Barker and Watts [11]. The reaction-field method makes use of the spherical-cutoff convention, but the region beyond the cutoff is now treated as a dielectric continuum. The presence of the continuum gives rise to a term in the potential energy equal to $-(1/2)\boldsymbol{m}_i \cdot \boldsymbol{R}(i)$, where $\boldsymbol{m}_i$ is the dipole-moment vector of molecule i and $\boldsymbol{R}(i)$ is the "reaction field", i.e. the field that acts on i as a result of the polarization of the continuum by the net dipole moment of the truncation sphere centred on i; $\boldsymbol{R}(i)$ is given explicitly by eq. (17) of the paper by Neumann (Reprint V.7). Inclusion of the reaction-field term has a dramatic effect on the orientational correlations, as illustrated in fig. 2 of Neumann's paper.

Both the Ewald (or Ewald–Kornfeld) and reaction-field methods have been in use for more than a decade, but for much of that time there has been considerable controversy over the question of how precisely the dipole-moment fluctuations of the sample are linked to the dielectric constant. The latter is an intensive property of the system and is therefore independent of the choice of boundary conditions, but the fluctuations, as we have seen, are not. Thus the relation between the fluctuations and the dielectric constant must itself depend on the way in which the dipolar interaction is handled. Neumann's paper treats this problem in a general way that is applicable to all the boundary conditions commonly used in simulations and is also relevant to the interpretation of results on frequency and wavenumber-dependent fluctuations.

References

[1] G.D. Harp and B.J. Berne, Phys. Rev. A 2 (1970) 975.
[2] M. Fixman, Proc. Natl. Acad. Sci. USA 71 (1974) 3050.
[3] S.G. Brush, H.L. Sahlin and E. Teller, J. Chem. Phys. 45 (1966) 2102.
[4] J.P. Hansen, I.R. McDonald and P. Vieillefosse, Phys. Rev. A 20 (1979) 2590.
[5] L.V. Woodcock and K. Singer, Trans. Faraday Soc. 67 (1971) 12.
[6] J.P. Hansen, in: Molecular Dynamics Simulation of Statistical-Mechanical Systems, eds G. Ciccotti and W.G. Hoover (North-Holland, Amsterdam, 1986).
[7] P. Turq, F. Lantelme and H.L. Friedman, J. Chem. Phys. 66 (1977) 3039.
[8] E. Dickinson, Chem. Soc. Rev. 14 (1985) 421.
[9] J.P. Hansen, D. Levesque and J.J. Weis, Phys. Rev. Lett. 43 (1979) 979.
[10] D.J. Adams and I.R. McDonald, Mol. Phys. 32 (1976) 931.
[11] J.A. Barker and R.O. Watts, Mol. Phys. 26 (1973) 789.

THE JOURNAL OF CHEMICAL PHYSICS VOLUME 55, NUMBER 7 1 OCTOBER 1971

Molecular Dynamics Study of Liquid Water*

ANEESUR RAHMAN
Argonne National Laboratory, Argonne, Illinois 60439
AND
FRANK H. STILLINGER
Bell Telephone Laboratories, Incorporated, Murray Hill, New Jersey 07974

(Received 6 May 1971)

A sample of water, consisting of 216 rigid molecules at mass density 1 gm/cm³, has been simulated by computer using the molecular dynamics technique. The system evolves in time by the laws of classical dynamics, subject to an effective pair potential that incorporates the principal structural effects of many-body interactions in real water. Both static structural properties and the kinetic behavior have been examined in considerable detail for a dynamics "run" at nominal temperature 34.3°C. In those few cases where direct comparisons with experiment can be made, agreement is moderately good; a simple energy rescaling of the potential (using the factor 1.06) however improves the closeness of agreement considerably. A sequence of stereoscopic pictures of the system's intermediate configurations reinforces conclusions inferred from the various "run" averages: (a) The liquid structure consists of a highly strained random hydrogen-bond network which bears little structural resemblance to known aqueous crystals; (b) the diffusion process proceeds continuously by cooperative interaction of neighbors, rather than through a sequence of discrete hops between positions of temporary residence. A preliminary assessment of temperature variations confirms the ability of this dynamical model to represent liquid water realistically.

I. INTRODUCTION

Although water occupies a preeminent position among liquids, this substance has not enjoyed the attention of a rapidly developing body of statistical mechanical theory devoted specifically to its own properties.[1] One obvious reason for this retardation is the internal structure of the water molecule, which at the very least requires considering orientational degrees of freedom. In addition, the potentials of interaction for water molecules have until very recently[2–5] been imperfectly known. Furthermore, it now appears that these interactions are nonadditive to a significant degree.[6–9] Finally, the maximum cohesive binding between pairs of molecules, in units of k_BT at the triple point, is roughly an order of magnitude greater than the same quantity for the theoretically popular liquified noble gases.

This combination of complications renders impractical a large part of conventional liquid state theory for studying water. One must forego reliance on the integral equation approaches to static pair correlation functions on the one hand, while it is clear on the other hand that the fundamental theory of kinetic processes becomes even more complex than usual.

Under these circumstances, the most promising approach at present seems to be the direct simulation of liquid water by electronic computer. Both the "Monte Carlo" method[10] and the technique of "molecular dynamics"[11] are available for this purpose. The former offers the possibility of generating canonical ensembles of given temperature, but it is entirely restricted to a study of static structural properties. Molecular dynamics (which is nominally microcanonical) however can probe both static and kinetic behavior, so in principle it is the more powerful tool. We have therefore chosen to utilize this more powerful approach. This paper provides details of computational strategy, and initial results, in our molecular dynamics investigation of liquid water.

The following Sec. II specifies the Hamiltonian used for our dynamical water model. The individual molecules are treated as rigid asymmetric rotors, i.e., their internal bond lengths and bond angles are invariant.

Classical mechanics describes the temporal evolution of our model system. The coupled differential equations for translational and rotational motion are considered in Sec. III for the model water system. Special choices are introduced there for system size (216 molecules), boundary conditions (periodic unit cell corresponding to liquid at 1 gm/cm³), and time increment for numerical integration of the coupled dynamical equations ($\Delta t = 4.355 \times 10^{-16}$ sec). Discussed as well in Sec. III is a force truncation scheme.

Section IV presents a body of results thus far accumulated which specifically bears on the static molecular structure for our water model. Separate radial correlation functions are reported for the three distinct types of nuclear pairs present (O–O, O–H, and H–H), and these are used to synthesize the hypothetical x-ray scattering pattern for the model liquid. Several aspects of the elaborate local orientational order are presented in Sec. IV by examining dipole direction correlation in successive concentric shells about a given molecule and by analyzing the oxygen–oxygen pairs in separate icosahedral sectors about a fixed molecule. The character of hydrogen bonding in the liquid has also been examined using the distribution function for pair interaction energy.

Having thus described the main features of equilibrium molecular order, we pass on to kinetic properties of the water model in Sec. V. Several autocorrelation functions are presented which reveal distinctive characteristics of translational and rotational motion. These autocorrelation functions permit one in principle to calculate the self-diffusion constant, the dielectric relaxation spectrum, neutron inelastic scattering, and NMR spin–lattice relaxation.

In order to supplement these conventional molecular dynamics quantities, we have also produced stereoscopic photographs (of a cathode-ray display) which visually present instantaneous configurations of the 216 molecules during the system's temporal evolution. Unfortunately, a printed paper such as this one does not provide an effective direct way to communicate these elaborate stereoscopic pictures. However, we have attempted verbally to summarize their contribution to our own understanding of liquid water at the appropriate points in Secs. IV and V. In particular, these pictures allow one to perceive the global features of liquid water hydrogen-bond patterns, and to appreciate details of local cooperativity in molecular Brownian motion.

The results in Secs. IV and V refer to a single computer "run" corresponding to water at a fixed temperature. Some early results for a substantially lower temperature are mentioned in Sec. VI. Although we reserve most of the details concerning temperature variations for a later publication, these few observations strengthen our conviction that the model used is a relatively faithful representation of real water.

Several items are taken up for discussion in the final Sec. VII. We list there some extensions of the present project that appear to us to have relatively high scientific merit. Included among these are possible modifications of the Hamiltonian that could well be required at the next precision level of computer simulation for aqueous fluids. In particular, we stress a simple energy rescaling that may be applied to the present results, which seems to produce substantial improvement in agreement with experiment.

II. WATER MODEL HAMILTONIAN

Neutron diffraction studies on heavy ice[12] have confirmed earlier reasoning by Bernal and Fowler[13] and by Pauling[14] that water molecules maintain their identity in condensed phases with very little distortion of their molecular geometry. In other words, the forces of interaction between neighbors tend to be largely ineffective in perturbing the stiff covalent bonds within the molecules. From our point of view this offers the distinct advantage that the water molecules may be treated as rigid asymmetric rotors (six degrees of mechanical freedom), rather than explicit triads of nuclei (nine degrees of freedom).

The classical Hamiltonian H_N for a collection of N rigid-rotor molecules consists of kinetic energy for translational and rotational motions, plus interaction potential energy V_N:

$$H_N=\tfrac{1}{2}\sum_{j=1}^{N}(m\,|\mathbf{v}_j|^2+\boldsymbol{\omega}_j\cdot\mathbf{I}_j\cdot\boldsymbol{\omega}_j)+V_N(\mathbf{x}_1\cdots\mathbf{x}_N). \quad (2.1)$$

The molecules all have mass m. Their linear and angular velocities are, respectively, denoted by $\mathbf{v}_j$ and $\boldsymbol{\omega}_j$, while the inertial moment tensors (whose elements depend on the molecular orientation) are symbolized by $\mathbf{I}_j$. The configurational vectors $\mathbf{x}_1\cdots\mathbf{x}_N$ for the rigid molecules each comprise six components: three specify the center-of-mass position and three Euler angles fix the spatial orientation about the center of mass.

The potential energy function for any substance may always be resolved systematically into pair, triplet, quadruplet, $\cdots$, contributions[15]:

$$V_N(\mathbf{x}_1\cdots\mathbf{x}_N)=\sum_{i<j=1}^{N}V^{(2)}(\mathbf{x}_i,\mathbf{x}_j)+\sum_{i<j<k=1}^{N}V^{(3)}(\mathbf{x}_i,\mathbf{x}_j,\mathbf{x}_k)$$
$$+\sum_{i<j<k<l=1}^{N}V^{(4)}(\mathbf{x}_i,\mathbf{x}_j,\mathbf{x}_k,\mathbf{x}_l)+\cdots+V^{(N)}(\mathbf{x}_1\cdots\mathbf{x}_N). \quad (2.2)$$

The component subset functions $V^{(n)}$ occurring here have unique definitions that are generated by successive reversion of expressions (2.2) for two, three, four, $\cdots$, molecules:

$$V^{(2)}(\mathbf{x}_i,\mathbf{x}_j)\equiv V_2(\mathbf{x}_i,\mathbf{x}_j),$$
$$V^{(3)}(\mathbf{x}_i,\mathbf{x}_j,\mathbf{x}_k)=V_3(\mathbf{x}_i,\mathbf{x}_j,\mathbf{x}_k)-V^{(2)}(\mathbf{x}_i,\mathbf{x}_j)-V^{(2)}(\mathbf{x}_i,\mathbf{x}_k)-V^{(2)}(\mathbf{x}_j,\mathbf{x}_k),$$
$$V^{(n)}(i_1\cdots i_n)=V_n(i_1\cdots i_n)-\sum_{m=2}^{n-1}\sum_{j_1<\cdots<j_m=1}^{n}V^{(m)}(j_1\cdots j_m). \quad (2.3)$$

In the case of fluids which consist of simple nonpolar particles, such as liquid argon, it is widely believed that V_N is nearly pairwise additive. In other words, the functions $V^{(n)}$ for $n>2$ are small and hence exert insignificant influence on the local structure of the fluid. We have already noted though that water fails to conform to this sort of simplification in the strict sense. Local structure in liquid water and its solutions thus depends to a significant degree on the character of at least the three-body terms $V^{(3)}$ in Eq. (2.3), if not those of even higher order.

While it is thus unrealistic to terminate the exact water V_N after the pair terms in Eq. (2.3), it is still legitimate to employ the format of a pairwise additive potential, *provided* one understands that the pair functions are *effective* pair potentials, $V_{\text{eff}}^{(2)}(\mathbf{x}_i,\mathbf{x}_j)$. A variational criterion is available[16] which optimally

Notes on p. 354

assigns a sum of effective pair potentials to any given function $V_N(\mathbf{x}_1\cdots\mathbf{x}_N)$; the assignment causes $V_{\rm eff}^{(2)}$ to differ from $V^{(2)}$ in such a way that it creates essentially the same structural shifts at equilibrium that would be produced by the aggregate of triplet, quadruplet, $\cdots$, terms in Eq. (2.3). Specifically $V_{\rm eff}^{(2)}$ is to be chosen so as to minimize the nonnegative quantity[16]:

$$\int\cdots\int\Big[\exp[-\tfrac{1}{2}\beta V_N(\mathbf{x}_1\cdots\mathbf{x}_N)] - \exp\Big(-\tfrac{1}{2}\beta\sum_{i<j=1}^{N} V_{\rm eff}^{(2)}(\mathbf{x}_i,\mathbf{x}_j)\Big)\Big]^2 \times d\mathbf{x}_1\cdots d\mathbf{x}_N, \qquad \beta=(k_BT)^{-1}. \quad (2.4)$$

Our molecular dynamics calculations have been based upon a specific estimate for the liquid water $V_{\rm eff}^{(2)}$ that has been proposed by Ben-Naim and Stillinger.[17] This estimate consists of a part $v_{\rm LJ}$ depending only on oxygen–oxygen separation r_{ij}, plus a function $v_{\rm el}$ [modulated by a factor $S(r_{ij})$] that sensitively depends upon orientations about the oxygen nuclei:

$$V_{\rm eff}^{(2)}(\mathbf{x}_i,\mathbf{x}_j)=v_{\rm LJ}(r_{ij})+S(r_{ij})v_{\rm el}(\mathbf{x}_i,\mathbf{x}_j). \quad (2.5)$$

The quantity $v_{\rm LJ}$ is a potential of the Lennard-Jones (12–6) type:

$$v_{\rm LJ}(r_{ij})=4\epsilon[(\sigma/r_{ij})^{12}-(\sigma/r_{ij})^6]. \quad (2.6)$$

Since the water molecule and the neon atom are isoelectronic closed-shell systems, parameters ϵ and σ in Eq. (2.6) were chosen by Ben-Naim and Stillinger to be the accepted neon values[18]:

$$\epsilon=5.01\times10^{-15}\ \text{erg}=7.21\times10^{-2}\ \text{kcal/mole},$$
$$\sigma=2.82\ \text{Å}. \quad (2.7)$$

Four point charges, each exactly 1 Å from the oxygen nucleus, are imagined to be embedded in each water molecule in order to produce $v_{\rm el}$. These charges are arranged to form the vertices of a regular tetrahedron. Two of them are positive ($+0.19e$ each) to simulate partially shielded protons, while the remaining two ($-0.19e$ each) act roughly as unshared electron pairs. The set of 16 charge pair interactions between two molecules forms $v_{\rm el}$:

$$v_{\rm el}(\mathbf{x}_i,\mathbf{x}_j)=(0.19e)^2\sum_{\alpha_i,\alpha_j=1}^{4}(-1)^{\alpha_i+\alpha_j}/d_{\alpha_i\alpha_j}(\mathbf{x}_i,\mathbf{x}_j). \quad (2.8)$$

Here the indices α_i and α_j are even for positive charges, odd for negative charges. The distance $d_{\alpha_i\alpha_j}$ between the subscripted charges obviously depends on the full set of relative configurational variables for molecules i and j.

If the radial distance r_{ij} between oxygen nuclei were 2 Å or less, it would be possible for one of the distances $d_{\alpha_i\alpha_j}$ to be zero. The resultant divergence of $v_{\rm el}$ surely would be physically meaningless. The "switching function" $S(r_{ij})$ however suppresses this possibility by vanishing identically at these small separations. In fact S varies continuously and differentially between 0 (small r_{ij}) and 1 (large r_{ij}):

$$\begin{aligned} S(r_{ij})&=0 && (0\le r_{ij}\le R_L),\\ &=(r_{ij}-R_L)^2(3R_U-R_L-2r_{ij})/(R_U-R_L)^3 && (R_L\le r_{ij}\le R_U),\\ &=1 && (R_U\le r_{ij}<\infty), \end{aligned} \quad (2.9)$$

where[17]

$$R_L=2.0379\ \text{Å}, \qquad R_U=3.1877\ \text{Å}. \quad (2.10)$$

The effective pair potential (2.5) incorporates the tendency of water molecules to associate by hydrogen bonding. The absolute minimum of $V_{\rm eff}^{(2)}(\mathbf{x}_i,\mathbf{x}_j)$ is achieved when one molecule forms a linear H bond (of length $r_{ij}=2.76$ Å) to the rear of the other molecule, i.e., when a charge $+0.19e$ on one is lined up with (but 0.76 Å distant from) a charge $-0.19e$ on the other. In this minimum energy configuration, the fully formed H bond has energy[17]

$$V_{\rm eff}^{(2)}(\mathbf{x}_i,\mathbf{x}_j)|_{\rm min}=-4.514\times10^{-13}\ \text{erg} =-6.50\ \text{kcal/mole pairs}. \quad (2.11)$$

Figure 1 illustrates this pair configuration, which after identification of its positive charges as proton positions agrees rather well with the stable dimer geometry predicted by *ab initio* quantum mechanical calculations.[6–9]

The distribution of mass in each molecule follows the tetrahedral geometry utilized in $V_{\rm eff}^{(2)}$. The oxygen atom mass (2.6555×10^{-23} g) is concentrated at the center of the tetrahedron (the force center for $v_{\rm LJ}$). A mass equal to $\frac{1}{16}$ this value is placed at each of the two positions bearing charges $+0.19e$ that are 1 Å away from the tetrahedron center, to act as

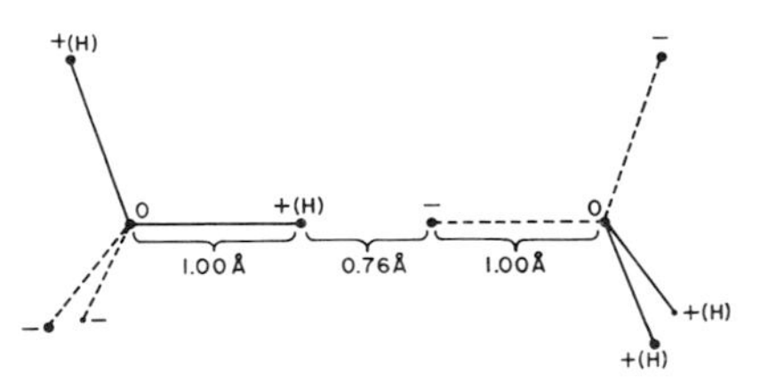

FIG. 1. Minimum energy configuration for two water molecules, according to potential (2.5). Each oxygen nucleus is symmetrically surrounded by a tetrad of four-point charges ($\pm0.19e$), the positive members of which represent partially shielded protons. The configuration shown has a plane of symmetry and incorporates a single linear H bond.

hydrogen atom masses.[19] The center of mass is therefore displaced along the molecular symmetry axis by 0.06415 Å, away from the oxygen mass. This rigid mass distribution of course fixes the inertial moment tensor for each molecule.

Investigation of the structure of ice crystals and the clathrate hydrates is not our primary objective in this paper. But the tetrahedral charge arrangements which underlie $V_{\rm eff}^{(2)}$ strongly favor the local tetrahedral pattern of H bonds about each molecule observed in these aqueous solids. Therefore there can be little doubt that $V_{\rm eff}^{(2)}$ in Eq. (2.8) will permit the existence of mechanically stable H-bond networks filling space in the ice and clathrate patterns. Our task now is to employ the effective pair potential version of the Hamiltonian:

$$H_N=\tfrac{1}{2}\sum_{j=1}^{N}(m\mid \mathbf{v}_j\mid^2+\boldsymbol{\omega}_j\cdot\mathbf{I}_j\cdot\boldsymbol{\omega}_j)+\sum_{i<j=1}^{N}V_{\rm eff}^{(2)}(\mathbf{x}_i,\mathbf{x}_j) \tag{2.12}$$

under liquid-phase conditions, to see what type of nonperiodic, strained, and defective H-bond networks spontaneously appear.

III. DYNAMICAL EQUATIONS AND METHOD OF SOLUTION

The configurational vector $\mathbf{x}_j$ for molecule j involves the following components:

$$\mathbf{x}_j=(X_j,\ Y_j,\ Z_j,\ \alpha_j,\ \beta_j,\ \gamma_j). \tag{3.1}$$

The first three are the Cartesian coordinates of the molecule's center of mass. The Euler angles α_j, β_j, and γ_j specify the orientation of the molecule's principal axes relative to a standard laboratory fixed orientation in the manner shown by Fig. 2. These angles have the following limits:

$$0\leq\alpha,\ \gamma<2\pi,\qquad 0\leq\beta\leq\pi. \tag{3.2}$$

The dynamical equations required to describe the temporal evolution of a set of rigid-rotor molecules are the coupled Newton–Euler equations. In the case of the center-of-mass position $\mathbf{R}_j=(X_j,\ Y_j,\ Z_j)$ for molecule j we have

$$m(d^2\mathbf{R}_j/dt^2)=\mathbf{F}_j, \tag{3.3}$$

where $\mathbf{F}_j$ is the total force exerted on that molecule by all others:

$$\mathbf{F}_j=-\nabla_{R_j}\sum_{k(\neq j)}V_{\rm eff}^{(2)}(\mathbf{x}_j,\mathbf{x}_k). \tag{3.4}$$

In an analogous fashion, rotational motion involves the torque $\mathbf{N}_j$ exerted on molecule j. The inertial moment tensor $\mathbf{I}_j$ is diagonal (I_1, I_2, I_3) in a Cartesian coordinate system affixed to the molecule as shown in Fig. 2. If we denote this molecule fixed system by primes, then the corresponding torque components are

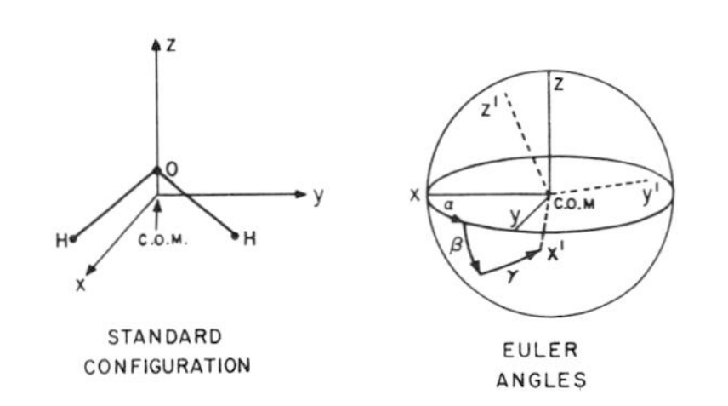

FIG. 2. Euler angle definition for water molecule orientation. The "standard configuration" relative to laboratory fixed coordinate system has $\alpha=\beta=\gamma=0$; the molecule lies in the yz plane, and z is the twofold molecular axis. The arcs denoted by α, β, and γ on the sphere surface are the path described by the molecule fixed axis x'. α is an angle of rotation about the initial $z=z'$ axis, β gives a rotation about the resulting y' axis, and finally, γ is a rotation angle about the displaced z' axis.

$$\mathbf{N}_j=(N_{jx'},\ N_{jy'},\ N_{jz'}). \tag{3.5}$$

In the same coordinate system, the angular velocity components must obey the Euler equations[20]:

$$\begin{aligned}I_1(d\omega_{jx'}/dt)-\omega_{jy'}\omega_{jz'}(I_2-I_3)&=N_{jx'},\\ I_2(d\omega_{jy'}/dt)-\omega_{jz'}\omega_{jx'}(I_3-I_1)&=N_{jy'},\\ I_3(d\omega_{jz'}/dt)-\omega_{jx'}\omega_{jy'}(I_1-I_2)&=N_{jz'}.\end{aligned} \tag{3.6}$$

The angular velocity components have the following representation in terms of the Euler angles of Fig. 2[21]:

$$\begin{aligned}\omega_{jx'}&=(d\alpha_j/dt)\sin\beta_j\sin\gamma_j+(d\beta_j/dt)\cos\gamma_j,\\ \omega_{jy'}&=(d\alpha_j/dt)\sin\beta_j\cos\gamma_j-(d\beta_j/dt)\sin\gamma_j,\\ \omega_{jz'}&=(d\alpha_j/dt)\cos\beta_j+(d\gamma_j/dt).\end{aligned} \tag{3.7}$$

From the computational point of view, it is convenient to cast the dynamical equations into dimensionless form. The parameters ϵ and σ, Eqs. (2.7), serve as units for energy and length. Then by invoking the total molecular mass m, the natural time unit becomes $\sigma(m/\epsilon)^{1/2}$; for water this is 2.179×10^{-12} sec.

Our calculations involve $N=216$ water molecules placed in a cubical container and subject to periodic boundary conditions so as to eliminate undesirable wall effects. This assembly is maintained at mass density 1 g/cm³ by fixing the cube side at $6.604\sigma=18.62$ Å. The corresponding number density ρ is 3.344×10^{22}/cm³.

Under periodic boundary conditions, each molecule would interact with an infinite number of others, including the remaining 215 in the same unit cell as well as all periodic images filling the entire space. In principle, then, Ewald sums would have to be carried out to evaluate the potential energy. In order

Notes on p. 354

to facilitate the molecular dynamics calculation,[22] it is advisable instead to include interactions only up to some finite limiting distance. In the present work, each molecule's oxygen nucleus is regarded as residing at the center of a sphere with radius 3.25σ, and only those other molecules having oxygens within this sphere are considered in computing the force and torque on the central molecule. By experimentation with various cutoff radii, it seems that the choice of cutoff at 3.25σ probably commits little error in predicted liquid structure, relative to a full Ewald sum (infinite cutoff radius). As with the full Ewald sum, the truncated dynamical equations maintain the initial periodicity at all times.

In the case of a substance such as Ar, with spherically symmetric particles, a force cutoff in the Newtonian equations of motion does not affect the role of energy as a constant of the motion, i.e., the system remains conservative. However with noncentral interactions as water requires, a truncation of force and torque terms in the coupled Newton–Euler equations leads in principle to a nonconservative situation.[23] Over a long dynamical run, this irreversibility would lead to a secular rise in temperature, as though the system were weakly coupled to a high-temperature heat reservoir. We have found, however, that this effect is manageable for the runs involved in the present investigation.

The calculations were carried out at the Argonne National Laboratory on an IBM 360-50-75 computer.[24] Details of the algorithm utilized to integrate the differential equations of motion are contained in Appendix A. By experimenting with two-molecule dynamics, an appropriate choice for the basic time increment for the numerical integration was found to be the following:

$$\Delta t = 2\times10^{-4}\times\sigma(m/\epsilon)^{1/2} = 4.355\times10^{-16}\ \text{sec}. \tag{3.8}$$

The smallness of this increment (relative to that required in liquid Ar calculations[22]) stems from the rapid angular velocity of the water molecules, which in turn derives from the small mass of the protons. To advance the system by Δt, the computer requires about 40 sec, so a time dilation factor of about 10^{17} applies in the relationship of real water molecule motions to those in our computation.

The 216 molecules were placed initially within the periodic cell at arbitrary positions, with random orientations, and with no translational or rotational velocities. Since this configuration corresponded to a very large potential energy, the velocities quickly increased to a distribution characteristic of about 2×10^4 °K within a few Δt steps. The system was then interrupted, and all velocities set to zero at the momentary configuration that obtained. This served to remove some of the excess energy. After several more Δt steps, the molecules again had achieved high velocities, so once again they were set equal to zero. Several repetitions of this energy reduction procedure were required to bring the ambient temperature near to the desired range. Thereafter, fractional uniform adjustment of all velocities permitted fine temperature control. In all, over 5000 steps of length Δt were expended in achieving the proper system energy and in allowing the system to "age." After this interval, the subsequent period of 5000 Δt was actually the time interval over which the molecular dynamics statistical averages reported below were calculated. We believe that the system had developed in time long enough to eliminate any undesirable remanent effects due to the choice of initial conditions.

Temperature is inferred from the average values of the translational and rotational kinetic energies over the molecular dynamics run; in a sufficiently long run:

$$\langle \tfrac{1}{2}\sum_{j=1}^{N} m\,|\mathbf{v}_j|^2\rangle = \langle \tfrac{1}{2}\sum_{j=1}^{N} \boldsymbol{\omega}_j\cdot\mathbf{I}_j\cdot\boldsymbol{\omega}_j\rangle = \tfrac{3}{2}Nk_BT. \tag{3.9}$$

Temperature variations for a new calculation can be implemented by suitably modifying linear and angular momenta at the end of a previous run, provided the system is allowed to "age" appropriately.

Integrating the dynamical equations in dimensionless form has the advantage that if length or strength rescaling of the potential ultimately proves to be required, results may be trivially renormalized to accommodate those changes. Thus if $V_{\text{eff}}^{(2)}$ were to be multiplied by the scalar factor $1+\zeta$, the energy unit would become $(1+\zeta)\epsilon$, and the unit of time would change to $\sigma[m/(1+\zeta)\epsilon]^{1/2}$. In view of Eq. (3.9), this would require that the temperature T originally assigned to a given molecular dynamics run be reinterpreted as $(1+\zeta)T$.

IV. STATIC STRUCTURE

From the average kinetic energy calculated for the 5000 step molecular dynamics run, the temperature of the system was determined to be 307.5°K (34.3°C). The force–torque cutoff irreversibility actually caused the total energy to drift upward by 1.5% during the run, with a mean value per molecule:

$$\langle H_N\rangle/N = -101.85\ \epsilon; \tag{4.1}$$

the potential energy contribution to this total was

$$\langle V_N\rangle/N = -127.3\ \epsilon = -9.184\ \text{kcal/mole}. \tag{4.2}$$

(The experimental value of this last quantity at 34.3°C is −9.84 kcal/mole.) Though the temperature effectively drifted upward during the computation, the structural features to be reported as run averages should correspond to the computed mean temperature 34.3°C for the run.

A. Radial Pair Correlation Functions

In water the three distinct types of nuclear pairs lead to three corresponding radial pair correlation functions, $g_{OO}^{(2)}(r)$, $g_{OH}^{(2)}(r)$, and $g_{HH}^{(2)}(r)$. These are defined for present purposes by the requirement that

$$\rho_\alpha \rho_\beta g_{\alpha\beta}^{(2)}(r)\,dv_1 dv_2 \qquad (\alpha, \beta = \text{O, H}) \qquad (4.3)$$

equal the fraction of time that differential volume elements dv_1 and dv_2 (separated by distance r) simultaneously and respectively contain nuclei of species α and β from distinct molecules. Here ρ_α and ρ_β stand for the number densities of the nuclei. In the large system limit, these radial pair correlation functions for a fluid phase all approach unity as $r \to \infty$.

Figure 3 exhibits the computed $g_{OO}^{(2)}(r)$, out to distance 3.25σ (9.165 Å), along with the "running coordination number":

$$n_{OO}(r) = 4\pi\rho_O \int_0^r s^2 g_{OO}^{(2)}(s)\,ds. \qquad (4.4)$$

The important features of the $g_{OO}^{(2)}(r)$ curve to note are the following:

(a) The relatively narrow first peak, with maximum at $r/\sigma = 0.975$, comprises an average of 5.5 neighbors out to the following minimum at $r/\sigma = 1.22$.

(b) The second peak is low and broad, with a maximum at about $r/\sigma = 1.65$. The ratio of second-peak to first-peak distances (1.69) is close to that observed for the ideal ice structure ($2\sqrt{2}/\sqrt{3} = 1.633$), where successive neighbor hydrogen bonds occur at the tetrahedral angle 109°28′.

(c) There is substantial filling between the first and second peaks.

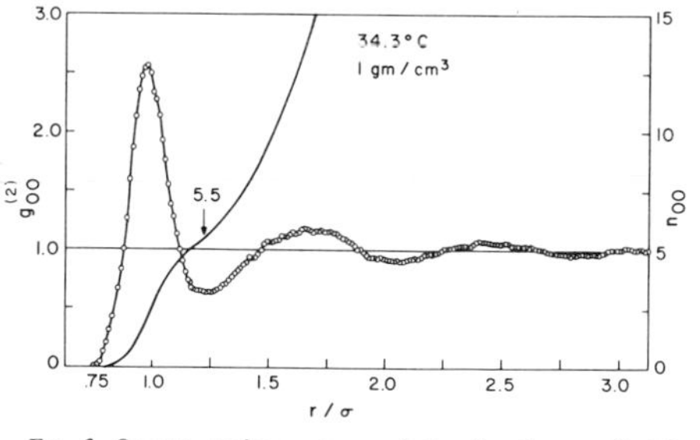

FIG. 3. Oxygen nucleus pair correlation function $g_{OO}^{(2)}$. The monotonically rising curve n_{OO} shows the average number of neighbor oxygens within any radial distance r. The liquid has temperature 34.3°C and density 1 g/cm³.

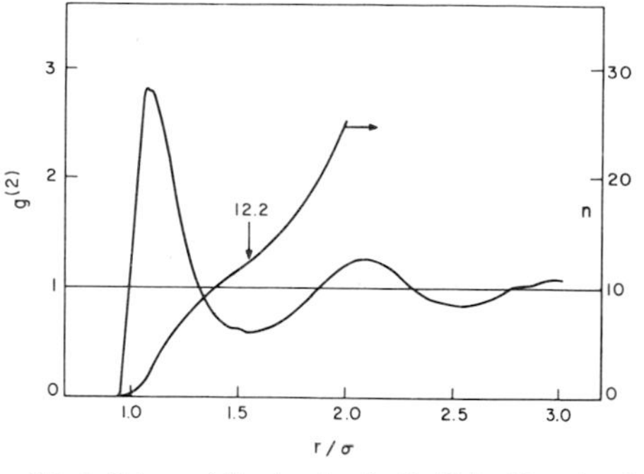

FIG. 4. Pair correlation function for liquid Ar. The reduced state parameters are $\rho^* = \rho\sigma^3 = 0.81$ and $T^* = k_B T/\epsilon = 0.74$. $n(r)$ gives the running neighbor count (right-hand scale).

(d) Although a weak third peak appears at $r/\sigma \cong 2.45$, the $g_{OO}^{(2)}$ curve has begun to damp rapidly to its asymptotic value unity. Beyond $r/\sigma = 3.00$ (8.46 Å), deviations from unity are apparently insignificant.

This liquid water pair correlation function stands in distinct contrast to its analog in liquid Ar, for which it is traditional to assume that the pair potential involves *only* a central Lennard-Jones interaction, and *no* directional forces. Figure 4 presents a liquid Ar $g^{(2)}(r)$, specially computed by the molecular dynamics procedure of Ref. 22, for comparison. Not only does the first peak encompass more neighbors than water (12.2 with a cutoff at the first minimum), but the distance ratio of second and first peaks (1.90) is considerably larger.

Evidently there is a substantial difference in the type of disordering attendant upon melting ice to liquid water on the one hand, and melting a face-centered cubic crystal of Ar to the corresponding liquid Ar on the other hand. The ice lattice second-neighbor peak, although broadened considerably, clearly remains after melting. The same is not the case for Ar, however, since Fig. 4 shows no persistence of the fcc lattice second-neighbor distance at $2^{1/2}$ times the first neighbor distance. Obviously the directionality of interactions in liquid water exerts a profound influence, not present in other liquids, on local order.

Figures 5 and 6, respectively, show $g_{OH}^{(2)}$ and $g_{HH}^{(2)}$ for water (these of course have no analogs in liquid Ar). The former of these functions plays a role in the theory of solute structural shifts in aqueous solutions.[25] The prominent first two peaks displayed by $g_{OH}^{(2)}$ evidently arise from neighboring molecules that are hydrogen bonded; the peak at smaller separation involves the proton along the bond and the acceptor oxygen toward which it points, while the larger dis-

Notes on p. 354

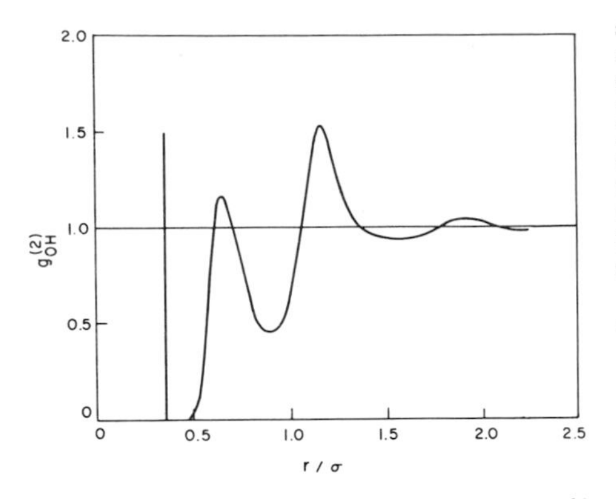

FIG. 5. Cross correlation function $g_{OH}^{(2)}(r)$ for water; 34.3°C and 1 g/cm³. The vertical line indicates the intramolecular O–H covalent bond length.

tance second peak comprises all remaining oxygen–proton distances across the bonded pair of molecules. The interpretation of $g_{HH}^{(2)}$ is less direct, owing to the multiplicity of possible proton pairs in hydrogen-bond networks, but it seems safe to assign the first peak to a proton along a hydrogen bond and a proton in the acceptor molecule. The very distinct shoulder in $g_{HH}^{(2)}$ in the region $r/\sigma \approx 0.5$ is especially interesting, since these very close proton pairs may in part arise from the situation in which three proton donors crowd together at the negative rear of an acceptor molecule.

Even in the absence of further information, the three water correlation functions lead to a picture of liquid water as a random, defective, and highly strained network of hydrogen bonds that fills space rather uniformly. The fact that each of the three correlation functions approaches its asymptotic value unity rather quickly as r increases indicates that large, low density clusters or "icebergs" are not present to a significant degree, though they have occasionally been advocated to explain the properties of water.[26,27] Recent small angle x-ray scattering measurements on liquid water appear to be consistent with this observation.[28]

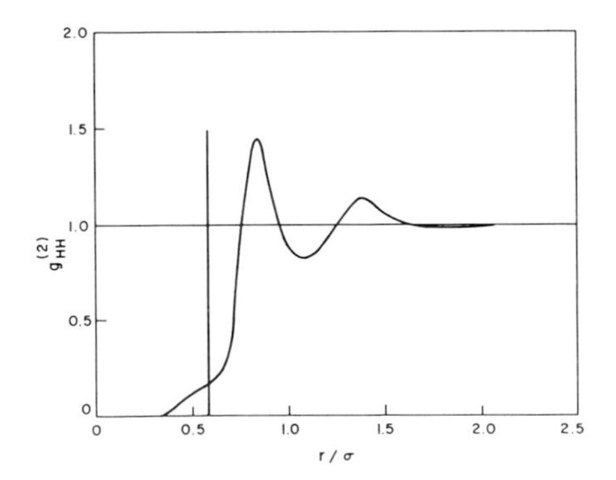

FIG. 6. Proton pair correlation function $g_{HH}^{(2)}(r)$ for water; 34.3°C and 1 g/cm³. The vertical line indicates the intramolecular H–H pair distance.

Unfortunately, experiments have not been carried out to provide measurements of the separate functions $g_{OO}^{(2)}$, $g_{OH}^{(2)}$, and $g_{HH}^{(2)}$ for direct comparison with the molecular dynamics results. In principle this could be done by combining x-ray scattering results with neutron scattering intensities for isotopically substituted waters. At present, though, only x-ray scattering has been carried out, which provides a weighted average of the three pair correlation functions.[29] We have therefore utilized our separate correlation functions, along with tabulated atomic scattering factors,[29] to synthesize the x-ray scattering intensity $I(s)$ [$s=(4\pi/\lambda)\sin\theta$, the magnitude of the scattering vector].

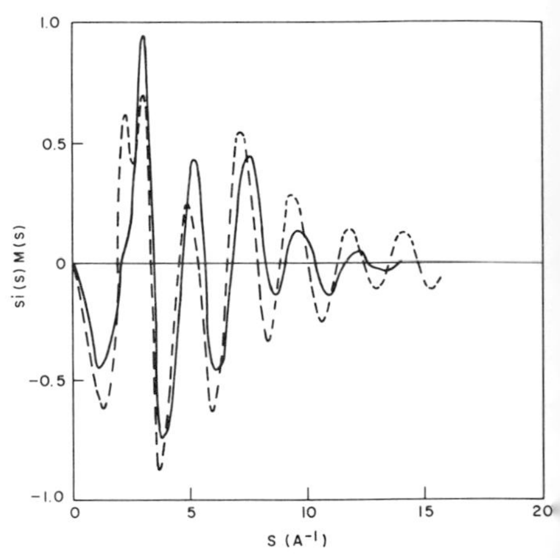

FIG. 7. Theoretical (solid line), and experimental (dotted line), x-ray scattering intensities for liquid water. The latter are taken from Narten, Ref. 29, and refer to 25°C.

Figure 7 presents computed values of $sI(s)$, along with Narten, Danford, and Levy's measurements of the same quantity at their nearest temperature, 25°C.[29] Agreement obtains only in modest degree, with the principal discrepancy occurring at the first peak ($s\approx 1.7$ to 3.0 Å^{-1}). At present it is not clear that the disagreement arises entirely from shortcomings in our basic water model. It may be, for example, that covalent intramolecular chemical bonds, and intermolecular hydrogen bonds, distort electron distributions sufficiently to invalidate the conventional assumption of independent spherically symmetric atomic scattering factors. In addition, we have treated the intramolecular geometry of nuclear positions as rigid, whereas the substantial zero-point motions should actually be taken into account to predict $I(s)$. More work is clearly required in this area, beyond the scope of the present paper.

B. Icosahedral $g_{OO}^{(2)}$ Resolution

The radial pair correlation function $g_{OO}^{(2)}(r)$ gives the mean density of oxygen nuclei over concentric spherical shells about a fixed oxygen nucleus. It thus provides little information about the angular distribution of oxygens on those shells if the molecule to which the central one belongs is held fixed in orientation. This angular information is important, however, if one is to understand the detailed architecture of hydrogen bond networks in liquid water.

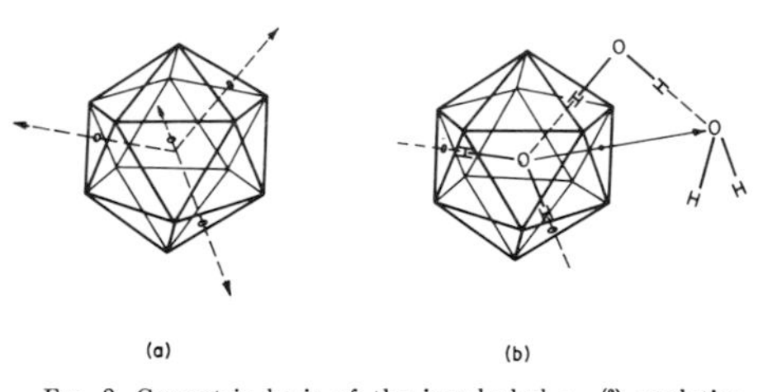

FIG. 8. Geometric basis of the icosahedral $g_{OO}^{(2)}$ resolution. In (a), a tetrahedrally directed quartet of directions passes simultaneously through the centroids (open circles) of four faces of a regular icosahedron. The correspondingly oriented water molecule is shown in (b). The four pierced faces in (a) form class I, the 12 faces sharing an edge with these four form class II, and the remaining four faces form class III.

The basic tetrahedral symmetry of our model water molecules suggests a convenient way to resolve some of the angular detail. Figure 8(a) shows that the four tetrahedral directions emanating from a point can be made to pass simultaneously through the centroids of four noncontiguous triangular faces of a regular icosahedron centered at that point. The four tetrahedral vectors in fact are perpendicular to the faces that they pierce. In Fig. 8(b), accordingly, a water molecule has been placed with its oxygen nucleus at the icosahedron center, and has been oriented so that the directions of the four undistorted hydrogen bonds in which it can engage pass through face centroids.

Let us denote the pierced triangular faces by I. Then first neighbors of a given water molecule which interact via undistorted hydrogen bonds with that molecule (as in ice) will invariably have their oxygen nuclei within those solid angles about the icosahedron center which are generated by class I faces.

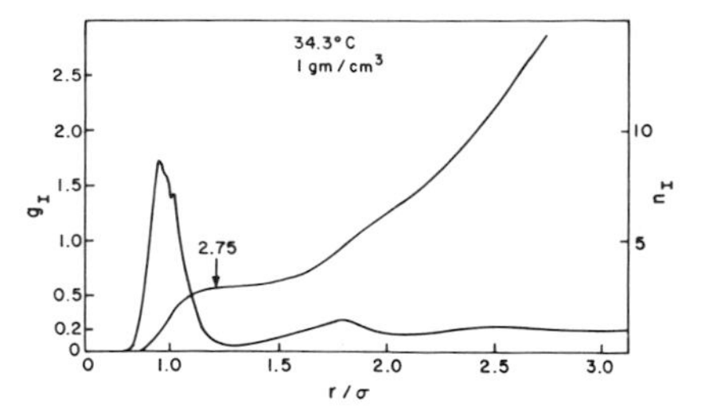

FIG. 9. $g_I(r)$ and $n_I(r)$ for water at 34.3°C and 1 g/cm³.

Twelve more icosahedron faces share edges with the four of class I. This new class of faces will be called class II. It is clear from Fig. 8(b) that second neighbor oxygens, located along a sequence of two undistorted hydrogen bonds, will occur within solid angles generated only by class II faces. Rotation around linear hydrogen bonds can cause the second-neighbor oxygen–oxygen direction to move across class II faces in an arc, but this arc will pass from one II face to another II face through a shared vertex.

The four I faces and twelve II faces leave four other faces to be accounted for. These are the "antitetrahedral faces," which lie directly opposite those of class I, across the icosahedron. We shall denote them by III. In a network of undistorted hydrogen bonds, oxygen nuclei from third- or higher-order neighbors only can occur in solid angles generated by class III faces.

The fact that the full solid angle 4π about the central oxygen in Figs. 8 has been split into three parts, I, II, and III, means that a corresponding resolution of $g_{OO}^{(2)}$ can be effected:

$$g_{OO}^{(2)}(r)=g_{I}(r)+g_{II}(r)+g_{III}(r), \quad (4.5)$$

where the three functions with Roman numeral sub-

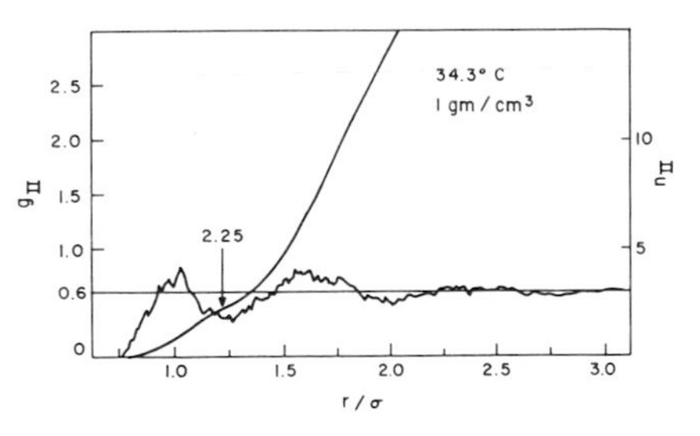

FIG. 10. $g_{II}(r)$ and $n_{II}(r)$ for water at 34.3°C and 1 g/cm³.

Notes on p. 354

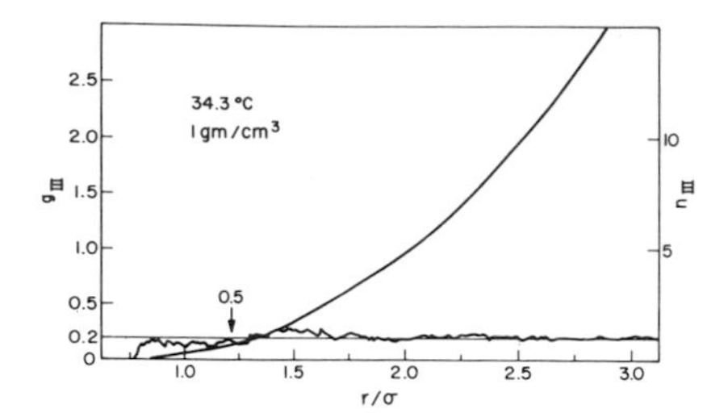

FIG. 11. $g_{III}(r)$ and $n_{III}(r)$ for water at 34.3°C and 1 g/cm³.

scripts represent the relative occurrence probabilities, at radial distance r, of oxygen nuclei in the respective solid angle classes. We will have

$$\lim_{r\to\infty} g_I(r) = \lim_{r\to\infty} g_{III}(r) = \tfrac{1}{5},$$

$$\lim_{r\to\infty} g_{II}(r) = \tfrac{3}{5}, \qquad (4.6)$$

reflecting merely the fraction of all 20 icosahedral faces in the respective classes.

Figures 9–11, respectively, present our computed functions g_I, g_{II}, and g_{III} for 34.3°C. Shown as well in each of these Figures are the corresponding running coordination numbers, defined in analogy to Eq. (4.4), e.g.,

$$n_I(r) = 4\pi\rho_0 \int_0^r s^2 g_I(s)\, ds. \qquad (4.7)$$

These results very clearly demonstrate that substantial deviations from ideal hydrogen-bonding directions are present. Although $g_I(r)$ exhibits a large peak at the position of the first $g_{OO}^{(2)}(r)$ maximum ($r/\sigma \cong 1$), $g_{II}(r)$ also has substantial weight there as well. Evidently some of the hydrogen bonds to neighbors have been bent out of class I solid angles into those contiguous solid angles of class II. On the basis of the curves $n_{OO}(r)$ and $n_{II}(r)$, evaluated at the first $g_{OO}^{(2)}(r)$ minimum, 2.25/5.50, or 41%, of the bonds suffer this fate. In addition, neither g_I nor g_{III} vanish at $r/\sigma = 1.633$, where ideally bonded second neighbors should appear only in g_{II}.

The function $g_{III}(r)$ shown in Fig. 11 is nearly flat from $r/\sigma = 0.75$ onward. For $r/\sigma \geq 2$ this is easy to understand merely in terms of the large number of ways that successive hydrogen bonds can link neighboring molecules so as to place ultimately an oxygen nucleus in a class III solid angle region. But when r/σ is near unity, the pairs which contribute to g_{III} are necessarily so seriously misaligned that no hope for a hydrogen bond between them exists. One possible explanation would be that an "interstitial" molecule were involved in such a pair, surrounded by, but only weakly interacting with, an enclosing network of hydrogen bonds. Alternatively, two molecules each incompletely hydrogen bonded to their surroundings, could by chance simply back into one another.

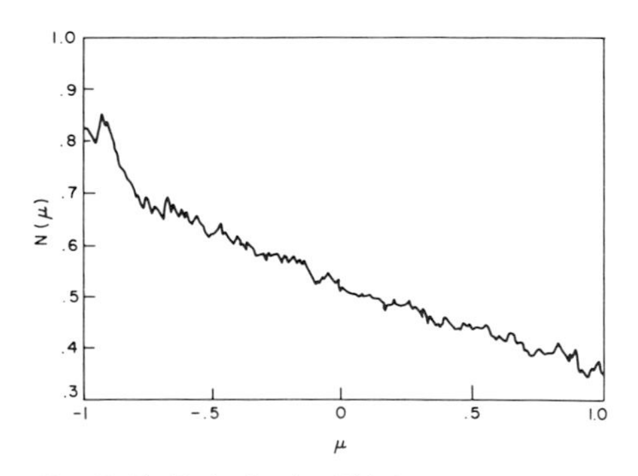

FIG. 12. Distribution function $N(\mu)$ for $\mu = \cos\theta_1$, defined by Eq. (4.13). The decline with increasing μ indicates a tendency for $\boldsymbol{\mu}_1^{(1)}$ and $\mathbf{N}_1$ to be antiparallel.

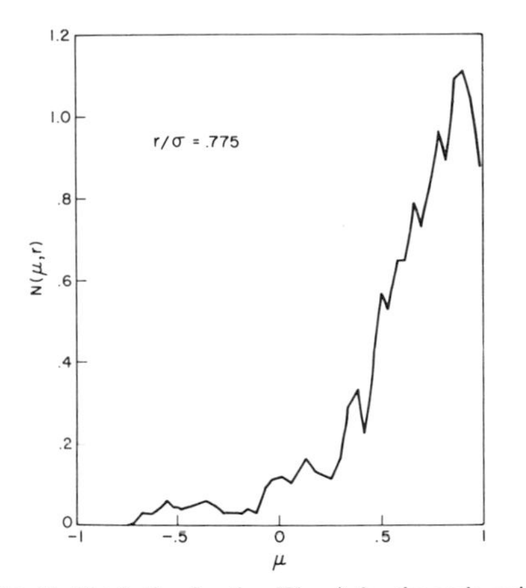

FIG. 13. Distribution function $N(\mu, r)$ for the angle cosine (μ) between a given molecule's dipole direction and the direction of the total moment of neighbors at distance r. Distances are reckoned in terms of oxygen atom positions. For this graph, $r = 0.775\sigma$.

C. Dipole Direction Correlation

The dipole moment of an undisturbed water molecule bisects its HOH bond angle. As another aspect of static orientational order in liquid water, it is interesting to see how these dipole directions are distributed for neighboring molecules.

Let $\boldsymbol{\mu}_i^{(1)}$ represent a unit vector along the dipole direction of molecule i, and set $\mathbf{M}$ equal to their vector sum for all $N=216$ molecules:

$$\mathbf{M}=\sum_{i=1}^{N}\boldsymbol{\mu}_i^{(1)}. \qquad (4.8)$$

If the dipole directions were entirely uncorrelated,

$$\langle M^2\rangle/N=1 \qquad \text{(uncorrelated)}. \qquad (4.9)$$

Instead, the average computed for the molecular dynamics run turns out to be

$$\langle M^2\rangle/N=0.171 \qquad \text{(molecular dynamics)}, \qquad (4.10)$$

so evidently the molecular interactions act in a way to quench the system's net moment.

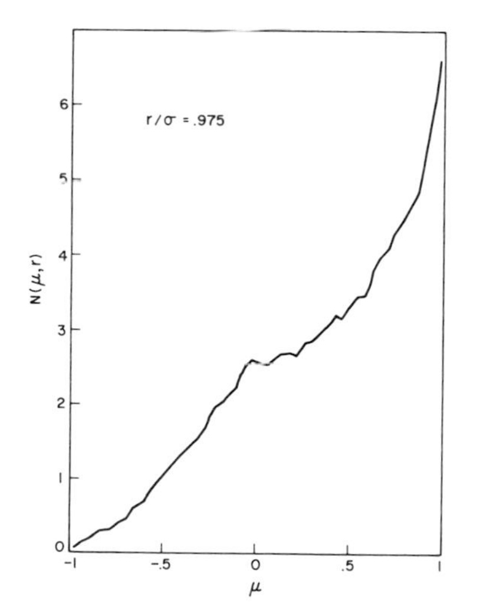

FIG. 14. Orientational distribution function $N(\mu,r)$ for $r=0.975\sigma$.

This quenching effect may be examined in a variety of alternative ways. One can, for instance, isolate one of the molecules (which for convenience we take to be the one numbered 1), and ask how its own dipole direction correlates with

$$\mathbf{N}_1=\mathbf{M}-\boldsymbol{\mu}_1^{(1)}, \qquad (4.11)$$

the total moment of all the other molecules. Of course the magnitude of $\mathbf{N}_1$ fluctuates, as revealed by the difference in computed averages:

$$\langle|\mathbf{N}_1|\rangle=5.69,$$
$$\langle N_1^2\rangle^{1/2}=6.13, \qquad (4.12)$$

as well as its direction relative to $\boldsymbol{\mu}_1^{(1)}$. The relevant

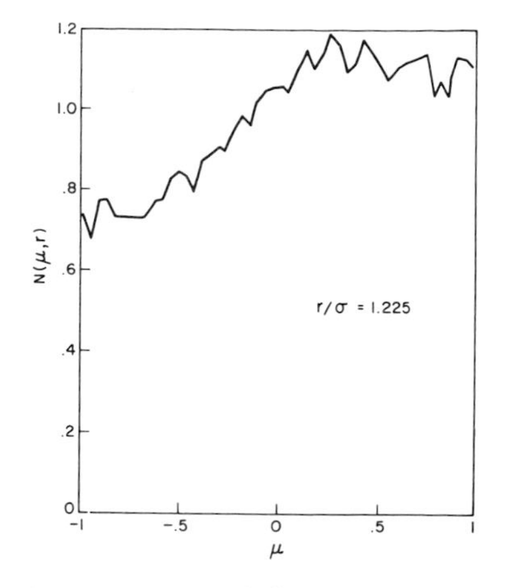

FIG. 15. Orientational distribution function $N(\mu,r)$ for $r=1.225\sigma$.

angle θ_1 is defined by

$$\cos\theta_1=\hat{\mathbf{N}}_1\cdot\boldsymbol{\mu}_1^{(1)}, \qquad \hat{\mathbf{N}}_1=\mathbf{N}_1/|\mathbf{N}_1|. \qquad (4.13)$$

For our molecular dynamics run:

$$\langle\cos\theta_1\rangle=-0.129, \qquad (4.14)$$

showing that on the average, $\mathbf{N}_1$ and $\boldsymbol{\mu}_1^{(1)}$ point in opposite directions.

Figure 12 shows the full distribution function $N(\mu)$ for $\mu=\cos\theta_1$, for which the average value shown in (4.14) applies. Although this distribution is nearly linear, a slight positive curvature beyond statistical uncertainty seems to be present.

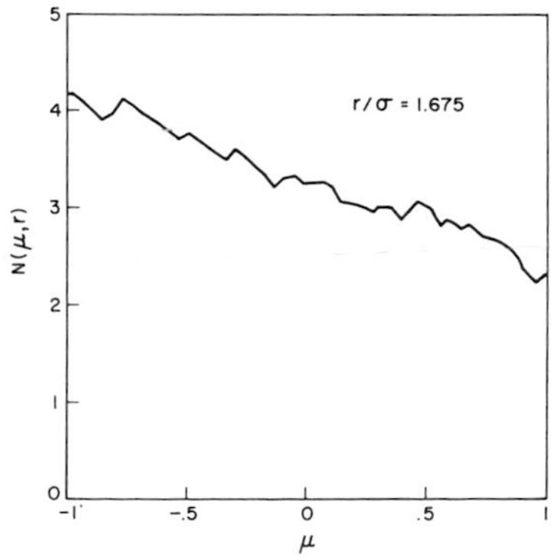

FIG. 16. Orientational distribution function $N(\mu,r)$ for $r=1.675\sigma$.

Notes on p. 354

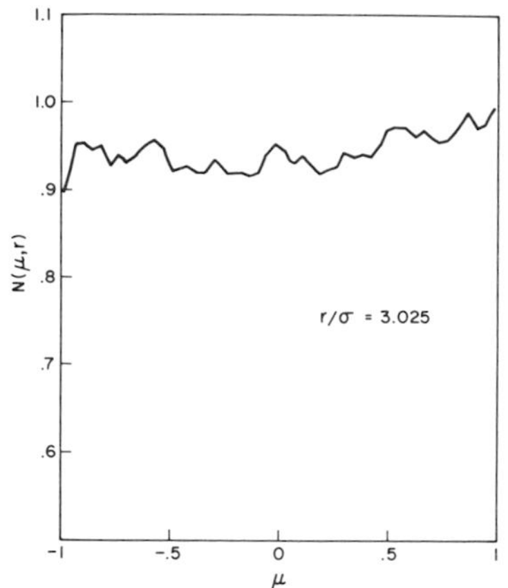

FIG. 17. Orientational distribution function $N(\mu, r)$ for $r=3.025\sigma$.

It is also instructive to analyze the separate contributions to $\mathbf{N}_1$ from distinct spherical shells centered about the oxygen nucleus of molecule 1. Figures 13–17 exhibit partial distribution functions $N(\mu, r)$ for μ, the cosine of the angle between $\boldsymbol{\mu}_1^{(1)}$ and the moment of a molecule contained within a shell of mean radius r, and width 0.05σ. The first two of the figures ($r/\sigma=0.755$ and 0.975, respectively) show that pairs of molecules close enough to hydrogen bond have a strong tendency toward dipole parallelism ($\mu=+1$), in spite of the over-all opposite tendency which is manifest in Fig. 12. At the somewhat larger distance $r/\sigma=1.225$ (Fig. 15), the alignment effect is less distinct, and it has reversed at $r/\sigma=1.675$ (Fig. 16). Figure 17 indicates a relatively flat distribution for $r/\sigma=3.025$.

The static dielectric constant ϵ_0 for polar fluids is intimately connected to the orientational correlations between neighboring molecules. The Kirkwood theory of polar dielectrics[30] expresses ϵ_0 in terms of the mean molecular polarizability α, the liquid phase dipole moment μ_l, and an orientational correlation function g_K:

$$(\epsilon_0-1)(2\epsilon_0+1)/3\epsilon_0=4\pi\rho[\alpha+(\mu_l^2 g_K/3k_BT)]. \quad (4.15)$$

The Kirkwood orientational correlation factor g_K may be expressed in terms of the full orientation and position dependent pair correlation function $g^{(2)}(\mathbf{x}_1, \mathbf{x}_2)$[17,31]:

$$g_K=1+(\rho/8\pi^2)\int d\mathbf{x}_2(\boldsymbol{\mu}_1^{(1)}\cdot\boldsymbol{\mu}_2^{(1)})g^{(2)}(\mathbf{x}_1, \mathbf{x}_2); \quad (4.16)$$

in this expression one must be careful to use only the infinite system limit function $g^{(2)}(\mathbf{x}_1, \mathbf{x}_2)$.

Our molecular dynamics calculation has not been set up in such a way that direct evaluation of ϵ_0 is possible. No external electric field is involved, of course. Furthermore, μ_l is unknown except to the extent that it should exceed the free molecule moment; even in ice, where the local structure is completely known, estimates of the mean molecular dipole moment vary considerably.[32,33]

Nevertheless, enough information is supplied by our calculation to evaluate g_K. The evaluation is not direct, however, since the infinite system limit required of $g^{(2)}$ in Eq. (4.16) is not available. Instead, only a finite system $g^{(2)}$ is at hand, and its finite limit integral analog of (4.16) gives an apparent Kirkwood orientational correlation factor G_K. The fundamental difference between g_K and G_K stems from inclusion only in the latter of a weak macroscopic polarization contribution associated with electrostatic boundary conditions at finite distance; in the present circumstance the "boundary" is generated by the interaction cutoff. One may show (Appendix B) that

$$G_K=[9\epsilon_0/(\epsilon_0+2)(2\epsilon_0+1)]g_K. \quad (4.17)$$

For the present calculations,

$$G_K=\langle M^2\rangle/N, \quad (4.18)$$

whose value has already been presented in Eq. (4.10). If we employ the measured dielectric constant for water at 34.3°C (75.25) to compute the conversion factor shown in Eq. (4.17), the implied result is

$$g_K=2.96. \quad (4.19)$$

This agrees roughly with values that have previously been proposed; Harris, Haycock, and Alder's[34] work, for instance, implies that $g_K\cong 2.6$ at 34.3°C. It should be pointed out however that the energy rescaling of our molecular dynamics results, which is discussed in Sec. VII, improves the agreement.

D. Bond Energy Distribution

Up to this point we have frequently invoked the concept of "hydrogen bonding" to interpret those aspects of water molecule correlation which stem from the characteristic tetrahedral directionality of the effective pair potential. We now require a precise definition of "hydrogen bond," so that precise statements may be formulated about the geometrical and topological character of the random networks that exist in liquid water.

The most obvious way to define hydrogen bonds in the present circumstance uses the effective pair interaction itself. Whenever the interaction energy for a given pair of molecules lies below a negative cutoff value V_{HB}, we shall say that the pair is hydrogen bonded; if their interaction equals or exceeds V_{HB}, they are by definition not hydrogen bonded:

$$V_{\text{eff}}^{(2)}(i,j)<V_{HB} \quad (i,j \text{ hydrogen bonded}),$$
$$V_{\text{eff}}^{(2)}(i,j)\geq V_{HB} \quad (i,j \text{ not hydrogen bonded}). \quad (4.20)$$

The cutoff parameter V_{HB} is arbitrary, at least within certain limits; the fullest understanding of the nature of hydrogen-bond networks in water would result by varying this parameter and observing the consequences.

In order to identify the range of values for V_{HB} of greatest chemical relevance and interest, it is valuable to examine the entire distribution of effective pair interactions in the liquid. This interaction density, p, may be expressed in terms of the following integral:

$$p(V)=(\rho/8\pi^2)\int d\mathbf{x}_j\delta[V-V_{\mathrm{eff}}^{(2)}(\mathbf{x}_i,\mathbf{x}_j)]g^{(2)}(\mathbf{x}_i,\mathbf{x}_j). \tag{4.21}$$

With this normalization,

$$n(V,V')=\int_V^{V'} p(V'')dV'' \tag{4.22}$$

will be the mean number of neighbors of any molecule whose instantaneous interaction with that molecule lies between V and V'. Naturally $p(V)$ will vanish if V declines below the absolute minimum of $V_{\mathrm{eff}}^{(2)}$ noted in Eq. (2.11). One easily establishes furthermore that $p(V)$ will diverge as V^{-2} near $V=0$, owing to the preponderance of weak dipolar pair interactions at large distances.

The distribution $p(V)$ has been calculated for the molecular dynamics run and the result is displayed in Fig. 18. The expected divergence at the origin is obvious in the Figure, but it is uninteresting since it conveys no specific structural information. The primary point of interest in the curve is the large, essentially flat region in the range of V from -4.5 kcal/mole to -2.0 kcal/mole. Evidently, this reveals a wide class of moderately strong "bonds," which often suffer considerable strain. The apparent relative maximum in $p(V)$ at about $+2.5$ kcal/mole we believe is a real effect, not a misleading statistical fluctuation. Probably it arises from pairs of molecules which are simultaneously bound to a third, but in such a relative configuration that they repel one another.[35]

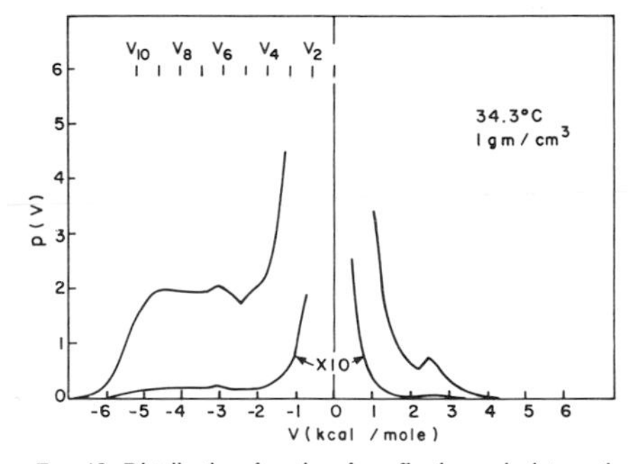

FIG. 18. Distribution function for effective pair interaction strength in water. The curves show relative numbers of molecules in successive energy intervals of width $\Delta V=4\epsilon=0.288$ kcal/mole.

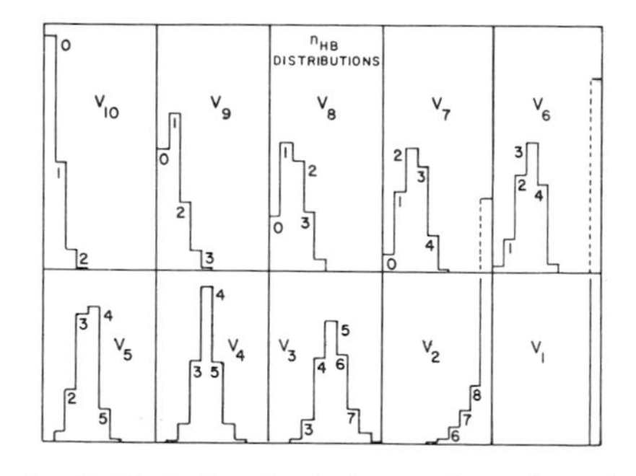

FIG. 19. Distribution of molecules according to the number of hydrogen bonds in which they engage. The set of cutoff energies V_{HB} used as alternative hydrogen-bond definitions is shown in Eq. (4.24).

On account of the negative-V plateau in $p(V)$, it seems plausible to select V_{HB} near the middle or upper limit of this range, for example,

$$V_{HB}=-2.9 \text{ kcal/mole}. \tag{4.23}$$

This choice would certainly be consistent with the chemical suggestion that a large number of hydrogen bonds are still present in the liquid after ice is melted, without at the same time being so permissive as to include pairs of molecules much more widely separated than nearest neighbors in the ice lattice.

Using several alternative choices for V_{HB}, including the range indicated by (4.23), the concentrations of molecules engaging in different numbers of hydrogen bonds simultaneously has been calculated. This set of coordination number distributions is displayed in Fig. 19. The alternative values selected for V_{HB} are equally spaced, and have the values

$$\begin{aligned} V_j &- -8(j-1)\epsilon \\ &= -0.577(j-1) \text{ kcal/mole}, \\ j &= 1, 2, \cdots, 10. \end{aligned} \tag{4.24}$$

It is clear from Fig. 19 that these choices span a wide range of hydrogen bond definitions, from an extremely permissive limit which assigns many more than four bonded neighbors to all molecules, to a very strigent limit which makes hydrogen bonding between neighbors a rare event.

The most significant conclusion to be drawn from Fig. 19 is that as V_{HB} is varied, the coordination number distribution shifts, but it always maintains a single-maximum character. This fact alone seems to rule out the class of two-state liquid water models[26,27]

Notes on p. 354

which postulate large side-by-side regions of bonded and of unbonded molecules. For such models, one would expect to observe for some V_{HB} choice a bimodal distribution with simultaneous maxima at zero and at four bonds.

It is clear from Fig. 19 that choice (4.23) for V_{HB} (essentially the value V_6) makes the most probable number of hydrogen bonds equal to three. At the same time a not insignificant number of molecules have fivefold coordination. From a detailed knowledge of $V_{eff}^{(2)}$ potential curves[17] it is possible to assert that essentially all molecules in ordinary ice will have precisely four bonded neighbors with choice (4.23), so the corresponding liquid phase distribution leads to a vivid picture of the nature of network disruption that accompanies melting.

It is instructive to place these bond distribution considerations for liquid water in the wider context of general liquid behavior. Thus, the anomalous character of water again stands out in comparison with the simple liquified noble gases. For liquid argon, the pair interaction distribution $p(V)$ may readily be expressed as an explicit, closed-form functional of the pair correlation function, provided V for that substance has the traditional Lennard-Jones (12–6) form. Using this hypothesis, and the argon correlation function of Fig. 4, the corresponding $p(V)$ was calculated. The result is presented in Fig. 20. Some of the obvious differences between this $p(V)$ and the one shown in Fig. 18 for water reflect the lack of rotational degrees of freedom for argon that can cause V to vary; the sharp peak at the minimum attainable V is one example. The most significant structural difference, though, is that no particularly suggestive features arise for negative V (such as a plateau region for water) that would imply a useful energy definition of "bond" between two argon atoms.

E. Stereoscopic Pictures

Intermediate configurations of the 216 water molecules, occurring every $500\Delta t = 2.1775\times10^{-13}$ sec, were placed on punched cards and then processed at Murray Hill to produce stereo slides. The computer program used for this purpose produces left and right eye views separately on a cathode-ray tube, whose display is then photographed with 35 mm film. The subsequently mounted slides can then be examined with a commercially available viewer,[36] to give a striking impression of the three-dimensional order.

The individual molecules in these pictures are rendered into stick figure form. The oxygen nucleus and the distinguishable hydrogen nucleus positions are indicated by small O, G, and H, respectively, with the covalent bonds between them (at the tetrahedral angle) drawn as straight lines. The time interval between successive pictures is sufficiently small that the same molecule can easily be identified and followed through the entire sequence, and the motion of its individual nuclei perceived.

The inevitable immediate impression conveyed by these pictures upon first viewing is that a high degree of disorder is present. Anything else of course would make one properly suspicious that a liquid was actually being simulated. Beyond this general feature, several more detailed observations can be listed:

(1) There is a very clear tendency for neighboring molecules to be oriented into rough approximations to tetrahedral hydrogen bonds, but the average degree of bending away from bond linearity and ideal approach directions is considerable.

(2) Except on the smallest scale, the random molecular configurations are rather homogeneous in density. No large "clusters" of anomalous density seem to occur.

(3) No recognizable patterns characteristic of the known ices or clathrates appear, beyond occasional polygons of hydrogen bonds. Such polygons occur with 4, 5, 6, 7 (and perhaps more) sides, but they tend to be distorted out of their most natural conformations.

(4) Dangling OH bonds exist, which are not included in hydrogen bonds. These entities persist far longer than water molecule vibrational periods, and hence may hold the key to the structurally sensitive band shapes that arise in infrared[37] and Raman[38,39] spectroscopy of water and its solutions.

(5) No obvious separation of molecules into "network" vs "interstitial" types suggests itself. This fact is consistent with the single-peak character of the hydrogen-bond coordination number distributions exhibited in Fig. 19. It also seems to diminish the validity of the interstitial models that have been proposed to explain liquid water.[40–42]

(6) In the case of moderately well-formed (i.e., undistorted) hydrogen bonds, all angles of rotation

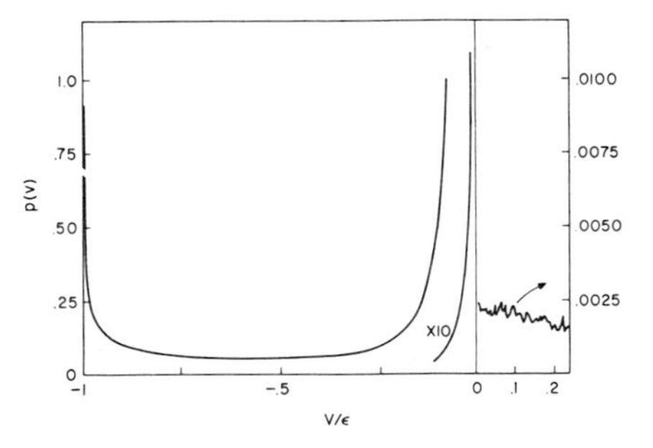

FIG. 20. Bond energy distribution for liquid Ar. The state parameters, and the correlation function employed, are those of Fig. 4.

of the molecules about the bond axis seem to be frequently represented. This behavior may have direct relevance in study of nonelectrolyte solvation, where the geometric requirements attendant upon formation of a hydrogen-bonded solvation cage forces the rotation angles into "eclipsed" configurations only,[43] thus lowering configurational entropy.

(7) No significant examples of network interpenetration were found, analogous to the interpenetration known to obtain in ice VII and ice VIII.[44]

V. KINETIC PROPERTIES

The internal structure of the water molecule requires that the static structure of the liquid be examined from many more independent points of view, for a given level of comprehension, than is required for liquid argon. This increased elaborateness persists into the regime of kinetic properties too. We now shall successively examine several aspects of the temporal development of our model water system. As in the static case, these aspects are not certainly the only informative ones that might have been chosen for study. Nevertheless, we believe that this selection serves to illuminate the dominant characteristics of water molecule motions in the liquid phase.

A. Self-Diffusion

The most obvious facet of the water molecule motions is their long time diffusion rate, which is measured by the self-diffusion coefficient D. This parameter may be related, in the long time limit, to the mean-square displacement of any fixed point in a given molecule. In the case of the molecular center of mass, at position $\mathbf{R}_j(t)$ at time t,

$$D=\lim_{t\to\infty}(6t)^{-1}\langle[\mathbf{R}_j(t)-\mathbf{R}_j(0)]^2\rangle; \qquad (5.1)$$

implicit in this limit expression is a previous infinite system size limit operation.

In the molecular dynamics calculations, one naturally is limited both in time interval, and by finite system size. However, as in the case of such studies for liquid Ar,[22] the computed mean-square center-of-mass displacement appears to approach a limiting slope sufficiently rapidly that D may conveniently be extracted from the calculations. But at the same time, the molecules do not diffuse far enough to span a periodicity cell edge length, which would invalidate use of (5.1) in the present context.

Figure 21 shows the computed curve for the water molecule center-of-mass mean-square displacement. Included as well is the mean-square displacement for a proton in the liquid. At sufficiently long times, these curves would become parallel straight lines, with the proton curve displaced upward from the center of mass curve by

$$2(0.3419\sigma)^2=0.2338\sigma^2, \qquad (5.2)$$

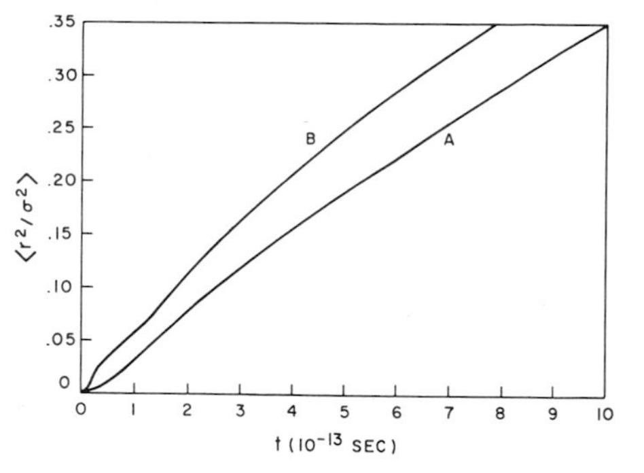

FIG. 21. Mean-square displacement of center-of-mass motion (Curve A), and of proton motion (Curve B). The apparent limiting slope of Curve A gives $D=4.2\times10^{-5}$ cm²/sec.

i.e., twice the OH bond length squared. In Fig. 21 though it is clear that the proton curve is still rising away from the essentially linear center-of-mass curve, due to the incompleteness of molecular rotation during the time allowed for computation. The motion of the center of mass therefore constitutes by far the more convenient means of estimating D.

The apparent limiting slope of the center-of-mass mean-square displacement curve in Fig. 21 implies

$$D=4.2\times10^{-5}\ \mathrm{cm^2/sec} \qquad (5.3)$$

for water at 34.3°C, and 1 g/cm³. This value is significantly larger than the value that may be inferred from recent spin–echo experiments[45,46]:

$$D=2.85\pm0.15\times10^{-5}\ \mathrm{cm^2/sec}. \qquad (5.4)$$

But in view of the rapid variation of the experimental D with temperature, the comparison should not be viewed as unfavorable. Evidently, a small change in the energy scale of $V_{\mathrm{eff}}^{(2)}$ could eliminate the discrepancy (see Sec. VII below).

In principle, the self-diffusion constant D could also be obtained from the velocity autocorrelation function for the molecular center-of-mass motion:

$$D=\tfrac{1}{3}\int_0^\infty\langle\mathbf{v}_j(0)\cdot\mathbf{v}_j(t)\rangle dt, \qquad \mathbf{v}_j(t)=d\mathbf{R}_j(t)/dt. \qquad (5.5)$$

This velocity autocorrelation function has been computed from the molecular dynamics run, and it is shown in normalized form in Fig. 22. Since only a limited average can be performed with the run of total length 2.1775×10^{-12} sec, this autocorrelation function necessarily represents an incomplete phase space average. The resulting statistical error is especially noticeable for "large" times ($6–7\times10^{-13}$ sec), where the exact autocorrelation function is apparently quite small. The "cutoff" indicated in Fig. 22 at

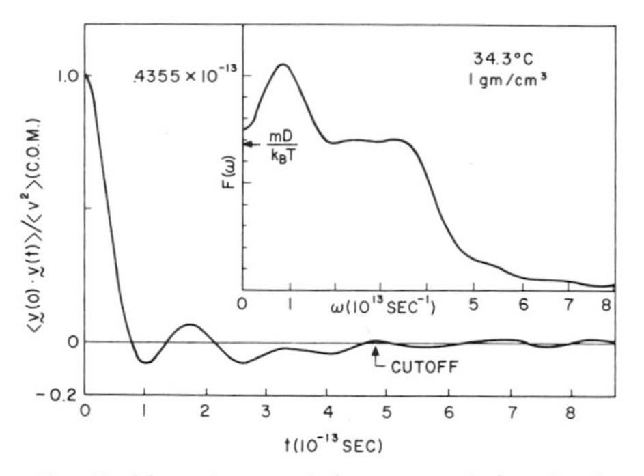

FIG. 22. Center-of-mass velocity autocorrelation function (normalized at $t=0$) for water and its Fourier cosine transform. The "cutoff" locates approximately the point beyond which statistical noise dominates the autocorrelation function.

4.78×10^{-13} sec represents our estimate of the point beyond which the curve is primarily statistical noise. At shorter times, though, the autocorrelation curve shown is probably a reasonably accurate approximation to the exact function.

Figure 22 also presents the Fourier transform[47]

$$F(\omega)=\int_0^\infty \frac{\langle \mathbf{v}(0)\cdot\mathbf{v}(t)\rangle}{\langle \mathbf{v}(0)\cdot\mathbf{v}(0)\rangle}\cos(\omega t)\,dt. \tag{5.6}$$

In principle, one should have

$$F(0)=mD/k_BT; \tag{5.7}$$

however, this identity is not quite obeyed using the D value in Eq. (5.3), as the arrow on the $F(\omega)$ scale shows. The discrepancy reflects the magnitude of errors in the incomplete phase averages involved in both methods of evaluating D. This suggests that the D value shown in Eq. (5.3) may be in error by

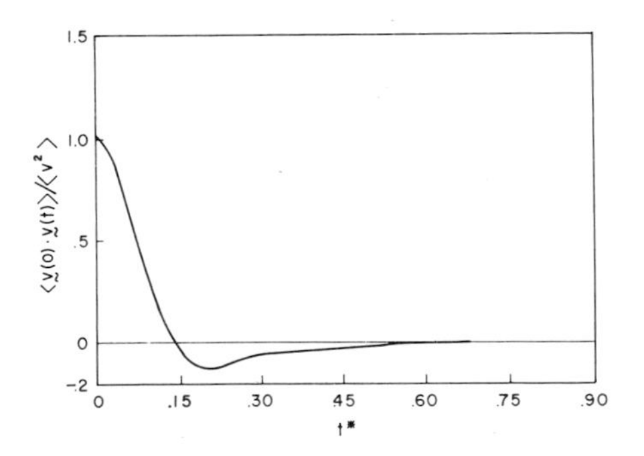

FIG. 23. Normalized velocity autocorrelation function for liquid argon; $\rho^*=0.81$, $T^*=0.74$. The reduced time t^* is measured in units $\sigma(m/\epsilon)^{1/2}$.

as much as 10%, compared to the model's precise self-diffusion constant at 34.3°C and 1 g/cm³.

Even accounting for the statistical uncertainty present in the center-of-mass velocity autocorrelation in Fig. 22, it is obvious that the average molecular motion is relatively oscillatory. Once again our water model stands in distinct contrast to liquid Ar, for which Fig. 23 presents the velocity autocorrelation function. For this comparatively simple liquid, the autocorrelation has only a single well-defined negative minimum, followed by a slow rise to zero. The hydrogen bonding in liquid water, however, produces a more persistently oscillatory motion, as a result of greater structural rigidity.

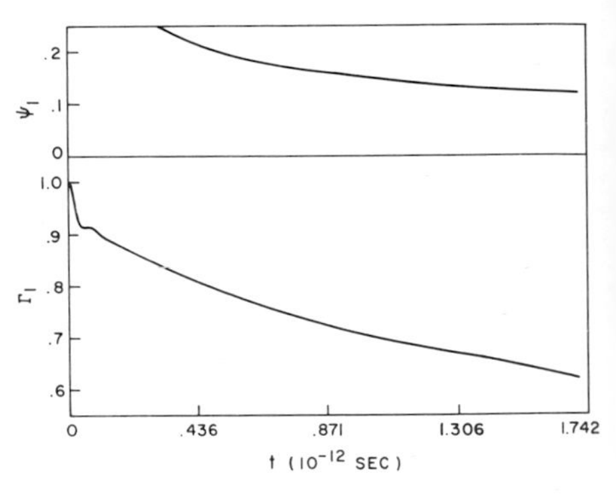

FIG. 24. Dipole direction relaxation function $\Gamma_1(t)$. The related $\psi_1(t)$ is defined in Eq. (5.9).

The stereo pictures of intermediate configurations are too widely separated in time (2.1775×10^{-13} sec) to resolve proton oscillations with great detail. However, they do suffice to give an idea of why and how translational diffusion proceeds. The hydrogen bonds between neighboring molecules are continually subject to varying degrees of strain. Accordingly, each molecule is tugged around in a random fashion by its imperfectly aligned neighbors, while at the same time those neighbors are also being forced to rearrange by *their* neighbors. In this circumstance, it is frequently the case that hydrogen bonds become strained to the breaking point; i.e., it is favorable for a molecule to reorient to form new hydrogen bonds to other nearby molecules. The amplitude of these very anharmonic motions is sufficiently high at the ambient temperature that settling down into a regular and relatively unstrained arrangement is overwhelmingly unlikely.

There are effectively so many available highly strained configurations near to one another, that their interconversion is a continual, rather than a discrete, process. There is no evidence in the stereo pictures for a hopping process between alternative positions of mechanical stability. Instead, translational diffusion proceeds via individual molecule participation in the continual restructuring of the labile random hydrogen-bond network.

B. Dipole Direction Relaxation

The forces of hydrogen bonding between molecules prevent rapid turning over of those molecules. The autocorrelation function for the dipole direction of a given molecule

$$\Gamma_1(t)=\langle \boldsymbol{\mu}_j^{(1)}(0)\cdot\boldsymbol{\mu}_j^{(1)}(t)\rangle\equiv\langle P_1[\cos\theta_j(t)]\rangle \quad (5.8)$$

clearly shows this rotational retardation. It is plotted in Fig. 24 both in direct form, as well as logarithmically in terms of the quantity

$$\psi_1(t)=-(10^3\Delta t/t)\ln\langle P_1[\cos\theta_j(t)]\rangle. \quad (5.9)$$

After a brief period of initial libration lasting roughly 10^{-13} sec, a long monotonic decay ensues which apparently persists well beyond the time limit imposed by the computation. By fitting a simple exponential function to Γ_1 in the monotonic range, one infers a longest relaxation time τ_1 equal to about 5.6×10^{-12} sec.

It has been claimed[48,49] that the macroscopic dielectric relaxation time (for polar liquids with large ϵ_0) should be $\frac{3}{2}$ or 2 times as large as τ_1. Our calculations would then imply for water at 34.3°C:

$$8.4\times10^{-12}\text{ sec}\leq\tau_d\leq11.2\times10^{-12}\text{ sec}. \quad (5.10)$$

The measured dielectric relaxation time at this temperature is[50]

$$\tau_d=6.7\times10^{-12}\text{ sec}, \quad (5.11)$$

so taking the various imprecisions into account, our water model seems not to be too far out of line.

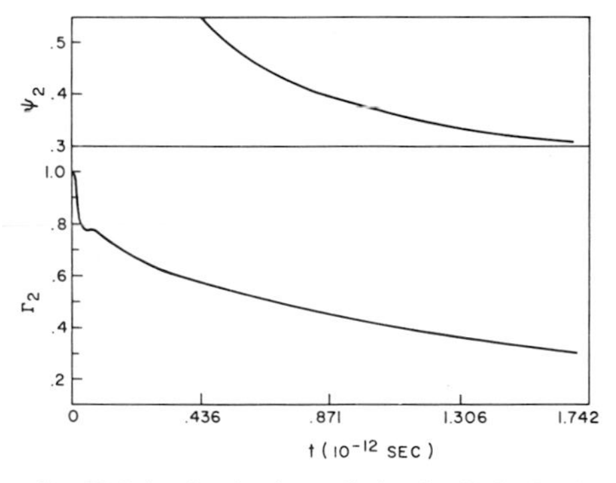

FIG. 25. Relaxation function $\Gamma_2(t)$ for the dipole direction second harmonic. The related function $\psi_2(t)$ is defined in Eq. (5.14).

The autocorrelation function $\Gamma_1(t)$ may be regarded as the leading member of an infinite sequence of autocorrelation functions for spherical harmonics of ascending order:

$$\Gamma_l(t)=\langle P_l[\boldsymbol{\mu}_j^{(1)}(0)\cdot\boldsymbol{\mu}_j^{(1)}(t)]\rangle, \quad (5.12)$$

where P_l is the lth Legendre polynomial. If the unit dipole direction vector $\boldsymbol{\mu}_j^{(1)}$ moved in time by a true rotational Brownian motion, the Γ_l would decay exponentially with relaxation times τ_l all simply related to τ_1:

$$\tau_l=2\tau_1/l(l+1). \quad (5.13)$$

In order partially to test this possibility, $\Gamma_2(t)$ was evaluated for the present water model, along with the $l=2$ analog of ψ_1:

$$\psi_2(t)=-(10^3\Delta t/t)\ln\Gamma_2(t). \quad (5.14)$$

These two functions are plotted in Fig. 25. Again an initial rapid libration shows up, followed by a longer monotonic decay. This monotonic portion is certainly not precisely characteristic of a single exponential decay, but indicates that the most persistent component exhibits a relaxation time

$$\tau_2\cong2.1\times10^{-12}\text{ sec}. \quad (5.15)$$

The ratio of $l=1$ to $l=2$ relaxation times for the water model computation is therefore somewhat less than the ideal Brownian motion ratio 3:

$$\tau_1/\tau_2\cong2.7. \quad (5.16)$$

This diminution should be expected however, since the water molecule rotational motion corresponds more closely to a sequence of finite stochastic jumps, rather than to the infinitely rapid infinitesimal jumps implied by classical Brownian motion (Wiener process[51]).

The nuclear magnetic resonance spin–lattice relaxation time T_1 contains a contribution, due to molecular rotation, that in principle measures $\Gamma_2(t)$ if the rotation is isotropic.[52] Krynicki[53] has inferred from his T_1 measurements how τ_2 should vary with temperature; his results imply that

$$\tau_2\cong1.9\times10^{-12}\text{ sec}. \quad (5.17)$$

In view of the several uncertainties involved in interpretation of the experiments, the τ_2 values (5.15) and (5.17) are in satisfactory agreement.

C. Dielectric Relaxation

The autocorrelation $\Gamma_1(t)$ is central to the frequency dependence of the liquid's dielectric constant. We have already noted (Sec. IV. C) that ignorance of the correct liquid-phase molecular dipole moment limits one's ability to predict the static dielectric constant ϵ_0. This ignorance naturally hinders full understanding of $\epsilon(\omega)$ as well, as does the present lack

Notes on p. 354

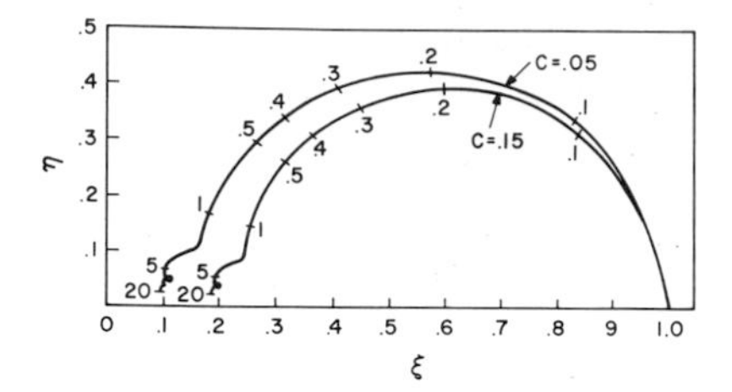

FIG. 26. Cole–Cole plots for the frequency-dependent dielectric constant $\epsilon(\omega)$. The curves are based upon the Nee–Zwanzig theory, Ref. 49. The reduced coordinates are defined by Eq. (5.19), and c defined by Eq. (5.20). Marks on the curves indicate frequency in units 10^{12} sec^{-1}.

of a fully general theory of time-dependent dielectric response in polar fluids.

Nevertheless, the recent approximate analysis of dielectric relaxation by Nee and Zwanzig[49] provides a convenient tentative basis on which to convert our molecular dynamics results into traditional Cole–Cole plots.[54] They derive the following relation:

$$\frac{\epsilon_0[\epsilon(\omega)-\epsilon_\infty][2\epsilon(\omega)+\epsilon_\infty]}{\epsilon(\omega)(\epsilon_0-\epsilon_\infty)(2\epsilon_0+\epsilon_\infty)} = -\int_0^\infty dt \exp(i\omega t)\frac{d\Gamma_1(t)}{dt}; \tag{5.18}$$

for present purposes the high-frequency dielectric constant ϵ_∞ corresponds to about the 10-cm^{-1} wavelength region, which has been reported[55] to yield an ϵ_∞ of 4.5 for water.

By using the previously evaluated $\Gamma_1(t)$ and a simple exponential extrapolation beyond the range shown in Fig. 24, it is possible to evaluate $\xi(\omega)$ and $\eta(\omega)$, the real and imaginary parts of $\epsilon(\omega)/\epsilon_0$:

$$\epsilon(\omega)/\epsilon_0 = \xi(\omega) + i\eta(\omega), \tag{5.19}$$

from expression (5.18). The precise value of

$$c = \epsilon_\infty/\epsilon_0 \tag{5.20}$$

for our specific model is unknown, but if real water at 34.3°C gives a reliable indication, it should be roughly 0.06.

Figure 26 shows the Cole–Cole plots obtained for the c values 0.05 and 0.15. The curves are rather close to the classic semicircular shape for $\omega < 2\times10^{-12}$ sec^{-1}. However for high frequencies, "curlicues" appear which may be attributed to the rapid initial librational motion in $\Gamma_1(t)$.

D. Proton Motion and Neutron Scattering

Owing to the fact that protons are strong incoherent scatterers, cold neutron scattering provides a convenient experimental tool for the study of single proton motions in water. It is therefore important to extract information from the molecular dynamics computation that bears specifically on these motions. There are in fact several independent dynamical quantities that deserve attention, beyond the mean-square displacement that has already been considered.

The angular velocity autocorrelations about the three principle axes of the molecule:

$$\langle\omega_\alpha(0)\omega_\alpha(t)\rangle/\langle\omega_\alpha^2\rangle, \qquad \alpha = 1, 2, 3, \tag{5.21}$$

illustrate one aspect of proton motion. Figure 27 presents these three rapidly decaying functions. All three indicate a substantial librational, or oscillatory, character. The rates of libration are in the same order as the reciprocals of the respective moments of inertia:

$$I_2^{-1} > I_3^{-1} > I_1^{-1}; \tag{5.22}$$

as the insert in Fig. 27 indicates, axis 1 is perpendicular to the molecular plane, axis 2 is in the molecular plane, and axis 3 is the twofold molecular symmetry axis.

A magnified view of the initial portion of $\Gamma_1(t)$ (shown previously in Fig. 24) is also included in Fig. 27. Only rotations about axes 1 and 2 affect the dipole direction $\boldsymbol{\mu}_j^{(1)}$ for molecule j, so it is a combination of these two motions which affects Γ_1. Since the librational rates are distinctly different about these two "dipole active" axes, we see why the com-

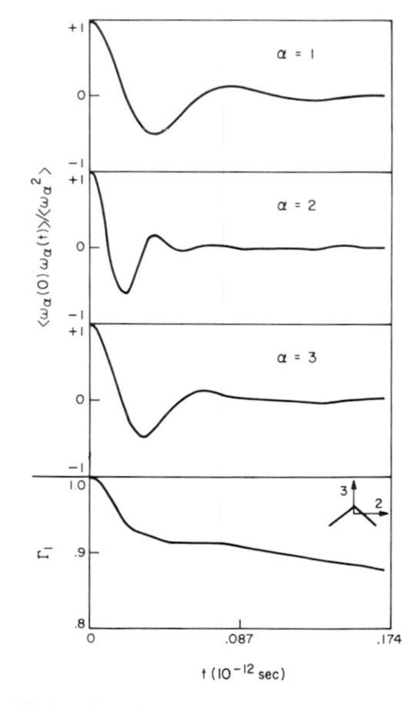

FIG. 27. Angular velocity autocorrelation functions. Included for comparison (bottom) is the initial behavior of $\Gamma_1(t)$, shown earlier in Fig. 24. The principal axis numbering is shown in the insert.

bination in Γ_1 exhibits the oscillatory character less vividly than the ω_α autocorrelations themselves.

The spectral resolutions, or Fourier transforms, of the ω_α autocorrelations are defined thus:

$$f_\alpha(\Omega) = \int_0^\infty \frac{\langle \omega_\alpha(0)\omega_\alpha(t)\rangle}{\langle \omega_\alpha^2\rangle} \cos(\Omega t)\,dt. \quad (5.23)$$

They are shown in Fig. 28. The positions of their respective maxima and centroids again reflect the ordering of librational rates according to the inertial moments.

The autocorrelation of the total angular velocity for a given molecule,

$$\langle \boldsymbol{\omega}(0)\cdot\boldsymbol{\omega}(t)\rangle / \langle |\boldsymbol{\omega}|^2\rangle, \quad (5.24)$$

may be obtained from a linear combination of the separate normalized autocorrelations (5.21), using weights that follow from the thermal equilibrium conditions,

$$\langle \omega_\alpha^2\rangle = k_B T / I_\alpha. \quad (5.25)$$

The resulting spectrum, $f_{tot}(\Omega)$, that follows from (5.24) is presented in Fig. 29.

If the vector $\boldsymbol{\varrho}$ denotes the position of a given proton relative to the center of mass of the molecule containing it, then the velocity of that proton measured relative to the center of mass will be

$$\boldsymbol{\omega}\times\boldsymbol{\varrho}, \quad (5.26)$$

where $\boldsymbol{\omega}$ is that molecule's angular velocity. Since our water model molecules are completely rigid, the

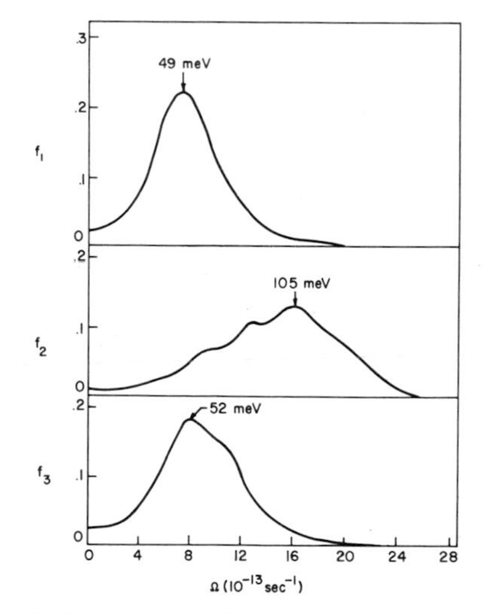

FIG. 28. Frequency spectra of the angular velocity autocorrelation functions.

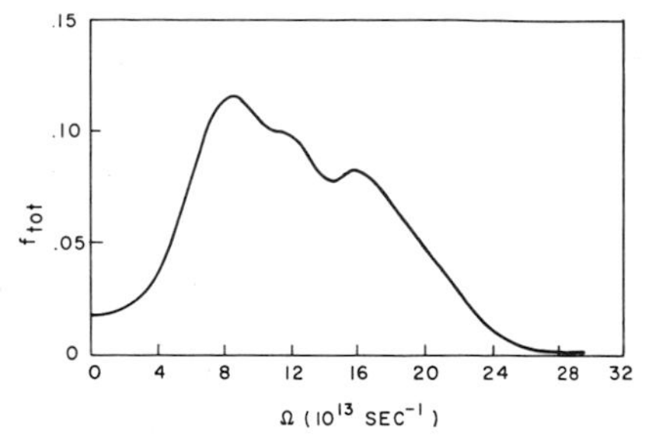

FIG. 29. Frequency spectrum of the total angular momentum autocorrelation function.

length $|\boldsymbol{\varrho}|$ remains fixed so that the proton seems to be moving on a spherical surface when viewed from the center of mass. The appropriate normalized velocity autocorrelation function for description of proton diffusion on the sphere is the following:

$$\phi(t) = \langle[\boldsymbol{\varrho}(0)\times\boldsymbol{\omega}(0)]\cdot[\boldsymbol{\varrho}(t)\times\boldsymbol{\omega}(t)]\rangle / \langle|\boldsymbol{\varrho}\times\boldsymbol{\omega}|^2\rangle, \quad (5.27)$$

and we shall represent its spectral resolution in the usual way:

$$\Phi(\Omega) = \int_0^\infty \phi(t)\cos(\Omega t)\,dt. \quad (5.28)$$

The Fourier transform Φ may be related to the time dependence of the mean square of $\mathbf{u}$, the proton displacement relative to the center of mass. Specifically, it is easy to establish that

$$\langle|\mathbf{u}(t)|^2\rangle = \left(\frac{4}{\pi}\right)\langle|\boldsymbol{\varrho}\times\boldsymbol{\omega}|^2\rangle \int_0^\infty \left(\frac{1-\cos(\Omega t)}{\Omega^2}\right)\Phi(\Omega)\,d\Omega. \quad (5.29)$$

Since the spherical surface upon which the proton must move is bounded, this last quantity must approach a finite limit as $t\to\infty$. Hence $\Phi(0)$ must vanish, and the integral

$$\int_0^\infty \Omega^{-2}\Phi(\Omega)\,d\Omega \quad (5.30)$$

is constrained to a value set by simple geometrical considerations.

The functions $\phi(t)$ and $\Phi(\Omega)$ obtained from the molecular dynamics are shown in Fig. 30. The former demonstrates once again the substantial oscillatory component of proton motion. To evaluate $\Phi(\Omega)$ numerically from $\phi(t)$, an integration cutoff time had to be imposed, which is indicated in Fig. 30 by an arrow. The resulting numerical $\Phi(\Omega)$ fails to vanish at the origin, due to absence of a long time negative tail in $\phi(t)$ that is associated with the eventual ($\approx 5\times$

Notes on p. 354

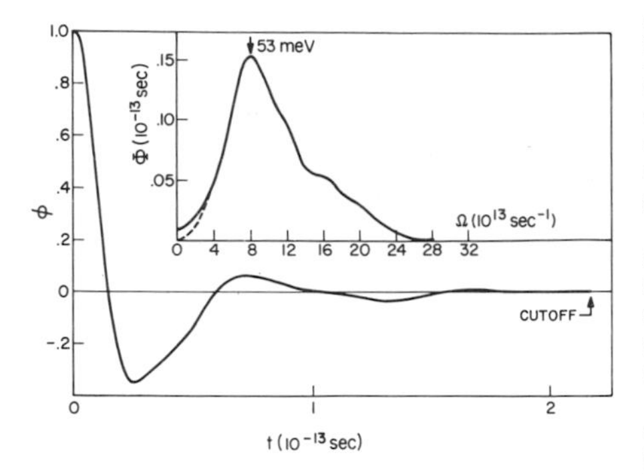

FIG. 30. Normalized velocity autocorrelation function (and its frequency spectrum) for proton motion relative to the center of mass.

10^{-12} sec) turning over of molecules. The dashed curve shown for Φ near the origin represents our estimate of how the accurate Φ (incorporating the effect of the negative ϕ tail) would have to behave, including the fact that integral (5.30) is fixed.

The central quantity probed by neutron inelastic scattering from water is the Van Hove[56] self-correlation function $G_s(r, t)$ for protons. This function gives the spatial distribution at time t (in an ensemble of identically prepared systems) for a proton initially at the measurement origin when $t=0$. Its spatial Fourier transform in the classical limit is given by

$$F_s(k, t) = \langle \exp\{i\mathbf{k}\cdot[\mathbf{r}(t) - \mathbf{r}(0)]\}\rangle. \qquad (5.31)$$

A substantial body of theoretical effort has been devoted to understanding these self-correlation functions in general liquids. A particularly popular approach, the "Gaussian approximation,"[57] has been motivated by the nature of the macroscopic diffusion process. This approximation requires $G_s(r, t)$ to be a Gaussian function in r at all times, with a width chosen to reproduce the correct microscopic mean-square displacement $\langle|\mathbf{r}(t)|^2\rangle$. The equivalent requirement is that $F_s(k, t)$ have the form

$$F_s(k, t) = \exp[-\tfrac{1}{6}k^2\langle r^2(t)\rangle]. \qquad (5.32)$$

Our molecular dynamics calculation enables us to test the validity of Eq. (5.32) directly, since independent calculations of $\langle r^2(t)\rangle$ and of expression (5.31) are possible. Figure 31 graphically shows the test of Eq. (5.32) for three wave vector choices that are consistent with the periodicity cube employed in the dynamics

$$k\sigma = 9.517,\ 14.276,\ 19.034. \qquad (5.33)$$

Although the specific Gaussian approximation (5.32) accounts qualitatively for the behavior of $F_s(k, t)$, it is obvious that substantial quantitative errors arise. For each wave vector, the Gaussian $F_s(k, t)$ decays too rapidly to zero with increasing t in Fig. 31. If one were to force neutron scattering measurements to fit expression (5.32) in this k range, the apparent $\langle r^2(t)\rangle$ would increase too slowly with t; i.e., the apparent self-diffusion constant would be anomalously small.

The failure of approximation (5.32) for intermediate k values is connected to the fact that the Van Hove function $G_s(r, t)$ is not itself Gaussian in r for intermediate times.[58] It is worth recalling that a distinctly non-Gaussian $G_s(r, t)$ has also been found in molecular dynamics calculations on liquid argon.[22]

The fact that the actual $F_s(k, t)$ curves in Fig. 31 decay in time more slowly than their Gaussian analogs is related to the narrowing that has been observed in water for neutron quasielastic scattering peaks.[59,60] One conceivable way in which this narrowing could be explained would be a jump diffusion mechanism, whereby molecules would execute occasional hops of considerable length between quasicrystalline sites of oscillation in the liquid. Indeed theories of precisely this character have been advanced to explain the neutron experiments by Singwi and Sjölander,[61] and Chudley and Elliot.[62] The former authors conclude for example that at 20°C each water molecule oscillates in place for about 4×10^{-12} sec before experiencing rapid diffusion (a "jump") to a new position of oscillation.

In confronting a phenomenon as complicated as proton motion in liquid water must surely be, one runs the risk that experimental data such as neutron scattering can be explained in a variety of ways. Thus, it may be that a jump diffusion mechanism is

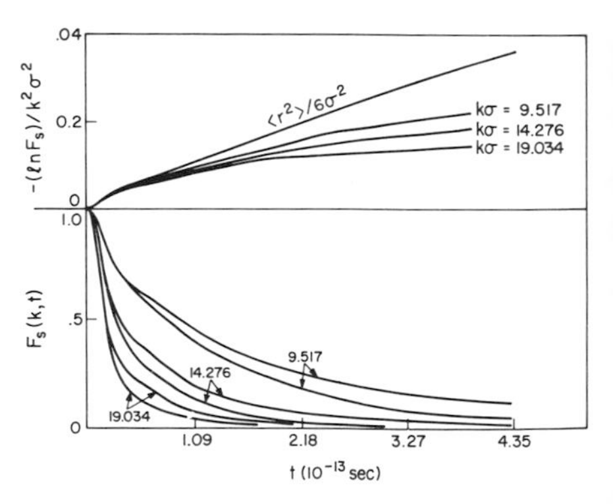

FIG. 31. Spatial Fourier transform $F_s(k, t)$ of the Van Hove self-correlation function for protons. Curves for the three values $k\sigma=9.517$, 14.276, 19.034 are shown. In the lower graph, the Gaussian approximation (5.32) gives the lower curve for each pair. In the upper logarithmic plot, the Gaussian approximation curves are coincident ($\langle r^2\rangle/6\sigma$).

sufficient to explain that data, but not logically *necessary*. In fact, the jump diffusion mechanism definitely conflicts with our molecular dynamics results. Both the temporal correlations, and the sequence of stereo pictures, show that well-defined quasicrystalline sites of residence do not exist in liquid water. The diffusive molecular motions are much more continuous and cooperative and apparently depend strongly upon the distinctive liquid-phase random hydrogen-bond network that is present and forever transforming its topological character.

Sakamoto *et al.*[63] have computed proton mean-square displacements vs time by Fourier-transforming neutron scattering data. Although their results have been interpreted as further support for the jump diffusion mechanism[64] such an interpretation is subject to exactly the same uniqueness criticism. For the diffusion times probed by the analysis of Sakamoto *et al.* (2×10^{-13} to 10^{-11} sec), the mean-square proton displacements are similar to those shown in Fig. 21 for the molecular dynamics. In this time range, the only clear distinctive feature exhibited by both approaches is that the proton displacement curve lies *above* the straight line passing through the origin with slope corresponding to the correct D. In order to effect a discriminating experimental test of diffusion mechanisms in liquid water, sufficient improvement in experimental technique is required to examine times of the order of 10^{-14} sec accurately.

As a final aspect of water molecule kinetics, we mention that the velocity autocorrelation for protons (within present accuracy) is found to be the sum of the molecular center-of-mass autocorrelation (Fig. 22), and the "motion on the sphere" autocorrelation (Fig. 30). Thus the translational and rotational motions of the molecules seem to proceed independently of one another, on the average. Figure 32 therefore exhibits the total proton motion spectrum that was obtained by linear combination of the spectra reported in Figs. 22 and 30.

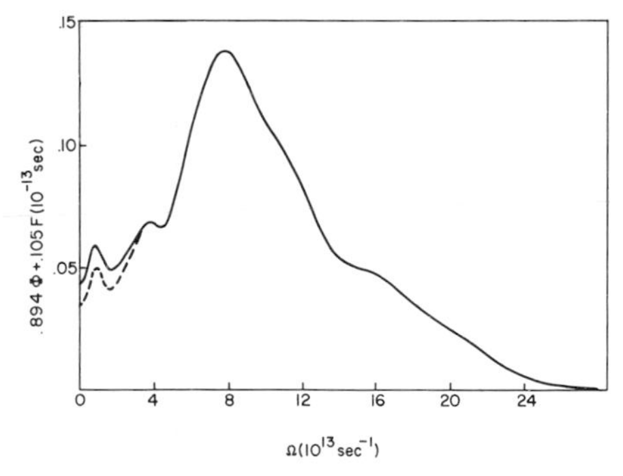

FIG. 32. Frequency spectrum for proton total velocity autocorrelation function. As the legend indicates, this was obtained as a linear combination of the spectra in Figs. 22 and 30.

VI. TEMPERATURE VARIATIONS

In addition to the one temperature studied at length in this paper, 34.3°C, our water model needs and deserves to be examined at several other temperatures to allow instructive comparisons to be carried out. These extensions are, in fact, underway and will be reported in due course. For the present, we shall only briefly mention a low temperature run to add perhaps some more credibility to our water model.

This new run involved the same condition as its predecessor ($N=216$, 18.62 Å cubical box with periodic boundary conditions), but the temperature was only 265°K (−8.2°C). The system therefore corresponded to liquid water in a state of moderate supercooling, but under the circumstances that prevail, essentially no chance existed for the liquid to nucleate and freeze into ice. The run lasted 4000 Δt.

The oxygen-nucleus pair correlation function $g_{OO}^{(2)}(r)$ computed for this lower temperature superficially resembles the one shown in Fig. 3. It approaches unity with increasing r rapidly enough again to exclude bulky "clusters" or "icebergs" from serious consideration. Furthermore, the distance ratio for second and first peaks is rather close to the ideal value for tetrahedral hydrogen bonding. However, the first maximum of $g_{OO}^{(2)}(r)$ is significantly larger (2.97, compared to 2.56 in Fig. 3) and the succeeding minimum deeper. The second maximum appears to be better developed and narrower. Evidently the random hydrogen-bond network has tightened up considerably, by discriminating against severe distortions of the ideal tetrahedral coordination geometry.

The stereo pictures that have been produced from this lower temperature run bear out that conclusion. The hydrogen bonds that form between neighboring molecules appear to be more nearly linear and to observe the tetrahedral angles of approach more frequently. Although the bonds in the random network tend to strengthen at lower temperature, this does not imply that it is easier then to identify a few "interstitial" molecules; as before we see no clear examples of any interstitials. Also, the liquid seems to show no marked tendency anywhere to organize the beginnings of ice nuclei.

To the extent that x-ray scattering experiments predominately reflect $g_{OO}^{(2)}(r)$, our structural shifts with temperature variation agree qualitatively with measurements reported by Narten, Danford, and Levy.[29]

Naturally the strengthening of the interactions tends to slow down molecular diffusive motions markedly. The self-diffusion constant D at 265°K is found to be approximately 1.5×10^{-5} cm²/sec for the molecular dynamics simulation. Once again this seems to be rather larger than the experimental value 0.75±

Notes on p. 354

0.03×10^{-5} cm²/sec,[65] but comparison of the ratio of values at -8.2 and 34.3°C (0.36 theoretical, 0.26 experimental) indicates roughly the necessary rapid temperature variation for D. Furthermore, the simple energy rescaling discussed in the following section induces far better agreement with the measured D values at both rescaled molecular dynamics temperatures.

In order to round out the study of temperature variations, we intend to study our model up to the neighborhood of the critical temperature (374°C). This should provide a comprehensive account of the thermal disruption of the hydrogen-bond network, and should permit a reasonably accurate evaluation of the constant volume heat capacity.

VII. DISCUSSION

In order to continue systematically applying the molecular dynamics method to water, two types of extensions of the present work are necessary. The first involves expanding the domain of application of the model employed here to include a wide range of densities and of temperatures (mentioned in the preceding section), as well as to include solvation and interface studies. The second type of extension is the analysis of ways in which the effective pair interaction $V_{eff}^{(2)}$ should be modified to yield a more accurate description of real water. These two aspects need to be carried forward in parallel, since accurate determination of the properties implied by a given Hamiltonian, in comparison with experimental results, provides the basis for modification.

With respect to future study of aqueous solutions, it should be mentioned that the model employed in this paper can be immediately adapted to simulation of a particularly simple solute. By retaining only v_{LJ} [see Eq. (2.5)] in the effective pair interaction between a chosen molecule and all the others, that molecule should reasonably well behave as a neon atom if in addition the full mass were placed at its center. This single solute particle of course could not hydrogen bond to its neighbors. One would then be particularly interested to see if the surrounding water molecules organized themselves into a sort of "cage" analogous to those present in clathrates. The results would be especially interesting in view of the "hydrophobic bond" concept that has been proposed to explain the interaction of nonpolar solutes in water.[66,67]

In order to broaden the scope of solution studies to include other solutes, information will be required about the interaction of water molecules with a variety of ions and molecules. Extensive quantum mechanical calculations would have considerable value in determining characteristic shapes of potential surfaces for water molecules in interaction with distinct chemical groupings, e.g., methyl groups, carbonyl groups, hydroxyls, conjugated double bonds between carbon atoms, amines, etc. With such information in hand, it would become possible to design molecular dynamics calculations to study the hydration of biological macromolecules and the interaction of water with membranes.

The simplest of all possible modifications that might be applied to the effective potential would be the strength rescaling mentioned briefly at the end of Sec. III. If one were to choose the interaction energy $\langle V_N\rangle/N$ and the self-diffusion constant D as having central importance, then a rescaling factor

$$1+\zeta=1.06 \tag{7.1}$$

would have the effect of changing the temperature to 52.8°C, while inducing much better agreement than previously between calculations and experiment both for mean interaction energy:

$$\begin{aligned}\langle V_N\rangle/N &= -9.735 \text{ kcal/mole (molecular dynamics rescaled)},\\ &= -9.63 \text{ kcal/mole (exptl, 52.8°C)}, \end{aligned} \tag{7.2}$$

and for the self-diffusion constant:

$$\begin{aligned}D &= 4.3\times10^{-5} \text{ cm}^2\text{/sec (molecular dynamics rescaled)},\\ &= 4.1\times10^{-5} \text{ cm}^2\text{/sec (exptl, 52.8°C)}. \end{aligned} \tag{7.3}$$

At the same time, of course, the rotational relaxation times τ_1 and τ_2 would increase by $(1+\zeta)^{1/2}$, making them apparently agree less well with experiment. But since the theoretical frameworks are weak in which τ_1 and τ_2 are related to dielectric and NMR experiments, this disagreement should probably be given relatively little weight at present.

Energy rescaling also affects the value predicted by molecular dynamics for the Kirkwood orientational correlation factor. When the measured ϵ_0 for water at 52.8°C is inserted in Eq. (4.17) to convert G_K to g_K, one obtains

$$g_K=2.72, \tag{7.4}$$

somewhat smaller than the "unscaled" result 2.96 in Eq. (4.19). The previously cited work of Harris, Haycock, and Alder,[34] which suggests g_K should be in the range 2.5–2.6 at this temperature, also supports the rescaling.

The same energy rescaling factor 1.06 seems tentatively also to produce considerable improvement in agreement with experiment for the low temperature run mentioned in the preceding Section. Although the nominal temperature 265°K (-8.2°C) for that run corresponded to supercooled water, the rescaled temperature 280.9°K (7.7°C) lies above the melting point of ice. Eventually, it would be very interesting to carry out a molecular dynamics simulation on a strongly supercooled water sample (even on the rescaled basis), to see what type of local order arises.

Beside the strength rescaling for $V_{eff}^{(2)}$, a length rescaling is also possible. This would produce a shift in density (rather than temperature) and would require reinterpretation of all quantities dependent upon the length unit. At present no compelling evidence motivates such a distance rescaling.

Probably the principal criticism which might be directed toward our choice for $V_{eff}^{(2)}$ is that it is somewhat too "tetrahedral." However, it is significant to have observed that even in the presence of this tetrahedral bias, the local structure produced in our model "water" still manifests very substantial randomness, and fails to mimic known crystal structures. It is therefore quite unlikely that a less tetrahedral $V_{eff}^{(2)}$, at a given temperature, would be more successful in building liquid-phase networks akin to the ice lattice or the clathrates at the local level.

Barker and Watts[68] have carried out a Monte Carlo calculation for $N=64$ water molecules (at 25°C), using the Rowlinson[69] pair potential. Like our own effective pair potential, the Rowlinson interaction is based on a four-point charge complex for each molecule. However, the positions of these charges do not lead to a natural tetrahedral arrangement of neighbors. As a result, the computed oxygen-nucleus pair correlation function tends to have a rather large number of nearest neighbors (6.4 out to distance 3.5 Å), and the positions of first and second maxima are far from the ideal tetrahedral ratio. Although the Rowlinson potential may provide a reasonable account of the interaction for isolated pairs of water molecules in the vapor, it probably deviates too far from the tetrahedral directionality required of a condensed phase $V_{eff}^{(2)}$ that must lead to essentially universal fourfold coordination in aqueous crystals.

A possible modification of our own $V_{eff}^{(2)}$, Eq. (2.5), which does not seriously compromise its directionality, would be to shorten the distance from the oxygen nucleus to the *negative* point charges, without changing any angles or the distance to the positive charges. At the same time it would be necessary to adjust the parameters η, R_L, and R_U, to maintain the strength and length of undistorted hydrogen bonds. This change would permit greater freedom in approach direction for formation of linear hydrogen bonds. The mean interaction energy in the liquid should thereupon increase in magnitude. It must be left for future investigation, however, to determine what the concomitant effects on static correlation functions and kinetic properties would be.

ACKNOWLEDGMENT

The authors are indebted to Dr. Seymour H. Koenig, IBM Watson Laboratory, for arranging a generous grant of computer time at the IBM Computation Center to carry out the low temperature run reported briefly in Sec. VI.

APPENDIX A

Given $\mathbf{x}_i$ and $\mathbf{x}_j$ [see Eq. (3.1)] and the positions of the oxygen and four point charges in each molecule, we can calculate the list of 17 distances: r_{ij}, the oxygen–oxygen distance; and 16 point charge distances $d_{\alpha_i\alpha_j}$. Then $V_{eff}^{(2)}$ gives rise to a force vector of magnitude:

$$24(\epsilon/r_{ij})[2(\sigma/r_{ij})^{12}-(\sigma/r_{ij})^6]$$
$$-6[(r_{ij}-R_L)(R_U-r_{ij})/(R_U-R_L)^3]v_{el}(\mathbf{x}_i,\mathbf{x}_j) \quad \text{(A1)}$$

acting between the two oxygen nuclei. The term with v_{el} [Eq. (2.8)] is present only for $R_L\leq r_{ij}\leq R_U$ [see Eq. (2.9)].

Each pair of point charges in the respective molecules gives a force vector of magnitude

$$S(r_{ij})(0.19e)^2(-1)^{\alpha_i+\alpha_j}/d_{\alpha_i\alpha_j}^2 \quad \text{(A2)}$$

acting between that pair.

The Cartesian components of these forces are obtained simply by multiplying the magnitudes by ratios of the type $(x_{0i}-x_{0j})/r_{ij}$ and $(x_{\alpha_i}-x_{\alpha_j})/d_{\alpha_i\alpha_j}$.

These 17 forces then give, by a summation of respective components, the total body force acting on a molecule. Using the transformation matrix which rotates xyz into $x'y'z'$ (Fig. 2), the components of the torque $\mathbf{N}_j$ [Eq. (3.5)] can also be calculated from the components of the force acting on each of the five points of a molecule (the positions of the oxygen nucleus and the four point charges). Adding the inertial terms [Eq. (3.6)] gives the derivatives of the angular velocities about the principal axes of inertia.

To indicate the scheme[70] we have used for solving the Newton–Euler equations of motion, we shall take the following three differential equations as prototypes for Eqs. (3.3), (3.6), and (3.7), respectively,

$$d^2x/dt^2=f(x,p),$$
$$dq/dt=g(x,p,q),$$
$$dp/dt=h(p,q). \quad \text{(A3)}$$

Here x typifies center of mass Cartesian coordinates, q is the angular velocities about the principal axes of inertia, and p is the Euler angles.

The problem is to get x, p, q at $t+\Delta t$ knowing the values at time t. Assume that the calculation has already been initiated so that, apart from x, p, q at t, we also know the first five derivatives of x and the first four of p and q. Let x_n, e.g., denote

$$[(\Delta t)^n n!]d^nx/dt^n.$$

Notes on p. 354

We predict the values of all these quantities at $t+\Delta t$ by using the Pascal triangle:

$$x_0' = x_0 + x_1 + x_2 + x_3 + x_4 + x_5,$$
$$x_1' = x_1 + 2x_2 + 3x_3 + 4x_4 + 5x_5,$$
$$x_2' = x_2 + 3x_3 + 6x_4 + 10x_5,$$
$$x_3' = x_3 + 4x_4 + 10x_5,$$
$$x_4' = x_4 + 5x_5, \qquad \text{(A4)}$$

and similarly for $p_0' \cdots p_3'$ and $q_0' \cdots q_3'$.

Using the predicted values x_0', p_0', q_0' we calculate $f(x_0', p_0')$, $g(x_0', p_0', q_0')$, and $h(p_0', q_0')$. Denote these values by f', g', h'.

The differences,

$$A = [(\Delta t)^2/2!]f' - x_2',$$
$$B = (\Delta t)g' - q_1',$$
$$C = (\Delta t)h' - p_1', \qquad \text{(A5)}$$

are then used to correct the predicted values in the following way:

$$x_n(t+\Delta t) = x_n' + f_{n2}A \qquad (n=0, 1, \cdots, 5),$$
$$\left.\begin{aligned} p_m(t+\Delta t) &= p_m' + f_{m1}B \\ q_m(t+\Delta t) &= q_m' + f_{m1}C \end{aligned}\right\} \qquad (m=0, 1, \cdots, 4), \qquad \text{(A6)}$$

where

$$f_{02}=3/16, \quad f_{12}=251/360, \quad f_{22}=1,$$
$$f_{32}=11/18, \quad f_{42}=1/6, \quad f_{52}=1/60, \qquad \text{(A7)}$$

and

$$f_{01}=251/720, \quad f_{11}=1, \quad f_{21}=11/12,$$
$$f_{31}=1/3, \quad f_{41}=1/24. \qquad \text{(A8)}$$

It will be seen from Ref. 70 that these coefficients depend on the "order" of the procedure used. In our molecular dynamics runs on water we have used a fifth-order procedure for the center-of-mass motion and a fourth-order one for angular motion.

At the start of the calculation the most convenient procedure is to take all the derivatives equal to zero. In the case of interest here (Newton–Euler equations), the angular velocities (typified by q in the schematic presentation) can also be equated to zero at the start of the calculation. The "aging" of the run then necessarily leads to a solution effectively unrelated to the specific starting procedure.

APPENDIX B

In the convention utilized for the molecular dynamics calculations, each molecule experiences forces and torques due only to those other molecules within a cutoff radius R. In the presence of a uniform, weak, external electric field $\mathbf{E}_{\text{ext}}$, the system will develop a uniform polarization $\mathbf{P}$. Classical electrostatics requires

$$4\pi\mathbf{P} = (\epsilon_0 - 1)\mathbf{E}, \qquad \text{(B1)}$$

where the electric field $\mathbf{E}$ is composed both of the imposed field $\mathbf{E}_{\text{ext}}$, and the internal field $\mathbf{E}_{\text{int}}$ due to the nonvanishing polarization

$$\mathbf{E} = \mathbf{E}_{\text{ext}} + \mathbf{E}_{\text{int}}. \qquad \text{(B2)}$$

The contribution of polarization in a small volume element δv at position $\mathbf{r}'$, to the internal field at position $\mathbf{r}$, is the following:

$$-\nabla_r[\delta v \mathbf{P}(\mathbf{r}')]\cdot\nabla_{r'}(|\mathbf{r}-\mathbf{r}'|^{-1}). \qquad \text{(B3)}$$

In writing an expression for the polarization $\mathbf{P}(\mathbf{r})$ for molecules at $\mathbf{r}$, that is consistent with the molecular dynamics, we clearly must integrate quantity (B3) only over the sphere of radius R surrounding position $\mathbf{r}$:

$$4\pi\mathbf{P}(r) = (\epsilon_0 - 1)\Big[\mathbf{E}_{\text{ext}} - \int_{|\mathbf{r}-\mathbf{r}'|\leq R} d\mathbf{r}'\nabla_r\mathbf{P}(\mathbf{r}')\cdot\nabla_{r'}(|\mathbf{r}-\mathbf{r}'|^{-1})\Big]. \qquad \text{(B4)}$$

The integral may easily be evaluated to yield:

$$4\pi\mathbf{P} = (\epsilon_0 - 1)[\mathbf{E}_{\text{ext}} - (4\pi/3)\mathbf{P}], \qquad \text{(B5)}$$

or

$$\mathbf{P} = [3(\epsilon_0 - 1)/4\pi(\epsilon_0 + 2)]\mathbf{E}_{\text{ext}}. \qquad \text{(B6)}$$

Not surprisingly, relation (B6) is exactly the same one that applies to the polarization of a spherical dielectric sample with a real boundary surface, placed in a vacuum region initially containing $\mathbf{E}_{\text{ext}}$ only. We may therefore call upon the Kirkwood theory of polar dielectrics,[30] which applies to those spherical specimen conditions.

Irrespective of boundary conditions, the polarization consists of a part $\mathbf{P}_\alpha$ due to induced molecular moments, plus a part due to reorientations that is proportional to $\langle M^2\rangle$, the mean-square system moment in the absence of external fields[71]:

$$\mathbf{P} = \mathbf{P}_\alpha + (\beta/3V)\langle M^2\rangle\mathbf{E}_{\text{ext}}, \qquad \text{(B7)}$$

where V stands for the system volume. The import of expressions (B6) and (B7) taken together is that for a given ϵ_0, the quantity $\langle M^2\rangle/V$ should be the same for the molecular dynamics situation as it is for a spherical dielectric sample in a vacuum.

Kirkwood[72] has shown that the local moment $\mathbf{m}$ of a fixed molecule and its immediate surroundings, in a dielectric sphere, is related to the average moment $\mathbf{M}$ of that sphere by

$$\mathbf{M} = [9\epsilon_0/(\epsilon_0+2)(2\epsilon_0+1)]\mathbf{m}. \qquad \text{(B8)}$$

It is precisely the distinction between these two moments which G_K and g_K for the dielectric sphere reflects, so we must have

$$G_K = [9\epsilon_0/(\epsilon_0+2)(2\epsilon_0+1)]g_K. \qquad \text{(B9)}$$

In view of the equivalence of the dielectric sphere to

the molecular dynamics system, the same relation applies to G_K and g_K for the latter, as shown by Eq. (4.17).

One should not take these considerations to mean that when $\mathbf{E}_{ext}$ vanishes, the polarization density surrounding fixed molecules in the two types of systems is the same (even if R for the molecular dynamics equals the dielectric sphere radius).

* Part of the work carried out at the Argonne National Laboratory was supported by the U.S. Atomic Energy Commission.

[1] A good review of the current situation is provided by D. Eisenberg and W. Kauzmann, *The Structure and Properties of Water* (Oxford U. P., New York, 1969).

[2] K. Morokuma and L. Pederson, J. Chem. Phys. **48,** 3275 (1968).

[3] P. A. Kollman and L. C. Allen, J. Chem. Phys. **51,** 3286 (1969).

[4] K. Morokuma and J. R. Winick, J. Chem. Phys. **52,** 1301 (1970).

[5] G. H. F. Diercksen, Chem. Phys. Letters **4,** 373 (1969).

[6] J. Del Bene and J. A. Pople, Chem. Phys. Letters **4,** 426 (1969).

[7] D. Hankins, J. W. Moskowitz, and F. H. Stillinger, Chem. Phys. Letters **4,** 527 (1970).

[8] J. Del Bene and J. A. Pople, J. Chem. Phys. **52,** 4858 (1970).

[9] D. Hankins, J. W. Moskowitz, and F. H. Stillinger, J. Chem. Phys. **53,** 4544 (1970).

[10] N. Metropolis, A. W. Rosenbluth, M. N. Rosenbluth, A. H. Teller, and E. Teller, J. Chem. Phys. **21,** 1087 (1953).

[11] B. J. Alder and T. E. Wainwright, J. Chem. Phys. **33,** 1439 (1960).

[12] S. W. Peterson and H. A. Levy, Acta Cryst. **10,** 70 (1957).

[13] J. D. Bernal and R. H. Fowler, J. Chem. Phys. **1,** 515 (1933).

[14] L. Pauling, J. Am. Chem. Soc. **57,** 2680 (1935).

[15] If interaction of the molecules with a container wall had to be explicitly considered, we would include single-molecule terms $V^{(1)}(\mathbf{x}_i)$ in this development. These extra potentials are unnecessary in the present context, however.

[16] F. H. Stillinger, J. Phys. Chem. **74,** 3677 (1970).

[17] A. Ben-Naim and F. H. Stillinger, "Aspects of the Statistical-Mechanical Theory of Water," in *Structure and Transport Processes in Water and Aqueous Solutions*, edited by R. A. Horne (Wiley–Interscience, New York, to be published).

[18] J. Corner, Trans. Faraday Soc. **44,** 914 (1948).

[19] The nuclear positions specified here do not conform exactly to the stable geometry of isolated water molecules. In particular, our O–H bonds are slightly too long (1 Å vs 0.957 Å measured length), and the bond angle somewhat too large (109° 28′ for the perfect tetrahedron vs 104° 31′ measured). However, there is experimental evidence that water molecule interactions in condensed phases may increase the average bond length (Peterson and Levy, Ref. 12) and theoretical evidence that they increase the average bond angle (Hankins, Moskowitz, and Stillinger, Ref. 9).

[20] H. Goldstein, *Classical Mechanics* (Addison-Wesley, Cambridge, Mass., 1953), p. 158.

[21] Reference 20, p. 134.

[22] A. Rahman, Phys. Rev. **136,** A405 (1964).

[23] It would be possible to restore energy conservation by forcing the cutoff to arise from a rapid decline in $S(r_{ij})$ from 1 to 0 when r_{ij} was close to 3.25σ. But this would entail unphysically large forces and torques at that large separation, which we consider to be undesirable.

[24] Some additional test runs were performed at the Bell Telephone Laboratories, Murray Hill, N.J., using a GE 635 computer.

[25] F. H. Stillinger and A. Ben-Naim, J. Phys. Chem. **73,** 900 (1969).

[26] H. S. Frank and W. Y. Wen, Discussions Faraday Soc. **24,** 133 (1957).

[27] G. Némethy and H. A. Scheraga, J. Chem. Phys. **36,** 3382 (1962).

[28] A. H. Narten and H. A. Levy, Science **165,** 447 (1969).

[29] A. H. Narten, ONRL Report No. ONRL-4578, July 1970. We are indebted to Dr. Narten for providing us with the atomic scattering factors used in the x-ray data treatment, whose original sources are identified in Ref. 29 of his report.

[30] J. G. Kirkwood, J. Chem. Phys. **7,** 911 (1939).

[31] The full pair correlation function $g^{(2)}(\mathbf{x}_1, \mathbf{x}_2)$ is normalized so as to approach $(8\pi^2)^{-1}$ at large separation r_{12}. In Eqs. (4.15) and (4.16), $\rho = \rho_0$ is the molecular number density.

[32] C. A. Coulson and D. Eisenberg, Proc. Roy. Soc. (London) **A291,** 444 (1966).

[33] L. Onsager and M. Dupuis, in *Electrolytes*, edited by B. Pesce (Pergamon, New York, 1962), p. 27.

[34] F. E. Harris, E. W. Haycock, and B. J. Alder, J. Chem. Phys. **21,** 1943 (1953); see also F. E. Harris and B. J. Alder, *ibid.* **21,** 1031 (1953).

[35] The members of the pair could simultaneously act either as proton donors to the third or as proton acceptors from the third.

[36] Realist Stereo Viewer, Realist, Inc., Menomonee, Wis.

[37] J. D. Worley and I. M. Klotz, J. Chem. Phys. **45,** 2868 (1966).

[38] G. E. Walrafen, J. Chem. Phys. **48,** 244 (1968).

[39] G. E. Walrafen, in *Hydrogen-Bonded Solvent Systems*, edited by A. K. Covington and P. Jones (Taylor and Francis, London, 1968), pp. 9–29.

[40] L. Pauling, *The Nature of the Chemical Bond* (Cornell U. P., Ithaca, N.Y., 1960), 3rd ed., p. 473.

[41] M. D. Danford and H. A. Levy, J. Am. Chem. Soc. **84,** 3965 (1962).

[42] O. Ya. Samoilov, *Structure of Aqueous Electrolyte Solutions and the Hydration of Ions* (Consultants Bureau, New York, 1965).

[43] These convex cages are illustrated by the clathrates in idealized form. Only by rotating its hydrogen bonds into eclipsed configurations can a cavity of sufficient size be built to accommodate the solute molecules.

[44] Reference 1, p. 89.

[45] R. Hausser, G. Maier, and F. Noack, Z. Naturforsch. **21a,** 1410 (1966).

[46] J. S. Murday and R. M. Cotts, J. Chem. Phys. **53,** 4724 (1970).

[47] The numerical calculation assumed that the autocorrelation function vanished beyond the "cutoff."

[48] J. G. Powles, J. Chem. Phys. **21,** 633 (1953).

[49] T.-W. Nee and R. W. Zwanzig, J. Chem. Phys. **52,** 6353 (1970).

[50] Reference 1, p. 207.

[51] S. G. Brush, Rev. Mod. Phys. **33,** 79 (1961).

[52] N. Bloembergen, E. M. Purcell, and R. V. Pound, Phys. Rev. **73,** 679 (1948).

[53] K. Krynicki, Physica **32,** 167 (1966).

[54] K. S. Cole and R. H. Cole, J. Chem. Phys. **9,** 341 (1941).

[55] E. H. Grant, T. J. Buchanan, and H. F. Cook, J. Chem. Phys. **26,** 156 (1957).

[56] L. Van Hove, Phys. Rev. **95,** 249 (1954).

[57] These Gaussian approximations are reviewed by A. Sjölander, in *Thermal Neutron Scattering*, edited by P. Egelstaff (Academic, New York, 1965), Chap. 7.

[58] It is however always Gaussian in both the short and long time limits.

[59] B. N. Brockhouse, Phys. Rev. Letters **2,** 287 (1959).

[60] D. J. Hughes, H. Palevsky, W. Kley, and E. Tunkelo, Phys. Rev. **119,** 872 (1960).

[61] K. S. Singwi and A. Sjölander, Phys. Rev. **119,** 863 (1960).

[62] C. T. Chudley and R. J. Elliott, Proc. Phys. Soc. (London) **77,** 353 (1961).

[63] M. Sakamoto, B. N. Brockhouse, R. G. Johnson, and N. K. Pope, J. Phys. Soc. Japan Suppl. B **17,** 370 (1962).

[64] Reference 1, p. 221.

[65] We are indebted to Dr. K. Gillen and Dr. D. C. Douglass for providing this value prior to publication.

[66] W. Kauzmann, Advan. Protein Chem. **14,** 1 (1959).

[67] G. Némethy, Angew. Chem. **6,** 195 (1967).

[68] J. A. Barker and R. O. Watts, Chem. Phys. Letters **3,** 144 (1969).

[69] J. S. Rowlinson, Trans. Faraday Soc. **47,** 120 (1951).

[70] C. W. Gear, ANL Report No. ANL-7126, 1966.

[71] F. E. Harris and B. J. Alder, Ref. 34.

[72] Reference 30, Eq. (12).

Notes on p. 354

Notes to Reprint V.1

1. The effects of temperature (p. 3337) were later discussed by Stillinger and Rahman (J. Chem. Phys. 57 (1972) 1281).

2. In solving the Euler equations (3.6), Rahman and Stillinger used the trick, described in the introduction to this chapter, whereby the coordinate frame attached to a molecule i is redefined whenever the angle β_i approaches the value 0 or π. Details of the method were first given in a paper on the simulation of liquid nitrogen by Barojas et al. (Phys. Rev. A 7 (1973) 1092).

3. Later molecular dynamics calculations on liquid water have used a time step larger by an order of magnitude than that defined by eq. (3.8).

4. The oxygen–oxygen pair distribution function derived from the X-ray scattering data of ref. [29] was published in a paper (Narten and Levy, J. Chem. Phys. 55 (1971) 2263) that appeared almost simultaneously with that of Rahman and Stillinger. The results are in fair agreement with those of fig. 3, but some improvement is obtained in simulations based on more recently proposed potential models (see, for example, Jorgensen et al., J. Chem. Phys. 79 (1983) 926).

5. The modification of the potential that is suggested in the last paragraph of Section VII was the source of the famous ST2 model of water (Stillinger and Rahman, J. Chem. Phys. 60 (1974) 1545).

6. Neumann (Reprint V.7) has proved that for a simulation carried out with the spherical cutoff convention the quantity $\langle |\mathbf{M}|^2 \rangle / N$ does have the same value as that for a sphere *in vacuo*, as anticipated in Appendix B, but only if $\mathbf{M}$ is calculated by summing over the full periodic cell, not merely over the truncation sphere.

MOLECULAR PHYSICS, 1982, VOL. 47, NO. 6, 1253–1264

Molecular dynamics of rigid systems in cartesian coordinates A general formulation

by G. CICCOTTI†
SRMP–CEN–Saclay, B.P. No 2, 91191 Gif-sur Yvette, Cedex, France

M. FERRARIO and J.-P. RYCKAERT
Pool de Physique, Universite Libre de Bruxelles, Campus Plaine, CP 223, Boulevard du Triomphe, 1050 Bruxelles, Belgium

(*Received* 23 *March* 1982 ; *accepted* 24 *May* 1982)

The dynamics of rigid polyatomic systems, either molecules or rigid portions of large molecules, is described by cartesian equations of motion for its atoms. In comparison with the original version of the method of constraints [1], the present approach has more general applicability, on account of a proper choice of holonomic constraints. Moreover, it always leads to a more efficient molecular dynamics simulation of polyatomic molecules. The dynamics of a rigid unit is provided by the motion of a ' basic ' subset of atoms : two for a linear molecule, three for a planar one, four for a tri-dimensional one. Explicit cartesian equations of motion of the ' basic ' atoms are derived from first principles. Verlet algorithm is shown to be particularly advantageous in their numerical integration. Illustrations are given for various molecular geometries, i.e. liquid CS_2, benzene and CCl_4.

1. INTRODUCTION

Among different molecular dynamics (MD) techniques appropriate to simulations of molecular liquids, the method of constraints [1] has been intensively exploited for rigid (or partly rigid) molecules interacting *via* atomic centres of forces. Originally applied to model n-alkanes, it has been used for a large variety of systems ranging from diatomic molecules to small proteins [2–5]. This method is based on the cartesian equations of motion of the individual atoms. The total force on a given particle appears as the sum of the force deriving from the potential energy and the force arising from the constraints. In the original method, [1] and in its recently modified versions, [6, 7] the rigidity always follows from the imposition of bond constraints between atomic pairs. As a result, the constraint force is in general a sum of restoring forces directed along the bonds.

Such constraints proved to be inadequate to assure the rigidness of some particular systems. The most notorious example is the linear triatomic molecule where it is impossible to define the four needed constraint relations by fixing interatomic distances. More generally, using bond constraints only, one can

† On leave of absence from : Ist. di Fisica ' G. Marconi ', Piazzale A. Moro 5, 00185, Rome, Italy.

0026–8976/82/4706 1253 $04·00

never recover the correct number of degrees of freedom for a linear molecule consisting of more than two atoms. In the case of planar molecules it is always possible to be left with six degrees of freedom after imposing bond constraint relations. However, as all bonds lie in the plane of the molecule, the component of the total constraint force orthogonal to the molecular plane vanishes for each atom. On the other side intermolecular forces on each atom are not subjected to any limitations. Therefore it is impossible to maintain the planar structure of molecules consisting of more than three atoms. Numerically one observes that the constraint equations cannot be satisfied for such systems because the constraint matrix is singular. We have also observed cases of singular constraint matrices for three-dimensional molecules having a group of more than three coplanar (or of more than two colinear) atoms.

Whenever it applies, the old constraint method presents further technical disadvantages. The use of bond constraints to maintain the rigidity of polyatomic molecules requires the explicit integration of three cartesian equations of motion per atom subjected to 3n–6 constraint relations. For large number of atoms the constraint matrix involves large computer memory and its inversion takes considerable computer time. These problems can be simplified but not circumvented by using the alternative procedure called SHAKE, described in [1].

In this paper we choose an appropriate set of holonomic constraints which are in part rigid bonds, in part linear relations in the atomic cartesian coordinates. The dynamics of the whole molecule can then be reduced to the motion of a basic subset of atoms : two for a linear molecule, three for a planar, four for a tridimensional one. The dynamical equations for the basic set have the structure of the constrained cartesian equations of motion for a diatomic, a non-linear triatomic and a tetrahedral molecule, respectively.

The main advantages of the present approach can be summarized as follows : (i) all kinds of rigid molecules can be treated by the method of constraints in a unified way ; (ii) the constraint relations to be numerically satisfied are always no more than six per rigid unit of atoms ; (iii) the cartesian equations of motion to be explicitly integrated are reduced to those of the basic set (no more than 12 scalar differential equations).

In § 2 and 3, we introduce the new constraint relations and the corresponding cartesian equations of motion for the basic set. § 4 presents their numerical implementation while in § 5 the method is adapted to molecular models including centres of force distinct from atomic positions. It might be useful for the impatient reader to read at first § 6 where some illustrative examples are discussed in detail. Some short comments are given in § 7.

2. A suitable set of holonomic constraints for rigid molecules

The use of partly rigid molecular models in MD simulations is now well-established. This follows from the possibility of decoupling the fast internal vibrations of polyatomic molecules from the rotational and translational degrees of freedom. Either generalized coordinates or cartesian coordinates of the individual atoms plus a suitable set of constraints can be used to describe the dynamics of such molecular systems. In the second case the description of a polyatomic unit (the whole molecule or one of its rigid parts) can be achieved by different sets of constraints. In this section we introduce a particularly simple set of (linear) constraints for K-dimensional ($K=1, 2, 3$) rigid units consisting of n point particles (atoms).

2.1. *Tridimensional molecules*

Consider a tridimensional molecule consisting of more than four particles. Choose four non-coplanar particles and connect them by six bond constraint relations, i.e. constraints fixing interparticle distances. With these four particles, we define a basic structure having six degrees of freedom, i.e. the number of degrees of freedom required to describe the whole molecule.

All the other particles can be linked to the basic structure by adding for each one three bond constraints connecting it to three out of the four particles of the basic structure. If then there are non-degeneracies, i.e. when none of the added particles lies in the plane of the three chosen basic particles, all constraints are well defined and one does not add, as expected, extra degrees of freedom. This description allows to apply the technique of [1] to integrate the equations of motion.

Alternatively, linear vectorial holonomic constraints can be used to connect to the basic structure the remaining particles. Let us consider the three (linearly independent) vectors $\mathbf{R}_{i1}=\mathbf{R}_i-\mathbf{R}_1, i=2,4$, where $\mathbf{R}_i, i=1, 4$, denotes the coordinates of the ith particle of the basic structure ; the position $\mathbf{r}_\alpha$ of any other particle of the molecule can be written as :

$$\begin{aligned}\mathbf{r}_\alpha &= \mathbf{R}_1+C_{\alpha 2}\mathbf{R}_{21}+C_{\alpha 3}\mathbf{R}_{31}+C_{\alpha 4}\mathbf{R}_{41} \\ &= \sum_{i=1}^{4} C_{\alpha i}R_i \quad \alpha=1,\ n-4,\end{aligned} \tag{1}$$

where the $\{C_{\alpha i}\}$, $i=1, 4$ follow from the geometry of the molecule and satisfy, for any α

$$\sum_{i=1}^{4} C_{\alpha i}=1. \tag{2}$$

Of course, this new set of constraints leaves us with the correct number of degrees of freedom, because, here again, any additional point introduces three new degrees of freedom and three new constraints.

2.2. *Planar molecules*

In the case of a planar molecule, three non-colinear particles define the plane of the molecule through the two basic vectors $\mathbf{R}_{21}$, $\mathbf{R}_{31}$. If the molecule contains more than three atoms, it is impossible to solve its dynamics by using bond constraints only. In fact these constraints do not support constraint forces orthogonal to the molecular plane while they are needed. Therefore, in this case, bond constraints alone are not sufficient while one can use the analogues of formula (1), i.e.

$$\mathbf{r}_\alpha=\mathbf{R}_1+C_{\alpha 2}\mathbf{R}_{21}+C_{\alpha 3}\mathbf{R}_{31}=\sum_{i=1}^{3} C_{\alpha i}\mathbf{R}_i \tag{3}$$

where again $\sum_{i=1}^{3} C_{\alpha i}=1$.

It is easily checked that the total number of constraints is again correctly given because, for this molecule, one has three bond constraints plus $3(n-3)$ linear constraints.

2 s 2

2.3. *Linear molecules*

For linear molecules of n atoms with $n>2$, only two particles need to be connected by bond constraints. The rigidity of such a molecule can be represented by linear constraints. The position of any other atom is easily expressed in terms of $\mathbf{R}_1$ and $\mathbf{R}_{21}$, i.e.

$$\mathbf{r}_\alpha = \mathbf{R}_1 + C_{\alpha 2}\mathbf{R}_{21} = \sum_{i=1}^{2} C_{\alpha i}\mathbf{R}_i, \tag{4}$$

with

$$C_{\alpha 1} + C_{\alpha 2} = 1.$$

The representation of linear molecules by bond constraints fails for the physical reason already described in the planar case. Again, the forces coming from bond constraints cannot give rise to components orthogonal to the molecular axis. Moreover, the correct number of degrees of freedom cannot be recovered.

2.4. *General formulation*

The general formulation of this new set of constraints for arbitrary rigid molecules is easily obtained. Let us divide the n atoms into (i) n_b primary particles, the basic structure, linked by $l_b = n_b(n_b-1)/2$ bond constraints σ_{ij} of length d_{ij} and (ii) $n_s = n - n_b$ secondary particles linked to the basic structure by $l_s = 3n_s$ linear constraints

$$\sigma_{ij} \equiv (\mathbf{R}_i - \mathbf{R}_j)^2 - d_{ij}{}^2 = 0, \quad i>j=1,\, n_b, \tag{5}$$

$$\boldsymbol{\tau}_\alpha \equiv \sum_{i=1}^{n_b} C_{\alpha i}\mathbf{R}_i - \mathbf{r}_\alpha = 0, \quad \alpha = 1,\, n_s, \tag{6}$$

To avoid confusion, primary and secondary particles will always be labelled by latin and greeks indices respectively. According to our previous discussion, linear, planar and three-dimensional molecules require basic structures of $n_b=2$, 3 or 4 particles.

The set $\{C_{\alpha i}\}$ is defined by the geometry of the rigid molecule and can be easily obtained given the positions of the n atoms in any reference frame : for each secondary particle, equations (6) and (2) give a non-singular system of linear equations in $C_{\alpha i}$.

3. Equations of motion

For a mechanical system composed of N rigid molecules, each consisting of (n_b+n_s) particles and endowed with the (l_b+l_s) constraints (5) and (6), the Lagrange equations of motion of the first kind are given by [8]

$$M_i\ddot{\mathbf{R}}_i = \mathbf{F}_i + \mathbf{G}_i = \mathbf{F}_i - \sum_{k=1}^{n_b-1}\sum_{j=k+1}^{n_b} \lambda_{jk}\boldsymbol{\nabla}_i\sigma_{jk} - \sum_{\beta=1}^{n_s} \boldsymbol{\nabla}_i(\boldsymbol{\mu}_\beta \cdot \boldsymbol{\tau}_\beta), \quad i=1,\, n_b, \tag{7 a}$$

$$m_\alpha\ddot{\mathbf{r}}_\alpha = \mathbf{f}_\alpha + \mathbf{g}_\alpha = \mathbf{f}_\alpha - \sum_{\beta=1}^{n_s} \boldsymbol{\nabla}_\alpha(\boldsymbol{\mu}_\beta \cdot \boldsymbol{\tau}_\beta), \quad \alpha=1,\, n_s \tag{7 b}$$

for each molecule. In (7), we kept implicit the molecular index to simplify notations. Moreover, $\mathbf{F}_i(\mathbf{f}_\alpha)$ is the force on particle $i(\alpha)$ resulting from the intermolecular interactions while $\mathbf{G}_i(\mathbf{g}_\alpha)$ is the total constraint force on the particle. $\lambda_{jk}(\boldsymbol{\mu}_\alpha)$ are the Lagrange multipliers associated to the constraints $\sigma_{jk}(\boldsymbol{\tau}_\alpha)$.

Equations (7 *b*) are in some way redundant to describe the motion of the rigid molecule. All we need indeed is to know the time evolution of the coordinates of the basic set $\{\mathbf{R}_i, i=1, n_b\}$. Then the time evolution of the secondary coordinates $\{\mathbf{r}_\alpha, \alpha=1, n_s\}$ follows from the constraint relations (6). This can be obtained solving $\{\boldsymbol{\mu}_\alpha\}$ in terms of the atomic forces $\{\mathbf{F}_i, \mathbf{f}_\alpha\}$ and the remaining Lagrange parameters λ_{jk}. That is equivalent, as we will see, to transport the atomic forces on secondary particles to the basic ones by equivalence operations.

Let us rewrite the equations of motion (7) in a more explicit form

$$M_i\ddot{\mathbf{R}}_i = \mathbf{F}_i - \sum_{k=1}^{n_b-1}\sum_{j=k+1}^{n_b} 2\mathbf{R}_{jk}(\delta_{ij}-\delta_{ik})\lambda_{jk} - \sum_{\beta=1}^{n_b} C_{\beta i}\boldsymbol{\mu}_\beta, \tag{8 a}$$

$$m_\alpha\ddot{\mathbf{r}}_\alpha = \mathbf{f}_\alpha + \boldsymbol{\mu}_\alpha. \tag{8 b}$$

As expected, the sum of constraint forces vanishes for each molecule. The constraint relations $\boldsymbol{\tau}_\alpha=0$, $\alpha=1, n_s$ for any time t, imply $\ddot{\mathbf{T}}_\alpha = \sum_{i=1}^{n_s} C_{\alpha i}\ddot{\mathbf{R}}_i - \ddot{\mathbf{r}}_\alpha = 0$. Substituting (8 *a*) and (8 *b*) in this relation, we obtain after some algebra

$$\sum_{\beta=1}^{n_s} A_{\alpha\beta}\boldsymbol{\mu}_\beta(t) = \mathbf{T}_\alpha(t) - 2\sum_{k=1}^{n_b-1}\sum_{j=k+1}^{n_b} B_{jk}{}^{\alpha}\,\mathbf{R}_{jk}(t)\lambda_{jk}(t), \tag{9}$$

where

$$A_{\alpha\beta} = \delta_{\alpha\beta}/m_\alpha + \sum_{i=1}^{n_b} C_{\alpha i}C_{\beta i}/M_i = A_{\beta\alpha},$$

$$\mathbf{T}_\alpha(t) = -\mathbf{f}_\alpha(t)/m_\alpha + \sum_{i=1}^{n_b} C_{\alpha i}\mathbf{F}_i(t)/M_i$$

and

$$B_{jk}{}^{\alpha} = C_{\alpha j}/M_j - C_{\alpha k}/M_k.$$

The $\boldsymbol{\mu}_\alpha(t)$ are now obtained by inverting the time independent matrix $A_{\alpha\beta}$

$$\boldsymbol{\mu}_\alpha(t) = \sum_{\beta=1}^{n_s}(A^{-1})_{\alpha\beta}\left\{\mathbf{T}_\beta(t) - 2\sum_{k=1}^{n_b-1}\sum_{j=k+1}^{n_b} B_{jk}{}^{\beta}\,\mathbf{R}_{jk}(t)\lambda_{jk}(t)\right\} \quad \alpha=1, n_s, \tag{10}$$

By inserting (10) in (8 *a*), we find the final form of the equation of motion for the basic particles

$$\ddot{\mathbf{R}}_i(t) = \mathscr{F}_i(t) - \left\{\sum_{p=1}^{l_b} \mathscr{R}_p{}^i(t)\lambda_p(t)\right\} \quad i=1, n_b, \tag{11}$$

where

$$\mathscr{F}_i(t) = \left(\mathbf{F}_i(t) - \sum_{\alpha=1}^{n_s}\sum_{\beta=1}^{n_s} C_{\alpha i}(A^{-1})_{\alpha\beta}\mathbf{T}_\beta(t)\right)\Big/ M_i \tag{12}$$

and

$$\mathscr{R}_p{}^i(t) = 2S_{jk}{}^i\,\mathbf{R}_{jk}(t), \tag{13 a}$$

$$S_{jk}{}^i = \left[(\delta_{ij}-\delta_{ik}) - \sum_{\alpha=1}^{n_s}\sum_{\beta=1}^{n_s} C_{\alpha i}(A^{-1})_{\alpha\beta}B_{jk}{}^{\beta}\right]\Big/ M_i. \tag{13 b}$$

In (11), (13 *a*) and followings, bond indices (jk) are denoted by a single index $p=1, l_b$.

These equations for the basic particles, supplemented by the conditions (6) for the positions of all other particles, provide the required representation of the motion of rigid molecules. Next section will provide the numerical implementation of the integration of the basic equations (11) supplemented by the constraint relations (5). Use will be made of the Verlet algorithm as it has proved particularly efficient and suitable for constraint dynamics. The general solution for arbitrary integration algorithm has been already discussed in [1].

4. Numerical implementation

If we adopt the usual central difference scheme [9] for the numerical integration of (11), the predicted coordinates at time $t+h$ are

$$\mathbf{R}_i(t+h)=\mathbf{R}'_i(t+h)-h^2\sum_{p=1}^{l_b}\mathscr{R}_p{}^i(t)\lambda_p,\quad i=1,n_b, \tag{14}$$

where h is the time step and

$$\mathbf{R}'_i(t+h)=-\mathbf{R}_i(t-h)+2\mathbf{R}_i(t)+h^2\mathscr{F}_i(t) \tag{15}$$

represents the coordinates of atom i in the absence of any bond constraint. The only unknown quantities are therefore the Lagrange multipliers $\{\lambda_p\}$ $p=1, l_b$ but the solution to equation (14) must satisfy the restriction that

$$|\mathbf{R}_{ij}(t+h)|^2=d_{ij}{}^2,\quad i>j=1,n_b. \tag{16}$$

Rewriting

$$\mathbf{R}'_{(p)}\equiv\mathbf{R}'_{ij}=\mathbf{R}'_i-\mathbf{R}'_j,\ i>j=1,n_b$$

and

$$\mathscr{R}_{pq}=\mathscr{R}_q{}^i-\mathscr{R}_q{}^j,\quad (i,j)\equiv p=1,l_b$$

to avoid cumbersome notations, the combination of (14) and (16) leads to a system of quadratic equations for $\{\lambda_p\}$. These equations may conveniently be written as

$$\lambda_p=\lambda_p{}^0+h^2\sum_{q'=1}^{l_b}\sum_{q''=1}^{l_b}\lambda_{q'}\lambda_{q''}\sum_{q=1}^{l_b}(C^{-1})_{pq}\mathscr{R}_{qq'}\cdot\mathscr{R}_{qq''}, \tag{17}$$

where

$$C_{pq}=2\mathbf{R}'_{(p)}\cdot\mathscr{R}_{pq} \tag{18}$$

and

$$\lambda_p{}^0=h^{-2}\sum_{q=1}^{l_b}(C^{-1})_{pq}(\mathbf{R}'_{(q)}{}^2-d_q{}^2) \tag{19}$$

is the first approximation to λ_p. Equation (17) is easily solved by iteration; two to four iterations are in general sufficient to maintain the bond length constant to within one part in 10^6 to 10^8. Each iteration requires only a single loop over the basic atoms of the molecule and therefore increases the total computing time negligibly. It has to be stressed that the $\{\lambda_p\}$ obtained are not the exact values of the Lagrange multipliers at time t. However it has been shown in reference [1], that the differences are consistent with the error introduced by the algorithm.

5. Inclusion of extra massless centres of force and/or inertial particles

We have implicitly assumed in the previous description that in the polyatomic system the point masses are coincident with the centres of force. However, it is well known that for some specific molecular models, e.g. HF [2], H_2O [10], NH_3 [3], extra massless coulombic centres of forces have been introduced to represent molecular charge distributions. More generally, the centres of force can be distinguished from the point-masses, as in the case of N [11] or C_6H_6 [12]. The present technique fits quite naturally with such situations.

Consider first the massless points. If they can be taken as secondary particles (in our previous classification), their effect can be easily taken into account. One simply writes for any such point (7 *b*)

$$0 = m_\alpha \ddot{\mathbf{r}}_\alpha = \mathbf{f}_\alpha + \boldsymbol{\mu}_\alpha \tag{20}$$

supplemented with the constraint relationship (6). Inserting $\boldsymbol{\mu}_\alpha = -\mathbf{f}_\alpha$ into equations (8 *a*), one gets rid of these extra degrees of freedom while the evolution of the point α is given by the constraint relationship. The physical meaning is clear. The Lagrange multipliers $\boldsymbol{\mu}_\alpha$ simply transfer the forces acting on point α to the basic points $\{\mathbf{R}_i\}$ $i=1, n_b$ by equivalence operations. In particular the force transferred on particle $\mathbf{R}_i$ reads $C_{\alpha i}\mathbf{f}_\alpha$.

In a special case the extra centre of force cannot be taken directly as a secondary particle. A planar molecule may have centres of force out of the plane, as in the ST2 model of water [10]. In such a case the position of the centre cannot be expressed as linear combination of the basic particles. Moreover it cannot be taken as basic particle as it does not carry mass. Then one can add to the basic vectors $\mathbf{R}_1$, $\mathbf{R}_2$, $\mathbf{R}_3$ (see (3)) the vector $\mathbf{R}_{21} \wedge \mathbf{R}_{31}$ orthogonal to the molecular plane. The constraint for an extra centre of force α is now given by

$$\boldsymbol{\tau}_\alpha = C_{\alpha 1}\mathbf{R}_1 + C\alpha_2\mathbf{R}_2 + C_{\alpha 3}\mathbf{R}_3 + C_{\alpha 4}(\mathbf{R}_{21} \wedge \mathbf{R}_{31}) - \mathbf{r}_\alpha = 0 \tag{21}$$

with

$$C_{\alpha 1} + C_{\alpha 2} + C_{\alpha 3} = 1.$$

The Lagrange multipliers $\boldsymbol{\mu}_\alpha$ are still given by (20), but the force $\mathbf{f}_\alpha$ is now transferred to the three basic particles in a more complicated way. Inserting (21) in (7 *a*), with $\boldsymbol{\mu}_\alpha = -\mathbf{f}_\alpha$ one finds the contribution of α to the force on particle i

$$C_{\alpha i}\mathbf{f}_\alpha + C_{\alpha 4}(\mathbf{R}_j - \mathbf{R}_k) \wedge \mathbf{f}_\alpha \tag{22}$$

($i=1$, 3 and (ijk) is a cyclic permutation of (1 2 3)). The evolution of the point α is given by the constraint relationship (21).

It remains to consider point masses which are not centres of force. They can be simply treated either as basic or as secondary particles in the approach developed in §§ 2 and 3. Subjected only to constraint forces, they play just an inertial role in the molecular motion.

6. Illustrations

In this section we give a detailed description of the technique in the case of the linear triatomic molecule and we compare it with integration methods based on generalized coordinates. Furthermore, in order to illustrate how the method works for other geometries, we outline the cases of a planar molecule, benzene, and of a tridimensional molecule, CCl_4. A comparison with the standard constraint method [1] is possible in this last case.

6.1. *Linear triatomic molecule* : CS_2

CS_2 is a linear triatomic molecule which has been widely investigated recently [13, 14, 15]. It is the simplest case where the cartesian equations of motion require the new formulation of the constraints. Let us take as primary particles the two sulphur atoms and call $\mathbf{R}_1$ and $\mathbf{R}_2$ their positions. The central carbon is then the secondary particle and its position will be denoted by $\mathbf{r}_1$. The molecule is subjected to the following constraints

$$\left.\begin{aligned} \sigma &\equiv (\mathbf{R}_1-\mathbf{R}_2)^2-d^2=0,\\ \boldsymbol{\tau} &\equiv (\mathbf{R}_1+\mathbf{R}_2)/2-\mathbf{r}_1=0, \end{aligned}\right\} \tag{23}$$

where d is the distance between sulphur atoms.

The equations of motion for each molecule read

$$\left.\begin{aligned} m_s\ddot{\mathbf{R}}_1 &= \mathbf{F}_1-2\lambda\mathbf{R}_{12}-\boldsymbol{\mu}/2,\\ m_s\ddot{\mathbf{R}}_2 &= \mathbf{F}_2+2\lambda\mathbf{R}_{12}-\boldsymbol{\mu}/2,\\ m_c\ddot{\mathbf{r}}_1 &= \mathbf{f}_1+\boldsymbol{\mu}, \end{aligned}\right\} \tag{24}$$

where $\mathbf{F}_1$, $\mathbf{F}_2$, $\mathbf{f}_1$ are the intermolecular forces on the atoms. Substituting the equations of motion in the second time derivatives of the constraints $\boldsymbol{\tau}$ one has

$$\boldsymbol{\mu}=\frac{m_c}{M}(\mathbf{F}_1+\mathbf{F}_2)-2\frac{m_s}{M}\mathbf{f}_1, \quad M=m_c+2m_s. \tag{25}$$

Inserting (25) in (24), we get the equations of motion for the basic particles

$$\left.\begin{aligned} m_s\ddot{\mathbf{R}}_1 &= \left(1-\frac{m_c}{2M}\right)\mathbf{F}_1+\frac{m_s}{M}\mathbf{f}_1-\frac{m_c}{2M}\mathbf{F}_2-2\lambda\mathbf{R}_{12}=\mathscr{F}_1-2\lambda\mathbf{R}_{12},\\ m_s\ddot{\mathbf{R}}_2 &= \left(1-\frac{m_c}{2M}\right)\mathbf{F}_2+\frac{m_s}{M}\mathbf{f}_1-\frac{m_c}{2M}\mathbf{F}_1+2\lambda\mathbf{R}_{12}=\mathscr{F}_2+2\lambda\mathbf{R}_{12} \end{aligned}\right\} \tag{26}$$

and for the secondary particles.

$$\mathbf{r}_1=(\mathbf{R}_1+\mathbf{R}_2)/2. \tag{27}$$

Equations (26) look like the equations of motion for a diatomic molecule but with a particular distribution of forces [15]. Equations (26) can be integrated in the usual way [15, 2].

We now compare our integration method with previous ones [13–15]. In recent investigations of rigid linear triatomic fluid by MD, two main methods have been used. Both separate the motion into a translation of the centre of mass and a rotation around it. Tildesley and Madden [14] used quaternions and a fifth order predictor corrector algorithm to simulate liquid CS_2. An extension of the Singer algorithm [16] for diatomic molecules has been exploited by Neumann and Steinhauser [13], again for CS_2 and by Murthy *et al.* [17] in the study of liquid CO_2.

In our calculation on CS_2 we describe the interactions between pairs of molecules with model A of [14]. It consists of three Lennard-Jones centres coincident with atomic sites with $\sigma_{cc}=3{\cdot}35$ Å, $\sigma_{ss}=3{\cdot}52$ Å, $\sigma_{cs}=3{\cdot}44$ Å and $\epsilon_{cc}/k_B=51{\cdot}2$ K,

Comparison of results for CS_2 molecular dynamics.

MD technique	Quaternions	Singer's algorithm	Constraints
Number of molecules (N)	256	64	64
Total time of MD run (10^{-12} s)	44·8	6·38	6·38
Cut-off distance (Å)	8·375	8·9	8·9
Time step (10^{-15} s)	4·48	12·8	12·8
Total energy conservation (rel. fluct.)	1×10^{-4}	$0{\cdot}9 \times 10^{-4}$	$0{\cdot}8 \times 10^{-4}$
Temperature (K)	193	187	186
Pot. energy/$(N\epsilon_{SS})$	−19·62	−19·60	−19·59
Pressure $\times \sigma_{SS}^3/\epsilon_{SS}$	0·08	0·41	0·46
C_V/R	2·37	2·36	2·40

$\epsilon_{cs}/k_B = 96{\cdot}8$ K, $\epsilon_{ss}/k_B = 183{\cdot}0$ K. The carbon–sulphur bond-length is 1·57 Å. As the performances of the quaternion algorithm are widely illustrated by the work of Tildesley and Madden we made a comparative test carrying out explicit MD runs using the two other methods, i.e. Singer's and the present constraint technique. Results for the three methods are compared in the table for the fixed density $\rho = 1{\cdot}42$ g cm^{-3}. Starting from the same equilibrated configuration we observe a close equivalence between the Singer method and the constraint one. Both algorithms require the same computer time per step (97 per cent being spent to evaluate pair interactions), and both conserve the total energy with the same accuracy, for the given value of the integration time step. No shift of total energy has been observed over length of 1000 steps, i.e. ~ 13 ps. This equivalence is not surprising as both methods are based on the Verlet integration scheme and treat intramolecular constraints in different fashion but with the same accuracy. Note that velocities were computed with an error of order h^3 by retaining atomic positions at time $(t-2h)$ as well. This provides a more accurate estimate of the temperature [18]. By comparing with the results of Tildesley and Madden, we see that the present constraint method allows a time step three times larger than their quaternion procedure. The gain in time should come mainly from the greater stability of the Verlet algorithm.

6.2. *Planar molecule : benzene*

Benzene is a highly symmetric planar molecule of considerable chemical interest. To our knowledge only two attempts have been made to investigate liquid benzene properties by computer simulation. Kushick and Berne [19] reported some dynamical properties of the rotational motion of benzene-like ellipsoids interacting through an angular dependent Lennard–Jones potential truncated at its minimum. Evans and Watts [12] developed a six-interaction site Lennard-Jones potential to investigate the liquid structure of benzene by Monte Carlo computer simulation.

Let us take as basic structure for a benzene molecule (see figure) three alternate carbon atoms $\mathbf{R}_1$, $\mathbf{R}_2$, $\mathbf{R}_3$ constrained to form an equilateral triangle of side equal to $3^{1/2} \times$ (carbon–carbon bond length).

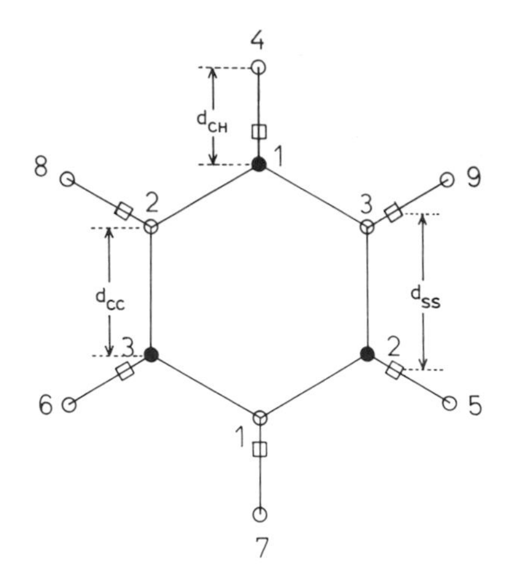

The molecular model for benzene ($d_{CC}=1{\cdot}395$ Å, $d_{CH}=1{\cdot}08$ Å, $d_{SS}=1{\cdot}756$ Å). ●, primary carbon atoms forming the basic structure ; ○, secondary atoms 1 to 3 are carbons and 4 to 9 are hydrogens ; □, Lennard-Jones interaction sites in the model of Evans and Watts [12].

The positions of the other atoms can be written as

$$\mathbf{r}_i=\tfrac{1}{3}(-\mathbf{R}_i+2\mathbf{R}_j+2\mathbf{R}_k),$$

for carbons and

$$\left.\begin{aligned}\mathbf{r}_{3+i}&=(1+\tfrac{2}{3}\eta)\mathbf{R}_i-\eta/3\mathbf{R}_j-\eta/3\mathbf{R}_k,\\ \mathbf{r}_{6+i}&=-(1+2\eta)/3\mathbf{R}_i+(2+\eta)/3\mathbf{R}_j+(2+\eta)/3\mathbf{R}_k,\end{aligned}\right\}\qquad(28)$$

for hydrogens where (i, j, k) is a cyclic permutation of (1, 2, 3) and η is the ratio between the carbon–hydrogen bond length (1·08 Å) and the carbon–carbon bond length (1·395 Å).

A preliminary run has been made using the Evans and Watts six LJ centre model ($\epsilon/_{k\text{B}}=77$ K, $\sigma=3{\cdot}5$ Å). In this model interactions sites are placed at a distance $d=1{\cdot}756$ Å from the molecular centre. At room temperature density a time step of 10^{-14} sec has been found to be appropriate.

Work is in progress to perform MD simulations of rigid benzene.

6.3. *Three-dimensional molecule : carbon tetrachloride*

The old version of constraint method has been employed by Van Hercke and Bellemans [20] for a five atomic model of CCl_4 and its methyl substituents. In this case nine interatomic distances (out of ten possible) were fixed by bond constraints. The present approach consists in fixing the four chlorine relative positions with six bond constraints, while the C position is naturally given by the centre of mass relationship.

$$\mathbf{r}_1=\tfrac{1}{4}(\mathbf{R}_1+\mathbf{R}_2+\mathbf{R}_3+\mathbf{R}_4)\qquad(29)$$

We have compared the efficiency of the two methods in the case of their model for liquid at room temperature. The constraint forces are computed twice as fast when the new formulation of constraint is used. On the whole there is a gain of 15 per cent on the total computer time. The most expensive part of the computation of constraints forces is the inversion the 6×6 matrix $\mathbf{C}$ equation (18). In the old formulation the order of the matrix to be inverted is equal to the total number of constraints, i.e. nine in the present case or $3n-6$ in the case of three-dimensional n-atomic molecule. Therefore greater savings of computer time must be expected for $n>5$.

7. Final remarks

In this paper we derived the equations of motion of a rigid polyatomic molecular system in terms of the cartesian coordinates of a basic set of up to four particles. In a unified way these equations describe the motion of molecules of arbitrary geometry with interaction models including centres of force distinct from particle positions (e.g. point charges). A very efficient MD technique follows when they are integrated with the Verlet algorithm. It is superior to the original method of constraints in many respects : (i) any rigid molecule can be handled ; (ii) only the dynamical evolution of basic particles has to be numerically integrated, saving both computer time and storage. In the respect of generalized coordinate integration schemes, it has all the advantages of the original constraint method, i.e. one has essentially a greater stability and can use larger time steps.

Finally it has to be stressed that in dealing with partly rigid molecules care must be taken in presence of linear or planar rigid atomic units. In fact a description in terms of bond constraints can fail while this kind of failure can be avoided by using the constraint relations introduced in this paper.

We thank André Bellemans for a very careful critical reading of the manuscript. One of us (G.C.) is pleased to thank Roger Impey for raising the problem of trilinear molecules and for many useful discussions ; he wants also to acknowledge some useful comments from Paul Madden. This work has been partly supported by the Nato Grant No 1865. One of us (M.F.) acknowledges financial support from the Italian Research Council (CNR).

References

[1] Ryckaert, J.-P., Ciccotti, G., and Berendsen, H. J. C., 1977, *J. comput. Phys.*, **23,** 327.
[2] Klein, M. L., and McDonald, I. R., 1979, *J. chem. Phys.*, **71,** 298.
[3] Klein, M. L., McDonald, I. R., and Righini, R., 1979, *J. chem. Phys.*, **71,** 3673.
[4] Ryckaert, J. P., and Bellemans, A., 1978, *Faraday Discuss. chem. Soc.*, **66,** 95.
[5] van Gunsteren, W. F., and Berendsen, H. J. C., 1977, *Molec. Phys.*, **34,** 1311.
[6] Stratt, R. M., Holmgren, S. L., and Chandler, D., 1981, *Molec. Phys.*, **42,** 1233.
[7] Memon, M. K., Hockney, R. W., and Mitra, S. K., 1982, *J. comput. Phys.*, **43,** 345.
[8] Bradbury, T. C., 1968, *Theoretical Mechanics* (Wiley), Chap. XI.
[9] Verlet, L., 1967, *Phys. Rev.*, **159,** 98.
[10] Rahman, A., and Stillinger, F. H., 1971, *J. chem. Phys.*, **55,** 3336.
[11] Murthy, C. S., Singer, K., Klein, M. L., and McDonald, I. R., 1980, *Molec. Phys.*, **41,** 1387.
[12] Evans, D. J., and Watts, R. O., (a) 1975, *Molec. Phys.*, **29,** 777; (b) 1976, *Ibid.*, **31,** 83; (c) 1976, *Ibid.*, **32,** 93.

Notes on p. 366

[13] Steinhauser, O., and Neumann, M., 1979, *Molec. Phys.*, **37**, 1921.
[14] Tildesley, D. J., and Madden, P. A., 1981, *Molec. Phys.*, **42**, 1137.
[15] Steinhauser, O., 1981, *Chem. Phys. Lett.*, **82**, 153. The equations of motion solved in this paper should be equivalent to our (26). However it is not clear if the central mass has been properly taken into account.
[16] Singer, K., Taylor, A., and Singer, J. V. L., 1977, *Molec. Phys.*, **33**, 1757.
[17] Murthy, C. S., Singer, K., and McDonald, I. R., 1981, *Molec. Phys.*, **44**, 135.
[18] Bellemans, A. (private communication).
[19] Kushick, J., and Berne, B., 1976, *J. chem. Phys.*, **64**, 1362.
[20] van Hercke, K., and Bellemans, A., 1981, Communication to the meeting of the Royal Society of Chemistry, Faraday Division, on 'The structure of molecular liquids', Cambridge.

Notes to Reprint V.2

1. The paper contains a number of misprints. (a) In eq. (8a), the sum over β should run from $\beta = 1$ to $\beta = n_S$. (b) The left-hand side of the equation in the second line after eq. (8b) should be "$\ddot{\boldsymbol{\tau}}_\alpha$", not $\ddot{\mathbf{T}}_\alpha$", and the sum over i on the right-hand side should run from $i = 1$ to $i = n_b$. (c) In the definition of τ_α in eq. (21), "$C\alpha_2$" should be replaced by "$C_{\alpha 2}$".

2. The method of constraints takes a very simple form in the case of a diatomic molecule; the relevant formulae are given in ref. [2].

3. The comparison made in the table (p. 1261) is complicated by the fact that results for the method of constraints were obtained with the Verlet algorithm (Reprint I.7), while those for the method of quaternions are based on the use of a predictor–corrector scheme. This in itself could account for the differences noted in the text.

4. The work on benzene mentioned on p. 1262 was later published by Claessens et al. (Mol. Phys. 50 (1983) 217).

5. The method described in this paper has subsequently been extended to a larger class of constraints (Ryckaert, Mol. Phys. 55 (1985) 549; Ciccotti and Ryckaert, Comput. Phys. Rep. 4 (1986) 345).

MOLECULAR PHYSICS, 1977, VOL. 34, NO. 2. 327–331

Singularity free algorithm for molecular dynamics simulation of rigid polyatomics

by DENIS J. EVANS and SOHAIL MURAD
School of Chemical Engineering, Olin Hall, Cornell University, Ithaca, New York 14853, U.S.A.

(*Received* 20 *April* 1977)

Using a transformation of the rigid body equations of motion due to Evans [4], a new algorithm is presented for the molecular dynamics simulation of rigid polyatomic molecules. The algorithm consists of solving the eulerian rigid body equations, using quaternions to represent orientations, by a fifth-order predictor corrector method. Compared to previous methods, it is shown that this algorithm leads to an order of magnitude increase in computing speed.

1. INTRODUCTION

The success of modelling monatomic fluids by molecular dynamics methods led to attempts to simulate rigid polyatomics. This was done by solving either Hamilton's, Lagrange's or Euler's equations of motion using Euler angles to represent the orientations. It was soon realized, however, that this procedure was very inefficient, since each of these equations possessed singularities at $\theta=0$ [1]. The singularities are first order for Euler's equations and second order for Lagrange's or Hamilton's equations. This problem was circumvented by Barojas *et al.* [2] in their simulation of nitrogen, by redefining the laboratory coordinate frame of any molecule for which $|\theta-n\pi|\leqslant\pi/10$ where $n=0, \pm 1$. Cheung has shown how to use the special properties of diatomic molecules to remove singularities from their equations of motion [3]. This paper presents an analogous improvement for arbitrary, rigid polyatomics.

Evans [4] has recently developed a set of four parameters (quaternions) that can be used in the equations of motion instead of Euler angles. The four quaternions are not mutually independent but give an orthogonal and euclidean representation of orientation space that makes the equations of motion singularity-free. We give an efficient algorithm for performing molecular dynamics calculations using quaternions and compare results using the new algorithm with those of the conventional Euler angle method, for otherwise identical simulations. This and other comparisons obtained using results given by previous workers, make the superiority of the quaternion algorithm apparent.

2. THEORY

The quaternion parameters, χ, η, ξ, ζ can be defined in terms of Goldstein Euler angles [1] by

$$\left.\begin{aligned}\chi&=\cos(\theta/2)\,.\,\cos(\psi+\phi)/2,\\ \eta&=\sin(\theta/2)\,.\,\cos(\psi-\phi)/2,\\ \xi&=\sin(\theta/2)\,.\,\sin(\psi-\phi)/2,\\ \zeta&=\cos(\theta/2)\,.\,\sin(\psi+\phi)/2.\end{aligned}\right\}\qquad(1)$$

From (1) it can be seen that the quaternions are not independent but satisfy the relation.

$$\chi^2+\eta^2+\xi^2+\zeta^2=1. \tag{2}$$

The rotation matrix **A** [1], defined by

$$\mathbf{V}_{\text{principal}}=\mathbf{A}\,.\,\mathbf{V}_{\text{lab}}, \tag{3}$$

is given by

$$\mathbf{A}=\begin{pmatrix} -\xi^2+\eta^2-\zeta^2+\chi^2, & 2(\zeta\chi-\xi\eta), & 2(\eta\zeta+\xi\chi) \\ -2(\xi\eta+\zeta\chi), & \xi^2-\eta^2-\zeta^2+\chi^2, & 2(\eta\chi-\xi\zeta) \\ 2(\eta\zeta-\xi\chi), & -2(\xi\zeta+\eta\chi), & -\xi^2-\eta^2+\zeta^2+\chi^2 \end{pmatrix} \tag{4}$$

The principal angular velocity $\boldsymbol{\omega}_{\text{p}}$, is related to the quaternions by the matrix equation

$$\begin{pmatrix} \dot{\xi} \\ \dot{\eta} \\ \dot{\zeta} \\ \dot{\chi} \end{pmatrix}=\tfrac{1}{2}\begin{pmatrix} -\zeta, & -\chi, & \eta, & \xi \\ \chi, & -\zeta, & -\xi, & \eta \\ \xi, & \eta, & \chi, & \zeta \\ -\eta, & \xi, & -\zeta, & \chi \end{pmatrix}\begin{pmatrix} \omega_{\text{p}x} \\ \omega_{\text{p}y} \\ \omega_{\text{p}z} \\ 0 \end{pmatrix} \tag{5}$$

The inverse of (5) is simple to find since the 4×4 matrix is orthogonal. The equations of motion for these parameters are thus free of singularities.

3. Method

A comparison was made of the relative efficiencies of quaternions and Euler angles for use in molecular dynamics programmes. A system of 108 model methane molecules was studied using the standard periodic boundary conditions [5]. Methane was modelled by an atom–atom potential due to Williams [6]. The details of this particular model and its predictions of the properties of methane are unimportant for our present purposes, but will be described fully in a later publication [7]. The intermolecular forces and torques were calculated in a laboratory coordinate system. The torques were then converted to principal torques $\mathbf{T}_{\text{p}}$, using the rotation matrix (equation (4)). Principal torques and forces $\mathbf{F}$, were used to evaluate the time derivatives of the principal angular velocities and cartesian positions $\mathbf{r}$, of the molecules using the equations,

$$\frac{d\omega_{\text{p}\alpha}}{dt}=T_{\text{p}\alpha}/I_{\text{p}\alpha}, \quad (\alpha=x,\ y,\ z), \tag{6}$$

$$\frac{d^2\mathbf{r}}{dt^2}=\mathbf{F}/m. \tag{7}$$

A fifth-order predictor method was used to integrate the centre or mass motion (equation (7)), while a fourth-order method was used for orientational equations of motion of the molecules (equation (6) and (5)).

It is apparent that our method treats the quaternions as if they were independent variables despite equation (2). The question therefore arises : in the limit of indefinitely small integration time steps, is the constraint equation (2), automatically satisfied ? If we use equation (5) to form the sum, $\chi\dot{\chi}+\eta\dot{\eta}+\zeta\dot{\zeta}+\xi\dot{\xi}$, it is easily seen to be zero for all values of principal angular velocities. However, this sum is just the time derivative of the equation of constraint (2). Thus we see that the constraint, if satisfied at time zero, will automatically be satisfied at all subsequent times. In practice, however, the equations of motion are not solved exactly, so that the constraint will only be approximately satisfied. Therefore, at each time step the modulus of the quaternion, $\sqrt{(\chi^2+\eta^2+\xi^2+\zeta^2)}$, was compared to one, and used to rescale the quaternions so that they satisfied equation (2) exactly at all times. For the levels of integration accuracy used in this work, such rescaling was observed to produce negligible changes in the particle trajectories over several hundred time steps.

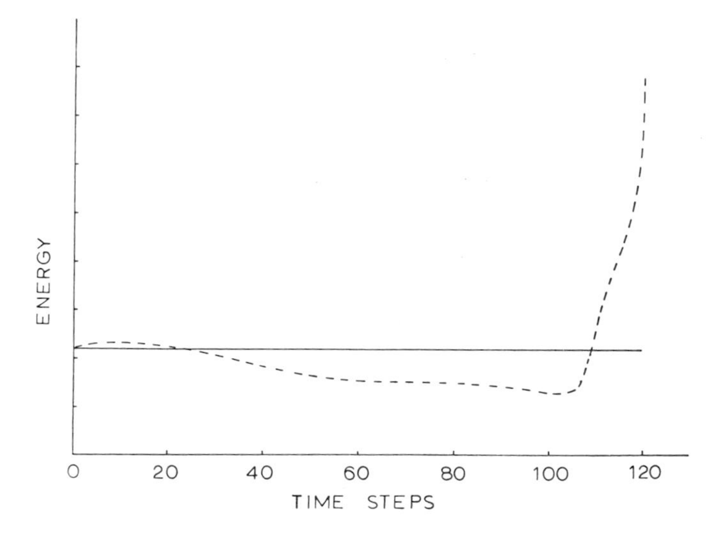

The figure compares total energy conservation as obtained from two simulations of the same system with identical starting configuration and time step length. (Dashed curve gives Euler angle algorithm and continuous curve gives quaternion results.) Units of energy are arbitrary.

4. Results

To determine the efficiency of the new algorithm, energy conservation was monitored for identical runs, one using equations of motion in quaternions and the other using Euler angles. The results of such a comparison are shown in the figure. Under identical conditions and integration step length, the figure shows that energy conservation for the quaternion programme, as revealed on a graph of the scale, is essentially exact. However, energy conservation for the Euler angle programme is quite poor. After just 110 time steps one molecule in the 108 molecule system approaches a singularity with $\theta=n\pi+\epsilon$ where $n=0, \pm 1, \pm 2, \ldots$, and $|\epsilon| \leqslant \pi/20$. The total energy of the system increases drastically. After the molecule moves out of the range of the singularity the energy will stabilize again, but at some other value. It is easily seen that in a

108 particle system molecules may be expected to come within the range of singularities relatively often. The run shown in the figure is typical in this respect.

The difficulties for the differential equation-solving algorithm do not arise because of round-off errors or floating point overflows, but because of the increased high-frequency structure of the differential equations themselves. In the neighbourhood of $\theta=0$, the equations for $\dot{\phi}$ and $\dot{\psi}$ [1] have the qualitative structure of sinusoids modulated by the cosecant of θ. Either side of $\theta=0$, $\dot{\phi}$ and $\dot{\psi}$ rapidly diverge to $\pm\infty$. If θ never actually becomes zero the singularities are never reached. However, if θ is within approximately 0·1 radians of the singularity, the equations are sufficiently rapidly changing to make the numerical integration very difficult with practical time steps.

Comparison of quaternions with other reported results.

Reference	Fluid studied and model used	Time step (s)	Energy conservation (percentage per 10 000 steps)
Stillinger and Rahman [9]	H_2O—S.T.2 model	2×10^{-16}	2·38
Rahman and Stillinger [10]	H_2O—B.N.S. model	4×10^{-16}	3·00
Cheung and Powles [11]	N_2—site site model	6×10^{-15}	0·10
Cheung and Powles [12]	N_2—site site plus quadrupole–quadrupole	7×10^{-15}	0·50
This study	CH_4—site site model	$2{\cdot}4\times10^{-15}$	0·15

In the table we give the time step length and the accuracy as determined by energy conservation, reported by a number of workers. The figures for the quaternion method are based on several 3000 step runs on the coexistence curve of methane between 100–200 K. Before we can compare relative efficiencies allowance must be made for differences in the characteristic time dependent behaviour of the different substances studied : nitrogen, methane and water.

As a first approximation we might assume that for the same level of accuracy, the time step lengths should be proportional to $\sqrt{I_p/T}$ where I_p is the smallest eigenvalue of the inertia tensor (6), and T is the temperature. This approximation becomes exact in the dilute gas limit and is obtained from the equipartition equation, $\frac{3}{2}kT=\frac{1}{2}I_p\omega_p{}^2$. Thus we would expect that the ratio of integration step lengths for nitrogen, methane and water would be approximately 2 : 1 : 0·5.

A direct comparison of the time steps given in the table is still not possible because the entries correspond to different degrees of accuracy. For instance, although the time steps used for the simulations of water are already much shorter than those used for the other simulations the energy conservation for water is also the poorest. Thus to achieve the degree of energy conservation of the other simulations, the time step used for water must inevitably become shorter still. Converting the time steps given in the table so that they all correspond to an integration accuracy of 0·15 per cent per 10 000 time steps we see that the actual ratio of time steps is approximately 2·3 : 1 : 0·05. This conversion was made using the known asymptotic variation of energy conservation with step length [8]. Comparing the actual ratio with the equipartition ratio we see that Cheung's method for nitrogen is of roughly the same efficiency as the quarternion method for methane. The Euler angle method for water is much more inefficient requiring roughly ten times as many function evaluations for the same accuracy.

It may be objected that our analysis has excluded consideration of the lack of conservation introduced by truncating the electrostatic intermolecular interactions [13]. However, comparing the relative accuracies of the simulations described in [3] and [11] shows that this is a relatively minor factor.

5. Conclusion

It has been demonstrated that under otherwise identical conditions, molecular dynamics programmes using quaternions to describe the orientations of molecules produce greatly superior results for energy conservation than do corresponding programmes using Euler angles. By analysing published results for the efficiency of molecular dynamics programmes developed by other workers, we have shown that the quaternion algorithm is roughly as efficient for rigid polyatomic molecules as is Cheung's algorithm for diatomics. Compared to the use of Euler angles both of these methods may be expected to decrease by an order of magnitude, the amount of computer time required for simulations of rigid molecules.

We thank the American Gas Association and the National Science Foundation for financial support of this work. It is a pleasure to thank the U.S. Military Academy, West Point, New York, for a generous allocation of computer time.

References

[1] Goldstein, H., 1971, *Classical Mechanics* (Addison-Wesley), Chap. 4.
[2] Barojas, J., Levesque, D., and Quentrec, B., 1973, *Phys. Rev.* A, **7,** 1092. Rahman and Stillinger [9, 10] do not comment on how they avoided this problem.
[3] Cheung, P. S. Y., 1976, *Chem. Phys. Lett.*, **40,** 19.
[4] Evans, D. J., 1977, *Molec. Phys.*, **34,** 317.
[5] Barker, J. A., Fisher, R. A., and Watts, R. O., 1971, *Molec. Phys.*, **21,** 657.
[6] Williams, D. E., 1966, *J. chem. Phys.*, **45,** 3770 ; 1967, *Ibid.*, **47,** 4680.
[7] Evans, D. J., Gubbins, K. E., Murad, S., and Streett, W. B. (to be submitted).
[8] Evans, D. J., and Watts, R. O., 1976, *Molec. Phys.*, **32,** 995.
[9] Stillinger, F. H., and Rahman, A., 1974, *J. chem. Phys.*, **60,** 1545.
[10] Rahman, A., and Stillinger, F. H., 1971, *J. chem. Phys.*, **55,** 3336.
[11] Cheung, P. S. Y., and Powles, J. G., 1975, *Molec. Phys.*, **30,** 921.
[12] Cheung, P. S. Y., and Powles, J. G., 1976, *Molec. Phys.*, **32,** 1383.
[13] See page 3340 of [10].

Notes to Reprint V.3

1. In addition to the method developed by Cheung (ref. [3]), an algorithm of a different type was worked out for the case of linear molecules (Singer et al., Mol. Phys. 33 (1977) 1957). The Singer algorithm can be shown to be a special case of the method of constraints (Reprint V.2).

2. The angles ϕ, θ and ψ of eq. (1) are equivalent, respectively, to the angles α, β and γ shown in fig. 2 of Reprint V.1.

3. The comparison made in the fig. (p. 329) relates to an Euler-angle algorithm in which no use is made of the device adopted by Barojas et al. (ref. [2]) and discussed on p. 327. In fact, all published accounts of work based on the Euler-angle approach make use of this trick; when this is done, the method is of comparable efficiency to that based on quaternions. For a recent application of the modified Euler-angle approach, see Levesque et al. (J. Chem. Phys. 79 (1983) 917).

Introduction of Andersen's demon in the molecular dynamics of systems with constraints

J. P. Ryckaert

Pool de Physique, Université Libre de Bruxelles, Campus Plaine, CP 223, Boulevard du Triomphe, 1050 Bruxelles, Belgium

G. Ciccotti[a]

S.R.M.P., C.E.N. Saclay, 91191 Gif Sur Yvette Cedex, France

(Received 4 October 1982; accepted 8 March 1983)

The method of constant pressure molecular dynamics (MD), developed by Andersen for monoatomic fluids is extended to the MD, in Cartesian coordinates, of molecular systems with constraints. Andersen's proof is easily generalized after decoupling internal degrees of freedom from the center-of-mass. Only these last degrees are directly affected by Andersen's transformation (demon). The Cartesian equations of motion of individual atoms are derived from a generalized Andersen's Lagrangian. The equations are quite similar to those of the usual MD simulation at constant volume apart from an additional term coupling the molecular center-of-mass and the volume of the sample. The volume appears now as a dynamical variable evolving from the imbalance between imposed external pressure and instantaneous values of the molecular stress tensor. Some numerical aspects are discussed and the technique is briefly illustrated for the case of rigid diatomic molecules.

I. INTRODUCTION

Various methods of computer simulation are now available for the study, within a classical mechanics approach, of condensed matter. They correspond in statistical mechanics to different ensembles according to the imposed thermodynamical and mechanical conditions.

Monte Carlo (MC) simulations in the NPT (isobaric–isothermal) ensemble have been exploited some years ago for simple liquids[2(a)] and their mixtures[2(b)]; recently, a regain of interest for simulations in NPH (isobaric–isoenthalpic) and NPT ensembles has followed the publication of Andersen's method of constant pressure molecular dynamics.[1]

As it has been beautifully illustrated by the recent work of Parrinello and Rahman on solid rubidium,[3] a slightly modified version of the constant pressure simulation technique seems a very promising tool for testing interatomic or intermolecular interaction in connection with the crystal structures and structural phase transitions. In fact, it seems reasonable to expect that, especially in solids, to allow greater density fluctuations in the small systems usually considered in simulations could lead to a faster relaxation of a given system toward its equilibrium state. On the contrary, usual constant volume simulations might well probe metastable states as a result of the too rigid boundary conditions.

In Andersen's method and in its subsequent applications,[3,4] only monoatomic systems have been considered. It is therefore natural to look for a generalization of the constant pressure simulations to molecular systems. If the molecules consist of a collection of atoms interacting through analytical potentials, both MC and MD methods apply equally well as a set of molecules can be viewed formally as an atomic system. However, the molecular model often consists of partially or totally rigid molecules. Then the methods need to be adapted in order to deal with the constraint relations.

Two different kinds of representation can be used when considering constrained systems. In the generalized coordinates approach, it seems natural to apply Andersen's demon to the center-of-mass of the molecules. With this choice, the volume fluctuations do not affect directly the internal structure of the molecule. This idea has been suggested by Brown[5] but it still needs a statistical mechanics justification, which in fact is easily found as a byproduct of the present work.

For rigid or semi-rigid molecules, the atomic description in terms of the Cartesian coordinates has proved to be very useful.[6,7] We have to adapt both the MC and Andersen methods to this atomic picture when the molecular model contains geometrical constraints. For the MC (in whatever ensemble), no recipe has been given so far to make a random move in terms of non-independent variables. On the contrary, Andersen's method can be generalized to yield a constant pressure version of the constant volume method of constraints.[6,7]

The object of the present paper is to present this dynamical representation of the isobaric–isoenthalpic ensemble for constrained systems represented by atomic Cartesian coordinates.

II. NVE AND NPH ENSEMBLES FOR MOLECULES WITH CONSTRAINTS

The Hamiltonian function H_1 associated to any molecular system can be written in terms of the Cartesian components of atomic positions $\{r_{i\alpha}\}$ and momenta $\{p_{i\alpha}\}$ (i and α indices denoting, respectively, specific atoms and molecules) as

$$H_1 = \sum_{\alpha=1}^{N} \sum_{i=1}^{n} \frac{\mathbf{p}_{i\alpha}^2}{2m_i} + V(\{r_{i\alpha}\}_{\substack{i=1,n \\ \alpha=1,N}}) , \qquad (1)$$

where V is the potential energy term which contains in general both inter and intramolecular contributions; N is the number of molecules, each of them being composed of n atoms.

[a]On leave of absence from: Ist. di Fisica "G. Marconi," Piazzale A. Moro 5, 00185 Roma (Italy).

 0021-9606/83/127368-07$02.10

We are interested, in rigid (or partly rigid) molecules where constraint relations are imposed between atomic positions and momenta. Such relations are essentially of two kinds[7]:

(i) rigid bond constraints, where two atoms $i_{i\alpha}$ and $j_{j\alpha}$ are connected by a bond of fixed length d_{ij}; this implies that, at any time

$$(\mathbf{r}_{i\alpha}-\mathbf{r}_{j\alpha})^2-d_{ij}^2=0\ , \qquad (2)$$

(ii) linear vectorial constraints, where the position of a given atom $\mathbf{r}_{i\alpha}$ follows at any time from the coordinates of a set of n_b basic atoms $\{r_{\nu\alpha}\}$ through a linear relation with constant coefficients C_ν, i.e., $\nu=1, n_b$,

$$\mathbf{r}_{i\alpha}-\sum_{1}^{n_b}{}_\nu C_\nu \mathbf{r}_{\nu\alpha}=0\ . \qquad (3)$$

(A simple example of this last constraints relation is provided by the rigid linear triatomic molecule where the coordinate of any atom can be written as a linear combination of the two other atomic positions.)

In classical mechanics, the constraint dynamics in Cartesian coordinates is most naturally obtained via the Lagrangian formalism[8]; standard MD simulations (at fixed NVE) have been implemented according to this atomic description.[6,7] The connection with statistical mechanics is the ergodic theorem which states that, given an arbitrary function $F(\{\mathbf{r}_{i\alpha}\},\{\mathbf{p}_{i\alpha}\})$, $\overline{F}$, the time average of F over the (constrained) trajectory is equal to a corresponding microcanonical average F_{NVE}. In this last ensemble average, we need to restrict the phase space according to the constraints connecting the variables. However, together with the constraints relations (2) and (3), we have also to include the constraints on the atomic momenta which follow from the time derivatives of relations (2) and (3), i.e.,

$$(\mathbf{r}_{i\alpha}-\mathbf{r}_{j\alpha})(\mathbf{p}_{i\alpha}/m_i-\mathbf{p}_{j\alpha}/m_j)=0\ , \qquad (4)$$

$$\mathbf{p}_{i\alpha}/m_i-\sum_{1}^{n_b}{}_\nu C_\nu \mathbf{p}_{\nu\alpha}/m_\nu=0\ . \qquad (5)$$

The vectorial constraints (3) or (5) represent three scalar constraints. In order to simplify the notation, for a molecule subject to l scalar constraints, the whole family of constraints (2) and (3) and (4) and (5) can be written in a unified way, respectively, as

$$C_k^\alpha \equiv C_k^\alpha(\{\mathbf{r}_{i\alpha}\})=0\ ,\quad k=1,l\ , \qquad (6)$$

$$\dot{C}_k^\alpha \equiv \dot{C}_k^\alpha(\{\mathbf{r}_{i\alpha},\mathbf{p}_{i\alpha}\})=0\ ,\quad k=1,l\ , \qquad (7)$$

where α denotes the molecule and k the constraint label. The microcanonical average F_{NVE} for a system of N identical molecules of n atoms subjected to l scalar constraints can be written in terms of Cartesian coordinates and momenta (see Appendix) as

$$F_{\mathrm{NVE}}(N,V,E)=[N!\,\Omega(N,V,E)]^{-1} \times \int_V d\mathbf{r}_{i\alpha}^{nN}\int d\mathbf{p}_{i\alpha}^{nN}\,\delta(H_1-E)\delta_c F\ , \qquad (8)$$

where

$$\Omega(N,V,E)=(N!)^{-1}\int_V d\mathbf{r}_{i\alpha}^{nN}\int d\mathbf{p}_{i\alpha}^{nN}\,\delta(H_1-E)\delta_c\ , \qquad (8a)$$

$$\delta_c=\delta_c(\{\mathbf{r}_{i\alpha},\mathbf{p}_{i\alpha}\}) =\prod_1^N{}_\alpha\prod_1^l{}_k\,\delta(C_k^\alpha)\,\delta\Big(\sum_1^l{}_j Z^{-1}_{\alpha,kj}\dot{C}_j^\alpha\Big)\ . \qquad (8b)$$

In Eq. (8b), $Z_{\alpha,kj}$ is the 1×1 matrix given by

$$Z_{\alpha,kj}=\sum_1^n{}_i\frac{1}{m_i}\frac{\partial C_k^\alpha}{\partial \mathbf{r}_{i\alpha}}\cdot\frac{\partial C_j^\alpha}{\partial \mathbf{r}_{i\alpha}}\ . \qquad (8c)$$

In Eq. (8), we adopt Andersen's notation for ensemble averages, here $F_{\mathrm{NVE}}(\mathrm{NVE})$, in which the subscript denotes the nature of the ensemble while the arguments represent numerical values of the ensemble parameters.

At this point, it is easy to introduce ensemble averages for the same mechanical system as defined by Eqs. (1)–(3). More specifically, we are presently interested to the NPH ensemble, the isobaric–isoenthalpic ensemble, for which we have

$$F_{\mathrm{NPH}}(N,P,H)=[N!\,\Gamma(N,P,H)]^{-1}\int_0^\infty dV\int_V d\mathbf{r}_{i\alpha}^{nN} \times\int d\mathbf{p}_{i\alpha}^{nN}\,\delta(H_1+PV-H)\delta_c F\ , \qquad (9)$$

where

$$\Gamma(N,P,H)=(N!)^{-1}\int_0^\infty dV\int_V d\mathbf{r}_{i\alpha}^{nN} \times\int d\mathbf{p}_{i\alpha}^{nN}\,\delta(H_1+PV-H)\delta_c\ . \qquad (9a)$$

The aim of this paper is to show how such ensemble averages F_{NPH} can be computed as time averages of $F(\mathbf{r}_{i\alpha},\mathbf{p}_{i\alpha})$ over a phase space trajectory produced by a molecular dynamics simulation relative to a closely related but different mechanical system, first proposed by Andersen in the case of atomic fluids. In this new molecular system, a dynamical variable Q representing the volume of the sample will be coupled to the molecular center-of-mass without directly affecting the intramolecular structure of the molecules. This procedure will be developed in detail and justified in the next section; however, we will anticipate somewhat and rewrite F_{NPH} in terms of variables which precisely decouple the center-of-mass coordinates from intramolecular coordinates.

Let us define the molecular center-of-mass $\{\mathbf{R}_\alpha\}$, their associated total momenta $\{\mathbf{P}_\alpha\}$, and a suitable set of $6nN$ internal variables $\{\mathbf{s}_{i\alpha},\mathbf{p}_{i\alpha}^s\}$ through the following transformation:

$$\mathbf{D}_\alpha\equiv\mathbf{R}_\alpha-(1/M)\sum{}_i m_i\mathbf{r}_{i\alpha}=0\ , \qquad (10)$$

$$\dot{\mathbf{D}}_\alpha\equiv(1/M)\Big(\mathbf{P}_\alpha-\sum{}_i\mathbf{p}_{i\alpha}\Big)=0\ , \qquad (11)$$

$$\mathbf{s}_{i\alpha}=\mathbf{r}_{i\alpha}-\mathbf{R}_\alpha\ , \qquad (12)$$

$$\mathbf{p}_{i\alpha}^s=\mathbf{p}_{i\alpha}-(m_i/M)\mathbf{P}_\alpha\ , \qquad (13)$$

where $M=\sum_i m_i$ is the total mass of the molecule. $\mathbf{D}_\alpha$ and $\dot{\mathbf{D}}_\alpha$ are symbols for the corresponding expressions.

Taking the defining relations (10) and (11) as linear vectorial constraints, the set of $6N(n+1)$ variables just defined gives an equivalent representation for the dynamical system given by the $6nN$ variables $\{\mathbf{r}_{i\alpha}, \mathbf{p}_{i\alpha}\}$ plus the Hamiltonian (1) and the constraints relations (6) and (7).

When performing this transformation, we find that:

(i) The Hamiltonian H_1 becomes

$$H_1=\sum_{1}^{N}{}_{\alpha}\left(\mathbf{P}_\alpha^2/2M+\sum_{1}^{n}{}_{i}(\mathbf{p}_{i\alpha}^s)^2/2m_i\right)+V(\{\mathbf{s}_{i\alpha}+\mathbf{R}_\alpha\}) . \quad (14)$$

(ii) It is easy to verify that the constraints relations (6), (7), and the matrix Z_α in Eq. (8c) are all independent from $\mathbf{R}_\alpha$, $\mathbf{P}_\alpha$, and maintain the same form with $\mathbf{s}(\mathbf{p}^s)$ replacing $\mathbf{r}(\mathbf{p})$

$$C_k^\alpha \equiv C_k^\alpha(\{\mathbf{s}_{i\alpha}\})=0 , \quad (6a)$$

$$\dot{C}_k^\alpha \equiv \dot{C}_k^\alpha(\{\mathbf{s}_{i\alpha}, \mathbf{p}_{i\alpha}^s\})=0 , \quad (7a)$$

$$\delta_c \equiv \delta_c(\{\mathbf{s}_{i\alpha}, \mathbf{p}_{i\alpha}^s\}) =\prod_{1}^{N}{}_{\alpha}\prod_{1}^{l}{}_{k}\delta(C_k^\alpha)\delta\left(\sum_{1}^{l}{}_{j}Z_{\alpha,kj}^{-1}(\{\mathbf{s}_{i\alpha}\})\dot{C}_j^\alpha\right) . \quad (8d)$$

(iii) Relations (10) and (11) read

$$\mathbf{D}_\alpha \equiv(-1/M)\sum_i m_i\mathbf{s}_{i\alpha}=0 , \quad (10a)$$

$$\dot{\mathbf{D}}_\alpha \equiv(-1/M)\sum_i \mathbf{p}_{i\alpha}^s=0 . \quad (11a)$$

(iv) Any ensemble average can be written in terms of this new set of variables. For example, for the NPH ensemble one finds:

$$F_{\mathrm{NPH}}(N,P,H)=[N!\,\Gamma(N,P,H)]^{-1}\int_0^\infty dV\int_V d\mathbf{R}_\alpha^N\int d\mathbf{P}_\alpha^N\int d\mathbf{s}_{i\alpha}^{nN}\int d(\mathbf{p}_{i\alpha}^s)^{nN} \times\delta(H_1+PV-H)\delta_c\,\delta_{\mathrm{c.m.}}F(\{\mathbf{R}_\alpha+\mathbf{s}_{i\alpha}, \mathbf{p}_{i\alpha}^s+(m_i/M)\mathbf{P}_\alpha\}) , \quad (15)$$

where

$$\delta_{\mathrm{c.m.}}=\prod_\alpha\delta(\mathbf{D}_\alpha)\delta[(1/M)\dot{\mathbf{D}}_\alpha] . \quad (16)$$

The set $\{\mathbf{R}_\alpha, \mathbf{P}_\alpha, \mathbf{s}_{i\alpha}, \mathbf{p}_{i\alpha}^s\}$ will prove particularly useful to devise a dynamical evolution such that, for any property, a time average over this trajectory is equivalent to a NPH ensemble average.

III. GENERALIZATION OF ANDERSEN'S METHOD TO MOLECULES WITH CONSTRAINTS

As in Andersen's original paper,[1] our starting point is the Lagrangian function $\mathcal{L}_1$ for the system of interest, i.e., a system of N identical molecules of n atoms with l constraints, already considered in the previous section. In terms of the variables $\{\{\mathbf{s}_{i\alpha}\}_{i=1,n}, \mathbf{R}_\alpha\}_{\alpha=1,N}$ introduced in Eqs. (10) and (12), the Lagrangian can be written

$$\mathcal{L}_1=\frac{1}{2}\sum_\alpha\left(\sum_i m_i\dot{\mathbf{s}}_{i\alpha}+M\dot{\mathbf{R}}_\alpha^2\right)-V(\{\mathbf{s}_{i\alpha}+\mathbf{R}_\alpha\}) \quad (17)$$

with the associated constraint relations (6a) and (10a).

We now parallel Andersen procedure by introducing the Lagrangian

$$\mathcal{L}_2=\frac{1}{2}\sum_\alpha\left(\sum_i m_i\dot{\sigma}_{i\alpha}^2+MQ^{2/3}\dot{\rho}_\alpha^2\right) -V(\{\sigma_{i\alpha}+Q^{1/3}\dot{\rho}_\alpha\})+\tfrac{1}{2}M_p\dot{Q}^2-P_{\mathrm{ext}}Q , \quad (18)$$

which is function of variables $\{\sigma_{i\alpha}, \rho_\alpha, Q\}$ to which, substituting $\sigma_{i\alpha}$ for $\mathbf{s}_{i\alpha}$, the constraint relations of the previous model are added. The Andersen transformation affects directly only the center-of-mass variables while adding the new variable Q. The range of variable ρ_α is now a cube of unit side, Q being an unrestricted positive variable. M_p is the inertial factor of the new variable Q and P_{ext} is a free parameter which will play in the following the role of the external pressure. This Lagrangian will generate the dynamics in which we are interested.

The isobaric–isoenthalpic average of any observable F of the dynamical system defined by Eq. (17) can be explicitly evaluated as a time average of the corresponding observable over the dynamical trajectory defined by the Lagrangian (18). To formulate properly this statistical mechanics equivalence, we need the Hamiltonian formalism. The momenta $\{\pi_{i\alpha}, \pi_\alpha, \pi\}$ conjugate to the variables $\{\sigma_{i\alpha}, \rho_\alpha, Q\}$ are given by

$$\pi_{i\alpha}=\frac{\partial\mathcal{L}_2}{\partial\dot{\sigma}_{i\alpha}}=m_i\dot{\sigma}_{i\alpha} ,$$

$$\pi_\alpha=\frac{\partial\mathcal{L}_2}{\partial\dot{\rho}_\alpha}=M\,Q^{2/3}\rho_\alpha , \quad (19)$$

$$\pi=\frac{\partial\mathcal{L}_2}{\partial\dot{Q}}=M_p\dot{Q}$$

in terms of which the Hamiltonian H_2 is

$$H_2=\frac{1}{2}\sum_\alpha\left(\sum_i\pi_{i\alpha}^2/m_i+\pi_\alpha^2/MQ^{2/3}\right) +V(\{\sigma_{i\alpha}+Q^{1/3}\rho_\alpha\})+\tfrac{1}{2}\pi^2/M_p+P_{\mathrm{ext}}Q . \quad (20)$$

We now define a correspondence between the variables of the new system and those of the original one, i.e.,

$$\mathbf{s}_{i\alpha}=\sigma_{i\alpha} , \quad \mathbf{p}_{i\alpha}^s=\pi_{i\alpha} ,$$

$$\mathbf{R}_\alpha=Q^{1/3}\rho_\alpha , \quad \mathbf{P}_\alpha=\pi_\alpha/Q^{1/3} , \quad V=Q . \quad (21)$$

Using this correspondence, any observable of the original system $F(\{\mathbf{r}_{i\alpha}, \mathbf{p}_{i\alpha}\}, V)=F(\{\mathbf{s}_{i\alpha}+\mathbf{R}_\alpha, \mathbf{p}_{i\alpha}^s+(m_i/M)\mathbf{P}_\alpha\}, V)$ identifies an observable

$$\hat{F}\equiv F[\{\mathbf{r}_{i\alpha}(\sigma_{i\alpha}, \rho_\alpha, Q), \; \mathbf{p}_{i\alpha}(\pi_{i\alpha}, \pi_\alpha, Q)\}, Q] \quad (22)$$

for the new system. We can now closely follow Andersen demonstration.[1]

Using the ergodic theorem, we have as in Eq. (15):

Notes on p. 379

$$\bar{\hat{F}} = \hat{F}_{NVE}(N, 1, H)$$

$$= [N!\Omega_2(N, 1, H)]^{-1} \int d\pi \int_0^\infty dQ \int_1 d\boldsymbol{\rho}_\alpha \int d\pi_\alpha \int d\sigma_{i\alpha}^{nN} \int d\pi_{i\alpha}^{nN}\, \delta(H_2 - H)\delta_c\, \delta_{c.m.}\, \hat{F} , \quad (23)$$

where [see Eq. (8b)] $\delta_c = \delta_c(\{\sigma_{i\alpha}, \pi_{i\alpha}\})$ and

$$\delta_{c.m.} = \Pi_\alpha \delta(\mathbf{D}_\alpha)\delta[(1/M)\dot{\mathbf{D}}_\alpha] .$$

Changing the variables of integration through Eq. (21), one obtains

$$\bar{\hat{F}} = \frac{\int_{-\infty}^{+\infty} d\pi \int_0^\infty dV \int_V d\mathbf{R}_\alpha \int d\mathbf{P}_\alpha \int d\mathbf{s}_{i\alpha}^{nN} \int d\mathbf{p}_{i\alpha}^{nN}\, \delta(H_1 + P_{ext} V + \pi^2/2M_p - H)\delta_c\, \delta_{c.m.}\, F}{\int_{-\infty}^{+\infty} d\pi \int_0^\infty dV \int_V d\mathbf{R}_\alpha \int d\mathbf{P}_\alpha \int d\mathbf{s}_{i\alpha}^{nN} \int d\mathbf{p}_{i\alpha}^{s,nN}\, \delta(H_1 + P_{ext} V + \pi^2/2M_p - H)\delta_c\, \delta_{c.m.}}$$

$$= \frac{\int_{-\infty}^{+\infty} d\pi \Gamma(N, P_{ext}, H - \pi^2/2M_p) F_{NPH}(N, P_{ext}, H - \pi^2/2M_p)}{\int_{-\infty}^{+\infty} d\pi \Gamma(N, P_{ext}, H - \pi^2/2M_p)} , \quad (24)$$

where in the last step use has been made of definitions (9) and (9a).

The final part of Andersen proof applies now without any change. Therefore, expanding $F_{NPH}(N, P_{ext}, H - \pi^2/M_p)$ in a power series of $\pi^2/2M_p$ up to second order, performing explicit integration on variable π and recombining the terms, one finds

$$\bar{\hat{F}} = F_{NPH}(N, P_{ext}, H - \overline{\pi}^2/2M_p) + O(N^{-2}) , \quad (25)$$

provided that F is an intensive quantity. Otherwise the correction term would be of order $O(N^{-1})$. As in the case of atomic fluids, the average $\overline{\pi}^2/2M_p$ is an intensive quantity because it is equal to $1/2\, kT$ where T is the temperature of the new system in the microcanonical ensemble at $V = 1$ and $E = H$.

IV. EQUATIONS OF MOTION AND MOLECULAR DYNAMICS IMPLEMENTATION

A. Equations of motion

The Lagrangian formulation (18) of the "scaled" mechanical system is described in terms of a volume dynamical variable Q, $3N$ scaled center-of-mass positions $\boldsymbol{\rho}_\alpha$, and $3nN$ intramolecular variables $\sigma_{i\alpha}$ subjected to constraints (6a) and (10a). As Q and $\boldsymbol{\rho}_\alpha$ are not involved in these constraints relations, their Lagrangian equations of motion are obtained using the standard procedure

$$\ddot{Q} = (1/M_p)\Big\{(1/3Q)\Big(\sum_\alpha M Q^{2/3} \dot{\boldsymbol{\rho}}_\alpha^2 + Q^{1/3} \sum_\alpha \boldsymbol{\rho}_\alpha \cdot \mathbf{F}_\alpha\Big) - P_{ext}\Big\} , \quad (26)$$

$$\ddot{\boldsymbol{\rho}}_\alpha = -(2/3)Q^{-1}\dot{Q}\boldsymbol{\rho}_\alpha + Q^{-1/3}\mathbf{F}_\alpha/M , \quad (27)$$

where

$$\mathbf{F}_\alpha = \sum_i \mathbf{F}_{i\alpha} = -\sum_i (\partial V/\partial \sigma_{i\alpha}) .$$

For variables $\sigma_{i\alpha}$ we derive the Lagrangian equation of the first kind with explicit constraint forces[8]; we find

$$m_i \sigma_{i\alpha} = \mathbf{F}_{i\alpha} - m_i \mu_\alpha - \sum_{k=1}^{l} \lambda_{k\alpha}(\partial C_k^\alpha/\partial \sigma_{i\alpha}) , \quad (28)$$

where $\{\mu_\alpha\}$ and $\{\lambda_{k\alpha}\}$ are Lagrangian multipliers associated to constraints $\mathbf{D}_\alpha$ (10a) and C_k^α (6a), respectively.

Time deriving $\mathbf{D}_\alpha$ [Eq. (10a)] twice and substituting Eq. (28) for $\ddot{\sigma}_{i\alpha}$ we get $\boldsymbol{\mu}_\alpha = \mathbf{F}_\alpha/M$. Equation (28) takes the form

$$m_i\ddot{\sigma}_{i\alpha} = \mathbf{F}_{i\alpha} - (m_i/M)\mathbf{F}_\alpha - \sum_{k=1}^{l} \lambda_{k\alpha}(\partial C_k^\alpha/\partial \sigma_{i\alpha}) . \quad (28a)$$

We can now easily come back to the atomic variables $\mathbf{r}_{i\alpha}$ which are most common in MD simulation with Cartesian coordinates. Using the correspondence (21) and then Eq. (12), Eqs. (26), (27), and (28a) can be rewritten

$$m_i\ddot{\mathbf{r}}_{i\alpha} = \mathbf{F}_{i\alpha} - \sum_k \lambda_{k\alpha}(\partial C_k^\alpha/\partial \mathbf{r}_{i\alpha}) + (m_i/3MV)[\ddot{V} - (2/3)V^{-1}\dot{V}^2]\sum_i m_i \mathbf{r}_{i\alpha} , \quad (29)$$

$$M_p\ddot{V} = (3V)^{-1}\sum_\alpha [M(\dot{\mathbf{R}}_\alpha - (3V)^{-1}\dot{V}\mathbf{R}_\alpha)^2 + \mathbf{R}_\alpha \cdot \mathbf{F}_\alpha] - P_{ext} , \quad (30)$$

where $\mathbf{R}_\alpha$ and $\mathbf{R}_\alpha$ are no longer extra variables but short symbols for their definition in terms of $\mathbf{r}_{i\alpha}$, $\dot{\mathbf{r}}_{i\alpha}$. Moreover, if the dynamics corresponding to the Lagrangian (18) is expressed in terms of the variables $\{\mathbf{r}_{i\alpha}\}$ [Eqs. (29) and (30)] one can need the generalized momenta corresponding to $\{\mathbf{r}_{i\alpha}\}$. Using Eqs. (18), (12), (13), and (21) one finds $\mathbf{p}_{i\alpha} = m_i(\dot{\mathbf{r}}_{i\alpha} - (1/3Q)\dot{Q}\mathbf{R}_\alpha)$. For example, the temperature is given by the usual relation $\langle \sum_i \mathbf{p}_{i\alpha}^2/2m_i\rangle = (f/2)kT$, where f is the number of degrees of freedom of the molecule. It is apparent from Eq. (30) that the change of the volume come out simply from the difference between the trace of the molecular stress tensor and the external pressure. Equation (29) can be further simplified by explicitly solving for the Lagrangian multipliers corresponding to linear constraint relations. This implies a reduction of the variables. The technique to achieve this reduction is discussed in details elsewhere.[7]

B. MD implementation

Apart from the additional Eq. (30) and the last term in their right-hand side, Eqs. (29) are the usual equations of motion in Cartesian coordinates of a molecular system with constraints. Their numerical implementation using the Verlet integration algorithm has been discussed in Refs. 6 and 7. This algorithm has proved to be very efficient. However, it applies directly only when the acceleration is independent from velocities. This is not true for the present equations but with a slight modification of the algorithm the simplicity of this integration still is useful. In the standard procedure, knowing the positions of all particles at time t and $t-h$ (where h is the time step and t the running time), one computes the positions at time $t+h$ and then the velocities at time t. The trajectory is obtained with an error of $O(h^4)$ while the velocities are computed with an error of $O(h^2)$. In the present case, with the knowledge of an extra configuration at time $t-2h$, one can predict velocities at time t, with a precision of order $O(h^2)$, using the formula

$$\mathbf{v}(t)=\frac{3\mathbf{r}(t)-4\mathbf{r}(t-h)+\mathbf{r}(t-2h)}{2h}+O(h^2)\ . \tag{31}$$

This prediction can be used to compute the acceleration. One recovers the standard procedure with the same precision. The reason is that the acceleration appears with a h^2 factor.

C. Illustration

MD experiments of molecular systems at constant pressure can be performed with slightly modified versions of existing standard MD programs. We tested the present technique on the fluid of rigid homonuclear diatomic molecules considered in Ref. 9. The potential energy between two molecules is a sum of four site–site shifted Lennard-Jones (LJ) interactions $[v^s(r)=v(r)-v(r^*)$ for $r\le r^*$ and $v^s(r)=0$ for $r>r^*]$, where $v(r)=4\epsilon[(\sigma/r)^{12}-(\sigma/r)^6]$ and $r^*=2.7\sigma$ is the adopted cutoff distance.

We first performed a standard MD experiment at a density of $\rho=0.41\sigma^{-3}$ and $kT=1.209\epsilon$ which corresponds to a dense liquid at normal temperature.

We then simulated the same system by a constant pressure–constant enthalpy MD according to Eqs. (29) and (30). The external pressure P_{ext} was fixed to the average pressure P_{NVE}(NVE) obtained in the first experiment and the total enthalpy was adjusted to

$$H=E+P_{ext}V+(1/2)kT\ , \tag{32}$$

where E, V, and T are, respectively, the total energy, volume, and average temperature relative to the NVE simulation (see Table I). The last term of the right-hand side of Eq. (32) is added to take into account the kinetic energy associated with the extra degree of freedom, i.e., the volume.

In both experiments, we considered a system of 216 molecules and adopted the time step $h=0.003\sigma(M/\epsilon)^{1/2}$ (where M is the total mass of the molecule). In the same units, the total time τ of integration was, respectively, $\tau=4.6$ and $\tau=8.7$ for NVE and NPH simulations. In the NPH MD experiment we fixed the inertial factor associated with the volume to $M_p=3.3\times10^{-4}\ M\sigma^{-4}$ as a similar value proved to be adequate for monoatomic LJ fluids.[4]

TABLE I. Comparison of NVE and NPH simulations of a liquid of homonuclear diatomic molecules at a similar thermodynamical state $\rho=0.41\sigma^{-3}$, $kT=1.2\epsilon$. If F is a fixed quantity its value is affected by the symbol *, while if F is a fluctuating quantity, δF is the rms deviation of F. (n_f is the total number of degrees of freedom, $H_s=E+PV$ is the total enthalpy of the system of interest while H is the total enthalpy including the kinetic energy associated to V.)

F	F_{NVE}	δF_{NVE}	F_{NPH}	δF_{NPH}
N	216*		216*	
n_f	1077*		1078*	
E/ϵ	−1281.09*	0.05[a]	−1282.6	b
kT/ϵ	1.209	0.029	1.201	0.043
$P/\epsilon\sigma^{-3}$	1.508	0.216	1.508*	
V/σ^3	526.8*		527.9	3.7
H_s/ϵ	−486.7	113.8	−486.6	b
H/ϵ			−485.98*	0.05[a]

[a]These numbers should theoretically be zero and are indicated here only to give the magnitude of the fluctuations due to the use of the numerical integration scheme.
[b]Quantities not evaluated.

In Table I, we list some fixed or averaged quantities relative to these experiments. The results suggest that both experiments are representative of the same thermodynamical state. Note finally the similar and good degree of conservation of the conserved quantities, i.e., the total energy in the NVE case and total enthalpy in the NPH case. Therefore, the modified version of the Verlet algorithm, introduced in our constant pressure simulation, does not introduce any numerical problem.

V. CONCLUSIONS

The extension of the constant pressure MD technique to molecular systems with constraints has been made possible by applying Andersen's scaling transformation to the molecular center-of-mass. This transformation was performed on the Lagrangian of the system of polyatomic molecules after substitution of atomic coordinates $\mathbf{r}_{i\alpha}$ by $\mathbf{s}_{i\alpha}+\mathbf{R}_\alpha$ where $\mathbf{s}_{i\alpha}$ is the position of atom i of molecule α relative to the center of mass $\mathbf{R}_\alpha$. Three additional variables per molecule have been introduced, and as a consequence, three constraint relations relating the $\mathbf{s}_{i\alpha}$ (but not $\mathbf{R}_\alpha$) have been included to the description [see Eqs. (17) and (10a)].

In the third section, we switched from the Lagrangian formalism to the Hamiltonian one to prove that time average on the scaled system corresponds to isobaric–isoenthalpic averages on the original system.

The equations of motion corresponding to the modified Lagrangian subject to all constraints were derived and written in terms of the usual Cartesian atomic coordinates in Sec. IV.

While these equations have been given within the framework of molecules with constraints treated in terms of the Cartesian atomic coordinates, they have wider implications.

(i) For unconstrained polyatomic molecules, they provide an alternative to the original Andersen's method

Notes on p. 379

where the volume couples directly to individual atoms.

(ii) Summing Eqs. (29) yields the corresponding equations of motion of the molecular center-of-mass in our constant pressure MD. In this way, Brown's suggestion[5] can be justified and therefore, the present method can also be performed in generalized coordinates if one simply adds to the center-of-mass equations, mentioned above, the equations of motion relative to the internal coordinates which are unaltered.

We are presently investigating the use of HPN techniques to test molecular interaction models on molecular crystal structures at nonzero temperature.

For such a purpose the generalized tensorial version of the HPN method due to Parrinello and Rahman is very useful. With the present technique it is easy to adapt this version to the case of constrained molecules.

ACKNOWLEDGMENTS

We are grateful for fruitful discussion with A. Rahman. We thank C. Moser for his hospitality at CECAM and for a critical reading of the manuscript. We thank the referee for pointing out a slight mistake in Eq. (8) which motivated the addition of the appendix. This work has been supported by the NATO Grant No. 1865.

APPENDIX

Let us introduce a set $\{q_{j\alpha}, p^q_{j\alpha}\}_{j=1,f}$ of $f = 3n - l$ independent generalized coordinates and conjugate momenta describing the mechanical state of molecule α ($\alpha = 1, N$) containing n atoms subject to the l geometrical constraints of Eq. (6).

Let us call $F^c(q, p^q)$ the dynamical variable equivalent to $F(\{r, p\})$ in Eq. (8),

$$F^c(\{q, p^q\}) = F(\{r(q), p(q, p^q)\}) \ . \tag{A1}$$

We now show that the familiar expression of the microcanonical average

$$F^c_{\mathrm{NVE}} = [N!\Omega(N, V, E)]^{-1} \times \int dq\, dp^q\, F^c(q\,p^q)\delta[H_c(q\,p^q) - E] \ , \tag{A2}$$

where

$$\Omega(N, V, E) = [N!]^{-1}\int dq\, dp^q\, \delta[H_c(q\,p^q) - E] \ , \tag{A3}$$

$$H_c(q\,p^q) = \frac{1}{2}\sum_{\alpha=1}^{N} \sum_{j=1}^{f} \sum_{k=1}^{f} \bar{A}^{-1}_{\alpha jk} p^q_{j\alpha} p^q_{k\alpha} + V_c(\{q\}) \ , \tag{A4}$$

and

$$\bar{A}_{\alpha jk} = \sum_{i=1}^{n} m_i \frac{\partial \mathbf{r}_{i\alpha}}{\partial q_{j\alpha}} \cdot \frac{\partial \mathbf{r}_{i\alpha}}{\partial q_{k\alpha}} \tag{A5}$$

is equivalent to our Eq. (8).

The system is described in Cartesian coordinates by the Lagrangian

$$\mathcal{L}(r\,\dot{r}) = \sum_{i,\alpha} \tfrac{1}{2} m_i \dot{\mathbf{r}}^2_{i\alpha} - V(\{r\}) \ , \tag{A6}$$

to which we add the constraints relations, Eqs. (6). The corresponding dynamics is given by the Lagrange equations of motion of first kind.[8] Adding to the fN generalized coordinates q_α, the set of $l \times N$ constraints relations C^α, Eqs. (6), as new coordinates which complete the set, we have[10,11]

$$\mathbf{r}_\alpha = \mathbf{r}_\alpha(\{q_\alpha, C^\alpha\}) \equiv \mathbf{r}_\alpha(\{u\}) \ , \tag{A7}$$

where $u = \{u_{\alpha s}\}$, $s = 1, 3n$, $\alpha = 1, N$ is a short symbol for $\{q, C\}$. The Lagrangian in the new variables is given, in matricial notation, by

$$\mathcal{L}'(u, \dot{u}) = \frac{1}{2}\sum_{\alpha=1}^{N} \dot{u}^T_\alpha M_\alpha \dot{u}_\alpha - V'(\{u\}) \ , \tag{A8}$$

where u^T is the transpose of vector u and

$$M_\alpha = \begin{pmatrix} A_\alpha & B_\alpha \\ B^T_\alpha & \Gamma_\alpha \end{pmatrix} \tag{A9}$$

is the metric matrix whose elements are given by

$$(M_\alpha)_{\mu\nu} = \sum_{i=1}^{3n} m_i \frac{\partial \mathbf{r}_{i\alpha}}{\partial u_\mu} \cdot \frac{\partial \mathbf{r}_{i\alpha}}{\partial u_\nu} \ . \tag{A10}$$

The submatrices A_α, B_α, Γ_α have as dimensions, respectively $(f\times f)$, $(f\times l)$, $(l\times l)$.

The corresponding Hamiltonian is given by

$$H'(u, p^u) = \frac{1}{2}\sum_{\alpha=1}^{N} p^{uT}_\alpha M^{-1}_\alpha p^u_\alpha + V'(u) \ , \tag{A11}$$

where

$$p^u_\alpha = \frac{\partial \mathcal{L}'}{\partial \dot{u}_\alpha} = M_\alpha \dot{u}_\alpha \tag{A12}$$

and

$$M^{-1}_\alpha \equiv \begin{pmatrix} \Delta_\alpha & E_\alpha \\ E^T_\alpha & Z_\alpha \end{pmatrix} = \left(\sum_{i=1}^{3n} \frac{1}{m_i} \frac{\partial u_{\alpha\mu}}{\partial \mathbf{r}_{i\alpha}} \cdot \frac{\partial u_{\alpha\nu}}{\partial \mathbf{r}_{i\alpha}}\right) . \tag{A13}$$

The Lagrangian of the system with constraints is recovered from $\mathcal{L}'$ by putting in Eq. (A8) $C^\alpha = \dot{C}^\alpha = 0$.

For the Hamiltonian description we have

$$\dot{C}^\alpha = E^T_\alpha p^q_\alpha + Z_\alpha p^c_\alpha \ . \tag{A14}$$

Therefore, $\dot{C}^\alpha = 0$ corresponds to $p^c_\alpha = -\bar{Z}^{-1}_\alpha \bar{E}^T_\alpha p^q_\alpha$, where $\bar{}$ means the matrices are evaluated at $C^\alpha = 0$. A direct calculation shows that the Hamiltonian of the constrained system is obtained for $C^\alpha = 0$, $p^c_\alpha = -\bar{Z}^{-1}_\alpha \bar{E}^T_\alpha p^q_\alpha$,

$$H_c(q_\alpha, p^q_\alpha) = H'(q_\alpha, p^q_\alpha, C_\alpha = 0, p^c_\alpha = -Z^{-1}_\alpha E^T_\alpha p^q_\alpha) \ . \tag{A15}$$

The explicit inversion of Eq. (A12) gives Eq. (A14), where $E^T_\alpha = Z_\alpha B^T_\alpha A^{-1}_\alpha$ and $Z_\alpha = (\Gamma_\alpha - B^T_\alpha A^{-1}_\alpha B_\alpha)^{-1}$. Then we have

$$p^c_\alpha + B^T_\alpha A^{-1}_\alpha p^q_\alpha = Z^{-1}_\alpha \dot{C}_\alpha \ . \tag{A16}$$

Finally we can write

$$F^c_{\mathrm{NVE}} = [N!\Omega(N, V, E)]^{-1}\int dq\, dp^q\, F^c(q\,p^q)\delta[H_c(q\,p^q) - E]$$

$$= [N!\Omega(N, V, E)]^{-1}\int du\, dp^u\, F'(u\,p^u)\delta[H'(u\,p^u) - E]\delta(C)\delta(p^c + B^T A^{-1} p^q)$$

$$= [N!\Omega(N, V, E)]^{-1} \int dr\, dp\, F(rp)\delta[H(rp) - E]\delta(C)\delta(Z^{-1}\dot{C}) . \quad \text{(A17)}$$

The last equality follows from the fact that the variable transformation $(u, p^u) \to (r, p)$ is a point transformation, therefore a canonical one. Moreover the Jacobian of a canonical transformation is ± 1 because the volume is a canonical invariant.[12]

One can easily be convinced that the matrix M_α does not depend on R_α. Writing

$$\mathbf{r}_{i\alpha} = \mathbf{R}_\alpha + \sum_{j \atop i} \frac{m_j}{M}(\mathbf{r}_{i\alpha} - \mathbf{r}_{j\alpha}) \quad \text{(A18)}$$

and expressing the $\mathbf{r}_{i\alpha}$ in terms of $l + (f-3)$ internal coordinates and constraint coordinates one immediately verifies that M_α, Eq. (A10), does not depend on R_α.

[1]H. C. Andersen, J. Chem. Phys. 72, 2384 (1980).
[2](a) W. W. Wood, J. Chem. Phys. 48, 415 (1968); 52, 729 (1970); (b) I. R. McDonald, Mol. Phys. 23, 41 (1972).
[3]M. Parrinello and A. Rahman, Phys. Rev. Lett. 45, 1196 (1980).
[4]J. M. Haile and H. W. Graben, J. Chem. Phys. 73, 2412 (1980).
[5]D. Brown, Information Quarterly for MD and MC Simulations 4, 32 (1982).
[6]J. P. Ryckaert, G. Ciccotti, and H. J. C. Berendsen, J. Comp. Phys. 23, 327 (1977).
[7]G. Ciccotti, M. Ferrario, and J. P. Ryckaert, Mol. Phys. 47, 1253 (1982).
[8]T. C. Bradbury, *Theoretical Mechanics* (Wiley–Interscience, New York, 1968), Chap. XI.
[9]J. P. Ryckaert, A. Bellemans, and G. Ciccotti, Mol. Phys. 44, 979 (1981).
[10]K. R. Symon, *Mechanics* (Addison–Wesley, Reading, Mass., 1960), Chap. 9.
[11]M. Fixman, Proc. Natl. Acad. Sci. U.S.A. 71, 3050 (1974).
[12]H. Goldstein, *Classical Mechanics*, 2nd ed. (Addison–Wesley, Reading, Mass., 1980), Chap. 9.

Notes to Reprint V.4

1. The following misprints should be noted. (a) Two lines before eq. (3), the coordinates of the basic atoms should be written as vectors. (b) In eq. (18), "$Q^{1/3}\dot{\boldsymbol{\rho}}_\alpha$" should be replaced by "$Q^{1/3}\boldsymbol{\rho}_\alpha$". (c) In eq. (19), "$\boldsymbol{\rho}_\alpha$" should be replaced by "$\dot{\boldsymbol{\rho}}_\alpha$". (d) In eqs. (23) and (24), the variables $\boldsymbol{\rho}_\alpha$, π_α, $\mathbf{R}_\alpha$ and $\mathbf{P}_\alpha$ should all carry a superscript N, and $\mathbf{p}_{i\alpha}$ in the numerator of eq. (24) should carry the same superscript as in the denominator. (e) In eq. (27), "$\boldsymbol{\rho}_\alpha$" should be replaced by "$\dot{\boldsymbol{\rho}}_\alpha$". (f) In eq. (28), "$\sigma_{i\alpha}$" on the left-hand side should be replaced by "$\ddot{\sigma}_{i\alpha}$". (g) In eq. (29), the sum on i should be a sum on j. (h) In the equation seven lines after eq. (30), "Q" and "$\dot{Q}$" should be replaced, respectively, by "V" and "$\dot{V}$". (i) In eq. (A15), the quantity E_α^T should carry a tilde.

2. The generalization mentioned on p. 7373 (top left) was carried through by Ferrario and Ryckaert (Mol. Phys. 54 (1980) 587), who also show how the method of constraints can be adapted to the constant-temperature case.

MOLECULAR PHYSICS, 1983, VOL. 50, NO. 5, 1055–1076

Constant pressure molecular dynamics for molecular systems†

by SHUICHI NOSÉ and M. L. KLEIN
Chemistry Division, National Research Council Canada,
Ottawa, Ontario, K1A 0R6, Canada

(*Received* 29 *April* 1983 ; *accepted* 15 *July* 1983)

Technical aspects of the constant pressure molecular dynamics (MD) method proposed by Andersen and extended by Parrinello and Rahman to allow changes in the shape of the MD cell are discussed. The new MD method is extended to treat molecular systems and to include long range charge–charge interactions.

Results on the conservation laws, the frequency of oscillation of the MD cell, and the equations which constrain the shape of the MD cell are also given. An additional constraint is introduced to stop the superfluous MD cell rotation which would otherwise complicate the analysis of crystal structures. The method is illustrated by examining the behaviour of solid nitrogen at high pressure.

1. INTRODUCTION

For many years, the molecular dynamics (MD) method was limited to the microcanonical (E, V, N) ensemble in which the volume and the total energy are conserved. On the other hand, the Monte Carlo (MC) method was long ago [1] extended from the normal canonical (T, V, N) ensemble to treat the constant pressure (T, P, N) and the grand canonical (T, V, μ) ensembles. It was Andersen [2] who first proposed the new constant pressure (NMD) method in which the volume was allowed to fluctuate, its average value being determined by the balance between the internal stresses in a system and the externally set pressure, P_{ex}. Andersen's method, which approximates a constant enthalpy–constant pressure (H, P, N) ensemble, introduces a parameter C, which determines the rate at which the volume fluctuates. At equilibrium, static quantities evaluated in this approximate (H, P, N) ensemble are independent of the value of this parameter [2, 3].

Parrinello and Rahman (PR) extended the method to allow the MD cell to change its *shape* and in this way they were able to explore the relationship between interaction potentials and crystal structures [4]. They demonstrated that Rb atoms interacting with a realistic potential and initially disposed in a f.c.c. structure changed spontaneously to a b.c.c. arrangement. The reverse transformation was observed for atoms interacting with a Lennard-Jones potential. The NMD method was then further extended to include external stresses and to discuss structural transformations in a Ni crystal [5]. More recently, NMD was used to investigate the temperature induced transformation between the non-conducting and superionic phases of AgI [6].

† Issued as N.R.C.C. No. 22766.

The first application of NMD to a molecular system was that of Nosé and Klein who carried out a simulation of a nitrogen crystal under pressure of about 70 kbar [7]. Structural transformations in carbon tetrafluoride were also investigated [8]. In both of these studies changes in the volume and shape of the MD cell occurred at the phase transitions. Clearly, the NMD method has a potentially wide application in the study of solid–solid and solid–fluid phase transitions.

The present article is devoted mainly to the technical aspects of the NMD method. The derivation of the dynamical equations and the conservation laws are discussed in § 2. For molecular systems, the possibility of an antisymmetric stress tensor means that the MD cell sometimes rotates as a whole. This rotation, while physically unimportant, leads to the difficulties in the analysis of molecular orientations and crystal structures. In § 3 we outline a procedure that avoids this problem. In § 4 an example of the NMD method is given. Finally, some comments on programming are collected in § 5.

2. Constant pressure molecular dynamics

2.1. *Equations of motion*

We first derive the equations of motion for a system of rigid molecules. To simplify the derivation, we make the assumption that the interactions between two molecules can be expressed as the sum of site–site interactions which depend only on the distance between the two sites. The extension to more general forms of interaction is straightforward.

Following PR, the coordinates of site k of molecule i are written as [4]

$$\mathbf{r}_i^k = \mathbf{u}_1 s_{i1}^k + \mathbf{u}_2 s_{i2}^k + \mathbf{u}_3 s_{i3}^k = \mathbf{h}\mathbf{s}_i^k = \mathbf{h}\mathbf{s}_i + \mathbf{p}_i^k,$$

where $\mathbf{u}_1$, $\mathbf{u}_2$, and $\mathbf{u}_3$ are vectors forming the edges of the MD cell. The value of $s_{i\alpha}^k$ is limited to the range $0 \leqslant s_{i\alpha}^k \leqslant 1$ ($\alpha = 1, 2, 3$) and we call $\mathbf{s}_i^k$ the *scaled frame* coordinate and $\mathbf{r}_i^k$ the *real frame* coordinate. The vector $\mathbf{r}_i$ (and $\mathbf{s}_i$) without an upper index locates the centre of mass of molecule i. The vector $\mathbf{p}_i^k$ is drawn from the centre of mass of molecule i to the site k, in the real frame, and $\mathbf{h} = (\mathbf{u}_1, \mathbf{u}_2, \mathbf{u}_3)$ is a transformation matrix from the scaled frame to the real frame, that acts only on the molecular centres of mass. The square of the distance between $\mathbf{r}_i^k$ and $\mathbf{r}_j^l$ is given by

$$\begin{aligned}(r_{ij}^{kl})^2 &= |\mathbf{r}_{ij}^{kl}|^2 = |\mathbf{r}_i^k - \mathbf{r}_j^l|^2 = (\mathbf{r}_i^k - \mathbf{r}_j^l)^t(\mathbf{r}_i^k - \mathbf{r}_j^l) = (\mathbf{s}_i^k - \mathbf{s}_j^l)^t\, \mathbf{h}^t\, \mathbf{h}(\mathbf{s}_i^k - \mathbf{s}_j^l) \\ &= (\mathbf{s}_i^k - \mathbf{s}_j^l)^t \mathbf{G}(\mathbf{s}_i^k - \mathbf{s}_j^l),\end{aligned}$$

where $\mathbf{G} = \mathbf{h}^t\mathbf{h}$ is a metric tensor. The volume of a MD cell is given by

$$V = \mathbf{u}_1 \,.\, (\mathbf{u}_2 \times \mathbf{u}_3) = \det \mathbf{h}.$$

Consider the following lagrangian,

$$\mathscr{L} = \tfrac{1}{2}\sum_i m_i \dot{\mathbf{s}}_i^t\, \mathbf{G}\dot{\mathbf{s}}_i + \tfrac{1}{2}\sum_i \boldsymbol{\omega}_i^t\, \mathbf{I}_i \boldsymbol{\omega}_i - \sum_i \sum_{j>i} \phi_{ij}(\mathbf{r}_{ij}, \boldsymbol{\alpha}_i, \boldsymbol{\alpha}_j) + \tfrac{1}{2} W \sum_{i,j} \dot{h}_{ij}^2 - P_{\text{ex}} V. \quad (2.1)$$

The first term is the kinetic energy for translational motion with the velocities of molecules written in a scaled form $\dot{\mathbf{r}}_i = \mathbf{h}\dot{\mathbf{s}}_i$. The second term is the kinetic energy for molecular rotation ; $\boldsymbol{\omega}_i$ and $\mathbf{I}_i$ are the angular velocity and the inertia tensor of the molecule i. The third term, the potential energy, is also dependent on the molecular orientations $\{\boldsymbol{\alpha}_i\}$. As the variable for $\boldsymbol{\alpha}_i$, we can use Euler angles or quaternions [9] or (in the case of a linear molecule) a vector parallel to the molecular axis. In this section, we assume the potential energy can be expressed by the sum of site–site interactions, hence

$$\phi = \sum_i \sum_{j>i} \phi_{ij}(\mathbf{r}_{ij}, \boldsymbol{\alpha}_i, \boldsymbol{\alpha}_j) = \sum_i \sum_k \sum_{j>i} \sum_l \phi_{ij}{}^{kl}(|\mathbf{r}_{ij}{}^{kl}|). \tag{2.2}$$

The fourth term is the kinetic energy associated with the MD cell deformation, W has the dimension of mass. The remaining term arises from the external pressure P_{ex}.

There are three distinct kinds of variables, $\mathbf{s}$ the coordinates of centre of mass, $\boldsymbol{\alpha}$, the orientation of the molecules, and $\mathbf{h}$ the shape of the MD cell. The equations of motion can be derived from the lagrangian equation

$$\frac{d}{dt}\left(\frac{\partial \mathscr{L}}{\partial \dot{Q}}\right) = \frac{\partial \mathscr{L}}{\partial Q}, \tag{2.3}$$

where Q stands for one of the variables mentioned above.

The equation for translational degrees of freedom is

$$\begin{aligned} \frac{d}{dt}\frac{\partial \mathscr{L}}{\partial \dot{\mathbf{s}}_i} &= \frac{d}{dt}(m_i \mathbf{h}^{\mathrm{t}}\,\mathbf{h}\dot{\mathbf{s}}_i) = \frac{d}{dt}(m_i \mathbf{G}\dot{\mathbf{s}}_i) = \frac{\partial \mathscr{L}}{\partial \mathbf{s}_i} = -\sum_{j\neq i} \frac{\partial \phi_{ij}(\mathbf{r}_{ij}, \boldsymbol{\alpha}_i, \boldsymbol{\alpha}_j)}{\partial \mathbf{s}_i} \\ &= -\mathbf{h}^{\mathrm{t}} \sum_{j\neq i} \frac{\partial \phi_{ij}}{\partial \mathbf{r}_i} = -\mathbf{h}^{\mathrm{t}} \sum_k \sum_{j\neq i} \sum_l \frac{\partial \phi_{ij}{}^{kl}(|\mathbf{r}_{ij}{}^{kl}|)}{\partial \mathbf{r}_{ij}{}^{kl}} = \mathbf{h}^{\mathrm{t}}\left(\sum_k \sum_{j\neq i} \sum_l \mathbf{f}_{ij}{}^{kl}\right) \\ &= \mathbf{h}^{\mathrm{t}}\left(\sum_k \mathbf{f}_i{}^k\right) = \mathbf{h}^{\mathrm{t}}\mathbf{f}_i, \end{aligned} \tag{2.4}$$

where the forces are defined by

$$\mathbf{f}_{ij}{}^{kl} = -\frac{\partial \phi_{ij}{}^{kl}}{\partial \mathbf{r}_{ij}{}^{kl}}, \quad \mathbf{f}_i{}^k = \sum_{j\neq i} \sum_l \mathbf{f}_{ij}{}^{kl}, \quad \text{and} \quad \mathbf{f}_i = \sum_k \mathbf{f}_i{}^k.$$

The result is the usual Newton's second law equation with a correction term arising from the change in shape of the MD cell.

$$m_i \ddot{\mathbf{s}}_i = \mathbf{h}^{-1}\,\mathbf{f}_i - m_i \mathbf{G}^{-1}\,\dot{\mathbf{G}}\dot{\mathbf{s}}_i. \tag{2.5}$$

The equations for angular coordinates are identical to those in the constant volume MD method. The rate of change of the angular momentum $\mathbf{M}_i{}^0$ of molecule i, is related to the torque $\mathbf{N}_i$ acting on the molecule.

$$\frac{d}{dt}(\mathbf{M}_i{}^0) = \frac{d}{dt}(\mathbf{I}_i \boldsymbol{\omega}_i) = \sum_k \sum_{j\neq i} \sum_l \mathbf{p}_i{}^k \times \left(-\frac{\partial \phi_{ij}{}^{kl}}{\partial \mathbf{p}_i{}^k}\right) = \sum_k \mathbf{p}_i{}^k \times \mathbf{f}_i{}^k = \mathbf{N}_i. \tag{2.6}$$

The equations of motion for the MD cell vectors are obtained from

$$\frac{d}{dt}\left(\frac{\partial \mathscr{L}}{\partial \dot{h}_{\alpha\beta}}\right) = \frac{\partial \mathscr{L}}{\partial h_{\alpha\beta}}. \tag{2.7}$$

Thus we find

$$W\ddot{h}_{\alpha\beta}=\frac{\partial \mathscr{L}}{\partial h_{\alpha\beta}}=\sum_i m_i(\mathbf{h}\dot{\mathbf{s}}_i)^{\mathrm{t}}_\alpha(\dot{\mathbf{s}}_i)_\beta-\sum_i\sum_k\sum_{j>i}\sum_l$$

$$\times\frac{\partial\phi_{ij}{}^{kl}}{\partial r_{ij}{}^{kl}}\frac{(\mathbf{r}_{ij}{}^{kl})_\alpha}{r_{ij}{}^{kl}}(\mathbf{s}_i-\mathbf{s}_j)_\beta-P_{\mathrm{ex}}\frac{\partial V}{\partial h_{\alpha\beta}}=\sum_\gamma(\Pi_{\alpha\gamma}-P_{\mathrm{ex}}\delta_{\alpha\gamma})\sigma_{\gamma\beta},\qquad(2.8)$$

where

$$\sigma_{\alpha\beta}=\frac{\partial V}{\partial h_{\alpha\beta}}=V(h^{-1\mathrm{t}})_{\alpha\beta}=V(h^{-1})_{\beta\alpha}\qquad(2.9)$$

and

$$\mathbf{h}^{-1}=\begin{pmatrix}\mathbf{k}_1{}^{\mathrm{t}}\\ \mathbf{k}_2{}^{\mathrm{t}}\\ \mathbf{k}_3{}^{\mathrm{t}}\end{pmatrix},\ \mathbf{k}_1=\frac{\mathbf{u}_2\times\mathbf{u}_3}{\mathbf{u}_1\cdot(\mathbf{u}_2\times\mathbf{u}_3)},\ \mathbf{k}_2=\frac{\mathbf{u}_3\times\mathbf{u}_1}{\mathbf{u}_1\cdot(\mathbf{u}_2\times\mathbf{u}_3)},\ \mathbf{k}_3=\frac{\mathbf{u}_1\times\mathbf{u}_2}{\mathbf{u}_1\cdot(\mathbf{u}_2\times\mathbf{u}_3)}.\qquad(2.10)$$

Here, $\mathbf{k}_1$, $\mathbf{k}_2$, and $\mathbf{k}_3$ are the reciprocal vectors of the MD cell. The internal stress tensor $\mathbf{\Pi}$ is defined by

$$\Pi_{\alpha\gamma}=\frac{1}{V}\left\{\sum_i m_i(\mathbf{h}\dot{\mathbf{s}}_i)^{\mathrm{t}}_\alpha(\mathbf{h}\dot{\mathbf{s}}_i)_\gamma+\sum_i\sum_k\sum_{j>i}\sum_l(\mathbf{f}_{ij}{}^{kl})_\alpha(\mathbf{h}(\mathbf{s}_i-\mathbf{s}_j))_\gamma\right\}.\qquad(2.11)$$

As defined above $\mathbf{\Pi}$ is symmetric only for a system of atoms with central force interactions.

Thus the MD cell changes its shape as the response to the imbalance of the internal stress $\mathbf{\Pi}$ and the external pressure P_{ex}. Parrinello and Rahman [5] have also indicated how to include an external strain term in the equations of motion but that will not concern us here.

In Appendix A, we indicate how to extend our equations to include the case of the long range charge–charge interactions so that the formalism can handle polar molecules and molecular ions [6].

2.2. *Conservation laws*

Three kinds of quantities, the total energy, the momentum, and the angular momentum are conserved in the usual constant volume MD method. (If we employ the periodic boundary condition, and if we reflect molecules back into the MD cell, the angular momentum is not conserved even in the constant volume case.)

The conservation laws in the NMD method can be derived in a fashion similar to the constant volume case.

The first conserved quantity is the hamiltonian of the system.

$$\mathscr{H}=\tfrac{1}{2}\sum_i m_i\dot{\mathbf{s}}_i{}^{\mathrm{t}}\,\mathbf{G}\dot{\mathbf{s}}_i+\tfrac{1}{2}\sum_i\boldsymbol{\omega}_i{}^{\mathrm{t}}\,\mathbf{I}_i\boldsymbol{\omega}_i+\sum_i\sum_{j>i}\phi_{ij}(\mathbf{r}_{ij},\boldsymbol{\alpha}_i,\boldsymbol{\alpha}_j)$$

$$+P_{\mathrm{ex}}V+\frac{W}{2}\sum\dot{h}_{ij}{}^2.\qquad(2.12)$$

For this ensemble the hamiltonian is the sum of the enthalpy and the kinetic energy of the MD cell vectors,

$$K_V = \frac{W}{2} \sum_{i,j} \dot{h}_{ij}^2.$$

As proved by Andersen [2] in the NMD method, the enthalpy is conserved only within the range of the fluctuation of the K_V term. The average of this term is $9/2\ kT$ in equilibrium. This sets an upper limit for deviations from the constant enthalpy–constant pressure ensemble.

The momentum conservation in the constant volume case is derived from Newton's action-reaction law, $\mathbf{f}_{ij}^{kl} = -\mathbf{f}_{ji}^{lk}$. As the result, we have

$$\sum_i \mathbf{f}_i = 0. \tag{2.13}$$

From (2.13) and (2.4), we obtain

$$\sum_i \mathbf{f}_i = \sum_i \left\{ \mathbf{h}^{-1t} \frac{d}{dt} (m_i \mathbf{h}^t \mathbf{h} \dot{\mathbf{s}}_i) \right\} = \mathbf{h}^{-1t} \frac{d}{dt} \left\{ \mathbf{h}^t \left(\sum_i m_i \mathbf{h} \dot{\mathbf{s}}_i \right) \right\} = 0. \tag{2.14}$$

The total momentum $\mathbf{P} = \sum_i m_i \mathbf{h} \dot{\mathbf{s}}_i$ can then be rewritten as

$$\mathbf{P} = \mathbf{h}^{-1t}\, \mathbf{c}, \tag{2.15}$$

where $\mathbf{c}$ is a constant vector.

The conserved quantity in this system is not the momentum $\mathbf{P}$ but $\mathbf{h}^t\mathbf{P}$. The momentum is only conserved in the special case $\mathbf{P} = 0$.

The conservation of the total angular momentum $\mathbf{M}$

$$\mathbf{M} = \sum_i (m_i \mathbf{h} \mathbf{s}_i \times \mathbf{h} \dot{\mathbf{s}}_i + \mathbf{I}_i \boldsymbol{\omega}_i)$$

is derived from the relation

$$\sum_i \mathbf{r}_i \times \mathbf{f}_i + \sum_i \mathbf{N}_i = 0. \tag{2.16}$$

From (2.16), (2.4) and (2.6) we have

$$\begin{aligned} \sum_i \mathbf{r}_i \times \mathbf{f}_i + \sum_i \mathbf{N}_i &= \sum_i \mathbf{h}\mathbf{s}_i \times \mathbf{h}^{-1t} \frac{d}{dt} (m_i \mathbf{h}^t \mathbf{h} \dot{\mathbf{s}}_i) + \frac{d}{dt} \left(\sum_i \mathbf{I}_i \omega_i \right) \\ &= \frac{d}{dt} \left[\sum_i (m_i \mathbf{h}\mathbf{s}_i \times \mathbf{h}\dot{\mathbf{s}}_i + \mathbf{I}_i \boldsymbol{\omega}_i) \right] \\ &\quad + \sum_i m_i \mathbf{h}\mathbf{s}_i \mathbf{h}^{-1t} \left(\frac{d\mathbf{h}^t}{dt} \right) \mathbf{h}\dot{\mathbf{s}}_i - \sum_i m_i \left(\frac{d\mathbf{h}}{dt} \right) \mathbf{s}_i \times \mathbf{h}\dot{\mathbf{s}}_i = 0. \end{aligned} \tag{2.17}$$

Since it does not seem possible to reexpress the second and the third terms in (2.17) as a time derivative, this means the total angular momentum generally is not conserved in the constant pressure method.

For the special case in which the MD cell changes its shape uniformly ($h_{\alpha\beta} = g_{\alpha\beta} L$; L is a parameter, and $g_{\alpha\beta}$ is a constant matrix), the second and the third terms of (2.17) cancel each other. Thus, the angular momentum is conserved in Andersen's uniform dilation case.

In the NMD method, the unit cell vector $\mathbf{h} = (\mathbf{u}_1, \mathbf{u}_2, \mathbf{u}_3)$ are not constants but variables for which we can formulate a conservation law.

We start with (2.8) and multiply both sides by $\mathbf{h}^{\mathrm{t}}$, so that $W\ddot{\mathbf{h}}\mathbf{h}^{\mathrm{t}} = (\boldsymbol{\Pi} - P_{\mathrm{ex}}\mathbf{1})V$. On subtracting the transposed form we then obtain

$$W(\ddot{\mathbf{h}}\mathbf{h}^{\mathrm{t}} - \mathbf{h}\ddot{\mathbf{h}}^{\mathrm{t}}) = W\frac{d}{dt}(\dot{\mathbf{h}}\mathbf{h}^{\mathrm{t}} - \mathbf{h}\dot{\mathbf{h}}^{\mathrm{t}}) = (\boldsymbol{\Pi} - \boldsymbol{\Pi}^{\mathrm{t}})V. \tag{2.18}$$

The following relationship holds :

$$(\dot{\mathbf{h}}\mathbf{h}^{\mathrm{t}} - \dot{\mathbf{h}}\mathbf{h}^{\mathrm{t}})_{\alpha\beta} = \sum_i (\dot{h}_{\alpha i}h_{\beta i} - h_{\alpha i}\dot{h}_{\beta i})$$

$$= \sum_i [(\dot{\mathbf{u}}_i)_\alpha(\mathbf{u}_i)_\beta - (\mathbf{u}_i)_\alpha(\dot{\mathbf{u}}_i)_\beta] = -\sum_i (\mathbf{u}_i \times \dot{\mathbf{u}}_i)_\gamma, \tag{2.19}$$

where (α, β, γ) is a cyclic rotation of (1, 2, 3). Hence only if $\boldsymbol{\Pi} = \boldsymbol{\Pi}^{\mathrm{t}}$ (i.e. an atomic system with central forces), does it follow that the angular momentum is conserved :

$$\frac{d}{dt}\left(W\sum_i \mathbf{u}_i \times \dot{\mathbf{u}}_i\right) = 0 \tag{2.20}$$

or

$$W\sum_i \mathbf{u}_i \times \dot{\mathbf{u}}_i = \mathbf{c} \quad (\mathbf{c}\text{, a constant vector}). \tag{2.21}$$

Thus if we select the initial value $\dot{\mathbf{u}}_i = 0$, the angular momentum is zero, and the MD unit cell does not rotate.

It was pointed out to us by Michiel Sprik that in the case of an antisymmetric stress tensor the appropriate conservation law is

$$\frac{d}{dt}\left(W\sum_i \mathbf{u}_i \times \dot{\mathbf{u}}_i + \sum_j \mathbf{I}_j\boldsymbol{\omega}_j\right) = 0.$$

The quantity $\sum_j \mathbf{I}_j\boldsymbol{\omega}_j$ is not conserved because this is always coupled with the term $\sum_j m_j\mathbf{r}_j \times \dot{\mathbf{r}}_j$. Hence, in general the MD cell will rotate as a whole. A method which can stop this rotation is proposed in § 3.

2.3. *Comments on the form of the dynamical equations*

There are several possible choices for the form K_{V}, the kinetic energy term associated with the volume changes in the NMD method. For example, consider the formulation used for the uniform dilation case of Andersen, where the appropriate variable is the volume V itself. The lagrangian in this case is

$$\mathscr{L} = \tfrac{1}{2}\sum_i m_i(V^{1/3}\dot{\mathbf{s}}_i)(V^{1/3}\dot{\mathbf{s}}_i) - \phi - P_{\mathrm{ex}}V + \tfrac{1}{2}C\dot{V}^2. \tag{2.22}$$

Only the term

$$K_{\mathrm{V}} = \tfrac{1}{2}C\dot{V}^2 \tag{2.23}$$

is necessary for the following discussion, and the equation of motion for V is

$$\frac{d}{dt}\left(\frac{\partial\mathscr{L}}{\partial\dot{V}}\right) = C\ddot{V} = \frac{\partial\mathscr{L}}{\partial V} = \tfrac{1}{3}\,\mathrm{Tr}\,(\boldsymbol{\Pi}) - P_{\mathrm{ex}}. \tag{2.24}$$

We can change the variable from V to $L=V^{1/3}$, then (2.23) and (2.24) become

$$K_V=\tfrac{9}{2}CL^4\dot{L}^2 \tag{2.25}$$

and

$$\ddot{L}=\frac{1}{3CL^2}(\tfrac{1}{3}\,\mathrm{Tr}\,(\boldsymbol{\Pi})-P_{ex})-\frac{4}{L}\dot{L}^2. \tag{2.26}$$

Equations (2.24) and (2.26) give exactly the same result for the time evolution of the volume.

However, if we choose the form

$$K_V=\tfrac{1}{2}W\dot{L}^2 \tag{2.27}$$

the equation of motion becomes

$$W\ddot{L}=3L^2[\tfrac{1}{3}\,\mathrm{Tr}\,(\boldsymbol{\Pi})-P_{ex}] \tag{2.28}$$

which is a totally different result. Equation (2.28) is the uniform dilation limit of (2.8). (See Appendix B for the derivation.)

The K_V term in (2.1) can be modified in an analogous way to

$$\tfrac{1}{2}\sum_{i,j} W_{ij}\dot{h}_{ij}^2 \quad \text{or} \quad \tfrac{1}{2}\,\mathrm{Tr}\,(\dot{\mathbf{h}}^t\mathbf{B}^t\mathbf{B}\dot{\mathbf{h}}),$$

where $\mathbf{B}$ is a matrix.

Any of these different formulations equally satisfy the basic requirement that the averaged values calculated in this method give the static quantities in the (H, P, N) ensemble.

We prefer (2.1) partly because of its simplicity and also because the discussion following (2.18–2.21) and that of § 3 is less straightforward for the other choices.

3. Equation of motion under the constraint of a symmetric h matrix

We can derive the equations of motion for a MD cell whose shape is constrained. The results for the special cases of uniform dilation, fixed-angle, and trigonal symmetry are given in Appendix B.

One of the difficulties arising from the use of (2.8) in molecular systems is illustrated in figure 1 (*a*). This figure was constructed using the results of a simulation on solid nitrogen in its high pressure cubic phase [7]. The MD cell vectors $\mathbf{u}_i$ have been projected on the surface of a sphere, and the sphere is viewed from the X, Y and Z directions respectively. During the course of the MD run, which spanned 7000 time steps, the shape of cell does not change much and the cell motion can be described as a rotation of the MD cell as a whole. The reason of this rotation can be understood in the following way. The matrix $\mathbf{h}$ has nine components, but only 6 parameters (3 lengths and 3 angles) are necessary to define the *shape* of the MD cell. The three remaining parameters, as a result, describe the orientation of a MD cell in space. The conservation law (2.21) is valid only in the atomic case. Although the cell rotation does not influence the dynamics of the system, it sometimes causes trouble in the analysis of the molecular orientations and crystal structures.

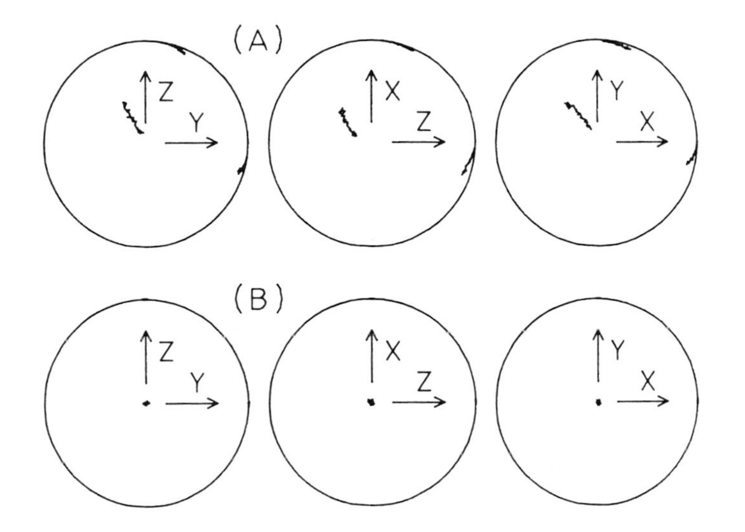

Figure 1. The upper portion of the figure shows the trajectories of the MD cell basis vectors derived using (2.8) while the lower portion is derived using (3.4) with a symmetrized **h** matrix. The cell vectors are projected onto the surface of a sphere and viewed from the X, Y and Z directions.

We derive here the equations eliminating the extraneous parameters responsible for the rotation. The transformation matrix **h** can be separated into a symmetric part **S** and an antisymmetric part **A**

$$\mathbf{h}=\mathbf{S}+\mathbf{A} \tag{3.1}$$

and the transpose of **h** becomes

$$\mathbf{h}^{\mathrm{t}}=\mathbf{S}-\mathbf{A}. \tag{3.2}$$

The kinetic energy term K_{V} is given by

$$K_{\mathrm{V}}=\tfrac{1}{2}W\sum_{i,j}\dot{h}_{ij}^{2}=\tfrac{1}{2}W\sum_{i,j}(\dot{S}_{ij}^{2}+\dot{A}_{ij}^{2}) \tag{3.3}$$

and differentiation by **S** and **A** are expressed as

$$\frac{\partial}{\partial S_{\alpha\beta}}=\sum_{\mu\nu}\frac{\partial h_{\mu\nu}}{\partial S_{\alpha\beta}}\frac{\partial}{\partial h_{\mu\nu}}=\frac{\partial}{\partial h_{\alpha\beta}}+\frac{\partial}{\partial h_{\beta\alpha}}$$

and

$$\frac{\partial}{\partial A_{\alpha\beta}}=\sum_{\mu\nu}\frac{\partial h_{\mu\nu}}{\partial A_{\alpha\beta}}\frac{\partial}{\partial h_{\mu\nu}}=\frac{\partial}{\partial h_{\alpha\beta}}-\frac{\partial}{\partial h_{\beta\alpha}}$$

The equation of motion for **S** and **A** are

$$\frac{d}{dt}\left(\frac{\partial\mathscr{L}}{\partial\dot{S}_{\alpha\beta}}\right)=2W\ddot{S}_{\alpha\beta}=\frac{\partial\mathscr{L}}{\partial S_{\alpha\beta}}$$

$$=\frac{\partial\mathscr{L}}{\partial h_{\alpha\beta}}+\frac{\partial\mathscr{L}}{\partial h_{\beta\alpha}}=[(\mathbf{\Pi}-P_{\mathrm{ex}}\mathbf{1})V\mathbf{h}^{-1\mathrm{t}}]_{\alpha\beta}+[(\mathbf{\Pi}-P_{\mathrm{ex}}\mathbf{1})V\mathbf{h}^{-1\mathrm{t}}]_{\beta\alpha}, \tag{3.4}$$

$$\frac{d}{dt}\left(\frac{\partial \mathscr{L}}{\partial \dot{A}_{\alpha\beta}}\right)=2W\ddot{A}_{\alpha\beta}=\frac{\partial \mathscr{L}}{\partial A_{\alpha\beta}}$$

$$=\frac{\partial \mathscr{L}}{\partial h_{\alpha\beta}}-\frac{\partial \mathscr{L}}{\partial h_{\beta\alpha}}=[(\mathbf{\Pi}-P_{\mathrm{ex}}\mathbf{1})V\mathbf{h}^{-1\mathrm{t}}]_{\alpha\beta}-[(\mathbf{\Pi}-P_{\mathrm{ex}}\mathbf{1})V\mathbf{h}^{-1\mathrm{t}}]_{\beta\alpha}. \quad (3.5)$$

If we initially select the $\mathbf{h}$ matrix in a symmetric form, the growth of the anti-symmetric component $\mathbf{A}$ is closely related to the rotation of the MD cell. The imposition of the constraint $\mathbf{A}\equiv 0$ reduces the number of independent parameters to six and can stop the rotation of MD cell.

The validity of (3.4) with the constraint $h_{\alpha\beta}=h_{\beta\alpha}$ is also justified by the fact that any $\mathbf{h}$ can be transformed into a symmetric form $\mathbf{S}$ by a rotation matrix R, $\mathbf{S}=R\mathbf{h}$. The imposition of this symmetry for $\mathbf{h}$ in the simulation has the same effect as the addition of a rotation that transforms $\mathbf{h}$ into a symmetric form at each time step. The rotation matrix R is an orthogonal matrix (Appendix C).

The procedure for using (3.4) is as follows : First, choose an initial configuration using an appropriate $\mathbf{h}$ matrix. Next, transform all the vectors in the real frame by the matrix R (Appendix C) and give symmetrical initial values for $\dot{\mathbf{h}}$, $\ddot{\mathbf{h}}$, ... Finally, carry out a simulation using (3.4).

At the second step, the transformation of the orientational variables of the molecules is also necessary. We adopt the same definition for the rotational matrix A as Evans [9]. The space fixed frame (real frame) vector $\mathbf{r}_{\mathrm{s}}$ is related to the molecular fixed frame vector $\mathbf{r}_{\mathrm{m}}$ by $\mathbf{r}_{\mathrm{m}}=A\mathbf{r}_{\mathrm{s}}$, and, the rotation R rotates $\mathbf{r}_{\mathrm{s}}$ to $\mathbf{r}'_{\mathrm{s}}=R\mathbf{r}_{\mathrm{s}}$. The rotational matrix in the new frame is defined by $\mathbf{r}_{\mathrm{m}}=A'\,\mathbf{r}'_{\mathrm{s}}$. Since $\mathbf{r}_{\mathrm{m}}=A\mathbf{r}_{\mathrm{s}}=AR^{-1}\,\mathbf{r}'_{\mathrm{s}}$, we have

$$A'=AR^{-1}. \quad (3.6)$$

The new quaternions $\tau_{A'}$ (corresponding to A') are obtained from the multiplication of the old quaternions τ_A by the quaternions $\tau_{R^{-1}}$. The quaternions $\tau_{R^{-1}}$, whose components τ_1, τ_2, τ_3, τ_4 correspond to ξ, η, ζ, χ of Evans [9], are expressed in terms of the elements of the R matrix, as follows :

$$\left.\begin{aligned}|\tau_1|&=[(1-R_{11}+R_{22}-R_{33})/4]^{1/2},\\ |\tau_2|&=[(1+R_{11}-R_{22}-R_{33})/4]^{1/2},\\ |\tau_3|&=[(1-R_{11}-R_{22}+R_{33})/4]^{1/2},\\ |\tau_4|&=[(1+R_{11}+R_{22}+R_{33})/4]^{1/2}.\end{aligned}\right\} \quad (3.7)$$

The sign of τ_i is determined by

$$\tau_1\tau_2=-\tfrac{1}{4}(R_{12}+R_{21}),\quad \tau_1\tau_3=-\tfrac{1}{4}(R_{23}+R_{32}),\quad \tau_1\tau_4=\tfrac{1}{4}(R_{31}-R_{13}). \quad (3.8)$$

The only difference between (3.4) and (2.8) is the symmetrization of the force term in the $\ddot{\mathbf{h}}$ equation.

4. Application to solid nitrogen at high pressure

We have employed both (2.8) and (3.4) to study the cubic high pressure orientationally disordered form of solid nitrogen [7]. A simple Lennard-Jones 12–6 atom–atom potential, $\phi(r)=4\epsilon[(\sigma/r)^{12}-(\sigma/r)^6]$ with $\epsilon=37{\cdot}3$ K, $\sigma=3{\cdot}31$ Å and the bond length $d = 1{\cdot}098$ Å was used. Initially, 64 rigid nitrogen molecules were arranged in the observed high pressure room temperature *Pm*3*n* structure with two different kinds of disordered sites. The structure is isomorphous with the high temperature, low pressure structure of solid oxygen, see figure 2 (*a*), (*b*).

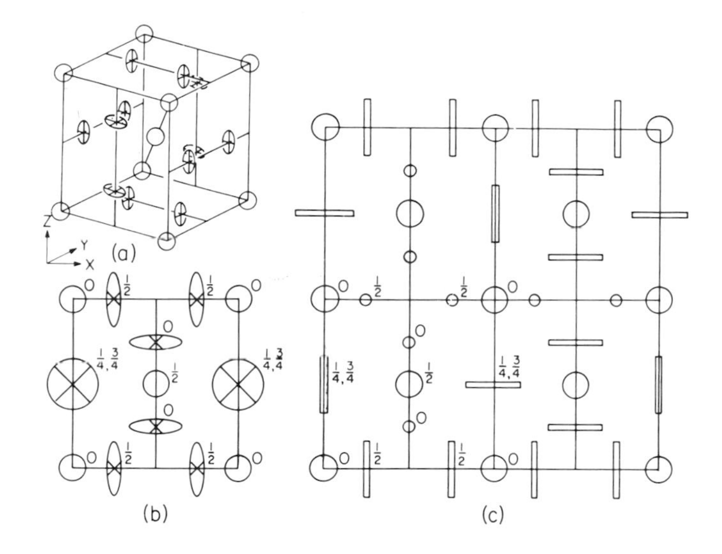

Figure 2. High pressure structures of solid nitrogen (*a*), (*b*) high temperature phase (*Pm*3*n*, $Z = 8$), (*c*) intermediate phase ($I2_13$, $Z = 64$). (*b*) and (*c*) are projections viewed from the *Z* axis, in (*c*) only the lower half of the unit cell is shown. Circles and ellipses show the disordered sites. A rod indicates the molecules that have a preferred orientation along *X*, *Y*, or *Z* directions. The fractions denote the *Z* coordinates of the centres of mass.

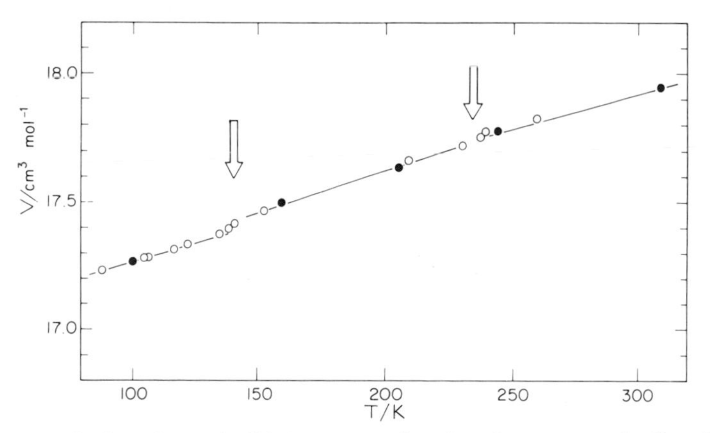

Figure 3. Molar volume of solid nitrogen as a function of temperature for $P_{ex} = 70{\cdot}0$ kbar. Solid circles indicate data reported in [7]. The line is a guide to the eye and arrows indicate the region of the phase transitions.

Notes on p. 401

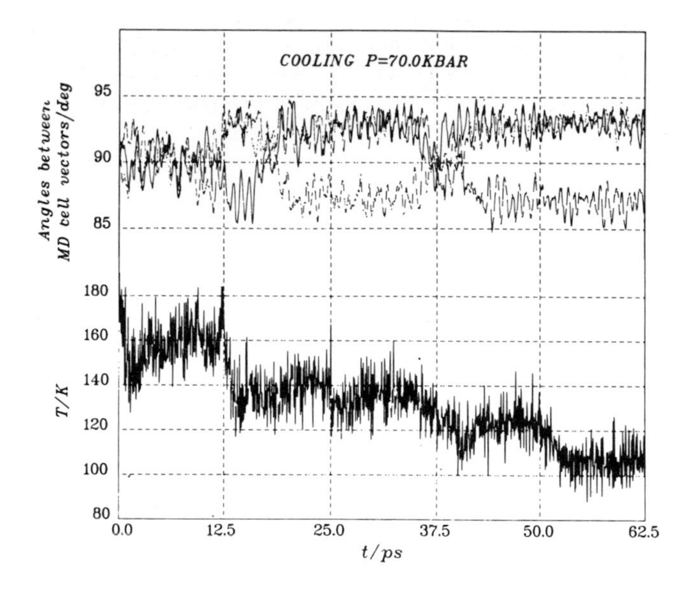

(*a*)

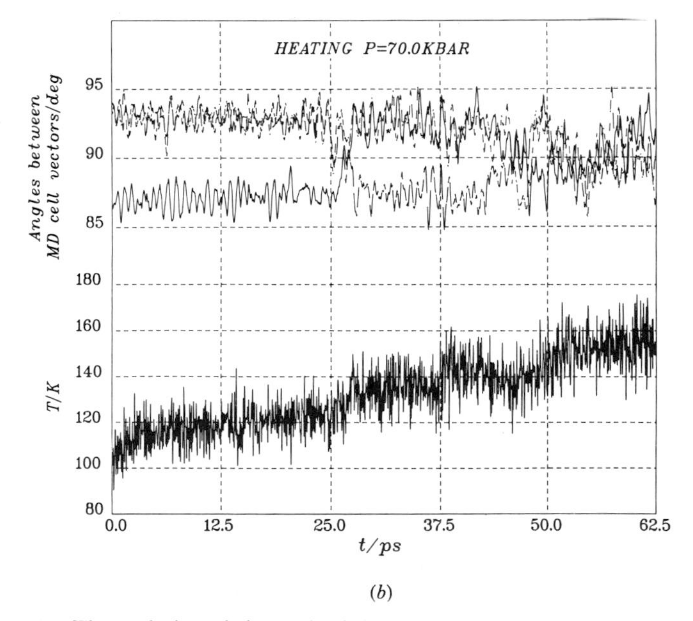

(*b*)

Figure 4. The evolution of the angles between MC cell vectors in cooling and heating processes.

In this structure, one quarter of the molecules rotate almost freely (T_d sites), while the remaining three quarters of the molecules have their motion confined within one plane (D_{2d} sites). To integrate the equations of motion, we used a time step 0·005 ps and a truncation length of 6·0 Å. A simulation of 2500 steps was carried at $P_{ex}=70{\cdot}0$ kbar and $T=300$ K, the first 500 steps being used for scaling, then the temperature was lowered in a step-wise fashion and the calculations were repeated. At about 230 K (see arrow in figure 3), the D_{2d} site molecules stop rotating and take preferred orientations parallel to unit cell vectors, but the T_d site molecules are still rotating and the crystal retains its cubic symmetry ($I2_13$, $Z=64$) (figure 2 (*c*)). Molecules in the D_{2d} sites running parallel to the crystal axes all have the same orientation. However, alternate rows of D_{2d} molecules have an antiferro type arrangement of their orientations. At about 140 K (see arrow in figure 3), all the T_d site molecules align along the same direction and the crystal spontaneously undergoes a shear distortion to a structure with trigonal symmetry ($R3c$, $Z=64$).

The molar volume is plotted versus the temperature in figure 3. The volume changes at both transition points is small but the three different phases are nevertheless distinguished by their different thermal expansion coefficients. The effect of cooling and heating the crystal in the neighbourhood of the lower phase transition around 140 K is shown in figure 4. In the transition region, the trigonal deformation was observed to change its direction. As shown in this example the transition is reversible. Additional calculations using a system of 512 molecules exhibited essentially the same phase transitions. However, in the intermediate and low temperature phases the MD cell separated into two domains, each of which had the same structure as in the 64 molecule system. Simulation with the symmetrical equation (3.4) reproduced exactly the same phase transitions as found previously using (2.8) [7].

The motion of the MD cell vectors for both cases is depicted in figure 1. The upper part contains the results of (2.8) and the lower part those given by (3.4). In constructing figure 1 only data from the first 7000 time steps have been used to show the rotation of MD cell. During this portion of the simulation, the MD cell retained its cubic symmetry. It should be noted that when (3.4) is used, the MD cell orientation undergoes small fluctuations around a mean value.

5. Comments on the programming and simulations

5.1. *Programming*

In the NMD constant pressure method, the three kinds of mutually coupled equations (2.5), (2.6), and (2.8) or (3.4) are integrated *simultaneously*. The force terms in these equations are dependent upon the velocity. Hence, neither the well known Verlet algorithm for integration [10], nor the method of constraints in molecular systems [11, 12] can be readily applied in this situation. Accordingly, we adopted the quaternion method [9] for the description of the molecular rotation and the predictor–corrector method [13–15] for integration of the equations. In practice, we used a 5th order algorithm for the second order differential equations (2.5), (2.8), and a 4th order one for the coupled first order differential equations for molecular rotation. It should be noted that the integration of equations of motion for the system with velocity dependent force terms is less accurate than in the usual velocity independent case.

The major changes needed to convert a constant volume MD program to the NMD constant pressure method are the introduction of two kinds of coordinate frames, the scaled frame and the real frame, and the calculation of the stress tensor components. It is useful to construct the MD program so as to be able to work under either constant volume or constant pressure by changing an optional parameter.

The use of the periodic boundary condition which reflects molecules into equivalent positions and integration of the equations of motion are carried in the scaled frame. On the other hand, the calculation of the forces, energy and other structural quantities must be carried out in the real frame. In molecular systems, a molecule fixed coordinate frame is convenient for describing the angular motion.

The flow of the program in the iteration process is as follows :

(1) Predict the variables $\mathbf{s}'$, $\boldsymbol{\alpha}'$, $\mathbf{h}'$, . . . at $t+\Delta t$.

(2) Determine the quantities $\mathbf{G}$, $\mathbf{h}^{-1}$, etc. relating to the shape of the MD cell.

(3) Obtain the coordinates of molecules and their interaction sites in the real frame. (The transformation from the scaled frame to the real frame $\mathbf{r}=\mathbf{hs}$ and from the molecule fixed frame $\mathbf{r}_{\mathrm{m}}$ to the space fixed real frame $\mathbf{r}_{\mathrm{s}}$ via $\mathbf{r}_{\mathrm{s}}=A^{\mathrm{t}}\mathbf{r}_{\mathrm{m}}$ [9] are necessary in this step.)

(4) Calculate the energy and forces.

(5) In the constant pressure case, add the correction term in (2.5) to the force and calculate the stress tensor ((2.8) or (3.4)).

(6) Correct the predicted values $\mathbf{s}'$, $\boldsymbol{\alpha}'$, $\mathbf{h}'$, . . . and obtain the final values for $\mathbf{s}$, $\boldsymbol{\alpha}$, $\mathbf{h}$, . . . corresponding to $t+\Delta t$.

Although the calculation of the energy, pressure, or other thermodynamic and structural quantities after step 6 (using the correct coordinates) gives the better result, we can save considerable computer time by doing these calculations after step 4 with predicted coordinates. The resulting error is usually very small.

In the NMD method, several new terms appear in equations of motion. We find the following method is useful to check for programming errors involving these terms.

In MD calculations, the hamiltonian of the system (2.12) is conserved. If we can avoid the influence of truncation effects, this conservation will be obeyed very precisely. Accordingly, we construct a configuration in which the molecules never cross the truncation boundary. We focus on a cluster of a few molecules contained within a sphere whose radius, r_{m}, that is much less than the truncation length r_{c}, which in turn satisfies the relationship $2r_{\mathrm{c}}<L$, the size of the MD cell. The cluster is placed in the central part of the MD cell. In this configuration, all the molecules are situated within the truncation sphere and never go out of this region, if we examine only a few time steps. Hence no truncation error can occur. With this device neither the periodic boundary condition nor the images of the molecules have to be considered at all.

Our program contains an option to construct such a configuration for testing purposes. It should be noted that only a small number of molecules and at most 30 time steps are necessary to check whether or not the hamiltonian is conserved. If the predictor–corrector method is used, at the beginning, the fluctuation in the hamiltonian is large, but after about 10 steps, the errors become extremely small. Non-conservation at this stage indicates some mistake in the program.

5.2. *Simulations*

In NMD method, the shape of MD cell changes in response to the imbalance between the internal stress and the external pressure P_{ex}. This is the consequence of (2.8) or (3.4). The MD cell geometry and the molecular coordinates are changing *simultaneoulsy*. This is a clear difference from the method employed by Pawley and Thomas [16]. They follow the integration for a certain number of steps in a fixed MD cell, then change the shape of the MD cell to fit the desired external pressure, and repeat this process.

If we define the instantaneous pressure value from virial equation (i.e. trace of stress tensor), the pressure value is not constant, but is always changing. However, if we take the average over a long time interval, the average value coincides with the external pressure P_{ex}. Therefore, the NMD method is a constant pressure simulation only in this sense. Also the instantaneous shape of MD cell is sometimes deformed far from the perfect symmetric structure. The symmetry of crystal structures should be determined by averaging over such fluctuations.

The uniform dilation method is recommended for simulation studies limited to the liquid state. If one uses the more general form of (2.8) or (3.4), the MD cell continually changes its shape in liquid phase. In fact, such deformations can be used as an indicator of liquid state, but it gives rise to difficulties in the analysis of the MD data.

The NMD method seems best suited to the study of solid state phase transitions. However, we must be careful about the inherent limitations of any simulation method which employs periodic boundaries [16]. Thus if there is a close relation between the two structures on both sides of a phase transition, such as the b.c.c. and f.c.c. structures that can change into each other by the expansion or shrinking of the one edge of the cell, the transformation may be reproducible. However, in the case of f.c.c. and h.c.p. structures, for example, the change of MD cell must be accompanied by the displacement of the molecules [17]. Such transformations may prove more difficult to obtain.

5.3. *Volume fluctuations*

In the constant pressure (NMD) method, the mass W that occurs in the expression of the kinetic energy of the MD cell vectors remains an additional adjustable parameter. Andersen proved that, in equilibrium, static quantities do not depend on W [2] but there is no rigorous proof concerning the effect of W on dynamical quantities. In a comparison of constant volume and constant pressure simulations for CF_4, carried out under similar state conditions, we found little difference between the velocity auto-correlation functions, so that this effect seems to be small for single particle properties [8]. However, the dynamics of the MD cell is governed by the value of W and the expression of the kinetic energy term K_V.

Here, we examine the relation between W and the time scale of the volume fluctuations. We assume the system is in equilibrium and that the shape of MD cell at any instant is not deformed far from its mean value. In this situation, using small deviations $\delta h_{\alpha\beta}=h_{\alpha\beta}-\langle h_{\alpha\beta}\rangle$ and (2.9), the volume fluctuation

Notes on p. 401

δV is expressed by

$$\delta V = \sum_{\alpha,\beta} \frac{\partial V}{\partial h_{\alpha\beta}} \delta h_{\alpha\beta} = \sum_{\alpha,\beta} V h_{\beta\alpha}{}^{-1} \delta h_{\alpha\beta}. \tag{5.1}$$

The equation of motion for δV is obtained from (2.8) if we ignore the higher terms in $\delta h_{\alpha\beta}$,

$$\frac{d^2}{dt^2}\delta V = \sum_{\alpha,\beta} V h_{\beta\alpha}{}^{-1} \frac{d^2}{dt^2}\delta h_{\alpha\beta} = \frac{V^2}{W}\sum_{\alpha,\beta}\sum_{\gamma} h_{\beta\alpha}{}^{-1}(\Pi_{\alpha\gamma} - P_{\text{ex}}\delta_{\alpha\gamma})h_{\beta\gamma}{}^{-1}$$

$$= \frac{V^2}{W}\,\text{Tr}\,\{(\mathbf{\Pi} - P_{\text{ex}}\mathbf{1})\mathbf{h}^{-1\text{t}}\,\mathbf{h}^{-1}\}. \tag{5.2}$$

For simplicity, we assume a cubic cell and let $L = V^{1/3}$. If we use the relationship

$$\delta P = P - P_{\text{ex}} = \frac{\partial P}{\partial V}\delta V = V\frac{\partial P}{\partial V}\frac{\partial V}{V} = -B\frac{\delta V}{V},$$

where $B = -V(\partial P/\partial V)$ is the bulk modulus, then we have the approximate equation

$$\frac{d^2}{dt^2}\delta V = \frac{V}{W}3L(P - P_{\text{ex}}) = -\frac{3LB}{W}\delta V. \tag{5.3}$$

This is an equation for harmonic oscillation from which the period of volume fluctuation T_0 is

$$T_0 = 2\pi\left(\frac{W}{3LB}\right)^{1/2}. \tag{5.4}$$

If we use the relation between W and the Andersen parameter C in (2.22), $W = 3CL^4$ the period T'_0 becomes

$$T'_0 = 2\pi\left(\frac{CL^4}{LB}\right)^{1/2} \sim 2\pi\left(\frac{L}{v_1}\right), \tag{5.5}$$

where v_1 is the sound velocity. Equation (5.5) is the same formula mentioned by Andersen [2]. We can check this relation from the data of Haile and Graben [3]. The results in figure 1 (*d*) of [3] are for the conditions $M^* = 0{\cdot}0001$, $\rho^* = 0{\cdot}627$, $T^* = 2{\cdot}46$, $X^* = 0{\cdot}052$ and $N = 108$ from which (5.5) gives $T'_0 = 0{\cdot}41$ ps. In the figure of [3] we count about 45 oscillations in 2000 time steps, hence the period is about 0·38 ps.

Some of our own data for simulation based on 256 carbon tetrafluoride molecules are shown in the table [8]. The bulk modulus reported there is obtained from the volume fluctuations via [3]

$$B_s = -V\left(\frac{\partial P}{\partial V}\right)_s = \frac{VkT}{\langle(\delta V)^2\rangle}. \tag{5.6}$$

In CF_4 the phase II solid is orientationally ordered while phase I is orientationally disordered, neither phase is cubic. However, the MD cells in both of phase I and phase II of CF_4 are approximately tetragonal.

Comparison between the calculated and observed period of the MD cell volume fluctuations for CF_4 using the NMD data from [8].

	W/g mol^{-1}	T/K	P_{ex}/GPa	V/cm^3 mol^{-1}	B/GPa	T_0/ps	T(obs)/ps
Phase II	100	70·7	0·12	41·64	5·4	0·40	0·45
	100	68·3	1·12	36·02	13·6	0·25	0·24
	100	72·3	2·12	33·87	16·1	0·24	0·23
Phase I	100	72·6	0·12	43·72	3·4	0·49	0·43
	400	74·1	0·12	43·20	5·1	0·80	0·80
	1600	75·5	0·12	42·85	2·9	2·15	1·74
Liquid	100	93·8	0·03	47·59	1·9	0·65	0·43
	100	127·7	0·03	51·52	1·5	0·71	0·49

We find that (5.4) is reasonably well fulfilled for these solids. In the liquid state, the shape of MD cell is deformed significantly from the assumed cubic form, and is always changing. In this situation, we must take the off-diagonal element of the stress tensor into consideration. This gives a smaller T_0 value than in (5.4) by at most factor $\sqrt{3}$. In any case either (5.4) or (5.5) give a good means for selecting an appropriate value of W.

6. Summary

We have indicated how the constant pressure molecular dynamics (NMD) method which allows a MD cell to change its shape can be applied to molecular solids and ionic crystals with molecular ions. In the application to systems that undergo phase transitions, the MD cell shape can serve as a useful order parameter to indicate the transition. The NMD method is therefore a powerful tool for the study of such transitions. Our article has been mainly devoted to technical aspects of the NMD method. We have discussed the usual conservation laws and find that some only hold under special conditions. We have proposed a method which can stop superfluous MD cell rotation, which in addition aids the analysis of molecular orientation and crystal structures. We have also indicated how to change the basic equations to include a constraint on the shape of the MD cell.

We thank Roger Impey, A. H. Louie, Anees Rahman, Ray Somorjai and Michiel Sprik, as well as the referee, for valuable comments and helpful discussions.

Appendix A

Formulae for coulombic interactions

In general, the charge distribution of a molecule or molecular ion can be represented by a set of charges. The long range charge–charge interactions that then arise in evaluation of the potential energy can be formally handled in the same way as described in § 2. However, as is well known, the direct pairwise summation of these interactions gives poor convergence. Several methods have

Notes on p. 401

been proposed for handling coulombic interactions. We employ the Ewald method [18, 19], in which the charge ζ_i^k at site k of molecule i (recall $\mathbf{r}_i^k = \mathbf{h}\mathbf{s}_i + \mathbf{p}_i^k$) is decomposed into two terms

$$\zeta_i^k \left(\frac{1}{\eta^3 \pi^{3/2}} \exp\left[-\frac{|\mathbf{r}_i^k|^2}{\eta^2} \right] - \frac{1}{V} \right)$$

and

$$\zeta_i^k \left(\delta(\mathbf{r}_i^k) - \frac{1}{\eta^3 \pi^{3/2}} \exp\left[-\frac{|\mathbf{r}_i^k|^2}{\eta^2} \right] \right).$$

The first is a gaussian distribution neutralized by a uniform background, and the second is a delta function neutralized by a gaussian distribution.

Using the periodicity of the lattice, the formula for the self-potential at the point of charge is given by

$$\psi(\mathbf{r}_i^k) = \frac{1}{4\pi\epsilon_0} \left[\frac{1}{\pi V} \sum_{\mathbf{n}}{}' S(-\mathbf{k}_\mathbf{n}) \exp\left[2\pi i \mathbf{k}_\mathbf{n} \cdot (\mathbf{h}\mathbf{s}_i + \mathbf{p}_i^k)\right] Q(\mathbf{k}_\mathbf{n}) \right.$$
$$\left. + \sum_{j \neq i}{}^{*} \sum_{l} \zeta_j^l \, 2\pi^{-1/2} \int_{|\mathbf{r}_{ij}^{kl}|/\eta}^{\infty} \exp(-t^2)\, dt / |\mathbf{r}_{ij}^{kl}| - \frac{2\zeta_i^k}{\pi^{1/2}\eta} \right] \quad \text{(A 1)}$$

where

$$S(\mathbf{k}_\mathbf{n}) = \sum_j \sum_l \zeta_j^l \exp\left[2\pi i \mathbf{k}_\mathbf{n} \cdot (\mathbf{h}\mathbf{s}_j + \mathbf{p}_j^l)\right], \quad \text{(A 2)}$$

and

$$\mathbf{r}_{ij}^{kl} = \mathbf{r}_i^k - \mathbf{r}_j^l = \mathbf{h}(\mathbf{s}_i - \mathbf{s}_j) + \mathbf{p}_i^k - \mathbf{p}_j^l,$$

$$Q(\mathbf{k}_\mathbf{n}) = \exp\left[-\pi^2 \eta^2 |\mathbf{k}_\mathbf{n}|^2\right] / |\mathbf{k}_\mathbf{n}|^2 \quad \text{(A 3)}$$

and the reciprocal vectors $\mathbf{k}_\mathbf{n}$ are defined as

$$\mathbf{k}_\mathbf{n} = \mathbf{k}_1 n_1 + \mathbf{k}_2 n_2 + \mathbf{k}_3 n_3 = \frac{1}{\mathbf{u}_1 \cdot (\mathbf{u}_2 \times \mathbf{u}_3)} \left[(\mathbf{u}_2 \times \mathbf{u}_3) n_1 + (\mathbf{u}_3 \times \mathbf{u}_1) n_2 + (\mathbf{u}_1 \times \mathbf{u}_2) n_3 \right]$$

$$= \mathbf{h}^{-1t} \begin{pmatrix} n_1 \\ n_2 \\ n_3 \end{pmatrix} = \mathbf{h}^{-1t}\, \mathbf{n}, \quad (n_1, n_2, n_3 \text{ integers}). \quad \text{(A 4)}$$

The summation $\sum_{\mathbf{n}}'$ excludes the $|\mathbf{n}| = 0$ and the summation $\sum_j^*$ is extended over the whole of space including the self-image site arising from the periodic boundary condition, $2\pi^{-1/2} \int_x^\infty \exp(-t^2)\, dt$ is an error function.

The potential energy of a coulomb system is thus

$$\phi = \tfrac{1}{2} \sum_{i,k} \zeta_i^k \psi(\mathbf{r}_i^k) = \frac{1}{4\pi\epsilon_0} \left[\frac{1}{2\pi V} \sum_{\mathbf{n}}{}' S(\mathbf{k}_\mathbf{n}) S(-\mathbf{k}_\mathbf{n}) Q(\mathbf{k}_\mathbf{n}) \right.$$
$$+ \tfrac{1}{2} \sum_i \sum_k \sum_{j \neq i}{}^{*} \sum_l \zeta_i^k \zeta_j^l \left(2\pi^{-1/2} \int_{|\mathbf{r}_{ij}^{kl}|/\eta}^{\infty} \exp(-t^2)\, dt \right) \Big/ |\mathbf{r}_{ij}^{kl}| - \sum_i \sum_k \frac{(\zeta_i^k)^2}{\pi^{1/2}\eta}$$
$$\left. - \tfrac{1}{2} \sum_i \sum_k \sum_{k' \neq k} \zeta_i^k \zeta_i^{k'} \left(2\pi^{-1/2} \int_0^{|p_i^k - p_i^{k'}|/\eta} \exp(-t^2)\, dt \right) \Big/ |\mathbf{p}_i^k - \mathbf{p}_i^{k'}| \right]. \quad \text{(A 5)}$$

The third term comes from the exclusion of the self energy and the fourth is the correction term for the subtraction of intramolecular interactions. These two are constant terms dependent only on the parameter η and the type of molecule, so we can ignore these terms in calculation of forces and stress tensor. Equation (A 5) does not depend on the value of η. The first term is an expansion which converges rapidly at large η. On the other hand, the second term converges fast near $\eta=0$. If we choose an appropriate value for η, then both expansions converge reasonably fast.

The force acting on site k of molecule i is

$$\mathbf{f}_i{}^k = -\frac{\partial \phi}{\partial \mathbf{r}_i{}^k} = \frac{\zeta_i{}^k}{4\pi\epsilon_0}\left[\frac{1}{2\pi V}\sum_{\mathbf{n}}{}' 2\pi i \mathbf{k}_{\mathbf{n}}\{S(\mathbf{k}_{\mathbf{n}})\exp[-2\pi i \mathbf{k}_{\mathbf{n}}\cdot\mathbf{r}_i{}^k]\right.$$
$$- S(-\mathbf{k}_{\mathbf{n}})\exp[2\pi i \mathbf{k}_{\mathbf{n}}\cdot\mathbf{r}_i{}^k]\}Q(\mathbf{k}_{\mathbf{n}})$$
$$+\sum_{j\neq i}{}^{*}\sum_{l}\zeta_j{}^l\,\mathbf{r}_{ij}{}^{kl}\left\{2\pi^{-1/2}\int_{|r_{ij}{}^{kl}|/\eta}^{\infty}\exp(-t^2)\,dt+\frac{|\mathbf{r}_{ij}{}^{kl}|}{\eta}2\pi^{-1/2}\right.$$
$$\left.\left.\times\exp\left[-\frac{|\mathbf{r}_{ij}{}^{kl}|^2}{\eta^2}\right]\right\}\bigg/|\mathbf{r}_{ij}{}^{kl}|^3\right]. \qquad \text{(A 6)}$$

The differentiation of ϕ by $h_{\alpha\beta}$ gives the result

$$-\frac{\partial\phi}{\partial h_{\alpha\beta}} = \sum_{\gamma}\Pi_{\alpha\gamma}Vh_{\beta\gamma}{}^{-1}, \qquad \text{(A 7)}$$

from which it follows that the stress tensor Π is given by

$$V\Pi_{\alpha\beta}=\frac{1}{4\pi\epsilon_0}\left[\frac{1}{2\pi V}\sum_{\mathbf{n}}{}' Q(\mathbf{k}_{\mathbf{n}})\left\{S(\mathbf{k}_{\mathbf{n}})S(-\mathbf{k}_{\mathbf{n}})\left(\delta_{\alpha\beta}-2\frac{(1+\pi^2\eta^2|\mathbf{k}_{\mathbf{n}}|^2)}{|\mathbf{k}_{\mathbf{n}}|^2}(\mathbf{k}_{\mathbf{n}})_\alpha(\mathbf{k}_{\mathbf{n}})_\beta\right)\right.\right.$$
$$-\sum_i\sum_k 2\pi i(\mathbf{k}_{\mathbf{n}})_\alpha(\mathbf{p}_i{}^k)_\beta(S(\mathbf{k}_{\mathbf{n}})\exp[-2\pi i\mathbf{k}_{\mathbf{n}}\cdot\mathbf{r}_i{}^k]$$
$$\left.-S(-\mathbf{k}_{\mathbf{n}})\exp[2\pi i\mathbf{k}_{\mathbf{n}}\cdot\mathbf{r}_i{}^k])\right\}+\tfrac{1}{2}\sum_i\sum_k\sum_{j\neq i}{}^{*}\sum_l\zeta_i{}^k\zeta_j{}^l\frac{(\mathbf{r}_{ij}{}^{kl})_\alpha(\mathbf{h}(\mathbf{s}_i-\mathbf{s}_j))_\beta}{|\mathbf{r}_{ij}{}^{kl}|^3}$$
$$\left.\times\left\{2\pi^{-1/2}\int_{|r_{ij}{}^{kl}|/\eta}^{\infty}\exp(-t^2)\,dt+\frac{|\mathbf{r}_{ij}{}^{kl}|}{\eta}2\pi^{-1/2}\exp\left[-\frac{|\mathbf{r}_{ij}{}^{kl}|^2}{\eta^2}\right]\right\}\right]. \qquad \text{(A 8)}$$

During the derivation of (A 8), the relations $(\partial V/\partial h_{\alpha\beta})=Vh_{\beta\alpha}{}^{-1}$, and $(\partial h_{\mu\nu}{}^{-1}/\partial h_{\alpha\beta})=-h_{\mu\alpha}{}^{-1}\,h_{\beta\nu}{}^{-1}$ are used. The latter is obtained from the differentiation of $\mathbf{hh}^{-1}=\mathbf{1}$. As usual the trace of stress tensor $\mathbf{\Pi}$ gives an expression for pressure.

Appendix B

The equations of motion for systems with a constrained $\mathbf{h}$ matrix.

(1) *Uniform dilation*

Here we assume

$$h_{\alpha\beta}=g_{\alpha\beta}L, \qquad \text{(B 1)}$$

where $g_{\alpha\beta}$ is a constant matrix which determines the shape of the MD cell, and L is a variable. The kinetic energy for the MD cell vectors is then

$$K_V=\tfrac{1}{2}A(L)\dot{L}^2. \qquad \text{(B 2)}$$

The derivative with respect to L can be calculated from the relationship

$$\frac{\partial}{\partial L}=\sum_{\alpha\beta}\frac{\partial h_{\alpha\beta}}{\partial L}\frac{\partial}{\partial h_{\alpha\beta}}=\sum_{\alpha\beta}g_{\alpha\beta}\frac{\partial}{\partial h_{\alpha\beta}}=\frac{1}{L}\sum_{\alpha\beta}h_{\alpha\beta}\frac{\partial}{\partial h_{\alpha\beta}}. \tag{B 3}$$

The equation of motion is

$$\begin{aligned}\frac{d}{dt}\left(\frac{\partial \mathscr{L}}{\partial \dot{L}}\right)&=\frac{d}{dt}(A(L)\dot{L})=A(L)\ddot{L}+\dot{A}(L)\dot{L}=A(L)\ddot{L}+\frac{\partial A}{\partial L}\dot{L}^2\\&=\frac{\partial \mathscr{L}}{\partial L}=\frac{1}{L}\sum_{\alpha\beta}h_{\alpha\beta}\frac{\partial \mathscr{L}}{\partial h_{\alpha\beta}}=\frac{1}{L}\sum_{\alpha\beta}\sum_{\gamma}h_{\alpha\beta}(\Pi_{\alpha\gamma}-P_{\mathrm{ex}}\delta_{\alpha\gamma})V(h^{-1^{\mathrm{t}}})_{\gamma\beta}\\&=\frac{V}{L}\operatorname{Tr}[\mathbf{h}^{\mathrm{t}}(\boldsymbol{\Pi}-P_{\mathrm{ex}}\mathbf{1})\mathbf{h}^{-1^{\mathrm{t}}}]=\frac{V}{L}\operatorname{Tr}(\boldsymbol{\Pi}-P_{\mathrm{ex}}\mathbf{1}).\end{aligned} \tag{B 4}$$

Thus the final result is

$$\ddot{L}=\frac{V}{A(L)L}\operatorname{Tr}(\boldsymbol{\Pi}-P_{\mathrm{ex}}\mathbf{1})-\frac{1}{A(L)}\frac{\partial A}{\partial L}\dot{L}^2. \tag{B 5}$$

If $A(L)=C(3L^2\det\mathbf{g})^2$, (B 2) and (B 5) become

$$K_{\mathrm{V}}=\frac{9L^4C}{2}(\det\mathbf{g})^2\dot{L}^2 \tag{B 6}$$

and

$$\ddot{L}=\frac{1}{9L^2C\det\mathbf{g}}\operatorname{Tr}(\boldsymbol{\Pi}-P_{\mathrm{ex}}\mathbf{1})-\frac{4}{L}\dot{L}^2. \tag{B 7}$$

Equation (B 7) is the identical equation with (2.26) (Andersen's case). If

$$A(L)=W,\quad K_{\mathrm{V}}=\frac{W}{2}\dot{L}^2 \tag{B 8}$$

and

$$\ddot{L}=\frac{L^2\det\mathbf{g}}{W}\operatorname{Tr}(\boldsymbol{\Pi}-P_{\mathrm{ex}}\mathbf{1}) \tag{B 9}$$

This one corresponds to the uniform dilation limit of (2.8).

(2) *Fixed angles*

In this situation, we assume

$$h_{\alpha\beta}=g_{\alpha\beta}L_{\beta}, \tag{B 10}$$

where $\mathbf{g}=(\mathbf{u}_1,\mathbf{u}_2,\mathbf{u}_3)$ is a constant matrix. Thus only a change in the length of the MD cell vector $\mathbf{u}_i$ is allowed. The kinetic energy for the MD cell vectors becomes

$$K_{\mathrm{V}}=\sum_{\alpha}\frac{W_{\alpha}}{2}\dot{L}_{\alpha}^2 \tag{B 11}$$

and the relationships

$$\frac{\partial}{\partial L_{\alpha}}=\sum_{\mu\nu}\frac{\partial h_{\mu\nu}}{\partial L_{\alpha}}\frac{\partial}{\partial h_{\mu\nu}}=\sum_{\mu\nu}g_{\mu\nu}\delta_{\nu\alpha}\frac{\partial}{\partial h_{\mu\nu}}=\sum_{\mu}g_{\mu\alpha}\frac{\partial}{\partial h_{\mu\alpha}},$$

enable us to derive the equations of motion

$$\frac{d}{dt}\left(\frac{\partial \mathscr{L}}{\partial \dot{L}_\alpha}\right)=W_\alpha \ddot{L}_\alpha=\frac{\partial \mathscr{L}}{\partial L_\alpha}=\sum_\mu g_{\mu\alpha}\frac{\partial \mathscr{L}}{\partial h_{\mu\alpha}}=\sum_{\mu\gamma} g_{\mu\alpha}(\Pi_{\mu\gamma}-P_{\rm ex}\delta_{\mu\gamma})V(h^{-1{\rm t}})_{\gamma\alpha}$$

$$=V[\mathbf{g}^{\rm t}(\boldsymbol{\Pi}-P_{\rm ex}\mathbf{1})\mathbf{h}^{-1{\rm t}}]_{\alpha\alpha}=\frac{V}{L_\alpha}[\mathbf{h}^{\rm t}(\boldsymbol{\Pi}-P_{\rm ex}\mathbf{1})\mathbf{h}^{1{\rm t}}]_{\alpha\alpha}. \qquad \text{(B 12)}$$

If $\mathbf{u}_i$ are orthogonal,

$$\mathbf{h}=\begin{pmatrix} L_1 & 0 & 0\\ 0 & L_2 & 0\\ 0 & 0 & L_3\end{pmatrix},$$

the equation of motion (B 12) becomes similar to (B 9)

$$\ddot{L}_\alpha=\frac{V}{W_\alpha L_\alpha}(\boldsymbol{\Pi}-P_{\rm ex}\mathbf{1})_{\alpha\alpha}. \qquad \text{(B 13)}$$

(3) *Trigonal symmetry*

Here we assume that the $\mathbf{h}$ matrix has the form

$$\mathbf{h}=\begin{pmatrix} L & A & A\\ A & L & A\\ A & A & L\end{pmatrix} \qquad \text{(B 14)}$$

or equivalently

$$h_{\alpha\beta}=\delta_{\alpha\beta}L+(\tau_{\alpha\beta}-\delta_{\alpha\beta})A, \qquad \text{(B 14')}$$

where τ is a matrix in which every element is unity. The length of the unit cell vector is $l=\sqrt{(L^2+2A^2)}$ and the cosine of the angle between the unit cell vectors is

$$\cos\alpha=\frac{(\mathbf{u}_1 \,.\, \mathbf{u}_2)}{|\mathbf{u}_1|^2}=\frac{2LA+A^2}{L^2+2A^2}.$$

The following relationships are needed in order to derive the equations of motion

$$\frac{\partial}{\partial L}=\sum_{\alpha\beta}\frac{\partial h_{\alpha\beta}}{\partial L}\frac{\partial}{\partial h_{\alpha\beta}}=\sum_{\alpha\beta}\delta_{\alpha\beta}\frac{\partial}{\partial h_{\alpha\beta}}=\sum_\alpha\frac{\partial}{\partial h_{\alpha\alpha}}$$

$$\frac{\partial}{\partial A}=\sum_{\alpha\beta}\frac{\partial h_{\alpha\beta}}{\partial A}\frac{\partial}{\partial h_{\alpha\beta}}=\sum_{\alpha\beta}(\tau_{\alpha\beta}-\delta_{\alpha\beta})\frac{\partial}{\partial h_{\alpha\beta}}=\sum_{\alpha\beta}\frac{\partial}{\partial h_{\alpha\beta}}-\sum_\alpha\frac{\partial}{\partial h_{\alpha\alpha}}.$$

The kinetic energy is

$$K_{\rm V}=\frac{W_L}{2}\dot{L}^2+\frac{W_A}{2}\dot{A}^2 \qquad \text{(B 15)}$$

and the equations of motion for L and A become

$$\frac{d}{dt}\left(\frac{\partial \mathscr{L}}{\partial \dot{L}}\right)=W_L\ddot{L}=\frac{\partial \mathscr{L}}{\partial L}=\sum_\alpha\frac{\partial \mathscr{L}}{\partial h_{\alpha\alpha}}={\rm Tr}\,[(\boldsymbol{\Pi}-P_{\rm ex}\mathbf{1})V\mathbf{h}^{-1{\rm t}}] \qquad \text{(B 16)}$$

and

$$\frac{d}{dt}\left(\frac{\partial \mathscr{L}}{\partial \dot{A}}\right)=W_A\ddot{A}=\frac{\partial \mathscr{L}}{\partial A}=\sum_\alpha\sum_{\beta\neq\alpha}[(\boldsymbol{\Pi}-P_{\rm ex}\mathbf{1})V\mathbf{h}^{-1{\rm t}}]_{\alpha\beta}. \qquad \text{(B 17)}$$

Appendix C

The rotation matrix R that transforms $\mathbf{h}$ into symmetric form is given formally by the equation

$$R = VU^{t}, \tag{C 1}$$

where U and V are orthogonal matrices. The proof follows from a theorem in linear algebra [20].

If $\mathbf{v}_i$ is an eigenvector of $\mathbf{h}^t\mathbf{h} = \mathbf{G}$ and λ_i^2 an eigenvalue, then since $\mathbf{G}$ is a symmetric and positive definite matrix, all its eigenvalues are positive,

$$G\mathbf{v}_i = \mathbf{h}^t\mathbf{h}\mathbf{v}_i = \lambda_i^2\, \mathbf{v}_i. \tag{C 2}$$

If we define

$$\mathbf{u}_i = \frac{1}{\lambda_i}\, \mathbf{h}\mathbf{v}_i, \tag{C 3}$$

then it can be shown that $\mathbf{u}_i$ is the eigenvector of $\mathbf{h}\mathbf{h}^t$. From (C 2) and (C 3), it follows that

$$(\mathbf{u}_i^{t}\, \mathbf{h}\mathbf{v}_j) = \frac{1}{\lambda_i}\,(\mathbf{v}_i^{t}\, \mathbf{h}^t\, \mathbf{h}\mathbf{v}_j) = \frac{\lambda_j^2}{\lambda_i}\,(\mathbf{v}_i^{t}\mathbf{v}_j) = \lambda_i\delta_{ij}. \tag{C 4}$$

Thus $\mathbf{h}$ can be diagonalized by the orthogonal matrices U^t and V.

It remains to demonstrate that $R\mathbf{h}$ is a symmetric matrix. Equation (C 4) can also be written as : $D = U^t\mathbf{h}V = V^{-1}(VU^t)\mathbf{h}V$, where D is a diagonal matrix. It follows that $VDV^{-1} = (VU^t)\mathbf{h} = R\mathbf{h}$, where VDV^{-1} is a symmetric matrix because V is an orthogonal matrix ($V^{-1} = V^t$), and hence

$$(VDV^{-1})_{\alpha\gamma} = \sum_{\beta} V_{\alpha\beta}D_{\beta}(V^{-1})_{\beta\gamma} = \sum_{\beta} V_{\alpha\beta}D_{\beta}V_{\gamma\beta} = (VDV^{-1})_{\gamma\alpha}.$$

We can therefore obtain a rotational matrix that transforms $\mathbf{h}$ into a symmetric form using the eigenvalues and eigenvectors of $\mathbf{h}^t\mathbf{h}$ and equations (C 1), (C 2), and (C 3).

References

[1] McDonald, I. R., 1972, *Molec. Phys.*, **23,** 42 ; 1972, *Ibid.*, **24,** 391.
[2] Andersen, H. C., 1980, *J. chem. Phys.*, **72,** 2384.
[3] Haile, J. M., and Graben, H. W., 1980, *J. chem. Phys.*, **73,** 2412 ; 1980, *Molec. Phys.*, **40,** 1433.
[4] Parrinello, M., and Rahman, A., 1980, *Phys. Rev. Lett.*, **45,** 1196.
[5] Parrinello, M., and Rahman, A., 1981, *J. appl. Phys.*, **52,** 7182 ; 1982, *J. chem. Phys.*, **76,** 2662.
[6] Parrinello, M., Rahman, A., and Vashishta, P., 1983, *Phys. Rev. Lett.*, **50,** 1073.
[7] Nosé, S., and Klein, M. L., 1983, *Phys. Rev. Lett.*, **50,** 1207.
[8] Nosé, S., and Klein, M. L., 1983, *J. chem. Phys.*, **78,** 6928.
[9] Evans, D. J., 1977, *Molec. Phys.*, **34,** 317.
[10] Verlet, L., 1967, *Phys. Rev.*, **159,** 98.
[11] Ryckaert, J.-P., Ciccotti, G., and Berendsen, H. J. C., 1977, *J. comput. Phys.*, **23,** 327.
[12] Ryckaert, J.-P., and Ciccotti, G., 1983, *J. chem. Phys.*, **78,** 7368.
[13] Nordsieck, A., 1962, *Math. Comput.*, **16,** 22.
[14] Gear, G. W., 1971, *Numerical Initial Value Problems in Ordinary Differential Equations* (Prentice Hall), Chap. 9.

[15] Rahman, A., and Stillinger, F. H., 1971, *J. chem. Phys.*, **55,** 3336. Appendix A contains a concise description of the predictor–corrector method. The appropriate coefficients, however, should be taken from [14], p. 154.
[16] Pawley, G. S., and Thomas, G. W., 1982, *Phys. Rev. Lett.*, **48,** 410.
[17] Parrinello, M., and Rahman, A., 1982, *Melting, Localization and Chaos,* edited by R. K. Kalia and P. Vashishta (Elsevier), p. 97.
[18] Tosi, M. P., 1964, *Solid St. Phys.*, **16,** 107.
[19] Rahman, A., and Vashishta, P., 1981, *Physics of Superionic Conductors,* edited by S. W. de Leeuw and J. W. Perram (Plenum) (to be published).
[20] Noble, B., and Daniel, J. W., 1977, *Applied Linear Algebra* (Prentice Hall), p. 327.

Notes to Reprint V.5

1. This paper uses the method of quaternions for the solution of the rotational equations of motion. The adaptation of the method of constraints to the new ensembles has also been achieved (see Reprint V.4 and Note V.4.2), and the remarks at the foot of p. 1066 are therefore unduly pessimistic.

2. For reviews of the use of both conventional and constant-pressure molecular dynamics in the study of rotator-phase crystals and solid–solid transformations, see the articles by Klein (Ann. Rev. Phys. Chem. 36 (1985) 525; and in: G. Ciccotti and W.G. Hoover, eds, Molecular Dynamics Simulation of Statistical-Mechanical Systems (North-Holland, Amsterdam, 1986).

PHYSICAL REVIEW A VOLUME 11, NUMBER 6 JUNE 1975

Statistical mechanics of dense ionized matter. IV. Density and charge fluctuations in a simple molten salt

Jean Pierre Hansen

Laboratoire de Physique Théorique des Liquides, Université de Paris VI, 4 Place Jussieu, 75005 Paris, France

Ian R. McDonald

Department of Chemistry, Royal Holloway College, Egham, Surrey TW20 OEX, England

(Received 24 February 1975)

The results of a molecular-dynamics study of a simple model of a molten salt are reported. The interionic pair potential which is used consists of the Coulombic term and an inverse-power repulsion which is assumed to be the same for all ions. The structure of the liquid is found to be dominated by charge-ordering effects and the calculated equilibrium properties are in good agreement with the predictions of the hypernetted-chain approximation. The relation between the self-diffusion coefficient and the electrical conductivity is discussed, and the observed deviations from the Nernst-Einstein relation in real molten salts are shown to have a natural explanation in terms of short-lived cross correlations. Data on the spectra of charge and particle density fluctuations are presented. At small wave numbers there is a propagating optic-type mode which shows a strong negative dispersion, but no Brillouin peak is seen even at the lowest wave number which is accessible. The data are analyzed in terms of a single-relaxation-time model incorporating the low-order spectral moments, for which we give explicit formulas. The fit achieved is fair, but the low-frequency behavior of the charge fluctuations at small wave numbers is incorrectly reproduced, and there is evidence for the necessity of introducing a second relaxation time. Comparison is made with results previously obtained for the classical one-component plasma.

I. INTRODUCTION

One of the most interesting results to emerge from recent research on simple liquids is the existence of well-defined collective excitations in a range of wave numbers k up to 2π times the inverse of the nearest-neighbor distance r_0. This information has come both from molecular-dynamics (MD) calculations and from neutron-scattering experiments. For example, the molecular-dynamics work of Levesque *et al.*[1] has shown that in the spectrum of density fluctuations in an argon-like liquid a three-peak structure characteristic of the hydrodynamic regime is still observable at wavelengths as short as $6r_0$. The position of the Brillouin peak in these calculations is at a frequency such that ω/k is very nearly equal to the macroscopic sound velocity. The persistence of propagating density fluctuations at short wavelengths is apparently closely related to the form of the interparticle potential. Whereas the work of Levesque *et al.*[1] was based on the familiar Lennard-Jones potential, a more recent computer "experiment" made on liquid rubidium by Rahman,[2] in which a long-range oscillatory potential due to Price[3] was used, has shown that in such a system density waves continue to propagate at wavelengths of order $2r_0$. The results obtained by Rahman[2] are in excellent agreement with the neutron-scattering measurements of Copley and Rowe.[4]

In an earlier paper of this series, to be referred to as I, we have, in collaboration with Pollock,[5] used the method of molecular dynamics to study the collective dynamical properties of the classical one-component plasma (OCP) in a uniform neutralizing background. Because of the assumed rigidity of the background, the dynamical structure factor $S(k,\omega)$ of the OCP describes simultaneously the spectrum both of density and charge fluctuations. At small wave numbers the dominant feature in $S(k,\omega)$ is a very sharp peak near the plasma frequency. The dispersion is typical of an optic-type mode, the peak being found at nonzero frequency even in the limit $k\to 0$.

The purpose of the present paper is to report the results of a molecular-dynamics simulation of a system of two components, differing only in the sign of their charges, which may be regarded as a simple model of a monovalent molten salt. The feature of molten salts which is of particular interest in the present context is the fact that the charge and density fluctuations are independent, at least in the long-wavelength limit. The possibility therefore exists of observing both the sound-wave propagation at low k which is characteristic of simple liquids and the plasma oscillations which we have previously studied in the case of the OCP. These two phenomena are the liquid-state analogs of the acoustic and optic modes of ionic crystals. In practice, we have been only partly successful, as we have been unable to detect any Brillouin peak in $S(k,\omega)$, even at the smallest k value which is accessible in the size of system we have studied.

Our interest lies primarily in elucidating the general character of short-wavelength excitations in molten salts rather than in making detailed quantitative calculations for a specific system. We wish in particular to avoid the complications which arise when the ions are of unequal size and mass. The model which we have used is one in which the masses of the cation and anion are equal, the ions are singly charged, and the interaction potential for two ions of either species is given by

$$\varphi(r_{12}) = \frac{e^2}{\lambda}\left[\frac{1}{n}\left(\frac{\lambda}{r_{12}}\right)^n + Q_1 Q_2\left(\frac{\lambda}{r_{12}}\right)\right]; \qquad (1)$$

Q_1 and Q_2 are the ionic charges in units of e, the electronic charge; λ is a characteristic length parameter which in fact is the separation at which the cation-anion potential is a minimum; and n is an exponent which we set equal to 9. Monte Carlo calculations[6] have shown that a model similar to (1) provides a fair description of the thermodynamic properties of the alkali-metal salts. The only difference between the potential used in Ref. 6 and that employed here is the fact that in the Monte Carlo work the short-range repulsive interactions between ions of like sign were ignored. From a practical point of view the distinction is unimportant because such interactions make a negligible contribution to the total potential energy of a typical configuration of ions. The thermodynamic properties of either model may be scaled on a corresponding-states basis, but the scaling of temperature is not independent of that of volume, as is the case in OCP.

The technical details of the molecular-dynamics calculations are similar to those in our work on the OCP, a few obvious changes being required in order to deal with a two-component system and short-range interionic repulsions. The algorithm adopted by Verlet[10] was used to integrate numerically the classical equations of motion of a system of 216 ions, 108 of each species, enclosed within a cube which is surrounded on all sides by periodically repeating images. The Coulombic force on each ion was computed by the Ewald method and the short-range interactions were truncated at a separation of one-half the length of the cube. As unit of length we choose the characteristic length λ, and as the unit of time we choose the inverse plasma frequency ω_p^{-1}, defined by

$$\omega_p^2 = 4\pi\rho e^2/m\ , \qquad (2)$$

where $\rho = N/V$ is the total number density of ions.

The molecular-dynamics "experiment" was made at a volume $V/N\lambda^3 = 2.72$ and the trajectories of the ions were followed for 2.5×10^4 time steps of length $\Delta t = 0.2$, extended to 5×10^4 steps for the calculation of electrical conductivity. The mean temperature of the run was $\lambda kT/e^2 = 0.0177$. To give some feeling for the magnitude of the quantities involved, we shall make a conversion to real units for the case of liquid NaCl. Taking the Monte Carlo calculations[6] as a guide, we choose $\lambda = 2.34$ Å and find $V = 41.9$ cm^3 mol^{-1}, $T = 1267$ K, and $\Delta t = 0.48\times10^{-14}$ sec. For comparison, the triple point[7] of NaCl is at $V = 37.52$ cm^3 mol^{-1}, $T = 1073$ K.

In order to simplify comparison with our results on the OCP, it is convenient to introduce two further quantities. These are in the ion-sphere radius a given by

$$a = (3/4\pi\rho)^{1/3}, \qquad (3)$$

and the (dimensionless) plasma parameter Γ defined as

$$\Gamma = e^2/ak_BT\ . \qquad (4)$$

Under the conditions of our "experiment" we find that $a = 0.866\lambda$ and $\Gamma = 64.6$.

The outline of the paper is as follows. Equilibrium properties are discussed in Sec. II and comparison made with results of the hypernetted-chain (HNC) theory. Section III is concerned with the phenomena of self-diffusion and electrical conductivity and the correlation of the two transport coefficients through the Nernst-Einstein relation; a brief account of this part of the work has previously appeared elsewhere.[8] Sections IV and V are devoted to the analysis, respectively, of density and charge fluctuations. Expressions for the low-order moments of the corresponding current correlation functions are derived and used to build a simple phenomenological model which provides a fair description of the observed spectra. Finally, in Sec. VI, we make some suggestions for future work.

II. EQUILIBRIUM PROPERTIES AND STRUCTURE

The equilibrium structure of a molten salt may be described in terms of the three partial radial distribution functions: $g_{++}(r)$, $g_{--}(r)$, and $g_{+-}(r)$. For our simple model there is some simplification because $g_{++}(r)$ and $g_{--}(r)$ are identical on grounds of symmetry. This makes it convenient to discuss the problem in terms of only two distribution functions:

$$g_1(r) = \tfrac{1}{2}[g_{++}(r) + g_{--}(r)], \qquad (5)$$

$$g_2(r) = g_{+-}(r)\ . \qquad (6)$$

Notes on p. 414

Corresponding to $g_1(r)$ and $g_2(r)$ we have two total correlation functions, $h_1(r)$ and $h_2(r)$, two direct correlation functions, $c_1(r)$ and $c_2(r)$, and two structure factors:

$$S_1(k) = 1 + \tfrac{1}{2}\rho^* \tilde{h}_1(k), \tag{7}$$

$$S_2(k) = \tfrac{1}{2}\rho^* \tilde{h}_2(k), \tag{8}$$

where $\rho^* = \rho\lambda^3$ and the tilde is used to denote a Fourier transform. Charge neutrality[9] requires that

$$\lim_{k\to 0} S_1(k) = \lim_{k\to 0} S_2(k), \tag{9}$$

and the perfect-screening condition[9] requires that

$$S_1(k) - S_2(k) \underset{k\to 0}{\sim} k^2/\kappa^2, \tag{10}$$

where κ is the inverse Debye length

$$\kappa = (4\pi\rho e^2/k_B T)^{1/2}. \tag{11}$$

Fluctuations in density are described in k space by the structure factor $S(k)$, given by

$$S(k) = S_1(k) + S_2(k), \tag{12}$$

and those in charge by the structure factor $S'(k)$:

$$S'(k) = S_1(k) - S_2(k). \tag{13}$$

The long-wavelength limits of (12) and (13) are determined by

$$\lim_{k\to 0} S(k) = \frac{1}{k_B T}\left[\left(\frac{\partial P}{\partial \rho}\right)_T\right]^{-1}, \tag{14}$$

$$\lim_{k\to 0} S'(k) = 0. \tag{15}$$

If we attempt to use the computed distribution functions to calculate the various structure factors, we encounter the usual problem, namely that the molecular-dynamics "experiment" yields values for the distribution functions only for interionic separations less than some cutoff distance r_c. In our case r_c is equal to one-half the length of the cube, i.e., $r_c = 4.185\lambda$. We have therefore extrapolated our results on $g_1(r)$ and $g_2(r)$ beyond $r = r_c$ by the method devised by Verlet,[10] except that we approximate the direct correlation functions at large r by means of the hypernetted-chain (HNC) approximation rather than the Percus-Yevick (PY) approximation. The justification for this change is the fact that HNC is known to be superior to the PY approximation in the case of long-range potentials, as evidenced by the work of Springer *et al.*[11] on the OCP.

To implement the method of Verlet,[10] we require a good initial guess for $c_1(r)$ and $c_2(r)$. In practice, we have found that this requires the solution to the full coupled HNC equations for our system. To obtain this we have adapted the very efficient method of Springer *et al.*[11] to the case of two components, but even then the calculations have proved very time consuming. We have been compelled to solve the HNC equations for a range of temperatures, starting at very high temperature and reducing in stages to the temperature of interest. At each new step in this procedure we used as input the direct correlation functions obtained at the previous stage. Approximately 500 iterations were needed at each step before satisfactory convergence was achieved and the extrapolation itself finally required 4000 iterations. The calculations, though lengthy, do have the advantage of yielding the HNC solution as a by-product. Though the HNC equation has often been used for electrolytes,[12] we are not aware of any previous calculations at charge densities typical of molten salts apart from some unpublished work by Larsen[13] on systems of hard-sphere ions.

The distribution functions $g_1(r)$ and $g_2(r)$ are plotted in Fig. 1, $c_1(r)$ and $c_2(r)$ in Fig. 2, and the structure factors $S(q)$ and $S'(q)$, where $q = \lambda k$, are shown in Fig. 3. What is very clear from Figs. 1 and 3 is the extent to which the structure of an ionic melt is determined by charge-ordering effects, a feature which has previously been discussed on the basis of Monte Carlo calculations by Woodcock and Singer.[14] The function $g_2(r)$ has a very sharp peak near $r = \lambda$ and the later oscillations differ in phase from those in $g_1(r)$ by almost exactly half a period. The result is that the overall radial distribution function has virtually no structure beyond its first peak. The function $g_1(r)$, in contrast to $g_2(r)$, has a rather weak and broad first peak near $r = 1.75\lambda$. However, the tail of $g_1(r)$ at small r extends inwards as far as the peak in $g_2(r)$, indicating that there is some small pene-

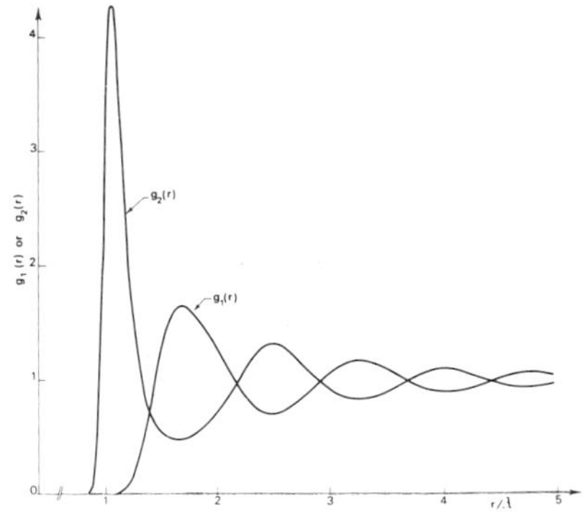

FIG. 1. Radial distribution functions.

tration of ions of like charge into the first coordination shell around a given ion. In q space the importance of charge ordering is equally obvious. The charge-charge structure factor $S'(q)$ has a very pronounced first peak and displays much more structure than the number-number structure factor $S(q)$. It should be noted that the main peak in $S'(q)$ occurs at a significantly smaller value of k than that in $S(q)$. The difference arises from the fact that the position of the peak in $S'(q)$ is determined by the period of oscillation of mean charge density around a reference ion, whereas that in $S(q)$ is determined by the value of the nearest-neighbor distance. As a check on the numerical accuracy we have confirmed that the conditions (8) and (9) are both very well obeyed.

The direct correlation function $c_1(r)$ is a monotonic function of r and qualitatively is very similar to that obtained for the OCP. The function $c_2(r)$, on the other hand, shows a break in slope near $r=\lambda$; this is a remnant of the hard-core type of discontinuity observed by Larsen.[13] Both $c_1(r)$ and $c_2(r)$ tend very rapidly to their Debye-Huckel limits. At $r=1.6\lambda$ the discrepancies in either case are already less than 1%.

For the sake of easy reference we list the results on the distribution functions in Table I and those on the structure factors in Table II. In Table I we also report the results of the HNC calculations. We have not plotted the HNC results in the figures simply because the differences are generally too small to make fair comparison possible, particularly in the steeply rising part of $g_2(r)$. In this case we find that

$$PV/NkT = 1+11.80-12.48 = 0.32 \quad \text{(MD)}$$
$$= 1+12.51-12.53 = 0.96 \quad \text{(HNC)}.$$

Overall the agreement with the molecular-dynamics calculations is surprisingly good, and better than might be expected on the basis of results obtained for the OCP.[11] The only major discrepancy is in the values of $S(k)$ at small k. In particular, we find in the long-wavelength limit that

$$S(0) = 0.16 \quad \text{(MD)}$$
$$= 0.50 \quad \text{(HNC)},$$

so that the HNC approximation yields a compressibility which is approximately three times too large.

For other thermodynamic properties the HNC results are much more satisfactory. For the excess internal energy we may divide the net result into contributions from short-range (SR) and Coulombic (C) interactions:

$$\frac{U}{Nk_BT} = \frac{U^{SR}}{Nk_BT} + \frac{U^C}{Nk_BT} . \tag{16}$$

On dividing the calculated values in the same way, we find that

$$U/Nk_BT = 3.93-37.45 = -33.52 \quad \text{(MD)}$$
$$= 4.17-37.59 = -33.42 \quad \text{(HNC)}.$$

A similar division may be made for the equation of state

$$\frac{PV}{Nk_BT} = 1 + \frac{P^{SR}V}{Nk_BT} + \frac{P^CV}{Nk_BT} . \tag{17}$$

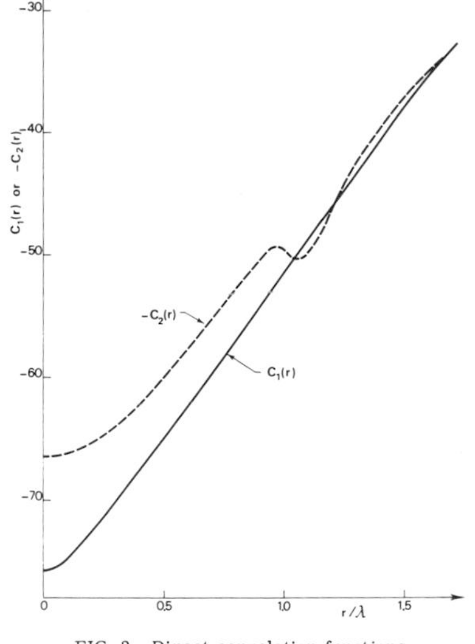

FIG. 2. Direct correlation functions.

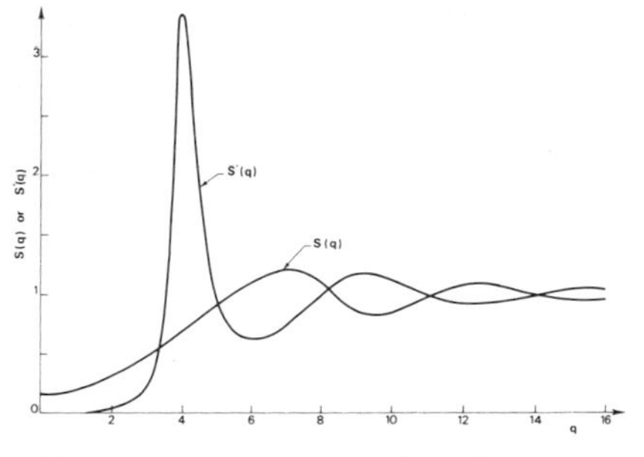

FIG. 3. Structure factors as functions of the dimensionless wave number $q=\lambda k$.

TABLE I. Radial distribution functions.

r	$g_1(r)$	$g_1^{\mathrm{HNC}}(r)$	$g_2(r)$	$g_2^{\mathrm{HNC}}(r)$	$c_1(r)$	$c_1^{\mathrm{HNC}}(r)$	$c_2(r)$	$c_2^{\mathrm{HNC}}(r)$
0.0	0.0	0.0	0.0	0.0	−75.76	−74.42	66.38	65.55
0.084	0.0	0.0	0.0	0.0	−75.08	−73.68	66.20	65.37
0.167	0.0	0.0	0.0	0.0	−73.40	−71.94	65.63	64.80
0.251	0.0	0.0	0.0	0.0	−71.34	−69.91	64.66	63.86
0.335	0.0	0.0	0.0	0.0	−69.20	−67.85	63.33	62.59
0.419	0.0	0.0	0.0	0.0	−67.03	−65.81	61.70	61.06
0.502	0.0	0.0	0.0	0.0	−64.84	−63.75	59.86	59.35
0.586	0.0	0.0	0.0	0.0	−62.62	−61.66	57.89	57.53
0.670	0.0	0.0	0.0	0.0	−60.37	−59.53	55.86	55.66
0.754	0.0	0.0	0.0	0.0	−58.10	−57.37	53.83	53.75
0.837	0.0	0.0	0.0	0.0	−55.81	−55.19	51.81	51.82
0.921	0.0	0.0	0.048	0.042	−53.52	−53.00	49.86	49.93
0.963	0.0	0.0	0.516	0.575	−52.37	−51.89	49.33	49.49
1.005	0.0	0.0	1.882	2.183	−51.23	−50.80	49.70	50.12
1.047	0.0	0.0	3.489	3.900	−50.09	−49.70	50.31	50.86
1.088	0.0	0.001	4.275	4.526	−48.96	−48.60	50.10	50.51
1.130	0.002	0.006	4.115	4.137	−47.84	−47.51	48.95	49.14
1.172	0.009	0.023	3.453	3.345	−46.73	−46.42	47.29	47.37
1.214	0.028	0.619	2.718	2.566	−45.62	−45.32	45.56	45.61
1.256	0.068	0.133	2.094	1.946	−44.50	−44.19	43.95	44.02
1.298	0.140	0.244	1.612	1.495	−43.36	−43.04	42.49	42.60
1.340	0.251	0.392	1.269	1.178	−42.21	−41.87	41.18	41.32
1.382	0.401	0.568	1.016	0.958	−41.03	−40.70	39.97	40.15
1.423	0.586	0.758	0.844	0.805	−39.84	−39.52	38.85	39.06
1.465	0.790	0.947	0.720	0.699	−38.66	−38.37	37.80	38.03
1.507	1.001	1.120	0.632	0.626	−37.49	−37.26	36.81	37.06
1.549	1.198	1.267	0.572	0.576	−36.35	−36.20	35.86	36.12
1.591	1.368	1.384	0.530	0.545	−35.28	−35.20	34.95	35.22
1.633	1.502	1.468	0.503	0.526	−34.26	−34.25	34.08	34.35
1.674	1.590	1.523	0.487	0.517	−33.32	−33.36	33.25	33.52
1.716	1.642	1.551	0.481	0.516	−32.45	−32.52	32.45	32.72
1.758	1.658	1.558	0.484	0.522	−31.64	−31.74	31.68	31.95
1.800	1.650	1.550	0.492	0.535	−30.89	−30.99	30.94	31.20
1.842	1.621	1.532	0.508	0.553	−30.19	−30.29	30.24	30.49
1.926	1.526	1.478	0.556	0.603	−28.92	−28.97	28.92	29.13
2.010	1.413	1.414	0.625	0.669	−27.78	−27.78	27.71	27.90
2.093	1.289	1.309	0.712	0.751	−26.73	−26.69	26.60	26.75
2.177	1.145	1.145	0.823	0.847	−25.73	−25.69	25.60	25.70
2.261	0.997	0.976	0.947	0.956	−24.79	−24.75	24.68	24.74
2.344	0.873	0.851	1.078	1.071	−23.90	−23.85	23.84	23.86
2.428	0.783	0.778	1.199	1.179	−23.08	−23.01	23.06	23.05
2.512	0.731	0.746	1.285	1.261	−22.32	−22.23	22.32	22.30
2.596	0.712	0.743	1.321	1.299	−21.61	−21.51	21.61	21.59
2.721	0.735	0.779	1.285	1.266	−20.64	−20.52	20.60	20.58
2.847	0.809	0.848	1.182	1.165	−19.75	−19.64	19.69	19.66
2.972	0.913	0.937	1.059	1.049	−18.92	−18.82	18.86	18.82
3.098	1.027	1.028	0.954	0.953	−18.14	−18.06	18.09	18.06
3.224	1.118	1.102	0.880	0.888	−17.42	−17.35	17.40	17.36
3.349	1.167	1.144	0.845	0.861	−16.75	−16.69	16.74	16.71
3.475	1.165	1.145	0.844	0.869	−16.14	−16.09	16.12	16.11
3.600	1.124	1.108	0.875	0.902	−15.56	−15.53	15.55	15.54
3.726	1.059	1.050	0.931	0.950	−15.03	−15.01	15.02	15.01
3.852	1.001	0.990	0.997	1.003	−14.52	−14.52	14.52	14.52

III. SELF-DIFFUSION AND ELECTRICAL CONDUCTANCE

The self-diffusion of ions of species α (where α may be + for cations, or − for anions) is usefully discussed in terms of the velocity autocorrelation function $Z_\alpha(t)$, defined as

$$Z_\alpha(t) = \langle \vec{v}_{i\alpha}(t) \cdot \vec{v}_{i\alpha}(0) \rangle , \qquad (18)$$

where the angular brackets denote an average over (a) all ions, labeled i, of species α, and (b) all choices of the time origin. For our simple model the autocorrelation functions $Z_+(t)$ and $Z_-(t)$ are identical and the subscripts may therefore be dropped. Then the coefficients of self-diffusion, D, is related to $Z(t)$ through the well-known expression

$$D = \frac{1}{3} \int_0^\infty Z(t)\, dt . \qquad (19)$$

From our calculations we find that $D = 0.0049$. For NaCl, taking the value of λ which we have already used, we find that our result corresponds to $D = 9.5 \times 10^{-5}$ cm^2 sec^{-1}. Experimentally,[7] for NaCl at $V = 38.2$ cm^3 mol^{-1} and $T = 1121$ K we find $D_+ = 9.99 \times 10^{-5}$ cm^2 sec^{-1}, $D_- = 7.21 \times 10^{-5}$ cm^2 sec^{-1}.

The electrical conductivity σ is related to the normalized autocorrelation function of the total current $J(t)$, defined as

$$J(t) = \left\langle \left(\sum_i \vec{v}_{i+}(t) - \sum_j \vec{v}_{j-}(t) \right) \times \left(\sum_i \vec{v}_{i+}(0) - \sum_j \vec{v}_{j-}(0) \right) \right\rangle . \qquad (20)$$

The principle of conservation of momentum may be used to rewrite this expression in several equivalent forms. The conductivity is given by

$$\sigma = \frac{e^2}{3VkT} \int_0^\infty J(t)\, dt . \qquad (21)$$

The electrical current is a collective property of the system and the averaging process labeled (a) above, which is used to improve the statistics on $Z(t)$, cannot be applied to the calculation of $J(t)$. Thus our estimate for σ is probably not accurate to better than 5%. In fact we find a value which for NaCl corresponds to 3.2 mho cm^{-1}. The experimental result[7] for NaCl at the state defined above is 3.7 mho cm^{-1}. We include this comparison with experiment both for D and σ primarily as evidence that our potential model is adequate to describe the main features of transport in molten salts.

A useful empirical link between the coefficients of self-diffusion and electrical conductivity is provided by the Nernst-Einstein relation. If we introduce a deviation parameter Δ, we may write this relation in a generalized form as

$$\sigma = \tfrac{1}{2}(Ne^2/VkT)(D_+ + D_-)(1 - \Delta) . \qquad (22)$$

From the data tabulated by Young and O'Connell[7] it is possible to compute values of Δ for eight alkali-metal salts at zero pressure and temperatures close to the respective triple points. We find that Δ is invariably positive, varying from 0.08 in the case of NaI to 0.43 for $LiNO_3$, with a mean of 0.26. Our own results on D and σ lead to a value of $\Delta = 0.19$. This compares particularly well with the experimental results for the chlorides: 0.15 (RbCl), 0.18 (NaCl), and 0.23 (CsCl).

The validity of the Nernst-Einstein relation, with $\Delta = 0$, can be deduced from Eqs. (18)–(21) if it is assumed that cross correlation terms of the type $\langle \vec{v}_{i-}(t) \cdot \vec{v}_{j+}(0) \rangle$ make zero contribution to $J(t)$. In such circumstances it follows that $J(t) = NZ(t)$ and the two normalized autocorrelation functions should be identical. The obvious explanation of deviations from the Nernst-Einstein relation lies in the formation of ionic complexes which contribute to the diffusive flux but not to the electrical current. Our results on $Z(t)$ and $J(t)$, normalized to unity at $t = 0$, are shown in Fig. 4. From the figure we see that there are some significant differences between the two functions. In particular,

TABLE II. Structure factors.

$q = \lambda k$	$S(q)$	$S'(q)$	$q = \lambda k$	$S(q)$	$S'(q)$
0.0	0.160	0.0	5.4	0.998	0.745
0.2	0.162	0.000 15	5.8	1.073	0.651
0.4	0.166	0.000 63	6.2	1.145	0.634
0.6	0.174	0.001 48	6.6	1.192	0.667
0.8	0.187	0.002 76	7.0	1.208	0.737
1.0	0.206	0.0046	7.4	1.205	0.829
1.2	0.228	0.0072	7.8	1.160	0.937
1.4	0.253	0.0108	8.2	1.049	1.045
1.6	0.277	0.0158	8.6	1.938	1.128
1.8	0.299	0.0228	9.0	0.866	1.168
2.0	0.321	0.0326	9.4	0.837	1.165
2.2	0.344	0.0464	9.8	0.844	1.133
2.4	0.371	0.0665	10.2	0.876	1.086
2.6	0.404	0.097	10.6	0.925	1.035
2.8	0.444	0.146	11.0	0.981	0.990
3.0	0.489	0.228	11.8	1.078	0.933
3.2	0.537	0.377	12.6	1.097	0.921
3.4	0.585	0.672	13.4	1.045	0.958
3.6	0.629	1.279	14.2	0.979	1.013
3.8	0.671	2.392	15.0	0.948	1.049
4.0	0.710	3.355	15.8	0.954	1.046
4.2	0.750	2.969	16.6	0.986	1.016
4.4	0.791	2.136	17.4	1.018	0.982
4.6	0.834	1.550	18.2	1.033	0.968
4.8	0.878	1.195	19.0	1.024	0.976
5.0	0.920	0.976	19.8	1.003	0.996

Notes on p. 414

the negative region is much more pronounced in $J(t)$ than in $Z(t)$. It is also clear that deviations from the Nernst-Einstein relation are not necessarily the result of a permanent association of ions of opposite charge. Figure 4 shows that the difference between the two normalized autocorrelation functions reaches a maximum near the first zero in each, and then rapidly disappears as t increases. In this case, at least, the Nernst-Einstein deviation has a natural explanation in terms of short-lived cross-correlation functions.

Finally, we mention that the velocity autocorrelation function resembles very closely that observed in an argonlike liquid at high density.[15] The work reported in I has shown that in the OCP the single-particle motion is very strongly coupled to the collective modes. This coupling is seen in the fact that the velocity autocorrelation function is characterized by well-defined oscillations, at a frequency close to ω_p, which are already evident at $\Gamma \approx 10$ and become increasingly more pronounced as the crystallization point ($\Gamma \approx 155$) is approached. In the present case, however, the power spectrum $\tilde{Z}(\omega)$ displays only a low-frequency diffusive peak and a weak shoulder at higher frequency which arises from the long tail in $Z(t)$. There is no evidence of any remnant of a plasma-type oscillation.

Self-diffusion in molten salts has also been studied by molecular dynamics by Lantelme *et al.*[16] and by Lewis and Singer[17]; the potential model used by these workers is considerably more complicated than that employed here.

IV. DENSITY FLUCTUATIONS

In order to discuss the collective longitudinal modes of the system under study, we first introduce the operators $\rho_{\vec{k}}^{+}(t)$ and $\rho_{\vec{k}}^{-}(t)$ which represent the Fourier components of the density of cations and anions

$$\rho_{\vec{k}}^{+}(t) = \sum_i e^{i\vec{k}\cdot\vec{r}_{i+}(t)}, \tag{23}$$

$$\rho_{\vec{k}}^{-}(t) = \sum_j e^{i\vec{k}\cdot\vec{r}_{j-}(t)}. \tag{24}$$

From these quantities we may construct a total number density operator $\rho_{\vec{k}}(t)$ given by

$$\rho_{\vec{k}}(t) = \rho_{\vec{k}}^{+}(t) + \rho_{\vec{k}}^{-}(t). \tag{25}$$

The time dependence of fluctuations in $\rho_{\vec{k}}(t)$ is described by the density-density correlation function $F(k, t)$

$$F(k, t) = (1/N)\langle \rho_{\vec{k}}(t)\rho_{-\vec{k}}(0)\rangle \tag{26}$$

The Fourier transform of $F(k, t)$ is the dynamical structure factor:

$$S(k, \omega) = \frac{1}{2\pi}\int_{-\infty}^{\infty} e^{i\omega t} F(k, t)\, dt \tag{27}$$

$S(k, \omega)$ represents the spectrum of number density fluctuations in the system.

It is also convenient to introduce the total particle current operators $\vec{j}_{\vec{k}}(t)$, defined as

$$\vec{j}_{\vec{k}}(t) = \vec{j}_{\vec{k}}^{+}(t) + \vec{j}_{\vec{k}}^{-}(t), \tag{28}$$

where

$$\vec{j}_{\vec{k}}^{+}(t) = \sum_i \vec{v}_{i+} e^{i\vec{k}\cdot\vec{r}_{i+}(t)}, \tag{29}$$

$$\vec{j}_{\vec{k}}^{-}(t) = \sum_j \vec{v}_{j-} e^{i\vec{k}\cdot\vec{r}_{j-}(t)}. \tag{30}$$

The current $\vec{j}_{\vec{k}}(t)$ may be separated into its components parallel and perpendicular to the wave vector $\vec{k}$. The time dependence of the parallel component is determined by the longitudinal current correlation function

$$C_1(k, t) = (1/N)\langle \vec{k}\cdot\vec{j}_{\vec{k}}(t)\vec{k}\cdot\vec{j}_{-\vec{k}}(0)\rangle. \tag{31}$$

We now form the Fourier-Laplace transforms of

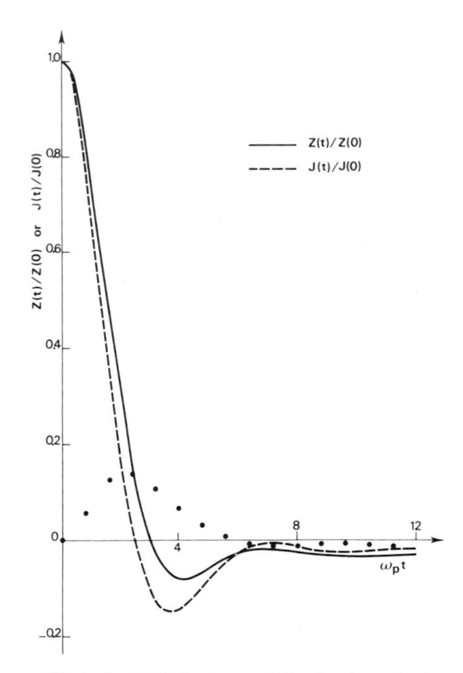

FIG. 4. Normalized autocorrelation functions of velocity and electrical current. The dots show the difference between the two curves.

(26) and (31)

$$\tilde{F}(k,\omega)=\int_0^\infty e^{i\omega t}F(k,t)\,dt\,, \tag{32}$$

$$\tilde{C}_1(k,\omega)=\int_0^\infty e^{i\omega t}C_1(k,t)\,dt\,, \tag{33}$$

from which we obtain two simple expressions for $S(k,\omega)$:

$$S(k,\omega)=(1/\pi)\,\Re\tilde{F}(k,\omega) \tag{34}$$

and

$$S(k,\omega)=(1/\pi\omega^2)\,\Re\tilde{C}_1(k,\omega)\,. \tag{35}$$

Equation (34) shows that the function $\omega^2 S(k,\omega)$ is proportional to the spectrum of fluctuations in the longitudinal current of particles.

We now introduce the dimensionless wave vector $q=\lambda k$ and make a short-time expansion of $F(q,t)$ in the form

$$F(q,t)=F(q,0)+\frac{t^2}{2!}\left.\frac{d^2F(q,t)}{dt^2}\right|_{t=0}+\frac{t^4}{4!}\left.\frac{d^4F(q,t)}{dt^4}\right|_{t=0}+\cdots\,. \tag{36}$$

The evaluation of the first two coefficients in (36) is straightforward and yields the same results as in the case of one-component uncharged fluids, namely

$$F(q,0)=(1/N)\langle\rho_{\vec q}\rho_{-\vec q}\rangle=S(q)\,, \tag{37}$$

$$\left.\frac{d^2F(q,t)}{dt^2}\right|_{t=0}=-\frac{1}{N}\langle\dot\rho_{\vec q}\dot\rho_{-\vec q}\rangle=-q^2k_BT/m=-\omega_0^2\,. \tag{38}$$

The evaluation of the coefficient of the term in t^4 is considerably more lengthy. After some tedious but straightforward algebra, following the classic method of de Gennes,[18] we find the following result, applicable only to the potential (1) with $n=9$ and with r expressed in units of λ:

$$\left.\frac{d^4F(q,t)}{dt^4}\right|_{t=0}=\frac{1}{N}\langle\ddot\rho_{\vec q}\ddot\rho_{-\vec q}\rangle=3\omega_0^4+\omega_0^2\omega_p^2\left(\int_0^\infty\frac{dr}{r^9}\,[g_1(r)+g_2(r)][F_2(qr)+\tfrac{4}{3}]-\int_0^\infty\frac{dr}{r}\,[g_1(r)-g_2(r)]\,F_1(qr)\right)\,, \tag{39}$$

where

$$F_1(x)=\frac{\sin x}{x}+\frac{3\cos x}{x^2}-\frac{3\sin x}{x^3}\,, \tag{40}$$

$$F_2(x)=\frac{11\sin x}{x^3}-\frac{11\cos x}{x^2}-\frac{5\sin x}{x}\,. \tag{41}$$

The properties of the Fourier transform ensure that the derivatives of $F(q,t)$ at $t=0$ are related to the spectral moments of $S(q,\omega)$. Specifically,

$$\langle\omega^{2n}S\rangle=\int_{-\infty}^{\infty}\omega^{2n}S(q,\omega)\,d\omega=(-1)^n\left.\frac{d^{2n}F(q,t)}{dt^{2n}}\right|_{t=0}. \tag{42}$$

A more general expression for the fourth moment of $S(k,\omega)$ in a molten salt has been given earlier by Abramo *et al.*[19] but is cast in a form which is unnecessarily complicated for our purposes.

In Fig. 5 we plot as a function of the dimensionless wave number q the theoretical values of the fourth root of the normalized fourth moment

$$[\omega_{LA}(q)]^4=\langle\omega^4S\rangle/S(q)\,, \tag{43}$$

where the subscript LA, denoting "longitudinal acoustic," is introduced by analogy with solid-state work. Equation (43) provides an approximate dispersion relation for longitudinal number density currents. There is much less structure in $\omega_{LA}(q)$ than in the corresponding curve for argonlike liquids. In particular, the curve is very flat in the neighborhood of the maximum in $S(q)$, which is a region where the dispersion of longitudinal currents in argon shows a well-defined double peak.

We have computed the density-density correlation function $F(q,t)$ for seven values of q, namely 0.751, 1.062, 1.678, 2.252, 2.809, 3.753, and 7.506. These wave numbers were chosen to facilitate comparison with the dynamical structure factor of the OCP which we discuss in I. The following conclusions emerge. First, $F(q,t)$ is a monotonically decreasing function of t at all values of q which we have studied. Even at the lowest q, therefore, there is no evidence of any Brillouin peak in $S(q,\omega)$. (Smaller values of q are inacces-

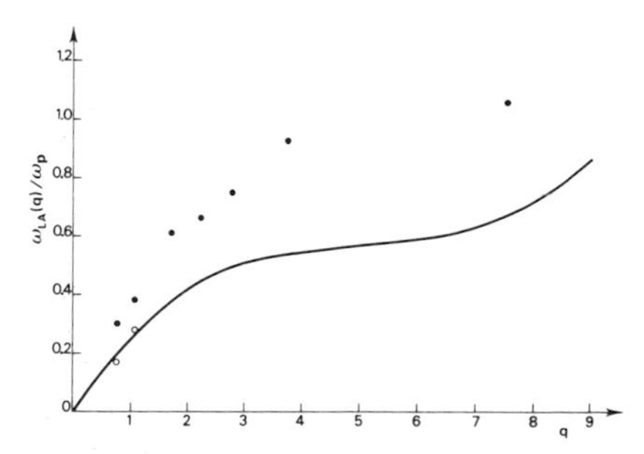

FIG. 5. Characteristic frequencies of the spectrum of density fluctuations. The curve is $\omega_{LA}(q)$ from Eq. (43), the open circles are the peaks in $\omega^2S(q,\omega)$, and the dots are the inverse relaxation time $1/\tau(q)$.

sible to us because of the periodic nature of our system.) Second, the spectrum of the longitudinal current of particles is very broad and flat at all except the smallest q values, a behavior which again is very different to that observed for the Lennard-Jones fluid. It is therefore impossible to draw an unambiguous dispersion curve for $q \gtrsim 1.5$.

Selected results on $S(q, \omega)$ are shown in Fig. 6. For the two lowest values of q it is possible to locate the peak in the current spectrum with reasonable precision; the resulting frequencies are plotted in Fig. 5 and agree quite well with the approximate dispersion relation, Eq. (43). Our results enable us to estimate the velocity of sound as 0.24. For NaCl this corresponds to 2.0×10^5 cm sec^{-1}, compared with the experimental[20] value (1100 K) of 1.7×10^5 cm sec^{-1}.

We now look for a more detailed phenomenological description of the density fluctuations, which we base on the memory function formalism of Mori.[21] We begin by writing a continued fraction expansion of $\tilde{F}(q, \omega)$ in the form

$$\tilde{F}(q, \omega)=\cfrac{\langle\rho_{\vec{q}}\rho_{-\vec{q}}\rangle}{-i\omega+\cfrac{\Delta_1}{-i\omega+\Delta_2\tilde{n}(q,\omega)}}, \qquad (44)$$

where

$$\Delta_1=\frac{\langle\dot{\rho}_{\vec{q}}\dot{\rho}_{-\vec{q}}\rangle}{\langle\rho_{\vec{q}}\rho_{-\vec{q}}\rangle}, \qquad (45)$$

$$\Delta_2=\frac{\langle\ddot{\rho}_{\vec{q}}\ddot{\rho}_{-\vec{q}}\rangle}{\langle\dot{\rho}_{\vec{q}}\dot{\rho}_{-\vec{q}}\rangle}-\frac{\langle\dot{\rho}_{\vec{q}}\dot{\rho}_{-\vec{q}}\rangle}{\langle\rho_{\vec{q}}\rho_{-\vec{q}}\rangle}, \qquad (46)$$

and $n(q, t)$ is an unknown memory function. We now choose a simple exponential form for $n(q, t)$, thereby introducing a single unknown relaxation time $\tau(q)$:

$$n(q, t)=e^{-t/\tau(q)}. \qquad (47)$$

The assumption of an exponential form for $n(q, t)$ has the advantage of leading to a closed form for $S(q, \omega)$ in terms of the moments of order 0, 2, and 4 and the unknown $\tau(q)$:

$$\frac{S(q,\omega)}{S(q)}=\frac{1}{\pi}\frac{\tau(q)\{\langle\omega^4S\rangle/S(q)-[\omega_0^2/S(q)]^2\}}{[\omega\tau(q)(\langle\omega^4S\rangle/\omega_0^2-\omega^2)]^2+(\omega_0^2-\omega^2)^2}. \qquad (48)$$

This single-relaxation-time approximation is known to work well for argonlike liquids, at least at large k, and a number of specific prescriptions for $\tau(k)$ have been proposed.[22-24] Equation (48) implies the assumption of a Maxwellian-type relaxation of the viscosity and the neglect of temperature fluctuations. The latter approximation is rather better justified for molten salts than for argon because the specific-heat ratio $\gamma=C_p/C_v$ is closer to 1, and the coupling between heat conduction and the stress tensor, which is determined by the quantity $(\gamma-1)$, is presumably smaller.

To evaluate $S(q, \omega)$ from Eq. (48) we use as input the calculated moments and adjust the parameter $\tau(q)$ so as to obtain a least-squares fit to the molecular-dynamics data. The results are shown in Fig. 6, from which we see that a very good fit is achieved. The inverse of the relaxation time which is obtained by this procedure is plotted in Fig. 5; the variation with q is roughly the same as that of $\omega_{LA}(q)$.

V. CHARGE FLUCTUATIONS

The formalism for the description of longitudinal charge currents may be developed by adapting that already laid down in Sec. IV for the case of number density fluctuations. We introduce a charge density operator $\rho_{\vec{k}}^{\,\ell}(t)$, defined as

$$\rho_{\vec{k}}^{\,\ell}(t)=e\sum_i Q_i e^{i\vec{k}\cdot\vec{r}_i(t)} = e[\rho_{\vec{k}}^{+}(t)-\rho_{\vec{k}}^{-}(t)]. \qquad (49)$$

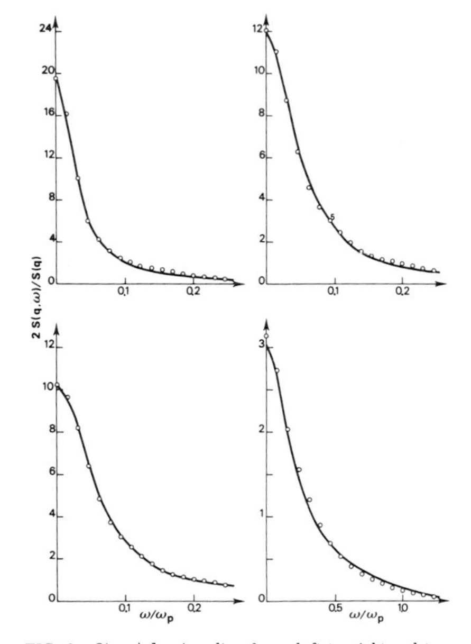

FIG. 6. $S(q, \omega)$ for (reading from left to right and top to bottom) $q=0.751$, 2.252, 2.809, and 7.506. The circles are the molecular-dynamics results and the curves show the single-relaxation-time approximation.

The charge-charge correlation function $F'(k,t)$ is defined as

$$F'(k,t) = (1/Ne^2)\langle \rho'_{\vec{k}}(t)\rho'_{-\vec{k}}(0)\rangle \tag{50}$$

with a Fourier transform, given by

$$S'(k,\omega) = \frac{1}{2\pi}\int_{-\infty}^{\infty} e^{i\omega t} F'(k,t)\,dt\ , \tag{51}$$

which describes the spectrum of charge density fluctuations.

We now make a short-time expansion of $F'(q,t)$ in a form entirely analogous to (36). The first two coefficients are again easy to evaluate

$$F'(q,0) = (1/Ne^2)\langle \rho'_{\vec{q}}\rho'_{-\vec{q}}\rangle = S'(q)\ , \tag{52}$$

$$\left.\frac{d^2F'(q,t)}{dt^2}\right|_{t=0} = \frac{-1}{Ne^2}\langle \dot{\rho}'_{\vec{q}}\dot{\rho}'_{-\vec{q}}\rangle = -\frac{q^2k_BT}{m} = -\omega_0^2\ . \tag{53}$$

Note that the coefficient of t^2 is the same for both $F(k,t)$ and $F'(k,t)$. For the coefficient of t^4 we find, by essentially the same methods used to obtain (39), that

$$\left.\frac{d^4F'(q,t)}{dt^4}\right|_{t=0} = \frac{1}{Ne^2}\langle \ddot{\rho}'_{\vec{q}}\ddot{\rho}'_{-\vec{q}}\rangle = 3\omega_0^4 + \omega_0^2\omega_p^2\left(\frac{2}{3} + \frac{4}{3}\int_0^{\infty}\frac{dr}{r^9}[g_1(r)+g_2(r)] + \int_0^{\infty}\frac{dr}{r^9}[g_1(r)-g_2(r)]\,F_2(qr) - \int_0^{\infty}\frac{dr}{r}[g_1(r)+g_2(r)-2]\,F_1(qr)\right)\ , \tag{54}$$

where F_1 and F_2 are the functions already defined in (40) and (41).

The moments of $S'(q,\omega)$ are given by a relation analogous to (42):

$$\langle\omega^{2n}S'\rangle = \int_{-\infty}^{\infty}\omega^{2n}S(q,\omega)\,d\omega = (-1)^n\left.\frac{d^{2n}F'(q,t)}{dt^{2n}}\right|_{t=0}\ . \tag{55}$$

However, the normalized moments, unlike those of $S(q,\omega)$, all tend to finite values in the limit $q \to 0$, a behavior which is characteristic of optic-type modes. In particular we find that

$$\langle\omega^2S'\rangle/S'(q) \underset{q\to 0}{\sim} \omega_p^2\ , \tag{56}$$

$$[\omega_{LO}(q)]^4 = \frac{\langle\omega^4S'\rangle}{S'(q)} \underset{q\to 0}{\sim} \tfrac{2}{3}\omega_p^4\left(1 + 4\int_0^{\infty}\frac{dr}{r^9}g_2(r)\right)\ , \tag{57}$$

where (57) also serves to define the "longitudinal optic" frequency, $\omega_{LO}(q)$. The latter is plotted as a function of q in Fig. 7. At small q it varies in a manner reminiscent of the longitudinal optic branches in an ionic crystal, but later passes through a minimum at a point almost coincident with the main peak in the static charge-charge structure factor.

In marked contrast to the featureless curves obtained for $F(q,t)$, those for $F'(q,t)$ display pronounced oscillations at small values of q. This behavior is illustrated in Fig. 8. An important point to note is the fact that the oscillations do not take place about the axis $F'(q,t)=0$, but about a small positive level which tends to zero only as the oscillations themselves disappear. This implies that at least two relaxation times are required to describe the underlying physical processes. The correlation function retains its oscillatory character to up to a critical wave number q_0 which we cannot locate precisely but is close to the first peak in $S'(q)$, hence also close to the minimum in $\omega_{LO}(q)$. At $q \approx q_0$, the form of $F'(q,t)$ changes dramatically; the oscillations disappear and the function decays very slowly with time. At still higher values of q the monotonic character is retained but the rate of decay increases with increasing q.

The variation in the form of $F'(q,t)$ is reflected in the q dependence of $S'(q,\omega)$, which we show in Figs. 9–11. For $q < q_0$ there is a well-defined optic-type peak at finite frequency, and another peak at $\omega = 0$, so that $S'(q,\omega)$ appears as the superposition of two almost symmetrical bell-shaped

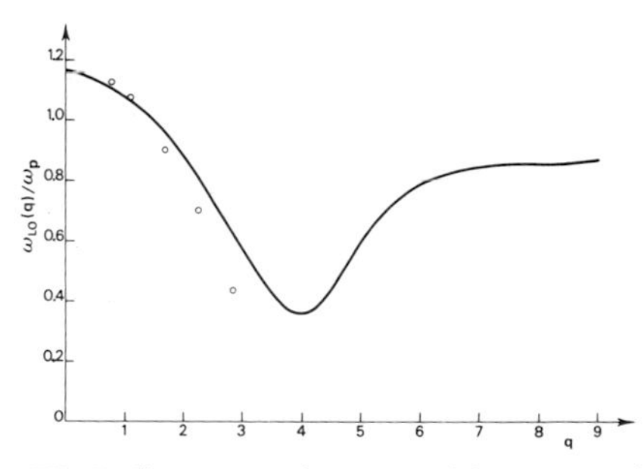

FIG. 7. Characteristic frequencies of the spectrum of charge fluctuations. The curve is $\omega_{LO}(q)$ from Eq. (57) and the circles are the peaks in $S'(q,\omega)$.

Notes on p. 414

curves. The two peaks correspond to the two different relaxation processes which we discussed above. As q increases the "optic" peak moves to lower frequency. This negative dispersion corresponds to what we have already observed[5] in the OCP at high charge densities, but there the rate of dispersion is significantly slower. For example, at both $\Gamma = 110$ and $\Gamma = 152$ the frequency of the plasmon mode of the OCP at a wave number $ak = 2.3$ is approximately 20% lower than its value at $k = 0$; in the present work the corresponding figure is 50%. The dispersion of the "optic" peak in $S'(q, \omega)$ is plotted in Fig. 7; it follows quite closely the curve of $\omega_{LO}(q)$.

We can adapt the phenomenological model used to describe the number density fluctuations by making a continued-fraction expansion of $\tilde{F}'(q, \omega)$, the Fourier-Laplace transform of $F'(q, t)$. The details are the same as before and need not be repeated here. The result is an expression for $S'(q, \omega)$ which is identical to (48) except that $S'(q)$ replaces $S(q)$, $\langle\omega^4 S'\rangle$ replaces $\langle\omega^4 S\rangle$, and a new relaxation time $\tau'(q)$ replaces $\tau(q)$. The results obtained for $S'(q, \omega)$ from a least-squares adjustment of $\tau'(q)$ are shown in Figs. 9–11. At high q the fit to the molecular-dynamics data is again very good. For $q < q_0$ the position of the "optic" peak is satisfactorily reproduced, particularly at the smallest values of q, but the use of a single-relaxation-time approximation necessarily means that the low-frequency behavior is incorrect. The situation that one encounters here is rather different than that which arises in the OCP.

In the latter case the spectrum of charge fluctuations at small q consists of a sharp peak at finite frequency and a low-frequency tail. The shape of the spectrum is therefore adequately described by a memory function with a single adjustable parameter.[5]

The values of $\tau'(q)$ which result from the fit to $S'(q, \omega)$ are listed in Table III, together with those of $\tau(q)$.

VI. CONCLUSIONS

We have carried out a molecular-dynamics calculation of the transport properties of an idealized molten salt, and have obtained in addition some rather precise information on the equilibrium properties of the system, particularly on the

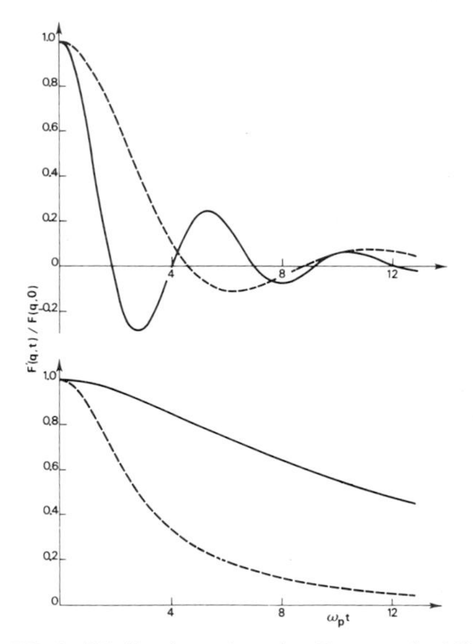

FIG. 8. $F'(q,t)$ at four values of q. Upper graph: full line, $q = 0.751$; dotted line, $q = 2.809$. Lower graph: full line, $q = 3.753$; dotted line, $q = 7.506$.

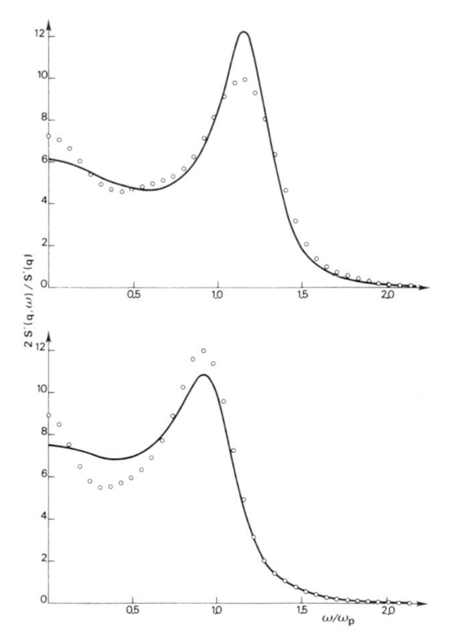

FIG. 9. $S'(q, \omega)$ for $q = 0.751$ (upper graph) and $q = 1.678$ (lower graph). The circles are molecular-dynamics results and the curves show the single-relaxation-time approximation.

static structure factors. By utilizing some of the equilibrium results, we have been able to build a simple phenomenological model to describe the spectra of fluctuations in number density and in charge, each mode being characterized by a single k-dependent relaxation time. The model works well at large values of k, but cannot reproduce satisfactorily the low-frequency part of the spectrum of charge fluctuations at small k. Its defects help to emphasize the features which a satisfactory theory must possess. To obtain a better description of the spectrum it is necessary, at least, to introduce a second relaxation time. We have not attempted a more elaborate parametrization because we believe that such a program must be carried through in parallel with a calculation of the transverse currents, which for practical reasons we have not attempted to compute. A meaningful discussion of the collective modes requires, in addition, that proper account be taken of the limiting hydrodynamic form of the correlation functions and of the coupling between the modes of optic and acoustic character. These are all obvious topics for future investigation. It would also be of great interest to extend the calculations to smaller values of k, where it should be possible to observe a propagating mode in the number density fluctuations. This requires a significant increase in the numbers of ions in the molecular-dynamics cube and is necessarily an expensive undertaking.

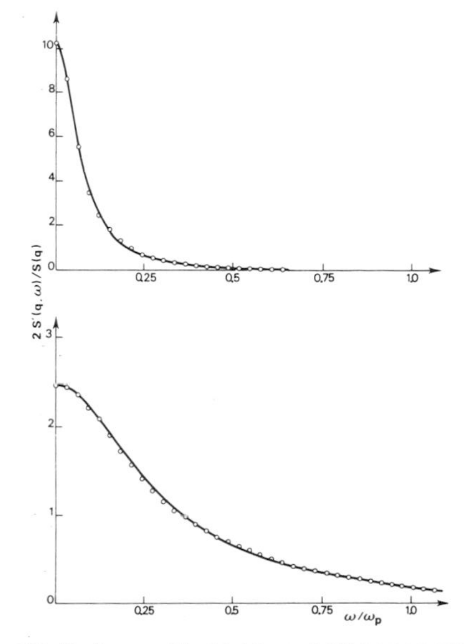

FIG. 11. Same as Fig. 9 but for $q = 3.753$ (upper graph) and $q = 7.506$ (lower graph).

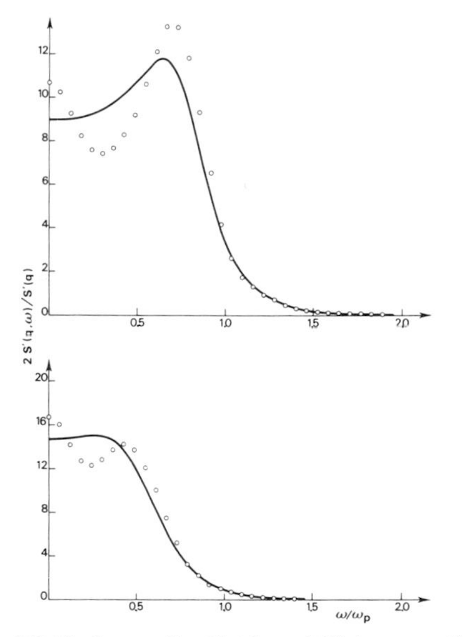

FIG. 10. Same as Fig. 9 but for $q = 2.252$ (upper graph) and $q = 2.809$ (lower graph).

ACKNOWLEDGMENTS

We thank Björn Larsen and Mario Tosi for a number of useful discussions. The work was initiated while I.R.McD. was a visitor in the Laboratoire de Physique Théorique et des Hautes Energies at Orsay, and he wishes to thank Loup Verlet and others for their hospitality. He is also indebted to the United Kingdom Science Research Council for financial support.

TABLE III. Values of the parameters $\tau(q)$ and $\tau'(q)$.

q	$\tau'(q)$	$\tau(q)$
0.751	1.089	3.386
1.062	1.081	2.639
1.678	0.939	1.641
2.252	0.785	1.526
2.809	0.690	1.308
3.753	0.498	1.078
7.506	0.525	0.942

Notes on p. 414

*Laboratoire associe au Centre National de la Recherche Scientifique.

[1]D. Levesque, L. Verlet, and J. Kürkijarvi, Phys. Rev. A 7, 1690 (1973).
[2]A. Rahman, Phys. Rev. Lett. 32, 52 (1974).
[3]D. L. Price, Phys. Rev. A 4, 358 (1971).
[4]J. R. D. Copley and J. M. Rowe, Phys. Rev. Lett. 32, 49 (1974).
[5]J. P. Hansen, I. R. McDonald, and E. L. Pollock, Phys. Rev. A 11, 1025 (1975).
[6]D. J. Adams and I. R. McDonald, Physica (to be published).
[7]R. E. Young and J. P. O'Connell, Ind. Eng. Chem. Fundam. 10, 418 (1971).
[8]J. P. Hansen and I. R. McDonald, J. Phys. C 7, L384 (1974).
[9]F. H. Stillinger and R. Lovett, J. Chem. Phys. 49, 1991 (1968).
[10]L. Verlet, Phys. Rev. 159, 98 (1967); 165, 201 (1968).
[11]J. F. Springer, M. A. Pokrant, and F. A. Stevens, J. Chem. Phys. 58, 4863 (1973).
[12]J. C. Rasaiah, J. Solution Chem. 2, 301 (1973).
[13]B. Larsen (private communication); see also Chem. Phys. Lett. 27, 47 (1974).
[14]L. V. Woodcock and K. Singer, Trans. Faraday Soc. 67, 12 (1971).
[15]D. Levesque and L. Verlet, Phys. Rev. A 2, 2514 (1970).
[16]F. Lantelme, P. Turq, B. Quentrec, and J. W. E. Lewis, Mol. Phys. 28, 1537 (1974).
[17]J. W. E. Lewis and K. Singer, JCS Faraday II. 71, 41 (1975).
[18]P. G. de Gennes, Physica 25, 825 (1959).
[19]M. C. Abramo, M. Parrinello, and M. P. Tosi, J. Nonmetals 2, 67 (1974).
[20]J. O'M. Bockris and N. E. Richards, Proc. R. Soc. A 241, 44 (1957).
[21]H. Mori, Prog. Theor. Phys. 33, 423 (1965).
[22]A. Z. Akcasu and E. Daniels, Phys. Rev. A 2, 962 (1970).
[23]C. J. Murase, J. Phys. Soc. Jpn. 29, 549 (1970).
[24]S. W. Lovesey, J. Phys. C 4, 3057 (1971).

Notes to Reprint V.6

1. Other papers in the series describe simulations of the classical, one-component plasma and of binary, ionic mixtures. See (Hansen, Phys. Rev. A 9 (1973) 3096; Pollock and Hansen, Phys. Rev. A 8 (1973) 3110; Hansen et al., Phys. Rev. A 11 (1975) 1025; Hansen et al., Phys. Rev. A 16 (1977) 2153; and Hansen et al., Phys. Rev. A 20 (1979) 2590). Much of this work is reviewed by Baus and Hansen (Phys. Rep. 59 (1980) 1).

2. Some sections of the text have been transposed in printing. The sentence that ends "... the steeply rising part of $g_2(r)$" at the top of the right-hand column of p. 2114 should be followed directly by the sentence "Overall the agreement...", while the material that at present comes between these two sentences should appear immediately after eq. (17).

3. Reference [6] was later published as Adams and McDonald (Physica B 79 (1975) 159).

4. Charge ordering of the type displayed in fig. 1, first observed by Woodcock and Singer (ref. [14]) in Monte Carlo calculations for molten potassium chloride, has subsequently been detected in neutron-scattering experiments on a number of molten salts (see Enderby and Neilson, Adv. Phys. 25 (1980) 323 and references therein).

5. Comparison of figs. 9 and 10 shows that as the wavenumber increases, the peak in the charge-fluctuation spectrum first sharpens and then becomes broader. It was later shown that this very unexpected behaviour is also predicted by mode-coupling theory (Bosse and Munakata, Phys. Rev. A 25 (1982) 2763).

6. Attempts have been made to detect a propagating charge-fluctuation mode by neutron-scattering techniques, but the results are inconclusive (McGreevy et al., J. Phys. C 17 (1984) 775). The most convincing evidence for the existence of such a mode in a molten salt comes from an analysis of infrared reflectivity data for lithium fluoride (Giaquinta et al., Physica 92A (1978) 185).

MOLECULAR PHYSICS, 1983, VOL. 50, NO. 4, 841–858

Dipole moment fluctuation formulas in computer simulations of polar systems

by MARTIN NEUMANN
Institut für Experimentalphysik der Universität Wien,
Strudlhofgasse 4, A-1090 Wien, Austria†,
and Vitreous State Laboratory, Keane Hall,
Catholic University of America, Washington, D.C. 20064, U.S.A.

(*Received* 24 *September* 1982 ; *accepted* 9 *May* 1983)

The fundamental equation of electrostatics, i.e. the integral equation for the polarization of a macroscopic dielectric in an arbitrary external field, is written in a form that allows explicit inclusion of the toroidal boundary conditions as well as the cutoff of dipolar interactions often used in the computer simulation of polar systems. The toroidal boundary conditions are accounted for in a natural way if the integral equation is formulated (and solved) in Fourier space. Rigorous expressions for the polarization induced by a homogeneous field as well as for the equilibrium dipole moment fluctuations, both as a function of the dielectric constant, are then easily derived for general systems in two and three dimensions. The equations obtained for spherical cutoff geometry with reaction field are identical to those valid for an infinite periodic system (Ewald sum plus reaction field).

When applied to the case of a highly polar Stockmayer system ($\mu^{*2}=3{\cdot}0$, $\rho^*=0{\cdot}822$, $T^*=1{\cdot}15$) the formulas lead to a consistent value of the dielectric constant that is independent of the precise value of the reaction field—if the number of particles is sufficiently large. In the thermodynamic limit the dielectric constant obtained with the reaction field method seems to be much higher than one would expect from the Ewald sum results published for the Stockmayer system at the same thermodynamic state.

1. Introduction

Computer simulations of polar systems are usually performed in one of the following geometries (a pair of boundary conditions and interaction truncation scheme will be called a ' geometry ') :

(*a*) Toroidal boundary conditions with minimum image truncation of dipolar interactions (MI).

(*b*) Toroidal boundary conditions with spherical interaction cutoff (SC).

(*c*) Toroidal boundary conditions with spherical cutoff and approximation of neglected (as compared to an infinite system) interactions by the reaction field (RF)‡.

† Permanent address ; also address for correspondence.

‡ ' RF ' will only be used as an abbreviation for reaction field *geometry* (*c*), as opposed to ' reaction field ' in the general sense.

(*d*) Periodic boundary conditions. Here interactions between particles in different cells of an infinite, though strictly periodic system are most often evaluated by the Ewald-Kornfeld or a similar lattice summation technique (EK)†.

In (*a*)–(*d*) we have deliberately used the terms 'toroidal boundary conditions' (TBC) and 'periodic boundary conditions' (PBC), in order to emphasize that the system being simulated is finite in cases (*a*)–(*c*) and infinite in case (*d*).

While SC is usually sufficient for systems with short-range interactions, the orientational correlations that are so important in polar systems, because of the long-range nature of dipolar interactions, are extremely sensitive to the type of geometry used in the simulation [1–12, 23, 35]. A well-known example is the equation relating the fluctuations of the total dipole moment $\mathbf{M}$ of a system to its dielectric constant ϵ. For an isolated spherical sample we have [13, 14]

$$\frac{\epsilon-1}{\epsilon+2}=\frac{4\pi}{3}\frac{\langle M^2\rangle}{3Vk_BT}, \tag{1}$$

while, if the sample is embedded in an infinite medium of its own dielectric constant, we have [13–15]

$$\frac{(2\epsilon+1)(\epsilon-1)}{9\epsilon}=\frac{4\pi}{3}\frac{\langle M^2\rangle}{3Vk_BT}. \tag{2}$$

Here V and T are volume and temperature of the sample; k_B is Boltzmann's constant. (1) is a Clausius–Mossotti (CM) type equation; (2) is known as the Kirkwood–Fröhlich (KF) equation. Both are quoted in gaussian units which are also used throughout this article; this is done in order to facilitate comparison of results in two and three dimensions.

Referring to an early paper by Rahman and Stillinger [16] it has been argued that (1) should also hold for computer simulations in SC [5, 17] and MI-geometry [5, 18], while (2) is assumed to apply for RF-geometry [2, 5, 9, 10, 19, 20]. Although CM and KF have been used to interpret simulation results for almost a decade, a rigorous proof for the applicability of (1) and (2) in computer simulations involving a cubical sample shape, toroidal boundary conditions, and a finite interaction range has yet to be given. This state of affairs is illustrated by the confusion that prevailed for some time, as to whether the volume entering (1) and (2) is that of the cutoff sphere or the whole sample [4, 5, 9–11, 17, 18, 20–22]. In a recent paper Patey *et al.* [23] have derived a formula for general RF-geometry that supports (1) and (2). They have also pointed out that the $\langle M^2\rangle$ and V to be used in their equation are those of the total sample. Although their formula *is* correct—it is identical with the one derived below—their proof includes neither the cubical sample shape nor the TBC explicitly.

Of course it is always possible to perform simulations in a geometry for which either CM or KF is known to hold, that is, one might try to simulate a spherical sample embedded in an infinite dielectric (as suggested by Friedman [24]) or a spherical sample surrounded by vacuum. The latter possibility

† 'EK' will be used for any one of these summation methods. That is, we would also call Ladd's [6, 39] technique an EK-geometry.

has been followed by Bossis and coworkers [25–28]. While these alternatives certainly remove some ambiguities they nevertheless destroy the homogeneity of the system by necessitating the introduction of walls.

In this article we will therefore give a rigorous, formal and straightforward derivation of general fluctuation formulas for toroidal boundary conditions. This renewed interest in TBC seems to be the more justified in the light of theoretical papers by de Leeuw *et al.* [29, 30], Felderhof [31, 32] and Smith [33], where exact fluctuation formulas for EK-geometry are given. In particular it is evident from [29, 30] and [33] that some kind of reaction field *has* to be used even in EK-geometry, if the system is not to obey CM. As will be shown below, the fluctuation formulas for RF-geometry are the same as those for EK with reaction field. Thus the inclusion of the reaction field seems to be more important for an adequate simulation of polar systems than whether the system is finite (TBC) or infinite (PBC). Which one of these geometries is to be ultimately preferred (for a given number of particles RF would be less time-consuming and easier to program than EK) can only be clarified by an empirical investigation of the N-dependence of the dielectric constant.

The formulas necessary for an unambiguous interpretation of RF-simulations in two and three dimensions will be derived below. Their validity and applicability is then examined in a series of simulations, in various toroidal geometries (MI, SC, RF), of a highly polar Stockmayer system ($\mu^{*2}=3\cdot0$, $\rho^*=0\cdot822$, $T^*=1\cdot15$) for which results under PBC are available in the literature [10].

2. Fluctuation formulas for toroidal boundary conditions

When deriving fluctuation formulas by the Kirkwood–Fröhlich method [13–15] one usually proceeds in two steps: First, a relation is established, with the aid of statistical mechanics, between the mean total dipole moment $\langle \mathbf{M}\rangle_{E_0}$ of the sample V in the presence of a homogeneous external field $\mathbf{E}_0$, and the equilibrium fluctuations $\langle M^2\rangle$ in the absence of any external perturbations. In the second step, classical electrostatics is used to calculate the macroscopic polarization $\mathbf{P}$ induced by $\mathbf{E}_0$ in the sample. After eliminating $\mathbf{E}_0$ from both equations one obtains a relation between $\langle M^2\rangle$ and the dielectric constant.

2.1. *The relation between* $\langle \mathbf{M}\rangle_{E_0}$ *and* $\langle M^2\rangle$

We will only briefly sketch the idea. A detailed derivation can be found in [13–15].

For a system of N particles with generalized coordinates $q^N=(q_1,\ldots,q_N)$ and interaction energy $U(q^N)$, enclosed in a volume V at temperature T, the mean value of the total dipole moment in the direction of a homogeneous external field $\mathbf{E}_0$,

$$\langle \mathbf{M}\rangle_{E_0}=\frac{\int dq^N \exp\{-[U(q^N)-\mathbf{M}(q^N)\,.\,\mathbf{E}_0]/k_BT\}\mathbf{M}(q^N)}{\int dq^N \exp\{-[U(q^N)-\mathbf{M}(q^N)\,.\,\mathbf{E}_0]/k_BT\}},\tag{3}$$

to first order in $\mathbf{E}_0$ is given by

$$\langle \mathbf{M}\rangle_{E_0}=\begin{cases}\dfrac{\langle M^2\rangle}{2k_BT}\,\mathbf{E}_0 & (4.2\mathrm{D})\\[2ex] \dfrac{\langle M^2\rangle}{3k_BT}\,\mathbf{E}_0. & (4.3\mathrm{D})\end{cases}$$

Notes on p. 432

In (4) use has been made of $\langle M_x{}^2\rangle = \langle M_y{}^2\rangle = \frac{1}{2}\langle M^2\rangle$ and $\langle M_x{}^2\rangle = \langle M_y{}^2\rangle = \langle M_z{}^2\rangle = \frac{1}{3}\langle M^2\rangle$ in 2D and 3D, respectively, which is true regardless of whether the sample is spherical, cubical or toroidal. Since there is no distinction between interior and exterior in TBC, the external field appearing in (3) and (4) is $\mathbf{E}_0$ itself, rather than a cavity field, as in (7.2) of [13].

2.2. *The relation between* **P** *and* $\mathbf{E}_0$

In the presence of an external field $\mathbf{E}_0$ a non-vanishing polarization (dipole moment density) **P** will be induced in the sample which may be either a cube V (in 3D) or a square A (in 2D). According to electrostatics the total field at any point **r** in the sample is given by

$$\mathbf{E}(\mathbf{r}) = \mathbf{E}_0(\mathbf{r}) + \int d\mathbf{r}'\, \mathbf{T}(\mathbf{r}-\mathbf{r}') \,.\, \mathbf{P}(\mathbf{r}'). \tag{5}$$

T is the dipole–dipole tensor, i.e.

$$T_{ab} = \begin{cases} \nabla_a \nabla_b(-\ln r) & \text{(6.2D)} \\ \nabla_a \nabla_b(1/r). & \text{(6.3D)} \end{cases}$$

Note that (5) is precisely the way in which the electric field would be calculated in a simulation, except that the integral would be replaced by a sum over particles.

With the aid of the constitutive relation $\mathbf{P}(\mathbf{r}) = \chi \mathbf{E}(\mathbf{r})$, with $\chi = (\epsilon - 1)/2\pi$ in 2D and $\chi = (\epsilon - 1)/4\pi$ in 3D, **E** may now be eliminated from (5), giving an integral equation for **P**

$$\mathbf{P}(\mathbf{r}) = \chi\{\mathbf{E}_0(\mathbf{r}) + \int d\mathbf{r}'\, \mathbf{T}(\mathbf{r}-\mathbf{r}') \,.\, \mathbf{P}(\mathbf{r}')\}. \tag{7}$$

Since **T** is singular at the origin it is convenient to split the integral on the right hand side into two parts by integrating either over $|\mathbf{r}-\mathbf{r}'| < \rho$ or $|\mathbf{r}-\mathbf{r}'| > \rho$ and letting $\rho \to 0$. For the part containing the singularity one obtains in 3D [13–15]

$$\lim_{\rho \to 0} \int_{|\mathbf{r}-\mathbf{r}'|<\rho} d\mathbf{r}'\, \mathbf{T}(\mathbf{r}-\mathbf{r}') \,.\, \mathbf{P}(\mathbf{r}') = -\frac{4\pi}{3}\mathbf{P}(\mathbf{r}). \tag{8.3D}$$

Equation (8) may also be derived explicitly by substituting the definition (6.3D) on the left hand side and rewriting the divergence as a surface integral. In the limit $\rho \to 0$ one then arrives at (8). The fastest way to find the 2D analogue is to formally replace the 3D volume of the unit sphere $(4\pi/3)$ by the 2D area of the unit circle (π)

$$\lim_{\rho \to 0} \int_{|\mathbf{r}-\mathbf{r}'|<\rho} d\mathbf{r}'\, \mathbf{T}(\mathbf{r}-\mathbf{r}') \,.\, \mathbf{P}(\mathbf{r}') = -\pi\mathbf{P}(\mathbf{r}). \tag{8.2D}$$

Thus the integral equation for **P** reads

$$\mathbf{P}(\mathbf{r}) = \lambda \left\{ \mathbf{E}_0(\mathbf{r}) + \lim_{\rho \to 0} \int_{|\mathbf{r}-\mathbf{r}'|>\rho} d\mathbf{r}'\, \mathbf{T}(\mathbf{r}-\mathbf{r}') \,.\, \mathbf{P}(\mathbf{r}') \right\}. \tag{9}$$

For the sake of brevity we shall omit the limit as well as the domain of integration in the following. λ is an abbreviation for

$$\lambda = \begin{cases} \dfrac{1}{\pi}\dfrac{\epsilon-1}{\epsilon+1} & \text{(10.2D)} \\[2ex] \dfrac{3}{4\pi}\dfrac{\epsilon-1}{\epsilon+2}. & \text{(10.3D)} \end{cases}$$

(9) and (10) are the basic equations of electrostatics. We have to solve them for the case of a homogeneous external field. Although (9) could easily be solved directly (for TBC $\mathbf{P}$ will be homogeneous too, if $\mathbf{E}_0$ is homogeneous), the method of solution best suited to an integral equation containing a convolution with toroidal boundary conditions is Fourier transformation. (This is illustrated for a one-dimensional example in the Appendix.)

We therefore write (9) in Fourier space,

$$\tilde{\mathbf{P}}(\mathbf{k}) = \lambda\{\tilde{\mathbf{E}}_0(\mathbf{k}) + \tilde{\mathbf{T}}(\mathbf{k}) \cdot \tilde{\mathbf{P}}(\mathbf{k})\}, \tag{11}$$

whose solution is

$$\tilde{\mathbf{P}}(\mathbf{k}) = \lambda[1 - \lambda\tilde{\mathbf{T}}(\mathbf{k})]^{-1} \cdot \tilde{\mathbf{E}}_0(\mathbf{k}). \tag{12}$$

Equation (12) is our central result. In a subsequent publication it will be used to calculate the distortions of the dipolar field of a particle that are created by various boundary conditions.

In the present case $\mathbf{E}_0$ is homogeneous, that is, if $E_{0,x} = E_{0,y} = 0$ and $E_{0,z} = E_0$ in 3D, we have

$$\left.\begin{aligned} &\tilde{E}_{0,x}(\mathbf{k}) = \tilde{E}_{0,y}(\mathbf{k}) = 0 \quad \text{for all } \mathbf{k}, \\ &\tilde{E}_{0,z}(\mathbf{k}) = \begin{cases} VE_0 & \text{for } \mathbf{k} = 0 \\ 0 & \text{otherwise.} \end{cases} \end{aligned}\right\} \tag{13}$$

Two typical components of $\mathbf{T}$ are $T_{xy} = 3xy/r^5$ and $T_{zz} = (3z^2 - r^2)/r^5$. Because of (13) it is sufficient to know $\tilde{\mathbf{T}}(\mathbf{k})$ for $\mathbf{k} = 0$, that is,

$$\tilde{\mathbf{T}}(0) = \int d\mathbf{r}\, \mathbf{T}(\mathbf{r}). \tag{14}$$

For MI-geometry the integral is over the whole cube, for SC-geometry it is only over a sphere of radius r_c, corresponding to the finite range of the dipole–dipole interaction. Since the singular part of $\mathbf{T}$, $r < \rho$, has been accounted for explicitly in (9), we may argue that T_{xy} is an odd function and therefore $\tilde{\mathbf{T}}_{xy}(0) = 0$, if T_{xy} is integrated over a cube or a sphere. Similarly, we have $\int d\mathbf{r}\, 3z^2/r^5 = \int d\mathbf{r}/r^3$ in both cases (cube and sphere), and therefore $\tilde{T}_{zz}(0) = 0$. Thus

$$\tilde{\mathbf{T}}(0) = 0 \tag{15}$$

for MI as well as SC-geometry.

(12) now simplifies to $\tilde{\mathbf{P}}(0)=\lambda\tilde{\mathbf{E}}_0(0)$, giving $M_z=VP_z=\tilde{P}_z(0)=\lambda VE_0$. Eliminating $\mathbf{E}_0$ from (4) we immediately conclude that CM, a Clausius–Mossotti-like equation, is valid for MI as well as SC-geometry

$$\frac{\epsilon-1}{\epsilon+1}=\pi\frac{\langle M^2\rangle}{2Ak_BT}, \tag{16.2D}$$

$$\frac{\epsilon-1}{\epsilon+2}=\frac{4\pi}{3}\frac{\langle M^2\rangle}{3Vk_BT}. \tag{16.3D}$$

From the above it is clear that A and V are the domain of integration used in the Fourier transforms, that is, area and volume, respectively, of the total system (rather than the cutoff sphere). The derivation of the 2D equation is completely analogous to the one given for 3D.

3. The reaction field

Here those interactions of every particle i with its more distant neighbours j $(r_{ij}>r_c)$ that are neglected in SC-geometry are replaced by an infinite continuum with dielectric constant ϵ_{RF} [2, 3]. In this way, it is hoped, one should be able to simulate an infinite system, provided $\epsilon_{RF}\approx\epsilon$. The total field acting on particle i now has to be corrected by the reaction field, in 3D

$$\mathbf{R}(i)=\frac{2(\epsilon_{RF}-1)}{2\epsilon_{RF}+1}\frac{1}{r_c^3}\mathbf{M}(i), \tag{17}$$

which is added to the field due to the immediate neighbours $(r_{ij}<r_c)$ of i. In (17) $\mathbf{M}(i)$ is the total dipole moment of the cutoff sphere around i. Equivalently [8], one may use a modified dipole–dipole interaction

$$u_{DD}(ij)=\begin{cases}-\boldsymbol{\mu}(i)\,.\,\mathbf{T}(ij)\,.\,\boldsymbol{\mu}(j)-\dfrac{2(\epsilon_{RF}-1)}{2\epsilon_{RF}+1}\dfrac{\boldsymbol{\mu}(i)\,.\,\boldsymbol{\mu}(j)}{r_c^3} & \text{if } r_{ij}<r_c\\ 0 \quad \text{otherwise.}\end{cases} \tag{18}$$

The way to write (18) that is best suited to our purposes is as $u_{DD}(ij)=-\boldsymbol{\mu}(i)\,.\,\mathbf{T}(ij,\epsilon_{RF})\,.\,\boldsymbol{\mu}(j)$. The modified **T**-tensor is thus defined by

$$T_{ab}(ij,\epsilon_{RF})=\begin{cases}T_{ab}(ij)+\dfrac{2(\epsilon_{RF}-1)}{2\epsilon_{RF}+1}\dfrac{1}{r_c^3}\delta_{ab} & \text{if } r_{ij}<r_c\\ 0 \quad \text{otherwise.}\end{cases} \tag{19.3D}$$

The corresponding expression in 2D is [11, 25, 33]

$$T_{ab}(ij,\epsilon_{RF})=\begin{cases}T_{ab}(ij)+\dfrac{\epsilon_{RF}-1}{\epsilon_{RF}+1}\dfrac{1}{r_c^2}\delta_{ab} & \text{if } r_{ij}<r_c\\ 0 \quad \text{otherwise.}\end{cases} \tag{19.2D}$$

In the following we will use the abbreviation

$$\mu = \begin{cases} \dfrac{1}{\pi}\dfrac{\epsilon_{\mathrm{RF}}-1}{\epsilon_{\mathrm{RF}}+1} & (20.2\mathrm{D}) \\[2ex] \dfrac{3}{4\pi}\dfrac{2(\epsilon_{\mathrm{RF}}-1)}{2\epsilon_{\mathrm{RF}}+1}. & (20.3\mathrm{D}) \end{cases}$$

The derivation of fluctuation formulas is now completely analogous to the case of SC, the only difference being that the integral

$$\tilde{\mathbf{T}}(\mathbf{k},\, \epsilon_{\mathrm{RF}}) = \int d\mathbf{r} \exp(-i\mathbf{k}\,.\,\mathbf{r})\mathbf{T}(\mathbf{r},\, \epsilon_{\mathrm{RF}})$$

does not vanish for $\mathbf{k}=0$. One rather has

$$\tilde{T}_{ab}(0,\, \epsilon_{\mathrm{RF}}) = \begin{cases} \pi^2 \mu \delta_{ab} & (21.2\mathrm{D}) \\[2ex] \left(\dfrac{4\pi}{3}\right)^2 \mu \delta_{ab}. & (21.3\mathrm{D}) \end{cases}$$

Combining (21) and (12) yields, in 3D,

$$\begin{aligned} \tilde{\mathbf{P}}(0) &= \lambda[1-\lambda\tilde{\mathbf{T}}(0,\, \epsilon_{\mathrm{RF}})]^{-1}\,.\,\tilde{\mathbf{E}}_0(0) \\ &= \lambda\left[1-\left(\frac{4\pi}{3}\right)^2 \lambda\mu\right]^{-1} \tilde{\mathbf{E}}_0(0) \end{aligned} \tag{22}$$

or

$$M_z = \lambda V E_0 \Big/ \left[1-\left(\frac{4\pi}{3}\right)^2 \lambda\mu\right]. \tag{23}$$

Eliminating $\mathbf{E}_0$ from (4) and substituting for λ and μ we obtain

$$\frac{\epsilon-1}{\epsilon+2}\left[1-\frac{\epsilon-1}{\epsilon+2}\frac{2(\epsilon_{\mathrm{RF}}-1)}{2\epsilon_{\mathrm{RF}}+1}\right]^{-1} = \frac{4\pi}{3}\frac{\langle M^2\rangle}{3Vk_{\mathrm{B}}T}. \tag{24.3D}$$

The result for 2D is

$$\frac{\epsilon-1}{\epsilon+1}\left[1-\frac{\epsilon-1}{\epsilon+1}\frac{\epsilon_{\mathrm{RF}}-1}{\epsilon_{\mathrm{RF}}+1}\right]^{-1} = \pi\frac{\langle M^2\rangle}{2Ak_{\mathrm{B}}T}. \tag{24.2D}$$

An alternative method to find the dielectric constant of a model system is to measure the response to a homogeneous external field $\mathbf{E}_0$, i.e. the polarization induced by that field [9, 10, 19, 22, 34]. Equation (23) immediately leads to

$$\pi\langle\mathbf{P}\rangle_{E_0} = \frac{\epsilon-1}{\epsilon+1}\left[1-\frac{\epsilon-1}{\epsilon+1}\frac{\epsilon_{\mathrm{RF}}-1}{\epsilon_{\mathrm{RF}}+1}\right]^{-1}\mathbf{E}_0, \tag{25.2D}$$

$$\frac{4\pi}{3}\langle\mathbf{P}\rangle_{E_0} = \frac{\epsilon-1}{\epsilon+2}\left[1-\frac{\epsilon-1}{\epsilon+2}\frac{2(\epsilon_{\mathrm{RF}}-1)}{2\epsilon_{\mathrm{RF}}+1}\right]^{-1}\mathbf{E}_0. \tag{25.3D}$$

$\langle\mathbf{P}\rangle_{E_0}$ is the polarization induced by, and parallel to, the external field.

4. Fluctuation and polarization formulas

Equations (24) and (25) of the preceding section are quite general in the sense that they permit the evaluation of the dielectric constant for arbitrary toroidal geometry. The formulas applicable to MI and SC-geometry are obtained by formally letting $\epsilon_{RF} \to 1$ in (24) and (25). Because of their importance some special cases are listed below.

4.1. *Fluctuation formulas*

Two dimensions

$$\pi \frac{\langle M^2 \rangle}{2Ak_BT} = \begin{cases} \dfrac{\epsilon-1}{\epsilon+1}\left[1-\dfrac{\epsilon-1}{\epsilon+1}\dfrac{\epsilon_{RF}-1}{\epsilon_{RF}+1}\right]^{-1} & \text{RF} \quad (26.1.2\text{D}) \\ \dfrac{\epsilon-1}{\epsilon+1} & \text{MI/SC} \quad (26.2.2\text{D}) \\ \dfrac{(\epsilon+1)(\epsilon-1)}{4\epsilon} & \epsilon_{RF}=\epsilon \quad (26.3.2\text{D}) \\ \dfrac{\epsilon-1}{2} & \epsilon_{RF}=\infty. \quad (26.4.2\text{D}) \end{cases}$$

Three dimensions

$$\frac{4\pi}{3}\frac{\langle M^2 \rangle}{3Vk_BT} = \begin{cases} \dfrac{\epsilon-1}{\epsilon+2}\left[1-\dfrac{\epsilon-1}{\epsilon+2}\dfrac{2(\epsilon_{RF}-1)}{2\epsilon_{RF}+1}\right]^{-1} & \text{RF} \quad (26.1.3\text{D}) \\ \dfrac{\epsilon-1}{\epsilon+2} & \text{MI/SC} \quad (26.2.3\text{D}) \\ \dfrac{(2\epsilon+1)(\epsilon-1)}{9\epsilon} & \epsilon_{RF}=\epsilon \quad (26.3.3\text{D}) \\ \dfrac{\epsilon-1}{3} & \epsilon_{RF}=\infty. \quad (26.4.3\text{D}) \end{cases}$$

As stated in the introduction, it is possible to prove rigorously that a Clausius–Mossotti-like equation actually holds for MI and SC-geometry while the Kirkwood–Fröhlich equation is correct for RF-geometry with $\epsilon_{RF}=\epsilon$. In contrast to [23] these equations are seen to be valid even for a cubical (square) volume (area) and toroidal boundary conditions. The volume (area) to be inserted on the left hand side of (26) in SC and RF-geometry is that of the total sample. This has been pointed out in [23] too. The formulas are in fact independent of r_c.

We also notice that (26) is identical to the formulas known to hold for EK-geometry with reaction field [29, 30, 33]. In particular (26.4) is just the formula for ‘ conducting boundary conditions ’

$$\epsilon = \begin{cases} 1+\pi\dfrac{\langle M^2 \rangle}{Ak_BT} & (27.2\text{D}) \\ 1+\dfrac{4\pi}{3}\dfrac{\langle M^2 \rangle}{Vk_BT}. & (27.3\text{D}) \end{cases}$$

We further notice that, in contrast to what one might expect, it makes quite a difference to replace ϵ_{RF} by ∞ in (17), even for systems with large ϵ; i.e. to replace $(\epsilon_{RF}-1)/(\epsilon_{RF}+1)$ or $2(\epsilon_{RF}-1)/(2\epsilon_{RF}+1)$ by unity: if (26.3) were used instead of (26.4) to extract the dielectric constant from a simulation with $\epsilon_{RF}=\infty$, a value of ϵ would result that is too large by a factor of 2 (in 2D) or 3/2 (in 3D). Thus it proves essential to work with the correct formula.

4.2. *Polarization formulas*

Two dimensions

$$\pi\langle P\rangle_{E_0}/E_0=\begin{cases}\dfrac{\epsilon-1}{\epsilon+1}\left[1-\dfrac{\epsilon-1}{\epsilon+1}\dfrac{\epsilon_{RF}-1}{\epsilon_{RF}+1}\right]^{-1} & \text{RF} \quad (28.1.2D)\\[2ex] \dfrac{\epsilon-1}{\epsilon+1} & \text{MI/SC} \quad (28.2.2D)\\[2ex] \dfrac{(\epsilon+1)(\epsilon-1)}{4\epsilon} & \epsilon_{RF}=\epsilon \quad (28.3.2D)\\[2ex] \dfrac{\epsilon-1}{2} & \epsilon_{RF}=\infty. \quad (28.4.2D)\end{cases}$$

Three dimensions

$$\frac{4\pi}{3}\langle P\rangle_{E_0}/E_0=\begin{cases}\dfrac{\epsilon-1}{\epsilon+2}\left[1-\dfrac{\epsilon-1}{\epsilon+2}\dfrac{2(\epsilon_{RF}-1)}{2\epsilon_{RF}+1}\right]^{-1} & \text{RF} \quad (28.1.3D)\\[2ex] \dfrac{\epsilon-1}{\epsilon+2} & \text{MI/SC} \quad (28.2.3D)\\[2ex] \dfrac{(2\epsilon+1)(\epsilon-1)}{9\epsilon} & \epsilon_{RF}=\epsilon \quad (28.3.3D)\\[2ex] \dfrac{\epsilon-1}{3} & \epsilon_{RF}=\infty. \quad (28.4.3D)\end{cases}$$

Again (28.4) is the formula for conducting boundary conditions

$$\epsilon=\begin{cases}1+\pi\langle P\rangle_{E_0}/E_0 & (29.2D)\\[1ex] 1+4\pi\langle P\rangle_{E_0}/E_0. & (29.3D)\end{cases}$$

$\mathbf{E}_0$ in (28) and (29) is the field actually added to the interaction of each particle, since this is how $\mathbf{E}_0$ has been defined in (5). The total field, including the contribution from the polarization of the system, is of course linked to $\mathbf{P}$ by the constitutive relation $\mathbf{P}=(\epsilon-1)/4\pi\mathbf{E}$, but generally not known in advance. In fact, knowledge of the correct relationship between $\mathbf{E}_0$ and $\mathbf{E}$ is essential for a successful interpretation of applied field simulations.

As an illustration, we consider the analysis by Adams and Adams [10] of their simulations in RF-geometry: Assuming that RF corresponds to a spherical sample embedded in an infinite medium with dielectric constant ϵ_{RF}, Adams and Adams used the appropriate cavity field $3\epsilon_{RF}/(2\epsilon_{RF}+1)\mathbf{E}$ as the $\mathbf{E}_0$ (in our notation) in their simulations. The dielectric constant was then determined from the constitutive relation

$$\epsilon = 1 + 4\pi\langle P\rangle/E. \tag{28.3.3D'}$$

Inserting Adams and Adams' $\mathbf{E}_0$ into (28.1.3D), however, shows that

$$\epsilon = 1 + \frac{2\epsilon_{RF}+\epsilon}{3\epsilon_{RF}} 4\pi\langle P\rangle/E \tag{28.1.3D'}$$

should have been used instead. Although (28.3.3D') is correct for $\epsilon \approx \epsilon_{RF}$, it will give a value for the dielectric constant that is too low if $\epsilon_{RF} < \epsilon$, and too high if $\epsilon_{RF} > \epsilon$ [41].

4.3. *Remarks*

From (26)–(29) it is evident that, regardless of whether the equilibrium fluctuations of the total dipole moment or the polarization in an applied field is used to calculate the dielectric constant, for highly polar systems only simulations including the reaction field can be successful : $\langle M^2\rangle$ and $\langle P\rangle_{E_0}/E_0$ are both bounded in the limit $\epsilon \to \infty$ for MI as well as SC-geometry (and also for EK without reaction field), but proportional to ϵ for RF-geometry (or EK with reaction field). Because of the limited statistical accuracy of simulations it will only be possible in the latter case to arrive at a reasonable value for the dielectric constant.

In order to remove the discontinuity introduced by the potential cutoff, in some simulations [10, 12, 20, 35] the dipole–dipole tensor (the modified **T**-tensor if the reaction field is used) is multiplied by a smoothing function $f(\mathbf{r})$, where $f(0)=1$, $f(\mathbf{r})=0$ for $r>r_c$, and $f(\mathbf{r})\mathbf{T}(\mathbf{r})$ is a continuously differentiable function. If **T** is replaced by $f\mathbf{T}$ throughout the derivation given in § 2, the formulas valid for MI and SC-geometry are found to apply to this case too : Since $f(\mathbf{r})$ usually has spherical or cubical symmetry one has $\int d\mathbf{r}\, f(\mathbf{r})\mathbf{T}(\mathbf{r})=0$, the only assumption needed in § 2. In RF-geometry, on the other hand, where **T** as well as the reaction field is modulated by $f(\mathbf{r})$, one just has to replace πr_c^2 or $4\pi r_c^3/3$ by

$$\int_{r<r_c} d\mathbf{r}\, f(\mathbf{r})$$

in (17)–(19). In this way (21) and eventually (24) and (25) are recovered again. This is the recipe suggested by Adams *et al.* [20] which now has been justified. Failing to do so means that one is working with the reaction field corresponding to a continuum with dielectric constant ϵ'_{RF}, where ϵ'_{RF} is given by

$$\frac{2(\epsilon'_{RF}-1)}{2\epsilon'_{RF}+1}\,\frac{4\pi/3}{\int_{r<r_c} d\mathbf{r}\, f(\mathbf{r})} = \frac{2(\epsilon_{RF}-1)}{2\epsilon_{RF}+1}\,\frac{1}{r_c^3}. \tag{30}$$

This is the value (generally $\epsilon'_{RF} < \epsilon_{RF}$) to be used in the analysis of the simulation results [12].

One might think that it is possible to use the reaction field in MI-geometry too : (21) remains unchanged if r_c^2 and r_c^3 in (17–19) are replaced by A/π and $3V/4\pi$, respectively. In this way (24) and (25) are obtained once more. While this is certainly true it can be shown [37] that the reaction field, when used in connection with MI-geometry, is not able to suppress, to the same degree as the reaction field used with a spherical cutoff, the influence of toroidal boundary conditions on the *local* (nearest neighbour) orientational structure of the system. Thus RF-geometry is to be preferred to MI in any case (with or without reaction field).

5. The dielectric constant of a highly polar Stockmayer system

In order to test the applicability and consistency of (26) in actual computer simulations, a typical model system has been studied under various geometries at a single thermodynamic state. The system chosen is the 3D Stockmayer system with reduced dipole moment $\mu^{*2}=3\cdot0$, at reduced density $\rho^*=0\cdot822$ and temperature $T^*=1\cdot15$ ($\mu^{*2}=\mu^2/\epsilon\sigma^3$, $\rho^*=\rho\sigma^3$, and $T^*=k_BT/\epsilon$, where ϵ and σ are the parameters of the Lennard-Jones part of the potential)†. The same system has also been investigated by Adams and Adams [10] in some detail, although mostly in EK-geometry. The simulations reported for RF-geometry in [10] have been analysed using $\langle M^2\rangle$ and V of the cutoff sphere and are not useful for our purposes.

The results reported below have been obtained from molecular dynamics (MD) simulations of a system of 256 particles. For the integration of the equations of motion the Singer–Verlet algorithm [38] with a (reduced) time step of $\Delta t^*=0\cdot0025$ was used. The moment of inertia (which has no effect on the purely static properties that are of sole interest here) was taken to be $I^*=0\cdot025$. With that system a simulation was performed each in MI, SC and RF (with $\epsilon_{RF}=50$ and $\epsilon_{RF}=\infty$) geometry. The potential cutoff in the latter simulations was always half the box length, i.e. $r_c=L/2=3\cdot389\sigma$ at that density. No smoothing function of the kind described at the end of § 4 was used. In order to compensate for the increase in energy that is to be expected in that case [20], the total energy of the system was artificially kept constant by rescaling the translational and rotational velocities after every 100 time steps. (The energy increase typically observed in that interval was of the order of 0·25–0·5 per cent, depending on boundary conditions.)

For the simulations in RF-geometry the desired statistical accuracy of 5 per cent for $\langle M^2\rangle$ (and thus for ϵ) was reached after 50 000 time steps. This is in agreement with [34]. The runs in MI and SC-geometry were terminated after 20 000 Δt since, for a system of only 256 particles, it would be impossible to obtain a reasonable value for the dielectric constant even after a much longer time (see discussion of figure 2). The number of time steps typically needed to equilibrate the system was 10 000. This may be interpreted as the time the system needs to readjust itself to a change of geometry.

In the table some properties of the Stockmayer system are listed for various geometries and compared with Adams and Adams' [10] results. The values of special interest are those of $\langle M^2\rangle$, G_K, g_K and ϵ. Here g_K is the Kirkwood g-factor, defined in the usual way [13–15] (i.e. for an infinite system); G_K is

† The system studied here is a singular case in the sense that, for a given polarity, its orientational correlations are of much longer range than those of a system with a more realistic interaction potential [12, 21, 35, 36].

Thermodynamic and dielectric properties of a Stockmayer system with $\mu^{*2}=3{\cdot}0$ at $\rho^*=0{\cdot}822$.

Geometry	N	T^*	$\langle U^*\rangle/N$	$\langle U_{DD}^*\rangle/N$	$\langle p^*\rangle$	$G_K=\langle M^2\rangle/N\mu^2$	g_K	ϵ	Time steps/ method †	Source
MI	108	1·15	−10·53	−5·25	−0·512	0·28	—	—	10 000 MC	[10]
MI	256	1·155	−10·430	−5·210	−0·551	0·29	—	—	20 000 MD	‡
SC	108	1·15	−9·97	−4·78	−0·085	0·32	—	—	10 000 MC	[10]
SC	256	1·148	−9·970	−4·777	−0·088	0·31	—	—	20 000 MD	‡
SC	500	1·15	−10·04	−4·85	−0·085	0·28	—	—	2000 MC	[10]
RF $\epsilon_{RF}=50$	108	1·148	−9·834	−4·664	0·100	3·56	3·51	47·9	100 000 MD	‡
RF $\epsilon_{RF}=50$	256	1·150	−9·896	−4·722	0·057	4·41	4·87	66·1	50 000 MD	‡
RF $\epsilon_{RF}=\infty$	108	1·149	−9·856	−4·692	0·119	5·78	3·89	53·0	100 000 MD	‡
RF $\epsilon_{RF}=\infty$	256	1·154	−9·906	−4·730	0·062	7·09	4·77	64·5	50 000 MD	‡
EK	108	1·15	−9·935	−4·737	−0·028	(4·29) §	(2·89)	39·5	50 000 MC	[10]
EK	256	1·15	−9·922	—	—	(3·48)	(2·36)	32·3	12 000 MC	[10]
EK	500	1·15	−9·902	−4·721	−0·020	(4·45)	(3·01)	41·0	20 000 MC	[10]
EK	500	1·155	−9·888	−4·705	0·056	(5·34)	(3·60)	48·8	30 000 MD	[10]

† For Monte Carlo simulations : order of magnitude of (number of moves)/N.

‡ This work. Estimated accuracy of G_K is 5 per cent. LJ energy and pressure include a cutoff correction. The LJ cutoff was always at $r_c=L/2$, even for MI. The electrostatic self-energy for RF-geometry, $\langle U_c^*\rangle/N=-(\mu^{*2}/r_c^{*3})(\epsilon_{RF}-1)/(2\epsilon_{RF}+1)$, is not included in $\langle U^*\rangle/N$, $\langle U_{DD}^*\rangle/N$ and $\langle p^*\rangle$.

§ The numbers in parentheses have been calculated from the dielectric constants given in [10].

the g-factor for a finite system [8], a cube with toroidal boundary conditions in our case. Thus, if $\boldsymbol{\mu}(i)$ is the dipole moment of particle i,

$$G_{\mathrm{K}}=\left\langle \sum_{j=1}^{N} \boldsymbol{\mu}(i) \cdot \boldsymbol{\mu}(j)\right\rangle \Big/ \mu^2=\langle M^2\rangle / N\mu^2. \tag{31}$$

Since G_{K} coincides with g_{K} for $\epsilon_{\mathrm{RF}}=\epsilon$, and because of (26), we have, for arbitrary reaction field, in 3D

$$g_{\mathrm{K}}=G_{\mathrm{K}}\frac{(2\epsilon+1)(\epsilon-1)}{9\epsilon}\left[1-\frac{\epsilon-1}{\epsilon+2}\frac{2(\epsilon_{\mathrm{RF}}-1)}{2\epsilon_{\mathrm{RF}}+1}\right]. \tag{32}$$

(Cf. also (14) of [16].) This is the way the g_{K}-values in the table have been obtained ; the values for the dielectric constant have been calculated with the aid of (26).

According to (26.2.3D) G_{K} is bounded for $\epsilon\to\infty$ in MI and SC-geometry:

$$G_{\mathrm{K,\,MI/SC}}<\frac{9k_{\mathrm{B}}T}{4\pi\sigma\mu^2}=0{\cdot}334. \tag{33}$$

Although this inequality is obeyed (our values are also identical with those of [10]), the ϵ-values for MI and SC-geometry are neither in agreement with one another nor with those calculated for RF-geometry.

If ϵ is determined in RF-geometry, however, the result is even independent of ϵ_{RF} (unless $\epsilon_{\mathrm{RF}}\ll\epsilon$)—provided the correct formula, namely (26.1), is used. We thus obtain $\epsilon=66$ for $\epsilon_{\mathrm{RF}}=50$ and $\epsilon=65$ for $\epsilon_{\mathrm{RF}}=\infty$, a consistent result. Yet this value is significantly higher than that reported by Adams and Adams ($\epsilon\approx 50$) for EK-geometry†. In principle one would expect RF and EK to give consistent results in the limit $N\to\infty$. Whether the discrepancy is due to a different N-dependence for both geometries remains to be seen. Given our computational facilities, a system of 256 was close to the limit of what can be done in a reasonable space of time. Some results obtained for a smaller system ($N=108$, $\epsilon_{\mathrm{RF}}=50$ and $\epsilon_{\mathrm{RF}}=\infty$, 100 000 Δt each) are also included in the table. Here the ϵ-values are in closer agreement with those of [10] ; they are also fairly consistent among themselves. However, preliminary results for a system of 500 particles seem to indicate that, in the thermodynamical limit, the dielectric constant is identical with that for $N=256$. The EK-values, on the other hand, increase much less with N : they are not even monotonous.

In this context it has to be pointed out once more (see also the discussion following (27)), how sensitive the fluctuations of the total dipole moment are to any change of the precise form of the dipole–dipole interaction. It might not be impossible that the approximations necessary for a full evaluation of the Ewald sum systematically distort the results‡. Ladd's [6, 39] multipole expansion of the Ewald sum, if augmented by the reaction field, would be conceptually more attractive : it can be shown [37] that the $\mathbf{k}=0$ component of the Fourier transform vanishes for the individual terms in the series expansion of the generalized $\mathbf{T}$-tensor (cf. (15)). Thus (26) is *exactly* obeyed, no matter where the series is truncated.

† Adams and Adams' EK-result is also the reason why $\epsilon_{\mathrm{RF}}=50$, rather than $\epsilon_{\mathrm{RF}}=\epsilon\approx 65$, was used in one of our simulations. They were run in parallel.

‡ *Note added in proof.*—We have recently been able to show that this is indeed the case [42].

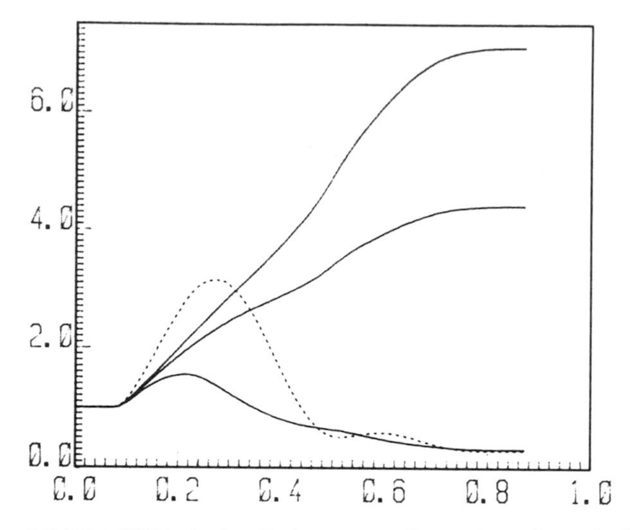

Figure 1. $\langle M^2(R)\rangle / \langle N(R)\rangle \mu^2$, the dipole moment fluctuations of a spherical volume with radius R, for a system of 256 Stockmayer particles with $\mu^{*2}=3{\cdot}0$ at $\rho^*=0{\cdot}822$ and $T^*=1{\cdot}15$. $N(R)$ is the number of particles in $V=4\pi R^3/3$. For $R>L/2$, V is the intersection of a sphere of radius R with the MD cell. R is in units of the box length L. Dashed curve: MI-geometry; solid curves (top to bottom): RF with $\epsilon_{\mathrm{RF}}=\infty$, RF with $\epsilon_{\mathrm{RF}}=50$, and SC-geometry.

For a dielectric sphere embedded in an infinite system, $\langle M^2\rangle/V$ should be independent of the radius of that sphere, according to (12). In figure 1 $\langle M^2(R)\rangle / \langle N(R)\rangle \mu^2$ is plotted as a function of R. Here

$$\langle M^2(R)\rangle = \left\langle \left(\sum_{|\mathbf{r}_i - \mathbf{r}_0(t)| < R} \boldsymbol{\mu}(i) \right)^2 \right\rangle \tag{34}$$

and

$$\langle N(R)\rangle = \left\langle \sum_{|\mathbf{r}_i - \mathbf{r}_0(t)| < R} \right\rangle. \tag{35}$$

$\mathbf{r}_0(t)$ is a point somewhere in the sample, chosen at random for every time step, and not necessarily coincident with the position of a particle. Thus $M(R)$ and $N(R)$ are the total dipole moment of, and the number of particles within, a sphere (or that part of a sphere that is within the cube-shaped sample) of radius R around $\mathbf{r}_0$. From figure 1 it is evident that it is impossible to find a reasonable value for the dielectric constant in MI or SC-geometry, using KF and the dipole moment fluctuations of a sphere that is small compared to the system as a whole: the resulting ϵ would be too small—at least for a system of our size. If CM is used instead, a negative value for ϵ is the most likely outcome [22]: $\langle M^2(R)\rangle / \langle N(R)\rangle \mu^2$ is everywhere greater than the maximum value permitted by (33). In RF-geometry, on the other hand, it is an increasing function of R, and ϵ can only be obtained from $\langle M^2\rangle$ for the total volume (and not from the cutoff sphere). For $R/L \to \sqrt{3}/2$ the curves in figure 1 approach the $\langle M^2\rangle$-values given in the table.

The R-dependent Kirkwood g-factor [8], the total dipole moment of a sphere of radius R around a reference particle,

$$G_K(R) = \left\langle \sum_{\substack{j \\ r_{ij} < R}} \boldsymbol{\mu}(i) \cdot \boldsymbol{\mu}(j) \right\rangle \Big/ \mu^2, \tag{36}$$

is plotted in figure 2. As before, the limits for $R/L \to \sqrt{3}/2$ are identical with the G_K-values in the table. The curve for SC-geometry shows the familiar pattern, a positive and a negative region, that is known to be caused by the boundary conditions [8]. For MI-geometry the situation is similar, with positive values being increased and negative values being decreased. For $R/L \to \sqrt{3}/2$, however, both curves meet at the same point—the one determined

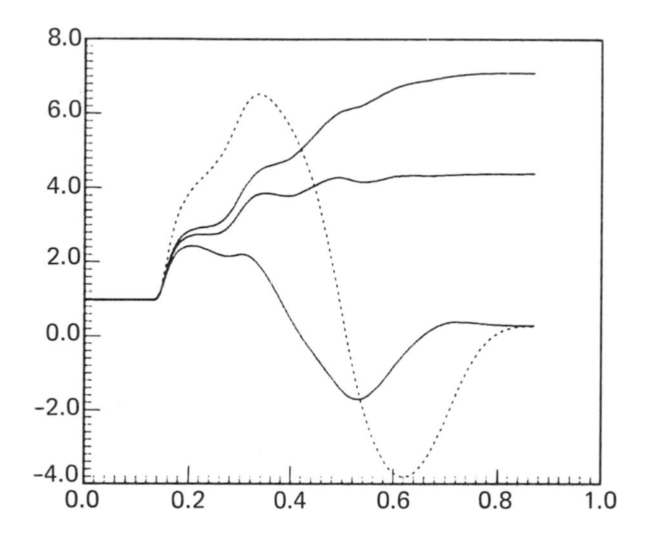

Figure 2. $G_K(R)$, the R-dependent Kirkwood g-factor (average total dipole moment of a sphere of radius R around a reference particle), for the system of figure 1. The symbols have the same meaning as in figure 1.

by (26.2.3D). The curve for $\epsilon_{RF} = \infty$ increases monotonously ; with a weaker reaction field ($\epsilon_{RF} = 50$) it remains virtually constant after three shells of neighbours whose dipole moments are positively correlated with the dipole of the reference particle. This is the behaviour one would expect for an infinite system. This is not entirely conclusive, though : first, the dielectric constant used for the reaction field ($\epsilon_{RF} = 50$) is lower than the ' true ' dielectric constant ($\epsilon \approx 65$). Second, for $R/L > 1/2$ (the corners of the cube) only incomplete shells of neighbours contribute to $G_K(R)$. We also conclude that, in contrast to what has been said in favour of MI-geometry [40], even local correlations differ more strongly from the infinite system behaviour than those resulting in SC-geometry. Thus SC is not only less time-consuming than MI, but also better.

Notes on p. 432

6. Discussion

It has been rigorously shown that—as has been tacitly assumed for a decade or so—the equations that connect the dipole moment fluctuations of a *spherical* system to its dielectric constant, (1) and (2), are also applicable for the *toroidal* boundary conditions (minimum image, spherical cutoff, reaction field geometry) often used in computer simulations of polar systems. The formula for reaction field geometry with arbitrary ϵ_{RF} is formally identical with that known to hold for periodic boundary conditions (Ewald sum plus reaction field).

The derivation presented above for two and three dimensions is a combination of statistical mechanics and electrostatics. It is completely analogous to the usual Kirkwood–Fröhlich treatment and completely general as well as straightforward. The crucial point is the solution, in Fourier space, of the integral equation for the polarization in the presence of an external field. Due to the close correspondence of that integral equation to the situation typical of computer simulations, the cubical shape of the sample and the toroidal boundary conditions as well as the interaction cutoff and the reaction field may now be treated exactly for the first time.

The only assumptions that had to be made are : (implicitly) that the system be sufficiently large to behave like a macroscopic dielectric (which is unavoidable if macroscopic properties are to be determined) ; and (explicitly) that the constitutive relation $\mathbf{P}=\chi\mathbf{E}$ be obeyed *locally*, i.e. at every point in space, rather than as a convolution

$$\mathbf{P}(\mathbf{r})= \int d\mathbf{r}'\, \chi(\mathbf{r}, \mathbf{r}')\mathbf{E}(\mathbf{r}'), \tag{37}$$

which form might be more appropriate for a small system.

When employed to interpret the results of actual simulations, the fluctuation formulas are found to give a consistent estimate of the dielectric constant for RF-geometry, while, for medium-size systems, the distortions of local correlations induced by MI or SC-geometry render those boundary conditions impractical. (This may apply to measurements of the polarization induced by an external field too.) In the thermodynamic limit the dielectric constant of the system studied here seems to be well above the values reported for EK-geometry in the literature.

The N-dependence of the dielectric constant, and the apparent inconsistency between its RF and EK estimates in particular, will be investigated in more detail. Because of the enormous amount of computer time involved comments from, and co-operation with, other laboratories would be most welcome.

The integral equation defining the macroscopic polarization may be used to study the influence of various geometries on the dipolar field of a particle. An investigation of this kind has been completed. Results will be reported in a separate publication [37].

The simulations reported in § 5 were carried out on a CDC CYBER 170-720 at the Interuniversitäres EDV-Zentrum and on a PDP 11/34 at the Prozeßrechenanlage Physik, both at Vienna. Many of the ideas presented in this article began to take shape while I was at the Vitreous State Laboratory. For making possible my stay there I would like to thank Professor T. A. Litovitz as well as Dr. H. A. Posch and Professor P. Weinzierl. It is a special pleasure to acknowledge the stimulating discussions I had with Dr. O. Steinhauser.

Appendix

Convolution with toroidal boundary conditions

Because of the importance of (15) in the text we will show here explicitly, for the one-dimensional case, that

$$\tilde{h}(k)=\tilde{f}(k)\tilde{g}(k) \tag{A 1}$$

is the Fourier transform of the convolution integral

$$h(x)=\int_{\text{TBC}} dx'\, f(x')g(x-x') \tag{A 2}$$

with toroidal boundary conditions. f, g, and h are assumed to be defined on the interval $(-1/2, 1/2)$; k is a point in Fourier space that is compatible with the length of this interval, i.e. k is a multiple of 2π.

Let $0<x<1/2$. f and g are only defined on $(-1/2, 1/2)$, so that (A 2), written explicitly, now reads

$$h(x)=\int_{-\frac{1}{2}}^{x-\frac{1}{2}} dx'\, f(x')g(x-x'-1)+\int_{x-\frac{1}{2}}^{\frac{1}{2}} dx'\, f(x')g(x-x'). \tag{A 3}$$

Inserting the Fourier expansion for f and g,

$$\begin{aligned} h(x)= &\int_{-\frac{1}{2}}^{x-\frac{1}{2}} dx' \left[\sum_{k'} \exp(ik'x')\tilde{f}(k')\right]\left[\sum_{k} \exp[ik(x-x'-1)]\tilde{g}(k)\right] \\ &+\int_{x-\frac{1}{2}}^{\frac{1}{2}} dx' \left[\sum_{k'} \exp[ik'x']\tilde{f}(k')\right]\left[\sum_{k} \exp[ik(x-x')]\tilde{g}(k)\right] \\ = &\int_{-\frac{1}{2}}^{x-\frac{1}{2}} dx' \sum_{k,k'} \exp i(k'-k)x' \exp(ikx) \exp(-ik)\tilde{f}(k')\tilde{g}(k) \\ &+\int_{x-\frac{1}{2}}^{\frac{1}{2}} dx' \sum_{k,k'} \exp[i(k'-k)x'] \exp(ikx)\tilde{f}(k')\tilde{g}(k), \end{aligned} \tag{A 4}$$

and since $\exp(-ik)=1$ for every k, both integrals may be combined :

$$h(x)=\sum_{k,k'} \exp(ikx)\tilde{f}(k')\tilde{g}(k)\int_{-\frac{1}{2}}^{\frac{1}{2}} dx' \exp[i(k'-k)x']. \tag{A 5}$$

The remaining integral is just $\delta_{k,k'}$, so that

$$h(x)=\sum_{k} \exp(ikx)\tilde{f}(k)\tilde{g}(k), \tag{A 6}$$

which proves (A 1).

For $-1/2<x<0$ the proof is similar.

References

[1] Stell, G., Patey, G. N., and Høye, J. S., 1981, *Adv. chem. Phys.*, **48**, 183. Wertheim, M. S., 1979, *A. Rev. phys. Chem.*, **30**, 471.

[2] Barker, J. A., and Watts, R. O., 1973, *Molec. Phys.*, **26**, 789.

[3] Watts, R. O., 1974, *Molec. Phys.*, **28**, 1069.

[4] Patey, G. N., 1977, *Molec. Phys.*, **34**, 427.

[5] Levesque, D., Patey, G. N., and Weis, J. J., 1977, *Molec. Phys.*, **34**, 1077.

Notes on p. 432

[6] LADD, A. J. C., 1978, *Molec. Phys.*, **36,** 463.
[7] PATEY, G. N., LEVESQUE, D., and WEIS, J. J., 1979, *Molec. Phys.*, **38,** 219.
[8] NEUMANN, M., and STEINHAUSER, O., 1980, *Molec. Phys.*, **39,** 437.
[9] ADAMS, D. J., 1980, *Molec. Phys.*, **40,** 1261.
[10] ADAMS, D. J., and ADAMS, E. M., 1981, *Molec. Phys.*, **42,** 907.
[11] CAILLOL, J. M., LEVESQUE, D., and WEIS, J. J., 1981, *Molec. Phys.*, **44,** 733.
[12] STEINHAUSER, O., 1983, *Ber. Bunsenges. phys. Chem.*, **87,** 128.
[13] FRÖHLICH, H., 1958, *Theory of Dielectrics*, 2nd edition (Clarendon Press).
[14] BÖTTCHER, C. J. F., 1973, *Theory of Electric Polarization*, Vol. 1 (Elsevier Scientific Publishing Company).
[15] KIRKWOOD, J. G., 1939, *J. chem. Phys.*, **7,** 911.
[16] RAHMAN, A., and STILLINGER, F. H., 1971, *J. chem. Phys.*, **55,** 3336.
[17] PATEY, G. N., and VALLEAU, J. P., 1974, *J. chem. Phys.*, **61,** 534.
[18] PATEY, G. N., and VALLEAU, J. P., 1976, *J. chem. Phys.*, **64,** 170.
[19] ADAMS, D. J., and MCDONALD, I. R., 1976, *Molec. Phys.*, **32,** 931.
[20] ADAMS, D. J., ADAMS, E. M., and HILLS, G. J., 1979, *Molec. Phys.*, **38,** 387.
[21] PATEY, G. N., LEVESQUE, D., and WEIS, J. J., 1979, *Molec. Phys.*, **38,** 1635.
[22] WATTS, R. O., 1981, *Chem. Phys.*, **57,** 185. Note that an incorrect formula has been used in this reference. Reanalysing Watts' results with (28.1–2.3D) instead of his (3) one now obtains $\epsilon \approx 20$ instead of $\epsilon \approx 8$.
[23] PATEY, G. N., LEVESQUE, D., and WEIS, J. J., 1982, *Molec. Phys.*, **45,** 733.
[24] FRIEDMAN, H. L., 1975, *Molec. Phys.*, **29,** 1533.
[25] BOSSIS, G., 1979, *Molec. Phys.*, **38,** 2023.
[26] BOSSIS, G., QUENTREC, B., and BROT, C., 1980, *Molec. Phys.*, **39,** 1233.
[27] BROT, C., BOSSIS, G., and HESSE-BEZOT, C., 1980, *Molec. Phys.*, **40,** 1053.
[28] BOSSIS, G., and BROT, C., 1981, *Molec. Phys.*, **43,** 1095.
[29] DELEEUW, S. W., PERRAM, J. W., and SMITH, E. R., 1980, *Proc. R. Soc.* A, **373,** 27.
[30] DELEEUW, S. W., PERRAM, J. W., and SMITH, E. R., 1980, *Proc. R. Soc.* A, **373,** 57.
[31] FELDERHOF, B. U., 1979, *Physica* A, **95,** 572.
[32] FELDERHOF, B. U., 1980, *Physica* A, **101,** 275.
[33] SMITH, E. R., 1982, *Molec. Phys.*, **45,** 915.
[34] POLLOCK, E. L., and ALDER, B. J., 1980, *Physica* A, **102,** 1.
[35] STEINHAUSER, O., 1982, *Molec. Phys.*, **45,** 335.
[36] STEINHAUSER, O., 1982, *Molec. Phys.*, **46,** 827.
[37] NEUMANN, M., 1982, *Molec. Phys.* (to be published).
[38] SINGER, K., TAYLOR, A., and SINGER, J. V. L., 1977, *Molec. Phys.*, **33,** 1757.
[39] LADD, A. J. C., 1977, *Molec. Phys.*, **33,** 1039.
[40] VALLEAU, J. P., and WHITTINGTON, S. G., 1977, *Modern Theoretical Chemistry*, Vol. 5, edited by B. J. Berne (Plenum Press), Part A.
[41] From (28.1.3D′) and (28.3.3D′) it is easily seen that the best values Adams and Adams could have obtained for the dielectric constant of the system considered in § 5 are 43 (for $\epsilon_{RF}=25$) and 53 ($\epsilon_{RF}=39$), which are both significantly lower than the ' true ' value $\epsilon \approx 65$.
[42] NEUMANN, M., and STEINHAUSER, O., 1983, *Chem. Phys. Lett.*, **95,** 417.

Notes to Reprint V.7

1. In a later paper, Neumann and Steinhauser (ref. [42]) showed that for the Ewald method of summation the correct fluctuation formula to use is indeed eq. (27.3), but only in an ideal case. The ideal situation is not always realized in a simulation, but Neumann and Steinhauser show how the necessary correction can be applied; this accounts for the discrepancies, discussed on p. 853, between the results of the present paper and the earlier calculations of Adams and Adams (ref. [10]).

2. The conclusion (see Note 1) that the Ewald method and the reaction-field method with $\epsilon_{RF} = \infty$ require the use of the same fluctuation formula can be understood as follows. At each

step in a simulation based on the usual periodic boundary conditions, the system as a whole is uniformly polarized, since every cell in the periodic array has the same net dipole moment. Associated with the uniform polarization is a contribution to the potential energy, but in general the latter is of no physical interest, since its value is shape dependent. In conventional applications of the Ewald method, the polarization contribution is ignored; in the notation of Nosé and Klein, it corresponds to the $\mathbf{n} = 0$ term in the reciprocal-space part of the Ewald sum (see eq. (A1) of Reprint V.5). So far as the fluctuations in dipole moment are concerned, neglect of the $\mathbf{n} = 0$ term is equivalent to supposing that the infinite, periodic system, whatever its shape, is embedded in a conducting medium, with the consequence that the same expression for ϵ applies in both cases. In other respects, the two methods are not equivalent, and do not represent the same physical situation.

3. The denominator in eq. (33) should be $4\pi\rho\mu^2$.

4. Equation (16) does not provide a practical route to the calculation of ϵ for materials of high dielectric constant because the left-hand side saturates as $\langle |\mathbf{M}|^2 \rangle \to \infty$. This difficulty, present also in the non-equilibrium molecular dynamics method of Evans and Morriss (Phys. Rev. A 33 (1986) 1408) is avoided when either the Ewald (see above) or usual ($\epsilon_{RF} = \epsilon$) reaction-field methods is used, since $\langle |\mathbf{M}|^2 \rangle$ is then linear in ϵ.

5. The methods described in this paper have also been applied to polarizable dipolar systems (Neumann and Steinhauser, Chem. Phys. Lett. 106 (1984) 563) and extended to include both frequency (Neumann and Steinhauser, Chem. Phys. Lett. 102 (1983) 508) and wavenumber-dependent (Neumann, Mol. Phys. 57 (1986) 97) fluctuations.

6. Some of the controversy mentioned in the introduction to this chapter has recently been revived: see (Perram and Smith, J. Stat. Phys. 46 (1987) 179) for a slightly different fluctuation formula for reaction-field boundary conditions.

CHAPTER VI

Trends and Prospects

Computer simulation is now a long-established area of research, but the level of activity continues to rise and the methods described in earlier chapters are finding new applications in many different branches of science. Although the interest in simple models remains high, a feature of recent work has been an increasing emphasis on the study of complex systems. Attempts have also been made, some very successful, to treat the quantum-mechanical problem by methods similar in spirit, if not in detail, to those used in the classical case. Our final group of papers have been chosen as illustrations of these developments; they should also give an indication of how the subject is likely to evolve in the future.

Reprints VI.1 to VI.5 highlight the attention currently paid to large-scale problems and to the simulation of real physical and chemical processes. The papers we have selected are taken from the fields of interfacial science, computational fluid dynamics, chemical reaction dynamics and enzyme catalysis. The choice is, of necessity, an arbitrary one and equally good examples can be found that address a wide range of other problems: the properties of polyelectrolyte solutions [1], the structure and dynamics of membranes [2], the onset of surface melting in solids [3], the behaviour of confined plasmas [4], the phase diagrams of fast-ion conductors [5] and other ionic crystals [6], the free energies of macromolecules [7], and so on.

The paper by Abraham, Rudge, Auerbach and Koch (Reprint VI.1) is concerned with the nature of the incommensurate phase formed by krypton atoms adsorbed on graphite. The work is of particular note for the fact that the dimensions of the graphite substrate ($\cong$ 2,000 Å) are comparable with the sample sizes used in present-day laboratory experiments. Reprint VI.2 (Landman et al.) also deals with a problem in surface physics, but the emphasis here is on the structure of the crystal-melt interface. The paper describes a molecular dynamics simulation of the pulsed heating of a silicon surface and the subsequent growth of a crystal-melt interface. The calculations are unusual insofar as the interatomic potential contains a three-body term. Most simulations are based on pairwise-additive potentials, but this would not be a realistic approximation for a covalently bonded material such as silicon. The article by Meiburg (Reprint VI.3) shows how molecular dynamics can be used to study the flow of a hard-sphere gas around a flat, inclined plate. These calculations, and the closely related work of Rapaport and Clementi [8], demonstrate that the flow patterns characteristic of real fluids at high Reynolds numbers are readily simulated by molecular dynamics, but only if the system is sufficiently large: typically, at least 100,000 particles are required. Reprint VI.4 (Bergsma et al.) describes a molecular dynamics algorithm for the study of chemical reactions in solution and applies the method to the case of the hypothetical reaction $Cl + Cl_2 \rightarrow Cl_2 + Cl$ in a rare-gas solvent. The simplest approach is to compute the trajectories of the reacting particles for all possible initial conditions, including those in which the particles are very far apart. This does not work well in practice, because successful reactions occur only rarely. It proves to be much more efficient to start the trajectories of the reactants from an equilibrium distribution over the potential-energy surface of the transition-state complex and to run them both forwards and backwards in time; trajectories that lead to successful reactions are then easily identified. The method [28] is similar to that devised earlier by Bennett [9] for the study of defect migration in solids. Finally, in Reprint VI.5, Brünger, Brooks and Karplus employ a combination of molecular dynamics and brownian dynamics techniques in an investigation of the behaviour of water molecules in the vicinity of the activated site of an enzyme in aqueous solution. As the authors point out, the catalytic events of interest in such a problem occur on a picosecond timescale, and hence are open to analysis by molecular dynamics, even though the overall rate of reaction is much slower.

The scope of Reprints VI.1 to VI.5, like that of papers discussed in earlier parts of the book, is limited to the extent that they rely on an almost completely classical description of the system under study. * On the other hand, there has been much effort over many years directed at the simulation of quantum-mechanical many-body systems. The methods used in quantum simulations are inevitably different in character from those that have been applied successfully to classical problems; they are also less highly developed and some of the technical problems have yet to be fully resolved.

The first reported quantum simulations were based on a variational approach [10]. Improvements came later with the introduction of the more sophisticated techniques of "Green's-function" [11] and "diffusion" [12] Monte Carlo. In the early papers, attention was focussed on the calculation of ground-state properties of systems such as liquid ^{4}He and spin-aligned hydrogen; this work has been well reviewed elsewhere [13]. Progress in the simulation of systems at non-zero temperature has centred largely on the use of a discretized, path-integral representation of the density matrix [14]. An ad-

* Thermodynamic properties obtained from classical simulations can be corrected for quantum effects. See, for example, Reprint IV.8 and Note II.3.1.

vantage of proceeding in this way is that it is possible to exploit the approximate isomorphism that can be established between the quantum particles of interest and a hypothetical classical system [14,15]. Since the two papers (Reprints VI.6 and VI.7) that we include as illustrations of the path-integral approach both presuppose some familiarity with the method, we give below an elementary treatment of the theory underlying it.

Consider the simple problem [16] of a single quantum particle in an external potential $\phi(\boldsymbol{r})$. The quantum-mechanical density operator of the system is

$$\rho = \exp(-\beta\mathcal{H}),$$

where $\mathcal{H} = -(\hbar^2/2m)\nabla^2 + \phi(\boldsymbol{r})$, the density matrix in the coordinate representation, is

$$\rho(\boldsymbol{r}_i, \boldsymbol{r}_j; \beta) = \langle \boldsymbol{r}_i | \exp(-\beta\mathcal{H}) | \boldsymbol{r}_j \rangle,$$

and the canonical partition function is

$$Q = \int \mathrm{d}\boldsymbol{r}_1\, \rho(\boldsymbol{r}_1, \boldsymbol{r}_1; \beta).$$

If we now put $\exp(-\beta\mathcal{H}) = [\exp(-\beta\mathcal{H}/P)]^P$, where P is a positive integer, and make repeated use of the rule that

$$\int \mathrm{d}\boldsymbol{r}_i\, |\boldsymbol{r}_i\rangle\langle\boldsymbol{r}_i| = 1$$

the partition function can be rewritten as

$$Q = \int \mathrm{d}\boldsymbol{r}_1 \cdots \mathrm{d}\boldsymbol{r}_P\, \rho(\boldsymbol{r}_1, \boldsymbol{r}_2; \beta/P)\, \rho(\boldsymbol{r}_2, \boldsymbol{r}_3; \beta/P) \cdots \rho(\boldsymbol{r}_P, \boldsymbol{r}_1; \beta/P).$$

If P is sufficiently large, we can make a high-temperature approximation * [19] of the form

$$\rho(\boldsymbol{r}_i, \boldsymbol{r}_j; \beta/P) \cong \rho_0(\boldsymbol{r}_i, \boldsymbol{r}_j; \beta/P) \exp\left[-\tfrac{1}{2}(\beta/P)\left(\phi(\boldsymbol{r}_i) + \phi(\boldsymbol{r}_j)\right)\right],$$

where

$$\rho_0(\boldsymbol{r}_i, \boldsymbol{r}_j; \beta/P) = \left(Pm/2\pi\beta\hbar^2\right)^{3/2} \exp\left[-\left(Pm/2\beta\hbar^2\right) |\boldsymbol{r}_i - \boldsymbol{r}_j|^2\right]$$

is the density matrix of a free particle [19]. Then

$$Q \cong \left(Pm/2\pi\beta\hbar^2\right)^{3P/2} \int \mathrm{d}\boldsymbol{r}_1 \cdots \mathrm{d}\boldsymbol{r}_P \exp\left[-\beta V_{\mathrm{eff}}(\boldsymbol{r}_1, \ldots, \boldsymbol{r}_P)\right],$$

* This is not a useful approximation for liquid helium [17]. For other possibilities, see ref. [18].

with

$$V_{\mathrm{eff}}(\boldsymbol{r}_1,\ldots,\boldsymbol{r}_P)=\sum_{i=1}^{P}(Pm/2\beta^2\hbar^2)\,|\boldsymbol{r}_i-\boldsymbol{r}_{i+1}|^2+\sum_{i=1}^{P}P^{-1}\phi(\boldsymbol{r}_i),$$

where $\boldsymbol{r}_{P+1}=\boldsymbol{r}_1$. The original quantum-mechanical problem has therefore been reduced to a classical one in which the quantum particle is replaced by a classical ring polymer consisting of P identical particles coupled together through nearest-neighbour "spring" forces. This means that the configuration space of a single quantum particle, or of a system of interacting particles, can be sampled by classical Monte Carlo or molecular dynamics methods.

The isomorphism between the quantum particle and a classical ring polymer becomes exact in the limit $P\to\infty$. Unfortunately, use of large values of P can lead to serious ergodic difficulties, since the strength of the "spring" forces is proportional to P. Highly efficient sampling techniques are therefore required; in this respect, there appears to be a distinct advantage to the use of Monte Carlo methods rather than molecular dynamics * [20]. So far as the calculation of physical properties is concerned, the choice between Monte Carlo and molecular dynamics is not a crucial one, because a "dynamic" path-integral simulation does not yield the true dynamics of the quantum system.

A particularly simple type of path-integral calculation is one in which a single quantum particle is immersed in a bath of classical particles. The earliest work along these lines was that of Parrinello and Rahman [16], who used the molecular dynamics method to study the solvation of an electron in molten potassium chloride. The paper by Sprik, Impey and Klein (Reprint VI.6) describes a similar calculation for an electron in liquid ammonia, but is based on a specialized form of Monte Carlo sampling that allows the use of very large values of P. A number of other systems that are partly classical and partly quantum mechanical have also been studied by path-integral methods. Examples include muonium and hydrogen in water [21], an electron in a hard-sphere fluid [22], and a simulation of water in which only the rotational degrees of freedom are treated quantum mechanically [23]. At the opposite extreme, the paper by Ceperley and Pollock (Reprint VI.7) contains the results of calculations for liquid ^{4}He (a bose fluid). In this case, each particle must be treated quantum mechanically, and there is the further complication that the effects of quantum statistics cannot be ignored. The inclusion of Bose–Einstein statistics requires a further elaboration of the sampling scheme: in the classical analogy, the polymer chains must be allowed to cross-link.

* Discretization, i.e. the use of finite P, can in principle be avoided altogether with the help of Green's-function method of ref. [11], but the computational problems are formidable.

Three other approaches to quantum simulation have also been explored. Singer and collaborators [24] * have developed a molecular dynamics method based on a variational solution of the time-dependent Schrödinger equation for a trial function consisting of a product of gaussian wave packets. Although limited to use in situations where quantum effects are weak, as in the case of liquid neon **, the method has the advantage of allowing the calculation of time-correlation functions of the quantum particles. Selloni et al. [27] have used a simultaneous integration of the time-dependent Schrödinger equation and Newton's equations of motion to study the dynamics of a single electron in a classical liquid. Finally, Car and Parrinello (Reprint VI.8) describe a completely different scheme in which classical molecular dynamics is combined with conventional methods of electronic-structure theory. The combined scheme provides a means of avoiding the special difficulties associated with particles that obey Fermi–Dirac statistics [12]; its key ingredient is a lagrangian (eq. (3) of the paper) from which equations of motion are derived for the nuclear coordinates and electronic orbitals. At least in principle, these equations provide the basis for what the authors call an *ab initio* molecular dynamics simulation, i.e. one in which the form of the intermolecular potential need not be specified in advance.

References

[1] V. Vlachy and D.J. Haymet, J. Chem. Phys. 84 (1986) 5874.
[2] P. van der Ploeg and H.J.C. Berendsen, Mol. Phys. 49 (1983) 233.
[3] V. Rosato, G. Ciccotti and V. Pontikis, Phys. Rev. B 33 (1986) 1860.
[4] A. Rahman and J.P. Schiffer, Phys. Rev. Lett. 57 (1986) 1133.
[5] J. Tallon, Phys. Rev. Lett. 57 (1986) 2427.
[6] R.W. Impey, M. Sprik and M.L. Klein, J. Chem. Phys. 83 (1985) 3638.
[7] A. Di Nola, H.J.C. Berendsen and O. Edholm, Macromol. 17 (1984) 2044.
[8] D.C. Rapaport and E. Clementi, Phys. Rev. Lett. 57 (1986) 695.
[9] C.H. Bennett, in: Algorithms for Chemical Computations, ed. R.E. Christofferson (American Chemical Society, Washington, 1977).
[10] W.L. McMillan, Phys. Rev. A 138 (1965) 442.
[11] M.H. Kalos, D. Levesque and L. Verlet, Phys. Rev. A 9 (1974) 2178.
D.M. Ceperley and B.J. Alder, J. Chem. Phys. 81 (1984) 5833.
[12] D.M. Ceperley and B.J. Alder, Phys. Rev. Lett. 45 (1980) 566.
P.J. Reynolds, D.M. Ceperley, B.J. Alder and W.A. Lester, J. Chem. Phys. 77 (1982) 5593.
[13] D.M. Ceperley and M.H. Kalos, in: Monte Carlo Methods in Statistical Physics, ed. K. Binder (Springer-Verlag, Berlin, 1979).
[14] J.A. Barker, J. Chem. Phys. 70 (1979) 2914.
[15] D. Chandler and P.G. Wolynes, J. Chem. Phys. 74 (1981) 4078.

* The methods used in these papers derive from the approach developed earlier by Heller [25] for the study of scattering phenomena.

** Singer and Smith [24] give results for solid, liquid and gaseous neon. Liquid neon has also been studied by the path-integral Monte Carlo method: see ref. [26].

[16] M. Parrinello and A. Rahman, J. Chem. Phys. 80 (1984) 860.
[17] E.L. Pollock and D.M. Ceperley, Phys. Rev. B 30 (1984) 2555.
[18] K.S. Schweizer, R.M. Stratt, D. Chandler and P.G. Wolynes, J. Chem. Phys. 75 (1981) 1347.
D. Thirumalai and B.J. Berne, J. Chem. Phys. 79 (1983) 5029.
[19] R.P. Feynman and A.R. Hibbs, Quantum Mechanics and Path Integrals (McGraw-Hill, New York, 1965).
R.P. Feynman, Statistical Mechanics (Benjamin, Reading, 1972).
[20] R.W. Hall and B.J. Berne, J. Chem. Phys. 81 (1984) 3641.
M. Sprik, M.L. Klein and D. Chandler, Phys. Rev. B 31 (1985) 4234.
[21] B. De Raedt, M. Sprik and M.L. Klein, J. Chem. Phys. 80 (1984) 5719.
[22] M. Sprik, M.L. Klein and D. Chandler, J. Chem. Phys. 83 (1985) 3042.
[23] R.A. Kuharski and P.J. Rossky, Chem. Phys. Lett. 103 (1984) 357.
[24] N. Corbin and K. Singer, Mol. Phys. 46 (1982) 671.
K. Singer and W. Smith, Mol. Phys. 57 (1986) 761.
[25] E.J. Heller, J. Chem. Phys. 64 (1976) 63.
[26] D. Thirumalai, R.W. Hall and B.J. Berne, J. Chem. Phys. 81 (1984) 2523.
[27] A. Selloni, P. Carnevali, R. Car and M. Parrinello, Localization, Hopping and Diffusion of Electrons in Molten Salts, Phys. Rev. Lett., to appear.
[28] In the context of simulation of chemical reactions the rules for proper yet practical sampling of rare, activated, events were first discussed by D. Chandler (J. Chem. Phys. 68 (1978) 2959) and then applied by R.O. Rosenberg, B.J. Berne and D. Chandler (Chem. Phys. Lett. 75 (1980) 162) to MD simulation of an isomerization reaction.
For a detailed and interesting reconstruction of all the various contributions to the method see the remarks by Chandler on p. 341 in the general discussion which follows the paper: D. Chandler and R.A. Kuharski, Faraday Discussion Chem. Soc. 85 (1988), 329.

Molecular-Dynamics Simulations of the Incommensurate Phase of Krypton on Graphite Using More than 100 000 Atoms

Farid F. Abraham, William E. Rudge, Daniel J. Auerbach, and S. W. Koch[a]
IBM Research Laboratory, San Jose, California 95193
(Received 14 November 1983)

The incommensurate phase of krypton on graphite is studied by use of the molecular-dynamics simulation technique for systems with graphite substrate dimensions comparable to present-day laboratory capabilities. At low temperature and for all coverages, honeycomb networks of "heavy" domain walls are observed for the first time. With increasing temperature, distortion from the perfect honeycomb structure becomes more prevalent, characterized by significant fluctuations from the symmetry directions, wall thickening, and wall roughening.

PACS numbers: 68.20.+t

The nature of the incommensurate (IC) phase of krypton on graphite and its transition to the $\sqrt{3}\times\sqrt{3}\ R30°$ commensurate (C) phase are being extensively investigated theoretically,[1-9] experimentally,[10-13] and by computer simulation.[14,15] For Kr on graphite, the domain walls may be in three different directions because of the hexagonal substrate structure. If wall intersections are energetically unfavorable, a *striped* phase might be expected where walls are only in one direction.[4] However, Villain has noted that a *honeycomb* array of walls has a degeneracy in which the hexagons of the array can expand or contract without changing the total wall length or the number of nodes (i.e., wall crossings).[1,2] Hence, Villain argues that this additional contribution to the entropy stabilizes the honeycomb phase relative to the striped phase and that the IC-C transition is first order. Expanding upon the Villain picture, Coppersmith *et al.*[5,6] have predicted that the *slightly* incommensurate phase is unstable to dislocations and conclude that sufficiently close to the IC-C transition the hexagonal overlayer will be a fluid at all temperatures, provided that the IC-C transition is weakly first order. The order of the transition has not been resolved experimentally.[10-12] However, Moncton *et al.*[11] have found the *high-temperature*, weakly incommensurate phase to be disordered with correlation lengths on the order of 100 Å. Computer simulation has reproduced the experimental measurements of this high-temperature phase and has established the microstructure of this "domain-wall liquid."[15] Such a demonstration suggests that computer experiments should be valuable in determining the incommensurate phase microstructure for all temperatures and coverages.

In this present investigation, we report on the structure of the incommensurate phase of krypton on graphite as a function of temperature and coverage using the molecular-dynamics simulation technique for systems of 103 041 and 161 604 krypton atoms. Hence, graphite substrate dimensions up to 1700 Å are realized and are essentially identical to present-day laboratory capabilities. We observe that the incommensurate phase consists of commensurate islands separated by a interconnecting network of incommensurate domain walls, the structure of this network being a sensitive function of temperature and coverage. At low temperature, Villain's honeycomb network of domain walls is observed for all coverages. Following the definition of Huse and Fisher,[7] all of the walls are *heavy* and are in agreement with the prediction of Kardar and Berker.[8] A low-temperature fluid phase or an incommensurate striped phase is not seen. With increasing temperature, distortions from the perfect honeycomb structure become more prevalent. At high temperatures, the individual domain walls fluctuate significantly from the symmetry directions while possessing boundary roughness and a greater wall thickness. But before we present the details of our findings, we briefly outline certain aspects of the simulation procedure.

Our numerical simulation method for studying rare-gas adsorption on graphite has been presented[15,16] and will not be described here. However, an additional feature of employing Hockney and Eastwood's[17] chain-link method to set up Verlet tables[18] for all atoms was adopted,[19] the table update being executed every ten time steps, each time step being 0.025 ps. Similar to our previous study,[15] we use the Lennard-Jones 12:6 pair potential to represent the interaction between the various atoms of the krypton-graphite system. The Lennard-Jones parameters are taken to be $\epsilon/k = 170$ K and $\sigma = 3.6$ Å for the krypton-krypton interaction and $\epsilon/k = 64.83$ K and $\sigma = 3.22$ Å for the krypton-carbon interaction, respectively. Simple pairwise additivity of the atomic interactions is assumed, and the carbon atoms defining the semi-infinite solid are fixed at their lattice sites. In order to reduce the computer memory requirement, we constrain the movement of the krypton atoms to a plane parallel to the graphite surface. Thus, in contrast to our earlier studies, we are dealing with a two-dimensional system in the external field of the graphite substrate. However, a detailed comparison between a two-dimensional simulation and the previous, quasi-two-dimensional simulation shows no physically relevant differences.[19] The external field of the graphite is reproduced to a very good approximation by the expression[20]

$$\varphi_{\text{Kr-G}} = -V_g[\cos 2\pi s_1 + \cos 2\pi s_2 + \cos 2\pi(s_1+s_2)], \quad (1)$$

where $V_g = 0.08\epsilon_{\text{Kr-Kr}}$ and s_1, s_2 are the basis vectors of the graphite unit cell. The planar x-y geometry of the computational cell is the same as described in Ref. 16, with the exception that our Cartesian coordinate system is now rotated by 90° about the origin. The 161 604 Kr atom system requires approximately 13 megabytes of memory and executes 110 time steps per central-processing-unit hour on the IBM/370 3081 computer.

We have performed two series of simulations: one in which the coverage was fixed at $\theta = 1.013$ and the *reduced* temperature was varied from $T^* \equiv kT/\epsilon_{\text{Kr-Kr}} = 0.05$ to 0.9; the other in which the temperature was fixed at $T^* = 0.05$ and the coverage was varied from 1.013 to 1.086. At the lowest temperature, the Kr system was initialized in a perfect incommensurate triangular lattice, and the atoms were assigned a velocity distribution corresponding to a temperature $T^* = 0.05$. Equilibrium was determined by monitoring the energy, the ratio of commensurate to incommensurate atoms, the ratio of the commensurate atoms in the three degenerate ground states, and the domain microstructure as a function of time. For the higher-temperature simulations, we slowly heated the system over 2000 time steps to the next higher temperature and repeated the equilibration procedure. Equilibration times were typically less than 1000 time steps and therefore very rapid. For a given temperature and coverage, 4000 to 8000 total time steps were performed. We have adopted the criterion that a krypton atom is commensurate if its average position is inside a circle with a radius $\delta = 0.2$ around the nearest adsorption site, in units of the graphite lattice constant $a = 2.46$ Å.

Figure 1 shows the principal results of our simulations for the 103 041 atom system. The incommensurate and commensurate regions are shown as solid black and solid white areas, respectively; in actual fact, we plotted the incommensurate atoms as points but the lack of graphical resolution merged the points to make a solid region. We first consider the case where the temperature is fixed at the low value of 0.05 and the coverage is varied. We note that for all coverages a honeycomb network of domain walls is established, the network consisting of straight walls with smooth boundaries which are aligned to the three symmetry directions of the graphite substrate. The commensurate regions form an array of honeycomb domains, the individual hexagons not being identical in size or shape. This honeycomb domain structure with *breathing* freedom is direct conformation of Villain's[1] picture of the incommensurate phase, and this is the first direct observation of this structure. At fixed temperature, the percentage Kr atoms that are commensurate (%C) decreases linearly with increasing coverage (90, 80, 75, 60, and 37 percent, respectively), while the domain-wall thickness remains essentially constant at 18 Å. This low-temperature wall thickness is in agreement with recent calculations.[9,21] This decrease of the commensurate percentage is associated with an increase of the total length of domain walls, and this gives rise to smaller and more numerous commensurate domains. For krypton on graphite, there exist two configurationally distinct types of walls,[7,8] sometimes referred to as heavy and light walls.[7] All of the domain walls in the low-temperature honeycomb network are the heavy type, and this is in agreement with the theoretical calculation of Kardar and Berker.[8] One can visualize two extremes for the atomic microstructure of the incommensurate state: (1) the Kr atoms are near registry except for thin domain walls, or (2) the Kr monolayer is a lattice that is weakly modulated by the substrate field. In Fig. 2, we present the honeycomb picture for $T^* = 0.05$, $\theta = 1.025$ and for the three commensurate-radius cutoff criteria $\delta = 0.1$, 0.2, and 0.3. We conclude that at this low temperature the Kr lattice is significantly modulated. However, with increasing coverage, this modulation will de-

Note on p. 443

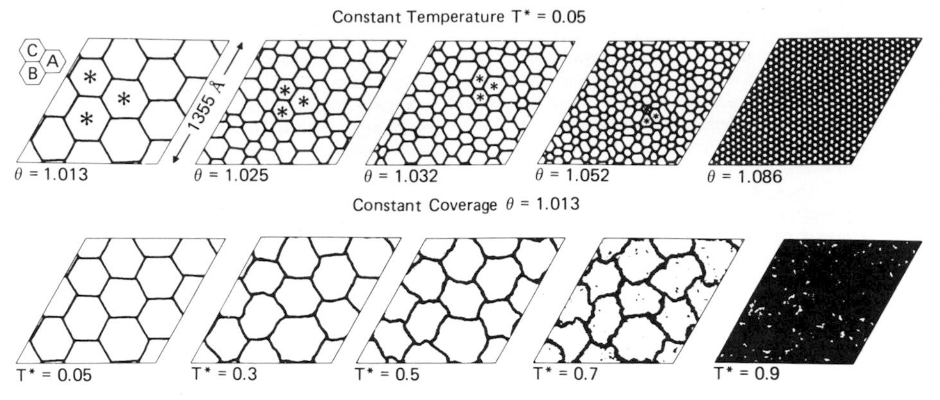

FIG. 1. Pictures of the domain-wall network for an equilibrium configuration of the incommensurate phase as a function of coverage θ at fixed temperature $T^* = 0.05$ and as a function of temperature at a fixed coverage $\theta = 1.013$. 103 041 Kr atoms on graphite.

crease; i.e., when the wall separation approaches the thickness of an individual wall.

In order to test the prediction of Coopersmith *et al.*[5,6] (i.e., the slightly incommensurate phase is unstable to dislocations sufficiently close to the IC-C transition and the hexagonal overlayer will be a *fluid* at all temperatures), we simulated as large a system as we felt was practical within the constraint of our computer resources—a 161 604 Kr atom system. The temperature and coverage are 0.05 and 1.005, respectively. In Fig. 3, we again see the incommensurate honeycomb structure and hence no evidence of a fluid phase at this low temperature. Furthermore, we do not observe the *striped* phase or two-phase coexistence between the commensurate and incommensurate phases, which would be indicative of a first-order transition when the total coverage is held constant in the two-phase region. Of course, the nonexistence of these features may be a consequence of our system not being sufficiently weakly incommensurate; i.e., a coverage may have to be much lower in order to observe these features. This same limitation exists for laboratory experiments.

Returning to Fig. 1, we now consider the series of simulations where the coverage was held fixed at 1.013. With increasing temperature, the overall appearance of the incommensurate phase remains that of the domain-wall network which becomes increasingly distorted, the walls becoming broader and the wall boundaries roughening considerably. One who feels compelled to classify the walls into the heavy- and light-wall catagories will note that the wall orientations favor the three symmetry directions, and those walls which can

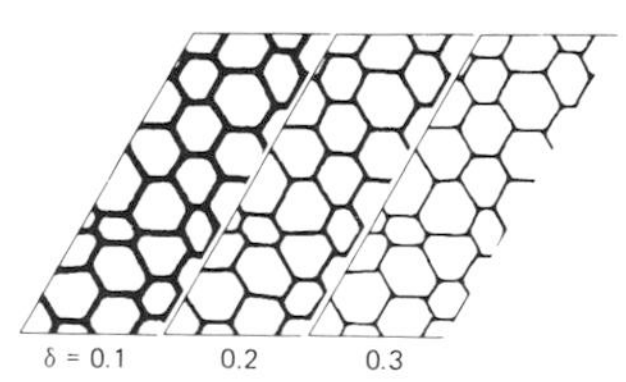

FIG. 2. Pictures of the domain-wall network for an equilibrium configuration of the incommensurate phase at a coverage $\theta = 1.025$ and temperature $T^* = 0.05$, for various values of the commensurate-radius cutoff criterion δ.

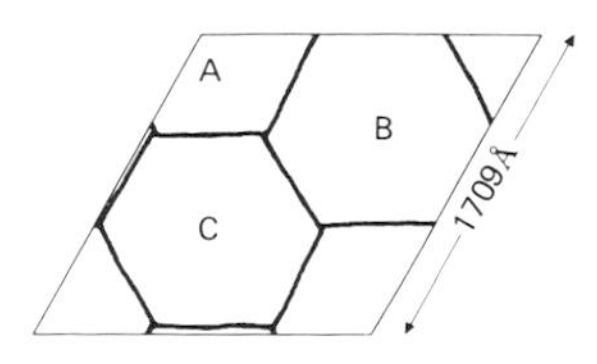

FIG. 3. A picture of the domain-wall network for an equilibrium configuration of the incommensurate phase simulated by 161 604 Kr atoms on graphite at a coverage $\theta = 1.005$ and temperature $T^* = 0.05$. The percentage of commensurate atoms is 96%.

be unambiguously classified are essentially heavy walls. Also, we note the marked increase in the wall thickness with increasing temperature, the wall thickness at $T^*=0.7$ being approximately twice the wall thickness at $T^*=0.05$. This is consistent with the gradual decrease of the percentage of commensurate atoms (90, 88, 85, and 81 percent, respectively). At the highest temperature, the system is mainly incommensurate since it is principally a liquid coexisting with very small patches of commensurate islands (%C = 26). This liquid is the familiar fluid state and not the fluid of Coopersmith *et al.*

Direct experimental confirmation of Villain's picture for the incommensurate phase of krypton on graphite has been achieved by simulating over 100 000 krypton atoms on a graphite substrate, thereby enabling us to study systems comparable to present-day laboratory experiments. A more detailed presentation including simulations of a smaller system (20 000 atoms) is forthcoming, where the variation of important quantities such as the wall thickness and average wall separation with temperature and coverage is presented.[19]

[a]Present address: Institut für Theoretische Physik, Universität Frankfurt, Frankfurt am Main, Federal Republic of Germany.

[1]J. Villain, in *Ordering in Strongly Fluctuating Condensed Matter Systems*, edited by T. Riste (Plenum, New York, 1980), p. 221.

[2]J. Villain and M. B. Gordon, Surf. Sci. 125, 1 (1983).

[3]P. Bak, Rep. Prog. Phys. 45, 587 (1982).

[4]V. L. Pokrovskii and A. L. Talapov, Phys. Rev. Lett. 42, 65 (1979).

[5]S. M. Coppersmith, D. S. Fisher, B. J. Halperin, P. A. Lee, and W. F. Brinkman, Phys. Rev. Lett. 44, 549 (1981).

[6]S. M. Coppersmith, D. S. Fisher, B. J. Halperin, P. A. Lee, and W. F. Brinkman, Phys. Rev. B 25, 349 (1982).

[7]D. A. Huse and M. E. Fisher, Phys. Rev. Lett. 49, 793 (1982).

[8]M. Kardar and A. N. Berker, Phys. Rev. Lett. 48, 1552 (1982).

[9]M. Schoebinger and S. W. Koch, to be published.

[10]M. D. Chinn and S. C. Fain, Phys. Rev. Lett. 39, 146 (1977).

[11]D. E. Moncton, P. W. Stephens, R. J. Birgeneau, P. Horn, and G. S. Brown, Phys. Rev. Lett. 46, 1533 (1981).

[12]M. Nielsen, J. Als-Nielsen, J. Bohr, and J. P. McTague, Phys. Rev. Lett. 47, 582 (1981).

[13]J. P. McTague, J. Als-Nielsen, J. Bohr, and M. Nielsen, Phys. Rev. B 25, 7765 (1982).

[14]F. Hanson and J. P. McTague, J. Chem. Phys. 72, 6363 (1980).

[15]F. F. Abraham, S. W. Koch, and W. E. Rudge, Phys. Rev. Lett. 49, 1830 (1982).

[16]S. W. Koch and F. F. Abraham, Phys. Rev. B 27, 2964 (1983).

[17]R. W. Hockney and J. W. Eastwood, *Computer Simulations Using Particles* (McGraw-Hill, New York, 1981).

[18]L. Verlet, Phys. Rev. 159, 98 (1967).

[19]S. W. Koch, W. E. Rudge, and F. F. Abraham, to be published.

[20]W. A. Steele, Surf. Sci. 36, 317 (1973).

[21]R. J. Gooding, B. Joos, and B. Bergerson, Phys. Rev. B 27, 7669 (1983).

Note to Reprint VI.1

Reference [9] was later published as (Schoebinger and Koch, Z. Phys. B 53 (1983) 233) and ref. [19] was published as (Koch et al., Surf. Sci. 145 (1984) 329).

Faceting at the Silicon (100) Crystal-Melt Interface: Theory and Experiment

Uzi Landman, W. D. Luedtke, R. N. Barnett, C. L. Cleveland, and M. W. Ribarsky
School of Physics, Georgia Institute of Technology, Atlanta, Georgia 30332

and

Emil Arnold, S. Ramesh, H. Baumgart, A. Martinez, and B. Khan
Philips Laboratories, Briarcliff Manor, New York 10510

(Received 30 September 1985)

Molecular-dynamics simulations and *in situ* experimental observations of the melting and equilibrium structure of the crystalline Si(100)-melt interface are described. The equilibrium interface is structured, exhibiting facets established on (111) planes.

PACS numbers: 68.55.Rt, 64.70.Dv, 68.45.Kg

The study of the solid-vapor and solid-liquid interfaces has attracted a recent surge of interest because of improved experimental and theoretical techniques for probing phenomena such as surface melting, surface roughening, and interface morphology. In particular, it has been suggested that the structure of the solid-melt interface during zone-melting recrystallization of silicon critically determines the generation of the observed networks of low-angle grain boundaries.[1] Ample evidence exists that surfaces of crystalline silicon become faceted upon melting, and the solid-melt interface of a growing Si crystal establishes itself on the (111) crystal planes.[2] Several models have been proposed to understand the solid-liquid interface morphology at equilibrium, notably the simple two-layer model of Jackson.[3] Silicon marginally satisfies the Jackson criterion for facet formation. It could be expected that under appropriate conditions such material would exhibit facets both on solidification and in melting. In previous experimental[4,5] and theoretical[1] studies the silicon crystal-melt interface was investigated under nonequilibrium conditions during growth by laser-induced zone melting of silicon films on SiO_2. In this Letter we report on the first theoretical microscopic simulations and experimental *in situ* observations of the melting and *equilibrium* structure of that interface, which provide evidence for a (111)-faceting transformation at the interface.

In our theoretical studies we have used the molecular dynamics method,[6] which consists of a numerical solution of the equations of motion of a large ensemble of interacting particles on refined spatial and temporal scales. The structural, dynamical, and stability properties of homogenous phases and equilibrium interphase interfaces, as well as the kinetics and dynamics of phase transformations (such as melting and solidification), are governed by the various contributions to the total energy (or free energy, at finite temperatures).[6b] The potential energy of an interacting system of particles can be written in general as a sum of contributions of varying order in the number of particles (one-body, two-body, . . . terms). Because of the directional covalent bonding, characteristic of tetrahedral semiconductors, a model of the potential for these materials must go beyond the often-used pair interactions, via the inclusion of nonadditive, angle-dependent contributions (three-body and higher order). In our simulations we have employed optimized two- and three-body potentials, V_2 and V_3, respectively, which have yielded a rather adequate description of the structural properties of crystalline and liquid silicon[7]:

$$V_2(r_{ij}) = A\,[Br_{ij}^{-P} - 1]\,g_\beta(r_{ij}), \tag{1a}$$

$$V_3(r_i,r_j,r_k) = v_{jik} + v_{ijk} + v_{ikj}, \tag{1b}$$

$$v_{jik} = \lambda g_\gamma(r_{ij}) g_\gamma(r_{ik}) [\cos\theta_{jik} + \tfrac{1}{3}]^2, \tag{1c}$$

where r_{ij} is the distance between atoms i and j, and $g_\gamma(r) = \exp[\gamma/(r-a)]$ for $r < a$ and vanishes for $r \geq a$. In Eqs. (1) r is expressed in units of $\sigma = 0.209\,51$ nm, the unit of energy is $\epsilon = 50$ kcal/mole, and that of temperature is $T = \epsilon/k_B$ (to convert to T in Kelvins multiply the reduced temperature by 2.5173×10^4); $A = 7.044\,556\,277$, $B = 0.602\,224\,558\,4$, $P = 4$, $a = 1.8$, $\lambda = 21$, $\beta = 1$, and $\gamma = 1.2$. The time unit, t.u., is $\sigma(m/\epsilon)^{1/2} = 7.6634\times10^{-14}$ s and the integration time step Δt, with use of a fifth-order predictor-corrector algorithm, is taken to be 1.5×10^{-2} t.u.; with this choice and a frequent updating of the interaction lists, the total energy is conserved to at least six significant figures. As seen from Eq. (1c), the three-body contribution to the potential energy vanishes for the perfect tetrahedral angle. Therefore, the liquid is characterized by a higher magnitude of V_3 than the solid. While improvements to the potential functions are desirable, their general form and the degree of agreement with observed data which they provide[7] warrant their use in our study of the interface between condensed phases, i.e., at the solid-melt interface (conduction-electron-density–dependent potential-energy terms could be incorporated in future studies in a manner similar to that proposed by us recently).[8] Additionally, the molecular dynamics technique which we use allows for dynamical variations in particle density and structural changes via the *Ansatz* Lagrangean of Parrinello and Rahman,[9] extended to include three-body interactions and planar 2D periodic boundary conditions.

Since zone-melted Si films on SiO_2 tend to recrystallize with (100) texture,[4,5] we start with a silicon crystal consisting of N_L dynamic layers, with N_P particles per layer, exposing the (001) face. The z axis is taken parallel to the [001] direction, and the 2D cell is defined by the [110] and [$\bar{1}$10] directions. The bottom layer of the crystal (layer number 1) is positioned in contact with a static silicon substrate. Simulations for two systems were performed: (i) $N_L=28$ and $N_P=36$, for which results are shown, and (ii) $N_L=24$ and $N_P=144$, yielding similar results. The N_L values were chosen so as to minimize the static substrate effects at the interface.

Following equilibration of the total system at a temperature $T=0.064$ and zero external pressure, a portion of the system (about $N_L/2$ from the top, free surface) was heated via scaling of particle velocities. During the subsequent dynamical evolution towards equilibrium, melting initiated, with the melting front propagating from the exposed surface towards the bulk, exhibiting a tendency for facet formation upon melting. After a prolonged period (for the data presented below, runs in excess of $1.5\times10^5\Delta t$) an equilibrium crystal-melt coexistence was established, at an average kinetic temperature $T_m=0.0662\pm0.0016$, uniform throughout the system. This value may be compared to the experimental melting temperature of silicon, $T_m=0.0669\equiv1683$ K. The equilibrium crystal-melt interface exhibits a pronounced structure, demonstrated by the sample particle trajectories, shown in Fig. 1(a). These are recorded for $2000\Delta t$ and viewed along the [$\bar{1}$10] direction (denoted by a circle at the bottom left), where the breakup into alternating (111) and ($\bar{1}\bar{1}$1) crystalline planes is indicated. The melt region in the vicinity of the solid (facet) planes exhibits a certain degree of ordering due to the crystalline potential, resulting in a diffuseness of the interface at that region.[6b] To complement the picture, we show in Fig. 1(b) particle trajectories in the region of the seventh layer ($l=7$), projected onto the 2D plane, exhibiting solid and partial-liquid characteristics. In extended runs we observed that the morphology of the interface fluctuates (on a time scale of $\sim5\times10^3\Delta t$) between equivalent facet configurations, the one shown in Fig. 1(a) where the facet runs along the ($\bar{1}$10) direction and the other one where the facet runs along the (110) direction.

Further insights are provided by the equilibrium particle density profile [Fig. 2(a)] and per-particle potential energies [Figs. 2(b)–2(d)], which show both the crystal-melt and melt-vacuum interfaces. Focusing on

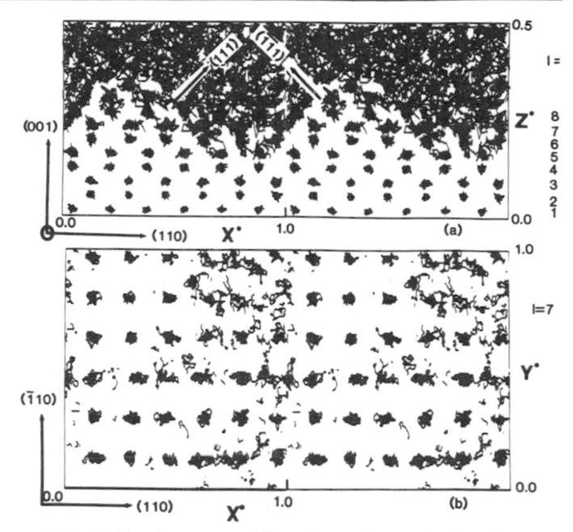

FIG. 1. Real-space particle trajectories at the interface region recorded at equilibrium. (a) Viewed along the [$\bar{1}$10] direction; (b) trajectories for particles at the region of the seventh layer ($l=7$) viewed from the [001] direction. $Z^*=1\equiv18.14\sigma$ and $X^*=Y^*=1\equiv10.9\sigma$. The 2D computational cell ($0\le X^*,Y^*\le1$) is replicated along the X^*[110] direction to aid visualization.

the former interface, we observe the opposing trends in the behavior of the two- and three-body potentials, V_2 and V_3, which, when added, yield the result shown in Fig. 2(b). The pronounced minima correspond to the locations of crystalline or partially crystalline layers. We observe that the variation in the total potential energy, V_2+V_3, upon transformation from the solid to the melted region, is smaller than the variation in the individual contributions V_2 and V_3. The behavior of the three-body potential can be used to distinguish the solid from the liquid and thus provide a vivid visualization of the solid component of the interface. In Fig. 2(e) contours of V_3 for $V_3\le0.45$ are shown. By restriction of the value of V_3, only solidlike regions are captured (the appearance of the V_3 contour map does not change significantly with small changes in the cutoff value). Figure 2(e), in conjunction with the real-space trajectories shown in Fig. 1, complements and corroborates our picture of the structure of the interface.

To affirm further our prediction of the (111) faceting of the equilibrium crystalline Si(100)-melt interface, we present in Fig. 3 results for the Si(111)-melt interface, simulated in a manner similar to that described above, with $N_L=20$ and $N_P=49$. The particle trajectories viewed along the (1$\bar{1}$0) direction [Fig. 3(a)] and in-layer trajectories for layers 10,11 and 8,9, shown in Figs. 3(b) and 3(c), respectively, along with the density and per-particle potential energy profiles in the (111) direction [Figs. 3(d) and 3(e), respectively],

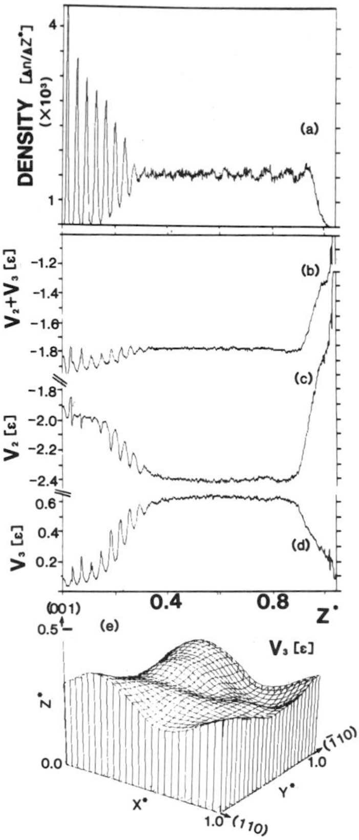

FIG. 2. (a) Equilibrium particle density $\Delta n/\Delta Z^*$, where Δn is the number of particles with Z^* coordinates between Z^* and $Z^*+\Delta Z^*$. (b)–(d) Per-particle potential-energy profiles vs distance along the [001] direction ($Z^*=1\equiv 18.14\sigma$). The total and the two- and three-body potential energies are shown in (b), (c), and (d), respectively. (e) A contour plot of V_3 for particles with $V_3 \leq 0.45\epsilon$, exhibiting the solidlike region of the sample, recorded at the same time as the trajectories given in Fig. 1.

provide clear evidence for a sharp, abrupt, and flat Si(111)-melt interface. Additionally, the melt region adjacent to the interface ($L=10$ and 11) possesses a residual degree of in-planar order.

To compare the predictions of the molecular-dynamics simulations with an experimental system, we have observed the solid-melt interface *in situ*, during melting with a shaped beam of a cw argon laser. The starting material was a 600-nm-thick film of polycrystalline silicon deposited on a quartz wafer by chemical vapor deposition and capped by a layer of silicon dioxide. The sample was mounted on a translation stage positioned upon a vibration-isolated table, and the liquid-solid interface was observed by means of a microscope equipped with a visible and near-infrared video camera. Two laser beams, shaped into elliptical cross sections, and positioned in close spatial proximity, were used to maintain a closely controlled temperature gradient across the molten zone. The size of the molten zone was approximately $25\times 100\ \mu m^2$. After an initial scan to convert the polycrystalline material into a monocrystalline ribbon, the scanning was stopped to establish quasiequilibrium temperature distribution across the molten zone. A view of the solid-melt interface is shown in Fig. 4. The recrystallized film has predominantly (100) texture, as has been confirmed by subsequent x-ray diffraction. In the direction normal to the original scan, small periodic deviations from the ⟨100⟩ axis occur, with a characteristic length of approximately 10 μm. The various sets of facets defined by alternating (111) and $(\bar{1}\bar{1}1)$ planes seen in Fig. 4 appear slightly tilted with respect to each other so as to preserve the relative alignment with the bulk texture. The structure shown in Fig. 4 is typical of many faceted configurations that were observed. The shape of the solid-melt interface fluctuates, as certain facets spontaneously disappear and new facets form. Individual facets typically persist for several video-frame times ($\frac{1}{30}$ s).

In comparison of the experimental and simulation results, the similarity in appearance is rather striking, particularly when one notices that the two differ vastly in spatial and temporal scales. On the foundation of the premise that the origins of all physical phenomena are microscopic in nature, we venture the hypothesis that above a certain size the energetics that govern the structure of the solid-melt interface in this system operate in a similar manner in both the microscopic (theory) and macroscopic (experiment) regimes, and that the dimensions of the morphological characteristics scale with system size. Clearly, the dimensions of the computational cell limit the size of facets that form. For a small facet, the relative number of edge and corner atoms to those located on the facet plane is high, thus affecting its stability. While analysis of the results obtained for the two system sizes which we simulated provides some support to this hypothesis, more theoretical and experimental work is needed. Such theoretical efforts would involve further, extended simulations and the development of improved microscopic theories of interfacial energetics, dynamics, and morphological stability.[10] On the experimental side, it would be desirable to observe the solid-melt interface on much smaller spatial and temporal scales to record the process of facet formation in greater detail.

The structure of the interface is governed by the en-

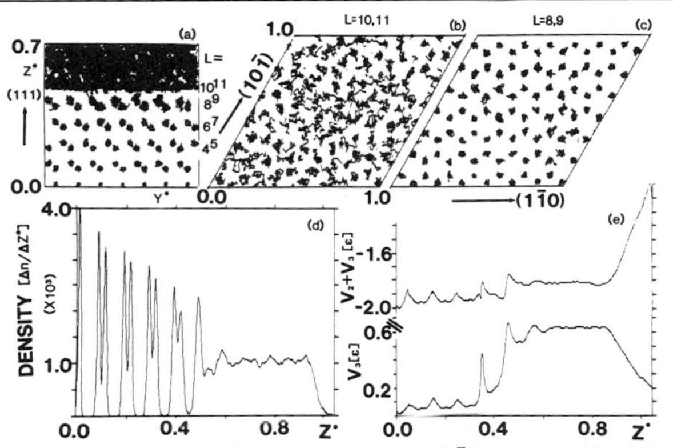

FIG. 3. (a) Particle trajectories recorded at equilibrium, viewed along the $[1\bar{1}0]$ direction. (b)–(c) In-plane trajectories, exhibiting solid and melt characteristics, in layers 10,11 and 8,9, viewed along the $[111]$, Z^*, direction. $Z^* = 1 = 14.96\sigma$ and the unit of length in the $[1\bar{1}0]$ and $[10\bar{1}]$ directions is 12.85σ. (d) Equilibrium particle density, $\Delta n/\Delta Z^*$, vs Z^*. (e) Per-particle total and three-body potential energies, $V_2 + V_3$ and V_3, vs Z^*.

ergetics, particularly the interplay between the two- and three-body contributions to the potential energy, and by entropic factors. Our observation that faceting initiates upon melting and then further refines upon achieving equilibrium is most likely a result of the fact that the closest-packed (111) planes of silicon are the directions of slowest growth.[11] In a dynamic equilibrium situation these faces are also the slowest to melt. An important issue in investigations of the recrystallization kinetics of zone-melted silicon is the role of the interface morphology in the generation of low-angle grain boundary defects, whose branching behavior was rather well described by a recent kinetic model,[1] with the assumptions of a faceted structure of the growth interface and that the growth rate is limited only by the nucleation rate of new monolayers. Having established the faceted equilibrium structure of the interface, including the partial ordering in the nearby melt region, we expect that further studies will allow us to elucidate the microscopic kinetics and dynamics of growth and subboundary formation.

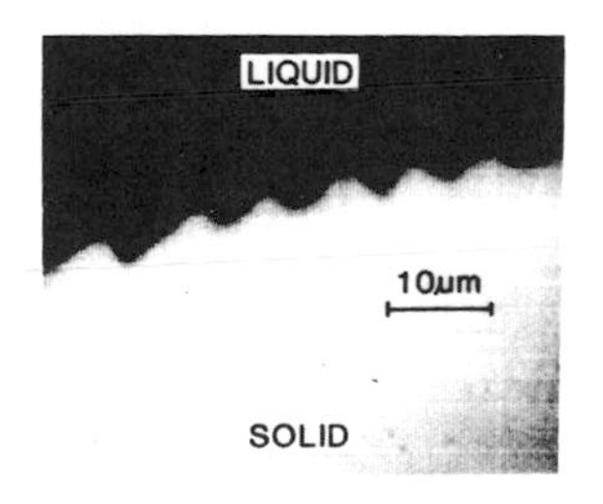

FIG. 4. *In situ* view (on a video monitor) of the solid-melt interface, in quasiequilibrium, exhibiting (111)-faceted structure.

The theoretical calculations were supported, in part, by U. S. Department of Energy Contract No. EG-S-05-5489 and were performed on the Cray-XMP at the the National Magnetic Fusion Energy Computer Center, Livermore, California.

[1]L. Pfeiffer, S. Paine, G. H. Gilmer, W. van Saarloos, and K. W. West, Phys. Rev. Lett. **54**, 1944 (1985).

[2]G. K. Celler, K. A. Jackson, L. E. Trimble, Mc. D. Robinson, and D. J. Lischner, in *Energy Beam–Solid Interactions and Transient Thermal Processing,* edited by J. C. C. Fan and N. M. Johnson (North-Holland, New York, 1984), p. 409.

[3]D. P. Woodruff, *The Solid–Liquid Interface,* (Cambridge Univ. Press, Cambridge, England, 1973), Chap. 3.

[4]M. W. Geis, H. J. Smith, B.-Y. Tsaur, J. C. C. Fan, D. J. Silversmith, and R. Mountain, J. Electrochem. Soc. **129**, 2813 (1982).

[5]K. F. Lee, T. J. Stultz, and J. F. Gibbons, in *Semiconductors and Semimetals,* edited by R. K. Willardson and A. C. Beer (Academic, New York, 1984), Vol. 17, p. 227.

[6(a)]F. F. Abraham, J. Vac. Sci. Technol. B **2**, 534 (1984).

[6(b)]U. Landman, R. N. Barnett, C. L. Cleveland, and R. H. Rast, J. Vac. Sci. Technol. A **3**, 1574 (1985).

[7]F. H. Stillinger and T. A. Weber, Phys. Rev. B **31**, 5262 (1985).

[8]R. N. Barnett, C. L. Cleveland, and U. Landman, Phys. Rev. Lett. **54** 1679 (1985).

[9]M. Parinello and A. Rahman, Phys. Rev. Lett. **45**, 1196 (1980).

[10]J. S. Langer, Rev. Mod. Phys. **52**, 1 (1980); R. F. Sekerka, Physica (Amsterdam) **12D**, 212 (1984).

[11]G. H. Gilmer, Mat. Res. Soc. Proc. **13**, 249 (1983); see also Ref. 6b.

ARTICLES

Comparison of the molecular dynamics method and the direct simulation Monte Carlo technique for flows around simple geometries

Eckart Meiburg[a)]
Institute for Theoretical Fluid Mechanics, Bunsenstrasse 10, D-3400 Göttingen, West Germany

(Received 15 May 1985; accepted 3 July 1986)

The molecular dynamics method (MD) and the direct simulation Monte Carlo technique (DSMC) are compared with respect to their capability of simulating vorticity distributions. The statistical assumptions underlying the evaluation of the collision term by the DSMC are analyzed. They lead to the nonconservation of angular momentum for the interaction of particles. Both methods yield equally good results for the Rayleigh–Stokes flow. Using the present parameters, however, only the MD simulation shows the generation of vortices for the flow past an inclined flat plate. This might indicate an effect on the computation of vortical flows. It is suggested that further systematic studies of the effect of the cell size and particle dimensions be carried out.

I. INTRODUCTION

The use of increasingly sophisticated satellites, launch vehicles, and structures in space has been stimulating research on efficient means of transportation among orbits of different altitudes and inclinations (for an overview see Walberg[1]). The payload capacity of such vehicles can be increased significantly by using aerodynamic forces during one or more passes through the atmosphere instead of all-rocket propulsion, leading to the concept of aero-assisted orbit transfer vehicles. As can be seen from Fig. 1, these vehicles will operate under free molecular flow conditions as well as under transitional and continuum flow conditions; the various regimes being characterized by the Knudsen number. Thus there is a need for efficient computational methods for the solution of Boltzmann's equation, which should be able to provide flowfield calculations from small Knudsen numbers through Knudsen numbers larger than unity. According to Bird,[2] for characteristic dimensions such as those that are of interest in flows around space vehicles, the continuum approach breaks down before significant fluctuations set in. This leads us to expect structures in the flowfield even where the Navier–Stokes equations are no longer valid, so we look for computational methods with which we can calculate collisionless flows in the high Knudsen number limit, as well as structures such as vortices in the low Knudsen number limit, together with the related forces and heat transfer rates. The present study investigates the applicability of both the direct simulation Monte Carlo technique described by Bird[2] and the approach first introduced by Alder and Wainwright[3] under the name of molecular dynamics. We want to elucidate the advantages and limitations of these computational methods by simulating flowfields around bodies of simple geometries for various Knudsen and Mach numbers. Our interest focuses on the question of whether or not basic fluid mechanical phenomena such as boundary layers or vortices can be described correctly.

II. MOLECULAR DYNAMICS METHOD VERSUS DIRECT SIMULATION MONTE CARLO TECHNIQUE

Both methods can be used for the simulation of a gas governed by the Boltzmann equation

$$\left(\frac{\partial}{\partial t} + \mathbf{c}\,\frac{\partial}{\partial \mathbf{r}} + \mathbf{F}\,\frac{\partial}{\partial \mathbf{c}}\right) f(\mathbf{r},\mathbf{c},t) = \frac{\partial f}{\partial t_{\text{coll}}},$$

which determines the particle distribution function f. Here $\mathbf{r}$ denotes the space vector, $\mathbf{c}$ the velocity vector, and $\mathbf{F}$ an external force acting on the particles. The term on the right-hand side takes the interaction among these particles into account. The simulation methods are based on the fact that this equation can also be formulated for a normalized distribution function, if at the same time the collision cross section of the particles is normalized correspondingly, so that solutions of the Boltzmann equation become independent of the number of particles. For a review see, for example, Derzko.[4]

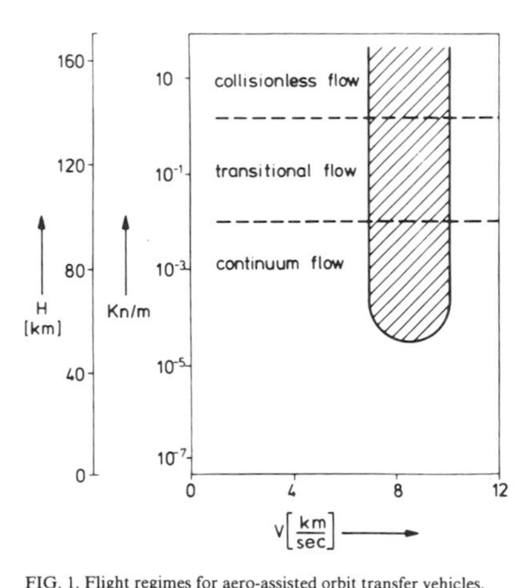

FIG. 1. Flight regimes for aero-assisted orbit transfer vehicles.

[a)] Present address: Department of Chemical Engineering, Stanford University, Stanford, California 94305.

 0031-9171/86/103107-07$01.90

This allows us to employ several thousand particles, all moving with their respective velocities and interacting according to the potential model chosen instead of the much higher number encountered in the real flow. In 1957, Alder and Wainwright[3] first applied their molecular dynamics method, and later, Bird[2] developed a more efficient statistical approach, which became known as the direct simulation Monte Carlo technique (DSMC) and has been applied to a wide variety of free molecular and transitional flows. In the following, a brief description of both methods is given.

A. Molecular dynamics simulation procedure

The calculation starts with the positioning of a given number of particles into the control volume. The initial positions of the particles depend on the physical problem under consideration, which means that they can be randomly distributed or prescribed explicitly. Similarly, the initial velocities are related to the physical problem; they can, for example, have a Maxwell distribution generated with the help of a random number generator. From this point on, the particles move with their individual velocities, subject to certain boundary conditions and to the law of interaction among one another. For an arbitrary particle potential, the calculation proceeds in very small time steps during which the particles move according to their velocities and the forces exerted upon them by the other particles. This force is calculated at the beginning of each time step, so that small steps are required. For the special case of a hard sphere potential, which has been used exclusively in the calculations reported here, particles do not interact except when their paths approach to within one diameter. This allows the time step to be as large as the time span between two subsequent collisions. Therefore, we must first calculate the time of the next collision in the flowfield by examining all particle pairs, then advance all particles up to this time, and finally compute the new velocities of the two particles involved in the collision by means of the laws of classical mechanics. When a particle approaches the boundaries of the control volume, different measures may be taken, varying according to the physical problem. The particle can be specularly reflected, which corresponds to a symmetry boundary condition, or it can be diffusely reflected, thus simulating a rigid wall. In this case, a new velocity would be calculated for the particle from the temperature of the wall. We can also apply an accommodation coefficient, which takes into account the extent to which the post-collision particle velocity depends on its pre-collision history. A periodic boundary condition can be simulated by placing the particle back into the flowfield close to the opposite boundary. Macroscopic quantities can be calculated at arbitrary times by sampling the properties of the particles.

As Alder and Wainwright already mentioned in their original paper, improving the computational procedure of selecting the collision partners and carrying out the collisions leads to a considerable speed-up of the calculation. Modern vector computers such as the CRAY-1S, on which all calculations reported here were performed, offer new opportunities in this regard (for details see Meiburg[5,6]). The molecular dynamics simulation procedure has the advantage of being a grid-free method. This allows the treatment of relatively complicated configurations, since in principle the only requirement to be satisfied by the geometry of the control volume or of a body in the flow is that it must allow us to calculate the time at which a particle will cross a boundary when its position and velocity are known. Thus it is easily feasible to calculate, for example, flows around satellites, as long as their shapes can be composed of cylinders, cones, planes, and other basic geometries.

B. Direct simulation Monte Carlo technique (DSMC)

This technique is similar to the molecular dynamics method in that it computes the trajectories of a large number of particles and calculates macroscopic quantities by sampling particle properties, but it has the advantage of being more efficient on electronic computers. It has been successfully applied to the simulation of a large variety of rarefied gas flows. A detailed description is given in Bird.[2] The DSMC differs from the molecular dynamics approach in the way interactions among the particles are treated. The place and time of a collision are no longer determined by comparing the trajectories of all particles, but by means of a statistical consideration. The principle steps of the DSMC are as follows: At the beginning of the calculation, the flowfield is divided into a net of cells. The particles are positioned into the cells in the same way as in the molecular dynamics method. The computation now proceeds in small discrete time steps Δt over which the motion of the particles and their interactions are uncoupled. Within one time step, all particles are first advanced according to their individual velocities, with the boundaries taken into account as in the molecular dynamics method. Then in each cell a certain number of statistical collisions is computed in the following way: From all the particles in the cell, two are selected randomly, without consideration of their actual positions. Then a "collision" between these two particles is computed as described below. It was shown by Bird[2] that this collision corresponds to a time increment

$$\Delta t_{\text{coll}} = 2/N\pi d^2 n c_{\text{rel}},$$

for hard sphere particles that have been used exclusively here. In this equation, N is the number of particles in the cell, πd^2 is the collision cross section of the particles, n represents the number density, and c_{rel} denotes the relative velocity of the particles involved in the collision. Further collisions are carried out until the sum of the time increments Δt_{coll} has reached the size of the time step Δt for each of the cells. For a sufficiently large number of particles per cell, this procedure yields the correct collision rate. The time step Δt should be small compared to the mean collision time, and the cell dimensions should be small compared to the mean free path, in order to yield accurate results.

A problem that arises from the calculation of "statistical collisions" will be discussed in the following. In general, the six post-collision velocity components u_1', v_1', w_1' and u_2', v_2', w_2' of two particles of mass m and pre-collision velocity components u_1, v_1, w_1 and u_2, v_2, w_2 undergoing an elastic collision are completely determined by:

(a) conservation of linear momentum per mass

Note on p. 455

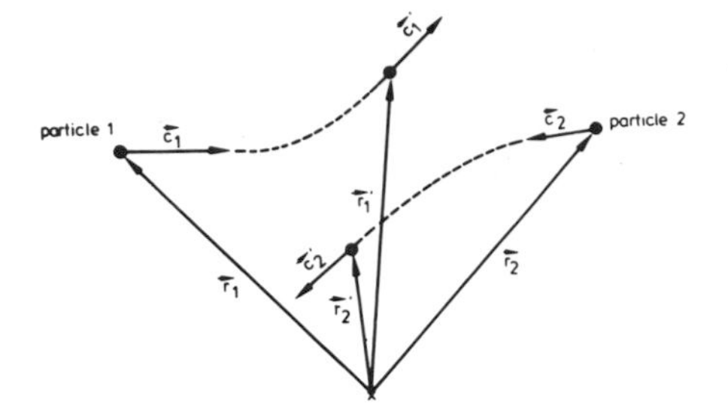

FIG. 2. Particle position and velocity vectors for a binary collision.

$$u_1' + u_2' = u_1 + u_2, \quad v_1' + v_2' = v_1 + v_2,$$
$$w_1' + w_2' = w_1 + w_2,$$

(b) conservation of kinetic energy per mass

$$u_1'^2 + v_1'^2 + w_1'^2 + u_2'^2 + v_2'^2 + w_2'^2$$
$$= u_1^2 + v_1^2 + w_1^2 + u_2^2 + v_2^2 + w_2^2,$$

(c) conservation of angular momentum per mass

$$\mathbf{r}_1' \times \mathbf{c}_1' + \mathbf{r}_2' \times \mathbf{c}_2' = \mathbf{r}_1 \times \mathbf{c}_1 + \mathbf{r}_2 \times \mathbf{c}_2,$$

(d) particle potential ϕ

$$\frac{\ddot{\mathbf{r}}_1}{m} = -\frac{\ddot{\mathbf{r}}_2}{m} = -\frac{d\phi}{ds},$$

where $\mathbf{r}_1, \mathbf{r}_2, \mathbf{r}_1', \mathbf{r}_2'$ are the pre- and post-collision particle position vectors (Fig. 2) and s is the distance between the particles. Since in the DSMC we try to calculate a collision between particles without considering their positions, the particle potential cannot be used to compute the post-collision velocities directly. But it can be shown that in a statistical sense, for the hard sphere particle potential, all directions are equally probable for the post-collision relative velocity of the two particles. This leads to the following method of computing collisions in the DSMC. First, the direction of the post-collision relative velocity is determined by randomly selecting an azimuthal and a polar angle. Then the four degrees of freedom remaining are used to satisfy the conservation of linear momentum and energy. Now the conservation of angular momentum can no longer be satisfied. Thus the DSMC yields both post-collision velocities sketched in Fig. 3

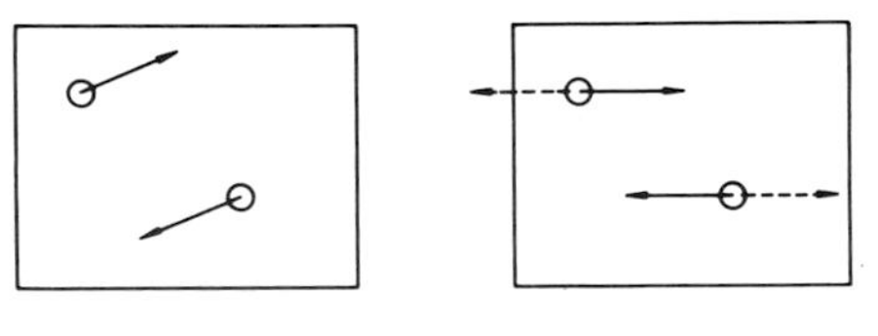

FIG. 3. Pre- and possible post-collisional velocities of two particles in a cell. The DSMC does not conserve angular momentum, since both the solid and the dashed velocity vectors present equally possible post-collisional particle velocities.

with equal probability. This is obviously incorrect, leading us to suspect that the DSMC might not correctly describe flowfields in which angular velocities are of importance, such as vortical flows. The nonconservation of angular momentum results from the fact that we are trying to calculate an interaction between two particles without considering their actual position. As long as this principle is employed, the effect of violating the law of conservation for angular momentum will therefore be seen, even for other particle potentials than the hard sphere. Obviously, the error depends on the distance between the interacting particles, and since these are always selected from the same cell, the error is expected to depend on the cell size. In principle, however, it will remain present even for small cells.

III. RESULTS

A. Impulsively started flat plate

As a test example for the two simulation methods described above we selected the flow induced by an infinite flat plate impulsively started from rest in its own plane in a compressible fluid, which had previously been considered by Bird.[2] Van Dyke[7] treated this problem by means of matched asymptotic expansions. Becker[8] generalized the boundary layer solution with respect to suction and blowing as well as to a change in the temperature of the plate. If ρ_0 is the density at the wall and y is the distance from the wall we can define the coordinate transformation

$$y' = \int_0^y \frac{\rho}{\rho_0} dy.$$

For u_w denoting the wall velocity, t the time, and ν_0 indicating the viscosity at the wall we obtain, for large Reynolds numbers, the similarity solution for the dimensionless velocity having the form

$$u/u_w = \text{erfc}\, \eta,$$

where

$$\eta = y'/2\sqrt{\nu_0 t}\ .$$

Figure 4 compares the numerical results from both a MD calculation and a DSMC simulation each employing

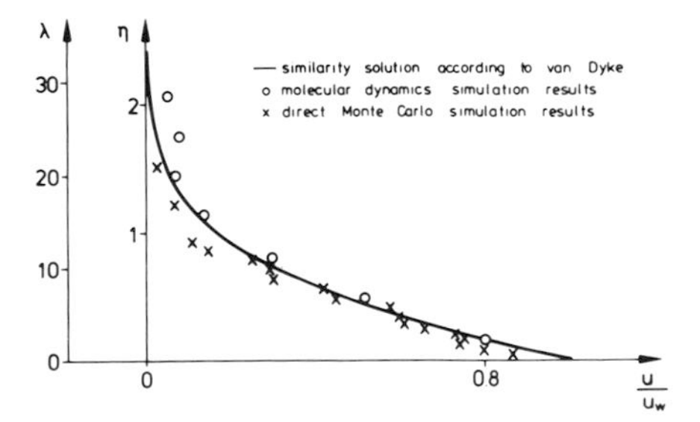

FIG. 4. Molecular dynamics and direct Monte Carlo simulation results compared to the exact solution for the impulsively started flat plate, Re = 85. Here η denotes the similarity variable and λ indicates the distance from the wall in mean free paths.

20 000 particles with the exact solution. We find that, apart from statistical scatter, the values agree quite well, provided that the dimension perpendicular to the plate of the cells in the DSMC simulation is small enough. As long as that is the case, then the fact that angular momentum is not conserved does not have a strong impact on the results. This can be explained by an analysis of the rate at which u-momentum is transferred from the wall to the flow. If the cell size is large compared to the mean free path of the flow, a particle that has just been reflected from the wall can be selected to collide with another particle in the same cell but relatively far away, and with a u-velocity considerably smaller than the velocity of the wall. Since angular momentum is not conserved, both particles will, on the average, have the same u-velocity after the collision. This results in an unrealistic deceleration of the particle close to the wall. As this particle is likely to hit the wall much earlier than the one farther away, the rate of transfer of u-momentum to the flow will be too high. If, on the other hand, the cell size close to the wall is small compared to the mean free path, a particle that has just been reflected from the wall will undergo the next collision four or five cells away from the wall. This in turn means that it is approximately the same distance from the wall as its collision partner, so that, on the average, both collision partners will hit the wall at the same time. Thus it is unimportant which one of them has the higher, and which one the lower u-velocity after the collision, so that in this case the conservation of angular momentum is without significance for the rate of transfer of u-momentum. With the rate of transfer of momentum being computed accurately, the DSMC can be expected to yield accurate results for the flowfield described above.

B. Vortex shedding behind an inclined flat plate

The process of vortex shedding is a classical problem in fluid mechanics. Experimentalists as well as theoreticians have carried out numerous investigations to shed light on different aspects of the problem, such as the generation and separation of vortices, the stability of vortex streets, the transition to turbulent wakes, the forces involved, or the possibilities of manipulating or suppressing the shedding process. For reviews of this topic, see for example, McCroskey,[9] Bearman and Graham,[10] and Bearman.[11] Most studies have been concerned with the incompressible flow around bodies of simple geometries, such as circular cylinders, thin flat plates, rectangular prisms, or cones. For example, Goldburg, Florsheim, and Washburn[12] have shown that vortex shedding also occurs in hypersonic flows, so the phenomenon seems to bear some relevance for the flow around space vehicles in the upper atmosphere. It is unknown, however, at which Knudsen numbers we can expect vortex street wakes. This gives rise to the question as to whether computational techniques for the solution of the Boltzmann equation—upon which a calculation of such flowfields must be based—are able to describe the generation and behavior of vortices. This in turn is a prerequisite for obtaining correct values for the related forces such as those acting upon the flaps with which the Space Shuttle is maneuvered.

As a model problem we studied the flow around a flat plate at 45° incidence, by means of an MD simulation as well as a DSMC simulation. The length of the plate d was 60 mean free paths, resulting in a Knudsen number of 0.017. The Mach number was 0.7, and the Reynolds number based on the length of the plate was 78. Both experiments and theoretical considerations suggest that the flow is unstable at this Reynolds number. The Knudsen number corresponds to the flow around a flat plate of 0.4 m at an altitude of 100 km. In order to avoid difficulties with the downstream boundary condition, the plate was impulsively started from rest and then towed through the three-dimensional control volume at constant speed, as shown in Fig. 5. Symmetry conditions were applied at all boundaries, so that no boundary layers form and the blockage effect is less severe than for a solid wall. Experiments have shown that blockage effects influence the onset of vortex shedding in incompressible low Re-number flows (Shair *et al.*[13]), but it is unknown so far how important blockage effects are in compressible low Re-number rarefied flows. When a particle collided with the plate, it was given a new velocity according to a Maxwellian distribution formed with the temperature of the plate. This corresponds to diffuse reflection of the particles at the plate, i.e., the accomodation coefficient was taken as 1.

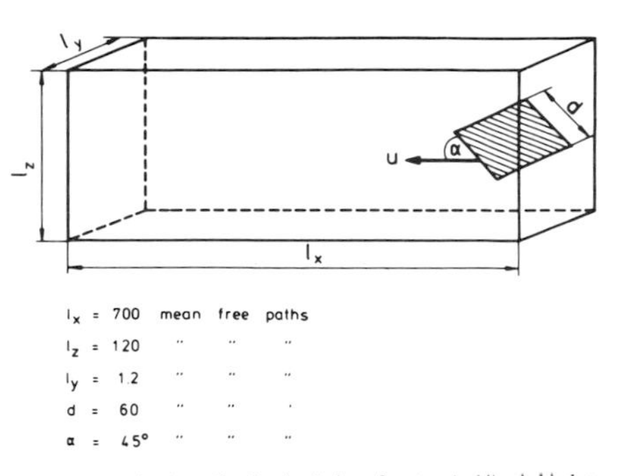

FIG. 5. Control volume for the simulation of vortex shedding behind an incline flat plate.

The MD calculation was carried out in a fully three-dimensional way. Because of CPU time and storage constraints, the number of particles had to be limited to 40 000 for this simulation, resulting in the relatively large molecular diameter of 0.753 mean free paths. This means that a fraction of approximately 8.9% of the control volume was occupied by the particles. In contrast to the MD simulation, which for the calculation of realistic collisions has to consider the positions of the particles in both x, y, and z direction, the DSMC simulation can be carried out in two dimensions as a consequence of the fact that it deals with a statistical set of collisions. The higher computational efficiency of the DSMC allowed us to employ 210 000 particles for the simulation of the flow past the inclined flat plate. The hard sphere potential was applied in both cases. The results had to be smoothed in order to be able to illustrate the flowfield by means of instantaneous streamlines and lines of constant

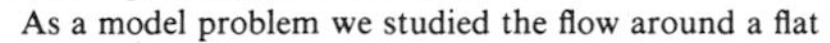

Note on p. 455

vorticity as well as plots of velocity vectors. These continuum–mechanical data were obtained by averaging the properties of particles in a net of cells, with the depth of the cells for the MD simulation being equal to the depth of the control volume. This means that, although the simulation was carried out in three dimensions, the results discussed in the following were obtained by integrating over the spanwise direction. In order to give an impression of the direction of the flow everywhere in the field, we integrated piecewise streamlines from a given grid of starting points. Their strong curvature in some regions indicates the presence of vortices, whereas their length does not have any physical meaning. The magnitude of the velocity is proportional to the length of the velocity vectors. Lengths have been nondimensionalized by the mean free path, and time has been made dimensionless with the length and the velocity of the plate.

First we will discuss the results of the MD simulation. Soon after the start of the plate, at time 0.36, the streamlines in the reference frame moving with the plate clearly follow the geometry of the plate (Fig. 6). Although the flow seems to have separated at the top, a vortex is not yet visible. Due to the inevitable statistical scatter, the exact position of the separation point cannot yet be recognized. The scatter is responsible for some streamlines' ending on or leaving from the plate. The separating streamline at the rear tip has not yet become tangential to the plate. The velocity vectors show a strong deceleration of the flow immediately behind the plate. At time 1.07, we can recognize for the first time a region of recirculating flow on the upper side of the plate (Fig. 7). The front stagnation point is close to the tip, and separation occurs immediately behind the upper tip. The vorticity distribution now shows a maximum close to the center of the vortex. The separating streamline at the lower tip is now almost tangential to the plate. A little later at time 2.31, we notice that the vortex has increased in strength with its cen-

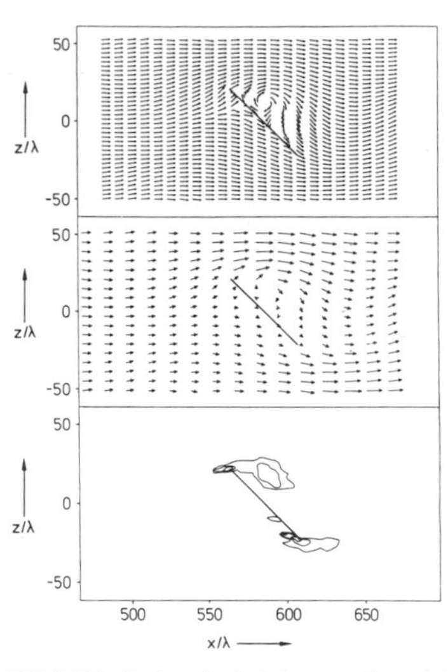

FIG. 7. Molecular dynamics simulation: streamlines, velocity vectors, and contours of constant vorticity at time 1.07.

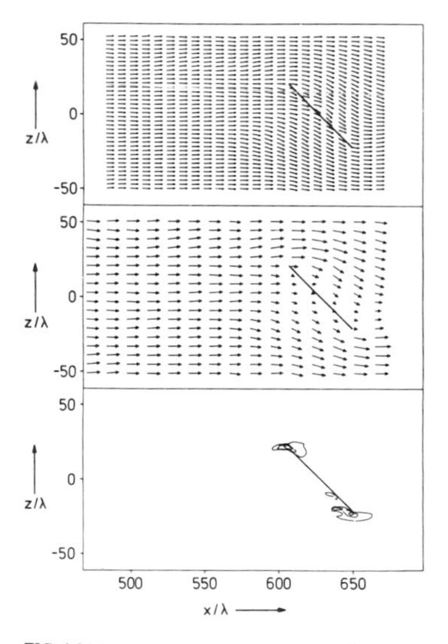

FIG. 6. Molecular dynamics simulation: streamlines, velocity vectors, and contours of constant vorticity at time 0.36.

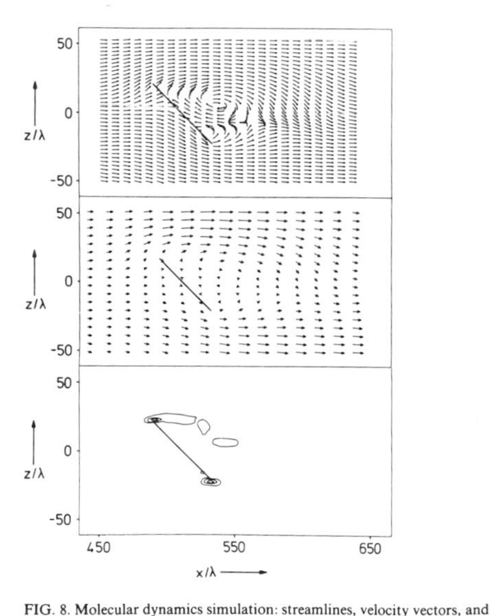

FIG. 8. Molecular dynamics simulation: streamlines, velocity vectors, and contours of constant vorticity at time 2.31.

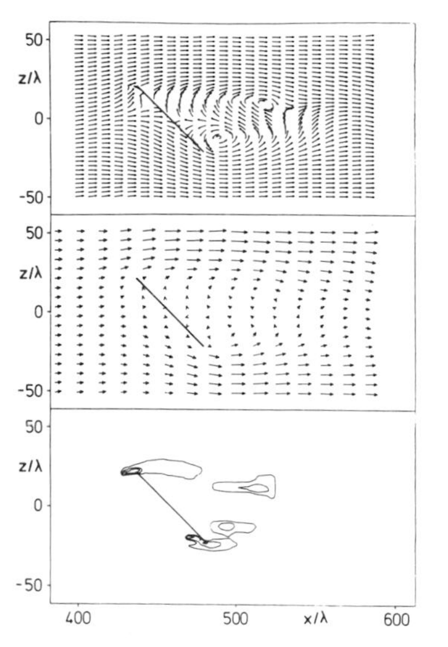

FIG. 9. Molecular dynamics simulation: streamlines, velocity vectors, and contours of constant vorticity at time 3.19.

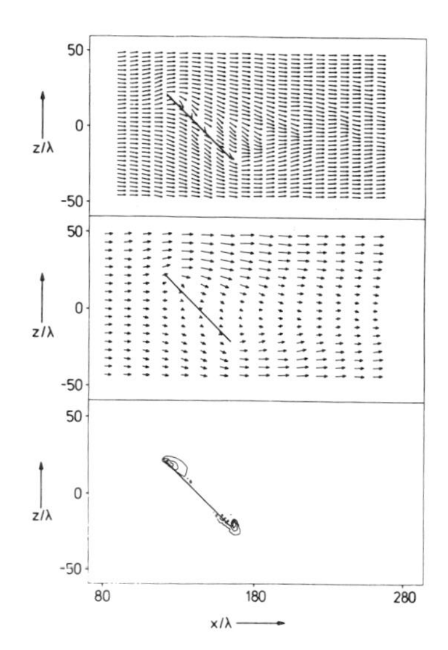

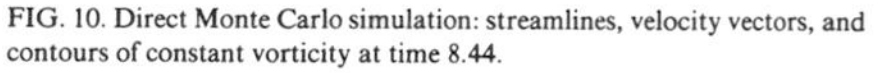

FIG. 10. Direct Monte Carlo simulation: streamlines, velocity vectors, and contours of constant vorticity at time 8.44.

ter having moved downstream, thus causing a second region of recirculating flow to form close to the lower tip (Fig. 8). At time 3.19, the region has grown to a full vortex, while at the same time the center of the initial vortex has moved further downstream. The maxima of the vorticity distribution again correspond to the centers of the vortices (Fig. 9). From this point on, periodic vortex shedding can be observed with alternating left-and right-rotating vortices forming behind the tips of the plate and moving downstream. The Strouhal number formed with the projection of the plate perpendicular to the oncoming flow is 0.25. For a Reynolds number as low as 78, this seems to be too high when compared with experimental results for incompressible flows. One reason for this discrepancy probably lies in the fact that the channel is fairly narrow, thus causing a strong acceleration of the flow above and below the plate, which in turn leads to separation of the vortices before they reach their full size.

The corresponding two-dimensional DSMC simulation yielded quite different results. The number of particles per cell was taken as 20 and the size of the cells came to approximately 2.9 mean free paths, which is small compared to the scale of the macroscopic vortices expected to be generated, but larger than the mean free path. Here too, at time 8.44 the flow follows the shape of the plate, and the maximum of the velocity production lies close to the tips of the plate (Fig. 10). However, it "diffuses" very quickly, so that no macroscopic vortices can form. The reason for this is believed to lie partly in the nonconservation of angular momentum. The vorticity that is continuously being generated at the surface of the plate as a consequence of the no-slip condition, i.e., diffuse reflection, decays in the process of the calculation of collisions. This means that close to the wall, there is still enough vorticity to cause local separation and recirculation of the flow, but the formation and separation of large scale vortices does not occur.

The difference between the MD and the DSMC results is further reflected in the streamline pattern of the whole flowfield at the end of each calculation. The pattern obtained from the MD calculation (Fig. 11) clearly shows the waviness of the wake indicating the presence of vortices moving downstream from the plate. The corresponding pattern of

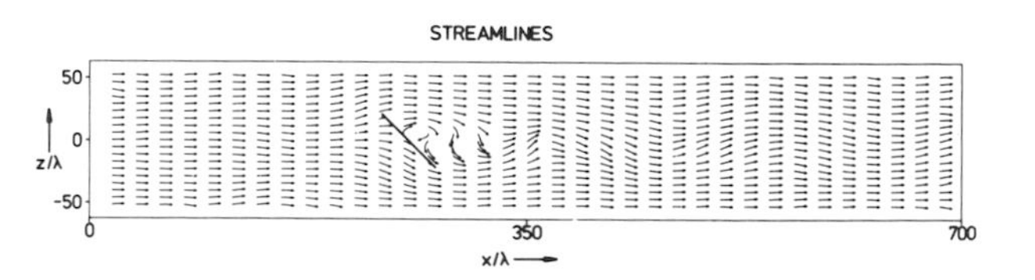

FIG. 11. Molecular dynamics simulation: streamline pattern at time 6.55.

Note on p. 455

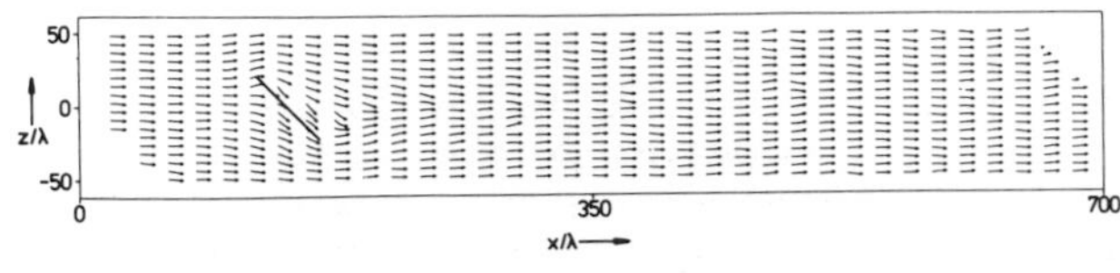

FIG. 12. Direct Monte Carlo simulation: streamline pattern at time 8.44.

the DSMC simulation (Fig. 12), on the other hand, does not show this waviness.

The degree to which angular momentum and vorticity are being "diffused" in the process of carrying out the collisions obviously can depend on the cell size as well as on the number and size of the particles used. The early destruction of angular momentum in the regions of high shear behind the tips of the plate might be reduced by using a locally refined mesh. The relatively large particle diameter in the MD simulation, on the other hand, results in the molecules' occupying a considerable fraction of the control volume, which could cause some dense gas effects and thus influence the diffusion of vorticity at the scale of the particle size. Therefore, a systematic study of the effect of the cell size in the DSMC on the nonconservation of angular momentum and vorticity should be carried out, and the influence of the number and size of particles on both the DSMC and the MD results presented here should be checked. It would be convenient to define a simpler test problem for this purpose, since on presently available computers it would be too costly to reduce the cell size used in the DSMC by a factor of, say, 10 and keep the number of particles per cell constant. This systematic study might help to clarify to what degree the loss of angular momentum depends on the numerical parameters such as the cell size and the number of particles. Even if the loss of angular momentum can be reduced by a large amount by using much smaller cells, the fact that a certain number of particles per cell has to be maintained could considerably reduce the numerical efficiency of the DSMC as compared to the MD for some transitional flows. This is of interest for future applications, which will certainly include calculations of flowfields around bodies in regimes where the mean free path is small compared to the dimensions of the body.

IV. CONCLUSIONS

Within the present study, two numerical methods for the simulation of Boltzmann's equation—the molecular dynamics method and the direct simulation Monte Carlo technique—have been compared with respect to their capability of calculating flows with strong gradients in the vorticity distribution. The goal was to study the effect of the statistical assumptions underlying the evaluation of the collision term in the Boltzmann equation by the DSMC technique, which lead to the nonconservation of angular momentum for the interaction of particles. Both methods yielded satisfactory results for the Rayleigh–Stokes problem. For the numerical parameters investigated here, however, the DSMC calculation for the flow past an inclined flat plate did not show periodic vortex shedding, whereas the MD simulation resulted in a vortex street, as was demonstrated by means of streamline-, vorticity-, and velocity-vector plots. The difference in the results is believed to be related to the loss of angular momentum on the particle level during the process of carrying out the collisions. Further work on a simpler test problem should be carried out in order to systematically investigate the effect of the cell size as well as the number and size of the particles on the loss of angular momentum and vorticity. Such a study would not only help to answer the question whether or not the DSMC in its present version has principle problems in simulating vortical flows, but would also allow conclusions with regard to the computational efficiency of both methods, which might favor the MD for some transitional flow problems if the DSMC requires cells that are too small. For large Knudsen number flows, however, there seems to be no doubt that, due to its higher numerical efficiency, the DSMC is to be preferred.

A possible next step may be the development of a hybrid scheme that is based on the MD approach where vortices are expected, and that works with the DSMC everywhere else. This would allow a combination of the advantages of both methods.

ACKNOWLEDGMENTS

I would like to express my thanks to Professor H. Oertel, who supervised this work, for his personal involvement and valuable advice that he provided during many helpful discussions, and to Dr. R. Kessler for giving a preliminary review of the manuscript.

[1]G. D. Walberg, AIAA Paper 82-1378, 1982.
[2]G. A. Bird, *Molecular Gas Dynamics* (Clarendon, Oxford, 1976).
[3]B. J. Alder and T. Wainwright, in *Proceedings of the International Symposium on Transport Processes in Statistical Mechanics* (Interscience, New York, 1957), p. 97.
[4]N. A. Derzko, UTIAS Rev. **35**, 1 (1972).
[5]E. Meiburg, in *Vectorization of Computer Programs with Application to Computational Fluid Dynamics* (Vieweg, Wiesbaden, 1984), p. 235.
[6]E. Meiburg, DFVLR Forschungsbericht **85**, 13 (1985).
[7]M. van Dyke, Z. Angew. Math. Phys. **3**, 343 (1952).
[8]E. Becker, Z. Angew. Math. Phys. **11**, 146 (1960).
[9]W. J. McCroskey, J. Fluid Eng. **99**, 8 (1977).
[10]P. W. Bearman and J. M. R. Graham, J. Fluid Mech. **99**, 225 (1980).
[11]P. W. Bearman, Annu. Rev. Fluid Mech. **16**, 195 (1984).
[12]A. Goldburg, B. H. Florsheim, and W. K. Washburn, AIAA J. **3**, 1332 (1965).
[13]F. H. Shair, A. S. Grove, E. E. Petersen, and A. Acrivos, J. Fluid Mech. **17**, 546 (1963).

Note to Reprint VI.3

The transport equation on p. 3107 is an exact result if the right-hand side is interpreted as the exact rate of change of $f(\mathbf{r}, \mathbf{c}, t)$ due to collisions between particles. The equation is then equivalent to the lowest-order member of the BBGKY hierarchy. In the literature of statistical mechanics, the name Boltzmann equation is usually reserved for the approximate, integro-differential equation obtained from the transport equation by assuming that collisions are strictly binary and invoking the assumption of "molecular chaos". In a molecular dynamics "experiment" on hard spheres, all collisions are treated exactly.

DYNAMICS OF THE A + BC REACTION IN SOLUTION

John P. BERGSMA, Pamela M. EDELSTEN, Bradley J. GERTNER, Kevin R. HUBER, Jeffrey R. REIMERS, Kent R. WILSON, Samuel M. WU
Department of Chemistry, University of California, San Diego, La Jolla, CA 92093, USA

and

James T. HYNES
Department of Chemistry, University of Colorado, Boulder, CO 80309, USA

Received 11 April 1985; in final form 31 October 1985

Molecular dynamics are computed for $A + BC \rightarrow AB + C$ in a rare gas solvent. Transition state theory is valid. For ± 0.02 ps about the barrier, reaction dynamics are essentially the same as without solvent. Reactive trajectories are translationally special over $\approx \pm 0.02$ ps, rotationally over $\approx \pm 0.5$ ps, and vibrationally over > 100 ps.

1. Introduction

The effects of the solvent on even the most simple chemical reactions are not well understood. Transition state theory (TST) [1,2] typically provides an excellent description of the kinetics of reactions in the gas phase but need not be generally applicable to reactions in the liquid phase [3]. For example, TST has been predicted to work well for high-barrier atom transfer reactions in a "weakly" interacting solvent, but not necessarily for charge transfer reactions (e.g. S_N2) for which there is strong interaction between the reactants and the solvent.

The thermodynamic formulation of TST clearly makes the fundamental assumption of the equilibrium nature of the transition state: the activated complex is supposed to exist in dynamic equilibrium with all possible important configurations of the system. This aspect at first sight appears to be easily satisfied in solution phase when frequent effective collisions between reactants and solvent occur. However, TST also (equivalently) assumes that an equilibrium distribution of particles crossing the energy barrier between reactants and products proceeds without any recrossings of the barrier top. If the solvent impedes the barrier passage through frequent collisions with reactants then recrossings can occur. Since TST regards every barrier crossing as being a successful reaction, it overestimates the reaction rate to an extent depending on the solvent-induced recrossing. Calculations of reaction rates in solution which take such recrossings into account can be based on a time correlation function approach [3–8]. In this approach the reactive flux through a transition state dividing surface is examined.

Molecular dynamics is a powerful and flexible tool for the study of solution reactions and is ideally suited to the evaluation of time correlation functions and other dynamical probes associated with reaction rates. For instance, the stable states picture (SSP) of Hynes, Northrup, and Grote [6,7] which is related to the treatments of Anderson [9] and Bennett [10], states that the reaction rate is determined by a net transition from within a stable reactant zone of phase space to a stable product zone of phase space. These trajectories pass through an intermediate zone which contains some saddle point on the potential surface and presents a barrier to the forward progression of the reaction. The transition state surface separates the reactant and product zones, and is located within this intermediate zone. To implement the SSP, molecular dynamics can be used to determine the history of any given reaction event through the intermediate zone and, although the barrier may be recrossed many times during a reactive trajectory, only the final outcome is counted toward the overall reaction rate. This approach also allows exploration of other aspects of the solvent's influence on the reaction dynamics. For example, reactants trapped in a solvent cage can undergo repeated collisions before finally reacting. In addition, the solvent may play a significant role in stabilizing the products by removing excess energy and preventing barrier recrossings, or it may accelerate the reaction by furnishing activation energy to the reactants.

The potential energy surface in the intermediate region is the gas phase potential, modified to include the interaction with the solvent. If the interaction is strong and the solvent density high, then the solvent can force the reactants together, favoring the more compact transition state over the separated reactants and, for example, lowering the activation energy [11].

To use molecular dynamics efficiently, trajectories are started from an equilibrium distribution on the transition state surface. They are run both forward and backward in time to determine the ensemble of trajectories which lead to successful reactions. This technique, first used in the gas phase by Keck [12] and by Anderson [9], is much more efficient than the straightforward approach of starting with an equilibrium distribution of stable reactants far from the transition state surface because successful reactions are rather infrequent when large activation energies are involved. It also has the advantage of revealing key post- and pre-reaction configurations and energetics of importance in the reactive process.

Our aim in this Letter is not to calculate absolute rates (although molecular dynamics can be used for this purpose) but rather to investigate the nature of the dynamics and corrections to TST rates. The motions of the atoms are calculated subject to the potential surface. The potential parameters can be varied at will to test the sensitivity of the dynamics to the details of the potentials. In this initial effort we consider one of the simplest systems possible, the weak solvent–solute interaction case with monatomic (rare gas) solvent. It is found that TST gives an accurate description. We find, however, that the liquid, even for this case, has a considerable effect on the reaction details: for example creating and destroying the excess translational, rotational and vibrational energy of the reactants and products respectively.

2. Calculation

We choose a very simple model reaction system with a high-energy reaction barrier and a weakly interacting solvent designed to test the qualitative features of solution reaction dynamics. In order that our calculations are performed for some chemically reasonable system, we choose the mass and separated molecule potential parameters that characterize a hypothetical reaction of Cl with Cl_2 in liquid argon, although the reaction, in particular the barrier height, is not intended as a model for a real $Cl + Cl_2$ reaction. For convenience we use the London–Eyring–Polanyi–Sato (LEPS) [13] potential surface for the $Cl + Cl_2 \rightarrow Cl_2 + Cl$ reaction, incorporating a 20.0 kcal/mol barrier height and the isolated molecule Cl_2 Morse potential [14]. The Ar–Ar and Cl–Ar interactions are described by the Lennard-Jones potential

$$V(r_{ij}) = 4\epsilon[(\sigma/r_{ij})^{12} - (\sigma/r_{ij})^{6}] , \quad (1)$$

where r_{ij} is the internuclear distance. The Lennard-Jones parameters used for Ar–Ar are ϵ/k = 119.8 K, σ = 3.405 Å. For Cl–Ar the Lennard-Jones parameters (calculated as the geometric mean of Cl–Cl parameters [15] and the Ar–Ar parameters [16]) are ϵ/k = 203 K and σ = 3.26 Å, where k is the Boltzmann constant. The solvent density used is 1.40 g/cm^3. Fifty solvent atoms are used in the molecular dynamics simulation.

We consider only the thermally relevant transition state near the lowest lying saddle point on the potential surface. In the gas phase the configuration at this saddle point is linear and symmetric with r_{Cl-Cl} = 2.254 Å for neighboring Cl atoms, and has an energy of −37.97 kcal/mol with respect to the ground electronic state dissociation energy of the isolated diatomic Cl_2 which is equal to 57.97 kcal/mol. We define the reaction coordinate as being the antisymmetric stretch vibrational normal mode a of the Cl_3 system, and the transition state surface is defined as the surface with a = 0. Note that the total potential energy is an even function of a for this symmetric reaction.

Notes on p. 460

Assuming that the reaction occurs at overall macroscopic thermal equilibrium, we may write the classical average value of any property A of the system as

$$\langle A\rangle = \frac{\int \mathrm{d}a\, \mathrm{d}p_a \int \mathrm{d}\boldsymbol{r}\, \mathrm{d}\boldsymbol{p}\, A \exp[-H(a,p_a,\boldsymbol{r},\boldsymbol{p})/kT]}{\int \mathrm{d}a\, \mathrm{d}p_a \int \mathrm{d}\boldsymbol{r}\, \mathrm{d}\boldsymbol{p} \exp[-H(a,p_a,\boldsymbol{r},\boldsymbol{p})/kT]}, \quad (2)$$

where $\boldsymbol{r}$ is a vector of all of the coordinates of the system except the reaction coordinate a, and $\boldsymbol{p}$ and p_a are the momenta conjugate to these coordinates, H is the full Hamiltonian of the system, and T is the temperature. We select p_a, $\boldsymbol{r}$, and $\boldsymbol{p}$ from a Boltzmann distribution but insist that all trajectories are initiated on the transition state surface, defining time such that at $t = 0, a = 0$. (This restriction is implied but not explicitly noted in eq. (2).) The algorithm used to do this makes some approximations which are not expected to be important for the case of a weakly interacting solvent. A discussion of this and other possible algorithms will be presented elsewhere [17]. An important feature of this algorithm is that it generates pairs of initial conditions which differ only in the sign of all the momenta and makes the greatest use of any other symmetry which may be present in the Hamiltonian. Trajectories are produced by propagating all atoms both forward and backward in time from the initial conditions which begin at time zero on the transition state surface. One hundred and twenty-six trajectories are used in the evaluation of eq. (2).

3. Results and discussion

One convenient and revealing way to characterize the dynamical role of the solvent in the reaction is to examine the details of how energy is exchanged between the Cl_3 reaction system and argon. Eq. (2) is used to calculate the average energy of the Cl_3 system at various times during the reaction. The average total energy of Cl_3 can be partitioned into the modes of Cl_2 vibration, rotation and translation and the translational energy of the Cl atom. This partitioning is not uniquely definable, and reflects the lack of chemical precision in defining reactants and products in the transition state region. We choose to conveniently define the two closest Cl atoms as being diatomic "Cl_2" and the other atom as being "free". The total vibrational energy is then taken to be the sum of the vibrational kinetic energy of these closest two atoms and the total LEPS potential energy of all three atoms. (There is of course interchange of vibrational energy with the solvent.) The zero of the LEPS potential energy is zero vibration in Cl_2. When one atom becomes well separated from the other two this definition becomes chemically meaningful and the LEPS potential reduces to just the bound Cl_2 Morse potential.

Fig. 1 shows the energy distribution of Cl_3 at T =

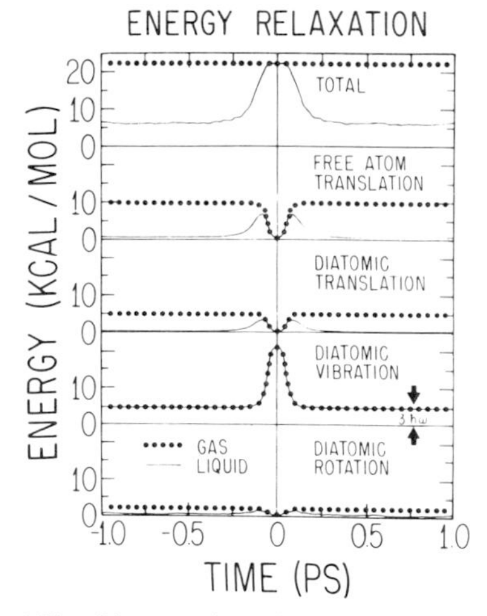

Fig. 1. Plot of the average internal energy of Cl_3 versus time for the reaction $Cl + Cl_2 \rightarrow Cl_2 + Cl$ for an ensemble of 126 reacting trajectories. Time zero is when the Cl_3 is first released at the saddle point. The solid lines show the average energy for the reaction in liquid argon solvent while the dots indicate the average energy for a reaction in the absence of solvent (gas phase). The translational energy arises from and decays to the solvent in 0.25 ps, rotational in about 1.0 ps and, although the figure does not show this, vibrational in >100 ps. The rise and fall on either side of the barrier crossing in vibrational energy is due to a loss of potential energy as the Cl_3 arises from and dissociates into the dimer Cl_2 and the free Cl atom. The excess vibrational energy of the Cl_2 is approximately $3\hbar\omega$, where ω is the ground state vibrational angular frequency of Cl_2. Note that the average energy for the liquid phase reaction is constant for the time period −0.02 ps to 0.02 ps during which the gas and liquid phase reaction trajectories are almost the same.

300 K for the reaction in both gas phase and in liquid argon solvent. From examination of the plot it is evident that translational energy arises from and is dissipated into the solvent from the Cl_3 in about 0.25 ps. On a somewhat longer time scale, about 0.5 ps, rotational energy arises from and is absorbed by the solvent, although it is hard to see on the scale of the figure. Finally, although this is not shown explicitly on the plot, vibrational energy is transformed on a much longer time scale, measuring at least one hundred picoseconds. The solvent is seen to provide the required activation energy in a manner analogous to the way it dissipates the energy after the reaction is complete.

Once the saddle point is reached, the reaction proceeds over the barrier top unimpeded by the solvent: in particular, recrossings are not observed in our sample. Two effects are important in this connection. First, the products are rapidly energetically stabilized by dissipating sufficient energy to the solvent and excess energy required for barrier recrossing is removed. Second, the reaction barrier is sharply peaked in energy as a function of the reaction coordinate and the reactants spend little time at the saddle point relative to the time scale for solvent motion. The solvent simply does not have enough time to interfere with the reaction progress once the barrier top is reached and recrossings are rare. Thus, we find TST to be a good description for this case of a sharp barrier reaction in a weakly interacting solvent. This is in accord with previous theoretical predictions [18,19].

As we have seen, the solvent does have considerable effect on the details of the trajectories, even for this weakly interacting case. This may be explored further using a method suggested by Anderson [20]. Eighty reactive trajectories are interrupted at a time τ before crossing the barrier and the solvent molecules are then given random momenta selected from a Boltzmann distribution while keeping all coordinates the same. The ability of such a solvent momentum perturbed trajectory to cross the barrier is a measure of the solvent momentum role in the reaction. Fig. 2 shows the probability of successful reaction versus time τ between momentum perturbation and original trajectory barrier crossing. After 0.14 ps the solvent is able to divert all trajectories away from reaction. As a check on numerical precision in these calculations the eighty reactive trajectories are interrupted at a time τ before crossing the barrier and a

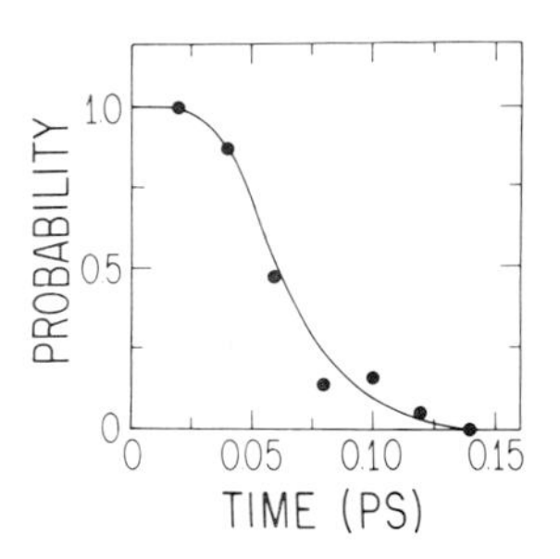

Fig. 2. Solvent momentum effect on reaction by taking an ensemble of 80 trajectories which react, going back a period in time before the barrier crossing given on the horizontal axis, and then while holding all coordinates constant, randomizing all solvent atom momenta with a Boltzmann distribution in velocities, and allowing the trajectory to resume. The probability on the vertical scale measures the fraction of trajectories which still lead to reaction. This gives a measure of time scale over which solvent momenta play a role in the reaction.

single random velocity increment is given to all the atoms in the system (i.e. Galilean transformation). All trajectories for all times τ are reactive after this transformation. Clearly, the solvent must participate actively in the reaction it is is to be succesful, and this suggests that the solvent distribution around the reactants must be considered when one analyzes the distribution of reactant states that leads to successful reactions. Note that the 100% diversion time is on the same time scale as the fastest translational energy relaxation process. Once the Cl_3 starts to lose energy to the solvent the associated entropy gain prevents the reaction from being reversed.

In this model reaction study a central dynamical role of the solvent is to furnish activation energy to the reactants and to similarly dissipate the excess energy of the products. The solvent interacts weakly with the short-lived transition state and, as shown in the numerical results and reflected in fig. 2, there is essentially unimpeded gas phase behavior for approximately the time period –0.02 ps to 0.02 ps during which the fate of the trajectory is decided. In fig. 1 one can see that the total energy of the Cl_3 remains constant during this time period for a reaction in both gas phase and liquid argon. It is in this region that TST is appropriately and successfully applied, even though

Notes on p. 460

for longer times the reaction in liquid phase bears little resemblance to a gas phase reaction. Thus, the essential qualities of TST for the rate constant are observed for this case of a high barrier and a weakly interacting solvent.

Acknowledgement

Acknowledgement is made to the donors of the Petroleum Research Fund, administered by the American Chemical Society, for partial support of this research. In addition, this work was supported by the National Science Foundation (Chemistry) and the Office of Naval Research (Chemistry).

References

[1] S. Glasstone, K.J. Laidler and H. Eyring, Theory of rate processes (McGraw-Hill, New York, 1941).
[2] D.G. Truhlar, W.L. Hase and J.T. Hynes, J. Phys. Chem. 87 (1983) 2664.
[3] J.T. Hynes, in: The theory of chemical reaction dynamics, ed. M. Baer (CRC Press, Cleveland, 1985), to be published.
[4] R.O. Rosenberg, B.J. Berne and D. Chandler, Chem. Phys. Letters 75 (1980) 162.
[5] D. Chandler, J. Chem. Phys. 68 (1978) 2959.
[6] S.H. Northrup and J.T. Hynes, J. Chem. Phys. 73 (1980) 2700.
[7] R.F. Grote and J.T. Hynes, J. Chem. Phys. 73 (1980) 2715.
[8] M.P. Allen and P. Schofield, Mol. Phys. 39 (1980) 207.
[9] J.B. Anderson, J. Chem. Phys. 58 (1973) 4684.
[10] C.H. Bennett, in: Algorithms for chemical computations: ACS Symposium Series, Vol. 46, ed. R.E. Christofferson (Am. Chem. Soc., Washington, 1977) p. 63.
[11] B.M. Ladanyi and J.T. Hynes, J. Chem. Phys., submitted for publication.
[12] J.C. Keck, Discussions Faraday Soc. 33 (1962) 173.
[13] P.J. Kuntz, in: Dynamics of molecular collisions, Part B, ed. W.H. Miller (Plenum Press, New York, 1976) p. 53.
[14] K.P. Huber and G. Herzberg, Constants of diatomic molecules (Van Nostrand Reinhold, New York, 1979).
[15] G. Heublein, R. Kühmstedt, P. Kadura and H. Dawczynski, Tetrahedron 26 (1970) 81.
[16] R.O. Watts and I.J. McGee, in: Liquid state chemical physics (Wiley–Interscience, New York, 1976).
[17] J.P. Bergsma, K.R. Wilson, J.R. Reimers and J.T. Hynes, to be submitted for publication.
[18] R.F. Grote and J.T. Hynes, J. Chem. Phys. 75 (1981) 2191.
[19] R.F. Grote, G. van der Zwan and J.T. Hynes, J. Phys. Chem. 88 (1984) 4676.
[20] H. Andersen, Stanford University, private communication.

Notes to Reprint VI.4

1. Reference [17], which contains a more complete account of the work, was later published as (Bergsma et al., J. Chem. Phys. 85 (1986) 5625).

2. For a discussion of the theoretical background to this paper, see the article by Hynes (in: Theory of Chemical Reactions, Vol. IV, ed. M. Baer (CRC Press, Boca Raton 1985)).

3. For subsequent work on more realistic model reactions, see Bergsma et al., J. Chem. Phys. 86 (1987) 1356; 86 (1987) 1377.

Proc. Natl. Acad. Sci. USA
Vol. 82, pp. 8458–8462, December 1985
Biophysics

Active site dynamics of ribonuclease

(enzyme kinetics/protein dynamics/solvation/stochastic boundary simulation/water structure)

AXEL T. BRÜNGER, CHARLES L. BROOKS III*, AND MARTIN KARPLUS

Department of Chemistry, Harvard University, Cambridge, MA 02138

Contributed by Martin Karplus, August 15, 1985

ABSTRACT **The stochastic boundary molecular dynamics method is used to study the structure, dynamics, and energetics of the solvated active site of bovine pancreatic ribonuclease A. Simulations of the native enzyme and of the enzyme complexed with the dinucleotide substrate CpA and the transition-state analog uridine vanadate are compared. Structural features and dynamical couplings for ribonuclease residues found in the simulation are consistent with experimental data. Water molecules, most of which are not observed in crystallographic studies, are shown to play an important role in the active site. Hydrogen bonding of residues with water molecules in the free enzyme is found to mimic the substrate–enzyme interactions of residues involved in binding. Networks of water stabilize the cluster of positively charged active site residues. Correlated fluctuations between the uridine vanadate complex and the distant lysine residues are mediated through water and may indicate a possible role for these residues in stabilizing the transition state.**

Bovine pancreatic ribonuclease A is an ideal system for a theoretical study of the structural and dynamic basis of enzyme catalysis. Although little is known about the physiological role of ribonuclease, a particularly wide range of biochemical, physical, and crystallographic data are available for this enzyme (1, 2). These have led to proposals for the catalysis of the hydrolysis of RNA by a two-step mechanism (3), in which a cyclic phosphate intermediate is formed and subsequently hydrolysed. Both steps are thought to involve in-line displacement at the phosphorus and to be catalyzed by the concerted action of a general acid and a general base. In spite of the accumulated experimental information, a full understanding of the mechanism and a detailed analysis of the rate enhancement produced by ribonuclease has not been achieved. One approach for supplementing the available experimental data is provided by molecular dynamics simulations. Such simulations are particularly useful for exploring aspects of the active site dynamics that are difficult to characterize by other techniques.

To achieve a realistic treatment of the solvent-accessible active site of ribonuclease, a new molecular dynamics simulation method has been implemented. It makes possible the simulation of a localized region, approximately spherical in shape, that is composed of the active site with or without ligands, the essential portions of the protein in the neighborhood of the active site, and the surrounding solvent. The approach provides a simple and convenient method for reducing the total number of atoms included in the simulation, while avoiding spurious edge effects. Using this method we have performed molecular dynamics simulations of the solvated active site of ribonuclease A, ribonuclease A with the dinucleotide substrate CpA, and ribonuclease A with the transition-state analog uridine vanadate (UVan). Analysis of the simulation provides new insights concerning the interactions between ligands and active site residues and permits elucidation of the role of water in stabilizing the structure and charges of the active site.

METHODS

The stochastic boundary molecular dynamics simulation method (4–6) for solvated proteins uses a known x-ray structure as the starting point; for the present problem we have made use of the refined high-resolution (1.5 to 2 Å) x-ray structures kindly provided by G. Petsko and co-workers (7, 8); the CpA structure was built from the crystal results for dCpA by introducing the 2′ oxygen into the deoxy sugar. The region of interest (here the active site of ribonuclease A) is defined by choosing a reference point (which is taken at the position of the phosphorus atom in the CpA complex) and constructing a sphere of 12 Å radius around this point. Space within the sphere not occupied by crystallographically determined atoms is filled by water molecules, introduced from an equilibrated sample of liquid water. A picture of the portion of the protein and the solvent included in the simulation is shown in Fig. 1. The 12 Å sphere is further subdivided into a reaction region (10 Å radius) treated by full molecular dynamics and a buffer region (the volume between 10 and 12 Å) treated by Langevin dynamics, in which Newton's equations of motion for the non-hydrogen atoms are augmented by a frictional term and a random-force term; these additional terms approximate the effects of the neglected parts of the system and permit energy transfer in and out of the reaction region. Water molecules diffuse freely between the reaction and buffer regions but are prevented from escaping by an average boundary force (5). The protein atoms in the buffer region are constrained by harmonic forces derived from crystallographic temperature factors (6). The forces on the atoms and their dynamics were calculated with the CHARMM program (9); the water molecules were represented by the ST2 model (10). The entire system for each simulation was equilibrated for 12 psec; this involved thermalization of the component atoms and adjustment of the number of waters in the active site region to obtain the normal liquid density in regions distant from the protein. After equilibration, 27 psec of simulation was performed for each of the three systems; it is the results of these simulations that are reported here. Details of the methodology and additional results will be given elsewhere.

RESULTS AND DISCUSSION

The behavior of the individual simulations and their comparison provide a number of insights into aspects of the active site region of ribonuclease that are important for an understanding of substrate binding and catalysis. Fig. 2*a* shows a

The publication costs of this article were defrayed in part by page charge payment. This article must therefore be hereby marked "*advertisement*" in accordance with 18 U.S.C. §1734 solely to indicate this fact.

Abbreviation: UVan, uridine vanadate.

*Present address: Department of Chemistry, Carnegie-Mellon University, Pittsburgh, PA 15213.

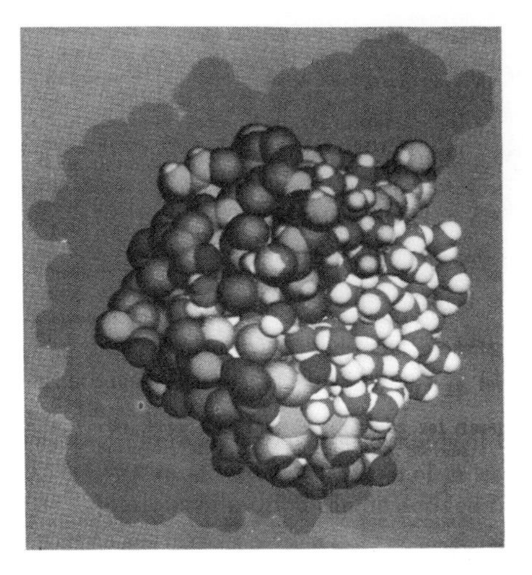

FIG. 1. Active site region of aqueous ribonuclease A within stochastic boundary; shown is a space-filling representation of the van der Waals spheres of protein and water atoms included in the stochastic boundary molecular dynamics simulation. The structure is obtained from a "snapshot" at 15 psec after equilibration. The shaded regions indicate the parts of the molecule that have been eliminated in the simulation (see text).

stereo picture of the average dynamics structure for ribonuclease, in the absence of substrate or other anions (sulfate, phosphate) bound to the active site (8). Residues that have been implicated in substrate binding (Thr-45 and Ser-123 in the "pyrimidine" site and Gln-69, Asn-71, and Glu-111 in the "purine" site) and in catalysis (His-12 and -119; Lys-7, -41, and -66; Phe-120, Asp-121, and Gln-11) are shown, with some additional residues that may be involved in maintaining the integrity of the active site; the average positions of water molecules are included. In what follows, we shall examine first the native results in the absence of bound anions and then compare them with those obtained in the presence of CpA and UVan.

One of the striking aspects of the active site of ribonuclease is the presence of a large number of positively charged groups, some of which may be involved in guiding and/or binding the substrate (11). The simulation demonstrated that these residues are stabilized in the absence of ligands by well-defined water networks. A particular example is shown in Fig. 3, which includes Lys-7, -41, and -66, Arg-39, and the doubly protonated His-119, as well as the water molecules within 3 Å of any charged site with their energetically most likely hydrogen bonds. The bridging waters, some of which are organized into trigonal bipyramidal structures, are able to stabilize the otherwise very unfavorable configuration of positive groups because the interaction energy between water and the charged $C\text{-}NH_n^+$ ($n = 1, 2,$ or 3) moieties is very large; e.g., at a donor–acceptor distance of 2.8 Å, the $C\text{-}NH_3^+\text{-}H_2O$ energy is -19 kcal/mol with the empirical potential used for the simulation (9), in approximate agreement with accurate quantum mechanical calculations (ref. 12; unpublished data) and gas-phase ion-molecule data (13). The average energy stabilization of the charged groups (Lys-7, -41, and -66, Arg-39, and His-119) and the 106 water molecules included in the simulation is -376.6 kcal/mol. This energy is calculated as the difference between the simulated system and a system composed of separate protein and bulk water. Unfavorable protein–protein charged-group interactions are balanced by favorable water–protein and water–water interactions. The average energy per molecule of pure water from an equivalent stochastic boundary simulation (5) is -9.0 kcal/mol, whereas that of the waters included in the active site simulation is -10.2 kcal/mol; in the latter a large contribution to the energy comes from the interactions between the water molecules and the protein atoms. It is such energy differences that are essential to a correct evaluation of binding equilibria and the changes introduced by site-specific mutagenesis (14).

During the simulation, the water molecules involved in the charged-group interactions oscillate around their average positions, generally without performing exchange. On a longer time scale, it is expected that the waters would exchange and that the side chains would undergo larger-scale displacements. This is in accord with the disorder found in the x-ray results for lysine and arginine residues (e.g., Lys-41 and Arg-39; see refs. 7 and 15), a fact that makes difficult a crystallographic determination of the water structure in this case. Water stabilization of charged groups has been observed (16, 17) or suggested (18) in other proteins. It is also of interest that Lys-7 and Lys-41 have an average separation of only 4 Å in the simulation, less than that found in the x-ray structure. That this like charged pair can exist in such a configuration is corroborated by experiments that have shown that the two lysines can be crosslinked (19); the structure of this compound has been reported recently (20) and is similar to that found in the native protein.

In addition to the role of water in stabilizing the charged groups that span the active site and participate in catalysis, water molecules make hydrogen bonds to protein polar groups that become involved in ligand binding. A particularly clear example is provided by the adenine-binding site in the CpA simulation (Fig. 2*b*). The NH_2 group of adenine acts as a donor, making hydrogen bonds to the carbonyl of Asn-67, and the ring N^{1A} acts as an acceptor for a hydrogen bond from the amide group of Glu-69 (see Table 1). Corresponding hydrogen bonds are present in the free ribonuclease simulation, with appropriately bound water molecules replacing the substrate. These waters and those that interact with the pyrimidine-site residues Thr-45 and Ser-123 (see below) help to preserve the protein structure in the optimal arrangement for binding. Similar substrate "mimicry" has been observed in x-ray structures of lysozyme (21) and of penicillopepsin (22). The presence of such waters in the active site has the consequence that water extrusion and its inverse must be intimately correlated with substrate binding and product release; e.g., there are 23 fewer waters in the active site of the CpA complex than for free ribonuclease in the present simulation.

A selected list of strong hydrogen bonds in the various x-ray and average dynamics structures is given in Table 1. Most of the interactions are similar in the crystal and simulated structures. Binding interactions in the pyrimidine-binding site are indicated by hydrogen bonds between Thr-45 and the cytidine (uridine) base in the CpA (UVan) complex. Comparisons of the x-ray structures of uridine and cytidine base inhibitor complexes (23) suggest that Ser-123 may be important for the binding of uridine. In the UVan simulation, Ser-123 shows persistent hydrogen bonding interaction with the uridine base complex (see Fig. 2*c*), whereas in the x-ray structure of Gilbert *et al.* (7), no bond between Ser-123 and the uridine base is indicated (see Table 1). Stabilization of the cytidine base in the dinucleotide substrate may also occur through perpendicular stacking interactions with Phe-120, as is apparent in Fig. 2*b* and in the x-ray structure (7). This type of interaction, which corresponds to that found in benzene

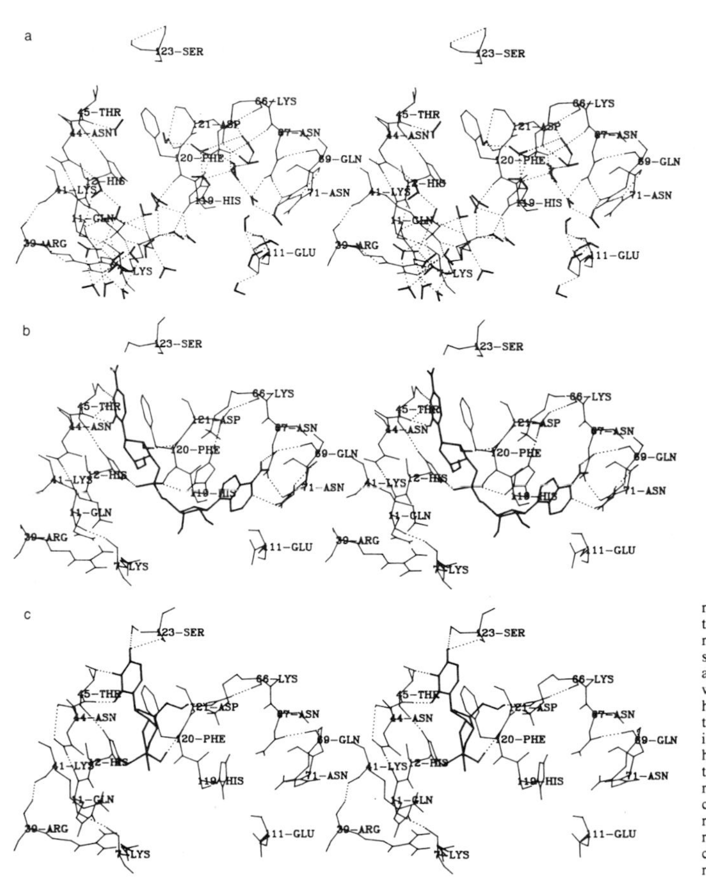

FIG. 2. (*a*) Stereo drawing of native ribonuclease A including active site residues and bound water molecules. The structure corresponds to the molecular dynamics average structure of the protein and water molecules. Protein·protein hydrogen bonds are indicated when the heavy donor–acceptor distance is less than 3.5 Å and the donor–hydrogen–acceptor angle is less than 65.0°. (b) Similar to *a*, for the ribonuclease–CpA complex, including substrate and active site residues. (*c*) Similar to *a*, for the ribonuclease–UVan complex, including substrate and active site residues.

crystals (24), has been discussed recently for proteins (25). His-12 forms a strong hydrogen bond via $N^{\epsilon 1}$ to the carbonyl of Thr-45 and is found to be relatively immobile and similarly positioned in all three structures. By contrast, His-119 is considerably more mobile and has different orientations in the various structures. Specifically, there is a hydrogen bond from $N^{\epsilon 2}$ to Asp-121 in the CpA complex but not in either of the other dynamics average structures; in the free ribonuclease simulation, His-119 participates in the water network already described. Also, His-119 undergoes dihedral-angle transitions in some of the simulations. This greater freedom of His-119, relative to His-12, which may be important for the catalytic mechanism, is in accord with the alternative conformations found in some x-ray structures (23, 26). In none

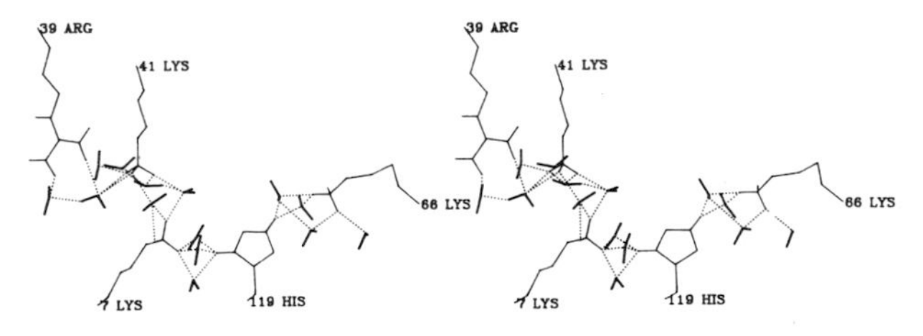

FIG. 3. Stabilization of positively charged groups by bridging water molecules in the active site of native ribonuclease A without anions. The stereo drawing corresponds to a snapshot of the molecular dynamics trajectory at 15 psec. The picture includes only the residues Lys-7, Arg-39, Lys-41, Lys-66, and His-119 and the bound water molecules. The hydrogen-bonding criteria are the same as used in Fig. 2.

Table 1. Comparison of hydrogen bonds in x-ray structures and in molecular dynamics (MD) average structures

Hydrogen bond		Distance,* Å					
		RNase		RNase–CpA		RNase–UVan	
Donor	Acceptor	x-ray	MD	x-ray	MD	x-ray	MD
His-12 ($N^{\delta1}$)	Thr-45 (O)	2.8	2.7	2.7	2.9	2.6	2.6
Lys-66 (N)	Asp-121 ($O^{\delta2}$)	2.8	2.8	2.7	2.8	3.0	2.8
His-119 ($N^{\epsilon2}$)	Asp-121 ($O^{\delta1}$)	2.8	—	2.7	2.6	—	—
His-12 ($N^{\epsilon2}$)	CpA (O^2)			3.2	2.8		
Lys-41 (N^{ζ})	CpA (O^2)			2.8	—		
Thr-45 (N)	CpA (N^{3C})			—	3.4		
Thr-45 (N)	CpA (O^{2C})			2.8	2.7		
Thr-45 ($O^{\gamma1}$)	CpA (N^{3C})			—	2.9		
Gln-69 ($N^{\epsilon2}$)	CpA (N^{1A})			3.4	3.0		
Asn-71 ($N^{\epsilon2}$)	CpA (N^{1A})			3.3	—		
His-119 ($N^{\delta1}$)	CpA (O^1)			3.2	2.8		
Phe-120 (N)	CpA (O^1)			2.9	3.1		
CpA (N^{6A})	Asn-67 ($O^{\delta1}$)			—	3.3		
CpA (N^{6A})	Asn-71 ($O^{\delta1}$)			3.1	—		
CpA ($O^{2'C}$)	Phe-120 (O)			—	2.7		
His-12 ($N^{\epsilon2}$)	UVan (O^8)					2.6	—
His-12 ($N^{\epsilon2}$)	UVan (O^1)					—	2.7
Thr-45 (N)	UVan (O^2)					2.8	2.7
Phe-120 (N)	UVan (O^8)					2.9	—
Ser-123 (N)	UVan (O^4)					—	3.5
Ser-123 (O^{γ})	UVan (O^4)					—	2.8
UVan (N^3)	Thr-45 ($O^{\gamma1}$)					2.6	2.7
UVan (O^8)	Phe-120 (O)					—	2.6
UVan ($O^{5'}$)	Asp-121 ($O^{\delta1}$)					—	2.8

Selected hydrogen bonds for the residues and substrates shown in Fig. 2 *a–c*; the criteria for the existence of a hydrogen bond are the same as in Fig. 2: the donor–acceptor distance has to be <3.5 Å and the donor–hydrogen–acceptor angle <65°.

*Between the donor and the acceptor atoms.

of the simulations do His-12 and His-119 get closer to each other than 5 Å; in the free ribonuclease simulation, they are separated by three water molecules. This confirms the suggestion of Campbell and Petsko (8), based on the x-ray structure, that the Witzel mechanism (27), which postulates a direct interaction between the two histidines, is not tenable.

In both the dCpA x-ray structure and the CpA simulation, $N^{\epsilon2}$ of His-12 is coupled to O^2 of the phosphate group (which is also hydrogen-bonded to two water molecules in the simulation) but is still well-positioned to interact with the 2′ hydroxyl of the sugar. $N^{\delta1}$ of His-119 is bonded to O^1 of the phosphate, as is the main-chain N of Phe-120 in both the x-ray and the simulation structure. For UVan, His-12 is in a similar position and the $N^{\epsilon2}$ group is hydrogen bonded to O^8 of the pentacoordinate vanadate. It is clear that His-12 moves to follow the negatively charged group of UVan which is displaced relative to that in CpA; such a "tracking" motion is also noted in ref. 7. His-119, which interacts with the $O^{5'}$ side chain of the sugar of UVan, has intervening water molecules between it and the vanadate ion, in accord with its role as a general base in the hydrolysis step of the reaction. In the x-ray structure (7) and the neutron structure (28), His-119 is somewhat closer to the vanadate group than in the simulation. As to Lys-41, neither in the x-ray structure (it is disordered) nor in the dynamic structure is it bonded to CpA. In the UVan complex, the crystal structure shows a direct interaction between the N^{ζ} groups of Lys-41 and O^2 of vanadate, whereas the dynamic structure shows a somewhat longer-range interaction with O^1. The reduced mobility for Lys-41 in the x-ray and neutron structures may originate from the use of a vanadate complex in the experiment; in the simulation, the smaller atomic charges appropriate for a phosphate group, rather than vanadate, were employed, and the Lys-41 mobility

is only slightly less in the UVan complex than in the free enzyme. Lys-7 is displaced toward the ligand on CpA and UVan binding, relative to the free enzyme, but has no direct interactions with the ligands in either case.

In the x-ray structure of the substrate analog dCpA (7), it was noted that the dinucleotide is in a *gauche, trans* configuration that would permit stereoelectronic assistance in phosphodiester hydrolysis (29). During the simulation, the phosphate backbone undergoes a transition to the more stable *gauche, gauche* configuration (30), where it remains for the remainder of the time. The significance of this transition requires further investigation, particularly because it may be related to the interaction with His-12 (see above), which is in the doubly protonated imidazolium ion form observed in the neutron crystal study. A singly protonated imidazole species is required for it to act as a general base in the catalytic mechanism. NMR measurements of the histidine protonation state (28) under the same conditions as in the neutron study suggest that His-12 and -119 are both present as a mixture of singly and doubly protonated species, although only the latter is observed in the crystal. Since the hydrogen-bonding behavior of the two protonation states is expected to be different, additional simulations need to be done to fully elucidate the role of the histidine residues (e.g., CpA with His-12 singly protonated and His-119 doubly protonated).

To supplement the positional results concerning important interactions obtained from the average structure, it is important to introduce *dynamical* information pertaining to the motions of different groups. This provides an indication of transient short-range interactions and longer-range correlated motions that may be important in interpreting chemical events but are not evident in the static average structure. The magnitudes of the correlated fluctuations between the atomic displacements of the protein and the ligand, relative to the dynamics average structure, are shown for the CpA and UVan complexes in Fig. 4 *a* and *b*, respectively. Although the time scale of the simulation is short compared with the observed rate constants, it is important to remember that the actual catalytic event in a given molecule occurs on a picosecond time scale. The overall time scale of the reaction can be orders of magnitude longer due to the fact that a reaction is rare, not that it is slow. Many of the large correlations (positive or negative) correspond to atoms in close

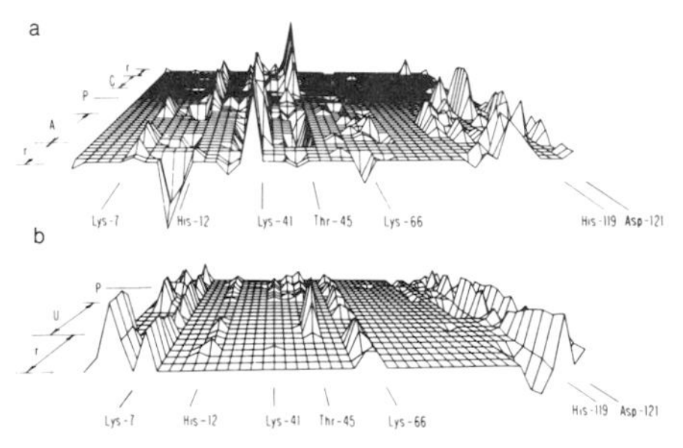

FIG. 4. (*a*) Normalized cross-correlations of the atomic fluctuations of the ribonuclease–CpA complex; only correlations with absolute values greater than 0.3 are shown, and an averaging over all atoms within each protein residue was performed. The 45 residues included in the simulation as well as in the drawing are Glu-2, Thr-3, Ala-4, Ala-5, Ala-6, Lys-7, Phe-8, Glu-9, Arg-10, Gln-11, His-12, Met-13, Met-30, Arg-33, Leu-35, Arg-39, Lys-41, Pro-42, Val-43, Asn-44, Thr-45, Phe-46, Val-47, Val-54, Cys-65, Lys-66, Asn-67, Asn-71, Cys-72, Gln-69, Cys-84, Ile-106, Ile-107, Val-108, Ala-109, Cys-110, Glu-111, Val-116, Pro-117, Val-118, His-119, Phe-120, Asp-121, Ala-122, and Ser-123 (in consecutive order, note the disconnectivities). r, Ribose ring atoms. (*b*) Same as *a*, for the ribonuclease–UVan complex.

proximity; an example is the correlation of Thr-45 and the pyrimidine, which have strong hydrogen-bonding interactions. His-119 has strong correlations with the phosphate group in CpA and the vanadate group in UVan; the latter is of particular interest because there are at least two water molecules, on the average, between the imidazole ring and the phosphate group. An analogous case is Asp-121, which has strong correlations with the substrate in both the CpA and the UVan complex; in the latter Asp-121 is hydrogen-bonded to the CH_2OH group of the sugar, but in the former it is coupled indirectly through His-119 and Phe-120. Lys-7 and -66 both are correlated significantly with the UVan ligand, although the N^{ζ} groups of the two lysines are more than 5 Å removed in both cases. The dynamic coupling appears to be mediated by an intervening water network, whose structural role was described above. By contrast Lys-41, which interacts directly with the vanadate group, shows only a very weak correlation. However, the results of a new simulation (unpublished data) of the cyclic cytidine monophosphate transition-state analog indicate a strong correlation between Lys-41 and the cyclic phosphate group. Such correlated fluctuations between the distant lysine residues and the phosphate group of the substrate may indicate a possible role for the lysine residues in stabilizing the transition state.

To further characterize catalysis by ribonuclease A, dynamical studies of substrate binding and product release, as well as of structural changes that occur along the reaction pathway, are required. Also, a detailed treatment of the reaction dynamics by activated dynamics techniques (31) is needed.

We are grateful to G. A. Petsko for making the x-ray coordinates available to us and for stimulating discussions of the results. We thank A. Wlodawer for kindly providing the neutron coordinates of the UVan complex. We have profited from discussions with L. C. Allen. A.T.B. wishes to thank W. Bode, G. M. Clore, A. M. Gronenborn, R. Huber, D. Oesterhelt, J. W. Pflugrath, and W. Steigemann for assistance during his stay at the Max-Planck-Institute for Biochemistry (Martinsried, F. R. G.). C.L.B. wishes to thank the National Institutes of Health for postdoctoral support. This work has been supported in part by a grant from the National Science Foundation.

1. Richards, F. M. & Wyckhoff, H. W. (1971) in *The Enzymes*, ed., Boyer, P. D. (Academic, New York), 3rd Ed., Vol. 4, pp. 647–907.
2. Blackburn, P. & Moore, S. (1982) in *The Enzymes*, ed., Boyer, P. D. (Academic, New York), 3rd Ed., Vol. 15, pp. 317–433.
3. Roberts, G. C. K., Dennis, E. A., Meadows, D. H., Cohen, J. S. & Jardetzky, O. (1969) *Proc. Natl. Acad. Sci. USA* **62,** 1151–1158.
4. Brooks, C. L. & Karplus, M. (1983) *J. Chem. Phys.* **79,** 6312–6323.
5. Brünger, A., Brooks, C. L. & Karplus, M. (1984) *Chem. Phys. Lett.* **105,** 495–500.
6. Brooks, C. L., Brünger, A. & Karplus, M. (1985) *Biopolymers* **24,** 843–865.
7. Gilbert, W. A., Fink, A. L. & Petsko, G. A. (1986) *Biochemistry*, in press.
8. Campbell, R. L. & Petsko, G. A. (1986) *Biochemistry*, in press.
9. Brooks, B. R., Bruccoleri, R. D., Olafson, B. O., States, D. J., Swaminathan, S. & Karplus, M. (1983) *J. Comp. Chem.* **4,** 187–217.
10. Stillinger, F. H. & Rahman, A. (1974) *J. Chem. Phys.* **60,** 1545–1557.
11. Matthew, J. B. & Richards, F. M. (1982) *Biochemistry* **21,** 4989–4999.
12. Desmeules, D. J. & Allen, L. C. (1980) *J. Chem. Phys.* **72,** 4731–4748.
13. Kebarle, P. (1977) *Annu. Rev. Phys. Chem.* **28,** 445–476.
14. Fersht, A. R., Shi, J.-P., Knill-Jones, J., Lowe, D. M., Wilkinson, A. J., Blow, D. M., Brick, P., Carter, P., Waye, M. M. Y. & Winter, G. (1985) *Nature (London)* **314,** 235–238.
15. Wlodawer, A. (1985) in *Biological Macromolecules and Assemblies: Volume 2–Nucleic Acids and Interactive Proteins*, eds., Jurnak, F. A. & McPherson, A. (Wiley, New York), pp. 394–439.
16. Baker, E. N. & Hubbard, R. E. (1984) *Prog. Biophys. Mol. Biol.* **44,** 97–179.
17. Sawyer, L. & James, M. N. G. (1982) *Nature (London)* **295,** 79–80.
18. Tabushi, I., Kiyosuke, Y. & Yamamura, K. (1981) *J. Am. Chem. Soc.* **103,** 5255–5257.
19. Marfey, P. S., Uziel, M. & Little, J. (1965) *J. Biol. Chem.* **240,** 3270–3275.
20. Weber, P. C., Salemme, F. R., Lin, S. H., Konishi, Y. & Scheraga, H. A. (1985) *J. Mol. Biol.* **181,** 453.
21. Blake, C. C. F., Pulford, W. C. A. & Artymiuk, P. J. (1983) *J. Mol. Biol.* **167,** 693–723.
22. James, M. N. G. & Sielecki, A. R. (1983) *J. Mol. Biol.* **163,** 299–361.
23. Borkakoti, N., Moss, D. S. & Palmer, R. A. (1982) *Acta Crystallogr. Sect. B* **38,** 2210–2217.
24. Wyckoff, R. W. G. (1969) *The Structure of Benzene Derivatives*, (Interscience, New York).
25. Burley, S. K. & Petsko, G. A. (1985) *Science* **229,** 23–28.
26. Borkakoti, N. (1983) *Eur. J. Biochem.* **132,** 89–94.
27. Witzel, H. (1963) *Prog. Nucleic Acid Res. Mol. Biol.* **2,** 221–258.
28. Borah, B., Chen, C.-W., Egan, W., Miller, M., Wlodawer, A. & Cohen, C.-W. (1985) *Biochemistry* **24,** 2058–2067.
29. Gorenstein, D. G., Findlay, J. B., Luxon, B. A. & Kar, D. (1977) *J. Am. Chem. Soc.* **99,** 3473–3479.
30. Alagona, G., Ghio, C. & Kollman, P. A. (1985) *J. Am. Chem. Soc.* **107,** 2229–2239.
31. Northrup, S. H., Pear, M. R., Lee, C.-Y., McCammon, J. A. & Karplus, M. (1982) *Proc. Natl. Acad. Sci. USA* **79,** 4035–4039.

Study of electron solvation in liquid ammonia using quantum path integral Monte Carlo calculations[a]

Michiel Sprik, Roger W. Impey, and Michael L. Klein
Chemistry Division, National Research Council of Canada, Ottawa, Ontario, Canada K1A 0R6

(Received 22 July 1985; accepted 16 August 1985)

The solvation of an electron in liquid ammonia has been studied using quantum path integral Monte Carlo calculations. In agreement with previous experimental and theoretical deductions the charge distribution of the electron is compact. Various distribution functions characterizing the structure around the solvated electron are presented and the surrounding solvent structure is compared to that around a classical atomic anion. A qualitative discussion is given of the absorption spectrum based upon the form of the complex time dependence of the electron mean squared displacement correlation function.

I. INTRODUCTION

The behavior of excess electrons in polar fluids has been the subject of extensive investigation and ammonia has been the focus of much of this work.[1-7] Depending upon the conditions, the charge density of an excess electron may be either extended, spanning many molecular diameters or highly compact, i.e., analogous to a solvated anion. Localization can even occur when the vapor density is 1000 times less than the normal liquid.[8,9] The optical absorption band of a solvated electron in liquid ammonia is asymmetric but peaks at about 0.8 eV. Moreover, the observation that isotopic substitution induces only a modest blue shift[10] suggests that an electron in liquid ammonia can be regarded as a quantum solute in a classical solvent.

The nature of the solvated electron has been the subject of a variety of theoretical studies, the solvent being treated as a static (rigid) structure or even as a dielectric continuum.[11-14] While considerable insight has been gleaned from such calculations, the treatment of the fluid/solvent was simplistic in the extreme. Recent advances in simulation techniques have exploited the isomorphism between a quantum particle and a classical flexible ring polymer (necklace) in order to study quantum solutes and bulk fluids.[15-28] The isomorphism becomes exact in the limit that the number of beads on the necklace $P = \infty$. Previous studies of quantum solutes that used molecular dynamics (MD) to sample the quantum paths of the solute in the coupled solute–solvent system used at most $P = 200$.[19,20] However, a number of arguments can be advanced which suggest that adequate sampling of the quantum paths of an electron, whose thermal wavelength at room temperature $\lambda = \hbar(\beta/m)^{1/2} = 17$ Å, may well require an order of magnitude larger value of P. In this regard, the path integral Monte Carlo approach offers a distinct advantage since integration algorithms have been devised to ensure efficient sampling of quantum fluctuations.[21,25] The present study employs the so-called staging algorithm[25] to sample the quantum paths of the electron. Through the use of this highly vectorized sampling algorithm it has proved possible to carry out calculations with $P = 2048$ without employing prodigious amounts of computing. Anticipating our results we find that the charge density of an electron in liquid ammonia is indeed compact, with a diameter of about 4 Å. The Monte Carlo calculations have been used to monitor the energy, second moment of the electron charge distribution, and certain other distribution functions which characterize the local structure around the solvated electron. In agreement with the accepted folklore, the coordination numbers in the solvent sheath of ammonia molecules resembles that around a large anion. However, the detailed structure is modified due to the quantum nature of the electron.

The outline of the article is as follows. In the next section we discuss our potential models for the solvent and the electron solvent interaction. In Sec. III we review the quantum path integral Monte Carlo method. Certain technical details of the Monte Carlo calculations are outlined in Sec. IV and the results are presented in Sec. V along with a qualitative discussion of the absorption spectrum. In the final section we consider the limitations of our calculations and the prospects for similar studies on related systems.

II. INTERACTION POTENTIALS

We have already alluded to the fact that isotopic substitution induces only a modest blue shift in the optical absorption spectrum of an excess electron in liquid ammonia.[10] Therefore, at least in an initial study, we can treat the ammonia as a classical fluid. (This is not an essential approximation since it is possible to carry out a fully quantum mechanical simulation of the solvent.[22]) Fortunately a simple effective pair potential already exists for liquid ammonia which gives a fair account of the liquid properties.[29] The potential is analogous to that of Bernal–Fowler for water. It consists of a Lennard-Jones (12-6) interatomic N–N potential with $\epsilon = 140$ K and $\sigma = 3.40$ Å plus electrostatic interactions between arrays of fractional charges, the latter being chosen to reproduce the molecular dipole moment ($\mu = 1.47$ D) and to yield a reasonable quadrupole moment. A suitable representation of the ammonia charge distribution consists of four charge sites; $Q^+ = 0.485e$ on each H atom plus a neutralizing charge $Q^- = -1.455e$ on the molecular C_{3v} axis 0.156 Å from the N atom, inside the molecule. The N–H

[a] Issued as NRCC Report Number 24743.

bond length was fixed at 1.0126 Å and the H–H distance at 1.625 Å.

The electron-ammonia interaction can in principle be evaluated using well defined procedures.[30] There are three essential ingredients in the potential: the electrostatic interaction, polarization terms, and the short range repulsion that arises from the requirement that the electron wave function be orthogonal to the electrons of the ammonia molecule. There is an extensive literature on electron scattering from polar molecules.[30–34] In some work the long range potential was simply truncated at short range, the truncation point being chosen by fitting to cross section data.[33] More recently consideration has been given to all three contributions to the potential.[34] In now appears that the polarization contribution is less important than previously supposed; it is severely damped at short range.[34]

In their work on electron solvation in molten KCl Parrinello and Rahman[19] used a local pseudopotential[35,36] to represent the electron–K^+ interaction; the charge–charge interaction being held constant inside the cation core whose radius, 1.5–2.5 Å, was an adjustable parameter. The electron–anion interaction was simply the bare Coulombic repulsion. The use of a pseudopotential for the cation has a number of advantages not the least of which is the smooth spatial variation which in turn reduces the value of P required to adequately sample the quantum paths of the electrons.[19] It is tempting to employ an analogous pseudopotential for the electron–NH_3 interaction.

This, in spirit is what we have done. However, we have used only the bare electron-site charge interactions and do not explicitly include either the polarization or short range interactions. This is, of course, a very crude approximation to the real electron–NH_3 potential. However, considering the NH_3 molecules as an array of charge sites has the advantage that we can then employ efficient high temperature (short time) approximations to the propagator of the pair Coulomb problem.[37–39] In effect we are borrowing techniques from the study of plasmas.[40] After all, the quantum mechanical solution of the hydrogen atom using path integral methods does not require a pseudopotential, but it is greatly facilitated if an improved propagator is employed. In fact, we will use the H atom as a means to estimate the degree of discretization (P) required for the electron–NH_3 problem.

III. PATH INTEGRAL MONTE CARLO

A. Path integral method

The simulation of the equilibrium properties of a quantum particle at finite temperature by means of the path integral method is well documented now.[15–28] The method employs the Feynman superposition principle which expresses the propagator (density matrix) as a sum of paths in configuration space. Each path is weighted by the exponent of the action functional which for the present problem takes the form

$$S(\mathbf{r}(t),\{\mathbf{R}_n\}) = -\hbar^{-1}\int_0^{\beta\hbar} dt\,[(\tfrac{1}{2}m|\dot{\mathbf{r}}(t)|^2 + U_I(\mathbf{r}(t),\{\mathbf{R}_n\})] - \beta U_S(\{\mathbf{R}_n\}), \quad (1)$$

where $2\pi\hbar$ is Planck's constant, β^{-1} is Boltzmann's constant times temperature, m is the electron mass, and $\mathbf{r}(t)$ is the position of the electron at imaginary time it; $\{\mathbf{R}_n\}$ denotes the set of position coordinates for the N_S charge sites in the fluid which interact with the electron. (These sites, are fixed in N_m rigid molecules, hence $\{\mathbf{R}_n\}$ is *not* to be considered as a set of independent molecular coordinates.) The motion of the solvent molecules is assumed to be classical so that their paths in imaginary time are reduced to points in configuration space. $U_S(\{\mathbf{R}_n\})$ and $U_I(\mathbf{r}(t),\{\mathbf{R}_n\})$ are, respectively, the site–site interaction potential for the solvent molecules and the interaction potential of the electron with the charge sites. The set of sites used in U_S and U_I are assumed to be the same but this is not an essential restriction.

In the path integral Monte Carlo method the quantum path variable $\mathbf{r}(t)$ is discretized by dividing the imaginary time period between 0 and $\beta\hbar$ into P intervals. The discretized path is represented by the set of coordinates $\{\mathbf{r}_i\}$, $i = 0,..P$, with $\mathbf{r}_i$ corresponding to the electron position at the ith time slice, subject to the condition $\mathbf{r}_0 = \mathbf{r}_p$. The weight functional of the path and fluid configuration is then factorized as

$$W(\{\mathbf{r}_i,\mathbf{R}_n\}) = \left[\prod_{i=1}^{P} K(\mathbf{r}_{i+1},\mathbf{r}_i;\{\mathbf{R}_n\};\hbar\beta/P)\right] \times \exp(-\beta U_S(\{\mathbf{R}_n\})), \quad (2)$$

where $K(\mathbf{r}',\mathbf{r};\{\mathbf{R}_n\};\hbar\beta/P)$ is the propagator for the electron to move from $\mathbf{r}$ to $\mathbf{r}'$ in time $\Delta\tau = \beta\hbar/P$. The simulation is implemented by taking the value of P sufficiently large and substituting for the factors K in Eq. (2) a convenient short time (high temperature) approximation $\widetilde{K}$. The integration over all configurations $\{\mathbf{r}_i,\mathbf{R}_n\}$ is then performed by a Monte Carlo sampling of the isomorphic classical system of a chain of P particles interacting with each other and the molecules by means of an effective potential derived from $\widetilde{K}$.

B. Pair product approximation

For an electron an efficient choice of $\widetilde{K}$ is of some importance since the thermal wavelength of an electron $\lambda = \hbar(\beta/m)^{1/2} \approx 17$ Å at 300 K is much larger than $l = (V/N_m)^{1/3}$, the average separation of molecules in the fluid, which is typically a few angstroms. Moreover, close to a molecule the electron potential U_I will be strongly varying over distances of order 1 bohr. This introduces a third length scale into the problem namely, the atomic length a_0. However, in this atomic regime, close to a particular interaction site in a molecule, the electron motion (path) is almost completely determined by this site and the pair product form[21] for K can be expected to be an accurate approximation

$$\widetilde{K}(\mathbf{r}_{i+1},\mathbf{r}_i;\{\mathbf{R}_n\};\beta\hbar/P) = K_0(\mathbf{r}_{i+1},\mathbf{r}_i;\hbar\beta/P)\prod_{n=1}^{N_S}\rho_2(\mathbf{r}_{i+1},\mathbf{r}_i;\mathbf{R}_n;\beta\hbar/P). \quad (3)$$

Here K_0 is the free propagator

$$K_0(\mathbf{r}_{i+1},\mathbf{r}_i;\hbar\beta/P) = \lambda^{-3}(P/2\pi)^{3/2}\exp[-(P/2\lambda^2)(\mathbf{r}_{i+1}-\mathbf{r}_i)^2], \quad (4)$$

and $\rho_2(\mathbf{r}_{i+1},\mathbf{r}_i;\mathbf{R}_n;\beta\hbar/P)$ is the ratio of the propagator as determined by interaction with the site $\mathbf{R}_n$ only and K_0. [For further justification of the pair product form (3) see Ref. 21.] Note that in a time $\Delta\tau = \beta\hbar/P$ a free particle moves over an average distance of in the order of $\Delta r = \lambda\,(3/P)^{1/2}$. From this argument it can therefore be concluded that application of the pair product form (3) requires an isomorphic system (chain) with an average nearest neighbor separation of almost an atomic unit. This amounts to a system with $P = (\lambda/a_0)^2 \approx 10^3$ particles.

C. Staging

The full length of the chain representing the electron path is of order λ, which is considerably larger than $l = (V/N_m)^{1/3}$. This implies that the chain, particularly in an extended configuration, may be wound in between and around several solvent molecules. On the other hand, a displacement by a Monte Carlo move of a vertex of a chain with average link size a_0 is also of the order a_0. Since a_0 is an order of magnitude smaller than l, it may be expected that a Monte Carlo procedure, that moves individual vertices, will sample the relevant configurations only very slowly and inefficiently. In previous work[25] dealing with an electron solvated in a hard sphere fluid, we have developed an improved procedure for sampling the fluctuations on the length scales l and a_0 simultaneously. This so-called staging algorithm moves the vertices of a primary (A) chain by an importance sampling procedure according to weights determined from direct sampling by a set of secondary (B) chain configurations constructed using nearest neighbor vertices of A chain as fixed end points. The number of particles in the primary chain P_A is chosen such that a displacement $\lambda/(P_A)^{1/2}$ is comparable to l. The numbers of vertices P_B of the secondary chains is obtained from the criterion that the size of a link $\lambda/(P_A P_B)^{1/2}$ of a B chain is reduced to the atomic length scale given by a_0, as discussed above. An obvious limitation of this combined importance and direct sampling procedure is that the accuracy of the calculation of the weights of the A chain links is affected by the attrition in the set of B chain configurations. In Sec. IV it will be argued that for systems under consideration here, the effective potential sampled by secondary chains of length $l \sim \lambda/(P_A)^{1/2}$ is sufficiently smooth and the attrition in the direct sampling is not too serious.

D. Correlation functions

Correlations in the positions of the particles in the isomorphic system and molecular fluid are determined from the statistics of the P_A position coordinates of the primary chain. The electron-site radial distribution function is then defined as

$$g_\alpha(r) = \langle \sum_{n\in\alpha} \delta[\mathbf{r}(t)]\delta(\mathbf{r} - \mathbf{R}_n)\rangle_A, \tag{5}$$

where the brackets denote averages over primary chain coordinates. The subscript α indicates the type of site considered, e.g., the positive charge sites in each molecule. It turns out that the radial distribution function (5) only reflects detail on an atomic length scale (a_0), i.e., structure that results from the interaction with a single charge site. The fluid structure in the vicinity of a solvated electron is best characterized by the center of mass distribution function

$$g_{\mathrm{cm},\alpha}(r) = \langle \sum_{n\in\alpha} \delta(\mathbf{r}_{\mathrm{cm}})\delta(\mathbf{r} - \mathbf{R}_n)\rangle_A, \tag{6}$$

where $\mathbf{r}_{\mathrm{cm}}$ is the center of mass of the P_A primary chain particles. Information on the thermal wave function (density matrix) of the electron can be obtained from the complex time correlation function[17]

$$\mathscr{R}^2(t-t') = \langle |\mathbf{r}(t) - \mathbf{r}(t')|^2\rangle_A. \tag{7}$$

In particular, the mean square displacement between particles halfway around the polymer chain yields the correlation length

$$\mathscr{R} = \mathscr{R}(\beta\hbar/2). \tag{8}$$

When there is ground state dominance, characteristic of a localized or compact state, the function $\mathscr{R}^2(t-t')$ starts from zero at $t - t' = 0$ and rapidly increases to its "saturation" value. The characteristic rise time τ of this initial (complex) time dependence is a measure of the energy difference ΔE between the ground state and the first manifold of excited states.

$$\Delta E = \hbar/\tau. \tag{9}$$

It should be pointed out that this gives only a very rough estimate of the excitation energy. For a more accurate evaluation real time methods are necessary.[27] But these have yet to be applied to systems of the type we consider here.

IV. TECHNICAL DETAILS

A. Regularization of the Coulomb potential

High temperature approximations to the density matrix of the two body Coulomb problem have been employed extensively in the Plasma Physics literature.[37–40] For example, a quantum effective pair potential[39] was used in a classical molecular dynamics simulation of a strongly coupled hydrogen plasma.[40] This effective potential V_q was obtained from a high temperature approximation to the diagonal element of the reduced two-body density matrix ρ_2 defined in Eqs. (3) and (4),

$$\rho_2(\mathbf{r},\mathbf{r};\mathbf{R};\beta\hbar/P) = \exp[-\beta V_q(\mathbf{r},\mathbf{R};\beta\hbar/P)/P]. \tag{10}$$

The quantum effective pair potential for the Coulomb problem given in Ref. 39 is in a particularly convenient form. For the case of an electron interacting with a classical charge Qe,

$$V_q(\mathbf{r};\beta\hbar/P) = -\frac{Qe^2}{r}(1 - \exp[-(2\pi P)^{1/2} r/\lambda]), \tag{11}$$

where we have set $R = 0$ for convenience. Assuming $\beta\hbar/P$ is small enough (high temperature) we can make the so-called "end point" approximation for the off-diagonal elements of ρ_2 (for further justification see Ref. 21).

$$\tilde{\rho}_2(\mathbf{r}_{i+1},\mathbf{r}_i;\mathbf{R};\beta\hbar/P) = \exp\{-\beta\,[V_q(\mathbf{r}_i,\mathbf{R};\beta\hbar/P) + V_q(\mathbf{r}_{i+1},\mathbf{R};\beta\hbar/P)]/2P\}. \tag{12}$$

If V_q of Eq. (11) is substituted into Eq. (12) we obtain a computationally simple approximation for ρ_2. The accuracy and the convergence with P was investigated by comparing path integral Monte Carlo results for the hydrogen atom with the well-known exact values. The results for the potential energy

TABLE I. Path integral Monte Carlo averages for the energy of a hydrogen atom at a temperature $\beta E_0 = -8$.

P^a	βU^b	$\beta V_q{}^b$	βK^b	βE^b
10	-12.3	-9.9	7.6	-4.7
20	-13.1	-11.4	7.8	-5.3
40	-14.1	-12.9	8.1	-6.0
80	-14.3	-13.5	7.8	-6.5
160	-14.7	-14.3	7.7	-7.0
320	-15.5	-15.2	8.3	-7.2

[a] P is the number of particles in the classical isomorphic system.
[b] U, V_q, K, E are, respectively, the average Coulombic potential energy, the quantum effective potential energy, virial kinetic energy, and total energy.

U and the total energy E are given in Table I for a temperature $-\beta E_0 = 8$ (E_0 is the ground state energy) or equivalently $\lambda = 4a_0$. For the kinetic energy, K, a modified virial estimator was employed.[19] The virial theorem has to be applied to the temperature dependent potential V_q given in Eq. (11), rather than to the bare Coulombic potential. Approximation (10) for the diagonal element of ρ_2 is virtually exact.[39] Thus, the underestimates of the absolute values of both U and E for small P are most likely due to the end point approximation (12). For the same reason the virial relation $K = -E = -\frac{1}{2}U$ for electrostatic interactions is only satisfied for the larger values of P in Table I. Structural properties of the ground state wave function such as the correlation length (second moment) defined in Eq. (8) are reproduced within a statistical error of about 10% and moreover do not show any clear systematic dependence on P. The convergence properties of V_q for like charges was not examined. However, from the form of Eq. (11) it can be expected that the results will be analogous to those for the attractive potential (Table I), i.e., an isomorphic system using a value of P which is too small will tend to underestimate the repulsion.

From the data in Table I it can be concluded that at room temperature ($\lambda \approx 32a_0$) a discretization of P equal to a few thousand underestimates the absolute value of potential energy by about 10%–20%. This error is most likely an upper bound for the error for the electron-fluid potential introduced in Sec. II. Here, the charges at the positive sites have Q values less than unity. Moreover, since the molecules are dipolar, the superposition of the contributions from the positive and negative charge sites will reduce the strength of the combined potential acting on the electron.

B. Staging and variable discretization

As outlined in Sec. III a two stage sampling procedure is used to simulate the classical isomorphic system. In addition to a more efficient sampling of the fluctuations on the molecular length scale l, the two stage algorithm has also a more technical advantage over the straightforward one stage algorithm. It allows for an easy implementation of a variable discretization scheme for interaction with molecules close to a chain vertex and molecules at larger distances. The beads on the secondary chains, sampled from the distribution of free chains with two nearest neighbor A chain vertices $\mathbf{r}_i^A$ and $\mathbf{r}_{i+1}^A$ as fixed end points are almost all located inside a ellipsoid with the points, $\mathbf{r}_i^A$ and $\mathbf{r}_{i+1}^A$ as foci. The volume of this ellipsoid is $\Omega_i \approx (\lambda / P_A^{1/2})^3$. If we choose $\lambda / P_A^{1/2} = l$, Ω_i will contain on average only a few molecules. The interaction with the sites on these molecules inside Ω_i is treated with the full discretization $P_A P_B$ by averaging over a set of B chain configurations. However, for the interaction with the molecules outside Ω_i, a discretization on the atomic scale $\lambda / (P_A P_B)^{1/2} = a_0$ is not necessary, since they are at distances l or greater, with respect to $\mathbf{r}_i^A$ and $\mathbf{r}_{i+1}^A$. The contribution to the propagator due to those molecules outside Ω_i (the majority) can be evaluated using the interaction with the primary chain vertices only, i.e., with a discretization P_A. The size of the nearest neighbor environment Ω_i can be taken larger or even somewhat smaller than $(\lambda / P_A^{1/2})^3$ and is in practice an adjustable parameter in the calculation. In the present application the size of nearest neighbor set is chosen to contain one molecule on average. The criterion for being part of a nearest neighbor set of the A chain link i is that if one site of a molecule is inside the boundary of the volume

TABLE II. Path integral Monte Carlo results for electron solvation in liquid NH_3 at $T = 260$ K.

Run	$P_A \times P_B{}^a$	$N_m{}^b$	V (cm³/mol)	NMC[c]	HISTORY[d]	$\mathscr{R}^e$ (Å)	U^f (eV)	ΔE^g (eV)
1	128×8	250	25.3	580	DELOC	4.23 ± 0.03	-4.5 ± 0.1	1.2 ± 0.2
2	128×8	107	25.3	400	LOC	4.05 ± 0.05	-4.0 ± 0.2	1.4 ± 0.2
3	128×8	107	23.5	400	LOC	3.70 ± 0.02	-4.2 ± 0.1	1.5 ± 0.2
4	128×16	107	23.5	725	LOC	3.76 ± 0.02	-4.3 ± 0.1	1.6 ± 0.2
5	64×16	107	23.5	300	LOC	3.68 ± 0.05	-4.5 ± 0.2	...
6	64×32	107	23.5	600	LOC	3.66 ± 0.02	-4.4 ± 0.1	...
7	64×16	108	23.3	420	DELOC	3.85 ± 0.04	-4.2 ± 0.1	...

[a] P_A is the number of vertices in the primary chain, P_B the number of vertices per link in the secondary chain.
[b] N_m is the number of solvent molecules.
[c] NMC is the number of Monte Carlo passes. One pass consists of P_A attempted moves of the primary vertices chain and $10 \times P_A$ attempted moves of solvent molecules.
[d] Starting configurations for the calculation: LOC (localized in a cavity with $\mathscr{R} \sim 2$ Å), DELOC (delocalized or extended with $\mathscr{R} \sim 7$ Å).
[e] The root mean square separation $\mathscr{R}(\beta\hbar/2)$ as defined in Eq. (8) of the text.
[f] Coulombic potential energy of the electron. The calculated kinetic energy is one half $|U|$, within its statistical error of about 15%; the total electron energy is thus essentially $U/2$.
[g] Mean electron excitation energy estimated from the complex time dependence of $\mathscr{R}^2$.

Ω_i, the entire molecule (i.e., all its sites) is considered inside. Hence, it is included in the direct sampling by B chains on the link between $\mathbf{r}_i^A$ and $\mathbf{r}_{i+1}^A$. This device of variable discretization leads to a considerable saving in computer time and is in practice the reason that simulations of isomorphic systems of $P \approx 10^3$–10^4 are feasible.

We now return to the problem of the accuracy of the evaluation of the weights for the primary chain links by means of direct sampling. For the degree of discretization discussed above the effective potential which is sampled by the B chains, is rather smooth; e.g., at the classical singularity $r = 0$, the effective potential (11) has an absolute value of $\beta V_q/P \approx 2.5$ for $P \approx 10^3$. Moreover, as with the case of two hard spheres,[16] it was found that the averages for the pair Coulomb problem are more efficiently determined by means of direct sampling from a distribution of free chain configurations. This was particularly evident for the larger values of P in Table I. Our conclusion is therefore that, provided $\lambda / P_A^{1/2} \sim l$, the attrition problem is not too serious. As a simple test we have examined the acceptance rates of zero displacements of primary chain particles, interacting with moving molecules. If the calculation of the weights used in the Metropolis algorithm is exact the acceptance rate is, of course, unity. Any inaccuracy reduces the acceptance rate. Our results, presented in the next section, were obtained using a set of 100 independent B chain configurations. For the systems listed in Table II, the zero displacement test gives acceptance rates in the range 0.95–0.99.

C. Simulation of the fluid

The polar fluid was simulated by means of the familiar Monte Carlo procedure of displacing the center of mass and rotating the orientation of a rigid molecule simultaneously.[41] Cubic periodic boundary conditions were applied to the molecular centers of mass. The electrostatic interactions between the solvent molecules ($N_m = 108 \sim 250$) were calculated with a spherical cutoff applied to the center of mass separation of a pair of molecules: If this distance is larger than 8 Å the interaction between the two molecules is neglected, if the distance is smaller the full electrostatic interaction is taken into account. Ten arbitrary solvent molecules are moved for each displacement of a primary chain vertex. This choice ensures that on average one molecule near the electron (i.e., inside the nearest neighbor maps discussed above) is moved for each move of a primary chain vertex. Unlike the solvent–solvent interaction, the electrostatic interaction between the isomorphic system and the solvent was evaluated by applying the minimum image convention to the difference coordinates of a chain vertex and the individual charge sites. The reason for this somewhat unusual truncation is technical and is not related to the physics of the problem. The effect of this procedure remains to be investigated. The simulation cell box length (see data Table II, next section) is comparable to the thermal wavelength of the electron. The fluid sample is therefore too small for a simulation of a completely extended or delocalized state. However, the thermal wave packet of a solvated electron in dense polar fluids is known to be considerably smaller than its free value. Accordingly, calculations commenced either with an electron thermal wave packet (diameter of primary chain) of about 7 Å, which is still considerably larger than the size of the solvent molecules, or in a compact state (diameter ~ 2 Å) localized in a cavity of the solvent (see the next section).

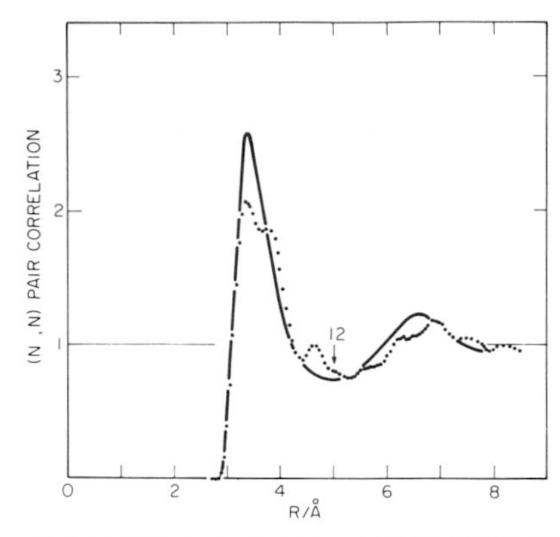

FIG. 1. Pair distribution function for the ammonia solvent (bold curve) and experimental room temperature x-ray data from Ref. 42 (dots).

The normal liquid range of ammonia is $T = 195$ K, $V = 23.3$ cm^3/mol at the triple point to $T = 269$ K, $V = 26.5$ cm^3/mol at the boiling point. Since the characteristics of the solvated electron are likely to depend more critically on the density than the temperature we have carried out all of our Monte Carlo calculations at one temperature ($T = 260$ K) but two molar volumes have been examined ($V = 23.3$ and 25.3 cm^3/mol). Figure 1 shows the pair distribution function of the solvent nitrogen atoms in the denser system. This is virtually indistinguishable from that reported previously for the pure liquid.[29] Moreover, the agreement with room temperature x-ray data[42] is reasonably good (see Fig. 1), especially since the latter refers to a higher temperature and larger molar volume.

V. RESULTS

A. Structure of solvated electron state

The first Monte Carlo study was performed on a system with $N_m = 250$ at $V = 25.3$ cm^3/mol. The initial state was a body centered cubic lattice with randomly oriented NH_3 molecules. The state of the electron was created by inserting a free chain ($P_A = 128$) appropriate to a particle with a mass five times larger; this procedure yields a value of $\mathscr{R} \sim 7$ Å. The electron mass was then immediately reset to its correct value and the calculation proceeded as outlined in the previous section using $P_B = 8$. After 200 passes through the A chain (i.e., 200×128, A-chain vertex moves) and ten times the number of moves of the solvent molecules, the value of $\mathscr{R}$ reduced to about 4 Å. During this period the Coulombic potential energy of the electron decreased from $\beta V \sim -70$ to -200. The system then appeared to be stable showing only modest fluctuations around these values. After an equilibration of 400 passes, averaging commenced. Figure 2

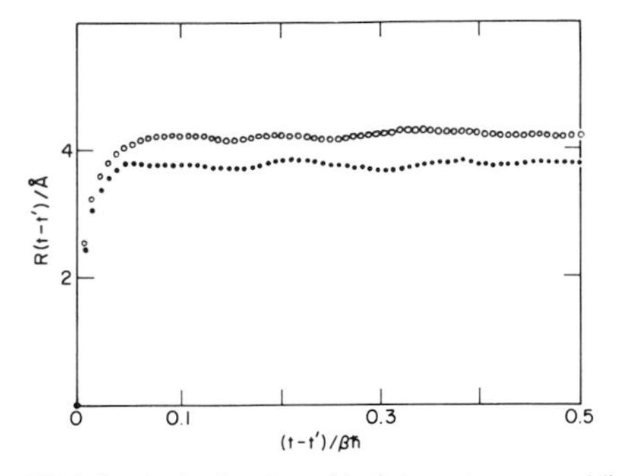

FIG. 2. Complex time dependence of the electron root mean squared displacement correlation function for an electron in liquid ammonia, open and full circles refer to runs 1 and 4 of Table II. The short time behavior is used to estimate ΔE, the excitation energy (see the text).

shows the function $\mathscr{R}(t-t')$ for this run, averaged over 600 further passes. A parabolic dependence on the complex time would imply an extended electron state but the observed behavior is clearly indicative of a compact equilibrium state[28] with $\mathscr{R}\sim 4$ Å. This finding is in excellent accord with previous experimental and theoretical deductions on the nature of the behavior of an electron in liquid ammonia.[1-14]

In order to better characterize the nature of this compact state the pair correlation functions for the electron center-of-mass (cm) and the solvent H and N atoms were examined. Apart from indicating that the electron resides in a cavity, these appeared to be rather featureless; the former having a weak maximum at about 2.1 Å and the latter a somewhat more enhanced peak at about 2.9 Å. Integration of the (cm–N) correlation function out to 4.3 Å gave a coordination number of about 8.

An attempt was made to enhance the structural features in the (cm–H) and (cm–N) pair correlation functions by carrying out additional Monte Carlo runs on a denser system ($V=23.3$ cm^3/mol). Also systematic studies were performed on the effect of varying both P_A and P_B. The results of all these calculations are summarized in Table II under the heading DELOC. Only modest changes were induced in the pair correlation functions and the coordination number remained essentially unchanged. The value of $\mathscr{R}$ decreased by about 10%. Since the solvated electron is clearly stable, i.e., we have observed the spontaneous formation of this compact state it is instructive to make a comparison with the solvation of a classical anion. To this end classical molecular dynamics calculations[43] were carried out on a system of 107 NH_3 molecules and one anion (X^-). The X–N (nonelectrostatic) potential was assumed to be identical to the N–N fluid ammonia potential, the anion carried a unit negative charge and no short range X–H potential was used. The (X–H) and (X–N) pair correlation functions (shown in Figs. 3 and 4) were very highly structured, in marked contrast to the results for a solvated electron. Integration of the (X–N) pair correlation function out to 4.3 Å yields a coordination number of about 9, which is very similar to that found for the

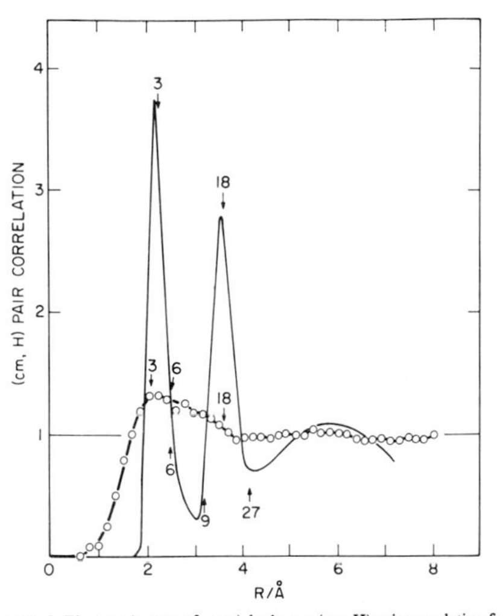

FIG. 3. Electron (center-of-mass)–hydrogen (cm–H) pair correlation function (circles) and corresponding function for a classical anion (thin curve) in liquid ammonia (run 4 of Table II). Coordination numbers are indicated by the arrows and Arabic numerals.

electron. Integration of the (X–H) pair correlation function out to the same distance yields a coordination number of about 27, which is entirely consistent with the (X–N) curve. The presence of two well resolved peaks in the (X–H) curve

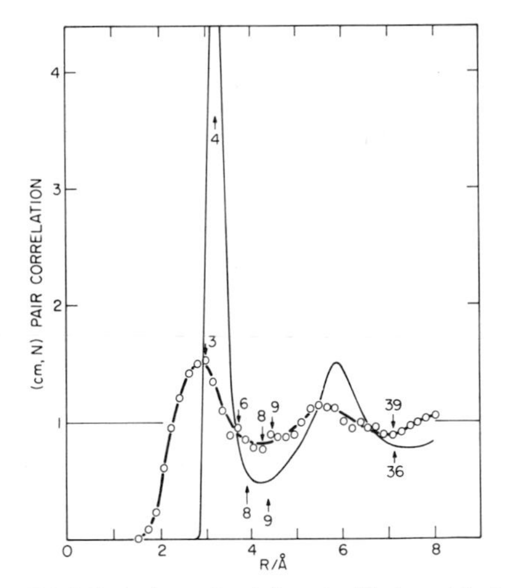

FIG. 4. Electron (center-of-mass)–nitrogen (cm–N) pair correlation function (circles) and corresponding function for a classical anion (thin curve) in liquid ammonia (run 4 of Table II). Coordination numbers are indicated by the arrows and Arabic numerals.

indicates that the anion solvation sheath structure is such that one H atom from each of the nine closest NH_3 molecules points directly at the anion, i.e., a bond-ordered structure.

Since the anion solvation sheath is highly ordered and has similar coordination numbers to that found for the solvated electron, it was decided to use this solvent configuration to initiate a path integral Monte Carlo calculation. Accordingly, the anion was removed and an electron with $\mathscr{R} \sim 2$ Å was inserted in its place. This initial state served as the genesis of the results quoted in Table II under the heading LOC. The molar volume for these runs was $V = 23.5$ cm^3/mol. Under these conditions the electron expanded in only 50 passes to $\mathscr{R} \sim 3.7$ Å. After suitable equilibration, averages were collected over 500 passes. The function $\mathscr{R}(t - t')$ for one of these runs is shown in Fig. 2 where it can be compared with the result for $V = 25.3$ cm^3/mol which was initialized from a delocalized state. Apart from a smaller asymptotic value the curves appear to be similar.

The pair correlation functions shown in Figs. 3 and 4 for the electron have a more clearly defined structure than obtained from those runs which were initiated from expanded states (DELOC in Table II). However, they are much less structured than those found for the classical anion but, as noted previously, the coordination numbers are very similar. The lack of a clear second peak in the electron cm–H distribution would seem to indicate a considerable degree of disorder in the bond orientation of the solvation sheath. Thus the diffuse nature of the electron charge density distribution enables the H atoms to achieve energetically favorable configurations without pointing directly at the center of mass of the electron. This would seem to be an important distinction between a classical and quantum solute.

Figure 5 shows the electron–positive charge site (Q^+) pair correlation function. The overall shape of this curve resembles quite closely one of the electron–potassium ion curves calculated for an electron in molten KCl (Fig. 4 in Ref. 19). Here, the electron density decays rather more rapidly but in a fashion consistent with that expected for a hydrogen-like atom carrying a nuclear charge $Q^+ \sim 0.5e$. In Fig. 6 the negative charge site (Q^-) and the N atom of the NH_3 molecule have been averaged together in constructing the pair correlation function. The curve contains a small core exclusion region but is otherwise quite featureless.

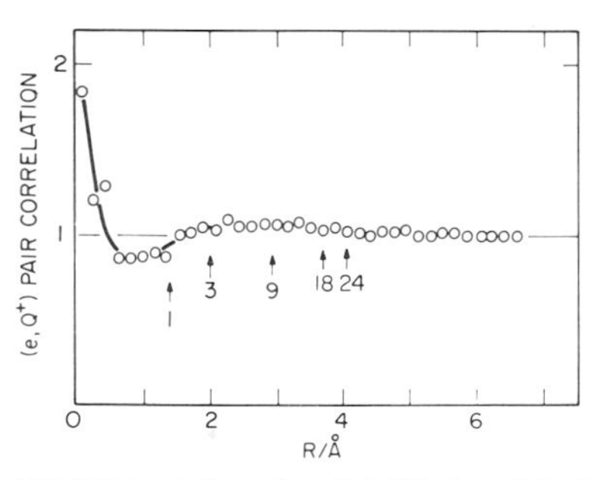

FIG. 5. Electron–hydrogen charge site (e–Q^+) pair correlation function for an electron solvated in liquid ammonia.

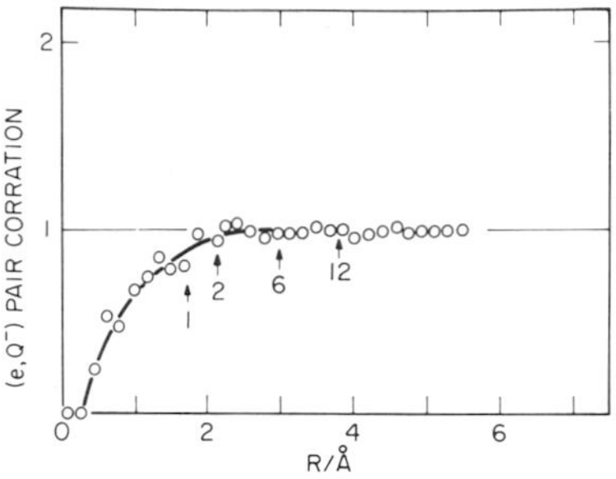

FIG. 6. The averaged electron–nitrogen and –negative charge site (e–Q^-) pair correlation function for an electron in liquid ammonia.

B. Absorption spectrum

The initial (complex) time dependence of the functions $\mathscr{R}^2(t - t')$ defined in Eq. (7) were analyzed to obtain the characteristic rise time τ. The time τ is obtained by subtracting the first five data points of $\mathscr{R}^2(t - t')$ from the asymptotic value of $\mathscr{R}^2$ at large $t - t'$ and then fitting the logarithm of this difference to a straight line. It was found that a linear relation is satisfied within the statistical error. The resulting excitation energies ΔE derived from τ by means of Eq. (9) are quoted in Table II. These energies are considerably greater than found experimentally (0.8 eV) at the same liquid densities.[10] However, it should be noted that the measured absorption spectrum is very asymmetric[10] and the procedure used here to estimate ΔE will likely correspond to an ill-defined mean excitation energy for the whole spectrum. Even allowing for the possible contribution from higher excited states, our calculated value ΔE would seem to be too large, the discrepancy being greater than likely systematic errors in the calculations (see Table II).

VI. DISCUSSION

The quantum path integral Monte Carlo technique has been applied to the problem of electron solvation in liquid ammonia. In agreement with previous theoretical and experimental evidence we find the minimum energy state of the electron to be highly compact or localized. This result was not imposed *a priori* since we have shown that an electron first prepared in an extended state with a diameter considerably larger than an ammonia molecule, spontaneously localized into a compact configuration. Complimentary MD calculations on the solvation of a classical anion have been carried out and yield a more highly structured configuration with bond order. The quantum nature of the electron gives rise to a considerable blurring of pair distribution functions compared with a classical solute. The peak in the electron (cm)–nitrogen atom distribution function occurs at about 2.9 ± 0.1 Å which is in good agreement with a value based upon a static model.[11] The coordination number is about 8, which is higher than previous estimates.[11,12]

The mean excitation energy ΔE is calculated to be considerably larger than the observed peak in the absorption spectrum. This discrepancy may well indicate that the simulated electron state is too compact. If true, this would imply that the electron pair correlation functions shown in Figs. 3 and 4 are too structured. This finding may be related to the fact that the electron–ammonia potential we employ is rather naive, the contributions of polarization and the short range electron repulsion having been ignored. Future work must necessarily focus on the consequences of using a more refined potential.

The staging algorithm we employ for the path integral Monte Carlo calculations has ensured an adequate sampling of the possible quantum paths of the solvated electron. This algorithm is particularly suited to a vector computer. Hence, our sampling is more efficient than in previous work based on MD codes; calculations having been performed with an effective discretization of $P = 2048$. Since the studies reported here did not involve prodigious amounts of computing, analogous calculations for other polar fluids are in hand.

ACKNOWLEDGMENTS

We thank Peter Rossky, Anees Rahman, and David Chandler, for freely communicating information concerning their own related but different calculations on solvated electrons and Barry Schneider and Richard LeSar for valuable discussions on electron–molecule interactions. The calculations were performed on the Cyber 205 of the Advanced Computing Technology Centre of the University of Calgary.

[1] *Metal–Ammonia Solutions*, edited by G. Lepoutre and M. J. Sienko (Benjamin, New York, 1964), Collogue Weyl I.
[2] *Metal–Ammonia Solutions*, edited by J. J. Lagowski and M. J. Sienko (Butterworths, London, 1970), Colloque Weyl II.
[3] *Electrons in Fluids*, edited by J. Jortner and N. R. Kestner (Springer, Berlin, 1973), Colloque Weyl III.
[4] J. Phys. Chem **79**, 2789–3079 (1975), Colloque Weyl IV.
[5] J. Phys. Chem **84**, 1065–1298 (1980), Colloque Weyl V, and references therein.
[6] J. Phys. Chem. **88**, 3699–3914 (1984), Colloque Weyl VI, Le Dernier Colloque Weyl, and references therein.
[7] Can. J. Chem. **55**, 1795–2279 (1977).
[8] J. Jortner and A. Gaathon, Can. J. Chem. **55**, 1801 (1977).
[9] P. Krebs, J. Phys. Chem. **88**, 3702 (1984).
[10] F. -Y. Jou and G. R. Freeman, J. Phys. Chem. **85**, 629 (1981).
[11] M. D. Newton, J. Phys. Chem. **79**, 2795 (1975), earlier references are quoted herein.
[12] N. R. Kestner, Can. J. Chem. **55**, 1937 (1977).
[13] P. R. Antoniewicz, G. T. Bennett, and J. C. Thompson, J. Chem. Phys. **77**, 4573 (1982).
[14] N. R. Kestner and J. Jortner, J. Phys. Chem. **88**, 3818 (1984).
[15] R. P. Feynman and A. R. Hibbs, *Quantum Mechanics and Path Integrals* (McGraw-Hill, New York, 1965); R. P. Feynman, *Statistical Mechanics* (Benjamin, Reading, MA, 1972).
[16] J. A. Barker, J. Chem. Phys. **70**, 2914 (1979); G. Jacucci and E. Omerti, *ibid.* **79**, 3051 (1983).
[17] D. Chandler and P. G. Wolynes, J. Chem. Phys. **74**, 4078 (1981); K. S. Schweizer, R. M. Stratt, D. Chandler, and P. G. Wolynes, *ibid.* **75**, 1347 (1981).
[18] M. F. Herman, E. J. Bruskin, and B. J. Berne, J. Chem. Phys. **76**, 5150 (1982).
[19] M. Parrinello and A. Rahman, J. Chem. Phys. **80**, 860 (1984).
[20] B. DeRaedt, M. Sprik, and M. L. Klein, J. Chem. Phys. **80**, 5719 (1984).
[21] E. L. Pollock and D. M. Ceperley, Phys. Rev. B **30**, 2555 (1984).
[22] R. A. Kuharski and P. J. Rossky, Chem. Phys. Lett. **103**, 357 (1984).
[23] D. Thirumalai, R. W. Hall, and B. J. Berne, J. Chem. Phys. **81**, 2523 (1984).
[24] R. W. Hall and B. J. Berne, J. Chem. Phys. **81**, 3641 (1984).
[25] M. Sprik, M. L. Klein, and D. Chandler, Phys. Rev. B **31**, 4234 (1985); **32**, 545 (1985); J. Chem. Phys. **83**, 3042 (1985).
[26] J. Bartholomew, R. W. Hall, and B. J. Berne, Phys. Rev. B **32**, 548 (1985).
[27] D. Thirumalai and B. J. Berne, Chem. Phys. Lett. **116**, 471 (1985).
[28] D. Chandler, Y. Singh, and D. M. Richardson, J. Chem. Phys. **81**, 1975 (1984); A. L. Nichols III, D. Chandler, Y. Singh, and D. M. Richardson, *ibid.* **81**, 5109 (1984).
[29] R. W. Impey and M. L. Klein, Chem. Phys. Lett. **104**, 579 (1984).
[30] D. G. Truhlar, K. Onda, R. A. Eades, and D. A. Dixon, Int. J. Quantum Chem. Symp. **13**, 601 (1979).
[31] Y. Itikawa, J. Phys. Soc. Jpn. **30**, 835 (1971); **31**, 1532 (1971).
[32] M. A. Morrison and L. A. Collins, Phys. Rev. A **17**, 918 (1978); **23**, 127 (1981); L. A. Collins and D. W. Norcross, *ibid.* **18**, 467 (1978); B. I. Schneider and L. A. Collins, *ibid.* **30**, 95 (1984).
[33] M. R. H. Rudge, J. Phys. B **13**, 1269 (1980); **11**, 1503, 2221 (1978).
[34] A. Jain and D. G. Thompson, J. Phys. B **16**, 1113, 2593 (1983); **15**, L631 (1982).
[35] V. Heine, Solid State Phys. **24**, 1 (1970).
[36] R. W. Shaw, Phys. Rev. **174**, 769 (1968).
[37] R. G. Storer, Phys. Rev. **176**, 326 (1968); J. Math Phys. **9**, 964 (1968).
[38] A. A. Barker, J. Chem. Phys. **55**, 1751 (1971).
[39] C. Deutsch and M. M. Gombert, J. Math. Phys. **17**, 1077 (1976); C. Deutsch, Phys. Lett. A **60**, 317 (1977); C. Deutsch, M. M. Gombert, and H. Minoo, *ibid.* **66**, 381 (1978).
[40] J. P. Hansen and I. R. McDonald, Phys. Rev. A **23**, 2041 (1981).
[41] J. A. Barker and R. O. Watts, Chem. Phys. Lett. **3**, 144 (1969).
[42] A. H. Narten, J. Chem. Phys. **66**, 3117 (1977).
[43] R. W. Impey, P. A. Madden, and I. R. McDonald, J. Phys. Chem. **87**, 5071 (1983).

Path-Integral Computation of the Low-Temperature Properties of Liquid ^{4}He

D. M. Ceperley and E. L. Pollock

Lawrence Livermore National Laboratory, University of California, Livermore, California 94550

(Received 28 October 1985)

Discretized path-integral computations of the energy and radial distribution function of ^{4}He in good accord with experiment are presented for temperatures down to 1 K at saturated vapor pressure. Results for the single-particle density matrix, momentum distribution, and condensate fraction agree at the lowest temperature with previous ground-state calculations.

PACS numbers: 67.40.–w

The unusual properties of liquid ^{4}He at low temperature were attributed to Bose-Einstein condensation by London in 1938.[1] The strength of the pair interaction between helium atoms, however, has so far prevented a first-principles study of this transition in ^{4}He. In this Letter we present a Monte Carlo discretized path-integral computation of the density matrix for liquid ^{4}He for temperatures spanning this transition which reproduces many of the experimental results and is in principle capable of arbitrary accuracy. We have assumed that the atoms interact via the Aziz pair potential.[2]

The calculations for the many-body density matrix,[3]

$$\rho(R,R';\beta) = \langle R | e^{-\beta H} | R' \rangle, \tag{1}$$

from which all equilibrium properties can be obtained, are based on the identity[3]

$$\rho(R,R';\beta) = \int \cdots \int \rho(R,R_1;\tau)\rho(R_1,R_2;\tau) \cdots \rho(R_{M-1},R';\tau)\,dR_1 \cdots dR_{M-1}, \tag{2}$$

where $\tau = \beta/M$, $M > 1$, and the R variables denote points in the $3N$-dimensional coordinate space. If an accurate many-body density matrix is known at some high temperature corresponding to τ then Eq. (2) allows its calculation at a lower temperature $T = 1/Mk\tau$. The density matrix for Bose systems is obtained by summing over all permutations of particle labels:

$$\rho_B(R,R';\beta) = (N!)^{-1} \sum_P \rho(R,PR';\beta). \tag{3}$$

Both the integral over paths and the sum over permutations are performed by a generalization of the Metropolis

TABLE I. Computed potential and kinetic energies for various temperatures at SVP. The statistical uncertainty in the potential energy is about 0.04 and 0.08 K in the kinetic energy. The densities used are based on Crawford.[a] The last six columns give the first three zeros and extremal values of the pair correlation function $h(r) = g(r) - 1$. The last row at 2.0 K is for distinguishable particles.

T(°K)	ρ(Å^{-3})	<U>/N (°K)	<K>/N (°K)	r_1(Å)	r_2(Å)	r_3(Å)	h_1	h_2	h_3
4.0	.01932	-18.91	15.65	2.973	4.583	6.271	.356	-.113	.040
3.333	.02072	-20.38	16.00	2.962	4.552	6.240	.369	-.126	.050
2.857	.02142	-21.14	15.99	2.950	4.537	6.234	.381	-.134	.055
2.50	.02179	-21.52	15.90	2.948	4.533	6.220	.383	-.136	.059
2.353	.02191	-21.60	15.75	2.945	4.526	6.205	.382	-.136	.060
2.222	.02197	-21.70	15.89	2.945	4.526	6.205	.382	-.139	.061
2.105	.02194	-21.57	15.10	2.940	4.522	6.205	.374	-.131	.055
2.0	.02191	-21.57	15.05	2.942	4.525	6.211	.377	-.132	.056
1.818	.02186	-21.44	14.71	2.939	4.526	6.209	.369	-.127	.053
1.600	.02183	-21.39	14.40	2.938	4.520	6.213	.368	-.124	.051
1.379	.02182	-21.35	14.23	2.937	4.515	6.206	.366	-.123	.050
1.1765	.02182	-21.35	14.17	2.938	4.520	6.212	.366	-.123	.050
2.0	.02191	-21.75	16.24	2.950	4.535	6.205	.385	-.140	.064

[a]Reference 5.

Monte Carlo method. A discussion of how this is implemented for distinguishable particles is given by Pollock and Ceperley.[4] In the extension of this work to bosons several new techniques were required which will be described in detail elsewhere, but before presenting our results we briefly mention two of the most important.

First (as in Ref. 4) the many-body density matrix at high temperature in Eq. (2) is taken as a product of one- and two-body density matrices which is exact in the high-temperature limit. Here we have used the full two-body density matrix rather than the "end point" approximation of earlier work. This is more accurate and allows larger values of τ (smaller M) to be used in Eq. (2). The high-temperature density matrix used was typically for a temperature of 40 K and thus paths of about twenty steps were needed for computations near T_λ. Had we been interested in only the structural properties rather than, for example, the kinetic energy, steps corresponding to 20 K or less would have sufficed. We have checked the adequacy of the step size by rerunning selected points using 80-K steps. A thorough convergence study of the earlier method was done in Ref. 4.

Secondly, a new method was used to construct trial paths for the multiparticle moves necessary in the sampling of the permutations of Eq. (3). The particular particles (here as many as four) for which permutation changes are attempted at one Monte Carlo move are initially selected on the basis of the free-particle density matrix. New trial paths are then generated by a "bisection method" which first generates new midpoints for paths and then new midpoints for the remaining halves and so on, with the possibility of rejecting the new paths at any stage in the construction. For permutations this is more efficient than the previous method of sequentially generating new paths step by step since now improbable paths may be rejected at an early stage in their construction, thus allowing many more trial moves for a given amount of computer time. The rejection step ensures that the accepted permutations and paths reflect the correct density matrix and not our initial guesses. Extensive tests of the convergence of the distribution of permutations were carried out.

Table I lists some of the temperatures and densities at saturated vapor pressure (SVP) along with the potential and kinetic energy and some structural properties where computations were done. The computed energy and specific heat as functions of temperature at SVP near T_λ are compared with experiment in Fig. 1. The simulations are for a periodic system of 64 atoms and each run takes about one hour on the Cray-1. The finite number of particles used in these simulations apparently depresses the computed energy in the temprerature region 2.1 K < T < 3 K.

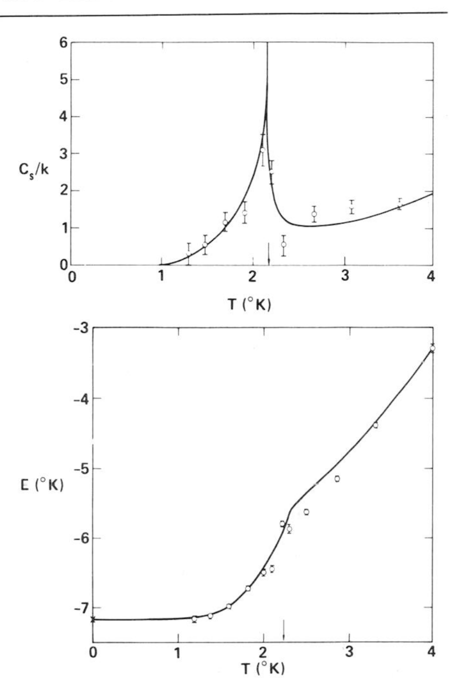

FIG. 1. Energy and specific heat at SVP near T_λ. The solid lines are the experimental values. [The energy was taken from Ref. 5 and the specific heat from Wilks (Ref. 6).] The simulation results for the specific heat were obtained by a differencing of the energy values. The energy computed from ground-state simulations (Ref. 7) is denoted by the cross. The experimental value for T_λ (2.17 K) is indicated by the arrow.

The effect of Bose statistics on the radial distribution function, $g(r)$, is small as shown at $T-2.0$ K SVP in Fig. 2(a) where the neutron-scattering results[8] are compared with the present simulation. The dashed line shown at the first peak and first minimum is for distinguishable particles [only the identity permutation is allowed in Eq. (3)] and shows the slightly increased spatial ordering attributed, via the uncertainty principle, to the decreased ordering in momentum space when the condensate is suppressed. Similar good agreement is obtained between the computed radial distribution and structure functions and the available neutron and x-ray[9] scattering data at other temperatures and pressures in the liquid phase.

The single-particle momentum distribution, $n(k)$, is the Fourier transform of the single-particle off-

Note on p. 477

diagonal density matrix, $n(r)$ [10]:

$$n(r)=\frac{\int \rho_B(r_1,r_2,\ldots,r_n,r_1+r,r_2,\ldots,r_n;\beta)dr_1\cdots dr_n}{\int \rho_B(r_1,r_2,\ldots,r_n,r_1,r_2,\ldots,r_n;\beta)dr_1\cdots dr_n}, \tag{4}$$

which in terms of path integrals corresponds to one open path beginning at r_1 and ending at r_1+r. (Here the r_i are the coordinates of atom i.) At temperatures well above T_λ where only the identity permutation is important, this open path involves only one particle and is restricted to a distance on the order of the thermal wavelength, $\hbar/(2mkT)^{1/2}$. This is primarily due to the free-particle part of the density matrix somewhat modified by many-body effects. Below T_λ this open path may involve a long cyclic permutation of many particles and the end-to-end distance will become macroscopic. The $n(r)$ in Fig. 2(b) shows this change in character on going through the transition.[12] The initial curvature is proportional to the kinetic energy and the value at large r is the fraction of particles in the zero-momentum state, the condensate. Figure 3(a) shows this condensate fraction as a function of temperature. The condensate fractions plotted there are obtained by our assuming $n(r)$ to be constant beyond 5 Å and averaging the values between 5 and 7 Å to obtain $n_0(T)$. Near T_λ $n(r)$ reaches its asymptotic value slowly and this procedure, because of the relatively small system simulated and periodic boundary effects, is not reliable. For example we find an $n_0(T)$ value of 1.4% at 2.5 K, significantly above the experimental transition temperature. Larger systems must be considered to determine the condensate fraction near T_λ. The momentum distribution of the noncondensed par-

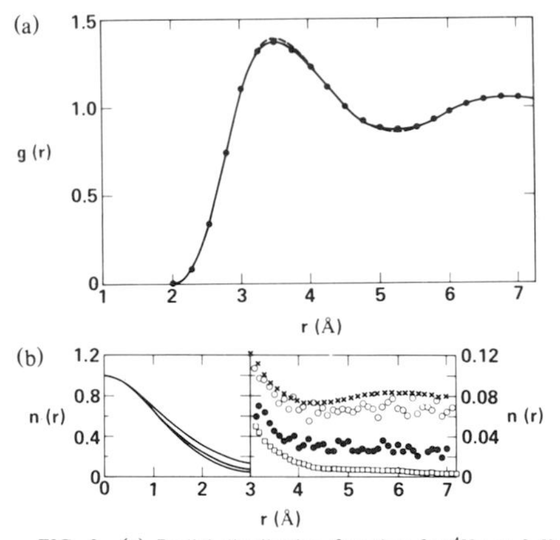

FIG. 2. (a) Radial distribution function for ^{4}He at 2 K and SVP. The solid line is the neutron-scattering result (Ref. 8). The circles are simulation results for bosons and the dashed line is for distinguishable particles. (b) Single-particle off-diagonal density matrix at 1.18 K (top curve and open circles), 2.22 K (middle curve and closed circles), and 3.33 K (lower curve and open squares). Beyond 3 Å the vertical axis is enlarged by 10 times and the interpolating curves are omitted. The crosses denote the ground-state results (Ref. 11) which are indistinguishable from the $T=1.18$ K results for $r<3$ Å on this graph.

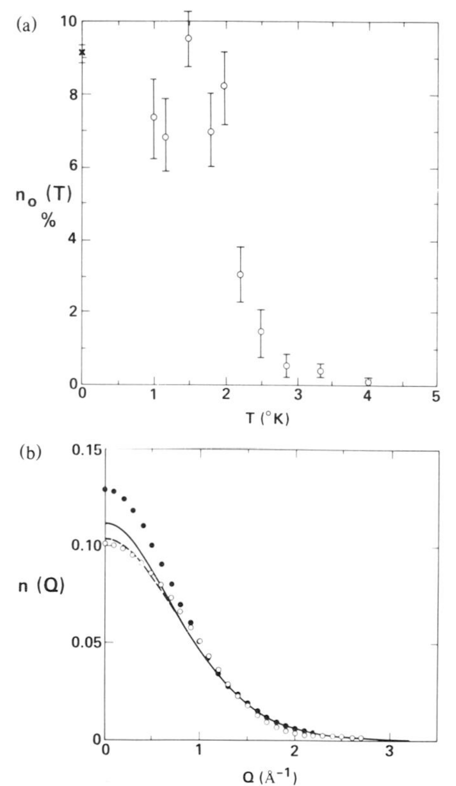

FIG. 3. (a) Percentage of atoms with zero momentum, $n_0(T)$, in ^{4}He at SVP. The indicated ground-state value is from Ref. 7. (b) Momentum distribution from simulations at temperatures of 3.33 K (solid curve), 2.22 K (dashed curve), 1.18 K (open circles), and for distinguishable particles at 2.22 K (solid circles).

ticles, Fig. 3(b), both is non-Gaussian and has a temperature-dependent shape. The present simulations are too noisy to test adequately the predicted low-momentum singularities in this distribution.[13]

In the past, estimates of the condensate fraction have been made[14] based on the hypothesis of Hyland, Rowlands, and Cummings[15] that the pair correlation function at large r has a constant shape below T_λ and is multiplied by $[1-n_0(T)]^2$ (intuitively speaking the probability is that neither atom in the pair is in the condensate and thus spatially uniform). We can only test this at moderate values of r but an estimate of $n_0(T)$ based on the second maximum, h_3, listed in Table I does not conflict with our results. Estimates based on the first minimum, h_2, at smaller r seem definitely too small. Another intuitive estimate[16] of $n_0(T)$ assumes the contribution of noncondensed atoms to the kinetic energy to be unchanged below T_λ where the condensate makes no contribution and thus the kinetic energy is proportional to $1-n_0(T)$. Figure 3(b) suggests that this assumption is only approximate; nevertheless this estimate, using Table I, also accords with our results of the still-sizable error bars in the $n_0(T)$ estimates.

Efforts are under way to extend these simulations to larger systems and to determine other properties of ^{4}He.

This work was performed under the auspices of the U.S. Department of Energy by the Lawrence Livermore National Laboratory under Contract No. W-7405-ENG-48.

[1]F. London, Nature (London) **141**, 643 (1938).

[2]R. A. Aziz, V. P. S. Nain, J. S. Carley, W. L. Taylor, and G. T. McConville, J. Chem. Phys. **70**, 4330 (1979).

[3]R. P. Feynman, *Statistical Mechanics* (Benjamin, Reading, Mass., 1972).

[4]E. L. Pollock and D. M. Ceperley, Phys. Rev. B **30**, 2555 (1984).

[5]R. K. Crawford, in *Rare Gas Solids,* edited by M. L. Klein and J. A. Venables (Academic, New York, 1976), Vol. 1, p. 663.

[6]J. Wilks, *Properties of Liquid and Solid Helium* (Clarendon, Oxford, 1967).

[7]M. H. Kalos, M. A. Lee, P. A. Whitlock, and G. V. Chester, Phys. Rev. B **24**, 115 (1981).

[8]E. C. Svensson, V. F. Sears, A. D. Woods, and P. Martel, Phys. Rev. B **21**, 3638 (1980).

[9]H. N. Robkoff and R. B. Hallock, Phys. Rev. B **24**, 159 (1981).

[10]O. Penrose and L. Onsager, Phys. Rev. **104**, 576 (1956).

[11]P. A. Whitlock, private communication of work in progress. For values of r less than the de Broglie thermal wavelength the $n(r)$ from the finite-temperature simulations is of comparable accuracy with the ground-state result. At larger r it proved necessary actually to simulate a system with one open path, as described in the text, and as seen in Fig. 2(b) the scatter in the finite-temperature results is greater there.

[12]It is interesting to note that a formal correspondence between Bose-Einstein condensation and polymerization was noted over 35 years ago. W. H. Stockmayer and B. H. Zimm, Annu. Rev. Phys. Chem. **35**, 1 (1984), and references therein.

[13]A. Griffin, Phys. Rev. B **30**, 5057 (1984).

[14]V. F. Sears and E. C. Svensson, Int. J. Quantum Chem. **14**, 715 (1980).

[15]G. J. Hyland, G. Rowlands, and F. W. Cummings, Phys. Lett. **31A**, 465 (1970).

[16]V. F. Sears, Phys. Rev. B **28**, 5109 (1983).

Note to Reprint VI.7

No use is made in this paper of the classical, ring-polymer analogy. The sampling method used in the work is describable in terms of a closed, random walk over a discrete set of points in configuration space (Pollock and Ceperley, Phys. Rev. B 30 (1984) 2555).

Unified Approach for Molecular Dynamics and Density-Functional Theory

R. Car
International School for Advanced Studies, Trieste, Italy

and

M. Parrinello
Dipartimento di Fisica Teorica, Università di Trieste, Trieste, Italy, and International School for Advanced Studies, Trieste, Italy
(Received 5 August 1985)

We present a unified scheme that, by combining molecular dynamics and density-functional theory, profoundly extends the range of both concepts. Our approach extends molecular dynamics beyond the usual pair-potential approximation, thereby making possible the simulation of both covalently bonded and metallic systems. In addition it permits the application of density-functional theory to much larger systems than previously feasible. The new technique is demonstrated by the calculation of some static and dynamic properties of crystalline silicon within a self-consistent pseudopotential framework.

PACS numbers: 71.10.+x, 65.50.+m, 71.45.Gm

Electronic structure calculations based on density-functional (DF) theory[1] and finite-temperature computer simulations based on molecular dynamics[2] (MD) have greatly contributed to our understanding of condensed-matter systems. MD calculations are able to predict equilibrium and nonequilibrium properties of condensed systems. However, in all practical applications MD calculations have used empirical interatomic potentials. This approach, while appropriate for systems like the rare gases, may fail for covalent and/or metallic systems. Furthermore, these calculations convey no information about electronic properties. On the other hand, DF calculations have provided an accurate, albeit approximate, description of the chemical bond in a large variety of systems,[1] but are computationally very demanding. This has so far precluded the application of DF schemes to the study of very large and/or disordered systems and to the computation of interatomic forces for MD simulations.

We wish to present here a new method that is able to overcome the above difficulties and to achieve the following results: (i) compute ground-state electronic properties of large and/or disordered systems at the level of state-of-the-art electronic structure calculations; (ii) perform *ab initio* MD simulations where the only assumptions are the validity of classical mechanics to describe ionic motion and the Born-Oppenheimer (BO) approximation to separate nuclear and electronic coordinates.

Following Kohn and Sham[3] (KS) we write the electron density in terms of occupied single-particle orthonormal orbitals: $n(\mathbf{r})=\sum_i|\psi_i(\mathbf{r})|^2$. A point of the BO potential energy surface is given by the minimum with respect to the $\psi_i(\mathbf{r})$ of the energy functional,

$$E[\{\psi_i\},\{R_I\},\{\alpha_\nu\}]=\sum_i\int_\Omega d^3r\,\psi_i^*(\mathbf{r})[-(\hbar^2/2m)\nabla^2]\psi_i(\mathbf{r})+U[n(\mathbf{r}),\{R_I\},\{\alpha_\nu\}]. \quad (1)$$

Here $\{R_I\}$ indicate the nuclear coordinates and $\{\alpha_\nu\}$ are all the possible external constraints imposed on the system, like the volume Ω, the strain $\epsilon_{\mu\nu}$, etc. The functional U contains the internuclear Coulomb repulsion and the effective electronic potential energy, including external nuclear, Hartree, and exchange and correlation contributions.

In the conventional formulation, minimization of the energy functional [Eq. (1)] with respect to the orbitals ψ_i, subject to the orthonormality constraint, leads to the self-consistent KS equations, i.e.,

$$\left\{-\frac{\hbar^2}{2m}\nabla^2+\frac{\delta U}{\delta n(\mathbf{r})}\right\}\psi_i(\mathbf{r})=\epsilon_i\psi_i(\mathbf{r}). \quad (2)$$

The solution of Eq. (2) involves repeated matrix diagonalizations with a computational effort rapidly growing with the size of the problem. Since the whole procedure has to be repeated for any new atomic configuration, the theoretical prediction of the equilibrium geometries, when these are not known from experiment, still remains an unsolved problem in most cases.

We adopt a quite different approach and regard the minimization of the KS functional as a complex optimization problem which can be solved by applying the concept of simulated annealing, recently introduced by Kirkpatrick, Gelatt, and Vecchi.[4] In this approach an objective function $O(\{\beta\})$ is minimized relative to the parameters $\{\beta\}$, by generation of a succession of $\{\beta\}$'s with a Boltzman-type probability distribution $\propto\exp(-O(\{\beta\})/T)$ via a Monte Carlo procedure. For $T\rightarrow 0$ the state of lowest $O(\{\beta\})$ is reached unless the system is trapped into some metastable state.

In our case the objective function is the total-energy functional and the variational parameters are the coefficients of the expansion of the KS orbitals in some convenient basis and possibly the ionic positions and/or the $\{\alpha_\nu\}$'s. We found that a simulated annealing strategy based on MD, rather than on the Metropolis Monte Carlo method of Kirkpatrick, Gelatt, and Vecchi,[4] can be applied efficiently to minimize the KS functional. This approach, which may be called "dynamical simulated annealing," not only is useful as a minimization procedure but, as we demonstrate here, it allows also the study of finite temperature properties.

In our method we consider the parameters $\{\psi_i\}$, $\{R_I\}$, $\{\alpha_\nu\}$ in the energy-functional [Eq. (1)] to be dependent on time and introduce the Lagrangean

$$L=\sum_i \tfrac{1}{2}\mu\int_\Omega d^3r\,|\dot\psi_i|^2+\sum_I \tfrac{1}{2}M_I\dot R_I^2+\sum_\nu \tfrac{1}{2}\mu_\nu\dot\alpha_\nu^2-E[\{\psi_i\},\{R_I\},\{\alpha_\nu\}], \qquad (3)$$

where the ψ_i are subject to the holonomic constraints

$$\int_\Omega d^3r\,\psi_i^*(\mathbf{r},t)\psi_j(\mathbf{r},t)=\delta_{ij}. \qquad (4)$$

In Eq. (3) the dot indicates time derivative, M_I are the physical ionic masses, and μ and μ_ν are arbitrary parameters of appropriate units.

The Lagrangean in Eq. (3) generates a dynamics for the parameters $\{\psi_i\}$'s, $\{R_I\}$'s, and $\{\alpha_\nu\}$'s through the equations of motion:

$$\mu\ddot\psi_i(\mathbf{r},t)=-\delta E/\delta\psi_i^*(\mathbf{r},t)+\sum_k\Lambda_{ik}\psi_k(\mathbf{r},t), \qquad (5a)$$

$$M_I\ddot R_I=-\nabla_{R_I}E, \qquad (5b)$$

$$\mu_\nu\ddot\alpha_\nu=-(\partial E/\partial\alpha_\nu), \qquad (5c)$$

where Λ_{ik} are Lagrange multipliers introduced in order to satisfy the constraints in Eq. (4). The ion dynamics in Eqs. (5) may have a real physical meaning, whereas the dynamics associated with the $\{\psi_i\}$'s and the $\{\alpha_\nu\}$'s is fictitious and has to be considered only as a tool to perform the dynamical simulated annealing. Equation (3) defines a potential energy E and a classical kinetic energy K given by

$$K=\sum_i \tfrac{1}{2}\mu\int_\Omega d^3r\,|\dot\psi_i|^2+\sum_I \tfrac{1}{2}M_I\dot R_I^2+\sum_\nu \tfrac{1}{2}\mu_\nu\dot\alpha_\nu^2. \qquad (6)$$

The equilibrium value $\langle K\rangle$ of the classical kinetic energy can be calculated as the temporal average over the trajectories generated by the equations of motion [Eqs. (5)] and related to the temperature of the system by suitable normalization. By variation of the velocities, i.e., the $\{\dot\psi_i\}$'s, $\{\dot R_I\}$'s, and $\{\dot\alpha_\nu\}$'s, the temperature of the system can be slowly reduced and for $T\to 0$ the equilibrium state of minimal E is reached. At equilibrium $\ddot\psi_i=0$, Eq. (5a) is identical within a unitary transformation to the KS equation [Eq. (2)], and the eigenvalues of the Λ matrix coincide with the occupied KS eigenvalues. Only when these conditions are satisfied does the Lagrangean in Eq. (3) describe a real physical system whose representative point in configurational space lies on the BO surface. For large systems our scheme is more efficient than standard diagonalization techniques.[5] Furthermore, in the present approach, diagonalization, self-consistency, ionic relaxation, and volume and strain relaxation are achieved *simultaneously*. The amount of classical kinetic energy is a measure of the departure of a system from the self-consistent minimum of its total energy.

It should be stressed that the dynamical simulated annealing technique introduced above is a method of quite general applicability in the context of functional minimization. As such it can be useful in many areas of physics. For instance, it can be applied to the study of classical field theories or to obtain the ground-state energy in Hartree-Fock or configuration interaction schemes. We also observe that, as far as functional minimization is concerned, Newtonian dynamics may be conveniently replaced by Langevin[6] or other types of dynamics.[7]

In order to illustrate how our method works in practice, we present results obtained for the ground-state electronic structure of Si as follows. We have considered a simple cubic supercell containing eight atoms subject to periodic boundary conditions. We have used a local pseudopotential[8] and a local-density approximation to the exact exchange and correlation functional.[9] The single-particle orbitals for the valence electrons have been expanded in plane waves with an energy cutoff of 8 Ry, which amounts to including 437 plane waves at the Γ point. For simplicity, only the Γ point of the Brillouin zone (BZ) of the supercell has been considered in the evaluation of the energy functional.[10] This leads to a total of 16×437 complex electronic variational parameters, since sixteen is the number of doubly occupied KS levels. A simulated annealing run is illustrated in Fig. 1. The lattice parameter was allowed to vary while the ions were kept in their perfect diamond arrangement. The total energy, the lattice parameter, and the eigenvalues of the matrix of the Lagrangean multipliers are plotted as functions of the simulation "time." The initial conditions for the electronic orbitals were fixed by filling the lowest available plane-wave states and giving a Maxwellian distribution of velocities to the components of the fields. The value of μ was chosen to be 1 a.u. The mass μ_Ω associated with variation in the volume was taken to be 10^{-5} a.u. The Verlet algo-

Note on p. 481

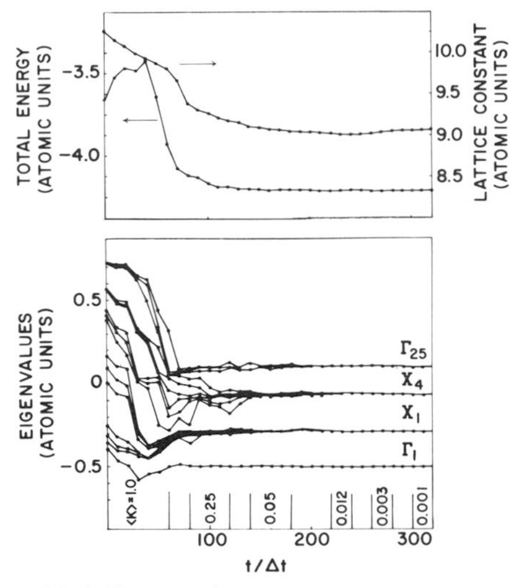

FIG. 1. Evolution of total energy per atom, lattice constant, and eigenvalues of the Λ matrix, during a typical dynamical annealing run. The partial averages of the classical kinetic energy K during each subsection of the run are indicated in the lower part of the picture. For $K \rightarrow 0$ the eigenvalues of the Λ matrix tend to the KS eigenvalues. The various multiplets are labeled according to the symmetry of the diamond lattice.

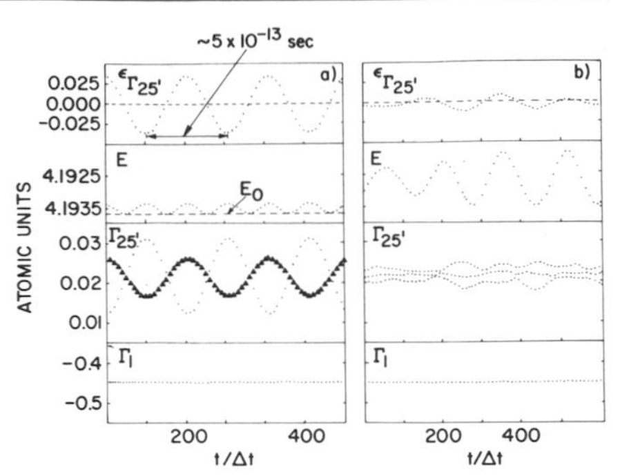

FIG. 2. From top to bottom, temporal evolution of average atomic displacement along $\epsilon_{\Gamma_{25'}}$, potential energy per atom, and $\Gamma_{25'}$ and Γ_1 multiplets for two different MD runs. The lattice constant was taken to be equal to the experimental value of 10.26 a.u.; Δt and μ were taken to be 10 and 300 a.u., respectively. The dashed line in the second panel from the top indicates the $T=0$ ground-state energy. The triangles indicate a doubly degenerate level.

rithm[11] with a time step of 0.1 a.u. was used and the values of the Lagrange multipliers were determined by the method of Ryckaert, Ciccotti, and Berendsen.[12] After some initial equilibration the temperature was progressively reduced to very small values. A satisfactory degree of convergence is seen to be achieved after ~ 200 time steps, when our results agree within numerical errors with those of a conventional self-consistent calculation for the same model.[13]

We can consider now a situation in which the ions, to which we associate their actual physical masses, are allowed to move at a given temperature, while the kinetic energy of the electronic variational parameters remains equal to zero. In this case the electrons are at any time in their ground state and the ions move under the action of BO forces. This can be achieved either by conveniently reoptimizing the electronic variational parameters or by realizing a metastable situation in which the kinetic energy associated with the ψ_i's remains always very small compared to the typical variations of the potential energy of the system. This is equivalent to giving the BO surface a finite thickness proportional to the temperature associated with the ψ_i's. If this temperature remains very small, the ion dynamics generated via Eqs. (5) provides a good representation of the actual dynamics of a physical system.

In Fig. 2 we report the results of two different sets of calculations in which we have performed dynamical simulations for our model. In these calculations it was not necessary to reoptimize the electronic variational parameters at each point along the trajectory, since the thickness of the BO surface never exceeded the value of 7×10^{-6} a.u. per atom, a rather small fraction of the potential energy variation. This is perfectly adequate to represent the interionic forces in this particular case. If the thickness of the BO surface were too large, not only would the forces be incorrectly estimated but also they might depend upon the path along which a given point of the potential energy surface is approached. On the other hand, for very small thicknesses the low velocities of the electronic variational parameters might lead to intolerably long relaxation times. In such a case a compromise would be necessary. In Fig. 2(a) the atoms were initially displaced from their perfect lattice position according to the eigenmode $\epsilon_{\Gamma_{25'}}$, corresponding to the optical phonon mode at the Γ point of the diamond lattice. The system undergoes slightly anharmonic oscillations whose frequency is 20 THz, in perfect agreement with the results of a static frozen-phonon calculation for the same model,[13] showing that the thickness of the BO surface was adequate. In Fig. 2 (a) we also report how the effect of the ionic oscillatory motion is reflected in some electronic properties. Notice that the threefold degenerate topmost $\Gamma_{25'}$ level splits, in perfect phase with the ionic motion, into a singlet and a doublet, whereas the low-lying Γ_1 state is hardly affected by the ionic motion. These results are contrasted with those reported in Fig. 2(b). Here the ions were first randomly displaced from their equilibrium position and a simulated annealing was performed in order to bring the electrons in the corresponding ground state. The ions were then allowed to move. After some equilibration the average kinetic energy associated with ionic motion had a value corresponding to ~ 250 K and the behavior of the system was as illustrated in Fig. 2(b). The projection of the ionic displacement along the $\epsilon_{\Gamma_{25'}}$ eigenvector and the electronic properties do not show any apparent correlation. The degeneracy of the $\Gamma_{25'}$ one-electron eigenstate is completely lifted by thermal disorder, while the Γ_1 state still remains hardly affected by the ionic motion.

The calculations presented here can all be performed on a VAX-like minicomputer. Access to supercomputers can make possible the simulation of larger systems and more realistic models. Because of the simplicity of Newton's equations the computer code can be fully vectorized with not much effort. However the main advantages of the present approach lie in its ability to perform a global minimization of the energy DF and, more importantly, in offering a convenient and, in principle, exact tool for studying finite temperature effects and dynamical properties.

We benefited from discussions with A. Baldereschi, P. Carnevali, A. Nobile, S. T. Pantelides, A. Selloni, E. Tosatti, and A. R. Williams. Special thanks are due to S. Baroni for precious suggestions and valuable help. This work has been supported by the Gruppo Nazionale di Struttura della Materia del Consiglio Nazionale delle Richerche and by the Ministero della Pubblica Istruzione.

[1]See, for instance, *Theory of the Inhomogeneous Electron Gas,* edited by S. Lundqvist and N. M. March (Plenum, New York, 1983).

[2]A. Rahman, in *Correlation Functions and Quasiparticle Interactions in Condensed Matter,* edited by J. Woods Halley, NATO Advanced Study Series Vol. 35 (Plenum, New York, 1977).

[3]W. Kohn and L. J. Sham, Phys. Rev. **140**, A1133 (1965).

[4]S. Kirkpatrick, C. D. Gelatt, Jr., and M. P. Vecchi, Science **220**, 671 (1983).

[5]For instance, if the N occupied single-particle orbitals are expanded into M plane waves ($M >> N$), a standard diagonalization requires $O(M^3)$ operations, whereas Eq. (5a) requires both $O(NM \ln M)$ and $O(N^2M)$ operations.

[6]P. Carnevali and A. Selloni, private communication.

[7]P. J. Rossky, J. D. Doll, and H. L. Friedman, J. Chem. Phys. **69**, 4628 (1978); C. Pangali, M. Rao, and B. J. Berne, Chem. Phys. Lett. **55**, 413 (1978).

[8]J. A. Appelbaum and D. R. Hamann, Phys. Rev. B **8**, 1777 (1973).

[9]J. P. Perdew and A. Zunger, Phys. Rev. B **23**, 5048 (1981).

[10]The sampling of the BZ and the pseudopotential used in this paper are not very accurate and should be replaced with better ones in more realistic calculations.

[11]L. Verlet, Phys. Rev. **159**, 98 (1967).

[12]J. P. Ryckaert, G. Ciccotti, and H. J. C. Berendsen, J. Comput. Phys. **23**, 327 (1977).

[13]When accurate pseudopotentials are used together with an accurate sampling of the BZ, local-density calculations agree very well with experiment [see M. T. Yin and M. L. Cohen, Phys. Rev. B **26**, 3259 (1982)].

Note to Reprint VI.8

For an application of the Car–Parrinello method to the study of a germanium surface, see Needels et al., Phys. Rev. Lett. 58 (1987) 1765.